The Biology of Gobies

The Biology of Gobies

Editors

Robert A. Patzner
Department Organismic Biology
University of Salzburg
Salzburg, Austria

James L. Van Tassell
Systematic Ichthyology
Biology Department
Hofstra University
Hempstead, NY, USA

Marcelo Kovačić
Prirodoslovni Muzej Rijeka
Lorenzov Prolaz 1
Rijeka, Croatia

B.G. Kapoor
Formerly Professor of Zoology
The University of Jodhpur
Jodhpur, India

CRC Press
Taylor & Francis Group
an informa business
www.crcpress.com

6000 Broken Sound Parkway, NW
Suite 300, Boca Raton, FL 33487
270 Madison Avenue
New York, NY 10016
2 Park Square, Milton Park
Abingdon, Oxon OX14 4RN, UK

Science Publishers
Jersey, British Isles
Enfield, New Hampshire

Published by Science Publishers, an imprint of Edenbridge Ltd.

- St. Helier, Jersey, British Channel Islands
- P.O. Box 699, Enfield, NH 03748, USA

E-mail: *info@scipub.net* Website: *www.scipub.net*

Marketed and distributed by:

CRC Press
Taylor & Francis Group
an **informa** business
www.crcpress.com

6000 Broken Sound Parkway, NW
Suite 300, Boca Raton, FL 33487
270 Madison Avenue
New York, NY 10016
2 Park Square, Milton Park
Abingdon, Oxon OX 14 4RN, UK

ISBN: 978-1-57808-436-4

Cover Illustrations: (from top to bottom)...

A pair of *Valenciennea strigata* (Gobiidae), Philippines, Pacific Ocean. Photo: Robert A. Patzner *Gobius cruentatus* (Gobiidae), Croatia, Mediterranean Sea. Photo: Robert A. Patzner Eggs of *Clamydogobius eremius* (Gobiidae), freshwater aquarium. Photo: Robert A. Patzner *Amblyeleotris randalli* (Gobiidae), Papua New-Guinea, Pacific Ocean. Photo: Robert A. Patzner

Library of Congress Cataloging-in-Publication Data

The biology of gobies / [edited by] Robert Patzner... [et al.]. -- 1st ed.
p. cm.
Includes bibliographical references and index.
ISBN 978-1-57808-436-4 (hardback)
1. Gobiidae. I. Patzner, Robert A.
QL638.G7.B56 2011
597.7--dc23

2011018817

The views expressed in this book are those of the author(s) and the publisher does not assume responsibility for the authenticity of the findings/conclusions drawn by the author(s). Also no responsibility is assumed by the publishers for any damage to the property or persons as a result of operation or use of this publication and/or the information contained herein.

Printed in the United States of America

Gobius niger by M.E. Bloch, 1782-84
from Bloch's Atlas Vol. 1, plate 107, Mergus Verlag, Melle, Germany

Preface

The Gobiiformes is one of the most diverse orders of the Percomorphs, with approximately 2000 species in 270 genera. They represent 5 to 10% of all teleosts, with species occurring in marine, fresh, and brackish water environments. Both one of the smallest extant vertebrates, *Trimmatom nanus*, with a maximum length of 8.0 mm SL for gravid females, and the lightest weight vertebrate, *Schindleria brevipinguis*, with a maximum weight of about 2 mg, belong to the order Gobioidei. Some gobioids are amphidromous returning to their freshwater streams by traversing the face of 350 m waterfalls. Gobioids occupy a variety of marine habitats including open water, sandy or rocky habitats, tidepools, caves, sponges, and mudflats. They are found at depths ranging from 0 to 1122 m (*Karsten totoyensis*) and in temperatures ranging from 7°C for *Leucopsarion petersi* to 40°C for *Chlamydogobius eremius*. Yet with all of this diversity and the important role they play in the ecology of regions a book about their biology has never been published. This book, the *Biology of Gobies*, contains the background information for others to join in on the study of this diverse group.

Systematics within the Gobiiformes and their placement within the teleosts is not resolved. Morphological trees based on synapomorphies differ from those proposed by molecular trees, with some molecular trees not supported by any morphological synapomorphies. The history of morphology and molecular systematics is provided to give readers an understanding of the problems faced in resolving monophyletic clades. The current status of the systematics of each of the families is presented by authors that have worked with systematics of those families for decades. The chapters on the study of morphological and molecular systematics, along with chapters on each of the families will help draw attention to the need for further studies in these areas.

Gobies are unparalleled among vertebrates in their ability to diversify and adapt. They excel in niche exploitation and microhabitat specialization which has led to rapid speciation and adaptive radiation. The section on zoogeography examines the genera and species of specific regions of the world including the Americas, north-eastern Atlantic, Mediterranean, southern Australia, New Zealand, Ponto-Caspian, and European freshwaters along with more generalized chapters on the diversity and distributions of marine and tropical freshwater gobies.

The reproduction of gobies is unusual compared to most fishes. There is a large investment in parental care with demersal eggs guarded by the male. They vary reproductively from gonochoric to hermaphrodite, monogamy to polygyny and promiscuity, some have short life spans and reproduce once only while others have longer life spans reproducing one or more times per year. Several chapters are devoted to reproductive strategies, reproductive organs, and the development of larvae and gobies.

Gobies form a major component of coral reef and coastal fish communities throughout the tropics and subtropics; in coral reef habitats gobies account for over 50 % of the energy flow. The short lifespans and high levels of reproductive output result in rapid generational turnover and the production of a significant component of the marine ichthyoplankton biomass. Symbiotic relationships of gobies as cleaners and in partnership with alpheid shrimp; paedomorphic species; mudskippers; and the role of gobies on coral reefs provide insight into the ecology of the group.

Thirty-three authors from 15 countries have contributed to the book to provide a synopsis of our current knowledge and areas for further studies.

Robert A. Patzner
James L. Van Tassell
Marcelo Kovačić
B.G. Kapoor

Contents

Section 2: Zoogeography

Section 3: General Biology

Section 4: Ecology

List of Contributors

Agorreta Ainhoa

Department of Zoology and Ecology, University of Navarra, Irunlarrea s/n, 31008 Pamplona, Spain and Department of Zoology. The Natural History Museum, Cromwell Road, SW7 5BD, London, UK.

E-mail: a.agorreta@nhm.ac.uk

Borges Rita

CCMAR, Centre of Marine Sciences, Universidade do Algarve, Campus de Gambelas 8005-139 Faro, Portugal and Eco-Ethology Research Unit, Instituto Superior de Psicologia Aplicada, Rua Jardim do Tabaco 34, 1149-041 Lisboa, Portugal.

E-mail: rborges@ispa.pt

Côté Isabelle M.

Department of Biological Sciences, Simon Fraser University, Burnaby BC, V5A 1S6, Canada.

E-mail: imcote@sfu.ca

Faria Cláudia

Instituto de Educação, Universidade de Lisboa, Campo Grande, Edificio C6, 1749-016, Lisboa, Portugal and Eco-Ethology Research Unit, Instituto Superior de Psicologia Aplicada, Rua Jardim do Tabaco 34, 1149-041 Lisboa, Portugal.

E-mail: cbmfaria@clix.pt

Freyhof Jörg

Leibniz-Institute of Freshwater Ecology and Inland Fisheries, Müggelseedamm 310, 12587 Berlin, Germany.

E-mail: freyhof@igb-berlin.de

Gil Fátima

Aquário Vasco da Gama, Rua Direita do Dafundo, 1495-718 Cruz Quebrada- Dafundo, Portugal.

E-mail: fatimagil@mail.telepac.pt

Gonçalves Emanuel J.

Eco-Ethology Research Unit, Instituto Superior de Psicologia Aplicada, Rua Jardim do Tabaco 34, 1149-041 Lisboa, Portugal.

E-mail: emanuel@ispa.pt

Gonzales Tomas T.

Center for General Education, AMA International University, Bahrain, P.O. Box 18041, Manama, Kingdom of Bahrain.

E-mail: t2gonzales@yahoo.com

Herler Juergen

Department of Theoretical Biology, Morphology Section, Faculty of Life Sciences, University of Vienna, Althanstrasse 14, 1090 Vienna, Austria.
E-mail: juergen.herler@univie.ac.at

Hernaman Vanessa

Centre for Marine Studies, University of Queensland, St Lucia, Queensland 4072, Australia.

E-mail: vanessa.hernaman@gmail.com

Ishimatsu Atsushi

Institute for East China Sea Research, Nagasaki University, Tairamachi 1551-7, Nagasaki 851-2213, Japan.

E-mail: a-ishima@nagasaki-u.ac.jp

Iwata Akihisa

Laboratory of Ecology and Environment, Division of Southeast Asian Area Studies, Graduate School of Asian and African Area Studies, Kyoto University Yoshida, Shimoadachi-cho 46, Sakyo-ku, Kyoto 606-8501 Japan.
E-mail: aiwata@asafas.kyoto-u.ac.jp

Karplus Ilan

Aquaculture Research Unit, the Institute of Animal Sciences, The Volcani Research Center, P.O. Box 6, Bet-Dagan 20250, Israel.
E-mail: Karplus@agri.gov.il

Keith Philippe

Muséum national d'Histoire naturelle, UMR 7208, ichtyologie, CP 26, 57 rue Cuvier, 75231 Paris Cedex 05, France.

E-mail: keith@mnhn.fr

Kovačić Marcelo

Prirodoslovni muzej Rijeka, Lorenzov prolaz 1, HR-51000 Rijeka, Croatia.

E-mail: marcelo@prirodoslovni.com

La Mesa Mario

ISMAR-CNR, Istituto di Scienze Marine, Sede di Ancona, Largo Fiera della Pesca, 60125 Ancona, Italy.

E-mail: m.lamesa@ismar.cnr.it

Larson Helen K.

65 Reid Road, Wongaling Beach, Queensland 4852, Australia.

E-mail: helen.larson@nt.gov.au

Lord Clara

Muséum national d'Histoire naturelle, UMR 7208, ichtyologie, CP 26, 57 rue Cuvier, 75231 Paris Cedex 05, France.

E-mail: claralord@mnhn.fr

Mazzoldi Carlotta

Department of Biology, University of Padova, via U. Bassi 58/B, 35121 Padova, Italy.

E-mail: mazzoldi@bio.unipd.it

Munday Philip L.

ARC Centre of Excellence for Coral Reef Studies / School of Marine and Tropical Biology, both: James Cook University, Townsville, Queensland 4811, Australia.

E-mail: philip.munday@jcu.edu.au

Murdy Edward

National Science Foundation, 4201 Wilson Blvd., Arlington, VA 22230, USA.

E-mail: emurdy@nsf.gov

Patzner Robert A.

Dept. Organismic Biology, University of Salzburg, Hellbrunnerstr. 34, 5020 Salzburg, Austria.

E-mail: robert.patzner@sbg.ac.at

Pezold Frank

Texas A&M University - Corpus Christi, Corpus Christi, TX 7841, USA.
E-mail: frank.pezold@tamucc.edu

Rasotto Maria B.

Department of Biology, University of Padova, via U. Bassi 58/B, 35121 Padova, Italy.

E-mail: rasotto@bio.unipd.it

Rüber Lukas

Department of Zoology, The Natural History Museum, Cromwell Road, SW7 5BD, London, UK.

E-mail: l.ruber@nhm.ac.uk

Schliewen Ulrich K.

Bavarian State Colletion of Zoology, Department of Ichthyology, Muenchhausenstr. 21, 81247 Munich, Germany.

E-mail: schliewen@zsm.mwn.de

Soares Marta C.

Unidade de Investigacão em Eco-Etologia, Instituto Superior de Psicologia Aplicada, Rua Jardim do Tabaco 34, 1149-041 Lisboa, Portugal and Department of Zoology, University of Neuchâtel, Emile-Argand 11, 2009 Neuchâtel, Switzerland.

E-mail: msoares@ispa.pt

Taylor Michael S.

Department of Biology, Southeast Missouri State University, Cape Girardeau, Missouri 63701, USA.

E-mail: mtaylor@semo.edu

Thacker Christine

Research and Collections, Section of Ichthyology, Natural History Museum of Los Angeles County, 900 Exposition Blvd., Los Angeles, CA 90007, USA.
E-mail: thacker@nhm.org

Thompson Andrew Richard

NOAA Fisheries Service, Southwest Fisheries Science Center, 8604 La Jolla Shores Drive, La Jolla, CA 92037-1508, USA.

E-mail: andrew.thompson@noaa.gov

Tornabene Luke

Department of Life Sciences, Texas A&M University-Corpus Christi, Corpus Christi, Texas 78412, USA.

E-mail: luke.tornabene@tamucc.edu

Van Tassell James L.

Systematic Ichthyology, Biology Department, Hofstra University, Hempstead, NY 11549, U.S.A.

E-mail: jvantassell@gobiidae.com

Zander C. Dieter

Biozentrum Grindel and Zoologisches Museum, University, Martin-Luther-King-Platz 3, 20146 Hamburg, Germany.

E-mail: cedezet@zoologie.uni-hamburg.de

SECTION 1

Systematics

CHAPTER

1.1

A History of Gobioid Morphological Systematics

James L. Van Tassell[1,*], Luke Tornabene[2] and Michael S. Taylor[3]

INTRODUCTION

Nelson (2006) divided the suborder Gobioidei into nine families, employing the six families (Rhyacichthyidae, Eleotridae, Kraemeriidae, Microdesmidae, Xenisthimidae, Gobiidae) recognized by Hoese (1984) and adding the Odontobutidae (Hoese and Gill, 1993), the Schindleriidae (Johnson and Brothers, 1993), and Ptereleotridae (Thacker, 2000). The nine families recognized by Nelson were Rhyacichthyidae (loach gobies, 1 genus, 2 spp.), Odontobutidae (freshwater sleepers, 5 genera, 15 spp.), Eleotridae (sleepers, about 35 genera, 155 spp.), Gobiidae (gobies, 210 genera, 1,950 spp.), Kraemeriidae (sandfishes, 2 genera, 8 spp.), Xenisthmidae (collared

Authors' addresses: [1]Systematic Ichthyology, Biology Department, Hofstra University, Hempstead, NY 11549, U.S.A.

[2]Department of Life Sciences, Texas A&M University-Corpus Christi, Corpus Christi, Texas 78412, USA. E-mail: luke.tornabene@tamucc.edu

[3]Department of Biology, Southeast Missouri State University, Cape Girardeau, Missouri 63701, USA. E-mail: mtaylor@semo.edu

**Corresponding author:* E-mail: jvantassell@gobiidae.com

wrigglers, 2 genera, 12 spp.), Microdesmidae (wormfishes, 5 genera, 30 spp.), Ptereleotridae (dartfishes, 5 genera, 36 spp.), and Schindleriidae (infantfishes, 1 genus, 3 spp.). A more up-to-date account of species and genera is found in Eschmeyer (2010), an online version listing species of fishes by family/subfamily. They cited the current number of species by families [as of April 2009] as: Rhyacichthyidae (3), Odontobutidae (20), Eleotridae (169), Gobiidae (1573), Kraemeriidae (9), Microdesmidae (80) with two subfamilies Microdesminae (32) and Ptereleotrinae (48), Xenisthmidae (13), and Schindleriidae (3). With the increasing use of DNA to illuminate cryptic species, the count in each group is likely to change considerably in the future. Pezold (1993) subdivided the Gobiidae of Hoese (1984) into five subfamilies, removing 56 genera from the subfamily Gobiinae and placing them in the Gobionellinae. With the development of molecular evolutionary techniques the classification presented above is continually changing. Thacker (2009) presented a new classification of the Gobioidei with six families Rhyacichthyidae, Odontobutidae (Odontobutidae + Milyeringa), Butidae, Eleotridae (Eleotrinae, Xenisthmidae), Gobionellidae (Gobionellinae, Oxudercinae, Amblyopinae, Sicydiinae), and Gobiidae (Gobiinae, Microdesmidae, Ptereleotridae, Kraemeridae, Ptereleotridae, Kraemeriidae, Schindleriidae). The molecular phylogeny of the Gobioidei is discussed in the following chapter.

HISTORY OF GOBIOID CLASSIFICATION

Until recently, the classification of gobioid fishes has been based largely on external morphology and osteology along with additional characters such as cutaneous papillae patterns, head canal system, and sensory pores. Finding reliable characters has been problematic due to miniaturization and, hence, reduction of features, further complicated by the lack of a clearly established sister group for character polarization.

In 1758, Linnaeus recognized one genus (*Gobius*) with seven species (*G. niger*, *G. paganellus*, *G. eleotris*, *G. aphya*, *G. jozo*, *G. pectinirostris*, and *G. anguillaris*). The genus was diagnosed as "...head with two simple pores between the eyes, another anterior to that, branchiostegal membrane with 4 elements; ventral fins united in an oval with 12 elements..." Gmelin (1789) increased the number of species to 25 and expanded the diagnosis of *Gobius* to include "...body small, compressed from side to side, and covered with thin scales, a small 'wart-like' structure located behind the anus, two dorsal fins..." Lacepède (1798) added six new genera: *Taenioides*, *Gobioides*, *Gobiomorus*, *Gobiomoroides*, *Bostrychus*, and *Bostrychoides*. He divided *Gobius* into two subgenera based on: 1) pectoral fins connected directly to the body or 2) pectoral fins with a fleshy base. *Taenioides* was diagnosed by "...a single anal fin; pelvic fins in the form of a disc, composed of a large number of rays; body and tail very elongated and compressed laterally; belly about the length of the head; the scales very small; the eyes barely noticeable..."; *Gobioides* by "...two pelvic fins united; a single dorsal fin; head small; branciostegal membrane united across the isthmus..."; *Gobiomorus* by "...two pelvic fins not united; two dorsal fins; head small; eyes close together; branchiostegal membrane attached across isthmus..."; *Gobiomoroides* by "...pelvic fins not united..."; and *Bostrychus* and *Bostrychoides* differing from the other genera by having barbels. The number of genera was further increased to 11 by Bloch and Schneider (1801), and then to 23 by Günther (1861), and to 99 by Bleeker (1874).

Günther (1861), Bleeker (1874), and Jordan (1923) proposed the earliest classification systems for gobioid interrelationships. These systems were based primarily on the structure of the fins, but also used numbers and type of scales (ctenoid *vs.* cycloid), teeth, and placement of eyes as diagnostic characters.

Günther (1861) recognized the family Gobiidae and divided it into groups: Gobiina, Amblyopina, Trypauchenina, Callionymina, and as an appendix the two genera *Luciogobius* and *Oxymetopon*. Günther's Gobiina contained species currently recognized as gobiids, eleotrids, and periophthalmids, but was divided into two subgroups based on whether the pelvic fins were united or separate. The Amblyopina, contained the genus *Amblyopus*; Trypauchenina, contained *Trypauchen* and *Trypauchenichthys*; and Callionymina had three genera: *Platyptera*, *Callionymus*, and *Vulsus*. Günther (1861) considered as synonyms of *Gobius* genera such as *Chaeturichthys*, *Gobionellus*, *Oxyurichthys*, *Ctenogobius*, *Acanthogobius*, *Rhinogobius*, *Chaenogobius*, and *Lepidogobius*. As he wrote:

> "...the attempts to divide the numerous species of this genus into smaller groups have failed to establish natural genera. The characters which may be used for that purpose as the form of the body, the structure and size of the scales, modifications of the dentition, extent of the fins, presence or absence of barbels either show gradual transitions between the most extreme forms, or such species are united by them as are otherwise entirely different from one another. Anyone who should extend his examination to a great number of species would soon become aware of the fruitlessness of such attempts, by which neither a natural arrangement is effected, nor the determination of the species facilitated. A generic separation of those species which are entirely devoid of scales, or which have teeth tricuspid or in a single series, appears to be the most justifiable."

Bleeker (1874) diagnosed four subfamilies: Eleotriformes, Gobiiformes, Amblyopodiformes (Amblyopina and Trypauchena of Günther), and Luciogobiiformes (gobioids without a first dorsal fin), but did not include Günther's Callionymina. Bleeker (1874) based his divisions primarily on whether the ventral fins were united or divided, and whether one or two dorsal fins were present; the latter could be united or divided. The subfamilies were further divided based on dentition. Herre (1927) commented that "...the character of teeth is a good one for limitation of genera, but has been relied upon too much by Bleeker, since many descriptions merely state 'wide bands of teeth' or 'bands of fine teeth,' or similar, indefinite statements." Koumans (1931) agreed with Herre's comments and noted that the descriptions given by Bleeker for each genus were too short in most cases to be useful. Koumans further noted that "...Bleeker himself afterwards altered the name of the genus to which a species belongs ... [in] his collection." Bleeker's publication (1874) was one of a few works that discussed the gobioid genera of the world and the first to give characters defining all genera known at that time. After Bleeker, publications appeared which discussed the gobioid fauna of regions e.g. *Fishes of India* (Day, 1878), *Fishes of North and Middle America* (Jordan and Evermann, 1898), and *Gobies of the Philippines* (Herre, 1927).

Regan (1911) developed the first classification based largely on osteology, e.g. bones of the cranium and pectoral girdle, and number of vertebrae. He recognized three families: the Eleotridae, Gobiidae, and Psammichthyidae. The Eleotridae was defined by: pelvic fins separated, palatine with an ascending stem articulating directly with a lateral ethmoid apophysis behind the origin of the maxillary process; mesopterygoid narrow, but well-developed; hypercoracoid and hypocoracoid well-developed; radials inserted on hypercoracoid, hypocoracoid, and on the ligamentous or cartilaginous tissue between these bones (Fig. 1.1.1A, B); vertebrae 25 to 28 (10 – 12 trunk vertebrae + 14 – 17 caudal vertebrae). Some of the genera included by Regan were *Ioglossus*, *Philypnus*, *Dormitator*, *Eleotris*, *Odontobutis* and *Rhiachichthys* (spelling of Regan). Regan's family Gobiidae was divided into two subfamilies, Gobiinae and Periophthalminae. The Gobiidae was defined by: pelvic fins, when well developed, united, usually forming an adhesive cup; palatine T-shaped, with a posterior process for articulation with the lateral ethmoid; mesopterygoid vestigial or absent; hypercoracoid absent; and radials inserting on cleithrum, only the ventralmost in contact with the hypocoracoid (Fig. 1.1.1C). The Gobiinae was further divided into two groups. The first group was defined by a spinous dorsal, when present, separate from second dorsal; second dorsal separate from caudal; pectorals large; vertebrae 25 to 28 (10 – 12 + 15 – 17). Most of the gobiid genera were in this group and included, *inter alia Gobius*, *Gobiodon*, *Cryptocentrus*, and *Sicydium*. The second group was defined by vertical fins confluent, pectoral fins small or moderate. This group was further divided based on vertebrae: 28 or 29 (11 + 17 – 18) and caudal vertebrae elongate (as in *Gobioides* and *Taenioides*); vertebrae 34 (10 + 24) and caudal vertebrae normal (as in *Trypauchen* and *Trypauchenichthys*). Regan's subfamily Periophthalminae was defined by the eyes being close together, prominent, erectile; base of pectoral fin very muscular; pectoral radials elongate, inserted on a broad laminar ridge of the cleithrum; hypocoracoid and cleithrum enclosing a large foramen (Fig. 1.1.1D); vertebrae 25 or 26 (10 – 11 + 14 – 16). Two genera were included, *Periophthalmus* and *Boleophthalmus*. Regan (1911) established the family Psammichthyidae (=Kraemeriidae according to Maugé, 1986 in Eschmeyer, 1990) and provisionally placed it within the Gobioidei. This family contained a single species, *Psammichthys nudus* (= *Kraemeria nuda)*.

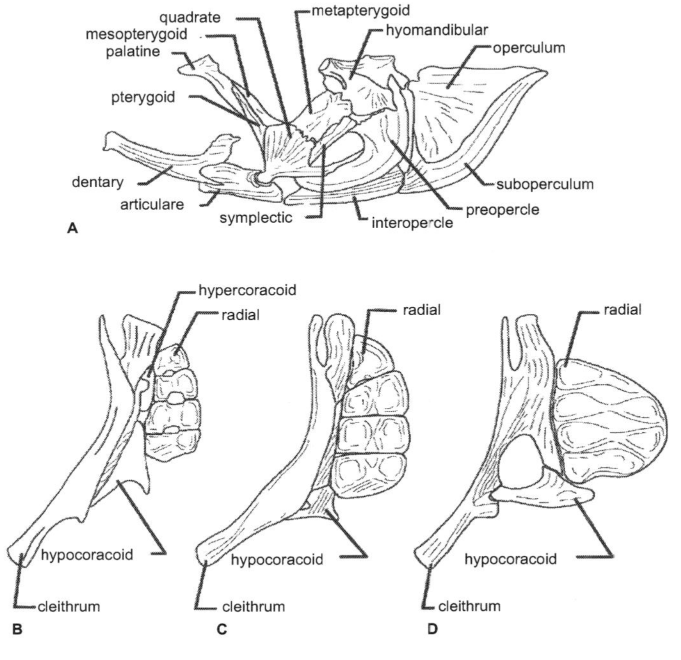

Fig. 1.1.1 Lateral views of the lower jaw, opercular and jaw suspension apparatus of *Eleotris marmorata* (*Oxyeleotris marmorata*), (A) and pectoral girdles of *E. marmorata* (*Oxyeleotris marmorata*) (B), *Gobius caputo* (*Gobius paganellus*) (C), and *Periophthalmus kaelreuteri* (*Periophthalmus barbarus*) (D). Modified from Regan (1911). Updated nomenclature in parentheses.

Sanzo (1911) was the first to study the gobioid lateralis system. This system of external neuromast organs (so-called 'sensory papillae') forms distinct patterns, primarily on the head region. Neuromast patterns can be divided into two groups, determined by whether the neuromasts are primarily arranged in longitudinal or transverse rows on the cheeks (Fig. 1.1.2). These patterns have been used to define some genera and species (Hoese 1971, 1983; Iljin 1930; Gill *et al.*, 1992; Pezold, 2004) and were used extensively by Miller (e.g., 1963, 1973, 1992a) to define taxa and arrange systems of classification. Hoese (1983) suggested that in *Acentrogobius* and *Glossogoboius* a longitudinal pattern may be more primitive, with a transverse papillae pattern evolving independently in members of each genus, though he cautioned the broad application of these results across other gobioid genera. Thus, the reliability of neuromast patterns as phylogenetically informative characters remains problematic, as assessment of homology and character polarization has not been determined. Wongrat and Miller (1991) proposed a classification of neuromast patterns based on the innervation of neuromast rows by the lateral-line nerve and nerve trunk or ramus. An attempt was also made to distinguish the rows according to their putative evolutionary history with respect to the reduction of the gobioid lateral line system. They recognized the following categories: 1) primary replacement neuromasts forming a linear series along the course of a deleted canal, and presumed to represent primary canal neuromasts; 2) secondary replacement neuromasts believed to be derived from superficial neuromasts once associated with canal pores and forming transverse rows along the former direction of the canal; and 3) secondary accessory neuromasts, which are those superficial neuromasts not linked with canal development. They did not clearly establish homology of the neuromast rows and because Wongrat and Miller (1991) used only six eleotridine species to assess innervation patterns, the systematic utility of their classification scheme for the gobioids is unclear and remains untested.

Jordan (1923) produced an extensive treatment of the gobioids. He divided the Gobioidei into eight families (Rhyacichthyidae, Eleotridae, Gobiidae, Periophthalmidae, Gobioididae, Trypauchenidae, Doliichthyidae, and Psammichthyidae) but did not provide diagnostic characters for each family. This was, however, the first work since Bleeker (1874) to name all genera of gobioid fishes then known. Herre (1928), following Jordan's (1923) system, gave the following description of the Order Gobioidea:

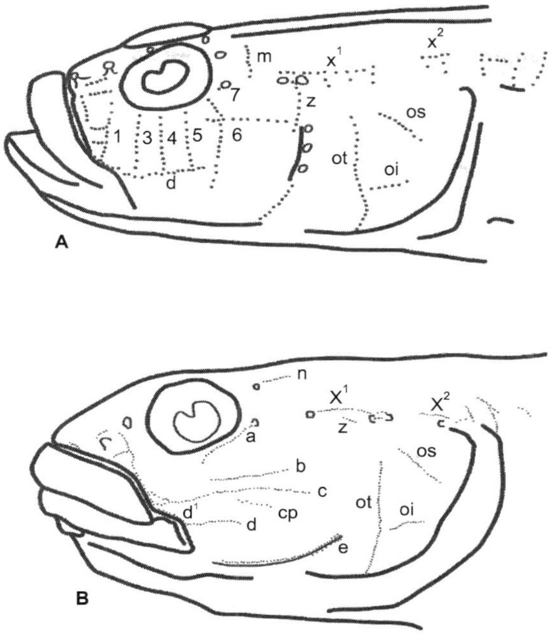

Fig. 1.1.2 Gobioid cephalic neuromast patterns. (A) typical transverse arrangement; (B) typical longitudinal arrangement. Letters indicate longitudinal neuromast rows, numbers indicate transverse neuromast columns. Modified from Sanzo (1911) and Van Tassell (1988).

" ...ventral fins are always present, thoracic in position, each having one spine and five, rarely four, soft rays. The ventrals are very close together, the inner rays the longer, and in by far the greater number, the two fins are completely joined to form a single fin, which may vary from very short and rounded to rather elongate; but, whether separated or fused into a functional unit, the ventrals are used as a sucking disk, or organ, for clinging to rocks or other objects. The spinous dorsal is nearly always present, short, and is made up of feeble spines, or is much less developed than the soft dorsal. There is no lateral line, the gill openings are confined to the sides, and there is an anal papilla. The papilla varies greatly in size and shape; in many species it is long, slender, and pointed in the males, and short, thick and cylindrical or subglobulose in the females. In other species it is alike in both sexes. The preopercle is without a bony stay; there are four gills. The gill membranes are attached to the isthmus, the gill openings thus restricted to the sides and sometimes very narrow. Pseudobranchiae are present, sometimes only in a rudimentary condition. Usually both air bladder and pyloric caecae are wanting..."

Table 1.1.1 Comparison of head pore terminology from unpublished work of Birdsong. See Fig. 1.1.5.

Sanzo (1911)	*Ginsburg (unpublished)*	*Wongrat (1977)*	*Lachner and McKinney (1978, 1979)*	*Akihito (1986)*	*Birdsong (unpublished)*	*Takagi (1967, 1988)*
--	--	σ^1	--	A	A	ethmoidal (ETH)
σ	σ	σ	nasal (NA)	B	B	nasal (NA)
λ	λ	λ	anterior interorbital (AITO)	C	C	anterior interorbital (ITOA)
κ	κ	κ	posterior interorbital (PITO)	D	D	posterior interorbital (ITOP)
ώ	ώ	ω	SOT		E	Esupraotic (SOT)
α	α	α	anterior otic (AOT)	F	F	postorbital (POR)
β	β	β	POT		G	Gintermediate otic(OTM)
ρ	ρ	β^1	intertemporal (IT)	H	H	extreme otic (OTE)
--	ζ	ρ(?)	--	--	H'	---
--	--	θ	--	I	I	supratemporal (SUT)
--	--	ρ	--	J	J	--
ρ^1	θ	ρ^1	anterior temporal (AT)	K	K	anterior cephalo -lateral (CELA)
ρ^2	τ	ρ^2	posterior tempora (PT)	L	L	posterior cephalo-lateral (CELP)
γ	γ	γ	preopercular (POP)	M	M	upper dorso -preopercular (PODU)
δ	δ	δ	POP	N	N	intermediate dorso -preopercular (PODM)
ε	ε	ε	POP	O	O	lower dorso -preopercular (PODL)
--	--	ε^1	--	P	P	upper ventro -preopercular (POVU)
--	--	ε^2	--	Q	Q	lower ventro -preopercular (POVL)

Duncker (1928) recognized four families (Eleotridae, Gobioididae, Periophthalmidae, and Gobiidae), primarily basing his work on Weber (1913) and others. Later, Berg (1940) rearranged the Gobioidei into two superfamilies. The Eleotrioidae contained the eleotrids and *Rhyacichthys*. The Gobioidae contained the remaining gobioids in three families: the Gobiidae (including taeniodids), the Periophthalmidae, and the Kraemeriidae.

Ginsburg (1933) and Koumans (1931) both pointed out the inadequacies of the classification system of gobioids based on primarily on pelvic fins being united or separated. The problem of separate pelvics was later addressed by Miller (1963; see below).

The next major osteological study of gobioid fishes was Gosline (1955). In this work, Gosline (1955) provided evidence for the placement of the microdesmids and kraemeriids among the gobioids and argued that the shape of the palatine and the presence or absence of the scapula did not adequately separate the Eleotridae and the Gobiidae. Gosline based his separation of these two families on the presence of six branchiostegal rays in eleotrids and five in gobiids; this character is still recognized today. Takagi (1967, 1988) (Table 1.1.1) made an extensive study of the cephalic sensory canal systems of gobies from Japan. Takagi examined 82 species of 54 genera from Japanese waters and amended the terminology of the cephalic sensory canals, head pores, and sensory papillae patterns. Takagi attempted to show that a sequence of cephalic sensory canals and associated pores could be used to represent a phylogenetic history of gobioids (Fig. 1.1.3). From his comparative analysis, Takagi divided the gobioids into two groups, those possessing both cephalic sensory canals and sensory papillae, which he considered as the plesiomorphic condition, and those possessing only sensory papillae, the derived condition.

Akihito (1963, 1967) (Fig. 1.1.4) conducted an extensive study on the scapula of gobioids, noting that the scapula is well-developed in primitive genera. Later, Akihito (1969) produced a detailed study of gobioid fishes based on the osteology of 85 species in 71 genera. Although Akihito (1969) did not suggest a classification of gobioids, he created four groups based on the presence or absence of the mesopterygoid (= endopterygoid *sensu* Harrison, 1989) and the upper postcleithrum, and the number of branchiostegal rays. The groups represented his interpretation of a graded series from most primitive (group 1) to most derived (group 4). Groups 1 and 2 consisted generally of eleotridids, group 3 consisted only of *Xenisthmus*, and group 4 comprised gobiids and microdesmids.

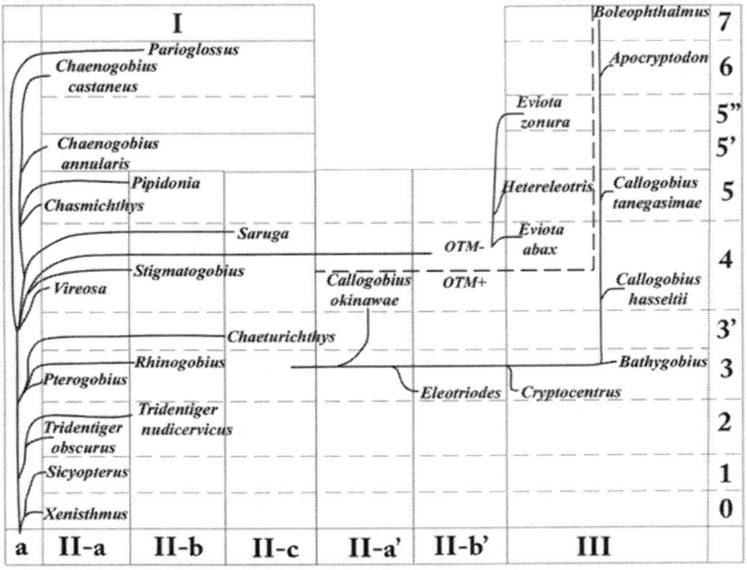

Fig. 1.1.3 Schematic dendrogram representing hypothesized graded change of oculoscapular canal and associated pore patterns for 21 genera of Japanese gobies. Roman numerals indicate supraorbital canal structure. Arabic numerals 0-7 indicate stages of canal and pore reduction to sensory papillae only. OTM+ = intermediate otic pore present; OMT- = intermediate otic pore absent. Graded change is relative to unspecified plesiomorphic ancestor (a). Current taxonomy - *Chaenogobius castaneus* = *Gymnogobius castaneus*, *Pipidonia* = *Gobiopsis*, *Chasmichthys* = *Chaenogobius*, *Vireosa* = *Ptereleotris*, *Eleotriodes* = *Valenciennea*. Modified from Takagi (1988).

Miller (1963) briefly outlined the major differences between gobiids and eleotrids, commenting that the position of separate pelvic fins was not sufficient to separate eleotrids from gobiids. He included the following characters as diagnostic of the Gobiidae: the number of epurals, hypural connection, the presence or absence of the endopterygoid, the number of branchiostegal rays, development of the scapula, the number of pectoral radials, presence or absence of the postcleithrum, a metapterygoid bridge to the quadrate, a preoperculum-symplectic bridge, and the extent of development of the oculoscapular and preopercular canals. Miller (1973) divided the suborder Gobioidei into two families: Rhyacichthyidae and Gobiidae. The monotypic Rhyacichthyidae was diagnosed by plesiomorphic characters, including three epurals and a well-developed cephalic lateralis system (Table 1.1.2), found also in percoid fishes. However, the

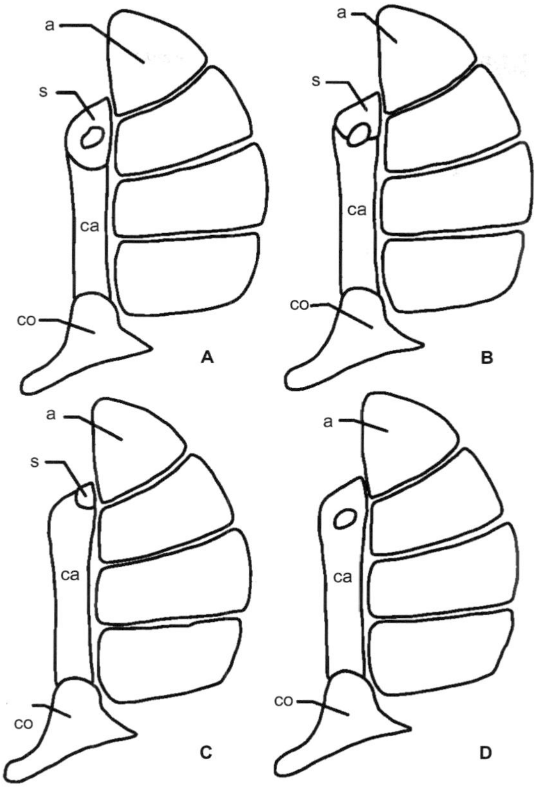

Fig.1.1.4 Gobioid scapula types. (A) enclosing; (B) forked; (C) blotched; (D) no scapula. a = actinost; ca = cartilage; co = coracoid; s = scapula. Modified from Akihito (1969).

combination of a lateral line system and free neuromast is, within the Gobioidei, a combination unique to Rhyacichthyidae (Ahnelt and Göschl, 2004), *Protogobius,* and *Terateleotris,* with the placement of the latter two genera being largely unresolved (Watson & Pöllabauer, 1998; Akihito *et al.*, 2000; Shibukawa *et al.,* 2001; Thacker and Hardman, 2005). The Gobiidae, with close to 2000 species, shared the derived character states of one or two epurals and a reduced cephalic lateralis system. Miller (1973) further divided the Gobiidae into seven subfamilies based on a graded distribution of the greatest number of primitive characters to the greatest number derived skeletal characters: Eleotrinae, Pirskeninae, Xenisthminae, Gobionellinae, Tridentigerinae, Gobiinae, and Kraemeriinae and included the Pholidichthyidae (currently placed in suborder Pholidichthyoidei, Springer and Johnson 2004) within the Gobiinae.

Table 1.1.2 Summary of Gobioidei classification schemes. Taxonomic levels indicated by the following prefixes in parentheses: superfamily (-oidea), family (-idae), subfamily (-inae). Modified from Pezold (1993).

Miller (1973)	*Hoese (1984)*	*Pezold (1993)*	*Nelson (2006)*	*Thacker (2009)*
Rhyacichthyidae	Rhyacichthyidae	Rhyacichthyidae	Rhyacichthyidae	Rhyacichthyidae
Gobiidae	Eleotridae	Rhyacichthyidae	Odontobutidae	Odontobutidae
Eleotrinae	Kraemeriidae	Gobioidea	Eleotridae	Butidae
Pirskeninae	Microdesmidae	Eleotridae	Butinae	Eleotridae
Xenisthminae	Microdesminae	Kraemeriidae	Eleotrinae	Eleotrinae
Gobionellinae	Ptereleotrinae	Microdesmidae	Xenisthmidae	Xenisthmidae
Tridentigerinae	Xenisthmidae	Xenisthmidae	Kraemeriidae	Gobionellinae
Gobiinae	Gobiidae	Gobiidae	Gobiidae	Gobionellinae
Kraemeriinae	Amblyopinae	Amblyopinae	Oxudercinae	Oxudercinae
	Gobiidae	Gobiinae	Amblyopinae	Amblyopinae
	Oxudercinae	Gobionellinae	Sicydiinae	Sicydiinae
	Sicydiinae	Oxudercinae	Gobionellinae	Gobiidae*
		Sicydiinae	Gobiinae	Gobiinae
			Microdesmidae	
			Ptereleotridae	
			Schindleriidae	

* Includes the families Microdesmidae, Ptereleotridae, Kraemriidae, and Schindleriidae but not formally designated as subfamilies

Birdsong (1975) criticized Miller's (1973) use of pleisiomorphic characters to define gobioid groups but, noting a scarcity of osteological studies among described gobioid taxa, did not suggest an alternative classification. Birdsong (1975) proposed a list of 26 osteological characters that "easily defined" the gobioids.

Springer (1983), and later Winterbottom (1993), used 23-25 characters suggested by Regan (1911), Gosline (1955), Miller (1973) and Birdsong (1975) as diagnostic for gobioids and compared them to an array of potential percomorph outgroup taxa. Springer (1983) identified at least four synapomorphies that unite the gobioids: parietals absent, pelvic intercleithral cartilage present, dorsal end of interhyal fails to meet the dorsal end of the symplectic, and basibranchial 1 cartilaginous. Springer (1983) noted that none of these putative synapomorphies are present in the Pholidichthyidae so he excluded this family from the Gobioidei. Miller (1984, 1992b) added the presence of a sperm duct gland as another gobioid synapomorphy, although the utility of this variable trait as a synapomorphy for the Gobioidei has been questioned (Mazzoldi *et al.*, 2005).

Springer (1983) also reviewed Miller's (1973) use of pleisiomorphic characters to diagnose Rhyacichthyidae from all other gobioids (Miller's Gobiidae) and proposed that the absence of coronomeckelian bones and the lack of a connection between the interhyal and the hyomandibula may be skeletal synapomorphies for Rhyacichthyidae. In addition, the rhyacichthyids appear to have a diagnostic dorsal-fin pterygiophore pattern (see Birdsong *et al.*, 1988).

Monophyly of the remaining gobioids is supported by three synapomorphies according to Springer (1983): (1) lateral line canal not extending onto the body, (2) ventral process of the hyomandibula broad with the dorsal tips of the interhyal and symplectic widely separated from each other; and (3) mandibular sensory canal absent. Hoese and Gill (1993) identified two additional synapomorphies to unite the non-rhyacichthyid gobioids. Firstly, the penultimate branchiostegal ray articulates with the anterior ceratohyal in all gobioids except for the rhyacichthyids. In the Rhyacichthyidae, the penultimate branchiostegal ray articulates with the posterior ceratohyal. Secondly, the rhyacichthyids have interconnected preopercular and oculoscapular canals. In all other gobioids, the preopercular canal is separate from the oculoscapular canal. Another potential synapomorphy to separate the Rhyacichthyidae from other gobioids is the caudal lateralis system composed entirely of free neuromasts in all gobioids except for Rhyacichthyidae (Ahnelt and Göschl, 2004).

Together, these putative synapomorphies support a basal sister relationship of the Rhyacichthyidae to a monophyletic group containing all other gobioids. Molecular data corroborate these morphological data (Thacker and Hardman, 2005). *Terateleotris* and *Protogobius* have affinities for both the Rhyacichthyidae and the remaining gobioids (i.e. both genera share a reduction in infraorbital and mandibular sensory-canals—a character also shared with most other gobioids except *Rhyacichthys*, however both *Terateleotris* and *Protogobius* possess a complete lateral line system as well as free neuromasts—a combination shared only with *Rhyacichthys* (Shibukawa *et al.*, 2001). Currently the placement of these two genera among the Rhyacichthyidae and other basal gobioids remains unresolved (Shibukawa *et al.*, 2001; Thacker and Hardman, 2005).

Subsequent morphological work has focused on the relationships within the non-rhyacichthyid gobioids. Hoese and Gill (1993) analyzed 16 morphological characters to derive a gobioid classification consisting of three families: Rhyacichthyidae, Odontobutidae, and Gobiidae, determined by the synapomorphies discussed above. The Gobiidae were separated from the Odontobutidae by four synapomorphies: (1) no autogenous middle radial in the first pterygiophore of the second dorsal fin; (2) upper proximal radial of the pectoral fin usually in contact with the cleithrum and extending well above the scapula; (3) anterior elongation of the procurrent caudal cartilage; and (4) scales without transforming ctenii. Springer and Johnson (2004), in examining dorsal gill-arch musculature, proposed that the passage of the internal levator-2 through the obliquus dorsalis may be a synapomorphy defining the Odontibutidae. Hoese and Gill (1993) provisionally divided the Gobiidae into three subfamilies (Butinae, Eleotridinae, and Gobiinae), but their character survey was not sufficient to resolve relationships among these three groups, nor did they include in their study representatives of Xenisthmidae, Microdesmidae, and Kraemeriidae.

Subsequently, others have provided a partial resolution of groups within the Eleotridae (Hoese and Gill, 1993), or given evidence for monophyly within the Gobiidae: Sicydiinae (Hoese, 1984; Harrison, 1989, 1993; Parenti and Maciolek, 1993; Parenti and Thomas, 1998); Oxudercinae (Murdy, 1989); Amblyopinae (Murdy and Shibukawa, 2001, 2002) and Gobiinae (Pezold, 1993).

Pezold (1993) studied the oculoscapular canals and associated pore patterns of 129 gobioid genera that represented all of the gobioid families recognized by Hoese (1984), with the exception of Rhyacichthyidae, (Fig. 1.1.5, Table 1.1.2). Pezold (1993) concluded that a monophyletic Gobiinae could be diagnosed by the presence of a single anterior interorbital pore, a fused interorbital canal, a single posterior pair of nasal pores, and one epural. Additional characters diagnostic for most species of Gobiinae include a 3-22110 first dorsal-fin pterygiophore formula, 26 or 27 vertebrae, and two prehemal pterygiophores. As a result of his study, Pezold (1993) recommended dividing Hoese's (1984) Gobiinae into a monophyletic Gobiinae and a smaller Gobionellinae, for which monophyly could not be established (Table 1.1.2).

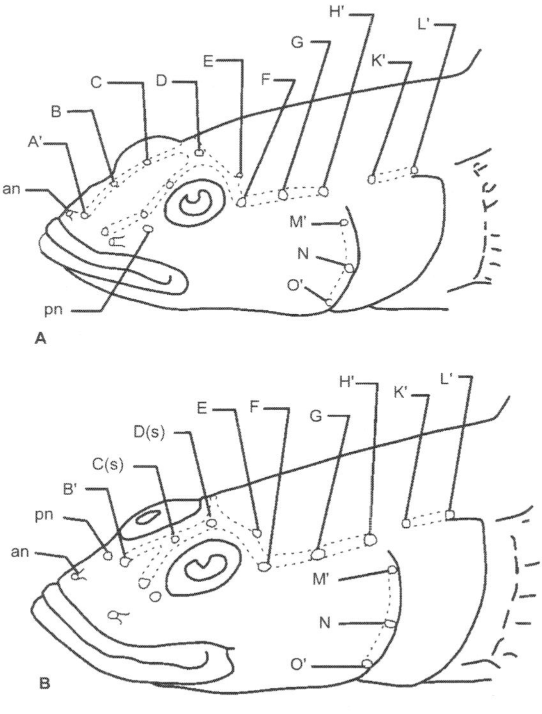

Fig. 1.1.5 Gobioid oculoscapular canals and associated pore patterns. (A) typical arrangement of Gobionellinae (sensu Pezold, 1993); (B) typical arrangement of Gobiinae (sensu Pezold, 1993). Upper case letters indicate sensory pores. Terminal pores indicated by ('). an = anterior nare. pn = posterior nare.

Genetic analyses have led to the rapid expansion of our understanding of the hypothesized relationships within the Gobioidei, as well as to other Acanthomorpha (see Chapter 1.2). The time-consuming techniques required to generate relatively few morphological characters for systematic comparison are quickly being overcome by molecular genetic techniques, which can more rapidly generate many phylogenetically informative characters for more taxa, and molecular phylogenies will continue to allow independent assessment of the possible gain and loss of morphological characters in and across the evolutionary history of the Gobioidei (Thacker, 2009). The significance of those "characters" (changing arrangements of four possible bases in a piece of sequence) however and their evolutionary context is not always clear. Despite the use of application of cladistic philosophies and approaches in some of the methods used, the phylogenies generated are by nature derived probabilistically. This statistical approach can have tremendous heuristic value in exploring evolutionary relationships, but should not be the sole foundation upon which relationships are understood and accepted. The continued identification and recognition of shared derived evolutionary novelties, whether morphological or molecular (not a bit of sequence but an actual gene or gene complex), is essential for the recovery of phylogenetic history.

References

Akihito, Prince. 1963. On the scapula of gobioid fishes. *Japanese Journal of Ichthyology* 11: 1-26.

Akihito, Prince. 1967. On four species of gobiid fishes of the genus *Eleotris* found in Japan. *Japanese Journal of Ichthyology* 14: 135-166.

Akihito, Prince. 1969. A systematic examination of the gobiid fishes based on the mesopterygoid, postcleithra, branchiostegals, pelvic fins, scapula and suborbital. *Japanese Journal of Ichthyology* 16: 93-114.

Akihito, A. Iwata, T. Kobayashi, K. Ikeo, T. Imanishi, H. Ono, Y. Umehara, C. Hamamatsu, K. Sugiyma, Y. Ikeda, K. Sakamoto, A. Fumihito, S. Ohno and T. Gojobori. 2000. Evolutionary aspects of gobioid fishes based upon a phylogenetic analysis of mitochondrial cytochrome *b* genes. *Gene* 259: 5-15.

Ahnelt, H. and J. Göschl. 2004. The pattern of lateral-line system on the caudal fin of *Perccottus glenii* Dybowski, 1877 (Teleostei: Odontobutidae), with comments on the arrangement of the lateral-line system on the caudal fin of Gobioidei. *Proceedings of the California Academy of Science* 55: 358-372.

Berg, L.S. 1940. *Classification of Fishes—Both Recent and Fossil.* Reprinted 1965. Applied Scientific Research Corporation of Thailand, Bangkok.

Birdsong, R.S. 1975. The osteology of *Microgobius signatus* Poey (Pisces: Gobiidae), with comments on other gobiid fishes. *Bulletin of the Florida State Museum, Biological Sciences* 19: 135-187.

Birdsong, R.S., E.O. Murdy and F.L. Pezold. 1988. A study of the vertebral column and median fin osteology in gobioid fishes with comments on gobioid relationships. *Bulletin of Marine Science* 42: 174-214.

Bleeker, P. 1874. Esquisse d'un système naturel des Gobioides. *Archives Néerlandaises des Sciences Exactes et Naturelles* 9: 289-331.

Bloch, M.E. and J.G. Schneider. 1801. *Systema Ichthyologia*. Berlin, Germany.

Day, F. 1878. *The Fishes of India; Being a Natural History of the Fishes Known to Inhabit the Seas and Fresh Waters of India, Burma, and Ceylon,* Vols. 1 and 2. Reprinted 1967. Today and Tomorrow's Book Agency, New Delhi.

Duncker, G. 1928. Teleostei Physoclisti 9. Gobiiformes In: *Die Tierwelt des Nord-und Ostsee*, Tiel 12, G. Grimpe and E. Wagler (eds.). Akademische Verlagsgesellschaft, Leipzig, pp. 121-148.

Eschmeyer, W.N. 1990. *Catalog of the Genera of Recent Fishes*. California Academy of Sciences, San Francisco.

Eschmeyer, W.N. (ed.) Catalog of Fishes electronic version (25 October 2010). http://research.calacademy.org/ichthyology/catalog/fishcatmain.asp

Gill, H.S., J.S. Bradley and P.J. Miller. 1992. Validation of the use of cephalic lateral-line papillae patterns for postulating relationships among gobioid genera. *Zoological Journal of the Linnean Society* 106: 97-114.

Ginsburg, I. 1933. Description of new and imperfectly known species and genera of gobioid and pleuronectid fishes in the United States National Museum. *Proceedings of the United States National Museum* 82: 1-23.

Gmelin, J.F. 1789. *Caroli a Linné...systema naturae per regna tria naturae, secundum classes, ordines, genera, species; cum characteribus, differentiis, synonymis, locis,* Vol. 1, Pt .3. Impensis Georg. Emanuel. Beer., Leipzig.

Gosline, W.A. 1955. The osteology and relationships of certain gobioid fishes, with particular reference to the genera *Kraemeria* and *Microdesmus*. *Pacific Science* 9: 158-170.

Günther, A. 1861. *Catalogue of the fishes in the British Museum. Catalogue of the acanthopterygian fishes in the collection of the British Museum.* 3. Gobiidae, Discoboli, Pediculati, Blenniidae, Labyrinthici, Mugilidae, Notacanthi. British Museum, London.

Harrison, I.J. 1989. Specialization of the gobioid palatopterygoquadrate complex and its relevance to gobioid systematics. *Journal of Natural History* 23: 325-353.

Harrison, I.J. 1993. The West African sicydiine fishes, with notes on the genus *Lentipes* (Teleostei: Gobiidae). *Ichthyological Exploration of Freshwaters* 4: 201-232.

Herre, A.W.C.T. 1927. *Gobies of the Philippines and the China Sea*. Bureau of Science (Manila), Monograph 23.

Herre, A.W.C.T. 1928. True fresh-water fishes of the Philippines, In: *Distribution of Life in the Philippines*, R.E. Dickerson, E.D. Merrill, R.C. McGregor, W. Schultze, E.H. Taylor and A.W. Herre (eds.). Bureau of Science (Manila), Monograph 21, pp. 242-247.

Hoese, D.F. 1971. *A revision of the eastern Pacific species of the gobiid fish genus* Gobiosoma, *with a discussion of relationships of the genus*. University of California, San Diego, Ph.D. Dissertation.

Hoese, D.F. 1983. Sensory papilla patterns of the cheek lateralis system in the gobiid fishes *Acentrogobius* and *Glossogobius*, and their significance for the classification of gobioid fishes. *Records of the Australian Museum* 35: 223-229.

Hoese, D.F. 1984. Gobioidei: relationships, In: *Ontogeny and systematics of fishes,* H.G. Moser, W.J. Richards, D.M. Cohen, M.P. Fahay, A.W. Kendall, Jr. and S.L. Richardson (eds.). Special Publication No. 1, American Society of Ichthyology and Herpetology, Allen Press, Lawrence Kansas. pp. 588-591.

Hoese, D.F. and A.C. Gill. 1993. Phylogenetic relationships of eleotrid fishes (Perciformes: Gobioidei). *Bulletin of Marine Science* 52: 415-440.

Iljin, B.S. 1930. Le systeme des Gobiides. *Trabajos del Institute Español de Oceanographia* 2: 1-63.

Johnson, G.D. and E.B. Brothers. 1993. *Schindleria*: a paedomorphic goby (Teleostei: Gobioidei). *Bulletin of Marine Science* 52: 554-626.

Jordan, D.S. 1923. A classification of fishes. *Stanford University Publications, Biological Sciences* 3: 79-243.

Jordan, D.S. and B.W. Evermann. 1898. The fishes of North and Middle America: a descriptive catalogue of the species of fish-like vertebrates found in the waters of North America, north of the Isthmus of Panama. Part III. *Bulletin of the United States National Museum* 47: 2183a-3136.

Koumans, F.P. 1931. *A Preliminary Revision of the Genera of the Gobioid Fishes with United Ventral Fins.* Proefschrift, Lisse.

Lacepède, B.G.E. 1798. *Histoire Naturelle des Poissons,* Vol. II. Plassan, Paris.

Linnaeus, C. 1758. *Systema Naturae, Regnum Animalie.* Guilielni Engelman, Leipzig.

Mazzoldi, C., C.W. Petersen and M.B. Rasotto. 2005. The influence of mating system on seminal vesicle variability among gobies (Teleostei, Gobiidae). Journal of Zoological Systematics and Evolutionary Research 43: 307-314.

Miller, P.J. 1963. Taxonomy and biology of the genus *Lebetus* (Teleostei-Gobioidea). *Bulletin of the British Museum of Natural History (Zoology)* 10: 207-256.

Miller, P.J. 1973. The osteology and adaptive features of *Rhyacichthys aspro* (Teleostei: Gobioidei) and the classification of gobioid fishes. *Journal of Zoology* 171: 397-434.

Miller, P.J. 1984. The tokology of gobioid fishes In: *Fish Reproduction: Strategies and Tactics,* G.W. Potts and R.J. Wootton (Eds.). Academic Press, London, pp. 119-153.

Miller, P.J. 1992a. A new species of *Didogobius* (Teleostei: Gobiidae) from the Adriatic Sea. *Journal of Natural History* 26: 1413-1419.

Miller, P.J. 1992b. The sperm duct gland: a visceral synapomorphy for the gobioid fishes. *Copeia* 1992: 253-256.

Murdy, E.O. 1989. A taxonomic revision and cladistic analysis of the oxudercine gobies (Gobiidae: Oxudercinae). *Records of the Australian Museum,* Supplement 11: 1-93.

Murdy, E.O. and K. Shibukawa. 2001. A revision of the gobiid fish genus *Odontamblyopus* (Gobiidae: Amblyopinae). *Ichthyological Research* 48: 31-43.

Murdy, E.O. and K. Shibukawa. 2002. Redescription of the gobiid fish genus *Pseudotrypauchen* (Gobiidae: Amblyopinae) and its significance in amblyopine phylogeny. *Marine & Freshwater Research* 53: 253-258.

Nelson, J.S. 2006. *Fishes of the World.* John Wiley and Sons, Hoboken, New Jersey.

Parenti, L.R. and J.A. Maciolek. 1993. New sicydiine gobies from Ponape and Palau, Micronesia, with comments on systematics of the subfamily Sicydiinae (Teleostei: Gobiidae). *Bulletin of Marine Science* 53: 945-972.

Parenti, L.R. and K.R. Thomas. 1998. Pharyngeal jaw morphology and homology in Sicydiine gobies (Teleostei: Gobiidae) and allies. *Journal of Morphology* 237: 257-274.

Pezold, F. 1993. Evidence for the monophyletic Gobiidae. *Copeia* 1993: 634-643.

Pezold, F. 2004. Redescriptions and synonymies of species of the American-West African genus *Gobionellus* (Teleostei, Gobiidae) with a key to species. *Copeia* 2004: 281-297.

Regan, C.T. 1911. The osteology and classification of the gobioid fishes. *Annals and Magazine of Natural History* 8: 729-733.

Sanzo, L. 1911. Distribuzione delle papille cutanee (organi ciatiformi) e suo valore sistematice nei Gobi. *Mitteilungen Zoologische Station zu Neapel* 20: 249-328.

Shibukawa, K., A. Iwata and S. Viravong. 2001. *Terateleotris*, a new gobioid fish genus from the Laos (Teleostei, Perciformes), with comments on its relationships. *Bulletin of the National Science Museum, Tokyo* Ser. A. 27: 229-257

Springer, V.G. 1983. *Tyson belos*, new genus and species of western Pacific fish (Gobiidae: Xenisthminae), with discussions of gobioid osteology and classification. *Smithsonian Contributions to Zoology* 390: 1-40.

Springer, V.G. and G.D. Johnson. 2004. Study of the dorsal gill-arch musculature of teleostome fishes, with special references to the Actinopterygii. *Bulletin of the Biological Society of Washington* 11: 1-260, plus 205 plates.

Takagi, K. 1967. Topologie du systeme sensoriel cephalique des Gobioidei. *La Mer (Bulletin de la Societe franco-japonaise d'Oceanographie)* 5: 37-51.

Takagi, K. 1988. Cephalic sensory canal system of the gobioid fishes of Japan: comparative morphology with special reference to phylogenetic significance. *Journal of Tokyo University of Fisheries* 75: 499-568.

Thacker, C.E. 2000. Phylogeny of the wormfishes (Teleostei: Gobioidei: Microdesmidae). *Copeia* 2000: 940-957.

Thacker, C.E. 2009. Phylogeny of gobioidei and placement within Acanthomorpha, with a new classification and investigation of diversification and character evolution. *Copeia* 2009: 93-104.

Thacker, C.E. and M.A. Hardman. 2005. Molecular phylogeny of basal gobioid fishes: Rhyacichthyidae, Odontobutidae, Xenisthmidae, Eleotridae (Teleostei: Perciformes: Gobioidei). *Molecular Phylogenetics and Evolution* 37: 858-871.

Van Tassell, J.L. 1988. A new species of *Didogobius* (Perciformes: Gobiidae) from the Canary Islands. *American Museum Novitates* 2906: 1-8.

Watson, R.R. and C. Pöllabauer. 1998. A new genus and species of freshwater goby from New Caledonia with a complete lateral line. *Senckenbergiana Biologica* 77: 147-153.

Weber, M. 1913. *Die Fische der Siboga-Expedition.* E.J. Brill, Leiden.

Winterbottom, R. 1993. Search for the Gobioid sister group (Actinopterygii: Percomorpha). *Bulletin of Marine Science* 52: 395-414.

Wongrat, P. 1977. *Systematics, comparative anatomy, and phylogeny of eleotrine gobies (Teleostei: Gobiodei).* University of Bristol, England, Ph.D. dissertation.

Wongrat, P. and P.J. Miller. 1991. The innervation of head neuromast rows in eleotridine gobies (Teleostei: Gobioidei). *Journal of Zoology* 225: 27-42.

CHAPTER

1.2

Molecular Systematics of Gobioid Fishes

Lukas Rüber[1,*] and Ainhoa Agorreta[1,2]

INTRODUCTION

The perciform suborder Gobioidei comprises nine families with about 270 genera, and approximately 2210 species (Nelson, 2006). Gobioids are small fishes normally between 4 to 10 cm in size that inhabit most freshwater, brackish, and marine habitats, and show a spectacular variety in morphology, behaviour, and ecology. Representative gobioid taxa are increasingly used as model organisms to address a wide variety of research topics a few of which are listed here: 1) early vertebrate development and embryology in *Leucopsarion petersii*, the neotenic ice goby "shiro-uo" (Nakatsuji *et al.*, 1997; Harada *et al.*, 2003; Miyake *et al.*, 2006), 2) gene expression studies under different environmental conditions in the eurythermal and euryhaline goby *Gillichthys mirabilis* (Gracey *et al.*, 2001;

Authors' addresses: [1]Department of Zoology, The Natural History Museum, Cromwell Road, SW7 5BD, London, UK.

[2]Department of Zoology and Ecology, University of Navarra, Irunlarrea s/n, 31008 Pamplona, Spain. E-mail: a.agorreta@nhm.ac.uk

[*]*Corresponding author*: E-mail: l.ruber@nhm.ac.uk

Lund *et al.*, 2006; Gracey, 2008; Bucciarelli *et al.*, 2009), 3) ecophysiological, functional morphological, and behavioural adaptations to terrestriality in mudskippers (Ishimatsu *et al.*, 1998; Kok *et al.*, 1998; Graham and Lee, 2004; Long and Gordon, 2004; Sayer, 2005) 4) sex determination and sex change in fishes including bidirectional sex change in *Gobiodon* and *Paragobiodon* (Kuwamura *et al.*, 1994; Nakashima *et al.*, 1996; Munday *et al.*, 1998) , *Lythrypnus* (St. Mary, 1994, 2000; Rodgers *et al.*, 2007), and *Trimma* (Sunobe *et al.*, 2005; Sakurai *et al.*, 2009), 5) parasite-host coevolution in the flatworm genus *Gyrodactylus* and their goby hosts *Pomatoschistus* (Huyse *et al.*, 2003, 2005; Huyse and Volckaert, 2005), 6) invasive species in Western Europe and North America such as round goby *Neogobius melanostomus*, tubenose goby *Proterorhinus marmoratus*, and Chinese sleeper *Perccottus glenii* (Dillon and Stepien, 2001; Stepien *et al.*, 2005; Stepien and Tumeo, 2006; Hegedis *et al.*, 2007; Brown and Stepien, 2009; Kovac *et al.*, 2009), 7) the role of sexual selection in the evolution of parental care in the sand goby *Pomatoschistus minutus* (Järvenpää and Lindström, 2004; Lindström *et al.*, 2006; Mobley *et al.*, 2009; Saaristo *et al.*, 2009), and 8) marine speciation in *Gobiodon* (Munday *et al.*, 2004), *Elacatinus*, (Taylor and Hellberg, 2003, 2005) and Gobiosomatini (Rüber *et al.*, 2003). While gobies belong to one of the most species rich fish groups—roughly 10% of all ray-fined fishes are gobioids—they are at the same time one of the least known. Despite their evolutionary and ecological importance, phylogenetic inter- and intrarelationships based both on morphological and molecular data of gobioids are still poorly understood.

Here, we provide a summary and overview of molecular systematic research of gobioid fishes conducted over the last two decades by reviewing the available data based on submitted nucleotide sequences to GenBank (www.ncbi.nlm.nih.gov), as well as a literature review of relevant publications retrieved from Web of Science and Zoological Record (www.isiknowledge.com).

Molecular Systematics: A General Introduction

Uncovering the genealogy of species remains a major challenge for evolutionary biology. Our understanding of the phylogenetic relationships among extant taxa has increased rapidly since molecular systematics started in the 1980s, when allozyme data became widely used in phylogenetic studies of closely related species. Subsequent advances in molecular biology, in particular the invention of the polymerase chain reaction (PCR; Saiki *et al.*, 1985) and direct sequencing of PCR-amplified DNA using universal primers

(Kocher *et al.*, 1989), had a great impact on the field. Reduced sequencing costs and next generation sequencing technology as well as developments in genomics are likely to further an increase in nucleotide sequence information for systematic studies focusing on phylogeny and species limits (Binladen *et al.*, 2007; Ellegren, 2008; Hauser and Seeb, 2008; Hudson, 2008). Similar methodological progress in phylogenetic reconstruction has been achieved (e.g. Felsenstein, 2004; Delsuc *et al.*, 2005).

DNA sequence data are potentially useful at any taxonomic and temporal scale, however genes should be chosen very carefully for the question at hand. Of the several genetic markers now available, mitochondrial DNA (mtDNA) sequences are still the most widely used for reconstructing the phylogeny of a variety of taxa across different taxonomic ranks; this is because mitochondrial genes are easy to amplify and sequence, and there is a wide range of universal fish primers available. Due to its predominantly maternal mode of inheritance, the effective number of mtDNA genomes is one quarter that of nuclear genes, leading to a more rapid rate of genetic differentiation through random drift (e.g. Moore, 1995). This smaller effective population size also means that mitochondrial genes are less subject to retention of ancient polymorphism. Hence, it is thought that mitochondrial genes generally are better estimators of species trees than nuclear genes (Moore, 1995) with the caveat that mitochondrial species trees may in some cases be confounded by introgressive hybridization. Genetic divergence is also enhanced by the higher rate of sequence evolution in vertebrate mtDNA compared to nuclear coding regions. The relatively fast mutation rate of mitochondrial genes allows resolving the phylogenetic relationships among closely related species as well as genealogical relationship among conspecific populations. This makes them ideal markers for phylogeographic studies. At deeper phylogenetic levels, the fast mutation rate of mitochondrial genes can be a disadvantage. For example, long terminal branches in a phylogeny can be placed erroneously as sisters groups due to the chance accumulation of shared character states, a phenomenon called long branch attraction (Huelsenbeck, 1997). Long branch attraction artefacts, however, are thought to be less severe with dense taxon sampling. It is believed that both dense taxon and nucleotide sampling combined with the utilization of both mitochondrial and nuclear loci will help to provide robust phylogenetic hypotheses. Several nuclear loci are now available for molecular systematic studies in fishes (e.g. Li *et al.*, 2007)

Molecular Systematics: Access to Nucleotide Sequence Information

The first sequence database, the Los Alamos Sequence Database, was established in 1979 at the Los Alamos National Laboratories and culminated in 1982 with the creation of the public GenBank initiative. From 1989 to 1992, the GenBank initiative transitioned to the newly created National Center for Biotechnology Information (NCBI) and in the mid-1990s it became part of the International Nucleotide Sequence Database Collaboration (INSD). The INSD comprises the DNA Data Bank of Japan (DDBJ), the European Molecular Biology Laboratory (EMBL), and GenBank. Together, they hold an annotated collection of all publicly available DNA sequence data. Each database has its own set of submission and retrieval tools, but the three databases exchange data daily so that they all contain the same set of sequences. Since 1982 the number of base pairs (bp) and sequences submitted has grown exponentially. For example, in 1982 GenBank contained 680,338 bp and 606 sequences, in 2008 these numbers were roughly 9.9×10^{10} and 9.8×10^{7} (information retrieved from NCBI and Wikipedia). The rapid accumulation of sequence data in GenBank (Fig. 1.2.1) allows combining data for the same sets of genes from different studies.

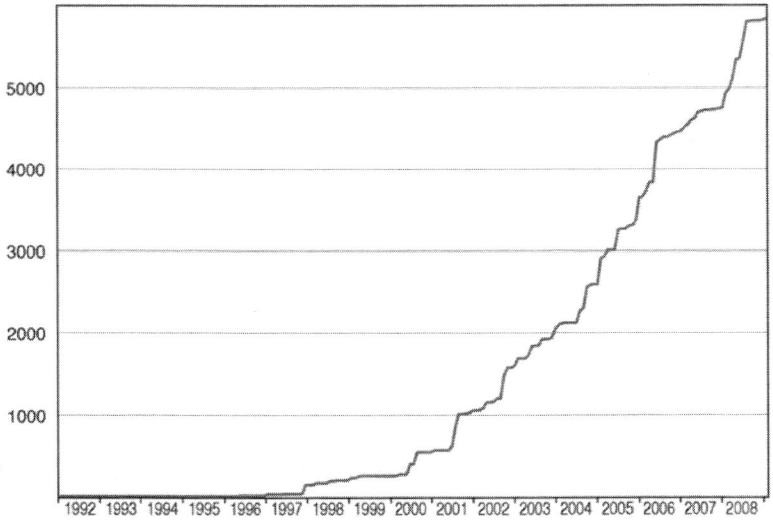

Fig. 1.2.1 Cumulative number of gobioid nucleotide sequences submitted to GenBank between December 1991 and January 2009.

GOBIOID MOLECULAR SYSTEMATICS: NUCLEOTIDE SEQUENCES REVIEW

For this review, information from GenBank was retrieved on 18 January, 2009. Up to that point a total of 5779 gobioid nucleotide sequences had been submitted (see Tables 1.2.1, 1.2.2). The first submission of a gobioid nucleotide sequence was a 1410 bp long fragment of the rhodopsin gene from the sand goby *Pomatoschistus minutus* in 1991 (Archer *et al.*, 1992; accession number X62405), followed by a 1440 bp fragment of DRP2, a novel human dystrophin homologue from an unidentified species of *Gobius* (Roberts *et al.*, 1996; accession number GSU43518). The third nucleotide submission had a systematic component and involved six accessions of partial cytochrome b (*cob*) sequences (365 bp) of *Proterorhinus marmoratus* and *Neogobius melanostoma* (Dougherty *et al.*, 1996; accessions NMU53673-NMU53677 and PMU53678).

The 5799 accession numbers corresponded to 178 individual submissions ranging from 281 accessions per submission in one case to just one accession in 21 occasions. For a total of 1190 accession numbers, information regarding their publication in a scientific journal was missing. Some of these might not have been published at the time of access, but a 'Title' search in ISI Web of Knowledge revealed that several were indeed published without showing the corresponding reference details in the GenBank entry. Although GenBank now routinely checks several major journals for accession numbers upon publication and then updates the corresponding 'Journal' field in the submission, authors should ensure that their GenBank entries are updated once the sequences are published or otherwise risk that their research is not readily accessible through GenBank. Most journals now require that sequences be deposited in GenBank prior to publication but it is still possible that a number of gobioid nucleotide sequences have been published without being deposited in GenBank.

Remarkably, the total number of nucleotide sequences for cyprinids and cichlids is 30 and 8 times higher, respectively, than those submitted for gobioids (see Table 1.2.1). The top five gobioid species represented in the nucleotide sequence data bank are (number of entries in brackets): *Gobiomorphus breviceps* (485), *Ctenogobiops feroculus* (318), *Gillichthys mirabilis* (227), *Gnatholepis scapulostigma* (192), and *Gnatholepis anjerensis* (158).

Table 1.2.1 GenBank entries for three different taxonomic groups (accessed January 18, 2009).

Taxon	*Nucleotide entries*	*Mitochondrial genome sequences*[(a)]	*EST*
Cyprinidae[(b)]	172,802	87	1,698,535
Cichlidae[(c)]	511,490	11	64,682
Gobioidei[(d)]	5,779	5	7,407

[(a)]included under nucleotide entries. Number of genera and species according to Nelson (2006): [(b)]220 genera and 2420 species, [(c)]390 genera and 1350 species (species number is certainly an underestimate since the massive species diversity found in the adaptive radiations of cichlids in the East African Great Lakes is not taken into account), [(d)]270 genera and 2210 species; the family Gobiidae alone comprises 210 genera and 1950 species.

Table 1.2.2 GenBank entries for gobioids by taxonomic group (accessed January 18, 2009).

Taxon	*Nucleotide entries*
Gobioidei	**5,799**
Eleotridae	**1,337**
Butinae	164
Eleotrinae	1,173
Gobiidae	**4,169**
Amblyopinae	4
Gobiinae	1,997
Gobionellinae	1,869
Oxudercinae	107
Sicydiinae	191
unclassified	1
Kraemeriidae	**7**
Microdesmidae	**17**
Odontobutidae	**62**
Ptereleotridae	**26**
Rhyacichthyidae	**4**
Schindleriidae	**173**
Xenisthmidae	**4**

Mitochondrial DNA: General

Almost 88% of the submitted nucleotide sequences are mitochondrial (i.e. 5091; Table 1.2.3). The three most common mitochondrial gene entries are *cob* (2212), control region (1098), and 12S and/or 16S rRNA (653).

Table 1.2.3 GenBank entries for gobioids by genome and genetic marker (accessed January 18, 2009).

Genome and genetic marker	*Nucleotide entries*
Mt DNA	**5,091**
Cytochrome *b* (*cob*)	2,212
Control region (D-loop)	1,098
12S rRNA and/or 16S rRNA	653
NADH dehydrogenase subunit 2 (*nad2*)	537
NADH dehydrogenase subunits 1 + 5 + 6 (*nad1, nad5, nad6*)	330
Cytochrome c oxidase subunit 1 (*cox1*)	192
ATP synthase F0 subunits 6 and 8 (*atp6, atp8*)	64
Complete mt genome	5
Nuclear DNA	**708**
Microsatellites	217
Recombination activating gene 1 (rag1)	207
Rhodopsin	74
Other	210
Total	**5,799**

Mitochondrial DNA: MT Genomes

Five complete mitochondrial genomes have been submitted to GenBank: *Acanthogobius hasta*, *Eleotris acanthopoma*, *Gymnogobius petschiliensis*, *Odontobutis platycephala*, and *Rhyacichthys aspro*.

Mitochondrial DNA: DNA Barcodes

DNA barcodes are short nucleotide sequences used for species identification. A 648 bp region of the cytochrome c oxidase subunit 1 gene (*cox1*) is now commonly used as universal DNA barcode although other mitochondrial gene regions might be better suited (Vences *et al.*, 2005). DNA barcoding has clearly polarized the systematic community but it is beyond the scope of our review to provide a detailed treatment of the discussion. While the primary aim and scope of DNA barcodes is to provide a DNA based species identification platform with wider applications such as in food identification, endangered species trade, or biodiversity inventories in hyperdiverse regions, evolutionary biologists increasingly use the short 650 bp long DNA barcode sequences to propose phylogenetic hypotheses even though the statistical power to reject alternative phylogenetic hypotheses might be weak if based on the very short *cox1* sequences alone.

Among the 5799 gobioid nucleotide sequence entries in GenBank only 24 have a positive hit with the search term "Barcode": 11 from Hubert *et al.* (2008), two from Ward and Holmes (2007), two from Victor (2007), one from James *et al.* (unpublished; accession number DQ521026 "DNA barcoding the fish species of the Hawaiian Islands"), and eight from Zhang (unpublished; EF607301- EF607306, EF607382, and EF607394 "DNA barcoding of fishes from South China Sea"). The two entries by Victor (2007) were not retrieved by searching for the string "/db_xref="BOLD:" that indicates an official Barcode of Life database entry. Surprisingly no gobioids were included in a recent barcoding project on Australian fish (Ward *et al.*, 2005) nor in a recent project on fish larvae identification with DNA barcodes in the southern Great Barrier Reef (Pegg *et al.*, 2006). The low number of submitted gobioid DNA barcodes is likely to increase shortly based on the progress for gobioid fishes reported by the Fish Barcode of Life Initiative (Fish-BOL; Table 1.2.4).

Table 1.2.4 Barcoding progress for selected fish clades and gobioid families. Source: http://www.fishbol.org/ (accessed January 18, 2009).

Taxon	*Number of species*	*Barcoded*[(a)]	*% Barcoded*
Actinopterygii	27,984	5,470	20
Perciformes	9,958	2,824	28
Eleotridae	162	34	21
Gobiidae	1,428	259	18
Kraemeriidae	9	1	11
Microdesmidae	26	7	27
Odontobutidae	20	2	10
Ptereleotridae	45	9	20
Rhyacichthyidae	3	2	67
Schindleriidae	3	2	67
Xenisthmidae	9	0	0

[(a)] 13 Barcode of Life (BoL) gobioid barcodes published until 18 January, 2009.

NUCLEAR DNA

Thus far only 708 gobioid nuclear sequences have been deposited in GenBank with microsatellites (217; see below), RAG1 (207), and rhodopsin (74) leading the list of entries (Table 1.2.3).

Nuclear DNA: Microsatellites

Microsatellites, or Simple Sequence Repeats (SSRs), are polymorphic loci that consist of short repeating units of 2-6 bp in length. Microsatellites show an exceptionally high rate of mutations due to the addition or subtraction of a small number of perfect repeats, most likely because of polymerase slippage during replication (Schlötterer and Tautz, 1992). They are typically neutral, co-dominant and are used as molecular markers which have wide-ranging applications in the field of genetics, including kinship and population studies (Jarne, 1996). Nucleotide sequences of the microsatellites themselves used to characterize are normally deposited in GenBank whereas data from applied studies that analyze allele size information of different individuals for several different microsatellite loci are not found in GenBank. Microsatellites have been isolated and characterized for the following gobioid genera: *Coryphopterus* (Hepburn *et al.*, 2006), *Ctenogobiops* (Thompson *et al.*, 2006), *Eucyclogobius* (Mendonca *et al.*, 2001), *Gnatholepis* (Thacker *et al.*, 2007), *Gobiusculus* (Mobley *et al.*, 2009), *Leucopsarison* (Kokita and Takahashi, 2008), neogobines (e.g. *Neogobius*, *Proterorhinus*; Dufour *et al.*, 2007; Vyskočilová *et al.*, 2007; Feldheim *et al.*, 2009), *Pomatoschistus* (Larmuseau *et al.*, 2008; Mobley *et al.*, 2009), *Pterogobius* (Nohara *et al.*, 2009), *Rhinogobius* (Ohara *et al.*, 2004), *Sicyopterus* (Hoareau *et al.*, 2009), and *Zosterisessor* (Gallini *et al.*, 2004).

Nuclear DNA: Expressed Sequence Tag (EST)

In addition to the sequences found in the nucleotide sequence database there are an additional 7407 gobioid expressed sequence tag (EST) sequences deposited in dbEST, a division of GenBank (Table 1.2.1). An EST is a short (300-500 bp) sequence read from mRNA or cDNA. ESTs play a vital role in identifying gene transcripts, gene discovery and gene expression studies (Nagaraj *et al.*, 2007). The identification of ESTs has proceeded rapidly, with approximately 62 million ESTs now available (accessed June 2009). The gobioid EST sequence entries are: *Gillichthys mirabilis* (5592), *Zosterisessor ophiocephalus* (1749), *Gillichthys seta* (62), *Periophthalmus modestus* (4) which reflect the importance of these species as model organisms in physiology and genomics (e.g. Gracey *et al.*, 2001; Sakamoto *et al.*, 2002).

GOBIOID MOLECULAR SYSTEMATICS: LITERATURE REVIEW

In order to review the existing literature we conducted a literature search in both Web of Science and Zoological Record for the following two search string in the "Topic" fields: "Gobioidei OR Goby OR Gobies OR Eleotridae OR Butinae OR Eleotrinae OR Gobiidae OR Amblyopinae OR Gobiinae OR Gobionellinae OR Oxudercinae OR Sicydiinae OR Kraemeriidae OR Microdesmidae OR Odontobutidae OR Ptereleotridae OR Rhyacichthyidae OR Schindleriidae OR Xenisthmidae" AND "systematic OR phylogeny OR phylogeography OR evolution OR relationship OR gene OR DNA OR sequence OR gene flow OR microsatellite OR population genetic OR population structure OR allozyme OR isozyme".

The search conducted on 29 April, 2009 retrieved 511 references in the Web of Science and 193 references in Zoological Record. A total of 69 duplicates were identified reducing the total number of references from 704 to 635. After manually filtering the remaining references a total of 132 were kept that can be broadly classified as dealing with molecular systematics of gobioids. Note, that articles that deal with broader systematic issues and that included some gobioid taxa could not be detected with the search criterion used. Table 1.2.5 summarizes which molecular markers are most commonly used in molecular systematic studies of gobioids with allozymes/isozymes (38 studies), mitochondrial *cob* gene (32 studies), and microsatellites (23 studies) ranked in the top three.

Table 1.2.5 List of the most commonly used molecular markers in gobioid systematics based on a review of 132 published studies (see text for more details). Note, not necessarily all nucleotide sequence data listed here were submitted to GenBank.

Genome	*Data type*	*Locus*	*Number of studies*[(a,b)]
Nuclear	Allozymes or isozymes		38
Mitochondrial	Nucleotide sequence data	*cob*	32
Nuclear	Microsatellites		23
Mitochondrial	Nucleotide sequence data	Control region	17
Mitochondrial	Nucleotide sequence data	12S rRNA	12
Mitochondrial	Nucleotide sequence data	16S rRNA	12
Mitochondrial	Nucleotide sequence data	*nad2*	6
Mitochondrial	Nucleotide sequence data	*cox1*	4
Nuclear	RFLP		4
Mitochondrial	Nucleotide sequence data	*nad1*	3
Mitochondrial	Nucleotide sequence data	*nad5*	2
Mitochondrial	Nucleotide sequence data	*atp6*	2

Contd...

Contd...

Nuclear	Nucleotide sequence data	*rag1*	2
Nuclear	Nucleotide sequence data	ITS	2
Mitochondrial	Nucleotide sequence data	*nad3, nad4L, nad4*	1
Nuclear	AFLP		1
Mitochondrial	Nucleotide sequence data	*atp8*	1
Nuclear	Intron length polymorphism		1
Nuclear	Nucleotide sequence data	Creatin kinase	1
Mitochondrial	Nucleotide sequence data	Mitochondrial genome	1

[a]Note that in several studies more than one molecular marker was employed; [b]for eight publications it was not possible to identify what nucleotide sequences were used based on the abstract alone.

References were classified into three categories (number of references in brackets): population genetics (57), phylogeography (32), and phylogenetics (43). Topics treated in these references are concerned with, for example, cryptic species and molecular identification (Kon *et al.*, 2007; Larmuseau *et al.*, 2008; Monteiro *et al.*, 2008; Neilson and Stepien, 2009a), invasive species (e.g. Stepien and Tumeo, 2006; Brown and Stepien, 2009; Neilson and Stepien, 2009b), hybridization (e.g. Wallis and Beardmore, 1980; Mukai *et al.*, 1997, 2000), and mating system and parentage analysis (e.g. Jones *et al.*, 2001; Ohara *et al.*, 2004; Takahashi and Ohara, 2006; Mobley *et al.*, 2009). The taxonomic scope of most articles is limited to a single genus, a species complex within a genus or a single species. The genera mostly covered in the literature are (number of references in brackets): *Pomatoschistus* (19, some include the closely related genus *Knipowitschia*), *Rhinogobius* (14), *Neogobius* (10, some include the closely related genus *Proterorhinus*), *Gnatholepis*, *Sicyopterus* and *Gobiomorphus* (4 each). Few studies deal with phylogenetic relationships above genus level. Some exceptions include Mediterranean gobies (McKay and Miller, 1991; Miller *et al.*, 1994; Penzo *et al.*, 1998; Sorice and Caputo, 1999; Giovannotti *et al.*, 2007), Gobiosomatini (Rüber *et al.*, 2003; Taylor and Hellberg, 2005; *Elacatinus* and related genera only), *Brachygobius* and related genera (McKay, 1993), and Red sea coral associated gobies, (*Gobiodon* and related genera; Herler *et al.*, 2009).

Gobioid Intrarelationships: An Overview and Reanalysis

Only five studies thus far have addressed gobioid intrarelationships based on molecular data (Akihito *et al.*, 2000; Wang *et al.*, 2001; Thacker, 2003, 2009; Thacker and Hardman, 2005). These studies use different taxon and nucleotide sampling, different outgroups as well as different methods of phylogeny inference (maximum parsimony, MP; neighbour joining, NJ; Bayesian inference, BI). Table 1.2.6 provides a summary of the most important findings of these five studies. To facilitate comparison among these studies we present a re-analyses of Akihito *et al.* (2000), Wang *et al.* (2001), and Thacker (2009) using a maximum likelihood (ML) framework and the gobioid genus *Rhyacichthys* to root the trees (Fig. 1.2.2-4). Again, for the sake of comparisons, we use the classification of Thacker (2000, 2003; see also Nelson, 2006) instead of the most recently proposed one (Thacker, 2009). The nine families and corresponding subfamilies in this classification are: Rhyacichthyidae, Odontobutidae, Eleotridae (Butinae, Eleotrinae), Xenisthmidae, Gobiidae (Oxudercinae, Amblyopinae, Sicydiinae, Gobionellinae, Gobiinae), Kraemeriidae, Microdesmidae, Ptereleotridae, and Schindleriidae.

The first phylogenetic analyses addressing gobioid intrarelationships was conducted by Akihito *et al.* (2000) based on complete *cob* sequences of 28 representative gobioid taxa. It is difficult to review the main results of this study since the authors only present unrooted trees. However, they note that "the phylogenetic trees obtained show that most species examined have diverged from each other almost simultaneously or during an extremely short period of time". In total six clusters were identified: 1) *Rhyacichthys* and *Protogobius* (Rhyacichthyidae), 2) *Calumia, Dormitator, Eleotris, Gobiomorphus, Hypseleotris* (all Eleotridae, Eleotrinae), 3) *Bostrychus, Butis, Ophiocara, Oxyeleotris* (all Eleotridae, Butinae), 4) *Micropercops* (Odontobutidae), *Acanthogobius* (Gobiidae, Gobionellinae), *Periophthalmus* (Gobiidae, Oxudercinae), *Taenioides* (Gobiidae, Amblyopinae), *Tridentiger* (Gobiidae, Gobionellinae), 5) *Kraemeria* (Kraemeriidae), *Ptereleotris* (Ptereleotridae), *Gunnelichthys* (Microdesmidae), and 6) *Mogurnda, Ophieleotris, Tateurndina* (all Eleotridae, Eleotrinae). The phylogenetic placement of two species (*Odontobutis obscura* (Odontobutidae) and *Xenisthmus sp* (Xenisthmidae) differs between the two analyses (ML and NJ) conducted. In the NJ analysis *Odontobutis obscura* was resolved as sister to cluster 6, and *Xenisthmus* as sister to cluster 4 and 5, whereas in the ML analyses the two taxa were found to be sister groups. The result of our reanalysis, shown in Fig. 1.2.2, differs in several aspects from the originally reported ML tree (see Table 1.2.6 and Fig. 1.2.2 for a comparison).

The original phylogenetic analyses of Wang *et al.* (2001) was based on 43 taxa and 1078 bp of 12S rRNA and $tRNA^{Val}$, and were analyzed using MP and NJ. The results of the original analyses are summarized in Table 1.2.6 and the results of our reanalyses are shown in Fig. 1.2.3.

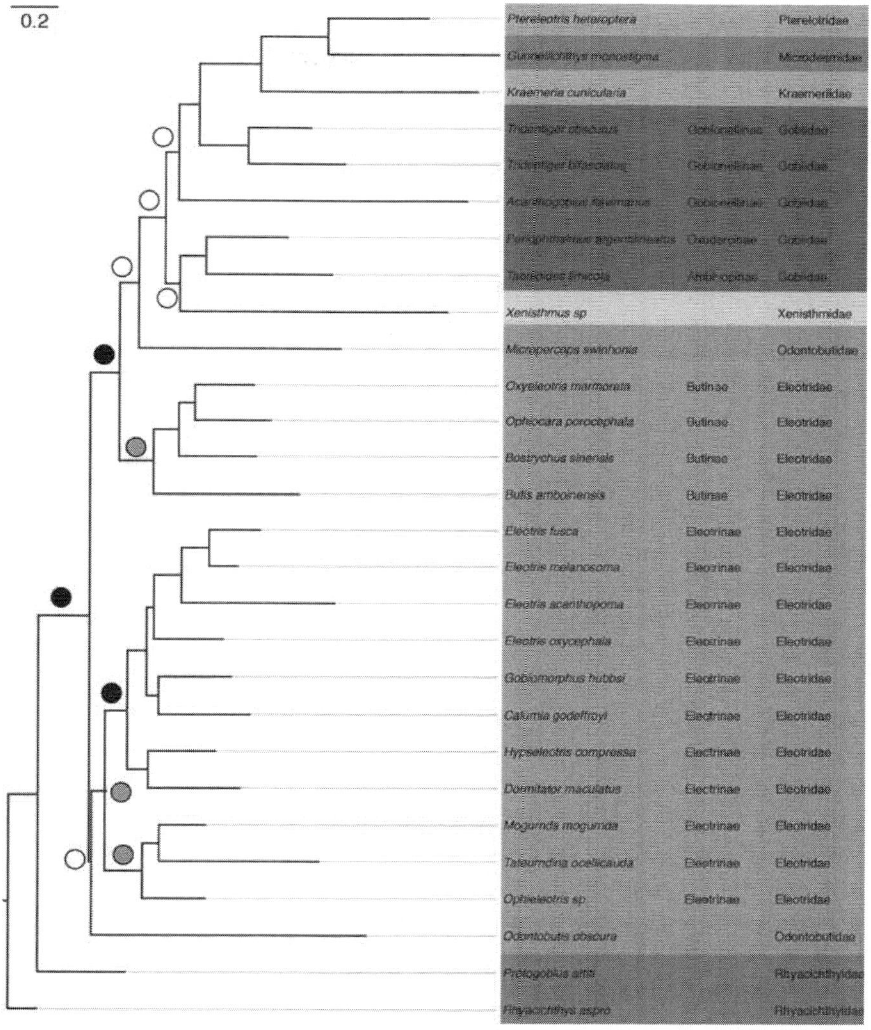

Fig. 1.2.2 Maximum likelihood reanalysis of the Akihito *et al.* (2000) data set using *Rhyacichthys* as outgroup. The analysis was based on a 1140 bp alignment of the complete *cob* gene and analyzed with Garli v0.96 (Zwickl, 2006) using default settings (-lnL = 14990.30) and visualized with FigTree v1.2.1 (Rambaut, 2006). Circles indicate bootstrap support (BS) for selected branches based on 1,000 replicates; white = BS < 50%, grey = 50 ≤ BS < 70, black = BS ≥ 70.

Color image of this figure appears in the color plate section at the end of the book.

Table 1.2.6 Summary of phylogenetic hypotheses regarding gobioid intrarelationships[a] recovered by Akihito *et al.* (2000), Wang *et al.* (2001), Thacker (2003), Thacker and Hardman (2005), and Thacker (2009). The classification follows Thacker (2000, 2003) and Nelson (2006).

Family	*Subfamily*	*Akihito et al. 2000*[b]	*Wang et al. 2001*[k]	*Thacker 2003*[o]	*Thacker and Hardman 2005*[v]	*Thacker 2009*[x]
Rhyacichthyidae		two species included[c]	not included	not included	one species included[w]	one species included[y]
Odontobutidae		polyphyletic[d]	one species included[l]	one species included[p]	monophyletic	monophyletic
Eleotridae		polyphyletic	paraphyletic	paraphyletic	polyphyletic	polyphyletic
Eleotridae	Eleotrinae	polyphyletic	monophyletic	paraphyletic[q]	paraphyletic[q]	paraphyletic[q]
Eleotridae	Butinae	monophyletic[e]	monophyletic	not included	polyphyletic	polyphyletic
Gobiidae		monophyletic	paraphyletic	paraphyletic	monophyletic	paraphyletic
Gobiidae	Gobiinae	not included	paraphyletic[m]	paraphyletic[r] included	one species included	paraphyletic[z]
Gobiidae	Gobionellinae	monophyletic[f]	paraphyletic[n]	paraphyletic[r]	one species included	paraphyletic[aa]
Gobiidae	Amblyopinae	one species included	not included	one species[s] included	not included	one species included[t]
Gobiidae	Oxudercinae	one species included	monophyletic	paraphyletic	not included	paraphyletic
Gobiidae	Sicydiinae	not included	monophyletic	monophyletic	not included	monophyletic
Kraemeriidae		one species included[g]	not included	one species included	not included	one species included
Microdesmidae		one species included[h]	not included	polyphyletic	not included	monophyletic

Contd...

Contd...

Pterelеotridae	one species included[i]	monophyletic	paraphyletic	not included	monophyletic
Schindleriidae	not included	not included	monophyletic	not included	monophyletic
Xenisthmidae	one species included[j]	not included	one species included[u]	one species included[u]	one species included[u]

[a]Note that some of sister-group relationships mentioned are not or only weakly supported by bootstrap analyses. Refer to the original references for information regarding bootstrap support; [b]based on unrooted ML tree presented by the authors which we converted into a rooted tree using *Rhyacichthys* as outgroup; [c]rooted with *Rhyacichthys*, *Protogobius* sister-group to remaining gobioids; [d]*Odontobutis* sister to Xenisthmidae and *Micropercops* sister to Gobiidae; [e]Butinae sister to *Micropercops* and Gobiidae; [f]sister-group to Oxudercinae and Amblyopinae; [g]sister-group to Microdesmidae and Ptereleotridae; [h]sister-group to Ptereleotridae; [i]sister-group to Microdesmidae; [j]sister-group to *Odontobutis*; [k]based on NJ tree presented by the authors; [l]sister-group to remaining gobioids; [m]Ptereleotridae nested within Gobiinae; [n]Sicydiinae and Oxudercinae nested within Gobionellinae; [o]based on MP tree presented by the author; [p]rooted with *Odontobutis*; [q]Xenisthmidae nested within Eleotrinae; [r]Microdesmidae, Ptereleotridae, and Schindleridae nested within Gobiinae; [s]Oxudercinae, Amblyopinae, Sicydiinae, and Kraemeriidae nested within Gobionellinae; [t]nested within Oxudercinae; [u]nested within Eleotrinae; [v]based on MP tree presented by the authors; [w]*Rhyacichtys* sister-group to remaining gobioids; [x]based on BI analysis presented by the author; [y]clade formed by *Rhyacichthys*, *Milyeringa* (Butinae) and Odontobutidae is sister-group to remaining gobioids; [z]Microdesmidae, Ptereleotridae, Schindleridae, and Kraemeriidae nested within Gobiinae; [aa]Sicydiinae, Oxucercinae, and Amblyopinae nested within Gobionellinae

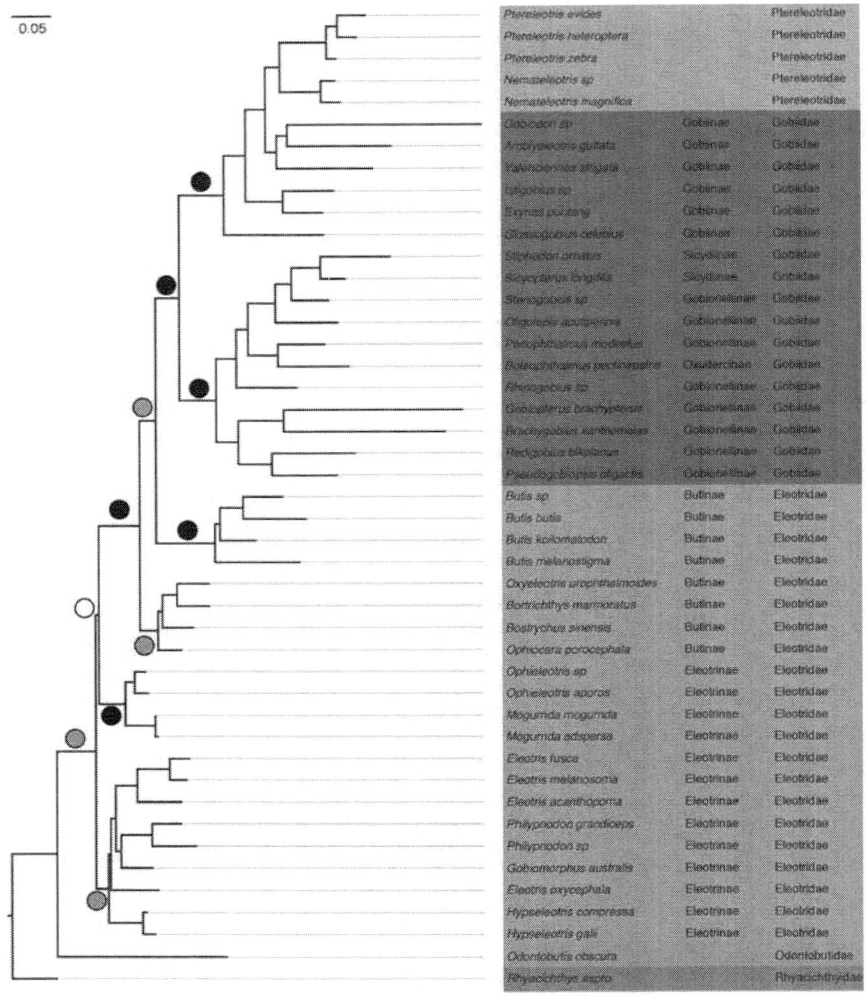

Fig. 1.2.3 Maximum likelihood reanalysis of the Wang *et al.* (2001) data set using *Rhyacichthys* as outgroup. The analysis was based on a 905 bp alignment of 12S rRNA and analyzed with Garli v0.96 (Zwickl, 2006) using default settings (-lnL = 11108.60) and visualized with FigTree v1.2.1 (Rambaut, 2006). Note, the original analyses by Wang *et al.* (2001) was based on 12S rRNA and tRNA^{Val}. However, the tRNA^{Val} data are not available from GenBank. Circles indicate bootstrap support (BS) for selected branches based on 1,000 replicates; white = BS < 50%, grey = 50 ≤ BS < 70, black = BS ≥ 70.

Color image of this figure appears in the color plate section at the end of the book.

While Thacker (2003) focused on derived gobioid lineages (79 specimens, 67 species, 3573 bp), Thacker and Hardman (2005) focused on basal gobioids (78 specimens, 45 species, 4397 bp). Both studies were based on NADH dehydrogenase subunits 1 and 2 (*nad1* and *nad2*), and *cox1* sequences and the latter also included *cob*. Thacker (2009) presented a combined analysis of her previous data including six additional taxa and *cob* sequences for some of the taxa of Thacker (2003; 153 specimens, 107 species, length of alignment not given, but probably the same as in Thacker and Hardman, 2005). The results of the original analyses are summarized in Table 1.2.6, and the result of our re-analysis of Thacker (2009) is shown in Fig. 1.2.4.

Some clear patterns seem to emerge from these phylogenetic studies. For example, Xenisthmidae is nested within Eleotrinae (Thacker, 2003, 2009; Thacker and Hardman, 2005); Oxudercinae and Sicydiinae are nested within Gobionellinae (Akihito *et al.*, 2000; Wang, 2001; Thacker, 2003, 2009) Ptereleotridae and Microdesmidae are nested within Gobiinae (Akihito *et al.*, 2000; Wang, 2001; Thacker, 2003, 2009). Other hypotheses, however, are less stable across different studies. In particular Ptereleotridae is monophyletic in Wang *et al.* (2001) and Thacker (2009), but non-monophyletic in Thacker (2003) and in our reanalysis (Fig. 1.2.4); Microdesmidae is monophyletic (Thacker, 2009) but non-monophyletic in Thacker (2003); Kraemeriidae is sistergroup to Microdesmidae and Ptereleotridae in a clade that includes some Eleotrinae (Akihito *et al.*, 2000), or Kraemeriidae sistergroup to Microdesmidae and Ptereleotridae in a clade with Gobionellinae (reanalysis of Akihito *et al.*, 2000; see Fig. 1.2.2), or Kraemeriidae is nested within Gobionellinae (Thacker, 2003; reanalysis of Thacker, 2009; see Fig. 1.2.4), or finally Kraemeriidae is nested within Gobiinae (Thacker, 2009).

Gobioid Interrelationships

Several molecular studies focusing on phylogenetic relationships within acanthomorphs have included gobioids. However, in our opinion, robust phylogenetic hypotheses regarding gobioid sister-group relationships have not yet been identified due to insufficient taxon and/or nucleotide sampling. In addition, a comparison between the proposed hypotheses is hindered by non-overlapping taxon sampling. The proposed sistergroups of gobioid based on molecular phylogenetic hypotheses are: 1) Dactylopteridae (Miya *et al.*, 2003, 2005), 2) a clade that includes amongst others

Synbranchiformes, Pleuronectiformes, Atherinomorpha, Lophiiformes, Tetraodontiformes (Dettaï and Lecointre, 2005), 3) Gobiesocoidei (Dettaï and Lecointre, 2008), 4) Apogonidae (Li *et al.*, 2009), and 5) a close relationship with Apogonidae, Pempheridae, and Kurtidae, and Leiognathidae (Thacker, 2009; Thacker and Roje, 2009).

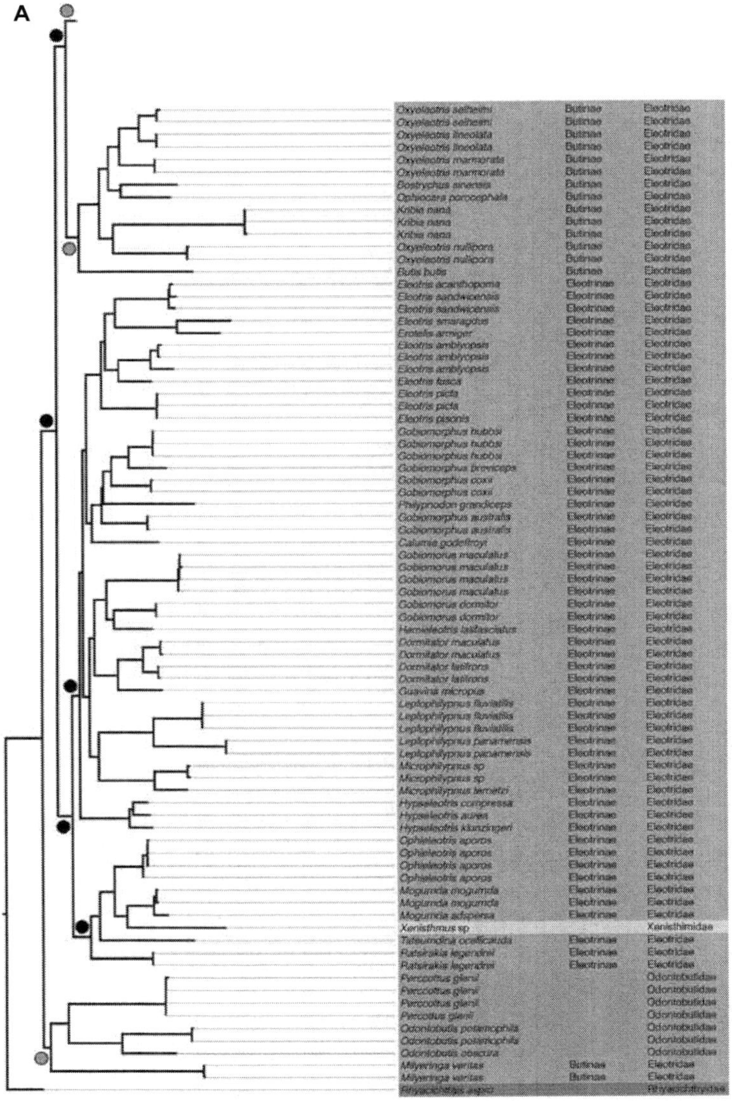

Fig. 1.2.4 contd...

Fig. 1.2.4 contd...

B

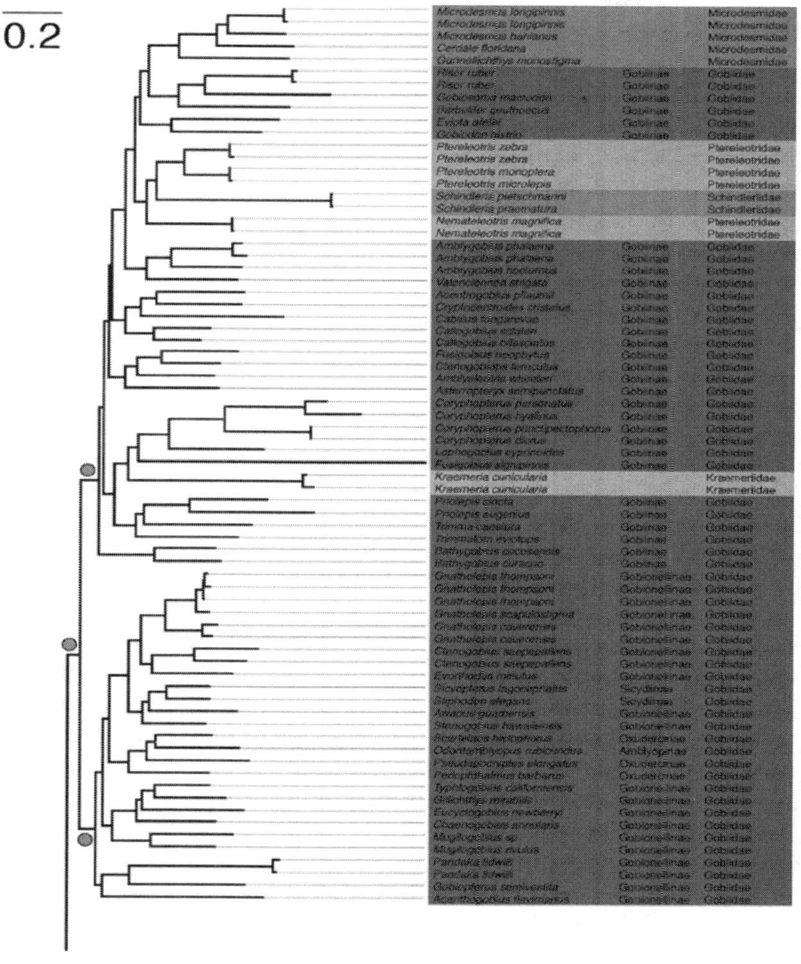

Fig. 1.2.4 Maximum likelihood reanalysis of the Thacker (2009) data set using *Rhyacichthys* as outgroup. The analysis was based on a 4647 bp alignment of concatenated *cox1*, *nad1*, *nad2*, and *cob* sequences and analyzed with Garli v0.96 (Zwickl, 2006) using default settings (-lnL = 196497.43) and visualized with FigTree v1.2.1 (Rambaut, 2006). Circles indicate bootstrap support (BS) for selected branches based on 1,000 replicates; white = BS < 50%, grey = 50 ≤ BS < 70, black = BS ≥ 70.

Color image of this figure appears in the color plate section at the end of the book.

GOBIOID MOLECULAR SYSTEMATICS: AN OUTLOOK

Clearly, the long list of topological discrepancies between different studies is not satisfactory. In order to better understand gobioid inter- and intrarelationships we need to address these issues and rigorously test whether insufficient taxon sampling, choice of molecular marker, method of phylogeny inference or a combination of these factors affected our hypotheses and how robust these hypotheses are against alternative ones. In addition, base compositional biases and/or long branch attraction phenomena are known to obscure the true phylogenetic signal in real world data sets and gobioids might be no exception.

While several studies have used molecular markers to tackle gobioid inter- and intrarelationships as shown in this review, they have just scratched the surface. Current phylogenetic hypotheses might be biased due to taxon sampling artefacts. In addition, the only five studies published thus far addressing gobioid intrarelationships were based on mitochondrial DNA alone. Undoubtedly, for an accurate understanding of gobioid phylogenetics large scale sequencing projects are needed to increase both taxon and nucleotide sampling (Zwickl and Hillis, 2002). Even then, it is possible that not all branches will be strongly supported and some clades may prove very difficult to resolve due to conflicting signal among genes or rapid speciation events (Rokas, 2005, 2006). Once a robust phylogenetic framework of gobioid intrarelationships has been achieved the major challenge will be to integrate it with the variety of phylogenetic studies focusing below the genus level that have already been published. This will allow us to better tackle comparative evolutionary questions ranging from biogeography, trait evolution, diversification patterns and mechanisms of speciation in these mesmerizing fishes.

Acknowledgments

We are grateful to Zeehan Jaafar and an anonymous reviewer for their comments on the first version of this chapter and to many colleagues for fruitful discussions on gobioids over the last couple of years: Akihito, Naomi Delventhal, Tony Gill, Ian Harrison, Doug Hoese, Yuji Ikeda, Zeehan Jaafar, Takeshi Kon, Marcelo Kovačić, Helen Larson, Peter Miller, Masaki Miya, Randy Mooi, Mutsumi Nishida, Frank Pezold, Gianluca Polgar, Uli Schliewen, Mike Taylor, Chris Thacker, Jim Van Tassell, and Rafael Zardoya. Ainhoa Agorreta was supported by a predoctoral fellowship (FPU AP2006-00608) of the Ministry of Science and Innovation of Spain.

References

Akihito, A.T. Iwata, K. Ikeo, T. Imanishi, H. Ono, Y. Umehara, C. Hamamatsu, K. Sugiyama, Y. Ikeda, K. Sakamoto, A. Fumihito, S. Ohno and T. Gojobori. 2000. Evolutionary aspects of gobioid fishes based upon a phylogenetic analysis of mitochondrial cytochrome genes. *Gene* 259: 5-15.

Archer, S., J. Lythgoe and L. Hall. 1992. Rod opsin cDNA sequence from the sand goby (*Pomatoschistus minutus*) compared with those of other vertebrates. *Proceedings of the Royal Society B: Biological Sciences* 248: 19-25.

Binladen, J., M. Gilbert, J. Bollback, F. Panitz, C. Bendixen, R. Nielsen, E. Willerslev and M. Hahn. 2007. The use of coded PCR primers enables high-throughput sequencing of multiple homolog amplification products by 454 parallel sequencing. *PLoS ONE* 2: e197.

Brown, J. and C.A. Stepien. 2009. Invasion genetics of the Eurasian round goby in North America: tracing sources and spread patterns. *Molecular Ecology* 18: 64-79.

Bucciarelli, G., M. Di Filippo, D. Costagliola, F. Alvarez-Valin, G. Bernardi and G. Bernardi. 2009. Environmental genomics: a tale of two fishes. *Molecular Biology and Evolution* 26: 1235-1243.

Delsuc, F., H. Brinkmann and H. Philippe. 2005. Phylogenomics and the reconstruction of the tree of life. *Nature Reviews Genetics* 6: 361-375.

Dettaï, A. and G. Lecointre. 2005. Further support for the clades obtained by multiple molecular phylogenies in the acanthomorph bush. *Comptes Rendus Biologies* 328: 674-689.

Dettaï, A. and G. Lecointre. 2008. New insights into the organization and evolution of vertebrate IRBP genes and utility of IRBP gene sequences for the phylogenetic study of the Acanthomorpha (Actinopterygii: Teleostei). *Molecular Phylogenetics and Evolution* 48: 258-269.

Dillon, A. and C.A. Stepien. 2001. Genetic and biogeographic relationships of the invasive round (*Neogobius melanostomus*) and tubenose (*Proterorhinus marmoratus*) gobies in the Great Lakes versus Eurasian populations. *Journal of Great Lakes Research* 27: 267-280.

Dougherty, J., W. Moore and J. Ram. 1996. Mitochondrial DNA analysis of round goby (*Neogobius melanostomus*) and tubenose goby (*Proterorhinus marmoratus*) in the Great Lakes basin. *Canadian Journal of Fisheries and Aquatic Sciences* 53: 474-480.

Dufour, B., T. Hogan and D. Heath. 2007. Ten polymorphic microsatellite markers in the invasive round goby (*Neogobius melanostomus*) and cross-species amplification. *Molecular Ecology Notes* 7: 1205-1207.

Ellegren, H. 2008. Sequencing goes 454 and takes large-scale genomics into the wild. *Molecular Ecology* 17: 1629-1631.

Feldheim, K., P. Willink, J. Brown, D. Murphy, M. Neilson and C.A. Stepien. 2009. Microsatellite loci for Ponto-Caspian gobies: markers for assessing exotic invasions. *Molecular Ecology Notes* 9: 639-644.

Felsenstein, J. 2004. *Inferring Phylogenies*. Sinauer Associates, Inc., Sunderland, MA.

Gallini, A., L. Zane and P. Bisol. 2004. Isolation and characterization of microsatellites in *Zosterisessor ophiocephalus* (Perciformes, Gobiidae). *Molecular Ecology Notes* 5: 24-26.

Giovannotti, M., P. Cerioni, M. La Mesa and V. Caputo. 2007. Molecular phylogeny of the three paedomorphic mediterranean gobies (Perciformes: Gobiidae). *Journal of Experimental Zoology* 308B: 722-729.

Gracey, A. 2008. The *Gillichthys mirabilis* Cooper array: a platform to investigate the molecular basis of phenotypic plasticity. *Journal of Fish Biology* 72: 2118-2132.

Gracey, A.Y., J.V. Troll and G.N. Somero. 2001. Hypoxia-induced gene expression profiling in the euryoxic fish *Gillichthys mirabilis*. *Proceedings of the National Academy of Sciences USA* 98: 1993-1998.

Graham, J.B. and H.J. Lee. 2004. Breathing air in air: In what ways might extant amphibious fish biology relate to prevailing concepts about early tetrapods, the evolution of vertebrate air breathing, and the vertebrate land transition? *Physiological and Biochemical Zoology* 77: 720-731.

Harada, Y., I. Kinoshita, T. Kaneko, S. Moriyama, M. Tanaka and M. Tagawa. 2003. Response of a neotenic goby, ice goby (*Leucopsarion petersii*), to thyroid hormone and thiourea treatments. *Zoological Science* 20: 877-882.

Hauser, L. and J. Seeb. 2008. Advances in molecular technology and their impact on fisheries genetics. *Fish and Fisheries* 9: 473-486.

Hegedis, A., M. Lenhardt, B. Mickovic, G. Cvijanovic, I. Jaric and Z. Gacic. 2007. Amur sleeper (*Perccottus glenii* Dubowski, 1877) spreading in the Danube River Basin. *Journal of Applied Ichthyology* 23: 705-706.

Hepburn, R., E. Mottillo, P. Bentzen and D. Heath. 2006. Polymorphic microsatellite loci for the masked goby, *Coryphopterus personatus* (Gobiidae). *Conservation Genetics* 6: 1059-1062.

Herler, J., S. Koblmüller and C. Sturmbauer. 2009. Phylogenetic relationships of coral-associated gobies (Teleostei, Gobiidae) from the Red Sea based on mitochondrial DNA data. *Marine Biology* 156: 725-739.

Hoareau, T., F. Barbisan, S. Dubois, L. Zane and P. Berrebi. 2009. Polymorphic microsatellite loci in the widespread amphidromous goby *Sicyopterus lagocephalus* and cross-genus amplification among Sicydiinae. *Molecular Ecology Notes* 9: 607-609.

Hubert, N., R. Hanner, E. Holm, N.E. Mandrak, E. Taylor, M. Burridge, D. Watkinson, P. Dumont, A. Curry, P. Bentzen, J. Zhang, J. April and L. Bernatchez. 2008. Identifying Canadian freshwater fishes through DNA barcodes. *PLoS ONE* 3: e2490.

Hudson, M. 2008. Sequencing breakthroughs for genomic ecology and evolutionary biology. *Molecular Ecology Resources* 8: 3-17.

Huelsenbeck, J.P. 1997. Is the Felsenstein zone a fly trap? *Systematic Biology* 46: 69-74.

Huyse, T. and F. Volckaert. 2005. Comparing host and parasite phylogenies: *Gyrodactylus* flatworms jumping from goby to goby. *Systematic Biology* 54: 710-718.

Huyse, T., V. Audenaert and F. Volckaert. 2003. Speciation and host–parasite relationships in the parasite genus *Gyrodactylus* (Monogenea, Platyhelminthes) infecting gobies of the genus *Pomatoschistus* (Gobiidae, Teleostei). *International Journal for Parasitology* 33: 1679-1689.

Huyse, T., R. Poulin and A. Theron. 2005. Speciation in parasites: a population genetics approach. *Trends in Parasitology* 21: 469-475.

Ishimatsu, A., Y. Hishida, T. Takita, T. Kanda, S. Oikawa, T. Takeda and K.K. Huat. 1998. Mudskippers store air in their burrows. *Nature* 391: 237-238.

Jarne, P. 1996. Microsatellites, from molecules to populations and back. *Trends in Ecology & Evolution* 11: 424-429.

Järvenpää, M. and K. Lindström. 2004. Water turbidity by algal blooms causes mating system breakdown in a shallow-water fish, the sand goby *Pomatoschistus minutus*. *Proceedings of the Royal Society B: Biological Sciences* 271: 2361-2365.

Jones, A.G., D. Walker, K. Lindstrom, C. Kvarnemo and J.C. Avise. 2001. Surprising similarity of sneaking rates and genetic mating patterns in two populations of sand goby experiencing disparate sexual selection regimes. *Molecular Ecology* 10: 461-469.

Kocher, T.D., W.K. Thomas, A. Meyer and S.V. Edwards. 1989. Dynamics of mitochondrial DNA evolution in animals: amplification and sequencing with conserved primers. *Proceedings of the National Academy of Sciences USA* 86: 6196-6200.

Kok, W., C. Lim, T. Lam and Y. Ip. 1998. The mudskipper *Periophthalmodon schlosseri* respires more efficiently on land than in water and vice versa for *Boleophthalmus boddaerti*. *Journal of Experimental Zoology Part A: Comparative Experimental Biology* 280: 86-90.

Kokita, T. and S. Takahashi. 2008. Isolation and characterization of polymorphic trinucleotide and dinucleotide microsatellite loci for the ice goby, *Leucopsarion petersii*. *Molecular Ecology Resources* 8: 305-307.

Kon, T., T. Yoshino, T. Mukai and M. Nishida. 2007. DNA sequences identify numerous cryptic species of the vertebrate: a lesson from the gobioid fish *Schindleria*. *Molecular Phylogenetics and Evolution* 44: 53-62.

Kovac, V., G.H. Copp and R.P. Sousa. 2009. Life-history traits of invasive bighead goby *Neogobius kessleri* (Gunther, 1861) from the middle Danube River, with a reflection on which goby species may win the competition. *Journal of Applied Ichthyology* 25: 33-37.

Kuwamura, T., Y. Nakashima and Y. Yogo. 1994. Sex change in either direction by growth-rate advantage in the monogamous coral goby, *Paragobiodon echinocephalus*. *Behavioral Ecology* 5: 434-438.

Larmuseau, T., J. Guelinckx, B. Hellemans, J. Van Houdt and F. Volckaert. 2008. Fast PCR-RFLP method facilitates identification of *Pomatoschistus* species from the North Atlantic. *Journal of Applied Ichthyology* 24: 342-344.

Li, B., A. Dettaï, C. Cruaud, A. Couloux, M. Desoutter-Meniger and G. Lecointre. 2009. RNF213, a new nuclear marker for acanthomorph phylogeny. *Molecular Phylogenetics and Evolution* 50: 345-363.

Li, C., G. Ortí, G. Zhang and G. Lu. 2007. A practical approach to phylogenomics: the phylogeny of ray-finned fish (Actinopterygii) as a case study. *BMC Evolutionary Biology* 7: 7-44.

Lindström, K., C. St. Mary and C. Pampoulie. 2006. Sexual selection for male parental care in the sand goby, *Pomatoschistus minutus*. *Behavioral Ecology and Sociobiology* 60: 46-51.

Long, J. and M. Gordon. 2004. The greatest step in vertebrate history: a paleobiological review of the fish-tetrapod transition. *Physiological and Biochemical Zoology* 77: 700-719.

Lund, S.G., M.R. Ruberté and G.E. Hofmann. 2006. Turning up the heat: the effects of thermal acclimation on the kinetics of hsp70 gene expression in the eurythermal goby, *Gillichthys mirabilis*. *Comparative Biochemistry and Physiology* 143: 435-446.

McKay, S. 1993. Genetic relationships of *Brachygobius* and related brackish Indo-west Pacific and Australasian genera (Teleostei: Gobiidae). *Journal of Fish Biology* 43: 723-738.

McKay, S. and P. Miller. 1991. Isozyme criteria in the testing of phyletic relationships between species of *Gobius* and related eastern Atlantic-Mediterranean genera (Teleostei: Gobiidae). *Journal of Fish Biology* 39: 291-299.

Mendonca, H., J. Smith and C. Brinegar. 2001. Isolation and characterization of four microsatellite loci in the tidewater goby (*Eucyclogobius newberryi*). *Marine Biotechnology* 3: 91-95.

Miller, P., M. Serventi, D. Soregaroli, P. Torricelli and G. Gandolfi. 1994. Isozyme genetics and the phylogeny of Italian freshwater gobies (Teleostei: Gobioidei). *Journal of Fish Biology* 44: 439-451.

Miya, M., T.P. Satoh and M. Nishida. 2005. The phylogenetic position of toadfishes (order Batrachoidiformes) in the higher ray-finned fish as inferred from partitioned Bayesian analysis of 102 whole mitochondrial genome sequences. *Biological Journal of the Linnean Society* 85: 289-306.

Miya, M., H. Takeshima, H. Endo, N.B. Ishiguro, J.G. Inoue, T. Mukai, T.P. Satoh, M. Yamaguchi, A. Kawaguchi, K. Mabuchi, S.M. Shirai and M. Nishida. 2003. Major patterns of higher teleostean phylogenies: a new perspective based on 100 complete mitochondrial DNA sequences. *Molecular Phylogenetics and Evolution* 26: 121-138.

Miyake, A., T. Saito, T. Kashiwagi, D. Ando, A. Yamamoto, T. Suzuki, N. Nakatsuji and T. Nakatsuji. 2006. Cloning and pattern of expression of the shiro-uo vasa gene during embryogenesis and its roles in PGC development. *International Journal of Developmental Biology* 50: 619-625.

Mobley, K., T. Amundsen, E. Forsgren, P. Svensson and A. Jones. 2009. Multiple mating and a low incidence of cuckoldry for nest-holding males in the two-spotted goby, *Gobiusculus flavescens*. *BMC Evolutionary Biology* 9: 6.

Monteiro, J., R. Borges, J. Robalo and V. Almada. 2008. Larval development of *Gobius xanthocephalus* with genetic validation of larval identification. *Journal of Fish Biology* 73: 123-138.

Moore, W.S. 1995. Inferring phylogenies from mtDNA variation: mitochondrial-gene trees versus nuclear-gene trees. *Evolution* 49: 718-726.

Mukai, T., T. Sato and M. Morisawa. 2000. Natural hybridization and gene flow between two *Tridentiger* gobies in Lake Hinuma (Ibaraki Prefecture, Japan). *Ichthyological Research* 47: 175-181.

Mukai, T., K. Naruse, T. Sato and A. Shima. 1997. Multiregional introgressions inferred from the mitochondrial DNA phylogeny of a hybridizing species complex of gobiid fishes, genus *Tridentiger*. *Molecular Biology and Evolution* 14: 1258-1265.

Munday, P., M. Caley and G. Jones. 1998. Bi-directional sex change in a coral-dwelling goby. *Behavioral Ecology and Sociobiology* 43: 371-377.

Munday, P., L. Vanherwerden and C. Dudgeon. 2004. Evidence for sympatric speciation by host shift in the sea. *Current Biology* 14: 1498-1504.

Nagaraj, S.H., R.B. Gasser and S. Ranganathan. 2007. A hitchhiker's guide to expressed sequence tag (EST) analysis. *Brief Bioinformatics* 8: 6-21.

Nakashima, Y., T. Kuwamura and Y. Yogo. 1996. Both-ways sex change in monogamous coral gobies, *Gobiodon* spp. *Environmental Biology of Fishes* 46: 281-288.

Nakatsuji, T., T. Kitano, N. Akiyama and N. Nakatsuji. 1997. Ice goby (hiro-uo), *Leucopsarion petersii*, may be a useful material for studying teleostean embryogenesis. *Zoological Science* 14: 443-448.

Neilson, M.E. and C.A. Stepien. 2009a. Evolution and phylogeography of the tubenose goby genus *Proterorhinus* (Gobiidae: Teleostei): evidence for cryptic species. *Biological Journal of the Linnean Society* 96: 664-684.

Neilson, M.E. and C.A. Stepien. 2009b. Escape from the Ponto-Caspian: evolution and biogeography of an endemic goby species flock (Benthophilinae: Gobiidae: Teleostei). *Molecular Phylogenetics and Evolution* 52: 84-102.

Nelson, J. 2006. *Fishes of the World*. John Wiley and Sons, Inc., New York. 4th edition.

Nohara, K.T., O. Tominga and T. Seikai. 2009. Isolation and characterization of 11 polymorphic microsatellite loci in the whitegirdled goby (*Pterogobius zonoleucus*) and cross-species amplification in the serpentine goby (*P. elapoides*). *Molecular Ecology Notes* 9: 610-612.

Ohara, K., D. Takahashi and M. Takagi. 2004. Isolation of microsatellite loci in the freshwater goby, *Rhinogobius* sp (Gobiidae). *Molecular Ecology Notes* 4: 449-451.

Pegg, G.G., B. Sinclair, L. Briskey and W.J. Aspden. 2006. MtDNA barcode identification of fish larvae in the southern Great Barrier Reef, Australia. *Scientia Marina* 70: 7-12.

Penzo, E., G. Gandolfi, L. Bargelloni, L. Colombo and T. Patarnello. 1998. Messinian salinity crisis and the origin of freshwater lifestyle in western Mediterranean gobies. *Molecular Biology and Evolution* 15: 1472-1480.

Rambaut, A. 2006. Fig Tree. Tree Figure Drawing Tool v 1.2.1. Institute of Evolutionary Biology, University of Edinburgh. Available at http://tree.bio.ed.ac.uk.

Roberts, R.G., T.C. Freeman, E. Kendall, D.L.P. Vetrie, A.K. Dixon, C. Shaw-Smith, Q. Bone and M. Bobrow. 1996. Characterization of DRP2, a novel human dystrophin homologue. Nature Genetics 13: 223-226.

Rodgers, E., R. Earley and M. Grober. 2007. Social status determines sexual phenotype in the bi-directional sex changing bluebanded goby Lythrypnus dalli. Journal of Fish Biology 70: 1660-1668.

Rokas, A. 2005. More genes or more taxa? The relative contribution of gene number and taxon number to phylogenetic accuracy. Molecular Biology and Evolution 22: 1337-1344.

Rokas, A. 2006. Animal evolution and the molecular signature of radiations compressed in time. Science 311: 1242-1242.

Rüber, L., J.L. Van Tassell and R. Zardoya. 2003. Rapid speciation and ecological divergence in the American seven-spined gobies (Gobiidae, Gobiosomatini) inferred from a molecular phylogeny. Evolution 57: 1584-1598.

Saaristo, M., J. Craft, K. Lehtonen, H. Björk and K. Lindström. 2009. Disruption of sexual selection in sand gobies (Pomatoschistus minutus) by 17α-ethinyl estradiol, an endocrine disruptor. Hormones and Behavior 55: 530-537.

Saiki, R., S. Scharf, F. Faloona, K. Mullis, G. Horn, H. Erlich and N. Arnheim. 1985. Enzymatic amplification of beta-globin genomic sequences and restriction site analysis for diagnosis of sickle cell anemia. *Science* 230: 1350.

Sakamoto, T., H. Yasungaga, S. Yokota and M. Ando. 2002. Differential display of skin mRNAs regulated under varying environmental conditions in a mudskipper. *Journal of Comparative Physiology B: Biochemical, Systemic, and Environmental Physiology* 172: 447-453.

Sakurai, M., S. Nakakoji, H. Manabe, S. Dewa, A. Shinomiya and T. Sunobe. 2009. Bi-directional sex change and gonad structure in the gobiid fish *Trimma yanagitai. Ichthyological Research* 56: 82-86.

Sayer, M. 2005. Adaptations of amphibious fish for surviving life out of water. *Fish and Fisheries* 6: 186-211.

Schlötterer, C. and D. Tautz. 1992. Slippage synthesis of simple sequence DNA. *Nucleic Acids Research* 20: 211.

Sorice, M. and V. Caputo. 1999. Genetic variation in seven goby species (Perciformes: Gobiidae) assessed by electrophoresis and taxonomic inference. *Marine Biology* 134: 327-333.

St. Mary, C. 1994. Sex allocation in a simultaneous hermaphrodite, the blue-banded goby (Lythrypnus dalli): the effects of body size and behavioral gender and the consequences for reproduction. *Behavioral Ecology* 5: 304.

St. Mary, C. 2000. Sex allocation in *Lythrypnus* (Gobiidae): variations on a hermaphroditic theme. *Environmental Biology of Fishes* 58: 321-333.

Stepien, C. and M. Tumeo. 2006. Invasion Genetics of Ponto-Caspian Gobies in the Great Lakes: A 'Cryptic' Species, Absence of Founder Effects, and Comparative Risk Analysis. *Biological Invasions* 8: 61-78.

Stepien, C.A., J. Brown, M.E. Neilson and M. Tumeo. 2005. Genetic diversity of invasive species in the Great Lakes versus their Eurasian source populations: Insights for risk analysis. *Risk Analysis* 25: 1043-1060.

Sunobe, T., M.Y. Nakamura, Y. Kobayashi, T. Kobayashi and Y. Nagahama. 2005. Aromatase immunoreactivity and the role of enzymes in steroid pathways for inducing sex change in the hermaphrodite gobiid fish. *Comparative Biochemistry and Physiology - Part A: Molecular & Integrative Physiology* 141: 54-59.

Takahashi, D. and K. Ohara. 2006. Parentage analyses of freshwater goby *Rhinogobius* sp. OR using microsatellite DNA markers and egg developmental stages of eggs in the nest. *Ichthyological Research* 53: 87-92.

Taylor, M. and M. Hellberg. 2003. Genetic evidence for local retention of pelagic larvae in a Caribbean reef fish. *Science* 299: 107-109.

Taylor, M. and M.E. Hellberg. 2005. Marine radiations at small geographic scales: speciation in neotropical reef gobies (*Elacatinus*). *Evolution* 59: 374-385.

Thacker, C. 2000. Phylogeny of the wormfishes (Teleostei: Gobioidei: Microdesmidae). *Copeia* 2000: 940-957.

Thacker, C. 2003. Molecular phylogeny of the gobioid fishes (Teleostei: Perciformes: Gobioidei). *Molecular Phylogenetics and Evolution* 26: 354-368.

Thacker, C. 2009. Phylogeny of Gobioidei and Placement within Acanthomorpha, with a New Classification and Investigation of Diversification and Character Evolution. *Copeia* 2009: 93-104.

Thacker, C. and M. Hardman. 2005. Molecular phylogeny of basal gobioid fishes: Rhyacichthyidae, Odontobutidae, Xenisthmidae, Eleotridae (Teleostei: Perciformes: Gobioidei). *Molecular Phylogenetics and Evolution* 37: 858-871.

Thacker, C. and D. Roje. 2009. Phylogeny of cardinal fishes (Teleostei: Gobiiformes: Apogonidae) and the evolution of visceral bioluminescence. *Molecular Phylogenetics and Evolution* 52: 735-745.

Thacker, C., A. Thompson, D. Roje and E. Shaw. 2007. New expansions in old clades: population genetics and phylogeny of *Gnatholepis* species (Teleostei: Gobioidei) in the Pacific. *Marine Biology* 153: 375-385.

Thompson, A., C. Thacker, E. Shaw and D. Roje. 2006. Characterization of 16 microsatellite loci for a common coral reef fish, the fierce shrimpgoby *Ctenogobiops feroculus*. *Molecular Ecology Notes* 6: 918-920.

Vences, M., M. Thomas, A. Van Der Meijden, Y. Chiari and D. Vieites. 2005. Comparative performance of the 16S rRNA gene in DNA barcoding of amphibians. *Frontiers in Zoology* 2: 5.

Victor, B. 2007. *Coryphopterus kuna*, a new goby (Perciformes: Gobiidae: Gobiinae) from the western Caribbean, with the identification of the late larval stage and an estimate of the pelagic larval duration. *Zootaxa* 1526: 51-61.

Vyskočilová, M., M. Ondračková, A. Šimková and J. Martin. 2007. Isolation and characterization of microsatellites in *Neogobius kessleri* (Perciformes, Gobiidae) and cross-species amplification within the family Gobiidae. *Molecular Ecology Notes* 7: 701-704.

Wallis, G. and J. Beardmore. 1980. Genetic evidence for naturally occurring fertile hybrids between two goby species, *Pomatoschistus minutus* and *P. lozanoi* (Pisces, Gobiidae). *Marine Ecology - Progress Series* 3: 309-315.

Wang, H.Y., M.P. Tsai, J. Dean and S.C. Lee. 2001. Molecular Phylogeny of Gobioid Fishes (Perciformes: Gobioidei) Based on Mitochondrial 12S rRNA Sequences. *Molecular Phylogenetics and Evolution* 20: 390-408.

Ward, R.D. and B. Holmes. 2007. An analysis of nucleotide and amino acid variability in the barcode region of cytochrome c oxidase I (*cox1*) in fishes. *Molecular Ecology Notes* 7: 899-907.

Ward, R.D., T.S. Zemlak, B.H. Innes, P.R. Last and P.D.N. Hebert. 2005. DNA barcoding Australia's fish species. *Philosophical Transactions of the Royal Society* B 360: 1847-1857.

Zwickl, D.J. 2006. Genetic algorithm approaches for the phylogenetic analysis of large biological sequence datasets under the maximum likelihood criterion. Ph.D. dissertation. The University of Texas, Austin.

Zwickl, D.J. and D. Hillis. 2002. Increased taxon sampling greatly reduces phylogenetic error. *Systematic Biology* 51: 588-598.

CHAPTER

1.3

Systematics of the Rhyacichthyidae

Helen K. Larson

The genus *Rhyacichthys* is the goby upon which everyone seems to agree—it is the most basal gobioid fish and sister to all other gobioids (Springer, 1983; Hoese 1984; Hoese and Gill, 1993; Akihito *et al.*, 2000; Thacker and Hardman, 2005; Mooi and Gill, 2008). It was first placed in its own family, the Rhyacichthyidae, by Jordan (1923) who created the family (with Platypteridae as alternate name) for *Rhyacichthys* and *Platyptera* (latter name preoccupied). Rhyacichthyids are easily distinguished externally by their possession of a complete lateral line along the side of the body (unlike the majority of gobioids). The type genus *Rhyacichthys* in particular possesses a mix of highly specialised and primitive features, which originally caused some uncertainty as to its systematic placement (Miller, 1973; Akihito, 1986).

The Rhyacichthyidae is currently represented by two genera, *Rhyacichthys* Boulenger (1901) and *Protogobius* Watson and Pöllabauer (1998). The family is known so far from the Indo-Australian Archipelago and particular localities in the Western Pacific, where suitable habitat exists.

Author's address: 165 Reid Road, Wongaling Beach, Queensland 4852, Australia. E-mail: helen.larson@nt.gov.au

The genus *Rhyacichthys* was created by Boulenger (1901) as a replacement name for *Platyptera* Valenciennes (1837) which is preoccupied by *Platyptera* Meigen, a genus of Diptera. As Roberts pointed out, there is an available older name which has been overlooked and never used: *Platypterus* Valenciennes (in Cuvier, 1836) (Roberts, 1993). As *Rhyacichthys* has been used since 1901 and nobody subsequent to 1836 has used *Platypterus* as a name, the author agrees with Roberts that *Rhyacichthys* should be retained. The other rhyacichthyid genus, *Protogobius*, was not discovered until 1996 (in New Caledonia) and was formally described in 1998.

Watanabe (1972) first described the appearance and distribution of *Rhyacichthys aspro* in some detail, while Miller (1973) provided the first (and so far, the only) detailed account of its morphology and osteology. Valenciennes (1837) appears to be the only person to have described the form of its internal organs. Springer (1983), Hoese and Gill (1993), Akihito *et al.* (2000) and Shibukawa *et al.* (2001) provided additional information on the characters of the Rhyacichthyidae.

There has been very little attempt to examine in detail the phylogenetic placement of the Rhyacichthyidae from a morphological viewpoint, other than Miller (1973), Springer (1983) and Mooi and Gill (2008). Springer and Orell (2004) carried out a phylogenetic analysis of 56 gill-arch muscle and skeletal characters, which placed *Rhyacichthys* (the only gobioid in their matrix of 168 taxa) as sister to percids (strict consensus and 50% majority rule consensus), in a clade that also included gasterosteids, gobiesocids, blenniids and kurtids (among many others). Using osteological and muscular characters, Mooi and Gill (2008) placed *Protogobius* and *Terateleotris* (presently considered to be an odontobutid (Nelson, 2006) as sisters together in a clade, with *Rhyacichthys* as sister to the entire gobioid clade.

Rhyacichthys has now been used as a "basal gobioid" in a number of molecular analyses of perciform relationships, such as that of the "higher teleosteans" (Miya *et al.*, 2003), the phylogeny of Batrachoidiformes (Miya *et al.*, 2005), the interrelationships of the gasterosteiformes (Kawahara *et al.* 2008), the phylogeny of the Apogonidae (Thacker and Roje, 2009), the placement of gobioids among the Acanthomorpha (Thacker, 2009) and an evaluation of facial nerve patterns (Yagishita *et al.*, 2009). These analyses placed gobioid fishes (including *Rhyacichthys*) as sister-group to several higher taxa (dactylopterids, gasterosteiforms, apogonids) or were unresolved (Thacker and Roje, 2009).

Akihito *et al.* (2000) presented an unrooted tree for 28 gobioid taxa, including *Rhyacichthys* and *Protogobius*, based on the mitochondrial cytB gene, with the two rhyacichthyids forming a distinct cluster. This study showed a close relationship between *Rhyacichthys* and *Protogobius*, despite their considerable difference in appearance. Indeed, it was found that the nucleotide substitution ratios for *Rhyacichthys* and *Protogobius* was very similar to the ratios for *Eleotris* (Eleotridae) and *Tridentiger* (Gobiidae), which fell out in different clades or 'clusters' (Akihito *et al.*, 2000). Wang *et al.* (2001) did not include *Rhyacichthys* in their molecular phylogeny of gobioid fishes; only one species of *Odontobutis* was used, with *Scomber* as outgroup.

Rhyacichthys does not look like a gobioid at all—species of this genus have a broad, depressed head and large wide pectoral fins set obliquely on the fin base, widely separated broad pelvic fins, the body is rounded anteriorly and laterally compressed posteriorly with a flat ventral surface and the small mouth is mostly under the rounded snout (resembling the balitorid loach genus *Gastromyzon*; Boulenger (1901) first compared it to *Homaloptera* and other similar fish). The genus has a number of primitive characters (that demonstrate its position as a basal gobioid) and some specialised characters (which give it the appearance of an atypical gobioid).

Other gobioids differ from *Rhyacichthys* in several features: the lateral line canal does not extend onto the body, the ventral portion of the hyomandibular is broad with the dorsal ends of the interhyal and symplectic widely separated from each other, the mandibular lateral canal is absent, the preopercular canal is separated from the oculoscapular canal and the penultimate branchiostegal ray articulates with the anterior ceratobranchial, not the posterior one as in *Rhyacichthys* (Springer, 1983; Hoese and Gill, 1993).

Protogobius, the other genus presently placed in the Rhyacichthyidae, has a much more typical 'goby-like' appearance, differing from *Rhyacichthys* in having the head and body slightly compressed and sub-cylindrical, with a somewhat pointed snout, small terminal mouth and large eyes, its pectoral fins are moderate and set on a mostly vertical fin base and the pelvic fins are separated (but not widely so) and long (reaching past the anus).

The primitive (plesiomorphic) characters of the Rhyacichthyidae are: a complete lateral line, with canal pores in grooved scales on the side of the body; infraorbital and supratemporal lateral canals and pores are present in one genus, *Rhyacichthys*, but not *Protogobius* (see Miller, 1973: Fig. 10; Watson and Pöllabauer, 1998; Fig. 1); the preopercular-mandibular lateral canal and its pores reaches far forward to below end of jaw in *Rhyacichthys* but not *Protogobius* (ending below eye); the cephalic portion of the lateral line canal is continuous with the body lateral line in *Rhyacichthys*, but not *Protogobius*, in which there is a break above the pectoral fin base (Miller, 1973: Fig. 10; Watson and Pöllabauer, 1998: Fig. 1; Shibukawa *et al.*, 2001); the lateral line canals on the head are continuous (with all pores joined by canals), a character shared with the odontobutid *Terateleotris* but not with *Protogobius* (which has the preopercular canal separate from the oculoscapular) (Shibukawa *et al.*, 2001); six branchiostegal rays are present (as in odontobutids and eleotrids (Miller, 1973); one supratemporal (or extrascapular) is present, as in some eleotrids and the odontobutid *Terateleotris* (Akihito, 1971; Springer, 1983; Shibukawa *et al.*, 2001); the preoperculum is broadened and reaches the symplectic in *Rhyacichthys* while the two bones barely meet in *Protogobius* (bones separate in other gobioids) (Springer, 1983; pers. obs.); two infraorbital bones are present, the lacrimal plus one infraorbital (the suborbital of Miller (1973)) (only the lacrimal is present in other gobioids) (Springer, 1983; Iwata in litt.); the metapterygoid reaches toward the quadrate and does not contact the endopterygoid (Miller, 1973); the endopterygoid is broad and stout (Miller, 1973); the scapula is well-developed (with a moderately large foramen) and extends dorsally to separate the uppermost pectoral radial from the cleithrum (shared with odontobutids); the dorsal postcleithrum is present (Miller, 1973; Springer, 1983); three epurals are present instead of two or one (shared with *Terateleotris*) (Miller 1973; Hoese and Gill, 1993); the dorsal and ventral procurrent cartilages in the caudal fin are quite small and do not support the anterior unsegmented rays, as in odontobutids (Hoese and Gill, 1993); the caudal fin is emarginate (shared with a few gobioids such as *Tyson* (Xenisthmidae) and *Ptereleotris* (Ptereleotrinae) (Springer, 1983); the body scales have up to 11 rows of cteni, with the outer row being of typical cteni in shape while the remaining inner rows are truncate (Hoese and Gill, 1993); and the pelvic fins are widely separated in *Rhyacichthys* (Miller, 1973) and less so in *Protogobius* (where they resemble the spacing seen in odontobutids and eleotrids).

Most of the more specialised (apomorphic) features of *Rhyacichthys aspro* were discussed by Miller (1973) and Akihito (1986); both workers being struck by the extreme adaptation of the species to its torrential habitat. These specialised features included: having an enlarged, thickened and wide pectoral fin set obliquely on the pectoral fin base, widely spaced broad fleshy pelvic fins, a depressed head and snout, relatively small dorsolaterally oriented eyes, a small ventrally-placed mouth, and accessory lateral line canals dorsal and ventral to the lateral line on the body, including the caudal fin base. As Shibukawa *et al.* (2001) and Ahnelt and Göschel (2004) have noted, many gobioids have rows of sensory papillae in various patterns along the side of the body and on the caudal fin, so it is difficult to be sure of this character's status. The sensory papillae on the head in *Rhyacichthys* are not arranged in any easily definable pattern, other than that those on the dorsal surface are arranged in lines and clusters, while those on the side of the head form scattered clusters (Larson, 2001: 20, Fig. 13). In other gobioids the sensory papillae can nearly always be classified as having a longitudinal or a transverse pattern (Hoese, 1983). *Protogobius* differs from *Rhyacichthys* in that it clearly has an abbreviated but distinct longitudinal papillae pattern reminiscent of the odontobutids *Micropercops* and *Perccottus* (Shibukawa *et al.*, 2001; pers. obs). *Protogobius* also has a row of firm fleshy pointed papillae around the posterodorsal edge of the eye (above the lateral line canal) that is absent in the more streamlined *Rhyacichthys* (Watson and Pöllabauer, 1998; pers. obs.). Hoese and Gill (1993) noted that *Rhyacichthys* had the adductor mandibulae tendon attached to the posterodorsal margin of the maxilla, unlike other gobioids while this tendon attaches just behind the anterior head of the maxilla (outer side) in *Protogobius* (Iwata in litt.) (the jaws of each genus clearly differ structurally).

Seven of the morphological characters listed above differ between *Rhyacichthys* and *Protogobius*. In *Rhyacichthys* the lateral line along body is continuous (broken in *Protogobius*), all lateral line canals on head are continuous (separate preopercular canal in *Protogobius*), the broad preopercular bone reaches the symplectic (hardly meeting in *Protogobius*), the pelvic fins are widely separated (in *Protogobius* these fins are separate but similar to the arrangement in eleotrids and odontobutids), the sensory papillae on the head are in scattered clusters (papillae arranged in an abbreviate longitudinal pattern in *Protogobius*), the posterodorsal edge of the orbit is smooth (*Protogobius* has a row of firm fleshy pointed papillae) and the adductor mandibulae tendon is attached to the posterodorsal margin of the maxilla (in *Protogobius* this tendon attaches just behind the anterior head of the maxilla). These differences plus the results shown in Mooi and Gill (2008) indicate that the Rhyacichthyidae is not monophyletic.

Rhyacichthys aspro (Valenciennes, 1837)

This is a freshwater goby that very much resembles a balitorid loach in form and habitat (its English common name is 'loach goby'). It lives in steep very swift to torrential rocky streams, inland or coastal, feeding on algae and diatoms (Miller, 1973; Allen, 1991). *Rhyacichthys aspro* clings to rocks and boulders very effectively with its broadened pelvic and pectoral fins and flattened head and snout, while the large caudal fin enables it to move swiftly—these are difficult fish to catch. They grow large, up to

250 mm SL (Allen, 1991). Good colour photographs of living fish can be seen in Senou *et al.* (2004) and Kawanabe and Mizuno (1989).

Koumans (1953) reported *R. aspro* from Indonesia (Ambon; Bali; Ceram; Java; Sulawesi (type locality); Sumatra), Papua New Guinea, Philippines and the Solomon Islands. It is also known from Japan (Iriomote; Masuda *et al.*, 1975), Taiwan (Watanabe, 1972) and the Republic of Palau (Jenkins in litt.). *Platyptera sinensis* (junior synonym of *R. aspro*) was described by Bleeker, based only on a Chinese painting, which does not mean the model for the fish was from China (Bleeker, 1872: 118). So far *R. aspro* is not known from mainland China; although Wu and Zhong (2008), in their summary of Chinese gobioid fishes, include Taiwan in 'China'.

Rhyacichthys guilberti Dingerkus and Seret, 1992

This species differs from *R. aspro* in having 12+15 vertebrae instead of 11+17 and in the head, body and fins being speckled and mottled with black instead of being shades of brown with darker rows of spots and lines along the body and fins (as well as some proportional differences). This species is known to reach 185 mm SL. Colour photographs of living fish are in Marquet *et al.* (2003) and Laboute and Grandperrin (2000). *Rhyacichthys guilberti* is now known only from the Tité and We Ina Rivers in north-eastern New Caledonia, having disappeared from the Ouaième River (Keith, 2002). It is considered by Keith (2002) to be critically endangered, but no rhyacichthyids appear on the International Union for the Conservation of Nature (IUCN) Red List of threatened species (gobioid fishes require assessment).

Protogobius attiti Watson and Pöllabauer, 1998

This genus and species were originally described without being assigned to a family. Shibukawa *et al.* (2001) first placed it in the Rhyacichthyidae based on its possession of the lateral line canal extending onto the body and the 12 morphological features first reported by Akihito *et al.* (2000). *Protogobius* resembles *Rhyacichthys* in possessing a complete lateral line along the body and short accessory canals on the body well above and below the lateral line as well as on the caudal fin base, but lacks an infraorbital lateral canal, the connection between the oculoscapular and preopercular canal and the extension of the preopercular canal on to the lower jaw (Watson and Pöllabauer, 1998). Akihito *et al.* (2000) found that it grouped next to *R. aspro* in their cytochrome *b* analysis and that the two species shared many morphological characters, indicating close relationship. However, it also differs from *Rhyacichthys* in a number of features. Shibukawa *et al.* (2001) discussed the relationships and characters of *Rhyacichthys, Protogobius* and the problematic *Terateleotris* Shibukawa, Iwata and Viravong, 2001 (the latter genus presently placed in the Odontobutidae).

Protogobius is known only from nine rivers and streams in New Caledonia, living over rock and gravel substrates, feeding on small crustaceans, especially atyid shrimps (Marquet *et al.*, 2003; Lord and Keith, 2006). These rivers do not have the torrential flow preferred by *Rhyacichthys*. *Protogobius* lives in an area close to nickel-mining activities, and is considered by Lord and Keith (2006) to be critically endangered (no rhyacichthyids have been assessed for the IUCN Red List). Colour photographs of living fish can be seen in Marquet *et al.* (2003) and Laboute and Grandperrin (2000). *Protogobius* is known to reach 80 mm SL.

While much has been made of this basal gobioid group, whatever their phylogenetic position, it is a pity that so little is known about them in the wild. After spending several hours with two colleagues in the swift Omijya River in Iriomote-jima, to capture just one *Rhyacichthys* alive, The author has great respect for these fascinating gobies and hopes that they continue to flourish.

References

Ahnelt, H. and J. Göschel. 2004. The pattern of the lateral-line system on the caudal fin of *Perccottus glenii* Dybowski, 1877 (Teleostei: Odontobutidae), with comments on the arrangement of the lateral-line system on the caudal fin of Gobioidei. *Proceedings of the California Academy of Sciences* 55: 358-372.

Akihito, Prince. 1971. On the supratemporals of gobiid fishes. *Japanese Journal of Ichthyology* 18: 57-64.

Akihito, Prince. 1986. Some morphological characters considered to be important in gobiid phylogeny. In: *Indo-Pacific Fish Biology: Proceedings of the Second International Conference on Indo-Pacific Fishes.* T. Uyeno, R. Arai, T. Taniuchi and K. Matsuura (eds.). Ichthyological Society of Japan, Tokyo, pp. 629-639.

Akihito, A. Iwata, T. Kobayashi, K. Ikeno, T. Imanishi, H. Ono, Y. Umehara, C. Hamamatsu, K. Sugiyama, Y. Ikeda, K. Sakamoto, F. Akishinonomiya, S. Ohno and T. Gojobori. 2000. Evolutionary aspects of gobioid fishes based upon a phylogenetic analysis of mitochondrial cytochrome *b* genes. *Gene* 259: 5-15.

Allen, G.R. 1991. *Field Guide to the Freshwater Fishes of New Guinea.* Publication No. 9; Christensen Research Institute, Madang, Papua New Guinea.

Bleeker, P. 1872. Mémoire sur la faune ichthyologique de Chine. *Nederlandsch Tijdschrift voor de Dierkunde* 4: 113-154.

Boulenger, G.A. 1901. Notes on the classification of teleostean fishes. I. On the Trachinidae and their allies. *Annals and Magazine of Natural History (Series 7)* 8: 261-271.

Cuvier, G. 1836. Le règne animal distribué d'après son organisation, pour servir de base à l'histoire naturelle des animaux et d'introduction à l'anatomie comparée. 3rd ed. 23 vols. Paris.

Dingerkus, G. and B. Seret. 1992. *Rhyacichthys guilberti*, a new species of loach goby from northeastern New Caledonia (Teleostei: Rhyacichthyidae). *Tropical Fish Hobbyist* July: 174-176.

Hoese, D.F. 1983. Sensory papilla patterns of the cheek lateralis system in the gobiid fishes *Acentrogobius* and *Glossogobius*, and their significance for the classification of gobioid fishes. *Records of the Australian Museum* 35: 223-229.

Hoese, D.F. 1984. Gobioidei: Relationships. In: *Ontogeny and Systematics of Fishes*. H.G. Moser (ed.). Special Publication of the American Society of Ichthyologists and Herpetologists No.1. Allen Press, Lawrence, Kansas, pp. 588-591.

Hoese, D.F. and A.C. Gill. 1993. Phylogenetic relationships of eleotridid fishes (Perciformes: Gobioidei). *Bulletin of Marine Science* 52: 415-440.

Jordan, D.S. 1923. A classification of fishes including families and genera as far as is known. *Stanford University Publications, University Series, Biological Sciences* 3: 77-243.

Kawahara, R., M. Miya, K. Mabuchi, S. Lavoué, J.G. Inoue, T. Satoh, A. Kawaguchi and M. Nishida. 2008. Interrelationships of the 11 gasterosteiform families (sticklebacks, pipefishes, and their relatives): a new perspective based on whole mitogenome sequences from 75 higher teleosts. *Molecular Phylogenetics and Evolution* 46: 224-236.

Kawanabe, H. and N. Mizuno. 1989. *Freshwater Fishes of Japan*. Yama-Kei Publishers, Tokyo [in Japanese].

Keith, P. 2002. Threatened fishes of the world: *Rhyacichthys guilberti* Dingerkus & Séret, 1992 (Rhyacichthyidae). *Environmental Biology of Fishes* 63: 40.

Koumans, F.P. 1953. X. Gobioidea. In: *The Fishes of the Indo-Australian Archipelago*. M. Weber and L.F. de Beaufort (eds.). E.J. Brill, Leiden.

Laboute, P. and R. Grandperrin. 2000. *Poissons de Nouvelle-Calédonie*. Editions Catherine Ledru, Nouméa.

Larson, H.K. 2001. A revision of the gobiid fish genus *Mugilogobius* (Teleostei: Gobioidei), and its systematic placement. *Records of the Western Australian Museum Supplement* 62: 1-233.

Lord, C. and P. Keith. 2006. Threatened fishes of the world: *Protogobius attiti* Watson & Pöllabauer 1998 (Rhyacichthyidae). *Environmental Biology of Fishes* 77: 101-102.

Marquet, G., P. Keith and E. Vigneux. 2003. *Atlas des poissons et des crustacés d'eau douce de Nouvelle-Calédonie. Patrimoines Naturels Volume 55*. Muséum National d'Histoire Naturelle, Paris.

Masuda, H., C. Araga and T. Yoshino. 1975. *Coastal Fishes of Southern Japan*. Tokai University Press, Tokyo.

Miller, P.J. 1973. The osteology and adaptive features of *Rhyacichthys aspro* (Teleostei: Gobioidei) and the classification of gobioid fishes. *Journal of Zoology, London* 171: 397-434.

Miya, M., T.P. Satoh and M. Nishida. 2005. The phylogenetic position of toadfishes (order Batrachoidiformes) in the higher ray-finned fish as inferred from partitioned Bayesian analysis of 102 whole mitochondrial genome sequences. *Biological Journal of the Linnean Society* 85: 289-306.

Miya, M., H. Takeshima, H. Endo, N.B. Ishiguro, J.G. Inoue, T. Mukai, T.P. Satoh, Y. Yamaguchi, A. Kawaguchi, K. Mabuchi, S.M. Shiair and M. Nishida. 2003. Major patterns of higher teleostean phylogenies: a new perspective based on 100 complete mitochondrial DNA sequences. *Molecular Phylogenetics and Evolution* 26: 121-138.

Mooi, R.D. and A.C. Gill. 2008. Exploring gobioid phylogeny using morphology—not a lost cause. Poster presentation, International Symposium on Systematics and Diversity of Fishes, 3-4 March 2008, National Museum of Nature and Science, Tokyo.

Nelson, J.S. 2006. *Fishes of the World*. John Wiley and Sons, Hoboken. Fourth edition.

Roberts, T.R. 1993. The freshwater fishes of Java, as observed by Kuhl and van Hasselt in 1820-23. *Zoologische Verhandelingen* 285: 1-94.

Senou, H., T. Suzuki, K. Shibukawa and K. Yano. 2004. *A Photographic Guide to the Gobioid Fishes of Japan*. Heibonsha, Tokyo [in Japanese].

Shibukawa, K., A. Iwata and S. Viravong. 2001. *Terateleotris*, a new gobioid fish genus from the Laos (Teleostei, Perciformes), with comments on its relationships. *Bulletin of the National Science Museum (Tokyo)* 27(4): 229-257.

Springer, V.G. 1983. *Tyson belos*, new genus and species of western Pacific fish (Gobiidae, Xenisthminae), with discussions of gobioid osteology and classification. *Smithsonian Contributions to Zoology* 390: 1-40.

Springer, V.G. and T. Orell. 2004. Phylogenetic analysis of the families of acanthomorph fishes based on dorsal gill-arch muscles and skeleton. Appendix. *Bulletin of the Biological Society of Washington* 11: 237-260.

Thacker, C.E. 2009. Phylogeny of Gobioidei and placement within Acanthomorpha, with a new classification and investigation of diversification and character evolution. *Copeia* 2009: 93-104.

Thacker, C.E. and M.A. Hardman. 2005. Molecular phylogeny of basal gobioid fishes: Rhyacichthyidae, Odontobutidae, Xenisthmidae, Eleotridae (Teleostei: Perciformes: Gobioidei). *Molecular Phylogenetics and Evolution* 37: 858-871.

Thacker, C.E. and D.M. Roje. 2009. Phylogeny of cardinal fishes (Teleostei: Gobiiformes: Apogonidae) and the evolution of visceral bioluminescence. *Molecular Phylogenetics and Evolution* 52: 735-745.

Valenciennes, A. 1837. Tome douzième. In: *Histoire Naturelle des Poissons*. G.L. Cuvier and A. Valenciennes. Levrault, Paris. Vol. 12.

Wang, H.-Y., M.-P. Tsai, J. Dean and S.-C. Lee. 2001. Molecular phylogeny of gobioid fishes (Perciformes: Gobioidei) based on mitochondrial 12S rRNA sequences. *Molecular Phylogenetics and Evolution* 22: 1-19.

Watanabe, M. 1972. First record of the gobioid fish, *Rhyacichthys aspro*, from Formosa. *Japanese Journal of Ichthyology* 19: 120-124.

Watson, R.E. and C. Pöllabauer. 1998. A new genus and species of freshwater goby from New Caledonia with a complete lateral line. *Senckenbergiana Biologia* 77(2): 147-153.

Wu, H. and J. Zhong. (eds.). 2008. *Fauna Sinica. Osteichthyes. Perciformes (V). Gobioidei*. Science Press, Beijing [in Chinese].

Yagishita, N., M. Miya, Y. Yamanoue, S. Shirai, K. Nakayama, N. Suzuki, T.P. Satoh, K. Mabuchi, M. Nishida and T. Nakabo. 2009. Mitogenomic evaluation of the unique facial nerve pattern as a phylogenetic marker within the perciform fishes. *Molecular Phylogenetics and Evolution* 53: 258-266.

CHAPTER

1.4

Systematics of Odontobutidae

Akihisa Iwata

The family Odontobutidae comprises at least six genera and about fifteen species, ranging in size from 3-20 cm. These fishes typically have a stout and cylindrical body and a benthic lifestyle (Fig. 1.4.1). However, a few odontobutid species are laterally compressed and nektonic. This family cannot easily be distinguished from eleotrids by external characters except for microscopic features of the scales. Almost all species occur in still or slow-flowing water in the middle and lower parts of rivers, ponds and lakes. A few species inhabit the upper reaches of mountain rivers. The distribution of the Odontobutidae is restricted to the sub-frigid, temperate, sub-tropical, and tropical zones of east and southeast Asia.

The family Odontobutidae was established by Hoese and Gill (1993), and can be distinguished from other gobioid fishes by the following combination of unspecialized characters (Table 3 of Hoese and Gill, 1993): (1) adductor mandibulae tendon attaches just posterior to, or on, the head of the maxilla; (2) dorsal and ventral procurrent caudal cartilages small and do not extend anteriorly to support the anterior unsegmented rays; (3) dorsal procurrent

Author's address: Laboratory of Ecology and Environment, Division of Southeast Asian Area Studies, Graduate School of Asian and African Area Studies, Kyoto University Yoshida, Shimoadachi-cho 46, Sakyo-ku, Kyoto 606-8501 Japan. E-mail: aiwata@asafas.kyoto-u.ac.jp

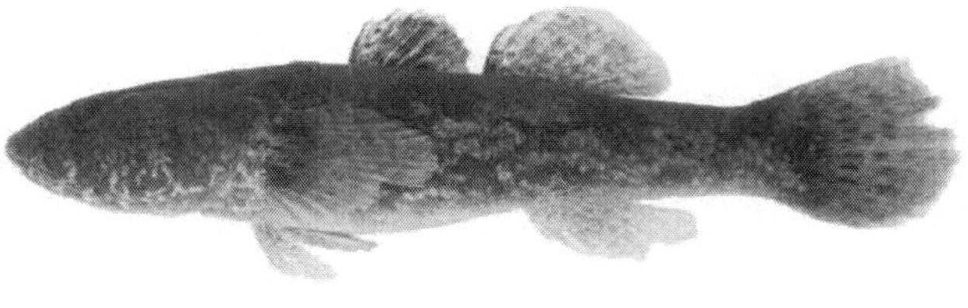

Fig. 1.4.1 *Odontobutis obscura*, 121.5 mm in standard length, Japan.

caudal cartilage short posteriorly and does not extend over the tip of the epural or epurals; (4) scapula well-developed and extends dorsally so that there is no contact between the upper proximal radial of the pectoral fin and the cleithrum; (5) autogenous middle radial present in the pterygiophore that supports the first element (spine) of the second dorsal fin; (6) bony support for the preopercular canal extends most of the length of the preoperculum; (7) transforming cteni are present on scales; (8) ventral transverse shelf on urohyal present; (9) first pterygiophore of second-dorsal fin begins immediately posterior to the last pterygiophore of the first-dorsal fin and there is no interneural space without a pterygiophore (interneural gap); (10) six branchiostegal rays; and (11) pelvic fins separate. Hoese and Gill (1993) hypothesized Odontobutidae as the sister group to the Eleotridae (butines and eleotrines) and gobies with *Odontobutis* Bleeker, 1874, *Micropercops* Fowler and Bean, 1920, and *Perccottus* Dybowski, 1877, included in the Odontobutidae. Subsequently, Shibukawa *et al.* 2001 described *Terateleotris,* which was based on *Odontobutis aspro* Kottelat, 1998. However, Shibukawa *et al.* (2001) did not classify their new genus in any gobioid family. Nelson (2006), however, placed *Terateleotris* in the Odontobutidae without explanation. Chen *et al.* (2002) described a new genus, *Neodontobutis* based on *Hypseleotris hainanensis* Chen, 1985 in Chen and Zheng (1985), and included their new genus within the Odontobutidae. In addition, *Sineleotris* Herre, 1940, and *Philypnus chalmersi* Nichols and Pope, 1927, were included by Chen *et al.* (2002) in Odontobutidae with the following justification: "Miller, pers. comm., pers. obs.". In Chen *et al.* (2002), the generic name of *Odontobutis aurarmus* Vidthayanon, 1995, was changed to *Neodontobutis*, and treated as a valid species, however, distinguishing characters between *N. hainanensis* and *N. aurarmus* were not provided. As mentioned above, the genera *Neodontobutis*, *Philypnus* and *Sineleotris* were included in Odontobutidae with limited or no justification by Chen *et al.* (2002). In particular, the taxonomic status of *Philypnus chalmersi* Nichols and Pope, 1927, is unstable. For example, some authors assign the genus *Perccottus* to this species (viz. Fowler, 1962; Hoese and Gill, 1993; Larson, 2001), whereas Chen *et al.* (2002) assigned this species to *Sineleotris*. Wu (2008) utilized the generic name *Philypnus* for this species (*P. chalmersi*).

This author examined clear and stained specimens of *Sineleotris saccharae*, *Philypnus chalmersi* and *Neodontobutis aurarmus*, and used information from Iwata (1994) and Shibukawa *et al.* (2001) for discussion of the genera *Odontobutis*, *Micropercops*, *Perccottus* and *Terateleotris* provided below. It was confirmed that the all materials examined possessed the characters listed in Table 3 of Hoese and Gill (1993) and all taxa were members of Odontobutidae.

Diagnostic characters of each genus and species of Odontobutidae are as follows:

Genus *Terateleotris* Shibukawa, Iwata and Viravong, 2001

Type Species: *Odontobutis aspro* Kottelat, 1998

This genus is distinguished from all other genera of Odontobutidae by the following unique characters among Odontobutidae (Table 1.4.1): complete lateral line present, connection between oculoscapular and preopercular canals present and shape of gill rakers on lower half outer surface of first gill arch short and stout with small spines. This genus is also distinguished from others by the following combination of characters (Table 1.4.1): furrow present on all mid-lateral line scales; cephalic sensory canals complete; sensory papillae on chin forming transverse pattern, right and left lines separate (Fig. 1.4.2); sensory papillae on tip of snout forming transverse pattern (Fig. 1.4.2); sensory papilla normal shape, not forming barbel-like projection; right and left gill membrane united at vertical from a point of near posterior edge of eye; gill opening wide, extending to a vertical from anterior margin of pupil; interorbital width narrower than eye diameter; predorsal scales ctenoid, extending forward to posterior part of interorbital space; transforming cteni of scales forming several rows; 5th pelvic-fin ray branched; distal radial of first-dorsal fin element present; suborbital bone present; three epurals; lateral extrascapula present and large; prevomerine teeth absent; ventral transverse shelf on posterior half of urohyal absent; gill rakers separated on inner part of 4th ceratobranchial; anterior projection on anterior tip of metapterygoid present (Fig. 1.4.3); connection between mesopterygoid and metapterygoid present (Fig. 1.4.3); overlap with lower part of ectopterygoid and quadrate absent (Fig. 1.4.3).

Table 1.4.1 Comparison of the conditions of selected external and osteological characters among the genera of Odontobutidae.

	Terateleotris	*Sineleotris*	*Micropercopus*	*Perccottus*	*Neodontobutis*	*Odontobutis*
External characters						
lateral line	present	absent	absent	absent	absent	absent
connection between oculoscapular and preopercular canals	present	absent	absent	absent	absent	absent
fullow on mid-lateral longitudinal scales	present	absent	absent	absent	absent	absent
cephalic sensory canals	complete	complete	moderate	absent	absent	absent or reduced
barbel-like projection on sensory papilla	absent	absent	absent	absent	present	absent
interorbital space	narrower than eye diameter	narrower than or equal to, eye diameter	nearly equal to eye diameter	nearly equal to or a little wider than eye diameter	narrower than eye diameter	wider than eye diameter
right and left gill membrane	united	united	separated, but very close	separated	separated	united
predorsal scales	ctenoid	cycloid	cycloid	cycloid	cycloid	ctenoid
extention of predorsal scales	posterior part of interorbital space	behind eye	posterior part of interorbital space	forward into of interorbital space	posterior part of interorbital space	forword to anterior part of interorbital space
transforming cteni	several rows	single row	single row	absent	single row	several rows
5th pelvic-fin ray	branched	branched	branched	unbranched	branched	branched

Contd...

Contd...

Osteological characters						
distal radial of first-dorsal fin	present .2nd-5th spines.	present .3rd-4th spines.	absent	absent	absent	absent
suborbital bone	present	present	present	absent	absent	absent
number of epural	3	3	2	2	2	2
lateral extrascapula	present(large)	present(small)	absent	absent	absent	absent
prevomerine teeth	absent	absent	absent	present	absent	absent
ventral transverse shelf on posterior half of urohyal	absent	present	present	present	present	absent
shape of gill rakers on lower half outer surface of first gill arch	short and stout with small spines	long and finger -like with small spines	long and ribbon -like without spine	disk-like with small spines	long and finger-like with small spines	disk-like with small spines
gill rakers on inner part of 4th ceratobranchial	separated	separated	separated	continuous and forming one row	continuous and forming one row	separated
anterior projection on anterior tip of metapterygoid	present	absent	absent	absent	absent	present
connection between mesopterygoid and metapterygoid	present	absent	absent	present	absent	present
overlap with lower part of ectopterygoid and quadrate	absent	absent	absent	absent	absent	present

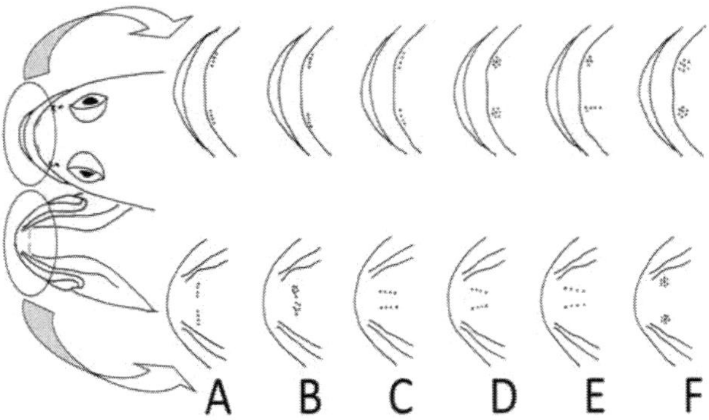

Fig. 1.4.2 Schematic illustration of sensory papilla patterns on snout (top) and chin (bottom) of the genera of Odontobutidae. A *Terateleotris*, B *Sineleotris*, C *Micropercops*, D *Perccottus*, E *Neodontobutis*, F *Odontobutis*.

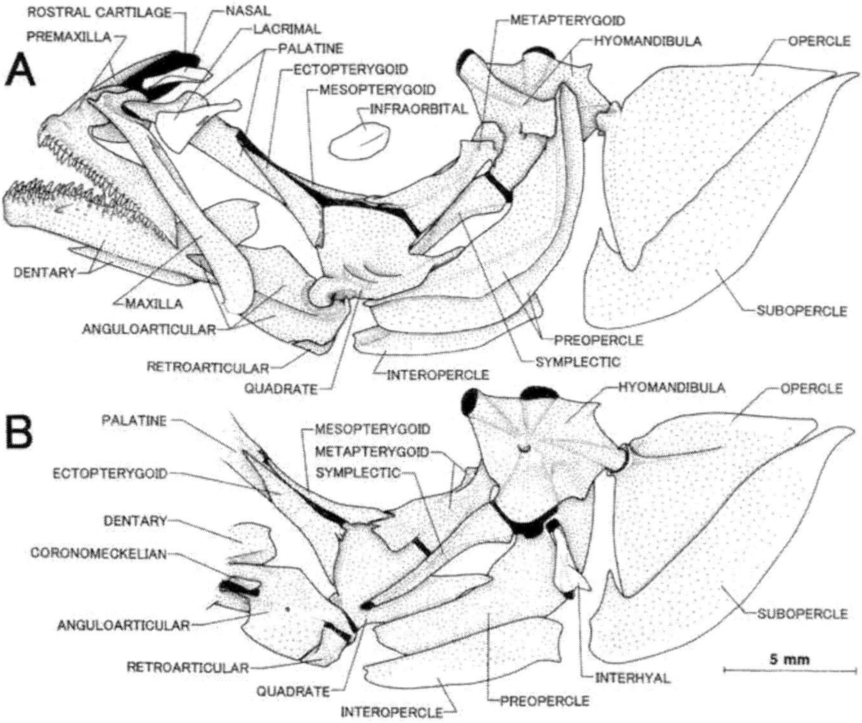

Fig. 1.4.3 Jaws, suspensorium, operculum and superficial bones (excepting lateral extrascapula) of *Trateleotris aspro*, from lateral (A) and medial (B) views (after Shibukawa *et al.*, 2001).

Terateleotris is known from only a single tributary of the Mekong River system of central Laos. Only one species, *Terateleotris aspro*, is known. Maximum body size is up to about 10 cm.

Genus *Sineleotris* Herre, 1940

Type Species: *Sineleotris saccharae* Herre, 1940

This genus is distinguished from all other genera of Odontobutidae by the following unique characters among Odontobutidae (Table 1.4.1): lateral extrascapula present, but small. This genus is also distinguished from others by the following combination of characters (Table 1.4.1): lateral line absent; connection between oculoscapular and preopercular canals absent; furrow absent from mid-lateral scales; cephalic sensory canals complete; sensory papillae on chin in two groups, often united (Fig. 1.4.2); sensory papillae on tip of snout forming transverse pattern (Fig. 1.4.2); sensory papilla normal shape, not forming barbel-like projection; right and left gill membrane united at vertical from mid-point of pupil; gill opening wide, extending to a vertical from mid-point of pupil; interorbital width narrower than, or nearly equal to, eye diameter; predorsal scales cycloid, reaching to behind eyes but not entering interorbital space; transforming cteni of scales forming single row; 5th pelvic-fin ray branched; distal radial of first-dorsal fin element present; suborbital bone present; three epurals; prevomerine teeth absent; ventral transverse shelf on posterior half of urohyal present; shape of gill rakers on lower half outer surface of first gill arch long and finger-like with small spines; gill rakers separated on inner part of 4th ceratobranchial; anterior projection on anterior tip of metapterygoid absent; connection between mesopterygoid and metapterygoid absent; overlap with lower part of ectopterygoid and quadrate absent.

Sineleotris is distributed in freshwaters of southeastern China. Only one species, *Sineleotris saccharae*, is known. Maximum body size is up to about 10 cm.

Genus *Micropercops* Fowler and Bean,1920

Type Species: *Micropercops dabryi* Fowler and Bean, 1920

This genus is distinguished from all other genera of Odontobutidae by the following unique characters among Odontobutidae (Table 1.4.1): cephalic sensory canals moderately developed; right and left gill membrane separated, but the attachment points of each side are very close together, gill opening extending to a vertical from posterior end of preoperculum, or midway between posterior end of preoperculum and posterior edge of eye; first and second epurals incompletely fused, consequently two epurals present; shape of gill rakers on lower half of outer surface of first gill arch long and ribbon-like without spine. This genus is also distinguished from others by the following combination of characters (Table 1.4.1): lateral line absent; connection between oculoscapular and preopercular canals absent; furrow absent from mid-lateral body scales; sensory papillae on chin in two longitudinal groups, right and left groups separate (Fig. 1.4.2); sensory papillae on tip of snout in transverse pattern (Fig. 1.4.2); sensory papilla normal shape, not forming barbel-like projection; interorbital width nearly equal to eye diameter; predorsal scales cycloid, reaching posterior part of interorbital space; transforming cteni of scales forming single row; 5th pelvic-fin ray branched; distal radial of first-dorsal fin element absent; suborbital bone present; lateral extrascapula absent; prevomerine teeth absent; ventral transverse shelf on posterior half of urohyal present; gill rakers separated on inner part of 4th ceratobranchial; anterior projection on anterior tip of metapterygoid absent; connection between mesopterygoid and metapterygoid absent; overlap with lower part of ectopterygoid and quadrate absent.

Micropercops dabryi Fowler and Bean, 1920, *Philypnus cinctus* Dabry de Thiersant, 1872, *Eleotris brachysoma* Bleeker, 1872, *Eleotris swinhonis* Günther, 1873, *Eleotris xanthi* Günther, 1888, and *Micropercops dabryi borealis* Nichols, 1930, may be members of *Micropercops*. However, the genus is in need of revision so some synonyms may exist.

The species of *Micropercops* are distributed in eastern Asia from Primorskii (Russia), some areas of Korea, to Fujian Province, in southern China. Maximum body size is less than around 10 cm.

Genus *Perccottus* Dybowski, 1877

Type Species: *Perccottus glenii* Dybowski, 1877

This genus is distinguished from all other genera of Odontobutidae by the following unique characters among Odontobutidae (Table 1.4.1): transforming cteni of scales absent (if present, only in large individuals); 5th pelvic-fin ray unbranched; prevomerine teeth present. This genus is also distinguished from others by the following combination of characters (Table 1.4.1): lateral line absent; connection between oculoscapular and preopercular canals absent; furrow absent from mid-lateral body scales; cephalic sensory canals absent; sensory papillae on chin in longitudinal pattern, right and left lines separate (Fig. 1.4.2); sensory papillae on tip of snout forming groups (Fig. 1.4.2); sensory papilla normal shape, not forming barbel-like projection; right and left gill membranes separate, gill opening extending to a vertical from a little behind mid-point between posterior end of preoperculum and posterior edge of eye; interorbital width nearly equal to or little wider than eye diameter; predorsal scales cycloid, reaching forward into interorbital space; distal radial of first-dorsal fin element absent; suborbital bone absent; two epurals; lateral extrascapula absent; ventral transverse shelf on posterior half of urohyal present; shape of gill rakers on lower half outer surface of first gill arch disk-like with small spines; gill rakers continuous and forming one row on inner part of 4th ceratobranchial; anterior projection on anterior tip of metapterygoid absent; connection between mesopterygoid and metapterygoid present; overlap with lower part of ectopterygoid and quadrate absent.

Perccottus is distributed in Primorskii (Russia) and northeastern China. Only one species, *Perccottus glenii*, is known. Maximum body size is up to about 20 cm.

Genus *Neodontobutis* Chen, Kottelat and Wu, 2002

Type Species: *Hypseleotris hainanensis* Chen, 1985

This genus is distinguished from all other genera of Odontobutidae by a unique character among the Odontobutidae (Table 1.4.1): each sensory papilla forming a barbel-like projection (Fig. 1.4.4). This genus is also distinguished from others by the following combination of characters (Table 1.4.1): lateral line absent; connection between oculoscapular and preopercular canals absent; furrow absent from mid-lateral body scales; cephalic sensory canals absent; sensory papillae on chin in longitudinal pattern, right and left line separate (Fig. 1.4.2); sensory papillae on tip of snout in groups or in longitudinal pattern(Fig. 1.4.2); right and left gill membrane separate, gill opening extending to a vertical from posterior end of eye, or midway between posterior end of preoperculum and posterior edge of eye; interorbital width narrower than eye diameter; predorsal scales cycloid, reaching posterior part of interorbital space; transforming cteni of scales forming single row; 5th pelvic-fin ray branched; distal radial of first-dorsal fin element absent; suborbital bone absent; two epurals; lateral extrascapula absent; prevomerine teeth absent; ventral transverse shelf on posterior half of urohyal present; shape of gill rakers on lower half outer surface of first gill arch long and finger-like with small spines; gill rakers continuous and forming one row on inner part of 4th ceratobranchial; anteriorly directed projection on anterior tip of metapterygoid absent;

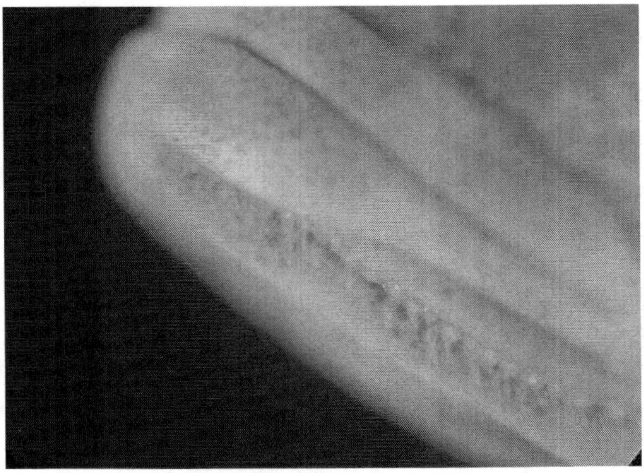

Fig. 1.4.4 Barbel-like projection of sensory papilla (ventral part of lower jaw) of *Neodontobutis hainanensis*.
Color image of this figure appears in the color plate section at the end of the book.

connection between mesopterygoid and metapterygoid absent; overlap with lower part of ectopterygoid and quadrate absent.

Neodontobutis hainanensis (Chen, 1985) is found in Hainan Island and the Pearl River system, mainland China. *Neodontobutis aurarmus* (Vidthayanon, 1995) is found in the Mekong River system of northeastern Thailand and central Laos.

Possible synonyms for *N. hainanensis* include *Micropercops macropectoralis* Mai, 1978, and *M. hotayensis* Mai, 1978 (the latter two species were described from northern Vietnam); this needs to be explored. Maximum body size does not exceed 10 cm.

Genus *Odontobutis* Bleeker, 1874

Type Species: *Eleotris obscura* Temminck and Schlegel,1845

This genus is distinguished from all other genera of Odontobutidae by the following unique characters among Odontobutidae (Table 1.4.1): sensory papillae on chin in groups, right and left group separate(Fig. 1.4.2); interorbital width wider than eye diameter; cycloid or ctenoid predorsal scales reaching forward to anterior part of interorbital space. This genus is also distinguished from others by the following combination of characters (Table 1.4.1): lateral line absent; connection between oculoscapular and preopercular canals absent; furrow absent from mid-lateral body scales; cephalic sensory canals absent or reduced; sensory papillae on tip of snout in groups (Fig. 1.4.2); sensory papilla normal shape, not forming barbel-like projection; right and left gill membranes united, gill opening extending to a vertical from posterior end of eye or middle of pupil; transforming cteni of scales forming several rows; 5th pelvic-fin ray branched; distal radial of first-dorsal fin element absent; ventral postcleithrum absent; suborbital bone absent; two epurals; lateral extrascapula absent; prevomerine teeth absent; ventral transverse shelf on posterior half of urohyal absent; shape of gill rakers on lower half outer surface of first gill arch disk-like with small spines; gill rakers separated on inner part of 4th ceratobranchial; anterior projection on anterior tip of metapterygoid present; connection between mesopterygoid and metapterygoid present; overlap with lower part of ectopterygoid and quadrate present.

This genus comprises the following eight species *Odontobutis obscura* (Temminck and Schlegel, 1845), *O. potamophila* (Günther, 1861), *O. interrupta* Iwata and Jeon, 1985, in Iwata *et al.* (1985), *O. platycephala* Iwata and Jeon, 1985, in Iwata *et al.* (1985), *O. haifengensis* Chen in Chen and Zheng (1985), *O. yaluensis* Wu, Wu and Xie, 1993, *O. hikimius* Iwata and Sakai, 2002, and *O. sinensis* Wu, Chen and Chong, 2002.

Odontobutis is found in western Japan, eastern Asia from the Yalu River (China), Jinlin Province to the Pearl River system, Guangxi Zhuang, and Hainan Island in China. Maximum body size is up to about 20 cm.

Philypnus Chalmersi Nichols and Pope, 1927

As mentioned above, *Philypnus chalmersi* is a member of Odontobutidae. However, *Philypnus* Valenciennes, 1837 in Cuvier and Valenciennes (1837), is a junior synonym of *Gobiomorus* Lacepède, 1800 (Reis *et al.*, 2003; Chen *et al.,* 2002), which belongs within the Eleotridinae (Hoese and Gill, 1993) and, thus, this generic name cannot be used for this species. Fowler (1962), Hoese and Gill (1993) and Larson (2001) assigned this species to *Perccottus,* and Chen *et al.* (2002) assigned it to *Sineleotris*. However, there are many morphological differences between this species and *Perccottus* as well as *Sineleotris*. The status of this species needs to be clarified.

This species is distributed from Hainan Island and southern China to northern Vietnam. Maximum body size is up to about 15 cm.

DISCUSSION

Terateleotris aspro has a complete lateral line. However, the condition in *Terateleotris aspro* does not correspond with that in *Rhyacichthys* and *Protogobius*. The lateral line of *Rhyacichthys* and *Protogobius* has two breaks at the body mid-point, with a single dorsal and ventral, and two caudal accessory canals. *Terateleotris aspro* possesses only a single uninterrupted lateral-line (Fig. 1.4.5) (Shibukawa *et al.*, 2001).

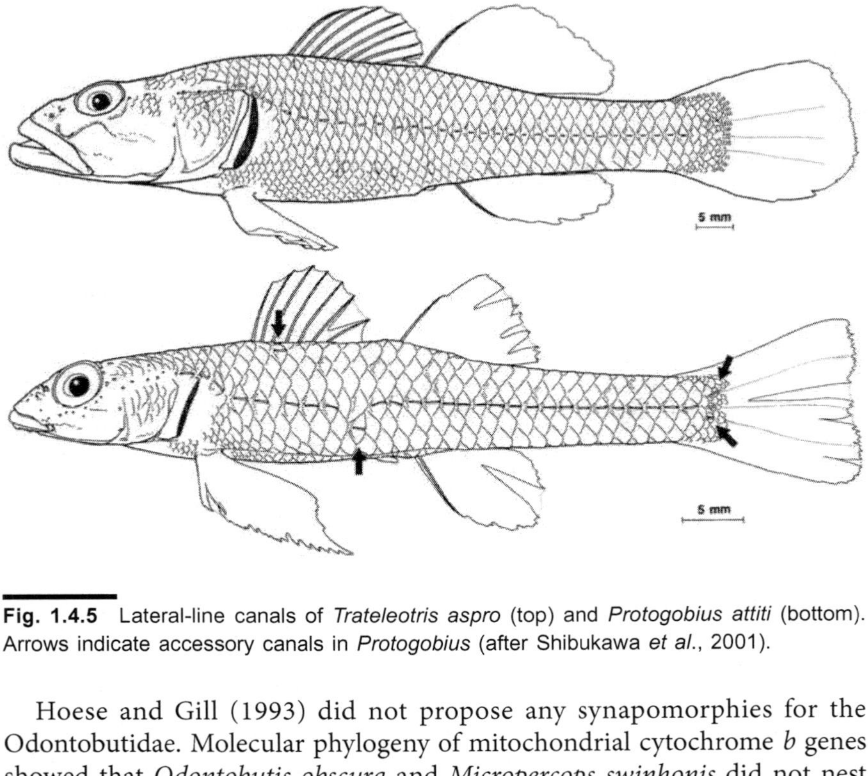

Fig. 1.4.5 Lateral-line canals of *Trateleotris aspro* (top) and *Protogobius attiti* (bottom). Arrows indicate accessory canals in *Protogobius* (after Shibukawa *et al.*, 2001).

Hoese and Gill (1993) did not propose any synapomorphies for the Odontobutidae. Molecular phylogeny of mitochondrial cytochrome *b* genes showed that *Odontobutis obscura* and *Micropercops swinhonis* did not nest in the same clade (Akihito *et al.*, 2000). Ahnelt and Göschl (2004) suggested that this family may not be monophyletic based on structures of the lateral-line system on the caudal fin.

Conversely, Iwata *et al.* (2001) suggested a possible synapomorphy between *Odontobutis obscura* and *Micropercops swinhonis* with respect to the developmental characteristics of the pelvic-fin bud position (fin buds widely separated). Springer and Johnson (2004) proposed a synapomorphy for *Odontobutis obscura*, *Perccottus glenii* and *Neodontobutis aurarmus* that involved dorsal gill-arch musculature.

The life history of odontobutid species is non-diadromous. *Micropercops swinhonis* is lacustrine with a planktonic period (Iwata *et al.*, 2001). *Odontobutis obscura, O. hikimius, O. interrupta* and *O. Platycephala* have a fluvial life history without a planktonic period (Iwata *et al.*, 1987, 1988a, 1988b; Doi and Aoyama, 2006). The other species of Odontobutidae are likely non-diadromous based on their distribution, where they occur inland far from the sea or not connected to rivers.

Species-level phylogeny among *Odontobutis*, *Micropercops* and *Perccottus* was discussed by Iwata *et al.* (1988b) and Iwata (1994). These workers argued that speciation occurred, changing from a lacustrine life history (with small egg and planktonic larval period) to a fluvial one (with large egg with direct development). A molecular phylogeny of *Odontobutis* based on the 12SrRNA gene was carried out by Ren and Zhang (2007).

A generic-level phylogeny of odontobutids based on morphology is still unknown. Thacker and Hardman (2005) reconfirmed that Odontobutidae is the most basal family next to Rhyacichthyidae among Gobioidei and showed that *Odontobutis* and *Perccottus* nested within the same clade. Chen *et al.* (2002) discussed the correct spelling of this family, i.e., Odontobutidae vs. Odontobutididae.

Perccottus tonkinensis Mai, 1978 (northern Vietnam and northern Laos according to Kottelat, 2001), "*Sineleotris namxamensis*" (northern Laos according to Kottelat, 2001) and *Odontobutis* sp. (Vietnam according to Serov *et al.* 2006), and *Odontobutis wui* Chen, 1934, appear to be the members of Odontobutidae, however, taxonomic and phylogenetic analyses remain to be conducted.

Milyeringa veritas Whitley 1945, a subterranean species from Western Australia was included in Odontobutidae based on molecular phylogeny by Thacker and Hardman (2005), however, decay indices were of very low value and the authors themselves noticed the weakness of their treatment. Clarification of the whole taxonomic and phylogenetic relationships among the species and genera of Odontobutidae remain to be completed.

Acknowledgments

I am grateful to Dr. Han-lin Wu (Shanghai Fisheries University, China) for the materials used in this study. I wish to express my gratitude to Dr. Akira Goto (Hokkaido University, Japan) for many considerations toward my research and Dr. Barbara Brown (American Museum of Natural History, U.S.A.) for an X-ray photograph of the holotype of *Philypnus chalmersi*. I wish to thank Dr. Chunguang Zhang, Dr. Yahui Zhao (Institute of Zoology, Chinese Academy of Sciences, China), Dr. Katsutoshi Watanabe (Kyoto University, Japan), Dr. Kei'ichiro Iguchi (National Research Institute of Fisheries Science, Japan), the staff of the Living Aquatic Resources Research Center (Lao P.D.R.) and the Faculty of Forestry, and Agriculture, National University of Laos (Lao P.D.R) and Dr. Koichi Shibukawa (Nagao Natural Environment Foundation, Japan) for their assistance and suggestions. This study was partly supported by Grants-in-Aid from the Ministry of Education, Culture, Sports, Science and Technology of Japan (no. 19405011).

References

Ahnelt, H. and J. Göschl. 2004. The pattern of the lateral-line system on the caudal fin of *Perccottus glehni* Dybowski, 1877 (Teleostei: Odontobutidae): with comments on the arrangement of the lateral-line system on the caudal fin of Gobioidei. *Proceedings of the California Academy of Sciences* 55: 358-371.

Akihito, A. Iwata, T. Kobayashi, K. Ikeno, T. Imanishi, H. Ono, Y. Umehara, C. Hamamatsu, K. Sugiyama, Y. Ikeda, K. Sakamoto, A. Fumihito, S. Ohno and T. Gojobori. 2000. Evolutionary aspects of Gobioid fishes based upon a phylogenetic analysis of mitochondrial cytochrome *b* genes. *Gene* 259: 5-15.

Bleeker, P. 1872. Mémoire sur la faune ichthyologique de Chine. *Nederlandsch Tijdschrift voor de Dierkunde*. 4: 113-154.

Bleeker, P. 1874. Esquisse d'un système naturel des Gobioïdes. *Archives Neerlandais de Sciences Naturelles, Haarlem* v. 9: 289-331.

Chen, I-S., M. Kottelat and H.-L. Wu. 2002. A new genus of freshwater sleeper (Teleostei: Odontobutididae) from southern China and mainland Southeast Asia. *Journal of the Fisheries Society of Taiwan* 29: 229-235. (In English with Chinese abstract).

Chen, J.T.F. 1934. Note sur les Gobioïdes de la collection du Muséum Métropolitain de Nankin. *Bulletin du Muséum National d'Histoire Naturelle (Série 2)* 6: 36-39.

Chen, W. and C.-Y. Zheng. 1985. On three new species of the family Eleotridae from Guangdong Province, China. *Journal of Science and Medicine of Jinan University* (1): 73-80. (In Chinese with English summary).

Cuvier, G. and A. Valenciennes. 1837. *Tome douzième. Suite du livre quatorzième. Gobioïdes. Livre quinzième. Acanthoptérygiens à pectorales pédiculées. Histoire naturelle des poissons. v. 12*: xxiv + 507pp + i., Pls. 344-368.

Dabry de Thiersant, P. 1872. *Nouvelles espèces de poissons de Chine. In: La pisciculture et la pêche en Chine*. Paris. Ix+ 196pp, 50pls.

Doi, T. and S. Aoyama. 2006. Ontogeny of the freshwater goby, *Odontobutis hikimius* in an aquarium. *Japanese Journal of Ichthyology* 53: 63-70. (In Japanese with English summary)

Dybowski, B.I. 1877. Fishes of the Amur water system. *Izvestiya Zapadno-Sibirskogo otdiela Imperatorskago russkago geograpficheskago obshchestva* 8 (1-2): 1-29.

Fowler, H.W. 1962. A synopsis of the fishes of China. Part X. The Gobioid fishes. *Quarterly Journal of the Taiwan Museum (Taipei)* 15: 1-77.

Fowler, H.W. and B.A. Bean. 1920. A small collection of fishes from Soochow, China, with descriptions of two new species. *Proceedings of the United States National Museum* to 58 (2338): 307-321.

Günther, A. 1861. *Catalogue of the fishes in the British Museum. Catalogue of the acanthopterygian fishes in the collection of the British Museum. 3. Gobiidae, Discoboli, Pediculati, Blenniidae, Labyrinthici, Mugilidae, Notacanthi.*: xxv + 586pp + x.

Günther, A. 1873. Report on a collection of fishes from China. *Annals and Magazine of Natural History (Series 4)* 12 (69): 239-250.

Günther, A. 1888. Contribution to our knowledge of the fishes of the Yangtsze-Kiang. *Annals and Magazine of Natural History (Series 6)* 1: 429-435.

Herre, A.W.C.T. 1940. Notes on fishes in the Zoölogical Museum of Stanford University, VIII. A new genus and two new species of Chinese gobies with remarks on some other species. *Philippine Journal of Science* 73: 293-299, 1pl.

Hoese, D.F. and A.C. Gill. 1993. Phylogenetic relationships of eleotridid fishes (Perciformes: Gobioidei). *Bulletin of Marine Science* 52: 415-440.

Iwata, A. 1994. *Systematics and speciation of the fishes of the genus* Odontobutis *(Gobioidei: Odontobutidae)*. Doctoral thesis, Tokyo University of Fisheries: ii+253pp. (In Japanese).

Iwata, A. and H. Sakai. 2002. *Odontobutis hikimius* n. sp.: A new freshwater goby from Japan, with a key to species of the genus. *Copeia* 2002: 104-110.

Iwata, A., S.R. Jeon, N. Mizuno and K.C. Choi. 1985. A revision of the eleotrid goby genus *Odontobutis* in Japan, Korea and China. *Japanese Journal of Ichthyology* 31: 73-388.

Iwata, A., S.R. Jeon, N. Mizuno and K.C. Choi. 1987. Embryonic development of a gobiid fish, *Odontobutis obscura interrupta. Science Report of the Yokosuka City Museum* 35: 29-36.

Iwata, A., S.R. Jeon, N. Mizuno and K.C. Choi. 1988a. Embryonic development of a gobiid fish, *Odontobutis platycephala. Korean Journal of Limnology* 21: 25-32. (In Korean with English summary).

Iwata, A., S.R. Jeon, N. Mizuno and K.C. Choi. 1988b. Larval development of a gobiid fish, *Odontobutis obscura obscura* in comparison with that of *O. o. interrupta* and of *O. platycephala. Japanese Journal of Ichthyology* 35: 371-381. (In Japanese with English summary).

Iwata, A., H. Sakai, K. Shibukawa and S.-R. Jeon. 2001. Developmental characteristics of a freshwater goby, *Micropercops swinhonis*, from Korea. *Zoological Science* 18: 91- 97.

Kottelat, M. 1998. Fishes of the Nam Theun and Xe Bangfai basins, Laos, with diagnoses of twenty-two new species (Teleostei: Cyprinidae, Balitoridae, Cobitidae, Coiidae and Odontobutidae). *Ichthyological Exploration of Freshwaters* 9: 1-128.

Kottelat, M. 2001. *Fishes of Laos*. WHT Publications (Pte) Ltd., Colombo.

Lacepède, B.G.E. 1800. *Histoire naturelle des poissons* 2: lxiv + 632pp., 20pls.

Larson, H.K. 2001. A revision of the gobiid fish genus *Mugilogobius* (Teleostei: Gobioidei), and its systematic placement. *Records of the Western Australian Museum Suppl.* 62: 1-233.

Mai, D.Y. 1978. *Identification of the Fresh-water Fishes of North Vietnam*. Ha Noi, Scientific & Technology, Publisher. (In Vietnamese).

Nelson, J.S. 2006. *Fishes of the World*. John Wiley & Sons. Inc., Hoboken. 4th edition.

Nichols, J.T. 1930. Some Chinese freshwater fishes. XXII. *Aphyocypris* describing a race from Shantung. XXIII. Gobies referable to the genus *Micropercops*. *American Museum Novitates* 402: 1-4.

Nichols, J.T. and C.H. Pope. 1927. The fishes of Hainan. *Bulletin of the American Museum of Natural History* 54 (art. 2): 321-394, 26pls.

Reis, R.E., S.O. Kullander and C.J. Ferraris, Jr. 2003. *Checklist of the Freshwater Fishes of South and Central America. CLOFFSCA*. Edipucrs, Porto Alegre, Brazil.

Ren, G. and Q. Zhang. 2007. Molecular phylogeny of the genus *Odontobutis* based upon partial sequences of mitochondrial 12SrRNA genes. *Acta Hydrobiologica Sinica* 31: 473-478 (In Chinese with English summary).

Serov, D.V., V.K. Nezdoliy and D.S. Pavlov. 2006. *The Freshwater Fishes of Central Vietnam*. KMK Scientific Press, Moscow.

Shibukawa, K., A. Iwata and S. Viravong 2001. *Terateleotris*, a new gobioid fish genus from the Laos (Teleostei, Perciformes), with comments on its relationships. *Bulletin of the National Science Museum (Tokyo)* 27: 229-257.

Springer, V.G. and G.D. Johnson. 2004. Study of the dorsal gill-arch musculature of teleostome fishes, with special reference to the Actinopterygii. *Bulletin of the Biological Society of Washington* 11: 1-260. (with Appendix by V.G. Springer and T.M. Orrell).

Temminck, C.J. and H. Schlegel. 1845. Pisces. In: *Fauna Japonica, sive descriptio animalium quae in itinere per Japoniam suscepto annis 1823-30 collegit, notis observationibus et adumbrationibus illustravit P. F. de Siebold. Parts 7-9*. pp. 113-172.

Thacker, C.E. and M.A. Hardman. 2005. Molecular phylogeny of basal gobioid fishes: Rhyacichthyidae, Odontobutidae, Xenisthmidae, Eleotridae (Teleostei: Perciformes: Gobioidei). *Molecular Phylogenetics and Evolution* 37: 858-871.

Vidthayanon, C. 1995. *Odontobutis aurarmus*, a new species of odontobutid goby from wetlands of the Mekong basin, Thailand. *Ichthyological Exploration of Freshwaters* 6: 235-242.

Whitley, G.P. 1945. New sharks and fishes from Western Australia. Part 2. *Australian Zoologist* 11: 1-42, 1pl.

Wu, H.-L. 2008. *Philypnus chalmersi.* In: *Fauna Sinica Ostichthyes Perciformes (V) Gobioidei*, H.L.Wu *et al.* (eds.). Science Press, Beijing, pp. 194-196.

Wu, H.-L., X.-Q. Wu and Y.-H. Xie. 1993. A revision of the genus *Odontobutis* from China with description of a new species. *Journal of Shanghai Fisheries University* 2: 52-61. (In Chinese with English summary).

Wu, H.-L., I-S. Chen, and D.-H. Chong. 2002 . A new species of the genus *Odontobutis* (Pisces, Odontobutidae) from China. *Journal of Shanghai Fisheries University* 11: 6-13. (In Chinese with English abstract).

CHAPTER

1.5

Systematics of Butidae and Eleotridae

Christine Thacker

Eleotridae and Butidae are two clades of gobioid fishes, both differing from relatives Gobiidae and Gobionellidae in that they are generally less morphologically simplified, without joined pelvic fins but with separate spinous and rayed dorsal fins. Eleotridae and Butidae share with Rhyacichthyidae and Odontobutidae the presence of six (rather than five) branchiostegal rays, a character commonly used to separate Gobiidae and Gobionellidae (with five rays) from the remainder of Gobioidei (Hoese, 1984; Hoese and Gill, 1993). Species of Eleotridae and Butidae tend to attain a larger size than most other gobioids (most are 10 – 25 cm in length, with some smaller and a few much larger), and also generally exhibit heavier scalation. Most species are known from fresh or brackish water, with exceptions in Eleotridae (*Calumia*, *Gobiomorphus*, *Grahamichthys*, *Thalasseleotris* and former xenisthmid genera *Allomicrodesmus*, *Paraxenisthmus*, *Rotuma*, *Tyson*, and *Xenisthmus*). Eleotridae also has a wider distribution than

Author's address: Research and Collections, Section of Ichthyology, Natural History Museum of Los Angeles County, 900 Exposition Blvd., Los Angeles, CA 90007 USA. E-mail: thacker@nhm.org

Butidae, found circumglobally in tropical and some temperate waters, while Butidae is restricted to the Old World tropics (Africa, Asia, Australia, and Oceania). Species of Butidae and Eleotridae tend to be robust and heavy bodied, particularly as compared to Gobiidae and Gobionellidae. In shape they range from stout and cylindrical (benthic) to laterally compressed (nektonic). Currently, 39 species of Butidae and 130 species of Eleotridae are described (data from Eschmeyer, 2008). Valid genera for Butidae and Eleotridae as well as other six-branchiostegal rayed gobioid families Rhyacichthyidae and Odontobutidae are listed in Table 1.5.1. As with other gobioids, new species of Butidae and Eleotridae continue to be discovered and identified (Hoese and Allen, 1983, 1987; Allen and Hoese, 1991; Allen and Jenkins, 1999a, 1999b; Allen, 2003; Hoese and Reader, 2006; Thacker *et al.*, 2006; Larson, 2007), and both butid and eleotrid systematics are active areas of research. In both families, just one or two genera contain the bulk of the species. The biggest fraction of the species in Butidae are *Oxyeleotris* (16 of 39) and similarly, the two largest genera in Eleotridae are *Eleotris* (31) and *Mogurnda* (26). Most of the remaining genera in both families have seven or fewer species, and many are monotypic.

Eleotrids and butids comprise a small but interesting portion of the coastal and occasionally inland freshwater fauna on all continents except Europe and Antarctica, but are particularly notable for their radiations in and Australia, New Guinea, and New Zealand (Butidae: *Oxyeleotris*, and Eleotridae including *Hypseleotris*, *Philypnodon*, *Allomogurnda*, *Mogurnda*, and *Gobiomorphus*). In those ecosystems, Ostariophysi, which are often dominent in freshwater, are mostly not present. Instead, endemic species and genera have evolved with a unique complement of freshwater fishes, many of which have secondarily invaded freshwater from marine environments (Allen, 1991; Allen *et al.*, 2002). Some of these eleotrid species exhibit vibrant breeding coloration in males, in conjunction with specific mating behaviors (Unmack, 2000); and the population genetics of some Australian species has been used to investigate the evolution of both the rivers they inhabit, as well as the species themselves (Hurwood and Hughes, 1998; Bertozzi *et al.*, 2000; Thacker *et al.*, 2007, 2008). Species-level phylogeny has also been investigated for one of these genera, *Hypseleotris* (Thacker and Unmack, 2005). Instances of dwarfism are scattered throughout both Eleotridae and Butidae, including freshwater taxa from Africa (*Kribia*), Australia (*Philypnodon macrostomus* and "*Oxyeleotris*" *nullipora*, a dwarf species not closely related to other *Oxyeleotris*, Thacker and Hardman, 2005), and the neotropics (*Microphilypnus* and *Leptophilypnus*) as well as the marine eleotrid genera *Allomicrodesmus*, *Calumia*, *Grahamichthys*,

Table 1.5.1 Valid genera of Rhyacichthyidae, Odontobutidae, Butidae and Eleotridae.

Rhyacichthyidae	***Odontobutidae***	***Butidae***	***Eleotridae***
Protogobius	*Micropercops*	*Bostrychus*	*Allomicrodesmus*
Rhyacichthys	*Milyeringa*	*Butis*	*Allomogurnda*
Terateleotris	*Odontobutis*	*Incara*	*Belobranchus*
	Percottus	*Kribia*	*Bunaka*
		Ophiocara	*Calumia*
		Oxyeleotris	*Dormitator*
		Prionobutis	*Eleotris*
		Typhleotris	*Erotelis*
			Gobiomorphus
			Gobiomorus
			Grahamichthys
			Guavina
			Hemieleotris
			Hypseleotris
			Kimberlyeleotris
			Leptophilypnus
			Microphilypnus
			Mogurnda
			Ophieleotris
			Paraxenisthmus
			Philypnodon
			Ratsirakea
			Rotuma
			Tateurndina
			Thalasseleotris
			Tyson
			Xenisthmus

Paraxenisthmus, *Rotuma*, *Thalasseleotris*, *Tyson*, and *Xenisthmus* (Allen *et al.*, 2002; Thacker and Hardman, 2005; Thacker *et al.*, 2006; Hoese and Reader, 2006).

Phylogenetic relationships of Butidae and Eleotridae, and of groups within those families, have rarely been assessed with morphological data. Hoese and Gill (1993) provided diagnostic characters for Eleotridae (then Eleotrinae), but not Butidae (then Butinae), and they did not include the former xenisthmid genera in their survey of eleotrid characters. Morphology and evolution of those genera has been most comprehensively documented by Springer (1983, 1988), and a phylogeny was provided by Gill and Hoese (1993). Butidae and Eleotridae have historically been grouped together as subfamilies within Eleotridae (Hoese, 1984; Hoese and Gill, 1993), but recent molecular phylogenetic analyses (Thacker, 2009) have demonstrated that Butidae is sister to a clade containing Gobiidae and Gobionellidae, with Eleotridae sister to that clade of three families. Thus, recognition of a group consisting of Butidae and Eleotridae would be paraphyletic, and they are recognized separately at the family level. Similarly, the five genera in the former family Xenisthmidae are included in Eleotridae so as not to render Eleotridae paraphyletic (Thacker and Hardman, 2005; Thacker, 2009). These five genera are all marine, small and elongate with little pigmentation and some reductions in the skeletal system, but with six branchiostegal rays (Gill and Hoese, 1993; Springer, 1983, 1988).

Within Eleotridae and Butidae, Birdsong *et al.* (1988) considered axial skeletal configuration as well as other morphological characters for several genera, placing most butids in their *Butis* group (which also contained the eleotrid *Gobiomorus*) except *Bostrychus* and *Kribia*, which were placed in their *Dormitator* group. The remaining eleotrids considered were classified into the *Eleotris*, *Gobiomorphus*, or *Hypseleotris* groups, with a separate *Xenisthmus* group for the former xenisthmid genera. These groups each share combinations of characters, but only in the case of the *Eleotris* group are the groupings confirmed with molecular phylogenetic analysis. Thacker (2009) did not recover any of Birdsong's groupings with the exception of demonstration of a close relationship between *Eleotris*/*Erotelis* and *Calumia*, a small, reef-dwelling eleotrid. However, molecular phylogenetic studies do confirm monophyly of most genera in both families; the most prominent exception is the butid genus *Oxyeleotris*, in which one species, the tiny *O. nullipora*, is recovered apart from the remainder of the genus (Thacker and Hardman, 2005; Thacker, 2009).

Miller (1998) and Pezold and Cage (2002) both considered the systematics of *Eleotris* and *Erotelis*. From surveys of morphology including sensory pores and papillae and scalation, Miller (1998) concluded that *Eleotris* and *Erotelis* were not distinct enough to warrant separate recognition. In contrast, Pezold and Cage (2002) documented additional morphological characters such as the caudal and median fin configurations, shape, and coloration, and postulated that *Eleotris* and *Erotelis* should both be recognized. Molecular phylogenetic hypotheses (Thacker and Hardman, 2005; Thacker, 2009) support the contention of Miller (1998) that *Erotelis* is nested within *Eleotris*. Not all *Eleotris* species were examined in those molecular studies, but of the species considered, the neotropical *Erotelis* were most closely related to two Pacific *Eleotris* species.

In Eleotridae, some genera (and some species within cosmopolitan genera) have invaded North and South America. These include some species of *Eleotris* (including all *Erotelis*), all species of *Dormitator* except one known from West Africa, and all species of *Gobiomorus*, *Guavina*, *Hemieleotris*, *Leptophilypnus*, and *Microphilypnus*. Invasion of the neotropics has occurred at least twice among these taxa; once in the neotropical *Eleotris* species, and as a separate event in the other genera (Thacker, 2009). Eleotridae also includes several instances of Neotropical transisthmian species pairs, in which the sister taxa inhabit different coasts of the isthmus of Panama. These species pairs exhibit different levels of genetic divergence, consistent with various ages for interruption of gene flow by uplift of the isthmus, with drainages becoming isolated from one another gradually (Thacker and Hardman, 2005). Taken together, morphological and molecular data have provided a range of complementary insights into the systematics, evolution, biogeography, and population-level processes in both Butidae and Eleotridae.

References

Allen, G.R. 1991. *Field guide to the freshwater fishes of New Guinea*. Publication No. 9 of the Christensen Research Institute, Madang, Papua New Guinea.

Allen, G.R. 2003. *Allomogurnda*, a new genus of gudgeon (Eleotridae) from freshwaters of New Guinea, with descriptions of seven new species. *Fishes of Sahul* 13: 978-997.

Allen, G.R. and D.F. Hoese. 1991. A review of the genus *Mogurnda* (Pisces: Eleotrididae) form New Guinea with descriptions of three new species. *Ichthyological Exploration of Freshwaters* 2: 31-46.

Allen, G.R. and A.P. Jenkins. 1999a. A review of the Australian freshwater gudgeons, genus *Mogurnda* (Eleotridae) with descriptions of three new species. *Aqua* 3: 141-156.

Allen G.R. and A.P. Jenkins. 1999b. Two new species of *Mogurnda* (Osteichthys: Eleotrididae) from the Etna Bay Region, Irian Jaya, Indonesia. *Ichthyological Exploration of Freshwaters* 10: 237-246.

Allen, G.R., S.H. Midgley and M. Allen. 2002. *Field guide to the freshwater fishes of Australia*. Perth: Western Australian Museum.

Bertozzi, T., M. Adams and K.F. Walker. 2000. Species boundaries in carp gudgeons (Eleotridae: *Hypseleotris*) from the River Murray, South Australia: evidence for multiple species and extensive hybridization. *Marine and Freshwater Research* 51: 805-815.

Birdsong, R.S., E.O. Murdy and F.L. Pezold. 1988. A study of the vertebral column and median fin osteology in gobioid fishes with comments on gobioid relationships. *Bulletin of Marine Science* 42: 174-214.

Eschmeyer, W.N. (Ed.). 2008. *Online Version of the Catalog of Fishes* www.calacademy.org/research/ichthyology/catalog/fishcatsearch.html; updated 18 September, 2008, accessed November, 2008.

Gill, A.C. and D.F. Hoese. 1993. *Paraxenisthmus springeri*, new genus and species of gobioid fish from the West Pacific, and its phylogenetic position within the Xenisthmidae. *Copeia* 1993: 1049-1057.

Hoese, D.F. 1984. Gobioidei: Relationships. In: *Ontogeny and Systematics of Fishes.* H.G. Moser (ed.). Special Publication of the American Society of Ichthyologists and Herpetologists No.1. Allen Press, Lawrence, Kansas. pp.588-591.

Hoese, D.F. and G.R. Allen. 1983. A review of the gudgeon genus *Hypseleotris* (Pisces: Eleotridae) of Western Australia, with descriptions of three new species. *Records of the Western Australian Museum* 10: 243-261.

Hoese, D.F. and G.R. Allen. 1987. New Australian Fishes Part 10. A new genus and two new species of freshwater eleotridid fishes (Gobioidei) from the Kimberly Region of Western Australia. *Memoirs of the Museum of Victoria* 48: 35-42.

Hoese, D.F. and A.C. Gill. 1993. Phylogenetic relationships of eleotridid fishes (Perciformes: Gobioidei). *Bulletin of Marine Science* 52: 415-440.

Hoese, D.F. and S. Reader. 2006. Description of a new species of dwarf *Philypnodon* (Teleostei: Gobioidei: Eleotridae) from south-eastern Australia. *Memoirs of the Museum of Victoria* 63: 15-19.

Hurwood, D.A. and J.M. Hughes. 1998. Phylogeography of the freshwater fish, *Mogurnda adspersa*, in streams of north-eastern Queensland, Australia: evidence for altered drainage patterns. *Molecular Ecology* 7: 1507-1517.

Larson, H.K. 2007. A new species of carp gudgeon, *Hypseleotris* (Pisces: Gobioidei: Eleotridae), from the Katherine River system, Northern Territory. *The Beagle, Records of the Museums and Art Galleries of the Northern Territory* 23: 111-117.

Miller, P.J. 1998. The West African species of *Eleotris* and their systematic affinities (Teleostei: Gobioidei). *Journal of Natural History* 32: 273-296.

Pezold, F. and B. Cage. 2002. A review of the spinycheek sleepers, genus *Eleotris* (Teleostei: Eleotridae) of the western hemisphere, with comparison to the West African species. *Tulane Studies in Zoology and Botany* 31: 19-63.

Springer, V.G. 1983. *Tyson belos*, new genus and species of Western Pacific fish (Gobiidae, Xenisthminae), with discussions of gobioid osteology and classification. *Smithsonian Contributions to Zoology* 390: 1-40.

Springer, V.G. 1988. *Rotuma lewisi*, new genus and species of fish from the Southwest Pacific (Gobioidei, Xenisthmidae). *Proceedings of the Biological Society of Washington* 101: 530-539.

Thacker, C.E. 2009. Phylogeny of Gobioidei and placement within Acanthomorpha, with a new classification and investigation of diversification and character evolution. *Copeia* 2009: 93-104.

Thacker, C.E. and M.A. Hardman. 2005. Molecular phylogeny of basal gobioid fishes: Rhyacichthyidae, Odontobutidae, Xenisthmidae, Eleotridae (Teleostei: Perciformes: Gobioidei). *Molecular Phylogenetics and Evolution* 37: 858-871.

Thacker, C.E. and P.J. Unmack. 2005. Phylogeny and biogeography of the eleotrid genus *Hypseleotris* (Teleostei: Gobioidei: Eleotridae) with redescription of *H. cyprinoides*. *Records of the Australian Museum* 57: 1-13.

Thacker, C.E., F. Pezold and R. Sutkuss. 2006. Redescription of the dwarf neotropical eleotrid genus *Leptophilypnus* (Teleostei: Gobioidei), including a new species and comments on *Microphilypnus*. *Copeia* 2006: 489-499.

Thacker, C.E., P. J. Unmack, L. Matsui and N. Rifenbark. 2007. Patterns of molecular evolution and phylogeography within and among *Hypseleotris* species in southeastern Australian drainages. *Journal of Biogeography* 34: 1518-1533.

Thacker, C.E., P.J. Unmack, L. Matsui, P. Duong and E. Huang. 2008. Phylogeography of *Philypnodon* species (Teleostei: Perciformes: Gobioidei: Eleotridae) across southeastern Australia: testing patterns of connectivity across drainage divides and among coastal rivers. *Biological Journal of the Linnean Society* 95: 175-192.

Unmack, P.J. 2000. The genus *Hypseleotris* of southeastern Australia: its identification and breeding biology. *Fishes of Sahul* 14: 647-657.

CHAPTER

1.6

Systematics of Gobionellidae

Frank Pezold

The classification adopted for this text (Thacker, 2009) is the culmination of recent molecular analyses comparing mitochondrial gene sequences of gobioid fishes. My observations and comments are largely focused on their relevance to the recognition of a family Gobionellidae including the previously recognized subfamily Gobionellinae (Pezold, 1993; Larson, 2001; Thacker, 2003) i.e. gobionellines *sensu stricto,* and the amblyopines, oxudercines and sicydiines (*sensu* Pezold, 1993). Additional observations are given on the Gobiidae clade (*sensu* Thacker, 2009), and the sister group relationship proposed between the Butidae and the two goby clades recovered by Thacker (2009). The clade-based classification proposed by Thacker (2009) names putative monophyletic groups, but no synapomorphy is given for the family Gobionellidae, or for any of the families recognized therein.

ORIGINS OF THE GOBIONELLINAE

Bleeker (1874) was the first person to use a derivative of *Gobionellus* for a grouping of taxa above the genus-level. The subphalanx Gobionelli

Author's address: Texas A&M University - Corpus Christi, Corpus Christi, TX 78412, USA.
E-mail: frank.pezold@tamucc.edu

identified by Bleeker contained the genera *Gobionellus* and *Synechogobius*, a synonym of *Acanthogobius*. Other species recognized as gobionellines today were scattered across several other named suprageneric groups along with non-gobionelline species (see Larson, 2001 for a synopsis). Miller (1973) first proposed a subfamily Gobionellinae as one of seven subfamilies of the Gobiidae. In his classification, gobioid fishes were simply split between two families - a monotypic Rhyacichthyidae and the Gobiidae which contained the roughly 2000 other species. Gobionellinae *sensu* Miller (1973) included typical gobiids like *Gobionellus*, *Stenogobius*, and *Oxyurichthys* (all three genera later recognized as members of a monophyletic *Stenogobius* group by Larson, 2001), as well as species generally recognized now as amblyopines and periophthalmines. Miller (1973) recognized another large group of gobies as the Gobiinae which was largely consistent with the Gobiinae of today, but also included microdesmids, *Rhinogobius*, sicydiines and ptereleotrines. Kraemeriidae formed another subfamily, as did *Tridentiger* and *Triaenopogon*. Subsequently, Birdsong *et al.* (1988) recognized 32 groups of gobioid fishes (based upon postcranial axial osteology), half of which were hypothesized as monophyletic. A *Gobionellus* Group was characterized by a shared spinous dorsal-fin pterygiophore insertion pattern of 3-12210, 2 epurals and 26 vertebrae. None of the characters could be identified as a synapomorphy. Included in the group were *Calamiana*, *Ctenogobius*, *Gnatholepis*, *Gobionellus*, *Mugilogobius*, *Oligolepis*, *Oxyurichthys*, *Pseudogobiopsis*, *Stenogobius* and *Tamanka*.

Harrison (1989) recognized a monophyletic *Ctenogobius* lineage including *Ctenogobius*, *Gnatholepis*, *Gobionellus*, *Evorthodus*, *Gobioides*, *Oligolepis*, *Stenogobius*, Sicydiinae and *Awaous* based upon shared derived features of the palatopterygoquadrate complex. A monophyletic *Oxyurichthys* lineage was also proposed based upon shared possession of a shared stubby palatine. This latter group included its gobionelline namesake *Oxyurichthys* and representative amblyopines and oxudercines.

Pezold (1993) recognized the subfamily Gobionellinae as one of five subfamilies of the Gobiidae. Although not diagnosed, the Gobionellinae included all species characterized by two anterior interorbital pores, two epurals and a 3-12210 spinous dorsal pterygiophore pattern. It also included some taxa in which one or more of these characters were modified to represent autapomorphies (e.g. *Gobioides* with a 3-12201 pterygiophore pattern) or convergences with taxa in other subfamilies (e.g. a single epural in various species of *Gobionellus*, *Ctenogobius*, *Evorthodus*, *Gobioides* and *Awaous* [Pezold 2004]). Pezold (1993) also included three groups of gobies previously, determined as monophyletic by Birdsong *et al.* (1988), (the *Acanthogobius, Astrabe* and *Chasmichthys* groups) and suggested they might themselves form a monophyletic temperate northern Pacific group. Monophyly was not proposed for the Gobionellinae; none of the three unifying characters was recognized as a synapomorphy because of limited understanding of character state distributions and polarities. Monophyly was hypothesized by Pezold (1993) for the revised subfamilies Gobiinae and Amblyopinae, in addition to the Sicydiinae and Oxudercinae proposed as monophyletic by earlier studies (Hoese, 1984; Harrison, 1989; Murdy, 1989).

The subfamily Gobionellinae was also recognized by Larson (2001). Adding to the characters observed by Pezold (1993), she indicated they also possess 25-28 vertebrae and 2 or 3 (occasionally 4) anal pterygiophores before the first hemal spine. The temperate northern Pacific gobionellines are characterized by a proliferation of vertebrae however, having counts as high as 42 in some species. In addition to the northern Pacific group, she recognized two other groups proposed as monophyletic: a *Mugilogobius* group and a *Stenogobius* group. The *Mugilogobius* group is characterized by a number of characters. These include the absence of an anterior nasal pore (reduction of the canals on the snout with a single pore on each canal anterior to the orbits), the presence of villi on the head, infraorbital pores present in those species with oculoscapular canals, and sensory papillae on the cheek primarily organized as longitudinal rows. Larson (2001) regarded all but the infraorbital pores as derived features. The *Stenogobius* species group was found to be characterized by canals on the snout including a pair of pores anterior to the orbits, no villi on the head, no infraorbital pores on the oculoscapular canal, and sensory papillae on the cheeks forming transverse rows. Only the loss of the infraorbital pore was regarded as evolutionarily derived. However, Larson (2001) refrained from naming the *Mugilogobius* and *Stenogobius* groups because no single derived feature was possessed by all members of either clade. Larson (2001) diagnosed 12 genera in the *Mugilogobius* group and provided a synoptic evaluation of relationships among the genera. The *Stenogobius* group identified by Larson (2001) is largely consistent with the *Ctenogobius* lineage of Harrison (1989).

In a phylogenetic analysis of *Gobionellus* species, Pezold (2004) examined relationships among genera of the *Stenogobius* group and delimited six monophyletic genera—*Ctenogobius, Evorthodus, Gobioides, Oligolepis, Oxyurichthys* and *Stenogobius*. The genus *Gobionellus* was differentially diagnosed, but evidence for monophyly not including *Gobioides* species was equivocal. Species historically assigned to *Gobionellus* (e.g. Gilbert and

Randall, 1979) fell into two groups—a *Gobionellus-Gobioides* clade sister to an *Awaous-Stenogobius* clade, and a *Ctenogobius-Oxyurichthys* clade which was in turn sister group to an *Evorthodus-Oligolepis* clade.

MOLECULAR ANALYSES AND THE EMERGENCE OF GOBIONELLIDAE

The first attempt to examine gobioid relationships using gene sequences was by Akihito *et al.* (2000). They compared partial sequences (380 bp) of the mtDNA gene cytochrome *b* for 28 gobioid fishes. Six species clusters, loosely joined by branches with low bootstrap values, were recovered. Of note was the high support for the grouping of the representative amblyopine species (*Taenioides limicola)* and the representative oxudercine species (*Periophthalmus argentilineatus*). In Akihito *et al.*'s study (2000) these two species were then clustered with three gobionelline species (*Acanthogobius flavigaster*, *Tridentiger bifasciatus* and *Tridentiger obscurus*), but with low support. These findings promised that more insight might be obtained from mtDNA sequences because they corroborated Harrison's (1989) placement of *Periophthalmus* and *Taenioides* in the Gobionellinae based on shared derived features of the suspensorium. Akihito *et al.*'s (2000) analysis also associated *Micropercops swinhonis*, an odontobutid sleeper, with this group, but again with low support. These results suggested that, while promising, the information obtained from mtDNA analyses would have to be interpreted cautiously and corroborated with other datasets, and that taxon sampling would need to be broadened and increased to more adequately represent the diversity comprised by gobioid fishes.

Following the Akihito *et al.* (2000) study, Wang *et al.* (2001) analyzed complete sequences of the 12s rRNA gene for 43 gobioid species spanning 32 genera in eight families or subfamilies. Their findings first identified some of the major groupings recognized in the classification followed here (Thacker, 2009). Significantly, a monophyletic Butinae was determined to be the sister group to a clade containing two monophyletic groups. One of these groups included the Gobionellinae, Sicydiinae and Oxudercinae (all *sensu* Pezold, 1993); the other monophyletic group consisted of the Gobiinae (*sensu* Pezold, 1993) plus the Microdesmidae (Ptereleotridae of Thacker, 2000). It was first noted by Wang *et al.* (2001) that the molecular phylogeny based on mtDNA was supported by the reduction of 6 branchiostegals to 5 for the clade containing these two monophyletic groups. They also observed that the lineage including the gobiines and microdesmids (ptereleotrids of Thacker, 2000) shared a derived trait, the possession of a single epural, while most species in the gobionelline clade have two epurals. A single epural also appears in some gobionellines and most sicydiines, but that was interpreted as homoplasious, representing a parallel evolutionary trend within the gobionelline clade. Subsequently, Kim *et al.* (2004) sequenced the complete mitochondrial genome of *Acanthogobius hasta*. Their phylogenetic analysis used the 12S rRNA gene sequences available from GenBank (from Wang *et al.* 2001) and the sequence for *Acanthogobius hasta* derived from their research. The four clades recognized by Wang *et al.* (2001) were supported and *A. hasta* was placed among the gobionellines.

Sixty-seven gobioid species from 51 genera in 12 families or subfamilies were examined by Thacker (2003) in an analysis using complete sequences of three other mitochondrial genes—ND1, ND2 and COI. As observed by Wang *et al.* (2001), two monophyletic clades of gobies were obtained—one including the gobionellines and another with gobiines (both *sensu* Pezold 1993). As observed by Wang *et al.* (2001), the clade including the Gobionellinae also contained the Amblyopinae, Sicydiinae and Oxudercinae. However, Thacker (2003) found the oxudercinae were paraphyletic if the representative amblyopine in the analysis, *Odontamblyopus rubicundus*, was excluded. The sicydiines were grouped with a paraphyletic set of gobionellines previously recognized by Larson (2001) as monophyletic (the *Stenogobius* group) due to the loss of an oculoscapular canal pore. *Kraemeria cunicularia*, the sole representative of the Kraemeriidae in the analysis, was grouped in another gobionelline clade with a mix of species from the putative monophyletic *Mugilogobius* group (Larson, 2001) and a northern Pacific gobionelline species.

Consistent with the findings of Wang *et al.* (2001), Thacker (2003) recovered the Ptereleotridae (and Microdesmidae *sensu* Thacker 2000) with the Gobiinae (*sensu* Pezold 1993). The Gobiinae were also recovered as paraphyletic (Thacker, 2003). Interestingly, the microdesmids were polyphyletic in the clade, with New World species grouped with New World gobiines and an Old World species, *Gunnellichthys monostigma*, with Old World gobiines. Ptereleotrids were sister to a clade containing *Gunnellichthys* and two species of Schindleriidae. The Microdesmidae and Ptereleotridae were not recovered as sister groups and neither was found to be monophyletic, contrary to the revision of the Microdesmidae by Thacker (2000) using morphological characters and other morphological studies (Randall and Hoese, 1985; Rennis and Hoese, 1987).

A study by Thacker and Hardman (2005) using complete sequences of ND1, ND2, COI and cytochrome *b* genes focused on basal gobioid relationships, but also included two gobiid species in the analysis. The phylogeny produced found the Butinae to be paraphyletic. This was due to *Butis butis* sharing a clade with the two gobiids instead of with other members of its subfamily.

In the most comprehensive molecular analysis of gobioid fishes to date, Thacker (2009) pulled together data from her earlier studies, work by Miya *et al.* (2003) and added information for a number of taxa to offer a phylogeny based on four mtDNA genes for 107 gobioid species and 41 outgroup species. Salient features of the resultant phylogeny are: a monophyletic Butidae as sister-group to two goby clades; a shift of the Kraemeriidae from the Gobionellidae/Gobionellinae clade (Thacker, 2003) to the Gobiidae/Gobiinae clade where it is now aligned with *Fusigobius* and *Coryphopterus* (instead of with *Pandaka*, *Gobiopterus* and *Acanthogobius*); sicydiines, amblyopines and oxudercines are part of the Gobionellidae clade; gobionellines outside of the *Stenogobius* Group are recovered as paraphyletic instead of monophyletic per Thacker (2003); the Microdesmidae and Ptereleotridae are each indicated as monophyletic instead of polyphyletic per Thacker (2003); and there is no division of gobiids into New World and Old World clades as recovered by Thacker (2003).

Some of the larger patterns of relationship (e.g. butine sleepers as sister group to gobies, and gobies comprising two major divisions—one including gobiines and the other gobionellines) are generally concordant with information from morphological studies (Hoese and Gill, 1993; Pezold, 1993). The devil is in the details however, even when comparing molecular studies as noted in the previous paragraph. Yet as more studies are attempted, the information accrued offers a hefty heuristic tool for testing hypotheses of monophyly and unraveling gobioid evolution. Another outcome presents more of a challenge than an aid—undiagnosed lineages determined by similarities of gene sequences. Classifications based upon morphology have been moving progressively towards the elucidation of monophyletic groups distinguished by features hypothesized to represent evolutionary novelties for the said groups, coupled with continuing analyses of different suites of features. This is embodied in all classifications since Miller (1973). The most recent classification presented by Thacker (2009) is particularly significant as it represents a departure from this philosophy, and it is not unique (Frost, 2006). The classification is clade-based as determined from analyses of mtDNA genes, instead of taxa being proposed based upon synapomorphies. There are some cautions advisable in this approach. There has been some notable instability in the progression of phylogenies that have been constructed (examples given in the previous paragraph). Broadening taxon sampling can reduce the impact of long branch attraction and is also needed to incorporate taxa critical to understanding the evolution of character states among gobioids, a group noted for its plasticity. Some attention is also necessary to which portions of the mtDNA genome are sequenced and the length of sequence read. In addition, although multiple genes are frequently used, presumably to avoid confusion of gobioid phylogeny with the reconstruction of a gene tree, the genes in all of the preceding molecular analyses are part of the mtDNA genome and thus are inherited together. In reality the addition of linked genes may be simply the addition of sequence. With those caveats in mind, the phylogenies constructed offer much insight, but demand greater corroboration through the addition of more taxa, additional independent genes and complementary morphological analyses.

ECOLOGY AND DISTRIBUTION OF THE GOBIONELLINAE

The Gobionellinae *sensu stricto* are found in tropical and temperate seas except for the northeastern Atlantic Ocean, the Mediterranean Sea and Ponto-Caspian. Most inhabit estuaries while some are freshwater species. The few species associated with coral reef habitats are typical of sand or fine sedimentary substrates. The *Mugilogobius* group predominates in the Indo-West Pacific and temperate Australia. Species in the *Stenogobius* group are also present in the Indo-West Pacific, and are the only gobionellines in the Western and East-Central Atlantic Ocean. The *Chasmichthys*, *Astrabe* and *Acanthogobius* groups identified by Birdsong *et al.* (1988) are endemic to the temperate northern Pacific and features of the post-cranial axial skeleton suggest that they may be part of a larger monophyletic clade that includes other genera endemic to the area. There are an estimated 370 species in 55 genera.

VALID GENERA OF THE SUBFAMILY GOBIONELLINAE

Northern Pacific Group

Acanthogobius Gill, 1859
Amblychaeturichthys Bleeker, 1874
Astrabe Jordan and Snyder, 1901

Chaenogobius Gill, 1859
Chaeturichthys Richardson, 1844
Clariger Jordan and Snyder, 1901
Clevelandia Eigenmann and Eigenmann, 1888
Eucyclogobius Gill, 1862
Eutaeniichthys Jordan and Snyder, 1901
Gillichthys Cooper, 1864
Gymnogobius Gill, 1863
Ilypnus Jordan and Evermann, 1896
Lepidogobius Gill, 1859
Lethops Hubbs, 1926
Leucopsarion Hilgendorf, 1880
Lophiogobius Günther, 1873
Luciogobius Gill, 1859
Polyspondylogobius Kimura and Wu, 1994
Pterogobius Gill, 1863
Quietula Jenkins and Evermann, 1895
Sagamia Jordan and Snyder, 1901
Suruga Jordan and Snyder, 1901
Tridentiger Gill, 1859
Typhlogobius Steindachner, 1880

Mugilogobius Group

Brachygobius Bleeker, 1874
Caecogobius Berti and Ercolini, 1991
Chlamydogobius Whitley, 1930
Eugnathogobius Smith, 1931
Gobiopterus Bleeker, 1874
Hemigobius Bleeker, 1874
Mistichthys Smith, 1902
Mugilogobius Smitt, 1899
Nesogobius Whitley, 1929
Paedogobius Iwata, Hosoya and Larson, 2001
Pandaka Herre, 1927
Papuligobius Chen and Kottelat, 2003

Pseudogobiopsis Koumans, 1934
Pseudogobius Popta, 1922
Redigobius Herre, 1927
Rhinogobius Gill, 1859
Schismatogobius Beaufort, 1912
Stigmatogobius Bleeker, 1874
Tamanka Herre, 1927
Tasmanogobius Scott, 1935
Tukugobius Herre, 1927

Stenogobius Group

Awaous Valenciennes, 1837
Ctenogobius Gill, 1858
Evorthodus Gill, 1859
Gnatholepis Bleeker, 1874
Gobioides Lacepède, 1800
Gobionellus Girard, 1858
Oligolepis Bleeker, 1874
Oxyurichthys Bleeker, 1857
Parawaous Watson, 1993
Stenogobius Bleeker, 1874

Acknowledgments

Doug Hoese provided information on genera and species of gobioid fishes. Kathleen Cole, Ryan Chabarria, Marcelo Kovacic and Luke Tornabene reviewed the manuscript and contributed helpful commentary and suggestions.

References

Akihito, A. Iwata, T. Kobayashi, K. Ikeo, T. Imanishi, H. Ono, Y. Umehara, C. Hamamatsu, K. Sugiyama, Y. Ikeda, K. Sakamoto, A. Fumihito, S. Ohno and T. Gojobori. 2000. Evolutionary aspects of gobioid fishes based upon a phylogenetic analysis of mitochondrial cytochrome *b* genes. *Gene* 259: 5-15.

Birdsong, R.S., E.O. Murdy and F.L. Pezold. 1988. A study of the vertebral column and median fin osteology in gobioid fishes with comments on gobioid relationships. *Bulletin of Marine Science* 42: 174-214.

Bleeker, P. 1874. Esquisse d'un systeme naturel des Gobioides. *Archives Nerlandaises des Sciences Exactes et Naturelles* 9: 289-331.

Frost, D.R., T. Grant, J.N. Faivovich, R.H. Bain, A. Haas, C.F.B. Haddad, R.O. De Sa′, A. Channing, M. Wilkinson, S.C. Donnellan, C.J. Raxworthy, J.A. Campbell, B.L. Blotto, P. Moler, R.C. Drewes, R.A. Nussbaum, J.D. Lynch, D.M. Green and W.C. Wheeler. 2006. The Amphibian Tree of Life. *Bulletin of the American Museum of Natural History* 297, 370 pp.

Gilbert, C.R. and J.E. Randall. 1979. Two new western Atlantic species of the gobiid fish genus *Gobionellus*, with remarks on characteristics of the genus. *Northeast Gulf Science* 3: 27-47.

Harrison, I.J. 1989. Specialization of the gobioid palatoquadrate complex and its relevance to gobioid systematics. *Journal of Natural History* 23: 235-353.

Hoese, D.F. 1984. Gobioidei: relationships In: *Ontogeny and Systematic of Fishes.* H.G. Moser, W.J. Richards, D.M. Cohen, M.P. Fahay, A.W. Kendall Jr. and S.L. Richardson (eds.). *American Society of Ichthyologists and Herpetologists Special Publication* 1, pp. 588-591.

Hoese, D.F. and A.C. Gill. 1993. Phylogenetic relationships of eleotridid fishes (Perciformes: Gobioidei). *Bulletin of Marine Science* 52: 415-440.

Kim, I-C., H.-S. Kweon, Y.J. Kim, C.-B. Kim, M.C. Gye, W.-O. Lee, Y.-S. Lee and J.-S. Lee. 2004. The complete mitochondrial genome of the jaeveline goby, *Acanthogobius hasta* (Perciformes: Gobiidae) and phylogenetic considerations. *Gene* 336: 147-153.

Larson, H.K. 2001. A revision of the gobiid fish genus *Mugilogobius* (Teleostei: Gobioidei), and its systematic placement. *Records of the Western Australian Museum Suppl.* 62: 1-233.

Miller, P.J. 1973. The osteology and adaptive features of *Rhyacichthys aspro* (Teleostei: Gobioidei) and the higher classification of gobioid fishes. *Journal of Zoology, London* 171: 397-434.

Miya, M., H. Takeshima, H. Endo, N.B. Ishiguro, J.G. Inoue, T. Mukai, T.P. Satoh, M. Yamaguchi, A. Kawaguchi, K. Mabuchi, S.M. Shirai and M. Nishida. 2003. Major patterns of higher teleostean phylogenies: a new perspective based on 100 complete mitochondrial DNA sequences. *Molecular Phylogenetics and Evolution* 26: 121-138.

Murdy, E.O. 1989. A taxonomic revision and cladistic analysis of the oxudercine gobies (Gobiidae: Oxudercidae). *Records of the Australian Museum Suppl.* 11: 1-93.

Pezold, F. 1993. Evidence for a monophyletic Gobiinae. *Copeia* 1993: 634-643.

Pezold, F. 2004. A phylogenetic analysis of the genus *Gobionellus* (Teleostei, Gobiidae). *Copeia* 2004: 260-280.

Randall, J.E. and D.F. Hoese. 1985. Revision if the Indo-Pacific dartfishes, genus *Ptereleotris* (Perciformes: Gobioidei). *Indo-Pacific Fishes* 7: 1-36.

Rennis D.S. and Hoese, D.F. 1987. *Aioliops*, a new genus of ptereleotrine fish (Pisces: Gobioides) from the tropical Indo-Pacific with descriptions of four new species. *Records of the Australian Museum* 39: 67-84.

Thacker, C. 2000. Phylogeny of the wormfishes (Teleostei: Gobioidei: Microdesmidae). *Copeia* 2000: 940-957.

Thacker, C.E. 2003. Molecular phylogeny of the gobioid fishes. *Molecular Phylogenetics and Evolution* 26: 354-368.

Thacker, C.E. 2009. Phylogeny of Gobioides and placement within Acanthomorpha, with a new classification and investigation of diversification and character evolution. *Copeia* 2009: 93-104.

Thacker, C.E. and M.A. Hardman. 2005. Molecular phylogenetics of basal gobioid fishes: Rhyacichthyidae, Odontobutidae, Xenisthmidae, Eleotridae (Teleostei: Perciformes: Gobioidei). *Molecular Phylogenetics and Evolution* 37: 858-871.

Wang, H-Y., M.-P. Tsai, J. Dean and S.-C. Lee. 2001. Molecular phylogeny of gobioid fishes (Perciformes: Gobioidei) based on mitochondrial 12S rRNA sequences. *Molecular Phylogenetics and Evolution* 20: 390-408.

CHAPTER

1.7

Systematics of Oxudercinae

Edward Murdy

The Oxudercinae is found on soft bottom intertidal areas and mangrove swamps in the Indo-West Pacific with one representative in tropical west Africa (Murdy, 1989). At present, this taxon comprises 10 genera and 40 species. Several oxudercine genera can be referred to as mudskippers, which describes their habits on exposed littoral surfaces. Adaptations of the axial skeleton, fins, and associated musculature enable mudskippers to move about on muddy surfaces with both speed and agility. Some oxudercines (e.g. *Scartelaos*) can even "tail-stand" and *Periophthalmus* spp. are able to "climb" rocks, logs, mangrove roots, and stems (Fig. 1.7.1A). Additionally, gill modifications and highly vascularized epithelial tissue facilitate aerial respiration in some oxudercines (Graham, 1997).

Oxudercines are elongate gobiid fishes that are compressed posteriorly. Scales are cycloid and vary in size from small to large, numbering from a few to as many as 275 in a longitudinal series. Similarly, the number of predorsal scales can be zero or as many as 113. The head is small to moderate (15-34% SL). The eyes are located dorsally on the head (Fig. 1.7.1B). The gape is moderate to wide. Teeth are caninoid, obtusely pointed, or bifid, and uniserial in both jaws (except in *Periophthalmodon*, which has two rows of

Author's address: National Science Foundation, 4201 Wilson Blvd., Arlington, VA 22230, USA.
E-mail: emurdy@nsf.gov

teeth in the upper jaw). A pair of recurved canine teeth are internal to the lower jaw symphysis (except in *Oxuderces dentatus*, *Periophthalmodon*, and *Periophthalmus*). The tongue is adnate. Meristics (Table 1.7.1) are as follows: spinous dorsal fin with 4 – 17 spines and 10 – 33 second dorsal-fin elements; anal fin with 9 – 31 elements; pectoral-fin rays 11 – 25; pelvic fin I, 5; vertebrae 10 + 15 + terminal vertebral element = 26; branchiostegals 5; epurals 2. One anal-fin pterygiophore precedes the first hemal spine. Oxudercine fishes are of small to moderate size with the largest members of the subfamily typically not exceeding 20 cm SL. They inhabit mangrove and mudflat areas from the west coast of Africa eastward to Samoa (Murdy, 1989).

Table 1.7.1 Means and ranges of dorsal- and anal-fin elements. Counts for second dorsal- and anal-fins include all elements.

Genus	*Spinous Dorsal Fin*		*Second Dorsal Fin*		*Anal Fin*	
	mean	*range*	*mean*	*range*	*mean*	*range*
Apocryptes	5.0	5	22.0	21–23	22.1	20–24
Apocryptodon	6.0	6	22.7	21–24	22.2	21–23
Boleophthalmus	5.0	5	25.3	23–28	25.0	22–27
Oxuderces	6.0	6	27.9	24–31	27.2	24–30
Parapocryptes	5.9	5–6	27.3	26–29	26.8	25–29
Periophthalmodon	7.7	4–15	12.8	11–14	12.0	10–14
Periophthalmus	11.2	4–17	12.5	10–14	11.6	9–14
Pseudapocryptes	5.0	5	31.2	29–33	29.1	27–31
Scartelaos	5.0	5	26.1	23–28	25.3	24–27
Zappa	5.0	5	28.1	27–30	27.3	26–28

Oxudercines are typically associated with soft bottom habitats, many are burrowers (Fig. 1.7.1C). *Periophthalmus* utilizes the mesozone (*sensu* Tomlinson, 1986) of the mangrove community whereas many of the other genera in the subfamily are typically found in the seaward zone that is oftentimes simply a soft, muddy tidal flat nearly or totally devoid of mangroves. This seaward zone is flooded at all high tides and exposed to some degree at low tide. Consequently, the distribution of non-*Periophthalmus* oxudercines is probably limited to availability of suitable mudflat habitat. Non-*Periophthalmus* oxudercines are distributed from the northwest Arabian Gulf (29° N latitude, 48° E longitude), eastward and southward to Queensland, Australia (20° S latitude, 150° E longitude) and northward to western Japan (34° N latitude, 134° E longitude) (Murdy, 1989).

In assessing the distribution of non-*Periophthalmus* oxudercines, only two genera (*Boleophthalmus*, *Scartelaos*) are found west of the southern tip of India and from there extend as far west as the Arabian Gulf. *Apocryptodon*, *Boleophthalmus*, and *Scartelaos* extend as far north as western Japan and these three genera as well as *Oxuderces* and *Periophthalmodon* reach as far south as northern Australia.

Many oxudercines are endemic to relatively restricted geographic areas. The Atlantic mudskipper, *Periophthalmus barbarus*, is found only on the west African coast (Murdy, 1989). One oxudercine from western Japan (*Apocryptodon punctatus*) is endemic there (Sato and Aizawa, 1992), whereas *Ps. magnuspinnatus* is only found in southern Korea and China (Baeck *et al.* 2007). In northern Australia, one species of *Boleophthalmus*, *B. birdsongi*, is endemic, whereas *Oxuderces wirzi* is known only from southern Papua New Guinea and northern Australia (Murdy, 1989). *Scartelaos cantoris* is known only from the Andamans, whereas *S. gigas* is known only from northern China (Murdy, 1989). *Zappa confluentus* is known only from Papua New Guinea (Polgar, 2007).

Mudskippers are the subject of much behavioral and physiological research due to their amphibious lifestyle and air-breathing abilities. Takita *et al.* (1999) reported on the distribution and habitat requirements of a number of oxudercine gobies from Malaysia and Indonesia. The seven genera and 12 species that Takita *et al.* (1999) studied had habitats that varied from sandy to muddy, elevated to level, exposed to immersed, and shoreline to tidal creeks. Similarly, Baeck *et al.* (2007) compared the ecology of two species of *Periophthalmus* in southern Korea; *Ps. magnuspinnatus* occupied elevated, vegetated mudflats whereas *Ps. modestus* inhabited level mudflats with no vegetation. Khaironizam and Norma-Rashid (2003) studied a population of *Periophthalmodon septemradiatus* in Peninsular Malaysia that inhabited a tidal mangrove area that was 15 km from the river mouth; at low tide the water was fresh and at high tide the salinity was 1-3 ppt. Graham (1997) reported that six oxudercine genera (*Periophthalmus*, *Periophthalmodon*, *Boleophthalmus*, *Scartelaos*, *Pseudapocryptes*, and *Apocryptes*) are amphibious. *Periophthalmus*, *Periophthalmodon*, *Boleophthalmus* are likely to be amphibious air-breathers according to Graham (1997) with *Pseudapocryptes* and *Apocryptes* known to be air-gulpers. Based on Ishimatsu *et al.* (1998), *Periophthalmodon schlosseri* gulps air at the surface of its burrow and then transports the air to the bottom of the burrow in order to aerate its eggs. According to Kok *et al.* (1998), *Periophthalmodon schlosseri* respires more efficiently, and has a higher metabolic rate, on land than in water; this observation is unique among fishes.

As oxudercines are found in shallow waters or on exposed surfaces, they are preyed upon by both terrestrial and aquatic predators including fishes, shorebirds, and water snakes (Clayton, 1993). Some oxudercines are themselves carnivorous (Fig. 1.7.1D). For instance, *Periophthalmus waltoni* feeds on a variety of ocypodid crabs (Clayton and Snowden, 2000) and *Periophthalmodon schlosseri* preys upon fiddler crabs (*Uca* spp.), insects, and worms (Ghaffar *et al.* 2006). Other oxudercines, such as *Boleophthalmus pectinirostris*, are herbivores that graze on diatoms on the mud surface during low tide (Yang *et al.* 2003). Many mudskippers are territorial and a number also exhibit elaborate courtship displays (Fig. 1.7.1E).

Fig. 1.7.1(A) *Periophthalmus chrysospilos* climbing log (image by E.O. Murdy)

(B) *Periophthalmodon schlosseri* at burrow entrance (image by E.O. Murdy)

(C) *Boleophthalmus dussumieri* entering burrow tail first (image by E.O. Murdy)

(D) *Periophthalmodon schlosseri* eating fiddler crab (image by E.O. Murdy)

(E) *Boleophthalmus dussumieri* in courting display at low tide on the mudflat (image by E.O. Murdy)

Color image of this figure appears in the color plate section at the end of the book.

A cladistic hypothesis of relationships was presented for the Oxudercinae by Murdy (1989) (Fig. 1.7.2). In contrast to Murdy's analysis, the monophyly of Oxudercinae was challenged by Thacker (2003) in her molecular analysis. Thacker (2003) presented a molecular phylogeny of the groups within the Gobioidei and her molecular hypothesis supported a sister taxon relationship between Oxudercinae + Amblyopinae, and several gobionelline and sicydiine genera. Based on Thacker's analysis, the Amblyopinae is nested within Oxudercinae and both are within a paraphyletic Gobionellinae.

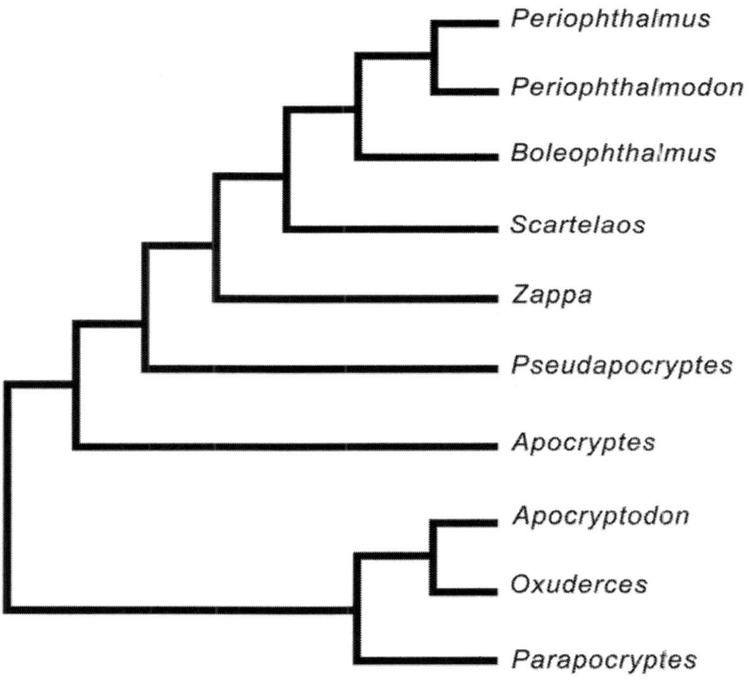

Fig. 1.7.2 A cladistic hypothesis of relationships presented for the Oxudercinae from Polgar (2007) which was modified from Murdy (1989).

All oxudercine genera, except *Periophthalmus*, are defined by at least one synapomorphy. Based on evidence provided in Murdy (1989), *Apocryptodon*, *Oxuderces*, and *Parapocryptes* represent one tribal-group subset of the Oxudercinae with *Apocryptes*, *Boleophthalmus*, *Periophthalmodon*, *Periophthalmus*, *Pseudapocryptes*, *Scartelaos*, and *Zappa* forming another.

LIST OF GENERA AND SPECIES

Subfamily Oxudercinae

Tribe Oxudercini

Genus	*Parapocryptes* Bleeker, 1874
	Parapocryptes rictuosus (Valenciennes, 1837)
	Parapocryptes serperaster (Richardson, 1845)
Genus	*Apocryptodon* Bleeker, 1874
	Apocryptodon madurensis (Bleeker, 1849)
	Apocryptodon punctatus (Tomiyama, 1934)
Genus	*Oxuderces* (Eydoux and Souleyet, 1848)
	Oxuderces dentatus (Eydoux and Souleyet, 1848)
	Oxuderces wirzi (Koumans, 1938)

Tribe Periophthalmini

Genus	*Apocryptes* (Valenciennes, 1837)
	Apocryptes bato (Hamilton, 1822)
Genus	*Pseudapocryptes* Bleeker, 1874
	Pseudapocryptes borneensis (Bleeker, 1855)
	Pseudapocryptes elongatus (Cuvier, 1816)
Genus	*Zappa* Murdy, 1989
	Zappa confluentus (Roberts, 1978)
Genus	*Scartelaos* Swainson, 1839
	Scartelaos cantoris (Day, 1871)
	Scartelaos gigas (Chu and Wu, 1963)
	Scartelaos histophorus (Valenciennes, 1837)
	Scartelaos tenuis (Day, 1876)
Genus	*Boleophthalmus* Valenciennes, 1837
	Boleophthalmus birdsongi Murdy, 1989
	Boleophthalmus boddarti (Pallas, 1770)
	Boleophthalmus caeruleomaculatus (McCulloch and Waite, 1918)
	Boleophthalmus dussumieri (Valenciennes, 1837)
	Boleophthalmus pectinirostris (Linnaeus, 1758)

Genus	*Periophthalmodon* Bleeker, 1874
	Periophthalmodon freycineti (Quoy and Gaimard, 1824)
	Periophthalmodon schlosseri (Pallas, 1770)
	Periophthalmodon septemradiatus (Hamilton, 1822)
Genus	*Periophthalmus* Bloch and Schneider, 1801
	Periophthalmus argentilineatus (Valenciennes, 1837)
	Periophthalmus barbarus (Linnaeus, 1766)
	Periophthalmus chrysospilos (Bleeker, 1852)
	Periophthalmus darwini (Larson and Takita, 2004)
	Periophthalmus gracilis (Eggert, 1935)
	Periophthalmus kalolo (Lesson, 1830)
	Periophthalmus magnuspinnatus (Lee, Choi and Ryu, 1995)
	Periophthalmus malaccensis (Eggert, 1935)
	Periophthalmus minutus (Eggert, 1935)
	Periophthalmus modestus (Cantor, 1842)
	Periophthalmus novaguineaensis (Eggert, 1935)
	Periophthalmus novemradiatus (Hamilton, 1822)
	Periophthalmus spilotus (Murdy and Takita, 1999)
	Periophthalmus takita (Jaafar and Larson, 2008)
	Periophthalmus variabilis (Eggert, 1935)
	Periophthalmus walailakae (Darumas and Tantichodok, 2002)
	Periophthalmus waltoni (Koumans, 1941)
	Periophthalmus weberi (Eggert, 1935)

References

Baeck, G.W., T. Takita and Y.H. Yoon. 2007. Lifestyle of Korean mudskipper *Periophthalmus magnuspinnatus* with reference to a congeneric species *Periophthalmus modestus*. *Ichthyological Research* 55: 43-52.

Clayton, D.A. 1993. Mudskippers. *Oceanography and Marine Biology Annual Review* 31: 507-577.

Clayton, D.A. and R. Snowden. 2000. Surface activity in the mudskipper, *Periophthalmus waltoni* Koumans 1941 in relation to prey activity and environmental factors. *Tropical Zoology* 13: 239-249.

Ghaffar, M., F. Yakob, S. Nor and A. Arshad. 2006. Foraging behavior and food selection of giant mudskipper (*Periophthalmodon schlosseri*) at Kuala

Gula, Matang Mangrove Reserve, Perak, Malaysia. *Coastal Marine Science* 30: 263-267.

Graham, J.B. 1997. *Air-Breathing Fishes*. Academic Press, San Diego.

Ishimatsu A., Y. Hishida, T. Takita, T. Kanda, S. Oikawa, T. Takeda and K.H. Khoo. 1998. Mudskippers store air in their burrows. *Nature* (*London*) 391: 237-238.

Khaironizam, M.Z. and Y. Norma-Rashid. 2003. First record of the mudskipper, *Periophthalmodon septemradiatus* (Hamilton) (Teleostei: Gobiidae) from Peninsular Malaysia. *The Raffles Bulletin of Zoology* 51: 97-100.

Kok, W.K., C.B. Lim, T.J. Lam and Y.K. Ip. 1998. The mudskipper *Periophthalmodon schlosseri* respires more efficiently on land than in water and vice versa for *Boleophthalmus boddarti*. *Journal of Experimental Zoology* 280: 86-90.

Murdy, E.O. 1989. A taxonomic revision and cladistic analysis of the oxudercine gobies (Gobiidae: Oxudercinae). *Records of the Australian Museum*, *Suppl.* 11: 1-93.

Polgar, G. (Ed.). 2007. The mudskipper. World Wide Web electronic publication. www.themudskipper.org (accessed 10/July/2008).

Sato, Y. and M. Aizawa. 1992. Record of the gobiid fish *Apocryptodon punctatus* from the Yoshino River estuary in Tokushima Prefecture, Japan, with comments on distribution of the species. *Bulletin of the Tokushima Prefectural Museum* 2: 43-50.

Takita, T., Agusnimar and A.B. Ali. 1999. Distribution and habitat requirements of oxudercine gobies (Gobiidae: Oxudercinae) along the Straits of Malacca. *Ichthyological Research* 46: 131-138.

Thacker, C.E. 2003. Molecular phylogeny of the gobioid fishes (Teleostei: Perciformes: Gobioidei). *Molecular Phylogenetics and Evolution* 26: 354-368.

Tomlinson, P.B. 1986. *The Botany of Mangroves*. Cambridge University Press, Cambridge.

Yang, K.Y., S.Y. Lee and G.A. Williams. 2003. Selective feeding by the mudskipper (*Boleophthalmus pectinirostris*) on the microalgal assemblage of a tropical mudflat. *Marine Biology* 143: 245-256.

CHAPTER

1.8

Systematics of Amblyopinae

Edward Murdy

Gobies of the gobiid subfamily Amblyopinae are elongate, mud-dwelling fishes of the Indo-West Pacific region that are commonly referred to as "eel gobies" or "worm gobies". Currently, 12 genera and 23 species of Amblyopinae are recognized. In life, these gobies are pink, purple, or red (Fig. 1.8.1). In contrast to the majority of gobies, amblyopine gobies have a continuous dorsal fin; the spinous (first) and soft (second) dorsal fins are connected by membrane. Because they live in turbid water habitats, the eyes of amblyopines are reduced in size and may have limited function. Based on axial skeletal features, Birdsong *et al.* (1988) subdivided the Amblyopinae into three units: 'Gobioides' Group, 'Taenioides' Group, and 'Trypauchen' Group. Birdsong *et al.* (1988) stated that the latter two groups shared the derived fin element to vertebra ratio of 2:1 whereas the 'Gobioides' Group possessed a ratio of 1:1 as is typical amongst gobioids. Subsequently, other characters have been found to ally the 'Gobioides' Group, comprising only *Gobioides*, with the Gobionellinae (Harrison, 1989; Pezold, 1993; Murdy, 1998). Murdy and Shibukawa (2001, 2002) recognized the following

Author's address: National Science Foundation, 4201 Wilson Blvd., Arlington, VA 22230, USA.
E-mail: emurdy@nsf.gov

two features as synapomorphies for the monophyletic Amblyopinae (i.e. 'Taenioides' Group + 'Trypauchen' Group): fin element to vertebra ratio of 2:1; and ultimate dorsal- and anal-fin pterygiophores supporting only a single ray.

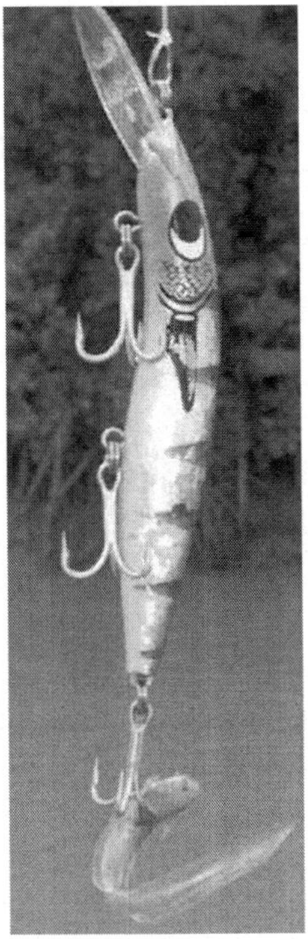

Fig. 1.8.1 *Caragobius rubristriatus* snagged on a fishing lure in Northern Territory, Australia, displaying the pinkish coloration common to many amblyopines. (Image by T. Robb and provided courtesy of M. McGrouther, Australian Museum).

Color image of this figure appears in the color plate section at the end of the book.

Birdsong *et al.* (1988) created the 'Trypauchen' Group as a putative monophyletic assemblage. All members of the 'Trypauchen' Group lack an interneural gap (space between neural arches lacking a pterygiophore), which is a synapomorphy amongst gobiid fishes; the 'Trypauchen' Group dorsal-fin pterygiophore formula is 3-1221 (= 3/I II II I/7·8 in Akihito *et al.* 1984). Birdsong *et al.* (1988) assigned all the remaining gobiid genera sharing the fin element to vertebra ratio of 2:1 to the 'Taenioides' Group, which is a phenetically united assemblage.

The 'Taenioides' Group of the Amblyopinae listed in Birdsong *et al.* (1988) comprised *Brachyamblyopus*, *Odontamblyopus*, and *Taenioides*. This group was phenetically united by the shared possession of a spinous dorsal-fin pterygiophore formula of 3-12210. To this group, Murdy and Shibukawa (2002) added *Pseudotrypauchen*. *Pseudotrypauchen* differs from the other group members and all other amblyopines by having the first two pterygiophores of the second dorsal fin (D2) separated by a neural spine (one pterygiophore resides in interneural space 8 and the other in interneural space 9) (Fig. 1.8.2). All other amblyopines have the first two pterygiophores of the second dorsal fin sharing the same interneural space (interneural space 7 or 8) (Fig. 1.8.3). The condition of the first two pterygiophores of the second dorsal fin in *Pseudotrypauchen* is considered the gobiid type of D2 anterior pterygiophore arrangement, whereas the condition shown by all other amblyopines is the eleotrid type of D2 anterior pterygiophore arrangement (gobiid and eleotrid types of 'interneural gap' of Hoese 1984). As the Oxudercinae has the gobiid type of D2 anterior pterygiophore arrangement and is considered the putative sister group to the Amblyopinae (Murdy and Shibukawa, 2001), the eleotrid type of D2 anterior pterygiophore arrangement is a reversal and is, thus, derived according to Murdy and Shibukawa (2002). With respect to the polarity of the D2 anterior pterygiophore arrangement in *Pseudotrypauchen*, Murdy and Shibukawa (2002) stated that it is primitive. As such, *Pseudotrypauchen* represents the basal member within the subfamily and is the sister group to all other amblyopines. (Thacker (2003) disputed a sister group relationship between the Oxudercinae and the Amblyopinae.)

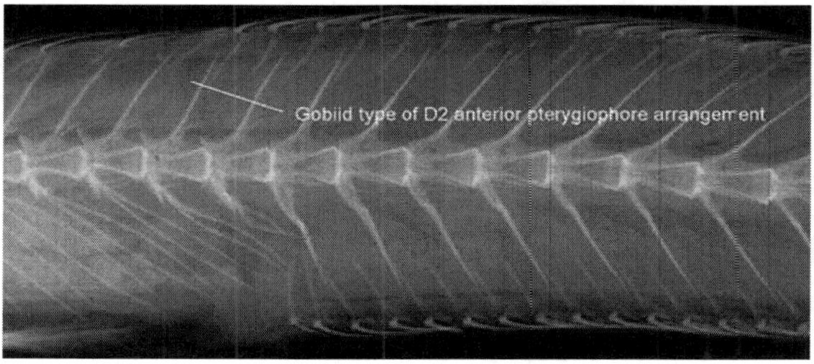

Fig. 1.8.2 Left lateral view of a radiograph of *Pseudotrypauchen multiradiatus* (USNM 341061) showing the first two pterygiophores of the second dorsal fin separated by a neural spine (gobiid type of D2 anterior pterygiophore arrangement). This is the typical condition in non-amblyopine gobies.

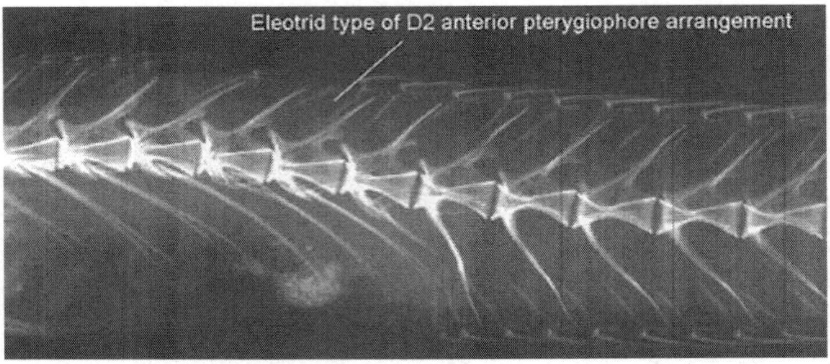

Fig. 1.8.3 Left lateral view of a radiograph of the holotype of *Brachyamblyopus brachysoma* (RMNH 4670) showing the first two pterygiophores of the second dorsal fin sharing the same interneural space (eleotrid type of D2 anterior pterygiophore arrangement). This is the typical condition in amblyopine gobies.

THE 'TAENIOIDES' GROUP

Pseudotrypauchen comprises a single species (*P. multiradiatus*) known from India, Malaysia, and Sumatra. The genus is unique within the Amblyopinae in having: pectoral fins longer than pelvic fins (113–205% of pelvic-fin length) and pointed posteriorly; short, compressed, and gently rounded teeth in the outermost tooth row in both jaws; and in having the first two pterygiophores of the second dorsal fin separated by a neural spine (Murdy and Shibukawa, 2002). Meristic counts are included in Table 1.8.1.

Table 1.8.1 Means and ranges of dorsal- and anal-fin elements in the Amblyopinae. Counts for dorsal- and anal-fins include all elements.

Genus	*Dorsal Fin Elements range*	*Anal Fin Elements range*	*Vertebral Count range*
Amblyotrypauchen	48 - 52	39 - 43	9 - 10 + 19, 23
Brachyamblyopus	35 - 39	28 - 34	10 + 16
Caragobius	36 - 52	31 - 45	10 + 18 - 22, 25 - 27
Ctenotrypauchen	53 - 57	41 - 46	10 + 23 - 25
Gymnoamblyopus	33	26	10 + 16
Karsten	46 - 51	40 - 44	9 + 19 - 22
Odontamblyopus	40 - 54	32 - 45	10 + 17, 20 - 24
Paratrypauchen	50 - 60	41 - 49	10 + 25 - 28
Pseudotrypauchen	36 - 39	30 - 32	10 + 17
Taenioides	41 - 66	34 - 65	10 + 17 - 23
Trypauchen	46 - 58	39 - 50	10 + 19 - 20, 23 - 24
Trypauchenichthys	45 - 48, 58 - 62	37 - 39, 46 - 52	10 + 31 - 32, 35 - 36, 40 - 41
Trypauchenopsis	33 - 42	27 - 33	10 + 16 - 18

Brachyamblyopus is little known and, thus, little studied. The holotype of *B. brachysoma* was collected in Sumatra and the species is also reported from Sulawesi, Papua New Guinea, the Persian Gulf, India, Thailand, and Hong Kong (Koumans, 1953). *Brachyamblyopus brachysoma* has a scaleless head, an enlarged, outer row of teeth in the jaws, and lacks head barbels (Murdy and Shibukawa, 2001).

Odontamblyopus comprises five species distributed from the west coast of India to northern China and Japan (Murdy and Shibukawa, 2001; Murdy and Shibukawa, 2003). *Odontamblyopus* has numerous fang-like teeth and shares with *Pseudotrypauchen* the condition of having the pectoral-fin rays free and silk-like. *O. lacepedii* is a marine and brackish-water goby that creates elaborate burrows in mud. Its burrows are tunnel-shaped, vertical, and extend to depths of 90 cm into the substrate; as many as six other smaller tunnels radiate from the primary tunnel and connect to the surface (Gonzales *et al.* 2008). According to Dotsu (1957), this species eats a variety of foods including bivalves, crustaceans, cephalopods, and small fishes. At low tide, this fish is a facultative air breather (Gonzales *et al.* 2006).

The genus *Taenioides* is in need of revision (Murdy and Randall, 2003). *Taenioides* is diagnosed amongst Gobiidae by the proximal radial of anterior second anal-fin pterygiophore being spatulate (this is equivalent to the Y-shaped, second anal-fin pterygiophore described by Birdsong *et al.* 1988). *Taenioides* spp. have raised dermal ridges bearing sensory papillae on the head and body as well as barbels on the underside of the head arranged in several different patterns (Fig. 1.8.4). The genus is distributed from the east coast of Africa, Arabian Gulf, throughout southeast Asia, Australia, China, and Japan.

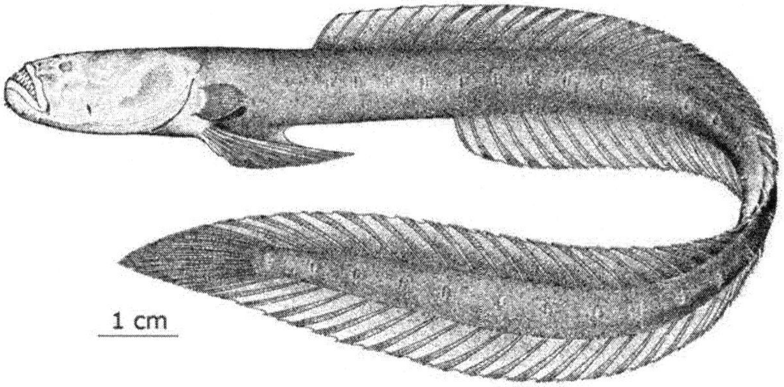

Fig. 1.8.4 *Taenioides snyderi* Jordan and Hubbs (1925). Originally figured as *Taenioides lacepedei* in Jordan and Snyder (1901).

Trypauchenopsis intermedia (Fig. 1.8.5) is known by the common name of bearded eel goby due to the presence of many barbel-like projections on the head. This taxon is characterized by the ultimate fin ray in either the dorsal or anal fins (or both) bearing a forked pterygiophore, the middle radial of which is bifurcated such that the proximal radial adjoins one hemal (or neural) spine and the bifurcated arm adjoins the hemal (or neural) spine posterior to it. This species has a very broad distribution from South Africa to Guam.

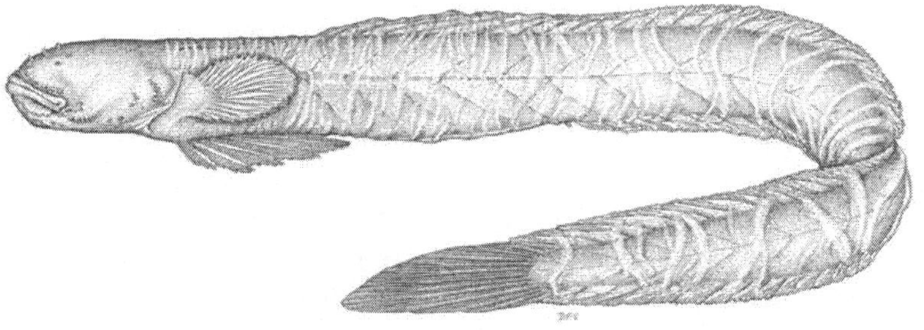

Fig. 1.8.5 *Trypauchenopsis intermedia*, SAIAB 14127, 106 mm SL, Mkomazi, Natal, South Africa. Figure drawn by D.P. Voorvelt of the J.L.B. Smith Institute and provided courtesy of E. Heemstra and J. Stapley.

The monotypic *Gymnoamblyopus* is the most recent addition to the 'Taenioides' Group (Murdy and Ferraris, 2003) and is known only from a single specimen collected in Fly River, Papua New Guinea. It differs from all other amblyopine gobies by the following combination of characters: only one anal-fin pterygiophore anterior to first hemal spine; no spatulate second anal-fin pterygiophore; no chin barbels; no scales; no raised dermal folds or ridges; pleural ribs short and posteriorly directed; and mouth vertical.

'Taenioides' Group

Genus	*Brachyamblyopus* Bleeker, 1874
	Brachyamblyopus brachysoma Bleeker, 1874
Genus	*Gymnoamblyopus* Murdy and Ferraris, 2003
	Gymnoamblyopus novaeguinea Murdy and Ferraris, 2003
Genus	*Odontamblyopus* Bleeker, 1874
	Odontamblyopus lacepedii (Temminck and Schlegel, 1845)
	Odontamblyopus rebecca Murdy and Shibukawa, 2003
	Odontamblyopus roseus (Valenciennes, 1837)

Odontamblyopus rubicundus (Hamilton-Buchanan, 1822)
Odontamblyopus tenuis (Day, 1876)

Genus *Taenioides* Lacepède, 1800
At least the following four species are considered valid:
Taenioides anguillaris (Linnaeus, 1758)
Taenioides gracilis (Valenciennes, 1837)
Taenioides kentalleni Murdy and Randall, 2002
Taenioides purpurascens De Vis, 1884

Genus *Trypauchenopsis* Volz, 1903
Trypauchenopsis intermedia Volz, 1903

The above listing of the 'Taenioides' Group comprises five genera and 12 species. In addition, two new genera await description. These new genera include at least three described species.

THE 'TRYPAUCHEN' GROUP

Among 'Trypauchen' Group members (*Amblyotrypauchen, Caragobius, Ctenotrypauchen, Karsten, Paratrypauchen, Trypauchen,* and *Trypauchenichthys*), all but *Caragobius* and *Karsten* possess a shallow pouch along the dorsal edge of the operculum (Fig. 1.8.6); the function of this opercular pouch is not known. In addition, all 'Trypauchen' Group members except *Karsten* have a pterygiophore formula of 3-1221 that is unique within the Gobioidei (Birdsong *et al.* 1988). Murdy (2002) proposed *Caragobius* as the sister group to all other 'Trypauchen' Group members and *Karsten* as the sister group to all 'Trypauchen' Group members that possess an opercular pouch.

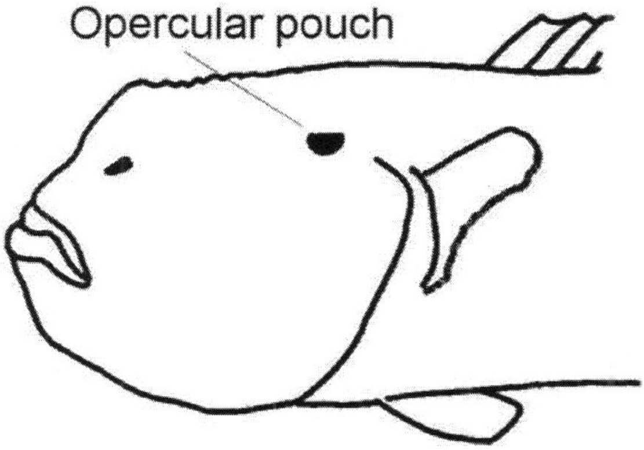

Fig. 1.8.6 Approximate relative size and location of the opercular pouch in those amblyopine gobies that possess one. (Modified from figure 313 in Kottelat *et al.*, 1993.)

Caragobius comprises two species, *C. rubristriatus* and *C. urolepis*; the former has a limited distribution, extending across northern Australia from the Prince Regent River to the Gulf of Carpentaria whereas the latter is known from the east coast of India eastward to Fiji. *Caragobius* is unique within the Amblyopinae in having: 3-7 anal-fin pterygiophores anterior to first hemal spine (AP); fifth hypural absent; and ribs lacking on 3rd precaudal vertebra. The other features useful in distinguishing it from all other 'Trypauchen' Group members are: pectoral fins broadly rounded, symmetrical dorsoventrally; head slightly depressed; no fang-like teeth; no opercular pouch; eyes rudimentary; and pelvic fins rounded (Murdy and Shibukawa, 2003).

The monotypic *Karsten* is distinguished from all other members of the subfamily in having a spinous dorsal-fin pterygiophore formula of 3 - 123, 3 - 132, or 3 - 1221 + 2. The genus is also distinct within the subfamily in having only nine precaudal vertebrae. *Karsten* lacks eyes as does *Amblyotrypauchen*. Of special note, one lot of *Karsten* (*K. totoyensis*, USNM 113201) was collected at a depth of 608 fathoms (1122 m) over a mud/sand bottom; this may represent the greatest depth from which any goby has ever been collected (Murdy, 2002). *Karsten* has only been collected from Indonesia, the Philippines, and Fiji.

Amblyotrypauchen is unique within the Amblyopinae in having 10+19 vertebrae, enlarged fang-like teeth in both jaws, partially separate pelvic fins, and scale patches on the opercle, cheeks, and dorsum of head. *Amblyotrypauchen* lacks eyes and has been collected off muddy bottoms from almost 17 m to over 183 m in depth. *Amblyotrypauchen* comprises a single species, *A. arctocephalus*, known from India, Malaysia, Hong Kong, the Philippines, Papua New Guinea, and Australia (Murdy, 2003).

Trypauchen is unique within the Amblyopinae in having the following combination of characters: 1) typically four, rarely three, anal-fin pterygiophores anterior to the first hemal spine; 2) pelvic fins small, united, and funnel-shaped with a well-developed interradial membrane; and 3) abdomen scaled. *Trypauchen* comprises two species: *T. pelaeos*, known from Myanmar, Malaysia, Thailand, Vietnam, and China; and *T. vagina*, distributed from Kuwait, along the coasts of India, ranging eastward to the Philippines, Taiwan, and China (Murdy, 2006).

The monotypic *Paratrypauchen* differs from *Trypauchen* in having a scaleless abdomen, emarginate pelvic fins, and in having three pterygiophores anterior to the first hemal spine (Murdy, 2008a). Akihito *et al.* (1984) reported that in Japan *Paratrypauchen microcephalus* inhabits soft mud bottoms (Fig. 1.8.7). Chen and Fang (1999) stated that *P. microcephalus* inhabits river mouth areas near mangrove forests and constructs burrows in the muddy substrate. This species is an omnivore feeding on benthic invertebrates such as crustaceans according to Chen and Fang (1999). *Paratrypauchen microcephalus* is known from Japan, Taiwan, the Philippines, Australia, Singapore, Malaysia, Vietnam, India, Mozambique, and South Africa. This species is one of the more common amblyopines in museum collections.

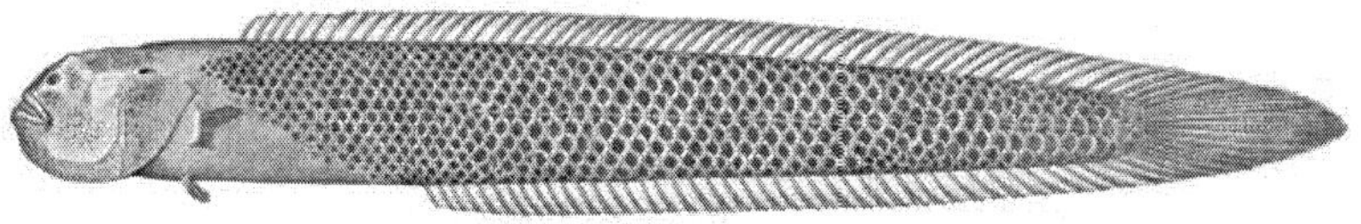

Fig. 1.8.7 *Trypauchen wakae* (= *Paratrypauchen microcephalus*) from Jordan and Snyder (1901).

Ctenotrypauchen is unique within the 'Trypauchen' Group in having a serrated frontal crest with united, but slightly emarginate pelvic fins. A specimen of *Ctenotrypauchen chinensis* from the Arafura Sea was collected at a depth of 62-65 meters. Based on the few specimens that exist in collections, it appears that this taxon may be confined to deeper waters than most other members of this subfamily excluding *Amblyotrypauchen* and *Karsten*.

Trypauchenichthys is unique within the Amblyopinae in having the pelvic fins separate (not joined by a connecting membrane) (Fig. 1.8.8). *Trypauchenichthys* comprises three species: *T. larsonae* known only from the Northern Territory, Australia; *T. sumatrensis* found in the Strait of Malacca with a reported occurrence in India (Fig. 1.8.9), and *T. typus* with specimens known only from the type locality in western Borneo and reports from Thailand, the Philippines, Sarawak, and the Riau Archipelago, Indonesia (Murdy, 2008b).

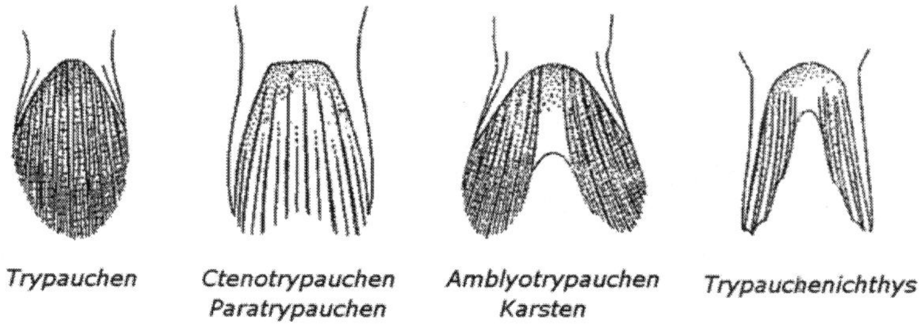

Fig. 1.8.8 Pelvic fin shapes of 'Trypauchen' group genera (modified from Hora, 1924).

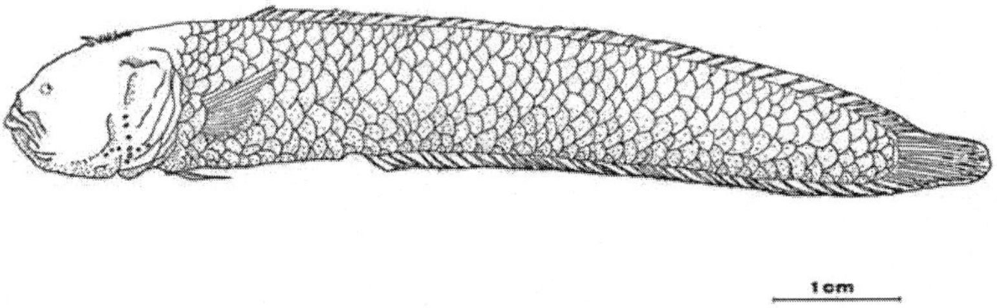

Fig. 1.8.9 *Trypauchenichthys sumatrensis* (drawn from a specimen removed from USNM 211295). The bony frontal crest is evident. Drawing by Christine Baer.

'Trypauchen' Group

Genus	*Amblyotrypauchen* Hora, 1924
	Amblyotrypauchen arctocephalus (Alcock, 1890)
Genus	*Caragobius* Smith and Seale, 1906
	Caragobius rubristriatus (Saville-Kent, 1889)
	Caragobius urolepis (Bleeker, 1852)
Genus	*Ctenotrypauchen* Steindachner, 1867
	Ctenotrypauchen chinensis Steindachner, 1867
Genus	*Karsten* Murdy, 2002
	Karsten totoyensis (Garman, 1903)
Genus	*Paratrypauchen* Murdy, 2008
	Paratrypauchen microcephalus (Bleeker, 1860)
Genus	*Trypauchen* Valenciennes, 1837
	Trypauchen pelaeos Murdy, 2006
	Trypauchen vagina (Bloch and Schneider, 1801)
Genus	*Trypauchenichthys* Bleeker, 1860
	Trypauchenichthys larsonae Murdy, 2008
	Trypauchenichthys sumatrensis Hardenberg, 1931
	Trypauchenichthys typus Bleeker, 1860

References

Akihito, Prince, M. Hayashi, T. Yoshino, K. Shimada and H. Senou. 1984. Suborder Gobioidei. In: *The Fishes of the Japanese Archipelago*, H. Masuda, K. Amaoka, C. Araga, T. Uyeno and T. Yoshino (eds.). Tokai University Press, Tokyo, pp. 236-238 and plates 235-258, 353-355.

Birdsong, R.S., E.O. Murdy and F.L. Pezold. 1988. A study of the vertebral column and median fin osteology in gobioid fishes with comments on gobioid relationships. *Bulletin of Marine Science* 42: 174-214.

Chen, I.-S. and L.-S Fang. 1999. *The Freshwater and Estuarine Fishes of Taiwan*. National Museum of Marine Biology and Aquarium, Pingtung. (In Chinese).

Dotsu, Y. 1957. On the bionomics and life history of the eel-like goby, *Odontamblyopus rubicundus* (Hamilton). *Science Bulletin of the Faculty of Agriculture, Kyushu University* 16: 101-110.

Gonzales, T.T., M. Katoh and A. Ishimatsu. 2006. Air-breathing of aquatic burrow-dwelling eel goby, *Odontamblyopus lacepedii* (Gobiidae: Amblyopinae). *Journal of Experimental Biology* 209: 1085-1092.

Gonzales, T.T., M. Katoh and A. Ishimatsu. 2008. Intertidal burrows of the air-breathing eel goby, *Odontamblyopus lacepedii* (Gobiidae: Amblyopinae). *Ichthyological Research* 55: 303-306.

Harrison, I.J. 1989. Specialization of the gobioid palatopterygoquadrate complex and its relevance to gobioid systematics. *Journal of Natural History* 23: 325-353.

Hoese, D.F. 1984. Gobioidei: Relationships. In: *Ontogeny and Systematic of Fishes*. H.G. Moser, W.J. Richards, D.M. Cohen, M.P. Fahay, A.K. Kendall, Jr. and S.L. Richardson (eds.), Allen Press, Lawrence, pp. 588-591.

Hora, S.L. 1924. Notes on fishes in the Indian Museum. VI. On a new genus of gobioid fishes (subfamily Trypaucheninae) with notes on related forms. *Records of the Indian Museum, Calcutta* 26: 155-163.

Jordan, D.S. and J.O. Snyder. 1901. A review of the gobioid fishes of Japan, with descriptions of twenty-one new species. *Proceedings of the United States National Museum* 24: 33-132.

Jordan, D.S. and C.L. Hubbs. 1925. Records of fishes obtained by David Starr Jordan in Japan, 1922. *Memoirs of the Carnegie Museum* 10: 93-346.

Kottelat, M., A.J. Whitten, S.N. Kartikasari and S. Wirjoatmodjo. 1993. *Freshwater Fishes of Western Indonesia and Sulawesi*. Wildlife Heritage Trust of Sri Lanka, Colombo.

Koumans, F.P. 1953. Gobioidea. In: *Fishes of the Indo-Australian Archipelago*, M. Weber and L.F. de Beaufort (eds.). E.J. Brill, Leiden, pp. 1-423.

Murdy, E.O. 1998. A review of the gobioid fish genus *Gobioides*. *Ichthyological Research* 45: 9-21.

Murdy, E.O. 2002. *Karsten*, a new genus of eel goby (Gobiidae: Amblyopinae) with a key to "Trypauchen" group genera. *Copeia* 2002(3): 787-791.

Murdy, E.O. 2003. A review of *Amblyotrypauchen* (Teleostei: Gobiidae), a genus of blind amblyopine gobies. *Proceedings of the Biological Society of Washington* 116: 330-336.

Murdy, E.O. 2006. A revision of the gobiid fish genus *Trypauchen* (Gobiidae: Amblyopinae). *Zootaxa* 1343: 55-68.

Murdy, E.O. 2008a. *Paratrypauchen*, a new genus for *Trypauchen microcephalus* Bleeker 1860 (Perciformes: Gobiidae: Amblyopinae) with redescription of

Ctenotrypauchen chinensis Steindachner 1867. *aqua, International Journal of Ichthyology* 14: 115-128.

Murdy, E.O. 2008b. *Trypauchenichthys larsonae*, a new species of amblyopine goby from Australia (Gobiidae: Amblyopinae) with a key to the species in the genus. *aqua, International Journal of Ichthyology* 14: 59-68.

Murdy, E.O. and K. Shibukawa. 2001. A revision of the gobioid fish genus *Odontamblyopus* (Gobiidae: Amblyopinae). *Ichthyological Research* 48: 31-43.

Murdy, E.O. and K. Shibukawa. 2002. A redescription of the gobiid fish genus *Pseudotrypauchen* (Gobiidae: Amblyopinae) and its significance in amblyopine phylogeny. *Marine and Freshwater Research* 53: 253-258.

Murdy, E.O. and C.J. Ferraris, Jr. 2003. *Gymnoamblyopus novaeguineae*, a new genus and species of worm goby from Papua New Guinea (Gobiidae: Amblyopinae). *Zootaxa* 150: 1-6.

Murdy, E.O. and J.E. Randall. 2003. *Taenioides kentalleni*, a new species of eel goby from Saudi Arabia (Gobiidae: Amblyopinae) *Zootaxa* 93: 1-6.

Murdy, E.O. and K. Shibukawa. 2003. A revision of the Indo-Pacific fish genus *Caragobius* (Gobiidae: Amblyopinae) *Zootaxa* 301: 1-12.

Pezold, F. 1993. Evidence for a monophyletic Gobiinae. *Copeia* 1993(3): 634-643.

Thacker, C.E. 2003. Molecular phylogeny of the gobioid fishes (Teleostei: Perciformes: Gobioidei). *Molecular Phylogenetics and Evolution* 26: 354-368.

CHAPTER

1.9

Systematics of Sicydiinae

Philippe Keith[1] and Clara Lord[2]

Sicydiins gobies occur in tropical, subtropical and some temperate insular river systems and they have an amphidromous life cycle (Keith, 2003) (see Chapter 2.5, "Amphidromy as a life cycle"). Their adult size varies from 2 cm to 15 cm.

In contrast to the majority of gobies, Sicydiinae pelvic fins are highly modified into a rounded sucking disc (Fig. 1.9.1). This disk has highly branched pelvic fin rays and thickened pelvic fin spines with a fleshy pad at the distal tip (Fig. 1.9.2). A thick interspinal frenum joins the posterior tips of left and right pelvic spines and these pelvic spines and the first ray are separated from the remaining four pelvic rays by a distinct gap (Fig. 1.9.2). The premaxilla is expanded dorsally (Fig. 1.9.3) and the rostral cartilage is small. The palatine bone is with a long dorsal process that articulates with the lateral ethmoid. Sicydiins also have 4 to 5 branchiostegal rays, one epural bone, and the dorsal pterygiophore formula is 3-12210.

Authors' address: Muséum national d'Histoire naturelle, UMR 7208, ichtyologie, CP 26, 57 rue Cuvier, 75231 Paris Cedex 05, France. E-mails: [1]keith@mnhn.fr, [2]claralord@mnhn.fr

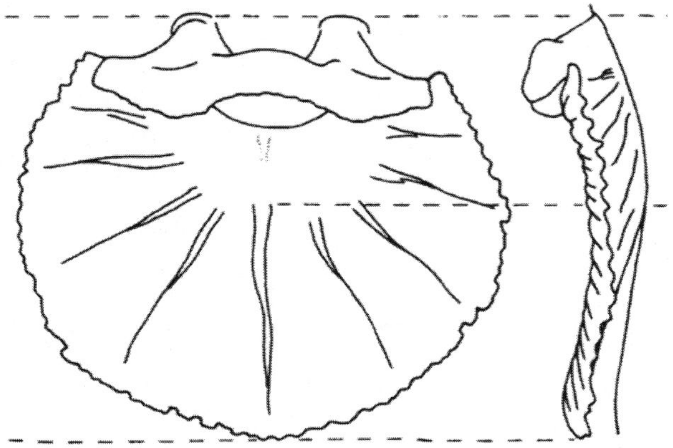

Fig. 1.9.1 Pelvic fins of *Sicyopus leprurus* in lateral (top) and ventral (bottom) views, adapted from Sakai and Nakamura (1979).

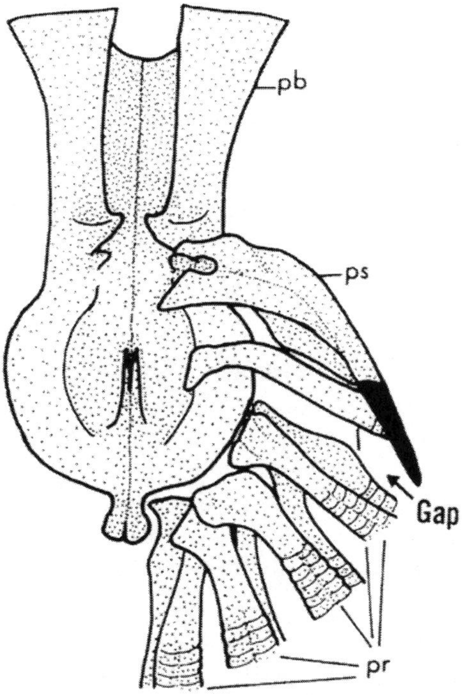

Fig. 1.9.2 Diagrammatic representation of pelvic girdle in ventral view of *Sicyopus felhmanni*, adapted from Parenti and Maciolek (1993). Abbreviations: pb, pelvic bones; ps, pelvic spine; pr, pelvic rays.

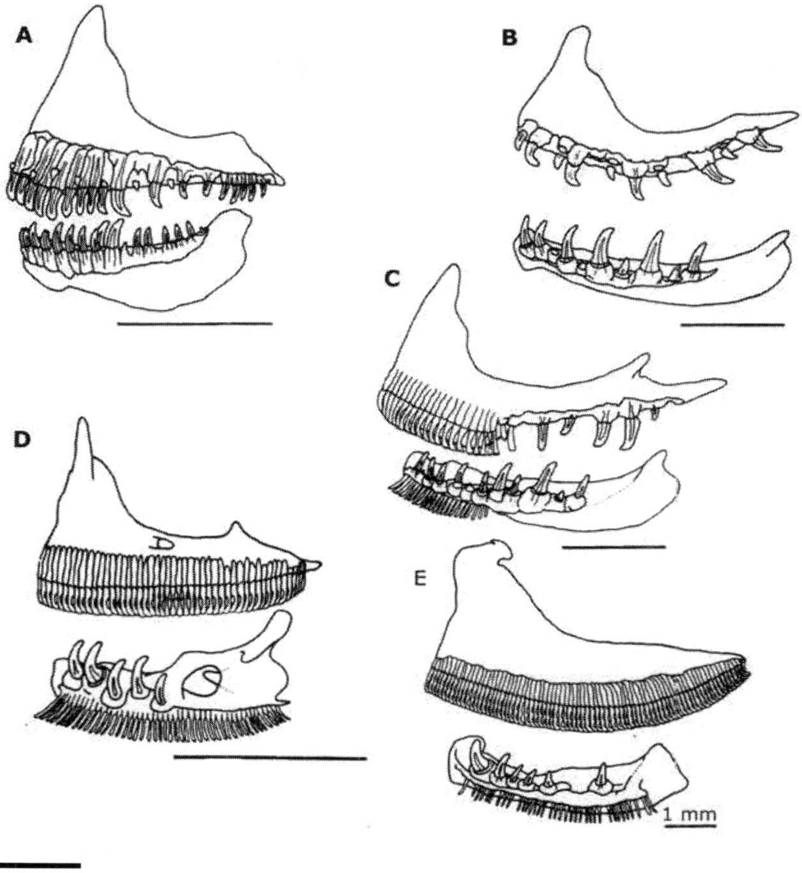

Fig. 1.9.3 Dentition of premaxilla and dentary of A = *Sicyopus leprurus*; B = *Sicyopus zosterophorum*; C = *Lentipes armatus*; D = *Stiphodon elegans*; E = *Sicyopterus japonicus*. Adapted from Sakai and Nakamura (1979).

The strongly effective pelvic suction cup of sicydiins allows them to rapidly access the upper reaches of the catchment area. They are carnivorous, omnivorous or herbivorous depending on the genus. Dentition (with replaceable teeth) is correlated with the feeding behaviour. Noticeable sexual dimorphism (or dichromatism) is characteristic of mature sicydiins (Fig. 1.9.4). In addition to genital papillae (usually pointed in males, rounded or lobed in females), differences may include relative size and shape of median fins, squamation, dentition, and especially bright live coloration.

The Sicydiinae subfamily Bleeker, 1874 comprises 8 genera (Fig. 1.9.4) (*Stiphodon* Weber, 1895 (nearly 30 species); *Sicyopus* Gill, 1863 (15 species); *Lentipes* Günther, 1861 (12 species); *Cotylopus* Guichenot, 1864 (2 species); *Sicyopterus* Gill, 1860 (nearly 30 species); *Sicydium* Valenciennes, 1837 (16 species); *Akihito* Watson, Keith and Marquet, 2007 (2 species) and *Parasicydium* Risch, 1980 (1 species), totalling nearly 110 species distributed in the Indo-Pacific, West African, Central American and Caribbean areas (Keith, 2003; Pezold *et al.*, 2006; Froese and Pauly, 2008).

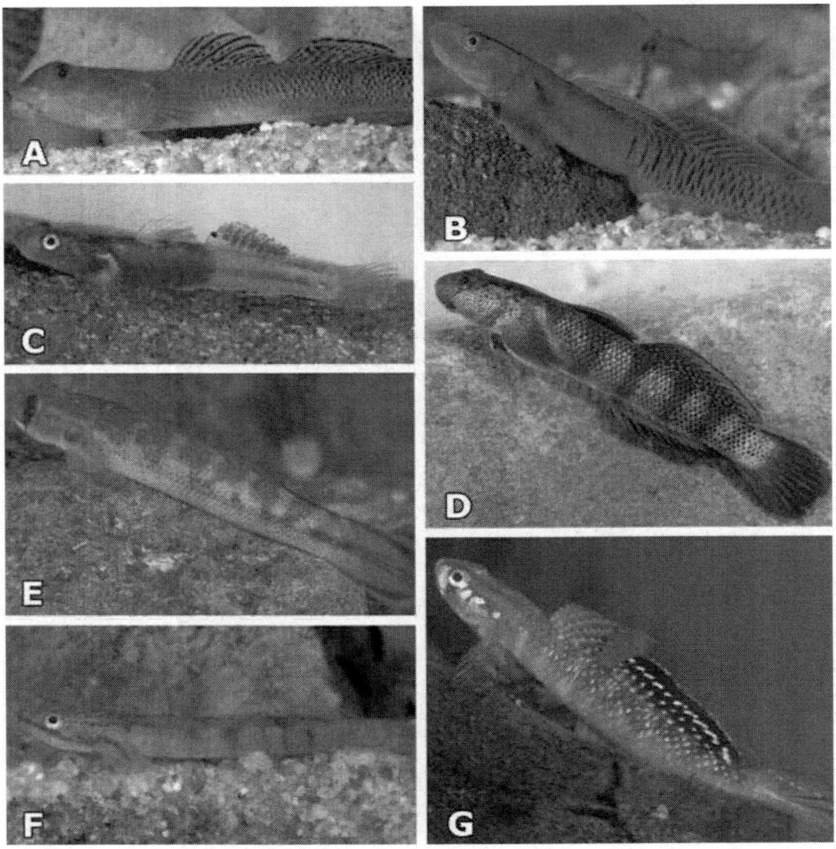

Fig. 1.9.4 Some Sicydiinae species (all male) showing the different genera. A = *Cotylopus rubripinnis*; B = *Akihito vanuatu*; C = *Lentipes kaaea*; D = *Sicydium plumieri*; E = *Sicyopterus lagocephalus*; F = *Sicyopus chloe*; G = *Stiphodon rutilaureus*. Photos P. Keith and E. Vigneux.

Color image of this figure appears in the color plate section at the end of the book.

In the Pacific area, *Sicyopterus* and *Stiphodon* are the most diversified genera and they were found in all the catchment areas from the lower to the upper reaches of rivers. The other genera (*Sicyopus, Lentipes, Akihito*) occur mainly in the middle and in the upper parts of the rivers.

All of these genera have a specific distribution. *Sicydium* is distributed throughout the Caribbean, Central America and West Africa (Pezold *et al.*, 2006); *Sicyopterus* is distributed in the Indo-Pacific from the Western Indian Ocean to the Eastern Pacific (Keith *et al.*, 2005a); *Stiphodon, Sicyopus,* and *Lentipes* are distributed from the Eastern Indian Ocean to the Eastern Pacific (Watson *et al.*, 2001, 2002; Keith and Marquet, 2007); *Cotylopus* is restricted to the West Indian Ocean (Keith *et al.*, 2005b); *Parasicydium* is restricted to Western Africa (Pezold *et al.*, 2006) and *Akihito,* a new recently described genus (Watson *et al.*, 2007; Keith *et al.*, 2007) appears to be restricted to the Western Pacific Ocean.

The main characters of sicydiins genera are summarized in Table 1.9.1.

Table 1.9.1 Main characters of sicydiins genera (D2: second dorsal fin; LS: lateral scales; PD: predorsal scales).

	D2	*Pectoral rays*	*Teeth (Upper Jaw)*	*LS*	*PD*
Sicyopterus	I, 10 - 11	18 - 22	35 - 120 Tricuspid	49 - 75	10 - 35
Sicydium	I, 9 - 10	17 - 22	25 - 83 Mono, Bi-or Tricuspid	49 - 74	0 - 31
Parasicydium	I, 10	16-18	Male 16 – 26 Tricuspid + 1 – 6 Conical Female 16 – 26 Tricuspid	50 - 57	0
Lentipes	I, 9 - 10	15 - 20	Male 6 – 27 Tricusp + 0 – 9 conical; Female 10-34 Tricuspid	20 - 34	0
Cotylopus	I, 9 - 10	15 - 17	7 - 23 Tricuspid	37 - 63	0
Stiphodon	I, 9 - 10	13 - 16	25 - 59 Tricuspid	12 - 39	0 - 17
Sicyopus	I, 9	13 - 16	12 - 25 Conical	12 - 47	0
Akihito	I 9 - 10	15 - 16	Male 4 – 17 Conical; Female 4 – 7 Tricuspid + 0 – 6 conical	14 - 32	0 - 2

Sicyopterus possesses numerous large tricuspid premaxillary teeth in both sexes (Fig. 1.9.3e). The ascending process on the premaxilla is broad at the dorsal tip (Fig. 1.9.3e). The tongue is fused to the floor of the mouth. The pelvic disc is adherent to the belly between all five rays. Scales are always present on the nape in adults. It has 18-22 pectoral rays. Unlike all the other known sicydiins, *Sicyopterus* is distinguished by a second dorsal fin I, 10-11 (*versus* generally I, 9-10) (Watson *et al.*, 2000).

Sicydium generally has numerous large tricuspid premaxillary teeth in both sexes (some species have mono or bicuspid teeth). The ascending process on the premaxilla is broad at the dorsal tip. The tongue is fused to the floor of the mouth. The pelvic disc is adherent to the belly between all five rays. Scales are generally present on the nape in adults. It has 17-22 pectoral rays (Harrison, 1993; Pezold *et al.*, 2006).

Parasicydium has tricuspid premaxillary teeth in both male and female but with 1-6 canine like teeth at the posterior tip of the premaxilla in males. The ascending process on the premaxilla is broad at the dorsal tip. The tongue is fused to the floor of the mouth. The pelvic disc is adherent to the belly between all five rays. Scales are generally absent on the nape in adults. It has 16-18 pectoral rays (Harrison, 1993; Pezold *et al.*, 2006).

Lentipes has tricuspid premaxillary teeth in both male and female but with 0-9 canine like teeth at the posterior tip of the premaxilla in males (Fig. 1.9.3c). The ascending process on the premaxilla is narrow at the dorsal tip. The tongue is fused to the floor of mouth. The pelvic disc is adherent to the belly between all five rays. Scales are never present on the nape and the belly in adults. It has 15-20 pectoral rays (Watson *et al.*, 2002).

Cotylopus possesses few large tricuspid premaxillary teeth on either side of the symphysis in both sexes. The ascending process on premaxilla is narrow at the dorsal tip. The tongue is fused to the floor of the mouth. The pelvic disc is adherent to the belly between all five rays. Scales are never present on nape in adults. It has 15-17 pectoral rays (Keith *et al.*, 2005b).

Stiphodon are unique among the Sicydiinae in having three anal pterygiophores prior to the first haemal spine (Birdsong *et al.*, 1988); in all the other genera belonging to the group, there are only two. *Stiphodon* possesses tricuspid premaxillary teeth in both sexes (Fig. 1.9.3d). The ascending process on the premaxilla is narrow at the dorsal tip (Fig. 1.9.3d). The tongue is fused to floor of the mouth. The pelvic disc is adherent to the belly between fifth rays only. Scales are generally present on the nape of adults and it has 13-16 pectoral rays (Keith and Marquet, 2007).

In the *Sicyopus* genus, premaxillary and dentary jaw teeth are unicuspid or caniniform, uniserial, and there are no horizontal teeth on the dentary.

Watson (1999) defined three subgenera of *Sicyopus* (*Juxtastiphodon*, *Smilosicyopus* and *Sicyopus*), based mainly on dental characteristics found in both jaws. *Juxtastiphodon* has conical teeth in both jaws crowded closely together, none recurved and without canine teeth. *Sicyopus* has no canine teeth but has widely spaced conical teeth in both jaws, most of them sharply recurved (Fig. 1.9.3b). *Smilosicyopus* has slightly recurved conical teeth anteriorly in both jaws, non-recurved needle-like teeth laterally, and between the anterior and lateral teeth, there is at least one (1-3) canine tooth, that is well-developed in males (Fig. 1.9.3a). *Sicyopus* genus is also characterized by a narrow ascending process at the dorsal tip of the premaxilla (Fig. 1.9.3). The tongue is free from the floor of the mouth. The pelvic disc is adherent to the belly between fifth rays only. Scales are sometimes present on the nape of adults and it has 13-16 pectoral rays (Watson, 1999).

Last, *Akihito* is characterized by premaxillary teeth in a single row, mostly tricuspid but with 0-6 canine-like teeth at the posterior tip of the premaxilla in females. Male only have conical premaxillary teeth. The dentary bears a single row of recurved conical to canine-like symphyseal teeth in both sexes. The ascending process on the premaxilla is narrow at the dorsal tip. The tongue is small, rounded and mostly free from floor of the mouth anteriorly. The pelvic disc is adherent to the belly between fifth rays only. Scales are generally absent on the nape of adults. It has 15-16 pectoral rays (Keith *et al.*, 2007).

RELATIONSHIPS AMONG SICYDIINAE

Several authors working on morphological characters of this subfamily (Sakai and Nakamura, 1979; Harrison, 1993; Pezold, 1993; Parenti and Maciolek, 1993; Parenti and Thomas, 1998; Watson *et al.*, 2007) have indicated that the buccal morphology of sicydiins provides valuable characters to assess this group's phylogeny. Synapomorphies of Sicydiinae include specializations of the pelvic fins, the jaws, the jaw suspensorium and of the soft anatomy that are all correlated with rock-climbing ability. The tongue is fused to the floor of the mouth or it is free only at its tip, the jaw suspension is highly modified, the pelvic rays are thickened and highly branched and there are fleshy pads at the tips of the pelvic spines. Harrison (1989) added another synapomorphy, which is the dorsal expansion of the premaxilla. Parenti and Thomas (1998) and Watson *et al.* (2007) stated that the Sicydiinae is a monophyletic group; they explained that monophyly is supported by the fact that the pelvic-fin spine articulates with the pelvic bone, that there is a fleshy pad on the tip of the pelvic-fin spine, that the ascending premaxillary and the articular processes are indistinct and, finally, that the rostral cartilage is small or absent.

Key of Sicydiinae genera

1a Pelvic disk adherent to belly between fifth rays only: **2**

1b Pelvic disk adherent to belly between all 5 rays: **4**

2a Tongue fused to the floor of mouth: ***Stiphodon***

2b Tongue free at its tip: **3**

3a Tricuspid teeth in upper jaw absent in male and female: ***Sicyopus***

3b Tricuspid teeth in upper jaw absent in male but present in female: ***Akihito***

4a Ascending process on the premaxilla broad at the dorsal tip : **5**

4b Ascending process on the premaxilla narrow at the dorsal tip : **7**

5a Scales on dorsal and ventral parts of body not extending anteriorly beyond origin of second dorsal and anal fins respectively; male with caniniform teeth in posterolateral parts of upper jaw: ***Parasicydium***

5b Scales on dorsal and ventral parts of body extending anteriorly to head; male without caniniform teeth in posterolateral parts of upper jaw: **6**

6a Upper jaw teeth in straight rows not meeting at the symphysis: ***Sicyopterus***

6b Upper jaw teeth in curved rows meeting at the symphysis: ***Sicydium***

7a Sac of replacement teeth in upper jaw present; labial teeth in a single band: ***Cotylopus***

7b Sac of replacement teeth in upper jaw absent; labial teeth in two short bands anteriorly: ***Lentipes***

Acknowledgments

Many colleagues have contributed to provide specimens, data and discussion that have led to the realisation of this work. We want to thank : G.R. Allen (CI), J. Aoyama (ORI), P. Béarez (MNHN), P. Bosc (ARDA), E. Feunteun (MNHN), R. Galzin (EPHE), P. Gerbeaux (IUCN), M. Iida (ORI), M. Jégu (MNHN), A. Jenkins (WI), D. Kalfatak (EU), M. Kottelat, M. Kovacic (NHMR), H. Larson (NTM), P. Lim (ENSAT), G. Marquet, F. Meunier (MNHN), L.R. Parenti (USNM), R.A. Patzner (Univ. Salzburg), F.

Pezold (UL), S. Planes (EPHE), G. Ségura (ISE), K. Tsukamoto (ORI), S. Watanabe (ORI), P. Valade (ARDA), E. Vigneux (HFC), R.E. Watson (UF), J. Williams (USNM).

The results obtained by the MNHN team were financially supported by the New Caledonian Government, the French Ichthyological Society, the French Ministery of Ecology, the French Foundation, the National Research Agency and the MNHN. We thank also the following for help and assistance during the various studies, trips and expeditions to Indian and Pacific Islands: the Environment Unit of Vanuatu, the Seychelles Environment Ministry, Conservation International from Madagascar and Pacific, the Provincial and Territorial authorities of New Caledonia and Wallis and Futuna, ARDA from La Réunion Island, and CNDRS from Comoros.

References

Birdsong, R.S., E.O. Murdy and F.L. Pezold. 1988. A study of the vertebral column and median fin osteology in gobioid fishes with comments on gobioid relationships. *Bulletin of Marine Science* 42: 172-214.

Froese, R. and D. Pauly. (eds.) 2008. FishBase. World Wide Web electronic publication. www.fishbase.org, version 12/2008.

Harrison, I.J. 1989. Specialisation of the Gobioid palatopterygoquadrate complex and its relevance to Gobioid systematics. *Journal of Natural History* 23: 325-353.

Harrison, I.J. 1993. The West African sicydiine fishes, with notes on the genus *Lentipes* (Teleostei: Gobiidae). *Ichthyological Exploration of Freshwaters* 4: 201-232.

Keith, P. 2003. Biology and ecology of amphidromous Gobiidae in the Indo-Pacific and the Caribbean regions. *Journal of Fish Biology* 63: 831-847.

Keith, P. and G. Marquet. 2007. *Stiphodon rubromaculatus,* a new species of freshwater goby from Futuna island (Teleostei: Gobioidei: Sicydiinae). *Cybium* 31: 45-49.

Keith, P., G. Marquet and R.E. Watson. 2007. *Akihito futuna*, a new species of freshwater goby from the South Pacific (Teleostei: Gobioidei: Sicydiinae). *Cybium* 31: 471-476.

Keith, P., T. Galewski, G. Cattaneo-Berrebi, T. Hoareau and P. Berrebi. 2005a. Ubiquity of *Sicyopterus lagocephalus* (Teleostei: Gobioidei) and phylogeography of the genus *Sicyopterus* in the Indo-Pacific area inferred from mitochondrial cytochrome *b* gene. *Molecular Phylogenetics and Evolution* 37: 721-732.

Keith, P., T. Hoareau and P. Bosc. 2005b. A new species of freshwater goby (Pisces: Teleostei: Gobioidei) from Mayotte island (Comoros) and comments about the genus *Cotylopus* endemic to Indian Ocean. *Journal of Natural History* 39: 1395-1405.

Parenti, L.R. and J.A. Maciolek. 1993. New sicydiine gobies from Ponape and Palau, Micronesia, with comments of systematics of the subfamily Sicydiinae (Teleostei: Gobiidae). *Bulletin of Marine Sciences* 53: 945-972.

Parenti, L.R. and K.R. Thomas. 1998. Pharyngeal jaw morphology and homology in sicydiine gobies (Teleostei: Gobiidae) and allies. *Journal of Morphology* 237: 257-274.

Pezold, F. 1993. Evidence for a monophyletic Gobiinae. *Copeia* 1993: 634-643.

Pezold, F., T. Iwamoto and I.J. Harrison. 2006. Multivariate analysis of sicydiines of São Tomé and Príncipe with redescription of *Sicydium brevifile* and *S. bustamantei* (Teleostei: Gobiidae) and a key to west African sicydiines. The California Academy of Sciences Gulf of Guinea Expedition (2201). *V. Proceedings of California Academy of Science* (Ser. 4) v. 57: 965-980.

Sakai, H. and M. Nakamura. 1979. Two new species of freshwater gobies (Gobiidae: Sicydiaphiinae) from Ishigaki Island, Japan. *Japanese Journal of Ichthyology* 26: 43-54.

Watson, R.E. 1999. Two new subgenera of *Sicyopus*, with a redescription of *Sicyopus zosterophorum* (Teleostei: Gobioidei: Sicydiinae). *Aqua: Journal of Ichthyology and Aquatic Biology* 3: 93-104.

Watson, R.E., G. Marquet and C. Pöllabauer. 2000. New Caledonia Fish Species of the genus *Sicyopterus* (Teleostei: Gobiodei: Sicydiinae). *Aqua: Journal of Ichthyology and Aquatic Biology* 4: 5-34.

Watson, R.E., P. Keith and G. Marquet. 2001. *Sicyopus (Smilosicyopus) chloe,* a new species of freshwater goby from New Caledonia (Teleostei: Gobioidei: Sicydiinae). *Cybium* 25: 41-52.

Watson, R.E., P. Keith and G. Marquet. 2002. *Lentipes kaaea*, a new species of freshwater goby from New Caledonia (Teleostei: Gobioidei: Sicydiinae). *Bulletin Français de pêche et de Pisciculture* 364: 173-185.

Watson, R.E., P. Keith and G. Marquet. 2007. *Akihito vanuatu*, a new genus and new species of freshwater goby from the South Pacific (Teleostei: Gobioidei: Sicydiinae). *Cybium* 31: 341-349.

CHAPTER

1.10

Systematics of Gobiidae

Christine Thacker

The family Gobiidae includes what are considered the typical gobies: small, usually marine fishes, with separate spinous and rayed dorsal fins and pelvic fins often joined into a disc, that live on or in the substrate and are generally inconspicuous. Color patterns may range from brightly patterned to drab and camouflaged. The composition of this family has changed since the description of the first *Gobius* by Linnaeus (1758), but it has always been the most speciose of gobioid taxa, and the one that contains the most well-known examples of goby biology and ecology. Various classifications based on morphology have delineated Gobiidae as the largest of six to nine gobioid families (Miller, 1973; Hoese, 1984; Harrison, 1989; Pezold, 1993; Nelson, 2006); reevaluations of gobioid systematics based on molecular data have shown that the smaller, distinctive families historically separated from the bulk of gobioid taxa are actually nested within the larger groups (Akihito *et al.*, 2000; Wang *et al.*, 2001; Thacker, 2003; Rüber and Agorreta, 2011, Chapter 1.2 of this volume). A revised six-family clade-based classification (Thacker, 2009) identifies a Gobiidae that includes all genera of the former taxa Gobiinae, Microdesmidae, Ptereleotridae, Schindleriidae,

---

Author's address: Research and Collections, Section of Ichthyology, Natural History Museum of Los Angeles County, 900 Exposition Blvd., Los Angeles, CA 90007 USA. E-mail: thacker@nhm.org and Kraemeriidae (*Kraemeria* recovered within Gobionellidae by Thacker [2003] but assigned to Gobiidae when analyzed in the context of more comprehensive sampling.) This monophyletic Gobiidae includes 1,107 species at this writing (calculated from an unpublished list of gobiid species compiled by Edward Murdy, augmented with data from the online version of the Catalog of Fishes [Eschmeyer, 2008]). The 167 genera included in Gobiidae are listed in Table 1.10.1.

Table 1.10.1 Valid genera of Gobiidae.

Acentrogobius	*Afurcagobius*	*Aioliops*
Akko	*Amblyeleotris*	*Amblygobius*
Amoya	*Anatirostrum*	*Aphia*
Arcygobius	*Arenigobius*	*Aruma*
Asterropteryx	*Aulopareia*	*Austrolethops*
Babka	*Barbulifer*	*Barbuligobius*
Bathygobius	*Benthophiloides*	*Benthophilus*
Bollmannia	*Bryaninops*	*Buenia*
Cabillus	*Caffrogobius*	*Callogobius*
Caspiosoma	*Cerdale*	*Chriolepis*
Chromogobius	*Clarkichthys*	*Corcyrogobius*
Coryogalops	*Coryphopterus*	*Cristatogobius*
Croilia	*Cryptocentroides*	*Cryptocentrus*
Crystallogobius	*Ctenogobiops*	
Deltentosteus	*Didogobius*	*Discordipinna*
Drombus		
Ebomegobius	*Echinogobius*	*Economidichthys*
Egglestonichthys	*Ego*	*Elacatinus*
Eleotrica	*Enypnias*	*Evermannichthys*
Eviota	*Exyrias*	
Favonigobius	*Feia*	*Flabelligobius*
Fusigobius		
Gammogobius	*Ginsburgellus*	*Gladiogobius*
Glossogobius	*Gobiodon*	*Gobiopsis*
Gobiosoma	*Gobitrichonotus*	*Gobius*
Gobiusculus	*Gobulus*	*Gorogobius*
Grallenia	*Gunnellichthys*	*Gymneleotris*
Hazeus	*Heterelotris*	*Heteroplopomus*
Hyrcanogobius		
Istigobius		
Kelloggella	*Knipowitschia*	*Kraemeria*
Larsonella	*Lebetus*	*Lesueurigobius*

Contd...

Contd...

Lobulogobius	*Lophiogobius*	*Lophogobius*
Lotilia	*Lubricogobius*	*Luposicya*
Lythrypnus		
Macrodontogobius	*Mahidolia*	*Mangarinus*
Mauligobius	*Mesogobius*	*Microdesmus*
Microgobius	*Millerigobius*	*Minysicya*
Myersina		
Nemateleotris	*Nematogobius*	*Neogobius*
Nes	*Nesogobius*	
Obliquogobius	*Odondebuenia*	*Ophiogobius*
Oplopomops	*Oplopomus*	*Opua*
Oxymetopon		
Padogobius	*Palatogobius*	*Palutrus*
Papillogobius	*Parachaeturichthys*	*Paragobiodon*
Paragunnellichthys	*Paratrimma*	*Pariah*
Parioglossus	*Parkraemeria*	*Parrella*
Pascua	*Phyllogobius*	*Platygobiopsis*
Pleurosicya	*Polyspondylogobius*	*Pomatoschistus*
Ponticola	*Porogobius*	*Priolepis*
Proterorhinus	*Psammogobius*	*Pseudaphya*
Psilogobius	*Psilotris*	*Ptereleotris*
Pycnomma		
Rhinogobiops	*Risor*	*Robinsichthys*
Schindleria	*Signigobius*	*Siphonogobius*
Silhouettea	*Speleogobius*	*Stonogobiops*
Sueviota	*Sufflogobius*	
Tasmanogobius	*Thorogobius*	*Tomiyamichthys*
Trimma	*Trimmatom*	*Tryssogobius*
Valenciennea	*Vanderhorstia*	*Vanneaugobius*
Varicus	*Vomerogobius*	
Wheelerigobius		
Yongeichthys		
Zebrus	*Zosterisessor*	

Tropical reef habitats are particularly rich in gobiids, but they are often inconspicuous due to their small size and ecology. Generally gobies will occupy crevices or interstices in the reef or rocky substrate, and many notable examples of mutualisms between gobies and invertebrates, particularly the burrow-digging snapping shrimps, have been documented (Karplus and Thompson, 2011, Chapter 4.4 of this volume). The reproductive strategy of gobiids is unusual compared to most fishes, and requires more investment and parental care. Typically, gobies will lay their eggs demersally, in a nest area prepared and guarded by the male; eggs bear adherent filaments and are fixed in an adhesive sperm trail. The clutch of eggs is then monitored and cared for by the male, until larvae hatch out and begin an extended planktonic larval period (Breder and Rosen, 1966; Ruple, 1984). Some of the reef-dwelling species exhibit protogynous hermaphroditism, with females transforming into males as needed in response to social and environmental cues (Cole, 1990, 2008). Gobiids are also notable for their long larval lives, both in absolute length and considered as a percentage of total lifespan. As indicated by otoliths, the larval life of a small gobiid (*Trimma nasa*) totaled 33.9 days, or 38.7% of the total lifespan of 87 days (Winterbottom and Southcott, 2008). Species of *Eviota* experience lifespans as short as 59 days total, with 41.8% of that time spent as larvae (Depczynski and Bellwood, 2005, 2006).

Finally, one of the most well-known characteristics of gobiids is their small adult size, including species that become mature at 8-10mm standard length (*Eviota*, *Trimma*, *Trimmatom*), and those that exhibit not just size reduction, but morphological simplification likely the result of paedomorphosis (such as *Schindleria*; Johnson and Brothers, 1993; Watson and Walker, 2004; Thacker and Grier, 2005; Nelson, 2006). The miniaturization, reduction, and simplification common among gobiid, gobies has hindered efforts to determine phylogeny, and although a fair amount of data has been gathered, more comprehensive phylogenetic surveys are clearly necessary to definitively determine relationships among gobiids.

Gobiidae are so diverse and such an important part of reef faunas that they have been studied as exemplars of a variety of evolutionary and ecological patterns. The coral-perching species of *Gobiodon* have provided one of the few clear examples of sympatric speciation by host shift, and the only such example from a marine system (Munday *et al.*, 2004). *Gobiodon* have also been studied to compare the impact of habitat degradation on ecological generalists as compared to specialists (Munday, 2004). Phylogenetic studies of the seven-spined cleaner gobies of the Caribbean have provided insights into larval transport and retention, as well as the tempo and ecological correlates of speciation (Rüber *et al.*, 2003; Rüber and Zardoya, 2005; Taylor and Hellberg, 2003, 2005). In each of these cases, understanding of phylogeny has been a crucial first step in exploring various hypotheses of evolutionary and ecological change.

Phylogeny of taxa and groups within Gobiidae has not been determined with morphological characters. Several surveys of morphological character patterns across Gobiidae have been performed, including osteology of the fins and axial skeleton (Birdsong *et al.*, 1988), suspensorium, branchial apparatus and pectoral girdle (Akihito, 1986; Harrison, 1989), and configuration of cephalic sensory pores (Pezold, 1993; Takagi, 1989). In particular, the study of Birdsong *et al.* (1988), although not phylogenetic, provided a breakdown of Gobiidae into groups of genera with common patterns of skeletal characters. These included temperate Pacific-rim *Acanthogobius*, *Astrabe*, and *Chasmichthys* groups; the New World tropical *Gobiosoma* and *Microgobius* groups; the Old World temperate *Gobius* and *Pomatoschistus* groups; and the highly diverse tropical marine *Priolepis* and *Bathygobius* groups. Pezold's (1993) study not only surveyed the cephalic pore configurations of a comprehensive sample of Gobiidae, but also proposed a diagnostic character for the group (former Gobiinae only): a single cephalic interorbital pore, rather than a pair of pores. He also noted that most of those taxa possess a common dorsal fin pterygiophore pattern and vertebral number, the same as that of the *Bathygobius* and *Priolepis* groups. Clearly, there are characters valuable for identifying gobiid groups; however, there are relatively few and they do not seem to show entirely consistent patterns, particularly when former families Microdesmidae, Ptereleotridae, Kraemeriidae, and Schindleriidae are included as part of gobiid diversity.

Given that Gobiidae is such a speciose group, it is not surprising that it is difficult to generalize about their morphology or biology. The most outstanding question still to be addressed in consideration of gobioid relationships is inference of the phylogeny of groups within Gobiidae. A robust phylogeny will enable evolutionary consideration of the wealth of interesting biological and ecological patterns documented among gobiids. Phylogenetic analysis of morphological characters has been performed for some subgroups (Gill, 1994; Thacker, 2000), others have been investigated with molecular data (Rüber *et al.*, 2003; Taylor and Hellberg, 2003, 2005), or a combination (Thacker and Cole, 2002; Harold *et al.*, 2008), and gobiids have been included in larger molecular phylogenetic hypotheses (Thacker, 2003, 2009; Thacker and Hardman, 2005). Patterns of morphological character distribution may be reconsidered in conjunction with molecular phylogenetic results, and generally it is the case that morphological character distribution correlates fairly well, but not completely, with clades proposed based on DNA sequence (Thacker, 2009). Within Gobiidae, the most notable results based on analysis of molecular data are that the large *Priolepis* group (Birdsong *et al.*, 1988) is not monophyletic, and that New World taxa are placed in several different clades (three in that hypothesis, comprising *Cerdale* plus *Microdesmus*; *Gobiosoma*, *Risor*, and *Barbulifer*; and *Coryphopterus* plus *Lophogobius*), nested within the more diverse Indo-

Pacific fauna (Thacker, 2003, 2009). Overall, the development of gobiid systematics seems to be progressing in much the same pattern as seen for the entire Gobioidei: distinctive groups are delineated and singled out from the majority of taxa, and as data accumulate, it is revealed that this perceived distinctiveness is simply a part of the range of diversity in the larger group. Comprehensive surveys of both morphology and molecular data will be required to further evaluate gobiid relationships, and to reveal the systematics of this diverse group.

References

Akihito, A. Iwata, K. Ikeo, T. Imanishi, H. Ono, Y. Umehara, C. Hamamatsu, K. Sugiyama, Y. Ikeda, K. Sakamoto, A. Fumihito, S. Ohno and T. Gojobori. 2000. Evolutionary aspects of gobioid fishes based on a phylogenetic analysis of mitochondrial cytochrome *b* genes. *Gene* 259: 5-15.

Akihito, P. 1986. Some morphological characters considered to be important in gobiid phylogeny. In: *Proceedings of the Second International Conference on Indo-Pacific Fishes*. The Ichthyological Society of Japan, Tokyo, pp. 629-639.

Birdsong, R.S., E.O. Murdy and F.L. Pezold. 1988. A study of the vertebral column and median fin osteology in gobioid fishes with comments on gobioid relationships. *Bulletin of Marine Science* 42: 174-214.

Breder, C.M. and D.E. Rosen. 1966. *Modes of Reproduction in Fishes*. Natural History Press, Garden City, New York.

Cole, K.S. 1990. Patterns of gonad structure in hermaphroditic gobies. *Environmental Biology of Fishes* 28: 125-142.

Cole, K.S. 2008. Transient ontogenetic expression of hermaphroditic gonad morphology within the Gobiosoma group of the neotropical seven-spined gobies (Teleostei: Gobiidae). *Marine Biology* 154: 943-951.

Depczynski, M. and D.R. Bellwood. 2005. Shortest recorded vertebrate lifespan found in a coral reef fish. *Current Biology* 15: R288-R289.

Depczynski, M. and D.R. Bellwood. 2006. Extremes, plasticity, and invariance in vertebrate life history traits: insights from coral reef fishes. *Ecology* 87: 3119-3127.

Eschmeyer, W.N. (ed.). 2008. *Online Version of the Catalog of Fishes* www.calacademy.org/research/ichthyology/catalog/fishcatsearch.html; updated 18 September, 2008, accessed November, 2008.

Gill, H.S. 1994. Phylogenetic relationships of the members of the *Bathygobius* and *Priolepis* groupings (sensu Birdsong *et al.*, 1988) which possess a longitudinal papillae pattern. In: *Proceedings of the Fourth Indo-Pacific Fish Conference: Systematics and Evolution of Indo-Pacific Fishes*. Bangkok, Thailand, pp. 40-59.

Harold, A.S., R. Winterbottom, P.L. Munday and R.W. Chapman. 2008. Phylogenetic relationships of Indo-Pacific coral gobies of the genus *Gobiodon* (Teleostei: Gobiidae) based on morphological and molecular data. *Bulletin of Marine Science* 82: 119-136.

Harrison, I.J. 1989. Specialization of the gobioid palatopteroquadrate complex and its relevance to gobioid systematics. *Journal of Natural History* 23: 325-353.

Hoese, D.F. 1984. Gobioidei: Relationships. In: *Ontogeny and Systematics of Fishes.* H.G. Moser (ed.). Special Publication of the American Society of Ichthyologists and Herpetologists No.1. Allen Press, Lawrence, Kansas. pp. 588-591.

Johnson, G.D. and E.B. Brothers. 1993. *Schindleria*: a paedomorphic goby (Teleostei: Gobioidei). *Bulletin of Marine Science* 52: 441-471.

Karplus, I. and A.R. Thompson. 2011. The partnership between gobiid fishes and burrowing alpheid shrimps. In: *The Biology of Gobies*, R.A. Patzner, J.L. Van Tassell, M. Kovačić and B.G. Kapoor (eds.), Science Publisher Inc., Enfield, pp. 559-607.

Linnaeus, C. 1758. Systema Naturae, Ed. X. (Systema naturae per regna tria naturae, secundum classes, ordines, genera, species, cum characteribus, differentiis, synonymis, locis. Tomus I. Editio decima, reformata.) Holmiae.

Miller, P.J. 1973. The osteology and adaptive features of *Rhyacichthys aspro* (Teleostei: Gobioidei) and the classification of gobioid fishes. *Journal of Zoology, London* 171: 397-434.

Munday, P.L. 2004. Habitat loss, resource specialization, and extinction on coral reefs. *Global Change Biology* 10: 1642-1647.

Munday, P.L., L. van Herwerden and C.L. Dudgeon. 2004. Evidence for sympatric speciation by host shift in the sea. *Current Biology* 14: 1498-1504.

Nelson, J.S. 2006. *Fishes of the World.* 4th edition. Wiley, New York.

Pezold, F. 1993. Evidence for a monophyletic Gobiinae. *Copeia* 1993: 634-643.

Rüber, L.J. and Zardoya, R. 2005. Rapid cladogenesis in marine fishes revisited. *Evolution* 59: 1119-1127.

Rüber, L.J. and A. Agorreta. 2011. Molecular systematics of gobioid fishes. In: *The Biology of Gobies*, R.A. Patzner, J.L. Van Tassell, M. Kovačić and B.G. Kapoor (eds.), Science Publisher Inc., Enfield, pp. 23-50.

Rüber, L., J.L. Van Tassell and R. Zardoya. 2003. Rapid speciation and ecological divergence in the American seven-spined gobies (Gobiidae, Gobiosomatini) inferred from molecular phylogeny. *Evolution* 57: 1584-1598.

Ruple, D. 1984. Gobioidei: Development. In: *Ontogeny and Systematics of Fishes.* H. G. Moser (ed.). Special Publication of the American Society of Ichthyologists and Herpetologists No. 1. Allen Press, Lawrence, Kansas. pp. 582-587.

Takagi, K. 1989. Cephalic sensory canal system of the gobioid fishes of Japan: comparative morphology with special reference to phylogenetic significance. *Journal of the Tokyo University of Fisheries* 75: 499-568.

Taylor, M.S. and M.E. Hellberg. 2003. Genetic evidence for local retention of pelagic larvae in a Caribbean reef fish. *Science* 299: 107-109.

Taylor, M.S. and M.E. Hellberg. 2005. Marine radiations at small geographic scales: speciation in neotropical reef gobies (*Elacatinus*). *Evolution* 59: 374-385.

Thacker, C.E. 2000. Phylogeny of the Wormfishes (Teleostei: Gobioidei: Microdesmidae). *Copeia* 2000: 940-957.

Thacker, C.E. 2003. Molecular phylogeny of the gobioid fishes (Teleostei: Perciformes: Gobioidei). *Molecular Phylogentics and Evolution* 26: 354-368.

Thacker, C.E. 2009. Phylogeny of Gobioidei and placement within Acanthomorpha, with a new classification and investigation of diversification and character evolution. *Copeia* 2009: 93-104.

Thacker, C.E. and K.S. Cole. 2002. Phylogeny and evolution of the gobiid genus *Coryphopterus*. *Bulletin of Marine Science* 70: 837-850.

Thacker, C.E. and H. Grier. 2005. Unusual gonad structure in the paedomorphic teleost *Schindleria praematura* (Teleostei: Gobioidei): a comparison with other gobioid fishes. *Journal of Fish Biology* 66: 378-391.

Thacker, C.E. and M.A. Hardman. 2005. Molecular phylogeny of basal gobioid fishes: Rhyacichthyidae, Odontobutidae, Xenisthmidae, Eleotridae (Teleostei: Perciformes: Gobioidei). *Molecular Phylogenetics and Evolution* 37: 858-871.

Wang, H.-Y., M.-P. Tsai, J. Dean and S.-C. Lee. 2001. Molecular phylogeny of gobioid fishes (Perciformes: Gobioidei) based on mitochondrial 12S rRNA sequences. *Molecular Phylogenetics and Evolution* 20: 390-408.

Watson, H. and H.J. Walker Jr. 2004. The world's smallest vertebrate, *Schindleria brevipinguis*, a new paedomorphic species in the family Schindleriidae (Perciformes: Gobioidei). *Records of the Australian Museum* 56: 139-142.

Winterbottom, R. and L. Southcott. 2008. Short lifespan and high mortality in the western Pacific coral reef goby *Trimma nasa*. *Marine Ecology Progress Series* 366: 203-208.

SECTION 2

Zoogeography

CHAPTER

2.1

Gobiiformes of the Americas

James L. Van Tassell

INTRODUCTION

The Gobiiformes (Wiley and Johnson, 2010) of the Americas comprise 308 species considered as valid in this chapter. An additional eight or more species are known and not currently described. Of the 308 species, 234 are in the family Gobiidae, 20 Microdesmidae, 4 Ptereleotridae, and 25 Eleotridae. Tables 2.1.1, 2.1.2, and 2.1.3 provide a list of all species in the Americas recognized herein. An initial species list was compiled using Catalog of Fishes (Eschmeyer, 2010), FishBase (Froese and Pauly, 2010), Shorefishes of the Tropical Eastern Pacific (Robertson and Allen, 2008), and the Global Invasive Species Database (http://www.issg.org/database/welcome/). That list was checked against the following field guides and faunal lists: Navarro and Pequeño, 1979; Menezes and Figueiredo, 1985; Courtenay *et al.*, 1986; Humann and Deloach, 1993; Cervigón, 1994; Grove and Lavenberg, 1997; Smith, 1997; De Moura *et al.*, 1999; Thomson *et al.*, 2000; Pezold and Cage, 2002; Floeter *et al.*, 2003; Kullander, 2003; Menezes, 2003; Moura, *et al.*, 2003; Moura and Menezes, 2003; Smith *et al.*, 2003; and my database. The final species tables were then reviewed by Carole Baldwin,

Author's address: Systematic Ichthyology, Biology Department, Hofstra University, Hempstead, NY 11549, U.S.A. E-mail: jvantassell@gobiidae.com

Table 2.1.1 All species of Gobiiformes in the Americas (excluding the Gobiosomatini), including distributions and habitats

Taxon	*Distribution*	*Habitat*
Gobiidae: Gobiinae		
Bathygobius Bleeker, 1878		
Bathygobius andrei (Sauvage, 1880)	Eastern Pacific: Guatemala to Peru	m
Bathygobius curacao (Metzelaar, 1919)	Western Atlantic: Bermuda, Florida, Bahamas to n. South America	m
Bathygobius lineatus (Jenyns, 1842)	Eastern Pacific: oceanic islands in tropical eastern Pacific, incl. Galápagos	m
Bathygobius mystacium Ginsburg, 1947	Western Atlantic: Bahamas, Caribbean, Brazil(N. of Espírito Santo)	m
Bathygobius ramosus Ginsburg, 1947	Eastern Pacific: n. Gulf of California to Peru	m
Bathygobius soporator (Valenciennes, 1837)	Western Atlantic: North Carolina to Brazil	m
Bathygobius geminatus Tornabene, Baldwin and Pezold, 2010	Western Atlantic: Florida, Puerto Rico	m
Bathygobius antilliensis Tornabene, Baldwin and Pezold, 2010	Western Atlantic: Central and South America, Caribbean	m
Bathygobius lacertus (Poey, 1860)	Western Atlantic: Florida, Central America, Caribbean	m
Coryphopterus Gill 1863		
Coryphopterus alloides Böhlke and Robins, 1960	Western Atlantic: s. Florida and Bahamas to Venezuela	m
Coryphopterus dicrus Böhlke and Robins, 1960	Western Atlantic: s. Florida to Brazil	m
Coryphopterus eidolon Böhlke and Robins, 1960	Western Atlantic: s. Florida to Lesser Antilles	m
Coryphopterus lipernes Böhlke and Robins, 1962	Western Atlantic: Florida Keys, Caribbean, Central America	m
Coryphopterus hyalinus Böhlke and Robins, 1962	Western Atlantic: Florida to Lesser Antilles	m
Coryphopterus kuna Victor, 2007	Western Atlantic: Florida, Caribbean	m
Coryphopterus personatus (Jordan and Thompson, 1905)	Western Atlantic: Florida to South America	m
Coryphopterus punctipectophorus Springer, 1960	Western Atlantic: South Carolina, Florida, Gulf of Mexico	m
Coryphopterus thrix Böhlke and Robins, 1960	Western Atlantic: Florida to Brazil	m
Coryphopterus tortugae (Jordan, 1904)	Western Atlantic: Florida to Venezuela	m
Coryphopterus glaucofraenum Gill, 1864	Western Atlantic: North Carolina to Brazil	m

Contd...

Contd...

Taxon	*Distribution*	*Habitat*
Coryphopterus urospilus Ginsburg, 1938	Eastern Pacific: n. Gulf of California to Colombia	m
Coryphopterus venezuelae (Cervigon, 1966)	Western Atlantic: s. Caribbean	m
Evermannia Jordan, 1895		
Evermannia erici Bussing, 1983	Eastern Pacific: Costa Rica	m
Evermannia longipinnis (Steindachner, 1879)	Eastern Pacific: w. central Gulf of California	m
Evermannia panamensis Gilbert and Starks, 1904	Eastern Pacific: Costa Rica, Panama	m
Evermannia zosterura (Jordan and Gilbert, 1882)	Eastern Pacific: s. Gulf of California to Peru	m
Lophogobius Gill, 1862		
Lophogobius cristulatus Ginsburg, 1939	Eastern Pacific: Costa Rica and Panama	m
Lophogobius cyprinoides (Pallas, 1770)	Western Atlantic: Bermuda, Florida, Caribbean to northern South America, Panama Canal	f,b,m
Lythrypnus Jordan and Evermann, 1896		
Lythrypnus alphigena Bussing, 1990	Eastern Pacific: Isla del Coco	m
Lythrypnus brasiliensis Greenfield, 1988	Western Atlantic: southern Brazil	m
Lythrypnus cobalus Bussing, 1990	Eastern Pacific: Isla del Coco	m
Lythrypnus crocodilus (Beebe and Tee-Van, 1928)	Western Atlantic: Bahamas, Antilles, Belize, Honduras, Nicaragua	m
Lythrypnus dalli(Gilbert, 1890)	Eastern Pacific: s. California and n. Gulf of California to Peru, Isla Malpelo, Galápagos	m
Lythrypnus elasson Böhlke and Robins, 1960	Western Atlantic: Bahamas, Cuba, Jamaica	m
Lythrypnus gilberti (Heller and Snodgrass, 1903)	Eastern Pacific: Galápagos	m
Lythrypnus heterochroma Ginsburg, 1939	Western Atlantic: Bahamas, Mexico, Belize	m
Lythrypnus insularis Bussing, 1990	Eastern Pacific: Isla Revillagigedo (off western Mexico)	m
Lythrypnus lavenbergi Bussing, 1990	Eastern Pacific: Isla del Coco	m
Lythrypnus minimus Garzón and Acero P., 1988	Western Atlantic: Bahamas, Isla de Providencia, Venezuela	m
Lythrypnus mowbrayi (Bean, 1906)	Western Atlantic: Bermuda	m

Contd...

Contd...

Taxon	*Distribution*	*Habitat*
Lythrypnus nesiotes Böhlke and Robins, 1960	Western Atlantic: Florida to n. South America	m
Lythrypnus okapia Robins and Böhlke, 1964	Western Atlantic: Bahamas	m
Lythrypnus phorellus Böhlke and Robins, 1960	Western Atlantic: Florida, Gulf of Mexico to Central America	m
Lythrypnus pulchellus Ginsburg, 1938	Eastern Pacific: n. Gulf of California to Panama	m
Lythrypnus rhizophora (Heller and Snodgrass, 1903)	Eastern Pacific: Isla del Coco and Galápagos	m
Lythrypnus solanensis Acero P., 1981	Eastern Pacific: Isla Gorgona (Colombia)	m
Lythrypnus spilus Böhlke and Robins 1960	Western Atlantic: Florida andBahamas to Greater Antilles	m
Lythrypnus zebra (Gilbert 1890)	Eastern Pacific: California to s. Baja California	m
Priolepis Valenciennes, 1837		
Priolepis dawsoni Greenfield, 1989	Western Atlantic: Trinidad to Brazil	m
Priolepis hipoliti (Metzelaar, 1922)	Western Atlantic: Florida to n. South America	m
Priolepis robinsi Garzón-Ferreira and Acero P., 1991	Western Atlantic: Colombia and Curaçao	m
Rhinogobiops Hubbs, 1926		
Rhinogobiops nicholsii (Bean, 1882)	Eastern Pacific: Canada to Baja California	m
Gobiidae: Gobionellinae		
Awaous Valenciennes, 1840		
Awaous banana (Valenciennes, 1837)	Western Atlantic: Florida to Venezuela; Eastern Pacific: s. Baja California to Peru	f,b,m
Awaous flavus (Valenciennes, 1837)	Western Atlantic: Colombia to n. Brazil	f,b,m
Awaous tajasica (Lichtenstein, 1822)	Western Atlantic: Florida to Brazil	f,b,m
Clevelandia Eigenmann and Eigenmann, 1888		
Clevelandia ios (Jordan and Gilbert, 1882)	Eastern Pacific: British Columbia to central. Baja California	f,b,m

Contd...

Contd...

Taxon	*Distribution*	*Habitat*
Ctenogobius Gill, 1858		
Ctenogobius boleosoma (Jordan and Gilbert, 1882)	Western Atlantic: North Carolina, Gulf of Mexico to Brazil	f,b,m
Ctenogobius claytonii (Meek, 1902)	Western Atlantic: North, Central, and South America	f,b,m
Ctenogobius fasciatus Gill, 1858	Western Atlantic: Florida to Venezuela	f,b,m
Ctenogobius manglicola (Jordan and Starks, 1895)	Eastern Pacific: central Gulf of California to Peru	b,m
Ctenogobius phenacus (Pezold and Lasala, 1987)	Western Atlantic: northern South America	b,m
Ctenogobius pseudofasciatus (Gilbert and Randall, 1971)	Western Atlantic: Florida, Central America to Suriname	f,b,m
Ctenogobius saepepallens (Gilbert and Randall, 1968)	Western Atlantic: Florida to South America	m
Ctenogobius sagittula (Günther, 1862)	Eastern Pacific: s. California (extremely rare) and n. Gulf of California to Peru	b,m
Ctenogobius shufeldti (Jordan and Eigenmann, 1887)	Western Atlantic: North Carolina to Gulf of Mexico	f,b,m
Ctenogobius smaragdus (Valenciennes, 1837)	Western Atlantic: South Carolina and Florida to Brazil	b,m
Ctenogobius stigmaticus (Poey, 1860)	Western Atlantic: South Carolina, Gulf of Mexico to Brazil	m
Ctenogobius stigmaturus (Goode and Bean, 1882)	Western Atlantic: South Carolina, Gulf of Mexico to Brazil	b,m
Ctenogobius thoropsis (Pezold and Gilbert, 1987)	Western Atlantic: n. South America	f,b,m
Eucyclogobius Gill, 1862		
Eucyclogobius newberryi (Girard, 1856)	Eastern Pacific: California	b,m
Evorthodus Gill 1859		
Evorthodus lyricus (Girard 1858)	Western Atlantic: Chesapeake Bay, Gulf of Mexico to northern South America	f,b,m
Evorthodus minutus Meek and Hildebrand 1928	Eastern Pacific: central Gulf of California to Peru	b,m
Gillichthys Cooper, 1864		
Gillichthys detrusus Gilbert and Scofield, 1898	Eastern Pacific: northernmost Gulf of California (Colorado River delta only)	f,b,m
Gillichthys mirabilis Cooper, 1864	Eastern Pacific: California, n. to central Gulf of California	f,b,m

Contd...

Contd...

Taxon	*Distribution*	*Habitat*
Gillichthys seta (Ginsburg, 1938)	Eastern Pacific: n. to central Gulf of California	m
Gnatholepis Bleeker, 1874		
Gnatholepis thompsoni Jordan, 1904	Western Atlantic and eastern Atlantic islands	m
Gobioides Lacepède, 1800		
Gobioides broussonnetii Lacepède, 1800	Western Atlantic: North, Central, South America, Caribbean	f,b,m
Gobioides grahamae Palmer and Wheeler, 1955	Western Atlantic: Guyana to n. Brazil	f,b,m
Gobioides peruanus (Steindachner, 1880)	Eastern Pacific: Mexico to Peru	f,b
Gobionellus Girard, 1858		
Gobionellus daguae (Eigenmann, 1918)	Eastern Pacific: Panama to Colombia	b,m
Gobionellus liolepis (Meek and Hildebrand, 1928)	Eastern Pacific: El Salvador to Panama	m
Gobionellus microdon (Gilbert, 1892)	Eastern Pacific: central Gulf of California to Ecuador	f,b,m
Gobionellus oceanicus (Pallas, 1770)	Western Atlantic: North Carolina, Gulf of Mexico, West Indies to Brazil	f,b,m
Gobionellus stomatus Starks, 1913	Western Atlantic: Brazil	b
Ilypnus Jordan and Evermann, 1896		
Ilypnus gilberti (Eigenmann and Eigenmann, 1889)	Eastern Pacific: California to s. Baja California, throughout Gulf of California	m
Ilypnus luculentus (Ginsburg, 1938)	Eastern Pacific: Gulf of California	m
Lethops Hubbs, 1926		
Lethops connectens Hubbs, 1926	Eastern Pacific: central California to n. Baja California	m
Lepidogobius Gill, 1859		
Lepidogobius lepidus (Girard, 1858)	Eastern Pacific: California to n. Baja California	m
Lophogobius Gill, 1862		
Lophogobius cristulatus Ginsburg ,1939	Eastern Pacific: Costa Rica and Panama	m
Lophogobius cyprinoides (Pallas, 1770)	Western Atlantic: Bermuda, Florida, Caribbean to northern South America, Panama Canal	f,b,m

Contd...

Contd...

Taxon	*Distribution*	*Habitat*
Oxyurichthys Bleeker, 1857		
Oxyurichthys stigmalophius (Mead and Böhlke, 1958)	Western Atlantic: Florida, Gulf of Mexico to Suriname, SW Brazil (Bahia, Espírito Santo)	m
Quietula Jordan and Evermann, 1895		
Quietula guaymasiae (Jenkins and Evermann, 1889)	Eastern Pacific: Gulf of California	b,m
Quietula y-cauda (Jenkins and Evermann, 1889)	Eastern Pacific: central California to Gulf of California	b,m
Typhlogobius Steindachner, 1879		
Typhlogobius californiensis Steindachner, 1879	Eastern Pacific: California to n. Baja California	m
Vomerogobius Gilbert, 1971		
Vomerogobius flavus Gilbert, 1971	Western Atlantic: Bahamas	m
Gobiidae: Sicydiinae		
Sicydium Valenciennes, 1837		
Sicydium adelum Bussing, 1996	Western Atlantic: Costa Rica	f
Sicydium altum Meek, 1907	Eastern Pacific: Costa Rica	f
Sicydium buscki Evermann and Clark, 1906	Western Atlantic: Dominican Republic	f
Sicydium cocoensis (Heller and Snodgrass, 1903)	Eastern Pacific: Isla del Coco	f,b,m
Sicydium fayae Brock, 1942	Eastern Pacific: Mexico	f,b,m
Sicydium gilberti Watson, 2000	Western Atlantic: Dominican Republic	f,b,m
Sicydium gymnogaster Ogilvie-Grant, 1884	Western Atlantic: Mexico, Honduras	f,b,m
Sicydium hildebrandi Eigenmann, 1918	Eastern Pacific: Colombia (Dagua River basin)	f
Sicydium montanum Hubbs, 1920	Western Atlantic: Venezuela	f
Sicydium multipunctatum (Regan, 1906)	Eastern Pacific: s. Mexico to Honduras	f
Sicydium plumieri (Bloch, 1786)	Western Atlantic: Antilles south of Cuba	f,b
Sicydium punctatum Perugia, 1896	Western Atlantic: South America, Caribbean	f,b,m
Sicydium rosenbergii (Boulanger, 1899)	Eastern Pacific: Ecuador	f
Sicydium salvini Ogilvie-Grant, 1884	Eastern Pacific: Nicaragua to Panama	f,b,m

Contd...

Contd...

Taxon	*Distribution*	*Habitat*
Microdesmidae		
Cerdale Jordan and Gilbert, 1882		
Cerdale fasciata Dawson, 1974	Western Atlantic: Brazil (Bahia)	b,m
Cerdale floridana Longley, 1934	Western Atlantic: Caribbean, Florida, Bahamas, Bermuda	b,m
Cerdale ionthas Jordan and Gilbert, 1882	Eastern Pacific: Costa Rica to Ecuador	b,m
Cerdale paludicola Dawson, 1974	Eastern Pacific: Costa Rica to Ecuador	b,m
Cerdale prolata Dawson, 1974	Eastern Pacific: Panama Bay	b,m
Clarkichthys Smith, 1958		
Clarkichthys bilineatus (Clark, 1936)	Eastern Pacific: s. Gulf of California to Colombia, Isla del Coco, Galápagos	m
Microdesmus Günther, 1864		
Microdesmus affinis Meek and Hildebrand, 1928	Eastern Pacific: n. Gulf of California to Ecuador	B,m
Microdesmus bahianus Dawson, 1973	Western Atlantic: Panama, Martinique, Brazil	m
Microdesmus carri Gilbert, 1966	Western Atlantic: SE Mexico, Costa Rica	b,m
Microdesmus dipus Günther, 1864	Eastern Pacific: central Gulf of California to Panama, Galápagos	b,m
Microdesmus dorsipunctatus Dawson, 1968	Eastern Pacific: s. Baja California and central Gulf of California to Panama	b,m
Microdesmus hildebrandi Reid, 1936	Eastern Pacific: Panama	m
Microdesmus intermedius Meek and Hildebrand, 1928	Eastern Pacific: Costa Rica to Panama	m
Microdesmus knappi Dawson, 1972	Eastern Pacific: Colombia	m
Microdesmus multiradiatus Meek and Hildebrand, 1928	Eastern Pacific: Panama Bay	b,m
Microdesmus lanceolatus Dawson, 1962	Western Atlantic: n. Gulf of Mexico	b,m
Microdesmus longipinnis (Weymouth, 1910)	Western Atlantic: North Carolina, n. Gulf of Mexico to Cayman Islands, Bermuda	m
Microdesmus luscus Dawson, 1977	Western Atlantic: Caribbean	m
Microdesmus retropinnis Jordan and Gilbert, 1882	Eastern Pacific: central Gulf of California (Mulegé) to Panama	b,m
Microdesmus suttkusi Gilbert, 1966	Eastern Pacific: central Gulf of California (Mulegé) to Colombia	b,m

Contd...

Contd...

Taxon	*Distribution*	*Habitat*
Ptereleotridae		
Ptereleotris Gill 1863		
Ptereleotris calliura (Jordan and Gilbert, 1882)	Western Atlantic: s. Florida and e. Gulf of Mexico	m
Ptereleotris carinata Bussing, 2001	Eastern Pacific: central Gulf of California to Colombia	m
Ptereleotris helenae (Randall, 1968)	Western Atlantic: s. Florida, Antilles, Brazil	m
Ptereleotris randalli Gasparini, Rocha and Floeter, 2001	Western Atlantic: Brazil to SE Venezuela	m
Eleotridae: Eleotrinae		
Dormitator Gill, 1861		
Dormitator cubanus Ginsburg, 1953 [Doubtful species]	Western Atlantic: Cuba	f
Dormitator latifrons (Richardson, 1844)	Eastern Pacific: s. California (rare), throughout Gulf of California to Peru, Galápagos	f,b,m
Dormitator levis (Eigenmann, 1918)	Eastern Pacific: Colombia	f
Dormitator maculatus (Bloch, 1792)	Western Atlantic: North Carolina to Brazil	f,b,m
Eleotris **Bloch and Schneider, 1801**		
Eleotris amblyopsis (Cope, 1871)	Western Atlantic: North Carolina to French Guyana, Cuba, Antilles (rare)	f,b,m
Eleotris perniger (Cope, 1871)	Western Atlantic: Western Atlantic: Caribbean Basin, Mexico, Bermuda	f,b,m
Eleotris picta Kner, 1863	Eastern Pacific: s. California and Gulf of California to Peru, Galápagos	f,b
Eleotris pisonis (Gmelin, 1789)	Western Atlantic: Brazil(including lower Amazon) to Orinoco delta	f,b,m
Eleotris tecta Bussing, 1996	Eastern Pacific: Costa Rica to Columbia	f,b
Eleotris tubularis Heller and Snodgrass, 1903	Eastern Pacific: Isla del Coco	f
Erotelis Poey, 1860		
Erotelis armiger (Jordan and Richardson, 1895)	Eastern Pacific: central Gulf of California to Colombia	f,b,m
Erotelis smaragdus (Valenciennes, 1837)	Western Atlantic: Bahamas, Florida, Gulf of Mexico to Brazil	f,b,m

Contd...

Contd...

Taxon	*Distribution*	*Habitat*
Gobiomorus Lacepède, 1800		
Gobiomorus dormitor Lacepède, 1800	Western Atlantic: Gulf of Mexico to Brazil	f,b,m
Gobiomorus maculatus (Günther, 1859)	Eastern Pacific: s. Gulf of California to Peru, Islas Revillagigedo, Isla del Coco, Galápagos	f,b,m
Gobiomorus polylepis Ginsburg, 1953	Eastern Pacific: Mexico to Costa Rica	f,b,m
Guavina Bleeker, 1874		
Guavina guavina (Valenciennes, 1837)	Western Atlantic: Florida (Port Canaveral), Mexico to Brazil	f,b,m
Guavina micropus Ginsburg, 1953	Eastern Pacific: Costa Rica and Panama	f,b
Hemieleotris Meek and Hildebrand, 1916		
Hemieleotris latifasciata (Meek and Hildebrand, 1912)	Eastern Pacific: Guatemala to Colombia	m
Hemieleotris levis Eigenmann, 1918	Eastern Pacific: Colombia	f
Leptophilypnus Meek and Hildebrand, 1916		
Leptophilypnus fluviatilis Meek and Hildebrand, 1916	Western Atlantic: Panama	f
Leptophilypnus guatemalensis Thacker and Pezold, 2006	Western Atlantic: Yucatan Peninsula	f
Leptophilypnus panamensis (Meek and Hildebrand, 1916)	Western Atlantic: Panama and Costa Rica	f
Microphilypnus Myers, 1927		
Microphilypnus ternetzi Myers, 1927	Western Atlantic: Venezuela (Orinoco River basin)	f
Microphilypnus amazonicus Myers, 1927	Western Atlantic: Brazil (lower Negro River basin)	f
Microphilypnus macrostoma Myers, 1927	Western Atlantic: Brazil (lower Negro River basin)	f

Pat Colin, Lloyd Findley, Jean-Christophe Joyeux, Ed Murdy, Frank Pezold, and Luke Tornabene. Special thanks to Lloyd Findley for contributing his unpublished distribution data for the eastern pacific species. Additional detailed information for each of the species, not provided here, is available at www.gobiidae.com.

The family Gobiidae is divided into five subfamilies: Amblyopinae (36 species worldwide); Gobiinae (1027); Gobionellinae (377); Oxudercinae (46); and Sicydiinae (109) (Eschmeyer and Fong, 2010). Members of the subfamilies Amblyopinae and Oxudercinae are absent from the Americas. The Gobiinae (Pezold, 1993) contain about 100 genera worldwide and are represented in the Americas by 31 genera, of which 27 are endemic. The Gobiinae tribe Gobiosomatini (= Birdsong's [1975] Gobiosomini — the proper spelling, Gobiosomatini, recognized by Smith and Baldwin [1999]) is endemic to the Americas and comprises 128 species — 73% of American Gobiinae and 55% of American Gobiidae.

The tribe, as originally proposed by Birdsong (1975), included what has been commonly called the American seven-spined gobies and several closely allied genera, all endemic to the Americas. The tribe was expanded by Hoese (1976) and again by Birdsong in 1988 and 1995. The genera currently included in the Gobiosomatini are: *Aboma*, *Akko, Aruma, Barbulifer, Bollmannia, Chriolepis, Elacatinus, Eleotrica, Enypnias, Evermannichthys, Ginsburgellus, Gobiosoma, Gobulus, Gymneleotris, Microgobius, Nes, Ophiogobius, Palatogobius, Pariah, Parrella, Psilotris, Pycnomma, Risor, Robinsichthys, Tigrigobius* and *Varicus*. The tribe was originally hypothesized to be monophyletic by Birdsong (1975); later work by Birdsong *et al.* (1988) inferred that a subset of the group, known as the "*Gobiosoma* Group" was monophyletic. The remaining genera (*Akko, Bollmannia, Microgobius, Palatogobius, Parrella*) were referred to as the "*Microgobius* Group." The monophyly of the seven-spined gobies, including the two major groups (*Gobiosoma* and *Microgobius* groups), is supported by the molecular phylogeny of Rüber *et al.* (2003).

The primary morphological characters used to unite the Gobiosomatini are a dorsal-fin pterygiophore formula of 3-221110 (in all genera except *Evermannichthys, Pariah*, and *Risor*), a vertebral count of 11+16-17 (in all but *Evermannichthys* and *Pariah*), and the fusion of hypurals 1-2 with 3-4 and the terminal vertebral element (in all genera except *Aboma*, *Bollmannia*, *Microgobius*, *Palatogobius*, and *Parrella*) (Birdsong, 1975; Van Tassell, 1998).

The Gobiosomatini display a great diversity in color, ecology, and behavior. They are found in fresh, brackish, and marine waters and in association with habitats of mud, shell, gravel, coral and/or rocky reefs. They occur in the intertidal and to depths of 500 meters (Rüber *et al.*,

2003; Colin, 2010). Differences in dentition indicate specialized diets, although little information is available for most species. The genera have diversified in behavior: for example, *Elacatinus* contains members that are involved in cleaning symbioses, removing ectoparasites from other fishes; *Nes* lives in association with snapping shrimp; *Evermannichthys* lives within sponges; and *Ginsburgellus* lives in association with urchins (Böhlke and Robins, 1968).

GOBIOSOMATINI

Gobiosoma Group

The "*Gobiosoma* Group" includes: *Aboma, Aruma, Barbulifer, Chriolepis, Elacatinus, Eleotrica, Enypnias, Evermannichthys, Ginsburgellus, Gobiosoma, Gobulus, Gymneleotris, Nes, Ophiogobius, Pariah, Psilotris, Pycnomma, Risor, Robinsichthys, Tigrigobius*, and *Varicus*. The primary difference between the "*Gobiosoma*" and "*Microgobius*" groups is that members of the former exhibit hypural fusion (1, 2 fused with 3, 4 and the terminal centrum) in all genera except *Aboma*. There is no hypural fusion in the "*Microgobius* Group."

Aboma Jordan and Starks, 1895

The genus comprises a single species, *A. etheostoma*. Some Indo-Pacific species have been placed in this genus, based on seven spines in the first dorsal, but they do not have the transverse papillae pattern diagnostic of the genus (Vasil'eva 1999, Larson *et al.*, 2008.) All have been removed and placed in other genera (see Eschmeyer (2010) for current status). *Aboma etheostoma* is a small goby, with large eyes, small mouth, ctenoid scales on body, 4 modified basicaudal scales on caudal peduncle, VII spines in first dorsal fin, the first 2 of which may be elongate in males, and mottled, light brown pigment on body, maximally reaching 3.4 cm SL. The genus is endemic to the tropical eastern Pacific and is found in Mexico from Sonora to Nayarit and south along the coast to Colombia. *Aboma etheostoma* inhabits shallow intertidal areas with sandy/mud substrate. (Thomson *et al.*, 2000) (Fig. 2.1.1A, Table 2.1.2).

Aruma Ginsburg, 1933

Aruma is a monotypic genus found only within the Gulf of California. *Aruma histro* is a small goby, maximally reaching a length of 6.3 cm SL

Table 2.1.2 Species of Gobiosomatini (*Gobiosoma* Group) including distributions and habitats.

Gobiidae: Gobiinae: Gobiosomatini: "Gobiosoma Group"	*Distribution*	*Habitat*
***Aboma* Jordan and Starks, 1895**		
Aboma etheostoma Jordan and Starks, 1895	Eastern Pacific: central Gulf of California (mainland side) to Colombia	b,m
***Aruma* Ginsburg, 1933**		
Aruma histrio (Jordan, 1884)	Eastern Pacific: Gulf of California	m
***Barbulifer* Eigenmann and Eigenmann, 1888**		
Barbulifer antennatus Böhlke and Robins, 1968	Western Atlantic: Bahamas to Antilles	m
Barbulifer ceuthoecus (Jordan and Gilbert, 1884)	Western Atlantic: Florida, Bahamas, south to Brazil (Santa Catarina); Eastern Pacific: Panama Canal area	m
Barbulifer enigmaticus Joyeux, Van Tassell and Macieira, 2009	Western Atlantic: Brazil: Pará (unconfirmed), Bahia, Espírito Santo, Rio de Janeiro, São Paulo	m
Barbulifer mexicanus Hoese and Larson, 1985	Eastern Pacific: Cape region of Baja peninsula, Islas Tres Marías to central Mexico (Guerrero)	m
Barbulifer pantherinus (Pellegrin, 1901)	Eastern Pacific: SW and NE Gulf of California	m
***Chriolepis* Gilbert, 1892**		
Chriolepis atrimelum Bussing, 1997	Eastern Pacific: Isla del Coco	m
Chriolepis benthonis Ginsburg, 1953	Western Atlantic: Arrowsmith Bank off Quintana Roo; deep water off Progreso, Yucatan	m
Chriolepis cuneata Bussing, 1990	Eastern Pacific: SW Gulf of California to Panama	m
Chriolepis dialepta Bussing, 1990	Eastern Pacific: Isla del Coco	m
Chriolepis fisheri Herre, 1942	Western Atlantic: Barbados, Belize, Honduras, Venezuela, Brazil: Rio Grande do Norte and Paraíba	m
Chriolepis lepidota Findley, 1975	Eastern Pacific: Isla Malpelo off Colombia	m
Chriolepis minutillus Gilbert, 1892	Eastern Pacific: Western Gulf of California	m
Chriolepis tagus Ginsburg, 1953	Eastern Pacific: Galápagos	m

Contd...

Contd...

Gobiidae: Gobiinae: Gobiosomatini: "Gobiosoma Group"	*Distribution*	*Habitat*
Chriolepis vespa Hastings and Bortone, 1981	Western Atlantic: NE Gulf of Mexico, possibly Guyana, Brazil: Rio Grande do Sul	m
Chriolepis zebra Ginsburg, 1938	Eastern Pacific: Central and southern Gulf of California (to Cape region of Baja peninsula, and Isla San Ignacio Farallón on mainland side)	m
***Elacatinus* Jordan, 1904**		
Elacatinus atronasus (Böhlke and Robins, 1968)	Western Atlantic: Exuma Sound, Bahamas	m
Elacatinus chancei (Beebe and Hollister, 1933)	Western Atlantic: southern Bahamas, Caicos Islands, Puerto Rico, Virgin Islands, Lesser Antilles, Aves Island (west of Guadeloupe), and some islands off coast of Venezuela.	m
Elacatinus colini Randall and Lobel, 2009	Western Atlantic: Belize and Islas de la Bahía, Honduras	m
Elacatinus evelynae (Böhlke and Robins, 1968)	Western Atlantic: Bahamas, Lesser Antilles, Venezuela and western Caribbean	m
Elacatinus figaro Sazima, Moura and Rosa, 1997	Western Atlantic: NE Brazil southward to Porto Belo, Santa Catarina, southern Brazil; probably north to Ceara	m
Elacatinus genie (Böhlke and Robins, 1968)	Western Atlantic: Bahamas and Grand Cayman Islands.	m
Elacatinus horsti (Metzelaar, 1922)	Western Atlantic: Curaço, Bonaire, Netherlands Antilles, Bahamas, Grand Cayman, Haiti, eastern Gulf of Mexico, Florida (SW Tampa Bay)	m
Elacatinus illecebrosus (Böhlke and Robins, 1968)	Western Atlantic: Panama and Colombia, also a few banks and islands in western Caribbean	m
Elacatinus jarocho Taylor and Akins, 2007	Western Atlantic: western Gulf of Mexico at Veracruz, Mexico	m
Elacatinus lobelia Randall and Colin, 2009	Western Atlantic: Belize and Islas de la Bahía, Honduras	m
Elacatinus lori Colin, 2002	Western Atlantic: Belize and Honduras	m
Elacatinus louisae (Böhlke and Robins, 1968)	Western Atlantic: Bahamas; Grand Cayman; Jamaica; Isla de Providencia; Belize (Glovers Reef) and (probably) Haiti.	m
Elacatinus oceanops Jordan , 1904	Western Atlantic: Florida	m
Elacatinus phthirophagus Sazima, Carvalho-Filho and Sazima, 2008	Western Atlantic: NE Brazil at Fernando de Noronha archipelago	m

Contd...

Contd...

Gobiidae: Gobiinae: Gobiosomatini: "Gobiosoma Group"	*Distribution*	*Habitat*
Elacatinus pridisi Guimarães, Gasparini and Rocha, 2004	Western Atlantic: SE Brazil (Trindade Island)	m
Elacatinus prochilos (Böhlke and Robins, 1968)	Western Atlantic: Lesser Antilles, Barbados, St. Croix, Jamaica, Belize, Yucatán, and (possibly) the N. Gulf of Mexico	m
Elacatinus puncticulatus (Ginsburg, 1938)	Eastern Pacific: Central Gulf of California to Ecuador	m
Elacatinus randalli (Böhlke and Robins, 1968)	Western Atlantic: Venezuela (Curacao, Bonaire, Gran Roque, Cubagua), the coast of Venezuela near Cumana, St. Vincent, Dominica, and Puerto Rico.	m
Elacatinus serranilla Randall and Lobel, 2009	Western Atlantic: Serranilla Bank (15°50'N, 79°50'W) and Jamaica	m
Elacatinus tenox (Böhlke and Robins, 1968)	Western Atlantic: Dominica, Nevis, Aves Is., and Panama (San Blas Islands)	m
Elacatinus xanthiprora (Böhlke and Robins, 1968)	Western Atlantic: Dry Tortugas northward to Miami and Gulf coast of Florida	m
***Eleotrica* Ginsburg, 1933**		
Eleotrica cableae Ginsburg, 1933	Eastern Pacific: Galápagos Islands	m
***Enypnias* Jordan and Evermann, 1898**		
Enypnias aceras Ginsburg, 1939	Eastern Pacific: El Salvador to Panama	m
Enypnias seminudus (Günther, 1861)	Eastern Pacific: SE Gulf of California to Colombia	m
***Evermannichthys* Metzelaar, 1919**		
Evermannichthys bicolor Thacker, 2001	Western Atlantic: Navassa Island, Haiti	m
Evermannichthys convictor Böhlke and Robins, 1969	Western Atlantic: Green Cay, Bahamas	m
Evermannichthys metzelaari Hubbs, 1923	Western Atlantic: Bahamas, Antilles southward to Curaçao and Venezuela	m
Evermannichthys silus Böhlke and Robins, 1969	Western Atlantic: Samana Cay, Bahamas and Curaçao	m
Evermannichthys spongicola (Radcliffe, 1917)	Western Atlantic: North Carolina to Florida, Dry Tortugas to Panama	m
***Ginsburgellus* Böhlke and Robins, 1968**		
Ginsburgellus novemlineatus (Fowler, 1950)	Western Atlantic: Bahamas, Cayman Islands, Antilles to Venezuela, Colombia	b,m

Contd...

Contd...

Gobiidae: Gobiinae: Gobiosomatini: "Gobiosoma Group"	*Distribution*	*Habitat*
***Gobiosoma* Girard, 1858**		
Gobiosoma bosc (Lacepède, 1800)	Western Atlantic: Massachusetts to Florida, (but not Florida Keys), Everglade City to Tampico, Mexico	b,m
Gobiosoma chiquita (Jenkins and Evermann, 1889)	Eastern Pacific: Northern, central and SW Gulf of California	b,m
Gobiosoma ginsburgi Hildebrand and Schroeder, 1928	Western Atlantic: Massachusetts to Jacksonville, Florida	b,m
Gobiosoma grosvenori (Robins, 1964)	Western Atlantic: SE Florida, Jamaica, Venezuela	m
Gobiosoma hemigymnum (Eigenmann and Eigenmann, 1888)	Western Atlantic: Brazil	m
Gobiosoma hildebrandi (Ginsburg, 1939)	Western Atlantic: Panama Canal and Panama Atlantic coast	b,m
Gobiosoma homochroma (Ginsburg, 1939)	Eastern Pacific: Coast of Panama and Pacific drainage of Canal Zone	b,m
Gobiosoma longipala Ginsburg, 1933	Western Atlantic: Gulf of Mexico from s. Florida to Texas	b,m
Gobiosoma nudum (Meek and Hildebrand, 1928)	Eastern Pacific: Southern Gulf of California to Colombia	b,m
Gobiosoma paradoxum (Günther, 1861)	Eastern Pacific: Central Gulf of California (mainland side) to Ecuador	b,m
Gobiosoma parri Ginsburg, 1933	Western Atlantic: Uruguay and Argentina	b,m
Gobiosoma robustum Ginsburg, 1933	Western Atlantic: Gulf of Mexico from Texas to Key West, Florida Atlantic coast from Miami to Cape Canaveral	b,m
Gobiosoma schultzi (Ginsburg, 1944)	Western Atlantic: Venezuela (Lake Maracaibo and surroundings)	b,m
Gobiosoma spes (Ginsburg, 1939)	Western Atlantic: Puerto Rico, Jamaica, Costa Rica, Panama and Venezuela	b,m
Gobiosoma spilotum (Ginsburg, 1939)	Western Atlantic: Panama (Colon to San Blas)	b,m
Gobiosoma yucatanum Dawson, 1971	Western Atlantic: Yucatán peninsula	b,m
***Gobulus* Ginsburg, 1933**		
Gobulus birdsongi Hoese and Reader, 2001	Eastern Pacific: Panama Canal, Miraflores Locks area	b,m
Gobulus crescentalis (Gilbert, 1892)	Eastern Pacific: Central outer coast of Baja California peninsula and throughout Gulf of California to Panama; Isla del Coco	m
Gobulus hancocki Ginsburg, 1938	Eastern Pacific: Central Gulf of California to Colombia	m

Contd...

Contd...

Gobiidae: Gobiinae: Gobiosomatini: "Gobiosoma Group"	*Distribution*	*Habitat*
Gobulus myersi Ginsburg, 1939	Western Atlantic: southern Florida, Bahamas to Venezuela, NE Gulf of Mexico, Antilles, Brazil (incl. Trindade)	m
***Gymneleotris* Bleeker, 1874**		
Gymneleotris seminuda (Günther, 1864)	Eastern Pacific: outer coast of southern Baja California peninsula (Bahía Magdalena) and central Gulf of California to Ecuador	m
***Nes* Ginsburg, 1933**		
Nes longus (Nichols, 1914)	Western Atlantic: Bermuda, Florida, Bahamas, s. Gulf of Mexico, Antilles, Yucatán to Venezuela and Colombia	m
***Ophiogobius* Gill, 1863**		
Ophiogobius jenynsi Hoese, 1976	Eastern Pacific: Chile	b,m
***Pariah* Böhlke, 1969**		
Pariah scotius Böhlke, 1969	Western Atlantic: Bahamas	m
***Psilotris* Ginsburg, 1953**		
Psilotris alepis Ginsburg, 1953	Western Atlantic: Bahamas, Cuba, Virgin Islands, Honduras	m
Psilotris amblyrhynchus Smith and Baldwin, 1999	Western Atlantic: Belize	m
Psilotris batrachodes Böhlke, 1963	Western Atlantic: Bahamas, Puerto Rico, Jamaica, Belize, Honduras, Isla de Providencia, Santa Marta, Punta de San Bernardo off Colombia	m
Psilotris boehlkei Greenfield, 1993	Western Atlantic: St. Barthelemy, Saba	m
Psilotris celsus Böhlke, 1963	Western Atlantic: Bermuda, Bahamas, Puerto Rico, Virgin Islands, Isla de Providencia, Santa Marta, Colombia, Brazil (Bahia)	m
Psilotris kaufmani Greenfield, Findley and Johnson, 1993	Western Atlantic: Jamaica, Belize, Honduras, Isla de Providencia	m
***Pycnomma* Rutter, 1904**		
Pycnomma roosevelti Ginsburg, 1939	Western Atlantic: Isla de Providencia, Belize, Puerto Rico, Honduras, Saba Bank	m
Pycnomma semisquamatum Rutter, 1904	Eastern Pacific: Central and southern Gulf of California	m
***Risor* Ginsburg, 1933**		
Risor ruber (Rosén, 1911)	Western Atlantic: southern Florida, Bahamas to NE Brazil	m

Contd...

Contd...

Gobiidae: Gobiinae: Gobiosomatini: "Gobiosoma Group"	*Distribution*	*Habitat*
***Robinsichthys* Birdsong, 1988**		
Robinsichthys arrowsmithensis Birdsong, 1988	Western Atlantic: Arrowsmith Bank (off NE Quintana Roo, Mexico)	m
***Tigrigobius* Fowler, 1931**		
Tigrigobius digueti	Eastern Pacific: Central Gulf of California to central Mexico	
Tigrigobius dilepis (Robins and Böhlke, 1964)	Western Atlantic: British West Indies, Bahamas, Cayman Islands. Belize, Jamaica, Lesser Antilles	m
Tigrigobius gemmatus (Ginsburg, 1939)	Western Atlantic: Lesser Antilles, Los Roques Island, Old Providence Island northward to British Honduras, Cayman Islands, Bahamas, Puerto Rico, Isla de Providencia, and Venezuela	m
Tigrigobius inornatus (Bussing, 1990)	Eastern Pacific: Costa Rica to Colombia	m
Tigrogobius janssi (Bussing, 1981)	Eastern Pacific: Mexico to Panama	m
Tigrigobius limbaughi (Hoese and Reader, 2001)	Eastern Pacific: Central Gulf of California to central Mexico	m
Tigrigobius macrodon Beebe and Tee-Van, 1928	Western Atlantic: southern Florida and Cuba to Haiti and Venezuela	m
Tigrigobius multifasciatus (Steindachner, 1876)	Western Atlantic: Bahamas, Cuba, Cayman Islands southward to Barbados and Curaçao and Colombia	m
Tigrigobius nesiotes(Bussing, 1990)	Eastern Pacific: Isla del Coco, the Galapagos and Gorgona Island (Colombia)	m
Tigrigobius pallens (Ginsburg, 1939)	Western Atlantic: Bahamas and throughout Caribbean Sea to northern South America	m
Tigrigobius redimiculus (Taylor and Akins, 2007)	Western Atlantic: Western Gulf of Mexico, off Veracruz (Isla de Lobos southward to Isla Verde)	m
Tigrigobius saucrus (Robins, 1960)	Western Atlantic: Florida Keys, Bahamas, Jamaica, Virgin Islands, Belize, Venezuela	m
Tigrigobius zebrella (Robins, 1958)	Western Atlantic: Trinidad, Venezuela	m
***Varicus* Robins and Böhlke, 1961**		
Varicus bucca Robins and Böhlke, 1961	Western Atlantic: Greater Antilles	m
Varicus imswe Greenfield, 1981	Western Atlantic: Eleuthera Island, Bahamas; Carrie Bow Cay, Belize	m
Varicus marilynae Gilmore, 1979	Western Atlantic: s. and w. coasts of Florida, Gulf of Mexico off Cape St. George	m

which has a depressed head, expanded cheeks, and no scales. Adults are benthic, hiding under rocks and in small crevices; they inhabit tidepools and rocky areas to depths of 14 m. Larvae are found in dense schools in association with sargassum weed (Thomson *et al.*, 2000) (Table 2.1.2).

Barbulifer Eigenmann and Eigenmann, 1888

A neotropical genus with five species, *Barbulifer* comprises two eastern Pacific species distributed from the Gulf of California south to central Mexico, and three western Atlantic species from southern Florida to southern Brazil. One species has passed through the Panama Canal and is now found in Pacific Panama. All species inhabit shallow water marine environments, generally in depths less than 1 m. *Barbulifer pantherinus,* a cryptic species found hiding under rocks or near the base of sponges or algae, reaches depths of 32 m. Species range in size from 2 to 5 cm SL. All species are characterized by the presence of barbels on the head and the complete absence of scales. Valid species include *B. antennatus*, *B. ceuthoecus*, *B. enigmaticus*, *B. mexicanus*, and *B. pantherinus* (Joyeux *et al.*, 2009) (Fig. 2.1.1B, Table 2.1.2).

Chriolepis Gilbert, 1892

This neotropical genus comprises 10 currently recognized species, seven in the Pacific and three in the western Atlantic. They range in the eastern Pacific from the Gulf of California to Colombia and in the western Atlantic from the Gulf of Mexico to northern South America. Several species from the Gulf of California are undescribed. Members of the genus are small, with maximum sizes ranging from 2.2 cm to 4.5 cm SL. They have VII spines in the first dorsal. The pelvic fins are completely separate, the inner rays connected at most by a very low, inconspicuous, fleshy ridge between the bases. There is no frenum between the spines. The first four pelvic rays are branched in adults, with the fourth ray generally the longest. The fifth ray is branched or unbranched and shorter than the fourth. Head pores are lacking, two-thirds (or less) of the body has scales, and modified basicaudal scales are present at the base of the caudal fin (Findley, 1983). Valid species include *C. atrimelum*, *C. benthonis*, *C. cuneata*, *C. dialepta*, *C. fischeri*, *C. lepidota*, *C. minutillus*, *C. tagus*, *C. vespa*, *C. zebra* (Fig. 2.1.1C, Table 2.1.2).

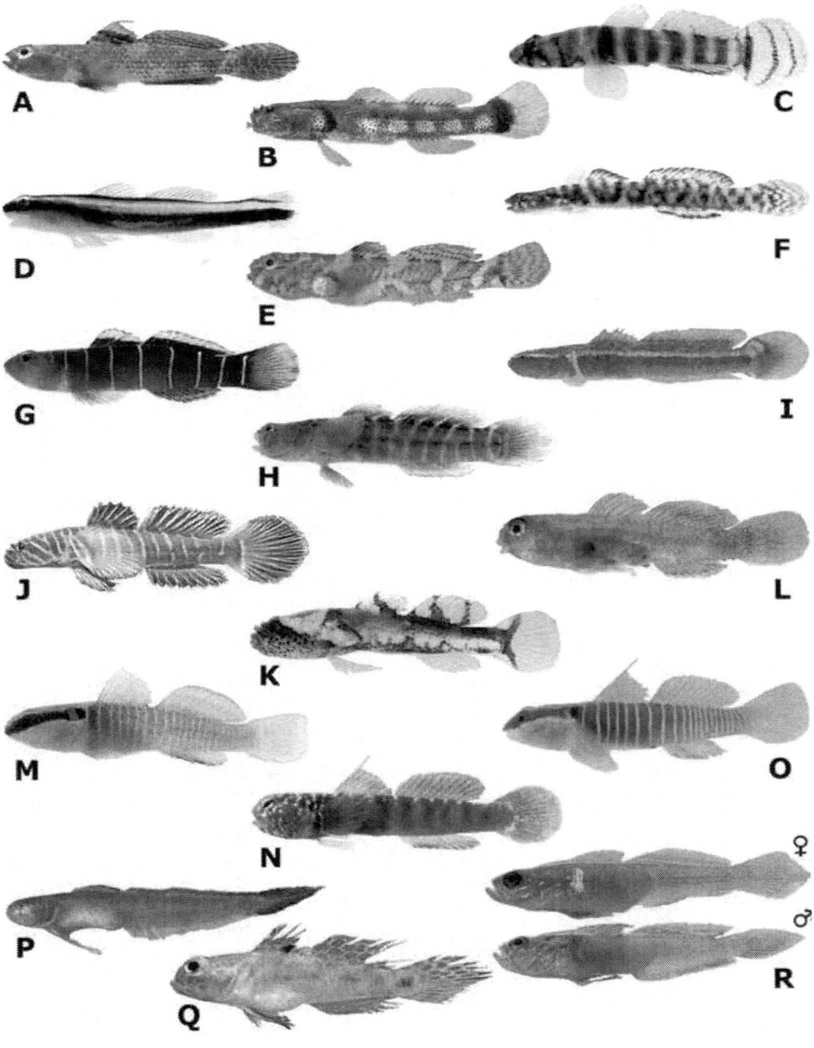

Fig. 2.1.1 Representative members of the Tribe Gobiosomatini. **(A)** *Aboma etheostoma* **(B)** *Barbulifer ceuthoecus* **(C)** *Chriolepis zebra* **(D)** *Elacatinus evelynae* **(E)** *Enypnias seminudus* **(F)** *Evermannichthys spongicola* **(G)** *Ginsburgellus novemlineatus* **(H)** *Gobiosoma bosc* **(I)** *Gobulus birdsong* **(J)** *Gymneleotris seminudus* **(K)** *Psilotris batracoides* **(M)** *Tigrigobius multifasciatus female* **(N)** *Tigrigobius gemmatum* **(O)** *Tigrigobius multifasciatus male* **(P)** *Akko brevis* **(Q)** *Bollmannia litura* **(R)** *Microgobius brevispinis*. Photographs by J. Van Tassell; plate composed by R. Patzner.

Color image of this figure appears in the color plate section at the end of the book.

Elacatinus Jordan, 1904

The inclusion of *Tigrigobius* in *Elacatinus,* as suggested by Hoese and Reader (2001a), is currently debated. In this chapter *Elacatinus* and *Tigrigobius* are treated as separate genera.

Tigrigobius was first proposed as a subgenus of *Gobiosoma* by Fowler (1931). When Ginsburg (1933) revised the species of *Gobiosoma* and *Garmannia,* he placed *Tigrigobius* as a subgenus of *Garmannia,* and *Elacatinus* as a subgenus of *Gobiosoma.* When Böhlke and Robins (1968) revised the Atlantic species of *Gobiosoma* they placed both *Tigrigobius* and *Elacatinus* as subgenera of *Gobiosoma.* They remained there until Hoese and Reader (2001a) made *Tigrigobius* a subgenus of *Elacatinus,* noting that *Elacatinus* differs from *Gobiosoma* in having 28 vertebrae (versus 27) and in several osteological characters of the skull. Recent molecular work by Rüber *et al.* (2003) indicates that *Tigrigobius* is not monophyletic. Two clades are recovered for species of *Tigrigobius:* one clade contains *T. macrodon* (type of *Tigrigobius*), *T. saucrus,* and *T. dilepis*; the other clade *T. gemmatum, T. pallens, T. multifasciatus, T. limbaughi* and the genera *Risor, Ginsburgellus,* and *Evermannichthys.* Futher studies are required to resolve the status of *Tigrigobius.* The *Elacatinus* clade of Rüber *et al.* (2003) includes two subclades, one containing fishes that dwell on sponges, the other that inhabit corals. The coral species all exhibit cleaning behavior.

Elacatinus comprises 21 species, all but one found in the western Atlantic. The single eastern Pacific species, *E. puncticulatus,* is basal to the rest of the *Elacatinus* clade (Rüber *et al.,* 2003) and exhibits a combination of characters found in both *Elacatinus* and *Tigrigobius.* The species all possess a bright stripe along the upper side of the body, from the eyes caudally, and are commonly known as the "neon" gobies (Colin, 1975). They lack scales, except for basicaudal scales on some species. Most species also possess a connection between the snout and upper lip, known as the rostral frenum.

The species and their differing color morphs have very restricted, non-overlapping, geographic distributions (Colin, 2010; Randall and Lobel, 2009). See Colin (2010) for a detailed list of color morphs and distributions. Most of the species are Caribbean, with others in the Gulf of Mexico and the offshore islands of Brazil. The following species are currently recognized: *E. atronasus, E. chancei, E. colini, E. evelynae, E. figaro, E. genie, E. horsti, E. illecebrosus, E. jarocho, E. lobelia, E. lori, E. louisae, E. oceanops, E. phthirophagus, E. pridisi, E. prochilos, E. puncticulatus, E. randalli, E. serranilla, E. tenox,* and *E. xanthiprora* (Fig 1D, Table 2.1.2).

Eleotrica Ginsburg, 1933

Eleotrica contains a single species, *E. cableae*, endemic to the Galapágos Islands. It inhabits sand-rubble bottoms near rocky reefs and is common in tidepools (Grove and Lavenberg, 1997). The species has a maximum length of 7.0 cm SL. The body lacks scales, except for several enlarged basicaudal scales at the base of the caudal fin, and the pelvic fins are almost completely separate (Humann and Deloach, 1993) (Table 2.1.2).

Enypnias Jordan and Evermann, 1898

Two species occur in this genus, both endemic to the tropical eastern Pacific. They occur from the Gulf of California to Colombia in coastal shallow environments to depths of 14 m. Generally found in habitats with mud or sand mixed with shell and/or rock, they may also be found in tidepools (Thomson *et al.*, 2000). *Enypnias* gobies are small (maximum sizes of 5 cm *E. aceras*, 7 cm *E. seminudus*) and have a mottled brown and white barred color pattern, small barbels near the anterior nostrils, a pair of barbels under the chin, and no scales on the head and body except for two enlarged basicaudal scales at base of caudal fin. Valid species include *E. aceras* and *E. seminudus* (Fig. 2.1.1E, Table 2.1.2).

Evermannichthys Metzelaar, 1919

Evermannichthys is a Caribbean genus with four species, all are obligate sponge dwellers. All are thin and elongate in form, with reduced scale cover and head pores. They range in length from 2 - 3 cm SL and have a body depth of about 2 mm. They live deep within the canals of the sponges, generally near the openings to the incurrent canals (personal observation). The sponges provide them with protection from predators and an available food source in the many invertebrates and their eggs found within the sponge or small plankton brought in through the incurrent pores (Tyler and Böhlke, 1972). Found from Florida and the Bahamas south to Curacao but rarely documented without specifically targeting sponges and using rotenone. Current valid species include *E. bicolor*, *E. convictor*, *E. metzelaari*, *E. silus*, and *E. spongicola* (Fig. 2.1.1F, Table 2.1.2).

Ginsburgellus Böhlke and Robins, 1968

This Caribbean genus comprises a single species, *G. novemlineatus*, found in association with sea urchins in shallow water with wave action. They are easily distinguished from other gobies by their dark brown body with nine, bright blue, thin, vertical lines, and the presence of a rostral frenum, maximally reaching 2.4 cm SL (Robins and Ray, 1986). They occur in the Bahamas, Cayman Islands, and Antilles south to Venezuela and Colombia (Smith, 1997; Acero and Garzón, 1987) (Fig. 2.1.1G, Table 2.1.2).

Gobiosoma Girard, 1858

Gobiosoma is a tropical to warm temperate genus with 16 described species. Species are small (< 7.5 cm) and found in fresh, brackish, and marine environments in association with rock or shell habitats. All inhabit shallow water with depth ranges of intertidal to 10 meters. They are light to dark brown with a combination of mottling and/or bars on the body. First dorsal with VII spines, pelvic fins united to form a complete disc, body with few or no scales, base of caudal fin with or without modified basicaudal scales, and cheeks puffy (Böhlke and Robins, 1968).

The 4 described eastern Pacific species range from the northern Gulf of California to Ecuador. One species is known only from the Panama Canal (*G. homochroma*), and a fifth species is a migrant through the Canal (*G. hildebrandi*) (Robertson and Allen, 2008). Three species await description. In the western Atlantic there are 12 species ranging from Massachusetts to Argentina. One Pacific species previously reported as a migrant through the Panama Canal (*G. nudum*) has not appeared in recent collections. *Gobiosoma nudum* reported in the literature as occurring in Brazil was recently described as new species of *Barbulifer* (Joyeux *et al.*, 2009) (Fig. 2.1.1H, Table 2.1.2).

Gobulus Ginsburg, 1933

This neotropical genus comprises four species, three in the tropical eastern Pacific and one in the western Atlantic. The tropical eastern Pacific species have restricted ranges: *Gobulus crescentalis* found only on the Gulf of California and the outer coast of Baja California; *Gobulus birdsongi*, 2001 is only known from a mangrove stream in the Panama Canal near the Miraflores Lock and in the lock area; *Gobulus hancocki* is widespread from the Gulf of California to Panama and Isla del Coco (Hoese and Reader, 2001). The western Atlantic species is distributed from Florida and the Bahamas south to Rio de Janeiro (Floeter *et al.*, 2003). All species are cryptic, benthic, and found on sandy bottoms near rocks or corals. In all species the head is depressed, and the body is counter shaded in light to dark browns and yellows. The first dorsal fin has VII spines; the pelvic fins are united but with only a small membrane between the spines forming a small flattened disc; and scales, head pores, and barbels are absent.

Gobulus attains maximum sizes range from 2.5 cm to 5.5 cm SL, and is generally a shallow coastal species occurring to maximum depths of 13 or 20 m. Atlantic species have been recorded at depths reaching 50 m (Hoese and Reader, 2001b) (Fig. 2.1.1I, Table 2.1.2).

Gymneleotris Bleeker, 1874

Gymneleotris is a tropical eastern Pacific genus with a single species, *G. seminuda*. It ranges from the Gulf of California south to Ecuador (Robertson and Allen, 2008). The species inhabits algae covered rocks and rubble reefs in waters up to 25m. The color pattern on the body is striking: thin white bars on a dark brown background. The head is broad, the first dorsal has VII spines, the pelvic fins are completely separate, and the body has cycloid scales extending from the base of the caudal fin forward to the third dorsal-fin spine (Böhlke and Robins, 1968) (Fig. 2.1.1J, Table 2.1.2).

Nes Ginsburg, 1933

Nes comprises a single western Atlantic species, *N. longus*, known from Bermuda, Florida, the Bahamas, southern Gulf of Mexico (arrecife Alacran), Antilles, and Yucatan to Venezuela (Smith, 1997). Garzón and Acero (1988) also reported it from Colombia. *Nes longus* is found on open silt-sand bottoms in channels between reef areas (Randall, 1996). *Nes* inhabits burrows with the alpheid shrimp *Alpheus floridanus*, with which it has an obligatory association (Randall *et al.*, 2005; Kramer *et al.*, 2009). Size ranges up to 10 cm TL (Randall, 1996). The body is slightly elongate and lacks scales, the ventral fins are completely united, and there are four pairs of orange blotches along the midside of the body (Smith, 1997) (Table 2.1.2).

Ophiogobius Gill, 1863

Ophiogobius contains a single species, *O. jenynsi*. The species was hypothesized by Hoese (1976) to be closely related to *Aruma histrio* based on similarities in papillae pattern and cranial osteology. Birdsong *et al.* (1988) placed it in their "*Gobiosoma* Group," along with *Pariah*, based on similarities in dorsal pterygiophore patterns. Molecular studies support its placement within the "*Gobiosoma* Group" but within the *Tigrigobius* clades, not closely aligned with *Aruma* (Rüber per. comm.). *Ophiogobius* has more dorsal-fin spines (VIII) and more vertebrae (13 + 19 = 32) than other members of the "*Gobiosoma* Group" (typically VII and 11 + 16 = 27).

Risor, *Pariah*, and *Evermannichthys* also show variations in patterns of dorsal-fin pterygiophores, numbers of dorsal-fin spines, and/or numbers of vertebrae that represent departures from the typical "*Gobiosoma* Group" conditions (Birdsong *et al.*, 1988).

Ophigobius jenynsi is distributed in Chile from Valparaiso to Punta Arenas (Navarro and Pequeño, 1979). It inhabits the rocky intertidal zone (Pequeño, 1989) where it feeds on small crustaceans (Muñoz and Ojeda, 1998) (Table 2.1.2).

Pariah Böhlke, 1969

Pariah is a Caribbean genus with a single species, *P. scotius,* which is known only from the Bahamas. A sponge-dwelling goby which inhabits loggerhead sponges, 1 - 2 ft in diameter, at depths of 12 - 28 m. *Pariah* generally co-occurs with *Evermannichthys metzelaari. Pariah scotius* is a small goby attaining a maximum size of 2.5 cm SL. It is without head pores, possess ctenoid scales only along the ventral edge and sides of the caudal peduncle, has two modified basicaudal scales at the base of caudal fin, a first dorsal with VIII spines and pelvic fins united forming a disc (Böhlke, 1969). The species possesses the dark internal bars also found in *Riscr* and several members of *Tigrigobius* (Table 2.1.2).

Psilotris Ginsburg, 1953

This Caribbean genus comprises six species that inhabit sandy bottoms in coral-reef habitats. Most species occur at depth ranges from 8 - 50 m, except for *P. alepis* which has only been collected in tidepools and sandy beaches in depths less than 1 m. Four of the species, *P. alepis, P. batrachodes, P. celsus,* and *P. kaufmani*, are broadly distributed geographically, but two species, *P. amblyrhynchus* and *P. boehlkei,* are known only from Belize and St. Barthelemy/Saba, respectively (Williams *et al.*, 2010). Additional collecting may reveal broader distributions for those species. All *Psilotris* species are small, with maximum sizes ranging from 1.8 - 4.0 cm SL. They lack scales, possess VII spines in the first dorsal, have no head or preopercle pores, and the ventral fins are completely separate; one species possesses a small frenum (Smith and Baldwin, 1999) (Fig. 2.1.1K, Table 2.1.2).

Pycnomma Rutter, 1904

Pycnomma is a neotropical genus with two species; *P. semisquamatum* is a tropical eastern Pacific species found within the Gulf of California and *P. roosevelti*, a Caribbean species known from Isla de Providencia, Belize, Puerto Rico, Honduras, and Saba bank. Both species live in shallow water

(< 20m) and inhabit sandy areas with rock or algae cover. They range in size from 2.5 - 6.1 cm SL, have VII spines in the first dorsal fin, lack scales on the head and anterior section of body, have modified basicaudal scales on the caudal peduncle, and have the pelvic fins completely separate. Body coloration is light to dark brown with darker bars on the sides (Robertson and Allen, 2008; Smith, 1997) (Table 2.1.2).

Risor Ginsburg, 1933

Risor is a Caribbean sponge dwelling genus with a single species. *R. ruber* is found in association with large barrel sponges, generally along the outside hiding in crevices, but also sometimes within the sponge tubules (Tyler and Böhlke, 1972). *Risor* has a maximum size of 2.5 cm SL, VII spines in the first dorsal fin, completely united pelvic fins, scales only on the posterior section of the trunk, and four modified basicaudal scales at the base of the caudal fin. The body color is light to dark brown, more or less solid, with faint internal bars visible on lighter specimens. *Risor* has a unique dentition: 2-4 enlarged re-curved canines in each jaw that project out of the mouth and point out and up in the upper jaw, down and back in the lower jaw. The teeth can be seen without the aid of a microscope, hence the common name "tusked" goby. The distribution includes the Bahamas, Florida, NE Gulf of Mexico, south through the Antilles to northeastern Brazil (Rocha *et al.*, 2000) (Fig. 2.1.1L, Table 2.1.2).

Robinsichthys Birdsong, 1988

A single species, *R. arrowsmithensis*, is known only from the type locality of Arrowsmith Bank in the western Caribbean. The only known specimens were trawled from deep water (92 – 596m), along the edge of the bank, which consists of hard substrate. The maximum size of this goby is 2.3 cm SL. The first dorsal fin possesses 6 – 7 spines, the body is completely scaled, the caudal-fin base has 2 basicaudal scales, the pelvic fins are completely separate, and head pores are lacking. Birdsong (1988) hypothesized that *Robinsichthys* forms a "bridge" between the "*Gobiosoma* Group" and the "*Microgobius* Group" of the Gobiosomatini; it shares with the *Microgobius* Group the absence of fused hypural 1 – 2 with 3 – 4 (present in the "*Gobiosoma* Group"), and it shares with the *Gobiosoma* Group the presence of basicaudal scales (not present in the "*Microgobius* Group"). He aligned *Robinsichthys* with *Chriolepis, Varicus*, and *Pycnomma* based on the absence of head pores and having separate pelvic fins, characters not found in the other members of the "*Gobiosoma* Group" (Table 2.1.2).

Tigrigobius Fowler, 1931

Currently there are 5 species in the eastern Pacific and 7 in the western Atlantic. The Pacific species reside on rock reefs with 4 (*T. digueti, T. inornatus, T. limbaughi, T. nesiotes*) of the 5 engaging in cleaning behavior (Thomson *et al.*, 2000). The western Atlantic species occupy a more diverse set of habitats. *Tigrigobius gemmatus* is reported from chiton burrows in red coralline algae at shallow depths (generally less than 1m); *T. pallens* is found at the base of small heads (10 – 30 cm) of *Montastrea* and *Siderastrea* corals (Taylor and Van Tassell 2002); *T. dilepis, T. macrodon, T. saucrus,* and *T. zebrella* are generally found in association with coral heads; and *T. multifasciatus* occurs in association with urchins on rocky coasts (pers. obs.). All species are small, reaching maximum sizes of 3 – 4.6 cm SL. They are brightly colored with reds, browns, blues, and yellows arranged in spots, bars, or stripes in varying combinations (Fig. 2.1.1M, N, O, Table 2.1.2).

Varicus Robins and Böhlke, 1961

This genus is restricted to the tropical western Atlantic and comprises three species: *Varicus bucca, V. imswe, V. marilynae. Varicus bucca* attains the largest size, 4.6 cm SL, with the remaining species maximally reaching 1.4 cm and 1.8 cm SL, respectively (Greenfield 1981, Williams and Gilbert, 1983). Due to their small size and deep depth preference (15 - 91 m), they are rare in collections and hence little is known about their distribution or ecology. As presently defined, *Varicus* is most closely related to *Chriolepis,* but differs in having the pelvic rays unbranched (except at the tips in some specimens); first four pelvic rays with expanded tips; fifth pelvic ray rudimentary; tongue bilobed; and scales under the first dorsal fin ctenoid. The exact limits of the two genera are under debate and with further investigation they may be combined (with *Chriolepis* having nomenclatural priority) (Table 2.1.2).

GOBIOSOMATINI

Microgobius Group

The "*Microgobius*" group includes *Akko, Bollmannia, Microgobius, Palatogobius,* and *Parrella* (Birdsong, 1975; Birdsong *et al.*, 1988; Birdsong, 1995; Van Tassell and Baldwin, 2004). The placement of *Akko* in the "*Microgobius*" group and the relationships of the genera within the group are supported by mtDNA studies (Rüber *et al.*, 2003).

Akko Birdsong and Robins, 1995

This neotropical genus consists of three species, *Akko dionaea* in the western Atlantic; *A. brevis* and *A. rossi* in the eastern tropical Pacific. All three species inhabit black mud, apparently living in burrows. They possess very small eyes and a light pink body color with darker red areas composed of blood capillaries close to the surface. The genus is characterized by an elongate body, first dorsal with VII spines, upper and lower jaws with enlarged, widely spaced teeth that overlap the lips of both jaws, and head pores greatly reduced in size and occasionally absent (Van Tassell and Baldwin, 2004) (Fig. 2.1.1P, Table 2.1.3).

Bollmannia Jordan, 1890

Twelve species of *Bollmannia* are recognized in this chapter as valid, four in the western Atlantic and nine in the eastern Pacific (two of the Pacific species are synonyms of *B. ocellata* according to an unpublished thesis). A review of the genus is required to delineate species limits and distributions. The eastern Pacific species range from the Gulf of California to northern Peru. The western Atlantic species occur from the Gulf of Mexico and Florida to the southern Caribbean. All of the species, except *B. boqueronensis*, inhabit muddy bottoms at depths ranging from 10 to 120 meters. *Bollmannia boqueronensis* occurs in sand-silt substrates. Characters that diagnose the genus include the following: head and body laterally compressed, mouth large and very oblique, body completely scaled with ctenoid scales, first dorsal with VII spines, pelvic fins fused forming a disc, and caudal fin lanceolate. Maximum sizes range from 6 to 14 cm SL. (Robertson and Allen 2008) (Fig. 2.1.1Q, Table 2.1.3).

Microgobius Poey, 1876

Microgobius contains 14 species, 8 in the eastern Pacific and 6 in the western Atlantic. The eastern Pacific species are found from the Gulf of California to northern Peru, the western Atlantic species from Florida and the Gulf of Mexico south to Brazil. Many of the species inhabit mangrove streams with very fine silt-mud substrate. Depths range from intertidal to 36 m, with most occurring in the shallow depths. Although the general body color is a light grey-brown, they typically have areas of very bright iridescent blue and/or yellow. The species are characterized by a laterally compressed body; large, nearly vertical mouth; body completely scaled; first

Table 2.1.3 Species of Gobiosomatini (*Microgobius* Group) including distributions and habitats.

Gobiidae: Gobiinae: Gobiosomatini: "Microgobius Group"	*Distribution*	*Habitat*
***Akko* Birdsong and Robins, 1995**		
Akko brevis (Günther, 1864)	Eastern Pacific: El Salvador, Panama, Peru	m
Akko dionaea Birdsong and Robins, 1995	Western Atlantic: Brazil (Amazon region) and Colombia	m
Akko rossi Van Tassell and Baldwin, 2004	Eastern Pacific: El Salvador	m
***Bollmannia* Jordan, 1890**		
Bollmannia boqueronensis Evermann and Marsh, 1899	Western Atlantic: Florida, Gulf of Mexico to Venezuela	m
Bollmannia chlamydes Jordan, 1890	Eastern Pacific: Nicaragua to northern Peru	m
Bollmannia communis Ginsburg, 1942	Western Atlantic: Gulf of Mexico to Guyana	m
Bollmannia eigenmanni (Garman, 1896)	Western Atlantic: Gulf of Mexico to Panama	m
Bollmannia litura Ginsburg, 1935	Western Atlantic: central and southern Caribbean	m
Bollmannia longipinnis Ginsburg, 1939 [=Syn. of *B. ocellata*]	Eastern Pacific: Gulf of California	m
Bollmannia macropoma Gilbert, 1892	Eastern Pacific: Mouth of the Gulf of California; Costa Rica	m
Bollmannia marginalis Ginsburg, 1939	Eastern Pacific: Southern Gulf of California to Ecuador	m
Bollmannia ocellata Gilbert, 1892	Eastern Pacific: Gulf of California to Ecuador	m
Bollmannia pawneea Ginsburg, 1939 [=Syn. of *B. ocellata*]	Eastern Pacific: Gulf of California to Panama	m
Bollmannia stigmatura Gilbert, 1892	Eastern Pacific: Gulf of California to northern Peru	m
Bollmannia umbrosa Ginsburg, 1939	Eastern Pacific: SW and eastern Gulf of California to Colombia	m
***Microgobius* Poey, 1876**		
Microgobius brevispinis Ginsburg, 1939	Eastern Pacific: outer coast southern Baja California, NE Gulf of California to Panama	b,m
Microgobius carri Fowler, 1945	Western Atlantic: Florida, NE Gulf of Mexico, Lesser Antilles to Tobago, Brazil (possible new species)	b,m

Contd...

Contd...

Gobiidae: Gobiinae: Gobiosomatini: "Microgobius Group"	*Distribution*	*Habitat*
Microgobius crocatus Birdsong, 1968	Eastern Pacific: El Salvador to Ecuador	b,m
Microgobius curtus Ginsburg, 1939	Eastern Pacific: El Salvador to northern Peru	b,m
Microgobius cyclolepis Gilbert, 1890	Eastern Pacific: outer coast southern Baja California, NE Gulf of California to Colombia	b,m
Microgobius emblematicus (Jordan and Gilbert, 1882)	Eastern Pacific: SE Gulf of California to northern Peru	b,m
Microgobius erectus Ginsburg, 1938	Eastern Pacific: Southern Baja California, northern and eastern Gulf of California to Panama	m
Microgobius gulosus (Girard, 1858)	Western Atlantic: Chesapeake Bay to Florida; Gulf of Mexico to Texas	b,m
Microgobius meeki Evermann and Marsh, 1899	Western Atlantic: Puerto Rico, Venezuela to Santos, Brazil	b,m
Microgobius microlepis Longley and Hildebrand, 1940	Western Atlantic: SE Florida, Dry Tortugas, Bahamas, Yucatán, Belize	b,m
Microgobius miraflorensis Gilbert and Starks, 1904	Eastern Pacific: outer coast southern Baja California (Bahía Magdalena), central eastern Gulf of California to northern Peru	b,m
Microgobius signatus Poey, 1876	Western Atlantic: Antilles, Venezuela	b,m
Microgobius tabogensis Meek and Hildebrand, 1928	Eastern Pacific: outer coast southern Baja California (n. Bahía Magdalena) to Peru (Puerto Pizarro)	b,m
Microgobius thalassinus (Jordan and Gilbert, 1883)	Western Atlantic: Chesapeake Bay to Florida (Cape Canaveral), n. Gulf of Mexico to Texas (Galveston)	b,m
Palatogobius **Gilbert, 1971**		
Palatogobius grandoculus Greenfield, 2002	Western Atlantic: Cozumel, Mexico	m
Palatogobius paradoxus Gilbert, 1971	Western Atlantic: Virgin Islands, NE Gulf of Mexico, Panama, Venezuela	m
Parrella **Ginsburg, 1938**		
Parrella fusca Ginsburg, 1939	Eastern Pacific: Gulf of Panama	m
Parrella ginsburgi Wade, 1946	Eastern Pacific: northern Gulf of California to Panama	m
Parrella lucretiae (Eigenmann and Eigenmann, 1888)	Eastern Pacific: El Salvador to Panama	m
Parrella macropteryx Ginsburg, 1939	Western Atlantic: Cuba (Isla de la Juventud), Panama, Puerto Rico, Brazil	m
Parrella maxillaris Ginsburg, 1938	Eastern Pacific: central Gulf of California to Panama	m

dorsal with VII spines; and ventral fins united. Maximum sizes range from 5 to 10 cm SL (Birdsong, 1981) (Fig. 2.1.1R, Table 2.1.3).

Palatogobius Gilbert, 1971

Two species are in this Caribbean genus. *Palatogobius paradoxus,* known from the Virgin Islands, NE Gulf of Mexico, Panama, and Venezuela, occurs on coral rubble and sand at depths of 23 – 78 m, most at the greater depths (Gilbert, 1971). *P. grandoculus* is known only from Cozumel, Mexico, where it was trawled from depths of 253 – 275 m (Greenfield, 2002). Both species are about 3 cm SL. The body has scales from the base of the caudal fin forward to the pectoral-fin base, the pelvic fins are united but lack a frenum, and most specimens possess vomerine teeth. They are rare in preserved collections, and little is known about their distribution or ecology (Table 2.1.3).

Parrella Ginsburg, 1938

Five species occur in the genus, four in the eastern Pacific and one in the western Atlantic. The eastern Pacific species are found from the Gulf of California to Panama, the western Atlantic species from Cuba south to Brazil. They are coastal species inhabiting areas with leaf, shell, and organic debris in depths of 1-10 m. Maximum sizes range from 5 to 8 cm SL. The body is slender and laterally compressed, the mouth is large and oblique, the pelvic fins are fused to form a disc, and the body is scaled from the base of the caudal fin anteriorly at least up to the pectoral-fin base. (Ginsburg, 1938; Hoese, 1973; Murdy and Hoese, 2003) (Table 2.1.3).

NON-NATIVE FAUNA

Additional information and updates on invasive species can be found on the Global Invasive Species Database <http://www.issg.org/database/welcome/>. That database focuses on invasive alien species that threaten native biodiversity and covers all taxonomic groups.

Acanthogobius Gill, 1859

Acanthogobius flavimanus (Temminck and Schlegel, 1845) is native to Japan, China, and Korea. It was first recorded in the Sacramento-San Joaquin Delta region in 1963 (Brittan *et al.*, 1963), and is now found in San Francisco Bay and other central and southern bays in California. In San Francisco Bay its population has grown such that it was the third most abundant fish in trawl catches in 1982 (Meng *et al.*, 1994). Maps on which its current distribution is plotted can be found in Brittan *et al.*, (1970), Courtenay and Hensley (1979a), and Lee *et al.*, (1980).

Butis Bleeker, 1856

Butis koilomatodon is native to the Indo-Pacific: Delagoa Bay, Mozambique, Madagascar, China, and the Philippines. It is also found in the eastern Atlantic at Port Harcourt (Niger delta in Nigeria) (Harrison and Miller, 1992). Its range has expanded to the Pacific coast of Panama and the delta of the Orinoco River, Venezuela (Lasso-Alcala *et al.*, 2005).

Neogobius Iljin, 1927

Neogobius melanostoma (Pallaas, 1814) is native in Eurasia, including the Black Sea, Caspian Sea, and Sea of Azov and tributaries (Miller, 1986). It was introduced into the St. Clair River (Michigan) in 1990, and is now found throughout the Great Lakes and many of the tributaries (Fuller *et al.*, 2010). It is found in fresh water habitats but prefers brackish (Stepien and Tumeo, 2006). See Stepien and Tumeo (2006) regarding the name change from *Neogobius melanostomus* to *N. melanostoma.*

Proterorhinus Smitt, 1900

Proterorhinus semilunaris Heckel, 1837 is native to Eurasia, including rivers and estuaries of the Black and Caspian Seas and Sea of Azov. It is also found in rivers of the northern Aegean and Aral Seas (Miller, 1986). This species is believed to be established but rare in the St. Clair River and in Lake St. Clair, Michigan (Jude, 1993).

Rhinogobius Gill 1859

Rhinogobius brunneus (Temminck and Schlegel, 1845) is native to Japan, Russian Far East, Taiwan, Korea, China and the Philippines (Froese and Pauly, 2010). It was first reported in the United States in East Fork Lewis River in western Washington in 2004. It is now expanded into the Columbia River and its estuary and in Portland, Oregon (Wonham *et al.*, 2000; Schofield, 2010).

Tridentiger Gill, 1859

Tridentiger barbatus (Günther, 1861) is native to marine and brackish waters in China, Japan, and Korea. It was first reported in San Joaquin River, California, in 1997, where it has possibly established itself (Matern *et al.*, 2002; Nico and Fuller, 2010a).

Tridentiger bifasciatus Steindachner, 1881 is native to fresh and brackish waters in Japan, Korea, China, Taiwan, and the former Soviet Union (Akihito and Sakamoto, 1989). It was first recorded in California in 1985 it is now found in the San Francisco Bay area and the Sacramento-San Joaquin estuary (Nico and Fuller, 2006b).

Tridentiger trigonocephalus (Gill, 1859) is a brackish and marine species that is native to China, Korea, and Japan (Courtenay *et al.*, 1986; Meng *et al.*, 1994). It was first recorded in 1960 in Los Angeles Harbor and has spread south along the coast to Mexico (Nico and Fuller, 2010c).

References

Acero, A. and J. Garzón. 1987. Peces arrecifales de la región de Santa Marta (Caribe colombiano). I. Lista de especies y comentarios generales. *Acta Biológica Colombiana* 1: 83-105.

Akihito, Emperor and K. Sakamoto. 1989. Reexamination of the status of the striped goby. *Japanese Journal of Ichthyology* 36: 100-112.

Birdsong, R.S. 1975. The osteology of *Microgobius signatus* Poey (Pisces: Gobiidae), with comments on other gobiid fishes. *Bulletin of the Florida State Museum, Biological Sciences* 19: 135-187.

Birdsong, R.S. 1981. A review of the gobiid fish genus *Microgobius* Poey. *Bulletin of Marine Science* 31: 267-306.

Birdsong, R.S. 1988. *Robinsichthys arrowsmithensis*, a new genus and species of deep-dwelling gobiid fish from the western Caribbean. *Proceedings of the Biological Society of Washington* 101: 438-443.

Birdsong, R.S. 1995. A new genus and species of seven-spined goby (Gobiidae: Gobiosomini) from the offing of the Amazon River, Brazil. *Copeia* 1995: 676-683.

Birdsong, R.S., E.O. Murdy and F.L. Pezold. 1988. A study of the vertebral column and median fin osteology in gobioid fishes with comments on gobioid relationships. *Bulletin of Marine Science* 42: 174-214.

Böhlke, J.E. 1969. *Pariah scotius*, a new sponge-dwelling gobiid fish from the Bahamas. *Notulae Naturae* 421: 1-7.

Böhlke, J.E. and C.R. Robins. 1968. Western Atlantic seven-spined gobies, with descriptions of ten new species and a new genus, and comments on Pacific relatives. *Proceedings of the Academy of Natural Sciences of Philadelphia* 120: 45-174.

Brittan, M., A. Albrecht and J. Hopkirk. 1963. An oriental goby collected in the San Joaquin River delta near Stockton, California. *California Fish and Game* 49: 302-304.

Brittan, M.R., J.D. Hopkirk, J.D. Conners and M. Martin. 1970. Explosive spread of the oriental goby *Acanthogobius flavimanus* in the San Francisco Bay-Delta region of California. *Proceedings of the California Academy of Sciences* 38: 207-214.

Cervigón, F. 1994. *Los peces marinos de Venezuela.* Volumen III. 2nd edition. Fundación Científica Los Roques.

Colin, P.L. 1975. *The Neon Gobies: Comparative Biology of the Gobies of the Genus Gobiosoma, Subgenus Elacatinus (Pisces: Gobiidae) in the Tropical Western Atlantic*, Tropical Fish Hobbyist Publications, Neptune, New Jersey.

Colin, P.L. 2010. Fishes as living tracers of connectivity in the tropical western North Atlantic: I. Distribution of the neon gobies, genus *Elacatinus* (Pisces: Gobiidae). *Zootaxa* 2370: 36-52.

Courtenay, W.R., Jr. and D.A. Hensley. 1979. *Survey of introduced non-native fishes. Phase I Report. Introduced exotic fishes in North America: status 1979.* Report submitted to National Fishery Research Laboratory, U.S. Fish and Wildlife Service, Gainesville, Florida.

Courtenay, W.R., Jr., D.A. Hensley, J.N. Taylor and J.A. McCann. 1986. Distribution of exotic fishes in North America. *The Zoogeography of North American Freshwater Fishes.* C.H. Hocutt and E.O. Wiley (eds.). John Wiley and Sons, New York, NY, pp. 675-698.

De Moura, R.L., J.L. Gasparini and I. Sazima. 1999. New records and range extensions of reef fishes in the western South Atlantic, with comments on reef fish distribution along the Brazilian coast. *Revista Brasileira de Zoologia* 16: 513-530.

Eschmeyer, W.N. (ed.) Catalog of Fishes. Electronic version (19 February 2010). http://research.calacademy.org/ichthyology/catalog/fishcatmain.asp.

Eschmeyer, W.N. and J.D. Fong. 2010. Species of Fishes by Family/Subfamily. On-line version dated 19 Feb 2010. http://research.calacademy.org/research/ichthyology/catalog/SpeciesByFamily.asp

Findley, L.T. 1983. *A revision of the eastern Pacific species of the gobiid fish genus* Chriolepis (*Teleostei: Gobioidei*). Ph.D. dissertation. University of Arizona, Tuscon.

Floeter, S.R., J.L. Gasparini, L.A. Rocha, C.E.L. Ferreira, C.A. Rangel and B.M. Feitoza. 2003Brazilian reef fish fauna: checklist and remarks (updated Jan. 2003). Brazilian Reef Fish Project: www.brazilianreeffish.cjb.net.

Fowler, H.W. 1931. Fishes obtained by the Barber Asphalt Company in Trinidad and Venezuela in 1930. *Proceedings of the Academy of Natural Sciences of Philadelphia* 83: 391-410.

Froese, R. and D. Pauly (eds.). 2010. Fishbase. World Wide Web Electronic publication. www.fishbase.org, version (01/2010).

Fuller, P., A. Benson and E. Maynard. 2010. *Neogobius* [= *Apollonia*] *melanostoma.* USGS Nonindigenous Aquatic Species Database, Gainesville, Florida.

<http://nas.er.usgs.gov/queries/FactSheet.asp?speciesID = 713> Revision Date: 3/7/2009.

Garzón F., J. and A. Acero P. 1988. Los peces Gobiidae de los arrecifes del caribe Colombiano. I. Especies, distribucion, frecuencia y abundancia. *Anales del Instituto de Investigaciones Marinas de Punta de Betín* 18: 59-81.

Gilbert, C.R. 1971. Two new genera and species of western Atlantic gobiid fishes with vomerine teeth. *Copeia* 1971: 27-38.

Ginsburg, I. 1933. A revision of the genus *Gobiosoma* (family Gobiidae) with an account of the genus *Garmannia. Bulletin of the Bingham Oceanographic Collection Yale University* 4: 1-59.

Ginsburg, I. 1938. Eight new species of gobioid fishes from the American Pacific coast. *Allan Hancock Pacific Expedition 1932-40*, Los Angeles 2: 109-121.

Greenfield, D.W. 1981. *Varicus imswe*, a new species of gobiid fish from Belize. *Copeia* 1981: 269-272.

Greenfield, D.W. 2002. *Palatogobius grandoculus*: a new deep-water western Caribbean goby (Teleostei: Gobiidae). *Copeia* 2002: 716-718.

Grove, J.S. and R.J. Lavenberg 1997. *The Fishes of the Galápagos Islands*. Stanford University Press, Stanford, California.

Harrison, I.J. and P.J. Miller. 1992. Eleotridae. In: *Faune des poissons d'eaux douces et saumâtres d'Afrique de l'Ouest* Tome 2. Coll. Faune Tropicale n° 28. C. Levêque, D. Paugy and G.G. Teugels (eds.). Musée Royal de l'Afrique Centrale, Tervuren, Belgique and O.R.S.T.O.M., Paris, France. pp. 822-836.

Hoese, D.F. 1973. *Gobius lucretiae* referred to the gobiid fish genus *Parrella,* with a review of the species. *Copeia* 1973: 817-819.

Hoese, D.F. 1976. Variation, synonymy and a redescription of the gobiid fish *Aruma histrio* and a discussion of the related genus *Ophiogobius*. *Copeia* 1976: 295-305.

Hoese, D.F. and S. Reader. 2001a. A preliminary review of the eastern Pacific species of *Elacatinus* (Perciformes: Gobiidae). *Revista de Biologia Tropical* 49 (Suppl. 1): 157-167.

Hoese, D.F. and S. Reader. 2001b. Revision of the eastern Pacific species of *Gobulus* (Perciformes: Gobiidae), with description of a new species. *Revista de Biologia Tropical* 49 (Suppl. 1): 169-176.

Humann, P. and N. Deloach. 1993. *Reef Fish Identification: Galapagos*. New World Publications, Jacksonville, Florida.

Jude, D.J. 1993. The alien goby in the Great Lakes Basin. *Great Lakes Information Network* (Online).

Joyeux, J.-C., J.L. Van Tassell and R.M. Macieira. 2009. *Barbulifer enigmaticus*, a new seven-spined goby (Pisces: Teleostei: Gobiidae) from the southwestern Atlantic. *Zootaxa* 2022: 58-68.

Kramer, A., J.L.Van Tassell and R.A. Patzner. 2009. A comparative study of two goby shrimp associations in the Caribbean Sea. *Symbiosis* 49: 137-141.

Kullander, S.O. 2003. Family Gobiidae (gobies). In: *Checklist of the Freshwater Fishes of South and Central America*. R.O. Reis, S.O. Kullander and C.J. Jr. Ferraris (Organizers) EDIPUCRS, Porto Alegre. pp. 657-665.

Larson, H.K., Z. Jaafar and K.K.P. Lim. 2008. An annotated checklist of the gobioid fishes of Singapore. *The Raffles Bulletin of Zoology* 56: 135-155.

Lasso-Alcala, O.M., C. Lasso A., F. Pezold and M.L. Smith. 2005. The mud sleeper *Butis koilomatodon* (Pisces: Eleotridae): first record from the western central Atlantic. *Revista de Biologia Tropical* 53: 211.

Lee, D.S., C.R. Gilbert, C.H. Hocutt, R.E. Jenkins, D.E. McAllister and J.R. Stauffer, Jr. 1980. *Atlas of North American Freshwater Fishes.* North Carolina State Museum of Natural History, Raleigh.

Matern, S.A., P.B. Moyle and L.C. Pierce. 2002. Native and alien fishes in a California estuarine marsh: twenty-one years of changing assemblages. *Transactions of the American Fisheries Society* 131: 797-816.

Meng, L., P.B. Moyle and B. Herbold. 1994. Changes in abundance and distribution of native and introduced fishes of Suisun Marsh. *Transactions of the American Fisheries Society* 123: 498-507.

Menezes, N.A. 2003. Família Eleotridae. In: *Catálogo das espécies de peixes marinhos do Brasil.* N.A. Menezes, P.A. Buckup, J.L. Figueiredo and R.L. Moura (eds.). São Paulo, Museu de Zoologia da Universidade de São Paulo. p. 97.

Menezes, N.A. and J.L. Figueiredo. 1985. *Manual de peixes marinhos do sudeste do Brasil. V. Teleostei.* Museu de Zoologia, Universidade de São Paulo, São Paulo.

Miller, P.J. 1986. Gobiidae. In: *Fishes of the North-Eastern Atlantic and the Mediterranean*, volume III. P.J.P. Whitehead, M.L. Bauchot., J.C. Hureau, J. Nielsen and E. Tortonese (eds.). United Nations Educational, Scientific and Cultural Organization, Paris, France. pp. 1019-1085.

Moura, R.L. and N.A. Menezes. 2003. Família Microdesmidae. In: *Catálogo das espécies de peixes marinhos do Brasil.* N.A. Menezes, P.A. Buckup, J.L. Figueiredo and R.L. Moura (eds.). São Paulo, Museu de Zoologia da Universidade de São Paulo pp. 100.

Moura, R.L., J.L. Figueiredo and N.A. Menezes. 2003. Família Gobiidae. In: *Catálogo das espécies de peixes marinhos do Brasil.* N.A. Menezes, P.A. Buckup, J.L. Figueiredo and R.L. Moura (eds.). São Paulo, Museu de Zoologia da Universidade de São Paulo. pp. 97-100.

Muñoz, A.A. and F.P. Ojeda. 1998. Guild structure of carnivorous intertidal fishes of the Chilean coast: implications of ontogenetic dietary shifts. *Oecologia* 114: 563-573.

Murdy, E.O. and D.F. Hoese. 2003. Gobiidae. In: *The Living Marine Resources of the Western Central Atlantic.* Carpenter 2003. Food and Agriculture Organization of the United Nations. Rome. p. 1781-1796.

Navarro, J. and R. Pequeño. 1979. Peces litorales de los Archipiélagos de Chiloé y Los Chonos, Chile. *Revista de Biologia Marina.* Departamento de Oceanología, Universidad de Chile. 16: 255-309.

Nico, L. and P. Fuller. 2010a. *Tridentiger barbatus.* USGS Nonindigenous Aquatic Species Database, Gainesville, Florida. <http://nas.er.usgs.gov/queries/FactSheet.asp? species ID = 715 > Revision Date: 4/24/2006

Nico, L. and P. Fuller. 2010b. *Tridentiger bifasciatus*. USGS Nonindigenous Aquatic Species Database, Gainesville, Florida. <http://nas.er.usgs.gov/queries/FactSheet.asp? species ID = 716 > Revision Date: 4/24/2006

Nico, L. and P. Fuller. 2010c. *Tridentiger trigonocephalus*. USGS Nonindigenous Aquatic Species Database, Gainesville, Florida. <http://nas.er.usgs.gov/queries/FactSheet.asp? species ID = 717> Revision Date: 4/24/2006

Pequeño, G. 1989. Peces de Chile. Lista sistematica revisada y comentada. *Revista de Biologia Marina y Oceanografia*, Valparaiso 24: 1-132.

Pezold, F. 1993. Evidence for a monophyletic Gobiinae. *Copeia*. 1993: 634-643.

Pezold, F. and B. Cage. 2002. A review of the spinycheek sleepers, genus *Eleotris* (Teleostei: Eleotridae), of the Western Hemisphere, with comparisons to the West African species. *Tulane Studies in Zoology and Botany* 31: 19-63.

Randall, J.E. 1996. *Caribbean Reef Fishes*. T.F.H. Publications, Inc. Ltd., Hong Kong. Third edition.

Randall, J.E. and P.S. Lobel. 2009. A literature review of the sponge-dwelling gobiid fishes of the genus *Elacatinus* from the western Atlantic, with description of two new Caribbean species. *Zootaxa* 2133: 1-19.

Randall, J.E., P.S. Lobel and C.W. Kennedy. 2005. Comparative ecology of the gobies *Nes longus* and *Ctenogobius saepepallens*, both symbiotic with the snapping shrimp *Alpheus florididians*. *Environmental Biology of Fishes* 74: 119-127.

Robertson, D.R. and G.R. Allen. 2008. Shorefishes of the Tropical Eastern Pacific online information system. Version 1.0. Smithsonian Tropical Research Institute, Balaboa, Panama. http://biogeodb.stri.si.edu/sftep/index.php.

Robins, C.R. and G.C. Ray. 1986. *A Field Guide to Atlantic Coast Fishes of North America*. Houghton Mifflin, Boston.

Rüber, L., J.L. Van Tassell and R. Zardoya. 2003. Rapid speciation and ecological divergence in the American seven-spined gobies (Gobiidae, Gobiosomatini) inferred from a moleuclar phylogeny. *Evolution* 57: 1584-1598.

Rocha, L.A., I.L. Rosa and B.M. Feitoza. 2000. Sponge dwelling fishes of northeastern Brazil. *Environmental Biology of Fishes* 59: 453-458.

Schofield, P.J. 2010. *Rhinogobius brunneus*. USGS Nonindigenous Aquatic Species Database, Gainesville, Florida.http://nas.er.usgs.gov/queries/FactSheet.asp?speciesID = 2612 Revision Date: 6/30/2009

Smith, C.L. 1997. *National Audubon Society Field Guide to Tropical Marine Fishes of the Caribbean, the Gulf of Mexico, Florida, the Bahamas and Bermuda*. Alfred Knopf, New York.

Smith, C.L., J.C. Tyler, W.P. Davis, R.S. Jones, D.G. Smith and C.C. Baldwin. 2003. Fishes of the Pelican Cays, Belize. *Atoll Reseach Bulletin* 497: 1-88.

Smith, D.G. and C.C. Baldwin. 1999. *Psilotris amblyrhynchus*, a new seven-spined goby (Teleostei: Gobiidae) from Belize, with notes on settlement-stage larvae. *Proceedings of the Biological Society of Washington* 112: 433-442.

Stepien, C.A. and M.A. Tumeo. 2006. Invasion genetics of Ponto-Caspian gobies in the Great Lakes: a 'cryptic' species, absence of founder effects and comparative risk analysis. *Biological Invasions* 8: 61-78.

Taylor, M.S. and J.L. Van Tassell. 2002. Observations on microhabitat utilization by three widely distributed neotropical gobies of the genus *Elacatinus*. *Copeia* 2002: 1134-1136.

Thomson, D.A., L.T. Findley and A.N. Kerstitch. 2000. *Reef Fishes of the Sea of Cortez: The Rocky-Shore Fishes of the Gulf of California.* University of Texas Press, Austin.

Tyler, J.C. and J.E. Böhlke. 1972. Records of sponge-dwelling fishes, primarily of the Caribbean. *Bulletin of Marine Science* 22: 601-642.

Van Tassell, J.L. 1998. *Phylogenetic relationships of species within the gobiid genus* Gobiosoma *sensu Bohlke and Robins (1968) with comments on their relationships to other genera in the tribe Gobiosomini.* PhD. dissertation. City University of New York, New York.

Van Tassell, J.L. 2009. Gobioidei in the Americas, an online information system. http://www.gobiidae.com.

Van Tassell, J.L. and C.C. Baldwin. 2004. A review of the gobiid genus *Akko* (Teleostei: Gobiidae) with description of a new species. *Zootaxa* 462: 1-15.

Vasil'eva, E.D. 1999. Various family accounts. In: *Annotated Check-List of Cyclotomata and Fishes of the Continental Waters of Russia.* Nauka Publishers, Moscow.

Wiley, E.O. and G.D. Johnson. 2010. A teleost classification based on monophyletic groups. In: *Origin and Phylogenetic Interrelationships of Teleosts.* J.S. Nelson, H.-P. Schultze and M.V.H. Wilson (eds.). Verlag Pfeil, München, Germany.

Williams, J.T. and C.R. Gilbert. 1983. Additional information on the gobiid fish *Varicus imswe*, with comments on the nominal species. *Northeast Gulf Science* 6: 185-189.

Williams, J.T., K.E. Carpenter, J.L. Van Tassell, P. Hoetjes, W. Toller, P. Etnoyer and M. Smith. 2010. Biodiversity Assessment of the Fishes of Saba Bank Atoll, Netherlands Antilles. PLoS ONE 5(5): e10676.doi: 10.1371/journal. pone 0010676.

Wonham, M.J., J.T. Carlton, G.M. Ruiz and L.D. Smith. 2000. Fish and ships: relating dispersal frequency to success in biological invasions. *Marine Biology* 136: 1111-1121.

CHAPTER

2.2

North-Eastern Atlantic and Mediterranean Gobies

Marcelo Kovačić[1] and Robert A. Patzner[2]

INTRODUCTION

The present review covers all currently known marine and brackishwater gobies in the eastern part of the amphiatlantic Temperate North Atlantic realm *sensu* Spalding *et al.* (2007) i.e. in its four eastern provinces: the Northern European Seas, the Lusitanian, the Mediterranean and the Black Sea. The data includes African coast to the Cape Vert peninsula in Senegal, but not the Cape Verde islands. In Briggs' (1974) classic biogeographic division of continental shelfs, temperate waters of the North-Eastern Atlantic and the Mediterranean were divided in cold-temperate Eastern Atlantic Boreal region and the warm-temperate Mediterranean-Atlantic region (Fig. 2.2.1A). The last one being further divided in the province Lusitania, including the Mediterranean, and the Black Sea and Caspian

Authors' addresses: [1]Prirodoslovni muzej Rijeka, Lorenzov prolaz 1, HR-51000 Rijeka, Croatia. E-mail: marcelo@prirodoslovni.com

[2]Dept. Organismic Biology, University of Salzburg, Hellbrunerstr. 34, 5020 Salzburg, Austria. E-mail: robert.Patzner@sbg.ac.at

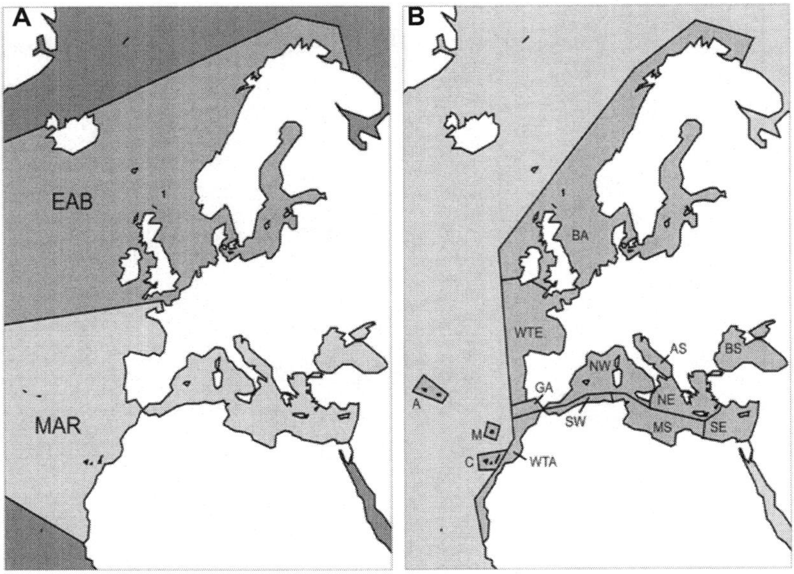

Fig. 2.2.1 A Zoogeographic regions according to Briggs (1974): EAB - cold temperate Eastern Atlantic Boreal region, MAR - warm temperate Mediterranean-Atlantic region. **B** The division of the zoogeographic areas of the North-Eastern Atlantic, the Mediterranean and the Black Sea. BA - the boreal Atlantic area, WTE - the warm temperate European Atlantic area, GA - the Gibraltar area, WTA - the warm temperate north African Atlantic area, A - Azores, M - Madeira islands, C - Canary and Salvage islands, NW - the northern westpart of the Mediterranean, SW - the southern westpart of the Mediterranean, AS - the Adriatic Sea, BS - the Black Sea and the Azov sea, NE - the northern eastpart of the Mediterranean, SE - the southern eastpart of the Mediterranean, MS - the middle south part of the Mediterranean.

Sea as separate provinces. Spalding *et al.* (2007) had similarly positioned boundaries in Atlantic between cold-temperate Northern European Seas and warm-temperate Lusitania as Briggs (1974).

GEOGRAPHY, HYDROGRAPHY AND GEOLOGICAL HISTORY

The recent geographic distribution of North-Eastern Atlantic and Mediterranean gobies was influenced by the geological history of the area, as well as by geographical, bathymetrical and hydrographical characteristics. Temperate waters of the North-Eastern Atlantic cover the large area of the Atlantic coast from the top of the Scandinavian Peninsula to the Cape Vert in Senegal in Africa. The boundaries of the temperate waters of the North-Eastern Atlantic are formed by the sea currents that originate from the North Atlantic current that comes from west and splits southwest of Ireland. The northern branch reaches Europe as the Norwegian current that carries warm and saline water into much higher latitudes than anywhere in the world (Briggs, 1974). The Canary current branches south from the North Atlantic current and flows toward the south-west about as far as Senegal. The boundary between cold-temperate and warm-temperate waters within this area also depends on the North Atlantic current, the relatively warm water of this current raises the sea temperature south of England before it turns southward and forms the Canary current. The coastline between the England and Cap-Vert, including the Mediterranean, is the largest warm-temperate area in the world (Briggs, 1974). Three enclosed seas are connected to the North-Eastern Atlantic, two by narrow straits, i.e. the Baltic Sea and the Mediterranean Sea; and the Black Sea indirectly, being opened only to the Mediterranean Sea through the Bosporus. The Baltic Sea is connected with cold-temperate Atlantic; it is fed by many large rivers, contributing a volume of 660 km^3 per year to the Baltic, and with low evaporation. Consequently, it is mostly characterised by the low salinity and high seasonal fluctuation in temperature (Briggs, 1974). The open surface water salinity is 6 to 8 ppt, deeper between 10 and 15 ppt, but at the semi-enclosed bays with major freshwater inflows the salinity is considerably lower. The Baltic Sea surface temperature varies seasonally from below 0°C in winter to 12 – 17°C in summer. On the contrary, the Mediterranean Sea is characterised by high salinity, much higher than the connected warm-temperate Atlantic. Evaporation is especially strong in the eastern end of the Mediterranean Sea and there is a clear northwest-southeast increase in salinity and temperature (northwest as minimum to southeast as maximum in surface salinity 36 ppt to 39.5 ppt, and in surface temperature: winter 9°C to 17°C, summer 20°C to 30°C). The waters of the Black Sea largely differ from the connected Mediterranean, receiving from many large rivers volume of 350 km^3 of freshwater per year and being much colder and with low salinity. The surface water has an average salinity of 18 to 18.5 ppt and temperature that varies seasonally from winter minimum about 0°C for the extreme north parts of the Black Sea, and about 10° for the south parts to the summer maximum of about 30°C.

The most important geological event that shaped today's gobiid fauna and fish biodiversity in general of the area was the desiccation of the Mediterranean known as "Messinian salinity crisis" (Hsü *et al.*, 1977; Quinard and Tomasini, 2000). The salinity crisis started at the same time over all the Mediterranean basin, about 5.96 million years ago, and the basin was isolated from the Atlantic Ocean between 5.59 and 5.33 million years ago (Krijgsman *et al.*, 1999). The marine Mediterranean palaeo-fauna was more or less destroyed. In the later stage of "Messinian salinity crisis", known as "Lago Mare" phase, both the Mediterranean and Paratethys seas were reduced to a network of lakes. After the connection to the Atlantic Ocean was restored, the Mediterranean Sea was recolonized by Atlantic species. However, the glacial-interglacial changes of temperature during the Pleistocene could also provide the mechanism of isolation and re-invasion of the Mediterranean by the warm water Atlantic species (Almada *et al.*, 2001). There are few, if any pre-messinian marine and brackish fishes present today in the Mediterranean, and those with ponto-caspic origin are equally rare (Quignard and Tomasini, 2000). The area of the present Black Sea, as the western part of the Paratethys, also became a normal sea regarding salinity for a short period in early Miocene. However, the later desalinization of the water in the Black Sea basin changed it to a freshwater lake environment until the late Quarternary, when Mediterranean seawater entered the Black Sea basin and transformed it into a semi-marine basin of low salinity (Hsü and Giovanoli, 1979; Yanko-Hombach *et al.,* 2007). The human influence must be added to the geological history that shaped the zoogeography of gobies of this area. After the opening of the Suez Canal in 1869, a migration of marine organisms started from the Red Sea into the Mediterranean, which are today known as Lessepsian immigrants.

HISTORY OF THE GOBIID RESEARCHES

The gobiid fishes, with 93 species, are the most species-rich fish family in the North-Eastern Atlantic, the Mediterranean and the Black Sea. These more or less European waters or waters close to Europe were supposed to be well researched, as several European ichthyologists had worked intensively on gobies at the end of 19th century and during the 20th century, i.e. J. Kolombatović, L. Sanzo, L. Fage, F. De Buen, B.S. Iljin and P.J. Miller. However, there had been no published list for the entire area until the synonymy of North-Eastern Atlantic and Mediterranean gobies was cleared by Miller (1973). The most recent updated list for the Mediterranean gobies was provided by Quignard and Tomasini (2000), for the North-Eastern Atlantic north of 30° by Miller (1986), and for the North-Eastern Atlantic south of 30° by Miller (1990). There is no published analysis on zoogeography of Gobiidae within the North-Eastern Atlantic, the Mediterranean and the Black Sea. The last published data with summary on distribution for each gobiid species is about 20 years old (Miller, 1986, 1990). Intensive activity including new records and new species descriptions in the recent years have quickly changed our knowledge on geographic presence and distribution of gobies (see references above). Taking in account just one year, 2007, eight papers were published that increase the gobiid diversity of different parts of this area (Boldyrev and Bogutskaya, 2007; Boltachev *et al.*, 2007; Engin *et al.*, 2007; Francour and Mangialajo, 2007; Freyhof and Naseka, 2007; Kovačić and Golani, 2007a, 2007b; Patzner, 2007). The present analysis of distribution patterns of Gobiidae in the North-Eastern Atlantic, the Mediterranean and the Black Sea is based on the inclusion of the most recent data. However, the new records and the species descriptions could soon make the data of the present analysis (last update in December 2008) out of date.

The most used work on Gobiidae from eastern Atlantic and Mediterranean Sea concerning taxonomy and distribution is the paper of Miller (1986). However, in the past 20 years there were several additions and changes. Seven gobiid species were described as new (*Chromogobius britoi, Didogobius kochi, D. schlieweni, D. splechtnai, Gobius kolombatovici, G. xanthocephalus, Vanneaugobius canariensis*), one invaded from the Red Sea (*Papillogobius =Favonigobius melanobranchus*), two have changed their names (*Monishia ochetica* to *Coryogalops ochetica, Gobius luteus* to *G. auratus*) and one was misidentified (*Oxyurichthys petersii* as *O. papuensis*). Most of the new species are rather small cryptobenthic gobies which have been detected by SCUBA diving and/or the use of anaesthetics.

C. britoi was described from Madeira and the Canarian Islands. Recently it was also found on the Atlantic coast of Portugal, south of Lisbon (Beldade *et al.*, 2006). This species is found from the intertidal and shallow subtidal to depths of up to 65 m. It occurs in small cracks and crevices on the vertical face of rock walls, under boulders covered with algae or under small rocks on sand as well as in tide pools.

D. kochi was described from the Canary Islands. Later is also was found at the Cape Verde Islands (Brito and Miller, 2001).It lives at rock outcroppings with large expanses of sand between.

D. schlieweni was described on the basis of a single immature specimen from Unije Island near Cres Island, Croatia (Miller, 1992). Later findings were made in the Adriatic Sea (Kovačić, 2005; Patzner, 2007), in southern France (Ballesta *et al.*, 1998) and in western Italy (Patzner, 2007). This species spends the day hidden to become active during the night. Swimming movements are slow, and when disturbed it swims approximately 20 cm above the bottom. *D. schlieweni* lives on a bottom substrate of pebbles or on sand among *Posidonia oceanica* or *Caulerpa racemosa*. It has been found at depths between 1.5 and 2.5 m in the Adriatic Sea, and at 10 to 14 m in the western Mediterranean (see Patzner, 2007).

D. splechtnai was described from Ibiza (Baleares)(Ahnelt and Patzner, 1995). Thereafter it was also found at Lampedusa (Italy) (Stefanni, 1999), in the northern Adriatic Sea and at Elba (Tyrrhenian Sea) (Herler and Patzner, 2002), at Sicily (Patzner, 2008) and in the southern Aegean Sea (Francour *et al.*, 2007). This species lives on rocky shores in deep caves in water depths from 5 to 56 m, usually resting directly in front of its hole (Herler *et al.*, 1999). It was also found in small cavities near sea grass (Herler and Patzner, 2002). In depths of more than 50 m it once was caught outside a cave (Ahnelt and Dorda, 2004).

G. kolombatovici was described in the northern Adriatic Sea. Later it was also found on several places in the north-western Mediterranean Sea (Francour and Mangialajo, 2007). Recently it was observed in the southern Tyrrhenian Sea and in the Ionian Sea (see Patzner, 2008). This species was observed in depths between 15 and more than 90 m. It inhabits localities with a combination of soft sediment and rock, usually at the foot of rock walls and among nearby boulders and rocks. They are usually resting on a soft substrate close to their shelter (Kovačić and Miller, 2000).

G. xanthocephalus was formerly misidentified as *Gobius auratus* or *G. fallax*. This species was described from southern France. It was later also found in northern Italy (Herler *et al.*, 2005), at Ibiza (Baleares)(Patzner, 2008), in the Atlantic Sea at the Canary Islands (Wirtz and Herrera,1995) and along the Portuguese coast (Almeida and Arruda, 1998) as well as in the Black Sea (Vasilyeva and Bogorodskii, 2004). It lives on rocky substrates, coralline grounds and also in the area of *Posidonia oceanica* meadows in water depth from 3 to 36 m (Heymer and Zander, 1993).

V. canariensis was described from the Canary Islands. Later it was also found in Guinea (Miller, 1990), at Madeira (Wirtz, 1998) and at the Cape Verde Islands (Brito and Miller, 2001). This species was observed in depths of 10 to 20 m on sandy bottoms (Wirtz, 1998).

BIODIVERSITY AND TAXONOMY

The 93 species of gobioid fishes present in the area of North-Eastern Atlantic, the Mediterranean and the Black Sea belong to 39 genera, 34 native and five being exotic immigrants (Tables 2.2.1 and 2.2.2). Currently 60 species of Gobiidae are known to occur in the Mediterranean Sea *sensu stricto* (Ahnelt and Dorda, 2004; Kovačić, 2005; Kovačić and Golani, 2007b), 33 gobiid species have been recorded in the Black Sea (Vasil'eva, 2003; Vasil'eva and Bogorodsky, 2004; Boldyrev and Bogutskaya, 2007; Boltachev *et al.*, 2007; Freyhof and Naseka, 2007; Kovačić and Engin, 2009), and 47 species in the North-Eastern Atlantic (Brownell, 1978; Brownell and Collignon; 1979; Miller, 1984, 1986, 1990; Dooley *et al.*, 1985; Van Tassell *et al.*, 1988; Gordo *et al.*, 1990; Nieto and Alberto, 1992; Alberto and Nieto, 1993; Wirtz, 1995, 1998; Santos *et al.*, 1997; Henriques *et al.*, 1999; Van Tassell, 2001; Araujo and Freitas, 2002; Ahnelt and Dorda, 2004; Beldade *et al.*, 2006). All these species are part of Gobiidae family *sensu* Nelson (2006) and the area is without representatives of other Gobioidei families *sensu* Nelson (2006). Pezold (1993) assigned the most of them to the subfamily Gobiinae, several to the subfamily Gobionellinae, including two exotic immigrants for this area, and a number of the Black Sea and Mediterranean endemic genera were not included in his study (Table 2.2.1). Thacker (this volume) listed the valid genera of the new defined family Gobiidae that includes only the subfamily Gobiinae among the previously recognized Gobiidae subfamilies. Thacker's list has not included four Gobionellinae genera and three other genera present in this area in her Gobiidae (Table 2.2.1).

SOME ECOLOGICAL ASPECTS

Unambiguously, gobies represent the most diverse fish family in the Mediterranean Sea and the eastern Atlantic. Currently, 93 species out of 39 genera (Tables 2.2.1 and 2.2.2) are known from throughout this region. However, the exact phylogenetic relationships of most of these species are unknown; the existence of many morphologically similar and endemic species in a restricted geographical may qualify the greater part of this diversity as one of the few known marine fish-species flocks. Although numerous studies on comparative morphology, on the ecology of single species, or on the ecology of gobiid communities in selected habitats have been published, very little is known about the comparative ecology and the diversity of life history of most Mediterranean and east-Atlantic gobies (Ahnelt and Patzner, 1996; Kovačić, 1997; Herler *et al.*, 1999; Patzner, 1999a; La Mesa *et al.*, 2006).

Habitat preferences

Mediterranean and east-Atlantic gobies (Gobiidae) inhabit many environments, from the shoreline down to a depth of more than 300 m (Miller, 1986; Patzner, 2008). The deepest records are: *Buenia jeffreysii* (330 m), *Lesueurigobius hetrofasciatus* (345 m), *Gobius roulei* (385 m) and

Table 2.2.1 The gobiid genera present in the area of North-Eastern Atlantic, the Mediterranean and the Black Sea. The assignments to the higher taxa from Pezold (1993) and Thacker (2011, Chapter 1.10 of this volume).

Genus	*Thacker (2011)*	*Pezold (1993)*
Aphia	Gobiidae	Gobiinellinae
Bathygobius	Gobiidae	Gobiinae
Benthophiloides	Gobiidae	/
Benthophilus	Gobiidae	/
Buenia	Gobiidae	Gobiinae
Caspiosoma	Gobiidae	/
Chromogobius	Gobiidae	Gobiinae
Corcyrogobius	Gobiidae	/
Coryogalops	Gobiidae	/
Crystallogobius	/	/
Deltentosteus	Gobiidae	Gobiinellinae
Didogobius	Gobiidae	Gobiinae
Gammogobius	Gobiidae	Gobiinae
Gnatholepis	/	Gobiinellinae
Gobioides	/	Gobiinellinae
Gobius	Gobiidae	Gobiinae
Gobiusculus	Gobiidae	Gobiinae
Gorogobius	Gobiidae	Gobiinae
Knipowitschia	Gobiidae	/
Lebetus	Gobiidae	/
Lesueurigobius	Gobiidae	Gobiinae
Mauligobius	Gobiidae	Gobiinae
Mesogobius	Gobiidae	Gobiinae
Millerigobius	Gobiidae	Gobiinae
Nematogobius	Gobiidae	Gobiinae
Neogobius	Gobiidae	/
Odondebuenia	Gobiidae	Gobiinae
Oxyurichthys	/	Gobiinellinae
Papillogobius	Gobiidae	/
Pomatoschistus	Gobiidae	Gobiinae
Proterorhinus	/	/
Pseudaphya	/	/
Silhouettea	Gobiidae	Gobiinae
Speleogobius	Gobiidae	/
Thorogobius	Gobiidae	Gobiinae
Tridentiger	/	Gobiinellinae
Vanneaugobius	Gobiidae	Gobiinae
Zebrus	Gobiidae	Gobiinae
Zosterisessor	Gobiidae	/

Table 2.2.2 Presence of gobiid species in the zoogeographic areas of the North-Eastern Atlantic, the Mediterranean and the Black Sea. bAt = the boreal Atlantic area, tAt = the warm temperate Atlantic European coast, Gi = the Gibraltar area, AtAf = the warm temperate Atlantic north African coast, Az = Azores, Ma = Madeira islands, Ca = Canaries and Salvage Islands, nwM = the northern westpart of the Mediterranean, swM = the southern westpart of the Mediterranean, Ad = the Adriatic Sea, Bs&Az = the Black Sea and the Azov sea, neM = the northern eastpart of the Mediterranean, seM = the southern eastpart of the Mediterranean, msM = the middle south part of the Mediterranean.

Area	*bAt*	*tAt*	*Gi*	*AtAf*	*Az*	*Ma*	*Ca*	*nwM*	*swM*	*Ad*	*Bs&Az*	*neM*	*seM*	*msM*
Species														
Aphia minuta mediterranea	+	+	+	+				+	+	+	+	+	+	
Bathygobius casamancus				+										
Bathygobius soporator				+										
Benthophiloides brauneri											+			
Benthophilus nudus											+			
Benthophilus magistri											+			
Benthophilus stellatus											+			
Buenia affinis								+		+		+		
Buenia jeffreysii	+	+												
Caspiosoma caspium												+		
Chromogobius britoi			+			+	+							
Chromogobius quadrivittatus								+		+	+	+	+	
Chromogobius zebratus			+					+		+		+	+	
Corcyrogobius liechtensteini								+		+				
Coryogalops ochetica													+	
Crystallogobius linearis	+	+		+				+		+		+		
Deltentosteus colonianus			+							+		+	+	
Deltentosteus quadrimaculatus		+	+	+				+		+		+	+	+
Didogobius bentuvii													+	
Didogobius kochi							+							
Didogobius schlieweni								+		+				

Contd...

Contd...

Area	*bAt*	*tAt*	*Gi*	*AtAf*	*Az*	*Ma*	*Ca*	*nwM*	*swM*	*Ad*	*Bs&Az*	*neM*	*seM*	*msM*
Species														
Didogobius splechtnai								+		+				+
Gammogobius steintzi								+		+				
Gnatholepis thompsoni						+								
Gobioides africanus				+										
Gobius ater								+		+		+		
Gobius auratus		+	+				+	+		+		+	+	
Gobius bucchichi			+	+				+	+	+	+	+	+	+
Gobius cobitis		+	+	+				+	+	+	+	+	+	+
Gobius couchi		+						+		+			+	
Gobius cruentatus		+	+	+				+	+	+	+	+	+	
Gobius fallax							+	+		+		+	+	
Gobius gasteveni		+	+			+	+	+						
Gobius geniporus								+		+		+	+	+
Gobius kolombatovici								+		+				
Gobius niger	+	+	+	+	+	+	+	+	+	+	+	+	+	+
Gobius paganellus	+	+	+	+	+	+	+	+	+	+	+	+	+	+
Gobius roulei			+					+		+			+	
Gobius rubropunctatus				+										
Gobius senegambiensis				+										
Gobius vittatus								+		+		+	+	
Gobius xanthocephalus			+				+	+			+			
Gobiusculus flavescens	+	+	+											
Gorogobius nigricintus				+										
Knipowitschia caucasica										+	+	+		
Knipowitschia longecaudata											+			

Contd...

Knipowitschia panizzae									+			
Lebetus guilleti	+					+			+			
Lebetus scorpioides	+	+										
Lesueurigobius friesii	+	+	+	+			+	+	+		+	+
Lesueurigobius heterofasciatus				+	+							
Lesueurigobius sanzoi			+	+				+				
Lesueurigobius suerii				+		+	+		+		+	+
Mauligobius maderensis					+	+						
Mesogobius batrachocephalus										+		
Millerigobius macrocephalus							+		+		+	+
Nematogobius brachynemus				+								
Nematogobius maindroni				+								
Neogobius cephalargoides										+		
Neogobius eurycephalus										+		
Neogobius fluviatilis										+		
Neogobius gymnotrachelus										+		
Neogobius kessleri										+		
Neogobius melanostomus										+	+	
Neogobius platyrostris										+		
Neogobius ratan										+		
Neogobius syrman										+		
Odondebuenia balearica							+		+		+	+
Oxyurichthys petersi											+	+
Papillogobius melanobranchus												+
Pomatoschistus bathi							+		+	+	+	
Pomatoschistus canestrini								+				
Pomatoschistus knerii							+		+		+	

Contd...

Contd...

Area	*bAt*	*tAt*	*Gi*	*AtAf*	*Az*	*Ma*	*Ca*	*nwM*	*swM*	*Ad*	*Bs&Az*	*neM*	*seM*	*msM*
Species														
Pomatoschistus lozanoi	+	+												
Pomatoschistus marmoratus			+				+		+	+	+	+	+	
Pomatoschistus microps	+	+	+	+			+	+	+					+
Pomatoschistus minutus	+	+	+	+				+	+	+	+	+		
Pomatoschistus norvegicus	+	+						+		+			+	
Pomatoschistus pictus	+	+	+		+	+	+	+		+	+	+		
Pomatoschistus quagga								+		+		+		
Pomatoschistus tortonesei								+						+
Proterorhinus marmoratus											+	+		
Pseudaphya ferreri								+		+		+	+	
Silhouettea aegyptia													+	
Speleogobius trigloides										+				
Thorogobius ephippiatus	+	+	+	+	+	+	+	+		+		+		
Thorogobius macrolepis								+		+		+		
Tridentiger trigonocephalus											+			
Vanneaugobius canariensis						+	+							
Vanneaugobius dollfusi				+						+				
Vanneaugobius pruvoti							+	+				+		
Zebrus zebrus			+					+		+	+	+	+	+
Zosterisessor ophiocephalus										+	+	+		+
Number of species per area	15	20	23	24	4	10	16	43	10	46	33	37	29	12

Crystallogobius linearis (400 m) (Miller, 1986). About 27 % have there preferred living range between the surface and a depth of 2 meters and 31% between 2 and 10 m. One third (29 %) lives mostly in depths of 10 to 50 m and only 13 % prefer deeper bottom (Fig. 2.2.2A.).

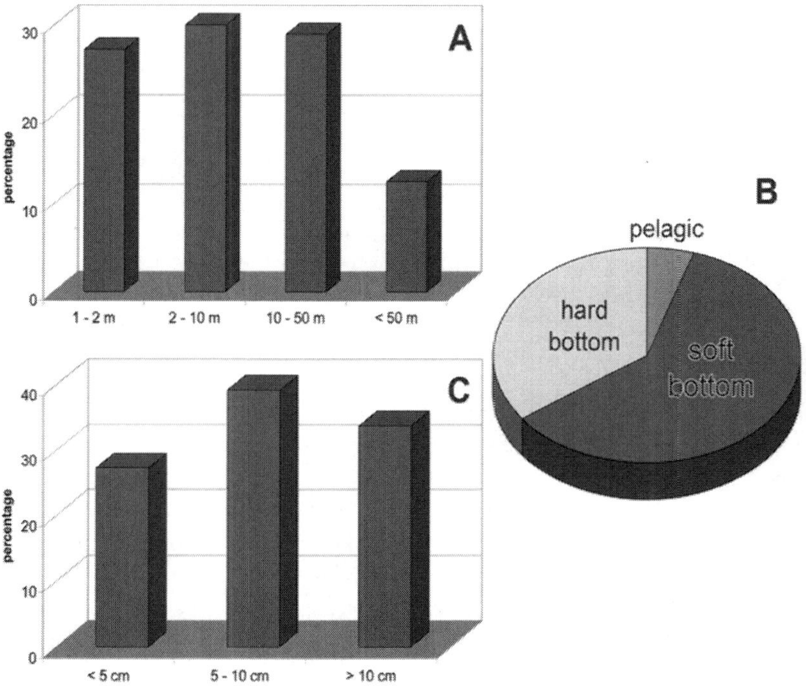

Fig. 2.2.2 Ecological aspects of gobies in the Mediterranean Sea and the eastern Atlantic. **A** prefered living range concerning water depth; **B** preferred habitats; **C** maximum size of animals.

Most of the species live a benthic life; however, *Aphia minuta*, *Crystallogobius linearis* and *Pseudaphya ferreri* are free swimming (nectonic). More than 60% of the benthic species prefer soft bottoms as substrate. About 40% live on hard bottoms and in caves (Fig. 2.2.2B). Several of the hard bottom dwellers show, more or less, a cryptic way of life. Some recently published papers have shed a little light on the living place of small cryptobenthic fishes (Ahnelt and Patzner, 1996; Kovačić, 1997; Herler *et al.*, 1999; Patzner, 1999a; La Mesa *et al.*, 2006). The term 'cryptobenthic' is here used according to the definition of Miller (1979, 1996): These species are always living inside different kinds of burrows (caves, cavities, holes, clefts) or below cover (stones, boulders, shells) and therefore can not be seen from above. A number of these species was believed to be extremely rare (Miller, 1986) but under intensified examination, they have been proved to be rather common.

Size

Gobies are small fish, only few species reach a length of more than 20 cm. Considering the species in the east-Atlantic and Mediterranean region, 27% of them do not reach a total length of 5 cm, 39 % get between 5-10 cm in total length and 34 % are longer than 10 cm (Fig. 2.2.2C). The smallest species in the area are *Speleogobius trigloides* (2.26 cm) and *Lebetus guilleti* (2.4 cm); the largest *Gobius cobitis* (27 cm) and *Neogobius cephalargoides* (25 cm) (Miller, 1986).

Associations

Gobius bucchichi is resistant against the strong poison of the snake-locks anemone *Anemonia viridis* (Pölzer, 1994; Puchner, 2002). It can move between the tentacles bearing the nettle cells and is therefore protected from predators. This association is not a true symbiosis, such as in tropical clown fishes, because the anemone gains no benefit from the presence of the goby. Usually the goby lives close to the anemone but a direct contact with the tentacles is avoided. Only in danger, the fish seeks a shelter between the tentacles. If the protecting mucous layer of the skin is experimentally removed (e.g. by hand) it takes several days to regenerate. During this time, the goby is vulnerable to the anemone's sting (Pölzer, 1994; Puchner, 2002). The association between fish and anemone is of variable intensity in different regions of the Mediterranean Sea. However, in the western part the partnership is obligatory, in the eastern part and in the Adriatic Sea it is much weaker and *G. bucchichi* is often found without an anemone (Velte, 1994).

In the Mediterranean Sea, beside some invertebrates (e.g. polychaets, shrimps, amphipods, ophiurids) juveniles of benthic teleost fishes seek shelter under sea urchins *Arbacia lixula*, *Paracentrotus lividus* and *Sphaerechinus granularis*. To date, the following species of gobies were found in this habitat: *Chromogobius zebratus*, *Gobius bucchichi*, *Millerigobius macrocephalus* and *Zebrus zebrus* (Patzner, 1999b).

BIOGEOGRAPHY

Relationships with other Zoogeographic Areas

The three zoogeographic areas with their gobiid fauna (the Ponto-Caspian and European freshwaters, the eastern tropical Atlantic coast along African coast, and the western Atlantic coast) adjoin the North-Eastern Atlantic, the Mediterranean and the Black Sea area. The cold Arctic waters north of the temperate waters of the North-Eastern Atlantic are without fish species of suborder Gobioidei. However, two gobiid species reach the south-east Iceland (*Buenia jeffreysii, Lebetus scorpioides*) (Miller, 1986).

The gobiid fauna of the North-Eastern Atlantic, the Mediterranean and the Black Sea shares nine native genera with the Ponto-Caspian and European freshwaters, about 1/4 of the North-Eastern Atlantic, the Mediterranean and the Black Sea total number of genera, and about 2/3 of the gobiid genera that occur in Ponto-Caspian and European freshwaters (Miller, 2003) (Table 2.2.3). All nine genera are present in the Black Sea, just three of them are present also elsewhere in the European temperate seas (*Knipowitschia, Pomatoschistus* and *Zosterisessor*). The zoogeographic connection between the European temperate seas and the Ponto-Caspian and European freshwater gobiid fauna is the area of the Black Sea, with its history of the reversals from marine to freshwater conditions. This probably enabled the adapting of the freshwater gobiid fauna to the highly different environment: brackish, semi-marine, and finally, a part of them, to the marine conditions.

The gobiid fauna of the North-Eastern Atlantic, the Mediterranean and the Black Sea is much richer in genera comparing to the eastern tropical Atlantic coast (Miller, 1990). Therefore, the 10 native genera present in the both areas are about 29% of genera from temperate waters and about 53% of genera from tropical waters (Table 2.2.3). These genera represent the mix of the clear tropical fauna that manage to reach tropical limit or shortly spread north of it (*Bathygobius, Gobioides, Gorogobius, Nematogobius*), with the taxa that are widespread in the temperate waters, but with representatives also in the tropical area (*Corcyrogobius, Gobius, Lesueurigobius, Thorogobius, Vanneaugobius*). *Gnatholepis thompsoni* (Jordan, 1904), goby from tropical western Atlantic and eastern Atlantic islands was recently recorded in Madeira Island (Araujo and Freitas, 2002). The gobiid fauna of the North-Eastern Atlantic shares three native genera with the Western Atlantic (Murdy and Hoese, 2008), which is about 9% of total number of genera in each of the areas (Table 2.2.3). Three amphiatlantic gobiid genera

Table 2.2.3 Presence of the native gobiid genera of the area of North-Eastern Atlantic, the Mediterranean and the Black Sea in the neighbouring areas (Miller, 1990, 2003; Murdy and Hoese, 2008).

North-Eastern Atlantic, Mediterranean and Black Sea	*Ponto-Caspian and European freshwaters*	*Eastern tropical Atlantic coast*	*Western Atlantic coast*
Aphia			
Bathygobius		+	+
Benthophiloides	+		
Benthophilus	+		
Buenia			
Caspiosoma	+		
Chromogobius			
Corcyrogobius		+	
Crystallogobius			
Deltentosteus			
Didogobius			
Gammogobius			
Gnatholepis		+	+
Gobioides		+	+
Gobius		+	
Gobiusculus			
Gorogobius		+	
Knipowitschia	+		
Lebetus			
Lesueurigobius		+	
Mauligobius			
Mesogobius	+		
Millerigobius			

Contd...

Contd...

Nematogobius		+	
Neogobius	+		
Odondebuenia			
Pomatoschistus	+		
Proterorhinus	+		
Pseudaphya			
Speleogobius			
Thorogobius		+	
Vanneaugobius		+	
Zebrus			
Zosterisessor	+		
Number of genera	9	10	3

(*Bathygobius*, *Gnatholepis* and *Gobioides*) that occur in the temperate North-Eastern Atlantic are all of tropical origin and present in the eastern tropical Atlantic. They could easily migrate north from the eastern tropical Atlantic. Therefore, there is no indication of any direct relationship between the warm temperate areas on each side of the North Atlantic, but the fauna of the Western and Eastern Atlantic is related through the tropical waters, what was already concluded for marine fauna in general by Briggs (1974).

The Zoogeographic Division of the North-Eastern Atlantic, the Mediterranean and the Black Sea

There is no published analysis on zoogeography of Gobiidae within the North-Eastern Atlantic, the Mediterranean and the Black Sea. Our division of the zoogeographic areas of the North-Eastern Atlantic, the Mediterranean and the Black Sea is based on the well-known biogeographic divisions of continental shelfs (Briggs, 1974; Spalding *et al.*, 2007); on geographical, hydrographical and geological characteristics of the areas; but it was also influenced to some degree by the available data on gobiid presence for some areas (Fig. 2.2.1). The division of North-Eastern Atlantic zoogeographic areas followed the border between warm temperate and cold temperate waters (North European Seas and Lusitanian provinces by Spalding *et al.*, 2007). Our further split of the warm temperate North-Eastern Atlantic included three areas around islands, and the three areas along the continental coast. The division on the continental coast was based on the possible influence of Mediterranean fauna in Gibraltar area and on the influence of tropical water fauna along African coast. The areas within the Mediterranean were delimited with the presumption that the distribution of small benthic littoral fishes like Gobiidae would follow a coastline, and limits of distribution of those species would also mostly depend on sea currents (spread of larvae), salinity and temperature. In the southern parts of Mediterranean the borders of the areas were adjusted to the state borders, due to the fact that the only recent sources of data were at best check-lists for each state.

The presented distribution patterns of Gobiidae in the North-Eastern Atlantic, the Mediterranean and the Black Sea are based on the inclusion of the most recent data (last update in December 2008). The following publications were source of data for species presence in the areas: the North-Eastern Atlantic (Brownell, 1978; Miller, 1984, 1986, 1990; Dooley *et al.*, 1985; Van Tassell *et al.*, 1988; Gordo *et al.*, 1990; Nieto and Alberto, 1992; Alberto and Nieto, 1993; Wirtz, 1995, 1998; Santos *et al.*, 1997; Henriques *et al.*, 1999; Van Tassell, 2001; Araujo and Freitas, 2002; Ahnelt and Dorda, 2004; Beldade *et al.*, 2006); the Mediterranean (Relini

Orsi and Fanciulli, 1977; Brownell, 1978; Brownell and Collignon, 1979; Zander, 1982; Županović and El Buni, 1982; Miller, 1986; Papakonstantinou, 1988; Ahnelt, 1990; Rakaj, 1995; Golani, 1996; Ibrahim and Soliman, 1996; Al Hassan and El-Silini, 1999; Stefanni and Mazzoldi, 1999; Mayol *et al.*, 2000; Louisy, 2002; Ahnelt and Dorda, 2004; Bradai *et al.*, 2004; Gysels *et al.*, 2004; Kovačić, 2005; La Mesa *et al.*, 2005; Kovačić and Golani, 2006; Francour and Mangialajo, 2007; Fischer *et al.*, 2007; Fricke *et al.*, 2007; Kovačić and Golani, 2007a, 2007b; Patzner, 2007; Mejri *et al.*, in press); the Black Sea and the Azov sea (Vasil'eva, 2003; Vasil'eva and Bogorodsky, 2004; Boldyrev and Bogutskaya, 2007; Boltachev *et al.*, 2007; Engin *et al.*, 2007; Freyhof and Naseka, 2007; Kovačić and Engin, 2009).

The species richness and the species composition reveal some grouping of the areas: within Mediterranean species rich areas of north Mediterranean and the Adriatic Sea, with the closely related Levant Sea differ from the west and middle south part of the Mediterranean; the Black Sea has the clearly distinct gobiid fauna composition among all areas; the north-eastern Atlantic islands have low number of species compared to the Atlantic continental coast areas and there is also a clear gradual change of species richness along the Atlantic continental coast with increase of number of species from north to south (Figs. 2.2.3. and 2.2.4A, Table 2.2.2).

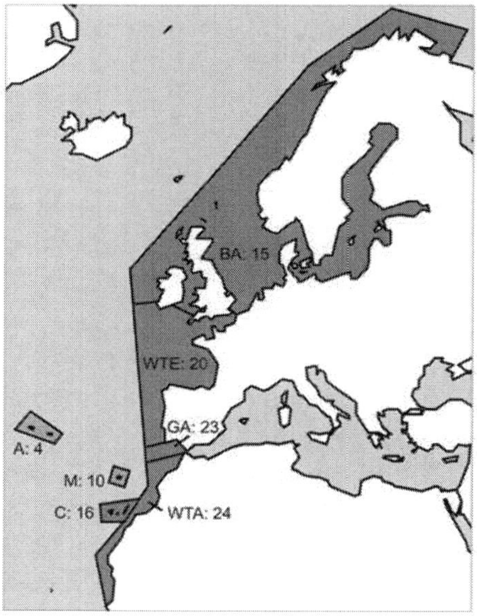

Fig. 2.2.3 Diversity of Gobiidae in the North-Eastern Atlantic areas (■ 15-24, ■ 4-10). BA - the boreal Atlantic area, WTE - the warm temperate European Atlantic area, GA - the Gibraltar area, WTA - the warm temperate north African Atlantic area, A - Azores, M - Madeira islands, C - Canary and Salvage islands.

The North-Eastern Atlantic and the Atlantic-Mediterranean Transition

In the north-eastern Atlantic, the reduced gobiid fauna of the Atlantic islands is zoogeographically distinct from the mainland gobiid fauna. At these islands a sharp decline in the number of gobiid species is proportional with the distance from the mainland (Fig. 2.2.3, Table 2.2.2). At the present stage of knowledge on species distributions, close relationship of the southern westpart of the Mediterranean with the Gibraltar area is clear, the first area is without the Mediterranean endemics in a narrow sense and all species being present there also inhabit the neighbouring Gibraltar area. However, there is a lack of data on gobiid distribution along the Atlantic coasts and at the southern westpart of the Mediterranean at present. Much more data are necessary for any final conclusion on the Atlantic-Mediterranean transition and on the question if the biogeographical separation of gobies between the Atlantic and the Mediterranean exist near the Strait of Gibraltar, west from the Strait or east from it. Additional data are also necessary for any conclusion about the justification of separate Gibraltar area and on the zoogeographic relationships between the warm temperate Atlantic coasts of Africa and Europe. The entire area of the north-eastern Atlantic shows very low level of endemism, with just 6 endemic gobies, two from the Atlantic islands (*Didogobius kochi, Mauligobius maderensis*), and four with more broad distributions in the temperate North-Eastern Atlantic (*Chromogobius britoi, Lebetus scorpioides, Lesueurigobius heterofasciatus, Pomatoschistus lozanoi*). A recent immigration of Ponto-Caspian species *Neogobius melanostomus* happened in the Baltic Sea. This highly invasive species was recorded for the first time near the Port of Hel in June 1990, and since than has spread along the coasts of Poland and Germany (Corkum *et al.*, 2004).

The Mediterranean Sea

The Mediterranean Sea *sensu stricto,* with 60 species is richer in gobies than the neighbouring large area of the north-eastern Atlantic and comparatively rich for its latitude. The north-eastern Atlantic has just six endemic gobies. On the other hand, 24 species (2/5 of Mediterranean gobies) are Mediterranean endemics in a narrow sense, i.e. their distribution is limited to in between the Strait of Gibraltar and the Bosphorus. In addition to this, five Mediterranean gobies extend their distribution into the Atlantic just to the Gibraltar area (*Chromogobius zebratus, Deltentosteus colonianus, Gobius xanthocephalus, Pomatoschistus marmoratus, Zebrus zebrus*). A similar situation was observed for blennies (Almada *et al.*, 2001). The most of today's Mediterranean endemic gobies probably evolved from the Atlantic species that recolonized the Mediterranean Sea after the connection to the Atlantic Ocean was restored. However, at the present state of knowledge on phylogeny of gobiids in the area, we could not be sure what part of the Mediterranean gobies were recolonized after the "Messinian salinity crisis" from brackish and freshwater species and what was the direction of invasion regarding Ponto-Caspian freshwater gobies and two freshwater species presently assigned to genus *Padogobius*. It could not be also rejected possibility that part of the pre-messinian gobiid fauna related to the Indo-Pacific gobies survived. McKay and Miller (1997) supposed that the Mediterranean sand gobies are sister group to the gobionellines widespread in the Indo-Pacific area. The gobiid species richness and the amount of endemics suggest that the Mediterranean Sea acted as the center of speciation and diversity of gobies. It was probably refuge for warm-temperate species during glaciations (Briggs, 1974; Almada *et al.*, 2001) and consequently may have influenced the present diversity of gobies of the temperate North-Eastern Atlantic.

The Mediterranean zoogeographic areas could be clearly divided in three groups according to species richness. The diversity is highest along the north Mediterranean coast, with 43 to 46 gobiid species in these areas, followed by the two Mediterranean eastern areas (29 to 37 species) and on the end two south areas along the African coast with just 10 to 12 gobiid species (Fig. 2.2.4A, Table 2.2.2). The increase of known gobiid diversity in the last 20 years in some areas (Vasil'eva and Bogorodsky, 2004; Kovačić, 2005; Kovačić and Golani, 2006; Boldyrev and Bogutskaya, 2007; Boltachev *et al.*, 2007; Engin *et al.*, 2007; Fischer *et al.*, 2007; Francour and Mangialajo, 2007; Freyhof and Naseka, 2007; Kovačić and Golani, 2007a, 2007b; Patzner, 2007; Kovačić and Engin, 2009) shows that the known diversity of Gobiidae along the Mediterranean and the Black Sea coasts is far from real species richness (Fig. 2.2.4B). To which degree does the known gobiid biodiversity mirror only sampling intensity? Could the eastern areas finally be as species rich as the northern ones, or could the southern areas reach the diversity of the eastern areas? The number of currently known gobiid species in the northern eastpart of the Mediterranean (the Ionian Sea, the Aegean Sea and the Marmara Sea) could easily be the result of poor research intensity in the last two decades (Figs. 2.2.4A and B, Table 2.2.2). On the other hand, it is hard to expect that the southern areas would triple the known number of species and reach the species richness of other Mediterranean areas (Figs. 2.2.4A and B, Table 2.2.2). The similarity between the Mediterranean areas by species composition matched well with the similarity between the areas in species number (Table 2.2.2). However, the two southern areas along the African coast with poor gobiid diversity differ in the species composition and share just five gobiid species (*Gobius bucchichi, G. cobitis, G. niger, G. paganellus, Pomatoschistus microps*). The western area has the Atlantic-Mediterranean species, while the middle southern part of the Mediterranean has also the Mediterranean endemics in a narrow sense (*Gobius geniporus, Pomatoschistus tortonesei*). Concerning the analysis of the endemism, about 2/5 of Mediterranean gobies (24 species) are the Mediterranean endemics in a narrow sense i.e. with distribution limited between strait of Gibraltar and Bosphorus strait. The differences in richness of Mediterranean endemics among areas are similar to the differences in total diversity of Gobiidae (Fig. 2.2.4C). About 2/3 of the Mediterranean endemics (14 species) are restricted just to the north Mediterranean areas (the northern west part of the Mediterranean, Adriatic, Ionian, Aegean seas). Only two Mediterranean endemics are restricted to the south Mediterranean areas: *Didogobius bentuvii* in the Levant and *Pomatoschistus tortonesei* in the middle south part of the Mediterranean, but the first species is known only from the single specimen and the distribution of the second species reaches the Sicilian coasts (Miller, 1986). Four exotic species are recorded until now in the Mediterranean, all known only from the Levant (*Coryogalops ochetica, Oxyurichthys petersi, Silhouettea aegyptia, Papillogobius melanobranchus*). The most recently found species, *P. melanobranchus* is not of the Red Sea origin, but its original distribution reaches just the western Indian Ocean (Kovačić and Golani, 2007b). Another exotic species, *Bathygobius soporator* (Valenciennes, 1837) was cited for Mediterranean in the check-list of Quignard and Tomasini (2000), but without any explanation. The intramediterranean regional specifics in the diversity of gobies match well with the results for the entire ichthyofauna. The west Mediterranean and the Adriatic Sea have high fish species richness compared to the south coasts of the Mediterranean (Quignard and Tomasini, 2000). The biodiversity of all fishes in the Levant has greatly increased in last decades due to the migratory flux of the lessepsian immigrants (Quignard and Tomasini, 2000). However, colonisations of the coldest areas of the Mediterranean by thermophilic fishes from the south has not yet been observed in gobies.

The Black Sea

The Black Sea has the most distinct gobiid fauna among areas, excluding the reduced gobiid fauna of the Atlantic islands (Table 2.2.2). That is the result of the specific hydrographic characteristics and geological history of this region. The history of the reversals from marine to freshwater conditions of the Black Sea probably enabled the mixing of faunas from highly different environments, the Mediterranean Sea and the Ponto-Caspian and European freshwaters. Therefore, the present gobiid fauna of Black Sea is a composite fauna, with a distinction between Mediterranean postglacial immigrants and endemic Ponto-Caspian species. However, the waters of the Black Sea, being much colder and with low salinity limited the colonisation of fishes from the Mediterranean. The winter minimum of the surface water, about 0°C for the extreme north parts of the Black Sea, and about 10°C for the south parts, match more the cold waters than the temperate waters (Briggs, 1974). Among 33 gobiid species recorded

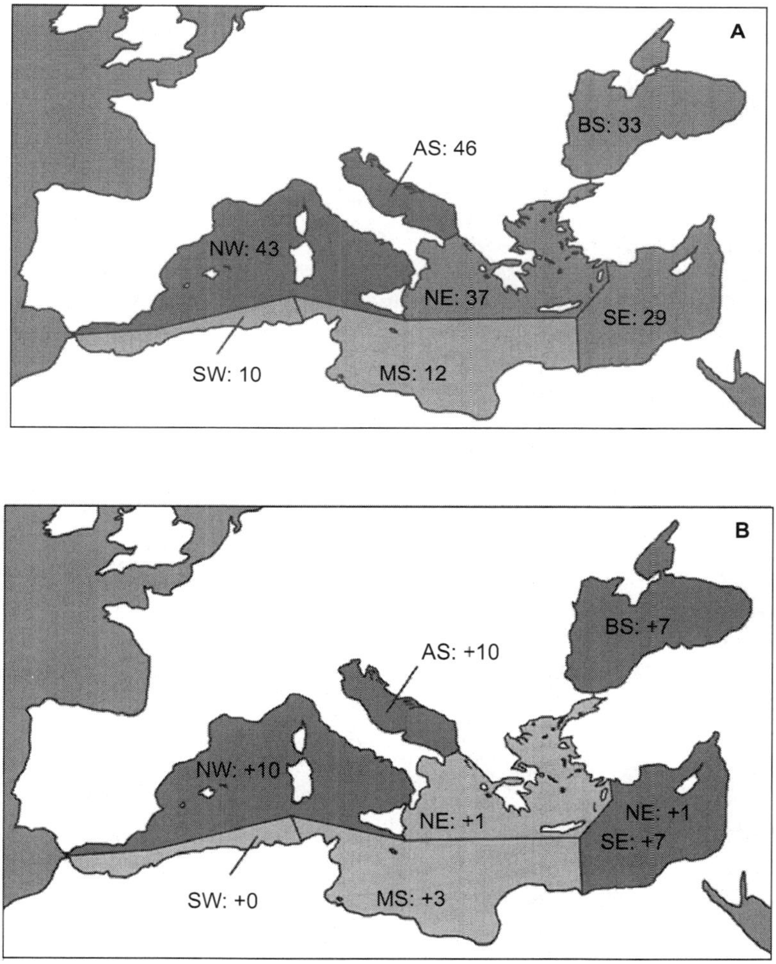

Fig. 2.2.4 Contd...

Fig. 2.2.4 Contd...

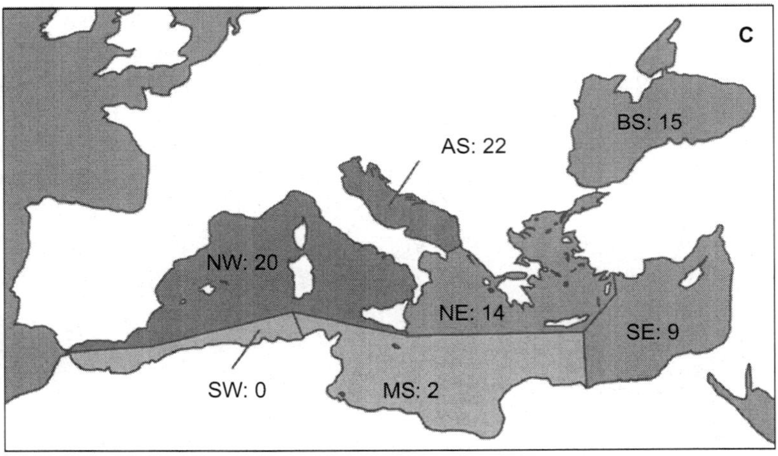

Fig. 2.2.4 A Diversity of Gobiidae in the Mediterranean areas and the Black Sea (■ 43-46, ■ 29-37, ■ 10-12). NW - the northern westpart of the Mediterranean, SW - the southern westpart of the Mediterranean, AS - the Adriatic Sea, BS - the Black Sea and the Azov sea, NE - the northern eastpart of the Mediterranean, SE - the southern eastpart of the Mediterranean, MS - the middle south part of the Mediterranean. **B** New records or species descriptions of Gobiidae for each of the Mediterranean areas and the Black Sea in the last 20 years (■ 7-10, ■ 0-3). **C** Number of endemic species of Gobiidae for each of the Mediterranean areas and the Black Sea (■ 20-22, ■ 9-15, ■ 0-2). The Mediterranean areas with the number of the Mediterranean endemics; the Black Sea and the Azov sea (BS) with the number of the endemics of Black Sea and the Caspian Sea.

in the Black Sea, 15 are not present in the Mediterranean. About half of these species are present also in the Caspian Sea, and therefore not Black Sea endemics in a narrow sense (Vasil'eva, 2003; Vasil'eva and Bogorodsky, 2004; Boldyrev and Bogutskaya, 2007; Boltachev *et al.*, 2007; Engin *et al.*, 2007; Freyhof and Naseka, 2007) (Figs. 2.2.4A and C). Recently, *Tridentiger trigonocephalus* (Gill 1859) (Boltachev *et al.*, 2007), a Pacific goby from Japan, China and Korea was recorded as the first alien goby in the Black Sea. The present gobiid fauna of Black Sea, with more than 1/2 of the species richness of the Mediterranean, has the higher relative biodiversity compared to the Black Sea ichthyofauna in general, that has less than 1/4 of the Mediterranean fish species richness (Quignard and Tomasini, 2000). It seems that the adapting and spreading of the freshwater gobiid genera to the brackish and semi-marine conditions mostly caused the relatively high biodiversity of gobies compared to other fishes in the Black Sea.

References

Ahnelt, H. 1990. *Chromogobius quadrivittatus, Chromogobius zebratus* und *Zebrus zebrus* (Pisces: Gobiidae): Erstnachweis für Korsika (Frankreich) und Sardinien (Italien). *Annalen des Naturhistorischen Museums in Wien. Serie B. Botanik und Zoologie* 91: 27-41.

Ahnelt, H. and R.A. Patzner. 1995. A new species of *Didogobius* (Teleostei: Gobiidae) from the western Mediterranean. *Cybium* 19: 95-102.

Ahnelt, H. and R.A. Patzner. 1996. Kryptobenthische Meergrundeln von den Balearen (Westliches Mittelmeer) mit Anmerkungen zum Unterartstatus von *Chromogobius zebratus levanticus* Miller, 1791 (Pisces: Teleostei: Gobiidae). *Annalen des Naturhistorischen Museums in Wien Serie B Botanik und Zoologie* 98B: 529-544.

Ahnelt, H. and J. Dorda. 2004. Gobioid fishes from the north eastern Atlantic and the Mediterranean: new records and rarely found species. *Annalen des Naturhistorischen Museums in Wien. Serie B. Botanik und Zoologie* 105: 5-19.

Al Hassan, L.A.J. and O.A. El-Silini. 1999. Checklist of bony fishes collected from the Mediterranean coast of Benghazi, Libya. *Revista de Biologia Marina y Oceanografia* 34: 291-301.

Alberto, L.J. and P. Nieto. 1993. Presence of *Chromogobius zebratus* (Kolombatovic, 1891) (Gobiidae) in the Atlantic. Comments on the subspecific characteristics and distribution. *Cybium* 17: 215-221.

Almada, V.C., E.J. Goncalves, R.F. Oliveira, A.J. Almeida, R.S. Santos and P. Wirtz. 2001. Patterns of diversity of the north-eastern Atlantic blenniid fish fauna (Pisces: Blenniidae). *Global Ecology & Biogeography* 10: 411-422.

Almeida, A.J. and L.M. Arruda. 1998. *Gobius xanthocephalus* Heymer & Zander, 1992 in Portuguese waters (Pisces: Gobiidae). *Arquivos do Museu Bocage Nova Serie* 3: 205-210.

Araujo, R. and M. Freitas. 2002. First record of the goldspot goby *Gnatholepis thompsoni* Jordan, 1904 (Pisces: Gobiidae) in Madeira Island (NE Atlantic Ocean). *Bocagiana-Funchal* 209: 1-4.

Ballesta, L., P.J. Miller and J.P. Quignard. 1998. First record of *Didogobius schlieweni* Miller, 1992 (Gobiidae) in the western Mediterranean Sea. *Cybium* 22: 290-292.

Beldade, R., J.L. Van-Tassell and E.J. Gonçalves. 2006. First record of *Chromogobius britoi* (Teleostei: Gobiidae) on the mainland European coast. *Journal of Fish Biology* 68: 608-612.

Boldyrev, V.S. and N.G. Bogutskaya. 2007. Revision of the tadpole-gobies of the genus *Benthophilus* (Teleostei: Gobiidae). *Ichthyological Exploration of Freshwaters* 18: 31-96.

Boltachev, A.R., E.D. Vasil'eva and O.N. Danilyuk. 2007. The first finding of the striped tripletooth goby *Tridentiger trigonocephalus* (Perciformes, Gobiidae) in the Black Sea (the estuary of the Chernaya River, Sevastopol Bay). *Journal of Ichthyology* 47: 802-805.

Bradai, M.N., J.-P. Quignard, A. Bouain, O. Jarboui, A. Ouannes-Ghorbel, L.B. Abdallah, J. Zaouali and S. Ben Salem. 2004. Ichtyofaune autochtone et exotique des cotes Tunisiennes: Recensement et biogéographie. *Cybium* 28: 315-328.

Briggs, J.C. 1974. *Marine Zoogeography*. McGraw-Hill, New York.

Brito, A. and P.J. Miller. 2001. Gobiid fishes from the Cape Verde Islands, including two new species of *Gobius* (Teleostei: Gobioidei). *Journal of Natural History* 35: 253-277.

Brownell, C. and J. Collignon. 1979. Sous-ordre des Gobioides, famille des Gobiidae. In: Catalogue raisonne des poissons des mers Marocaines, J. Collignon (ed.). 4. Partie. *Bulletin de l'Institut de Peches Maritimes* 24: 139-155.

Brownell, C.L. 1978. Sur quelques collections de poissons littoraux de l'Atlantique Marocain. *Bulletin de l'Institut de Peches Maritimes* 23: 111-133.

Corkum, L.D., M.R. Sapota and K.E. Skora. 2004. The round goby, *Neogobius melanostomus*, a fish invader on both sides of the Atlantic Ocean. *Biological Invasions* 8: 105-112.

Dooley, J.K., J. Van-Tassell and A. Brito. 1985. An annotated checklist of the shorefishes of the Canary Islands. *American Museum Novitates* 282: 41-49.

Engin, S., D. Turan and M. Kovačić. 2007. First record of the red-mouthed goby, *Gobius cruentatus* (Gobiidae), in the Black Sea. *Cybium* 31: 87-88.

Fischer, S., R.A. Patzner, C.H.G. Müller and H.M. Winkler. 2007. Studies on the ichthyofauna in the coastal waters of Ibiza (Balearic Islands, Spain). *Rostocker Meeresbiologische Beiträge* 18: 31-63.

Francour, P. and L. Mangialajo. 2007. *Gobius kolombatovici*, a common species of Gobiidae in the north-western Mediterranean Sea? *Cybium* 31: 389-390.

Freyhof, J. and A.M. Naseka. 2007. *Proterorhinus tataricus*, a new tubenose goby from Crimea, Ukraine (Teleostei: Gobiidae). *Ichthyological Exploration of Freshwaters* 18: 325-334.

Fricke, R., M. Bilecenoglu and H. Musa Sari. 2007. Annotated checklist of fish and lamprey species (Gnathostomata and Petromyzontomorphi) of Turkey, including a Red List of threatened and declining species. *Stuttgarter Beiträge zur Naturkunde Serie A (Biologie)* 706: 1-169.

Golani, D. 1996. The marine ichthyofauna of the eastern Levant—history, inventory and characterization. *Israel Journal of Zoology* 42: 15-55.

Gordo, L.S., L.M. Arruda and P.J. Miller. 1990. List of Portuguese fishes caught by the R/V Poseidon and in the collection of the Museu Bocage. *Arquivos do Museu Bocage, Nova Série, I* 42: 613-620.

Gysels, E.S., B. Hellemans, C. Pampulie and A.M. Volckaert. 2004. Phylogeography of the common goby, *Pomatoschistus microps*, with particular emphasis on the colonization of the Mediterranean and the North Sea. *Molecular Ecology* 13: 403-417.

Henriques, M., E.J. Gonçalves and V.C. Almada. 1999. The conservation of littoral fish communities: a case study at Arrábida coast (Portugal). In: *Behaviour and*

Conservation of Litoral Fishes, V.C. Almada, R.F. Oliveira and E.J. Gonçalves (eds.), Instituto superior de psicologia aplicada, Lisboa, pp. 473-519.

Herler, J. and R.A. Patzner. 2002. New records of *Didogobius splechtnai* Ahnelt & Patzner, 1955 (Pisces: Gobiidae) from the Tyrrhenian and northern Adriatic Sea. *Cybium* 26: 153-155.

Herler, J., R.A. Patzner and C. Sturmbauer. 2005. A preliminary revision of the *Gobius auratus* species complex with redescription of *Gobius auratus* Risso, 1810. *Journal of Natural History* 39: 1043-1075.

Herler, J., R.A. Patzner, H. Ahnelt and H. Hilgers. 1999. Habitat selection and ecology of two speleophilic gobiid fishes (Pisces: Gobiidae) from the western Mediterranean Sea. *Marine Ecology* 20: 49-62.

Heymer, A. and C.D. Zander. 1993. La discrimination phenotypique, meristique et eco-ethologique entre *Gobius auratus* Risso, 1810 et *Gobius xanthocephalus* Heymer et Zander, 1992 (Teleostei, Gobiidae). *Revue Francaise d'Aquariologie Herpetologie* 20: 81-92.

Hsü, K.J. and F. Giovanoli. 1979. Messinian event in the Black Sea. *Palaeogeography, Palaeoclimatology, Palaeoecology* 29: 75-93.

Hsü, K.J., L. Montadert, D. Bernoulli, M.B. Cita, A. Erickson, R.E. Garrison, R.B. Kidd, F. Melieres, C. Muller and R. Wright. 1977. History of the Mediterranean salinity crisis. *Nature* 267: 399-403.

Ibrahim, M.A. and I.A. Soliman. 1996. Checklist of bony fish species in the Mediterranean waters of Egypt. *Bulletin of the National Institute of Oceanography and Fisheries* 22: 43-57.

Kovačić, M. 1997. Cryptobenthic gobies (Pisces, Perciformes, Gobiidae) and clingfishes (Pisces, Gobiesociformes, Gobiesocidae) in the Kvarner area, Adriatic Sea. *Natura Croatica* 6: 423-435.

Kovačić, M. 2005. An annotated checklist of the family Gobiidae in the Adriatic Sea. *Annals for Istrian and Mediterranean Studies, Series Historiae Naturalis* 15: 1-24.

Kovačić, M. and P.J. Miller. 2000. A new species of *Gobius* (Teleostei: Gobiidae) from the northern Adriatic Sea. *Cybium* 24: 231-239.

Kovačić, M. and D. Golani. 2006. First record of the Roule's goby, *Gobius roulei* (Gobiidae), in the Levant. *Cybium* 30: 189-190.

Kovačić, M. and D. Golani. 2007a. First record of three gobiid species in the Levant. *Cybium* 31: 89-91.

Kovačić, M. and D. Golani. 2007b. The first record of *Papillogobius melanobranchus* (Fowler, 1934) in the Mediterranean Sea and the new data on geographic distributions, morphology and bathymetric ranges of several small benthic fishes in the Levant. *Cybium* 31: 417-425.

Kovačić, M. and S. Engin. 2009. First record of the zebra goby, *Zebrus zebrus* (Gobiidae), in the Black Sea. *Cybium* 33: 83-84.

Krebs, C.J. 1999. *Ecological Methodology*. Addison Wesley Longman, Menlo Park, California.

Krijgsman, W., F.J. Hilgen, I. Raffi, F.J. Sierro and D.S. Wilson. 1999. Chronology, causes and progression of the Messinian salinity crisis. *Nature* 400: 652-655.

La Mesa, M., E. Arneri, V. Caputo and M. Iglesias. 2005. The transparent goby, *Aphia minuta*: review of biology and fisheries of a paedomorphic European fish. *Reviews in Fish Biology and Fisheries* 15: 89-109.

La Mesa, G., S. Di Muccio and M. Vacchi. 2006. Structure of a Mediterranean cryptobenthic fish community and its relationships with habitat characteristics. *Marine Biology Berlin* 149: 149-167.

Louisy, P. 2002. Meeresfische, Westeuropa und Mittelmeer. Ulmer Verlag, Stuttgart.

Mayol, J., A. Grau, F. Riera and J. Oliver. 2000. *Lista Vermella dels Peixos de les Balears*. Documents tecnics de Conservacio. Il epoca num.7, Govern de les Illes Balears, Palma: 1-126. http://dgcapea.caib.es/pe/publicacions.es.htm.

McKay, S.I. and P.J. Miller. 1997. The affinities of European sand gobies (Teleostei: Gobiidae). *Journal of Natural History* 31: 1457-1482.

Mejri, R., M. Kovačić and O.K. Ben Hassine. 2009. First record of *Pomatoschistus marmoratus* (Pisces: Gobiidae) on the Tunisian coast. *Cybium* 33: 171-172.

Miller, P.J. 1973. Gobiidae. In: *Checklist of the Fishes of the North-Eastern Atlantic and of the Mediterranean (CLOFNAM)*, J.C. Hureau and T. Monod (eds.). UNESCO, Paris, pp. 483-515.

Miller, P.J. 1979. Adaptiveness and implications of small size in teleosts. *Symposia of the Zoological Society of London* 44: 263-306.

Miller, P.J. 1984. The gobiid fishes of temperate Macaronesia (eastern Atlantic). *Journal of Zoology, London* 204: 363-412.

Miller, P.J. 1986. Gobiidae. In: *Fishes of the North-eastern Atlantic and the Mediterranean* Vol. 3, P.J.P. Whitehead, M.-L. Bauchot, J.-C. Hureau, J. Nielsen and E.Tortonese (eds.). UNESCO, Paris, pp. 1019-1085.

Miller, P.J. 1990. Gobiidae. In: *Checklist of the Fishes of the Eastern Tropical Atlantic (CLOFETA) 2*, J.C. Quero, J.C. Hureau, C. Karrer, A. Post and L. Saldanha (eds.). JNICT, Lisbon; SEI, Paris; and UNESCO, Paris, pp. 925-951.

Miller, P.J., 1992. A new species of *Didogobius* (Teleostei: Gobiidae) from the Adriatic Sea. *Journal of Natural History* 26: 1413-1419.

Miller, P.J. 1996. The functional ecology of small fish: some opportunities and consequences. *Symposia of the Zoological Society of London* 69: 175-199.

Miller, P.J. 2003. Gobiidae. In: *The Freshwater Fishes of Europe,* Vol. 8/I, P.J. Miller (ed.). Aula Verlag Wiebelsheim, pp. 157-162.

Murdy, E.O. and D.F. Hoese. 2008. *FAO family account and key to WA gobies*. World Wide Web electronic publication. http://gobiidae.com/FAO_WA_gobies.htm

Nelson, J.S. 2006. *Fishes of the World.* John Wiley and Sons, Inc. Hoboken, New Jersey. 4th edition.

Nieto, P. and L.J. Alberto. 1992. The presence of *Zebrus zebrus* (Risso, 1826) (Gobiidae) on the Atlantic coast of Spain. *Cybium* 16: 137-144.

Papakonstantinou, C. 1988. *Checklist of marine fishes of Greece*. Fauna Graeciae IV: 1-257.

Patzner, R.A. 1999a. Habitat utilization and depth distribution of small cryptobenthic fishes (Blenniidae, Gobiesocidae, Gobiidae, Tripterygiidae) in Ibiza (western Mediterranean Sea). *Environmental Biology of Fishes* 55: 207-214.

Patzner R.A., 1999b. Sea urchins as hiding-place for juvenile benthic teleosts (Gobiidae, Gobiesocidae) in the Mediterranean Sea. *Cybium* 23: 93-97.

Patzner, R.A. 2007. A recent finding of *Didogobius schlieweni* (Gobiidae) with some notes on its ecology and distribution. *Acta Ichthyologica e Piscatoria* 37: 63-65.

Patzner, R.A. 2008. *Mediterranean gobies, Gobiidae, order Perciformes*. World Wide Web. Electronic publication. http://www.users.sbg.ac.at/~patzner/Gobiidae.htm.

Pezold, F. 1993. Evidence for a monophyletic Gobiinae. *Copeia* 1993: 634-643.

Pölzer, W. 1994. *Beitrag zur Biologie der mediterranen Anemonengrundel* Gobius bucchichii *Steindachner, 1870 (Teleostei, Gobiidae) mit besonderer Berücksichtigung der Nahrungs-, Fortpflanzungs- und Verhaltensbiologie*. MSc. Thesis, University of Salzburg.

Puchner, T. 2002. *Eine fakultative Symbiose zwischen der Anemonengrundel* Gobius bucchichi (*Teleostei, Gobiidae*) *und der Wachsrose* Anemonia sulcata (*Cnidaria, Actiniidae*). MSc. Thesis, University of Vienna.

Quignard, J.P. and J.A. Tomasini. 2000. Mediterranean fish biodiversity. *Biologia del mare Mediterraneo* 7: 1-66.

Rakaj, N. 1995. *Iktiofauna e Shqipërisë*. Shtëpia Boutese "Libri Universitar", Tiranë.

Relini Orsi, L. and G. Fanciulli. 1977. First record of *Crystallogobius linearis* in the Ligurian sea and identification of its young stages as "bianchetti di fondo" (Osteichthyes). *Natura* 68: 111-122.

Santos, R.S., F.M. Porteiro and J.P. Barreiros. 1997. Marine fishes of the Azores. Annotated checklist and bibliography. *Arquipelago Supplement* 1: 1-244.

Spalding, M.D., H.E. Fox, G.R. Allen, N. Davidson, Z.A. Ferdaña, M. Finlayson, B.S. Halpern, M.A. Jorge, A. Lombana, S.A. Lourie, K.D. Martin, E. McManus, J. Molnar, C.A. Recchia and J. Robertson. 2007. Marine ecoregions of the world: a bioregionalization of coastal and shelf areas. *BioScience* 573: 573-583.

Stefanni, S. 1999. A new record of *Didogobius splechtnai* Ahnelt & Patzner, 1995 (Gobiidae) from the central Mediterranean Sea. *Cybium* 23: 105-107.

Stefanni, S. and C. Mazzoldi. 1999. The presence of Couch's goby in the Mediterranean Sea. *Journal of Fish Biology* 54: 1128-1131.

Thacker, C. 2011. Systematics of Gobiidae. In: *The Biology of Gobies*, R.A. Patzner, J.L. Van Tassell, M. Kovačić and B.G. Kapoor (eds.). Science Publisher Inc., Enfield, pp. 129-136.

Van Tassell, J.L. 2001. *Chromogobius* (Teleostei: Gobiidae): a new species from the eastern Atlantic. *Copeia* 2001: 1073-1080.

Van Tassel, J.L., P.J. Miller and A. Brito. 1988. A revision of *Vanneaugobius* (Teleostei: Gobiidae), with description of a new species. *Journal of Natural History* 22: 545-567.

Vasil'eva, E.D. 2003. An annotated catalogue of fishes and fish-like organisms living in seas of Russia and adjacent countries: Part 4. Gobioidei. *Journal of Ichthyology* 43: 41-56.

Vasil'eva, E.D. and S.V. Bogorodsky. 2004. Two new species of gobies (Gobiidae) in the ichthyofauna of the Black Sea. *Journal of Ichthyology* 44: 555-562.

Velte, F. 1994. Ein Anemonenfisch im Mittelmeer—die Streifengrundel. *Die Aquarien- und Terrarienzeitschrift* 47: 300-302.

Wirtz, P. 1998. Twelve invertebrate and eight fish species new to the marine fauna of Madeira, and a discussion of the zoogeography of the area. *Helgoländer Meeresuntersuchungen* 52: 197-207.

Wirtz, P. and R. Herrera. 1995. The lobster *Enoplometopus antillensis* (Decapoda: Enoplometopidae), and the goby *Gobius xanthocephalus* (Pisces: Gobiidae)— new records for the marine fauna of the Canary Islands. *Arquipelago-Boletim da Universidade dos Acores Ciencias Biologicas e Marinhas* 13A: 115-118.

Yanko-Hombach, V., A.S. Gilbert and P. Dolukhanov. 2007. Controversy over the great flood hypotheses in the Black Sea in light of geological, paleontological, and archaeological evidence. *Quaternary International* 167-168: 91-113.

Zander, C.D. 1982. Zur Morphologie und Biologie einiger seltener Grundeln des Mittelmeeres (Pisces, Gobioidei, Gobiidae). *Senckenberg maritima* 14: 1-8.

Županović, S. and A.A. El Buni. 1982. A contribution to demersal fish studies off the Libyan coast. *Bulletin of the Marine Research Centre in Tripoli, Libya* 3: 77-122.

CHAPTER

2.3

Diversity and Distribution of Marine, Euryhaline and Amphidromous Gobies from Western, Central and Southern Africa

Ulrich K. Schliewen

INTRODUCTION

Although quite a few checklists of marine, brackish and freshwater species have been published over the last 25 years (Brito *et al.*, 2001; Hoese, 1986; Miller, 1990; Harrison *et al.*, 2003a, b, c; 2007a, b; Pezold *et al.*, 2007; Wirtz *et al.*, 2007), knowledge of the gobioid fauna of the western and southern african coastline and of the western african islands is very incomplete. The major reason is that many species are represented in collections only by few to very few specimens, which are often in bad condition. In addition, collection techniques in the often turbid waters along the West African coast typically neglect or destroy small to very small fishes and some species, e.g. sicydiines, are notoriously difficult to collect. Nevertheless, a number of recent publications have clarified several taxonomic problems in western and southern African gobies, several new species have been described, and

Author's address: Bavarian State Colletion of Zoology, Department of Ichthyology, Muenchhausenstr. 21, 81247 Munich, Germany. E-mail: schliewen@zsm.mwn.de limited phylogenetic information on relationships has become available over the years (see detailed citations in the species accounts). As complementary information to Kovačić and Patzner (2011, Chapter 2.2 of this volume), a short account on the diversity and distributional data of all currently valid marine and brackish water gobies is prepared for those species which are known from marine ecoregions and provinces (based on Spalding *et al.*, 2001) from Senegal south of the Cape Verde peninsula (including Goreé island) to the Natal ecoregion of the Agulhas Province in Southern Africa (only up to Coffee Bay, South Africa), including the Gulf of Guinea islands and Cape Verde islands (see Fig. 2.3.1) . The cut off point at Coffee Bay is chosen as a somewhat arbitrary point, as coastal fishes do not reflect a clear biogeographic break between Agulhas Natal and Benguela ecoregions according to Spalding *et al.* (2001), but show a gradual species turnover east of Cape Point (Turpie *et al.*, 2001). During the preparation of this chapter it became obvious, that north of Coffee Bay tropical Indo-Pacific goby species are increasingly common, which are dealt with in another chapter (Taylor, this volume). This small paper is not intended to replace previous taxonomic checklists. It rather shall serve as a preliminary update and starting point for the understanding of the goby diversity of one of largest and relatively unexplored coastlines worldwide. Currently, there are 87 gobioid species from 31 genera recorded from this area, out of which 51 species and 9 genera are endemic. Goby classification follows Thacker (2009; for a critical discussion of this relatively new classification see Rüber, this volume) and terminology and borders of biogeographic provinces and ecoregions follows Spalding *et al.* (2007) if not otherwise mentioned.

GOBIID GENERA WITH ENDEMIC SPECIES IN THE REGION

Family Eleotridae

Genus *Dormitator* Gill, 1862

Four *Dormitator* species are known from the western Atlantic and Eastern Pacific. The fifth species, *D. lebretonis* (Steindachner, 1870) is endemic to the West African coastline from Senegal to the Cunene River mouth in Namibia. A brackish water species, which sometimes enters freshwater (Harrison *et al.*, 2003c; Pezold *et al.*, 2007).

Genus *Eleotris* Bloch and Schneider, 1801

More than 30 species are distributed pantropically and pansubtropically in brackish and coastal freshwaters. Six species are known from the area, of

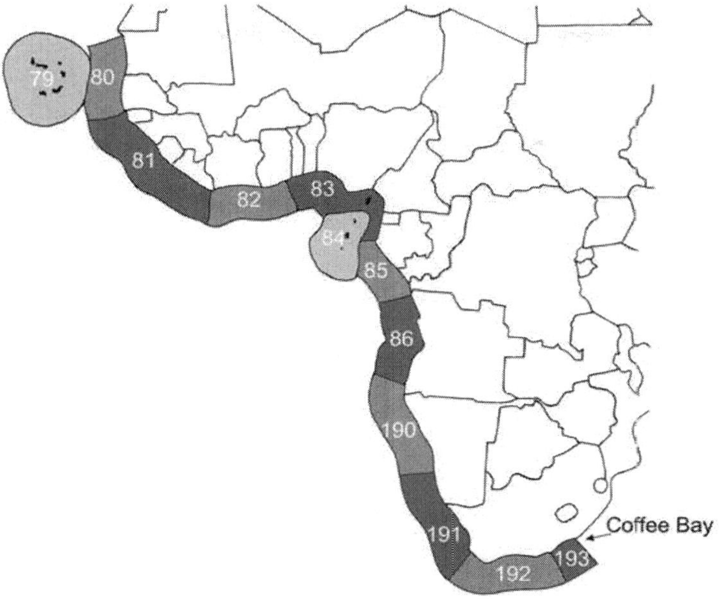

Fig. 2.3.1 Western, central and southern African marine biogeographic ecoregions according to Spalding *et al.* (2007): West African Transition Province: Cape Verde (79) and Sahelian Upwelling (80); Gulf of Guinea Province: West (81), Upwelling (82), Central (83), Islands (84), South (85) and Angolan (86); Benguela Province: Namib (190) and Namaqua (191); Temperate Southern Africa Province: Agulhas Bank (192) and Natal (193, only up to Coffee Bay, South Africa). Drawing by R. Kühbandner.

which three are recorded from the West African coastline between Senegal and the northern Namibia: *E. daganensis* Steindachner, 1870, *E. senegalensis* Steindachner, 1870, *E. vittata* Duméril, 1861 (see Table 2.3.1 for species specific distribution records). One species (*E. annobonensis* Blanc, Cadenat and Stauch, 1968) is endemic to Gulf of Guinea volcanic islands of Bioko, Sao Tome, Principe and Annobon (Miller, 1998; Harrison *et al.*, 2003c; Pezold and Cage, 2002, 2007; Wirtz *et al.*, 2007). The remaining two species (*E. fusca* (Forster, 1801) and *E. melanosoma* Bleeker, 1852) are widespread Indo-Pacific species, which also occur along the Indian Ocean coast in South Africa (Skelton, 2001). Pezold and Cage (2002) compared western hemisphere and West African *Eleotris*. They concluded that each of the four West African species appears close to the New World species: Carribean *E. amblyopsis* (Cope, 1871) are morphologically indistinguishable from *E. daganensis* and western Atlantic *E. perniger* (Cope, 1871) are extremely similar to *E. annobonensis*. They did not take taxonomic action, because of the "scant, subtle and overlapping morphological variation of diagnostic characters observed among the western Atlantic taxa".

Family Butidae

Genus *Bostrychus* Lacépède, 1802

A predominantly Indo-Pacific fresh- and brackishwater genus containing six species, of which *Bostrychus africanus* (Steindachner, 1879) is the only african species. It is endemic to brackish waters along Atlantic coastline from Senegal to Angola and the Gulf of Guinea islands, but may enter freshwater (Harrison *et al.*, 2003c, Pezold *et al.*, 2007; Wirtz *et al.*, 2007; Zarske, 1993).

Family Gobionellidae

Genus *Awaous* Valenciennes, 1837

Two species are known from the eastern Atlantic, the widespread *A. lateristriga* (Duméril, 1861) and *A. bustamantei* (Greef, 1882), which is an endemic to the Gulf of Guinea volcanic islands (Harrison *et al.*, 2003a, 2007c; Watson, 1992; for *A. bustamantei* see Wirtz *et al.*, 2007). The third species of the area is *A. aeneofuscus* (Peters, 1852) from the Indian Ocean of South Africa (Hoese, 1986a; Skelton, 2001). Adult *Awaous* are amphidromous fishes living on sandy substrates in predominantly clear coastal rivers, larger streams and sometimes lakes, i.e. crater Lake Barombi Mbo in Cameroon (author´s pers. obs.). About 11-14 species are recognized from pantropical areas and East African species are in the need of revision (Harrison, pers. comm.), especially with regard to the status of "*Gobius*" *scorteccii* Poll, 1961 and madagascan *Awaous*, which are closely related *A. aeneofuscus* (Poll, 1961).

Genus *Ctenogobius* Gill, 1858

The only african species of this otherwise Western Atlantic Eastern Pacific genus is *C. lepturus* (Pfaff, 1993), which is an inshore marine and brackish water species distributed from Senegal to Congo (Harrison *et al.*, 2003a, 2007a). Based on a parsimony analysis of morphological characters, *C. lepturus* is most closely related to a clade comprised of three Western

Atlantic and two Eastern Pacific *Ctenogobius*, i.e. *C. manglicola* (Jordan and Starks, 1895), *C. phenacus* (Pezold and Lasala, 1987), *C. saepepallens* (Gilbert and Randall, 1968), *C. sagittula* (Günther, 1862) and *C. smaragdus* (Valenciennes, 1837) (Pezold, 2004a).

Genus *Gobioides* Lacépède, 1800

Gobioides are very slender and microphthalmic. They are distributed with altogether five species in the genus both in the tropical Eastern Pacific as well as on both sides of the Atlantic (Murdy, 1998). They are fineparticle feeders and two species occur in muddy inshore marine habitats and estuarine habitats along the West African coast, *G. africanus* (Giltay, 1935) from Ghana to Congo and *G. sagitta* (Günther, 1862) from Senegal to Congo (Harrison *et al.*, 2003a, 2007a; Murdy, 1998).

Genus *Gobionellus* Girard, 1858

Gobionellus was revised recently (Pezold, 2004a, b) and contains six species. The single african species, *G. occidentalis* (Boulenger, 1909) shares with two atlantic species *G. oceanicus* (Pallas, 1770) and the eastern Pacific *G. liolepis* (Meek and Hildebrand, 1828) coloration similarities and an elongate bodyform (Pezold, 2004b). In a morphological parsimony analysis which did not include those coloration characters, these species form a monophyletic group together with *G. daguae*, *G. broussoneti* and *G. grahamae*. *G. occidentalis* is distributed over inshore muddy bottoms in purely marine and brackish water environments, where it may retreat in burrows (Harrison *et al.*, 2003a; 2007a).

Genus *Parasicydium* Risch 1980

The western african endemic *Parasicydium bandama* Risch, 1980 is a small sicydiine goby known only from few scattered coastal stream- and river localities along West African coastline between the Bandama River in Ivory Coast and the Koilou River in the Republic of Congo (Harrison, 2003a, 2007a). There is currently an unsettled debate on the validity of the genus *Parasicidyium* with respect to *Lentipes* (see Harrison, 1993; Harrison *et al.*, 2007a; Watson, 2002). We follow here Harrison (1993) and Harrison *et al.*, 2007a), who tentatively places three species in a valid genus *Parasicydium*, of which two are of Indo-Pacific origin.

Genus *Periophthalmus* Bloch and Schneider, 1801

The western african coast has one endemic mudskipper, *Periophthalmus barbarus* (Linnaeus, 1766), which is distributed from Senegal to Angola as well as on the Gulf of Guinea islands (Harrison *et al.*, 2003a, 2007a; Wirtz *et al.*, 2007). Only two species are apparently recorded from the east coast of South Africa as deduced from the key in Jaafar and Larson (2008) and SAIAB database entries (Froese and Pauly, 2009 (SAIAB entries only)): *P. kalolo* Lesson, 1830 and *P. argentilineatus* Valencinnes, 1837 (Jaafar and Larson, 2008). *P. kalolo* is distributed from the western Indian Ocean to Oceania and *P. argentilineatus* from east Africa to western Pacific (Jaafar and Larson, 2008). SAIAB data base entries for *Periophthalmus* from the region include also *P. sobrinus* Eggert, 1935, a synonym of *P. argentilineatus* according to Murdy (1989), and *P. koelreuteri* (Pallas, 1770), a synonym of *P. barbarus* (according to Murdy, 1989).

Genus *Redigobius* Herre, 1927

The tropical Indo-Pacific genus with about 11 species is represented in the southwestern Indian Ocean in South Africa with two species (Larson, 2001; Skelton, 2001). One is the wide spread Indo-Pacific *R. bikolanus* (Herre, 1927). The other one, *R. dewaali* (Weber, 1897), is endemic to the southwestern Indian Ocean along african coast from southern Mozambique to the Agulhas Province with isolated records in the cooler Agulhas Bank province (Skelton, 2001). They are brackish water species, which regularly enter freshwater.

Genus *Sicydium* Valenciennes 1837

With approximately 16 species in the genus *Sicydium* is the major western Atlantic and eastern Pacific sicydiine genus with three eastern Atlantic species: *S. crenilabrum* Harrison 1993, *S. brevifile* Ogilivie-Grant, 1884 and *S. bustamantei* Greef, 1884 (Pezold *et al.*, 2006). The two latter species seem to be amphidromous endemics living as adults in clearwater streams of the volcanic islands in the Gulf of Guinea and of Mount Cameroon streams in Western Cameroon (*S. brevifile*) and from rivers Lobe and Kribi, Southern Cameroon (*S. bustamantei*) (Harrison *et al.*, 2007a). One species, *S. bustamantei,* is known from a stream located above several hundred meter high cascade of Rio Ilady, Bioko Island at an altitude of over 100m (pers. obs.). In contrast, *S. crenilabrum* occurs in coastal streams between

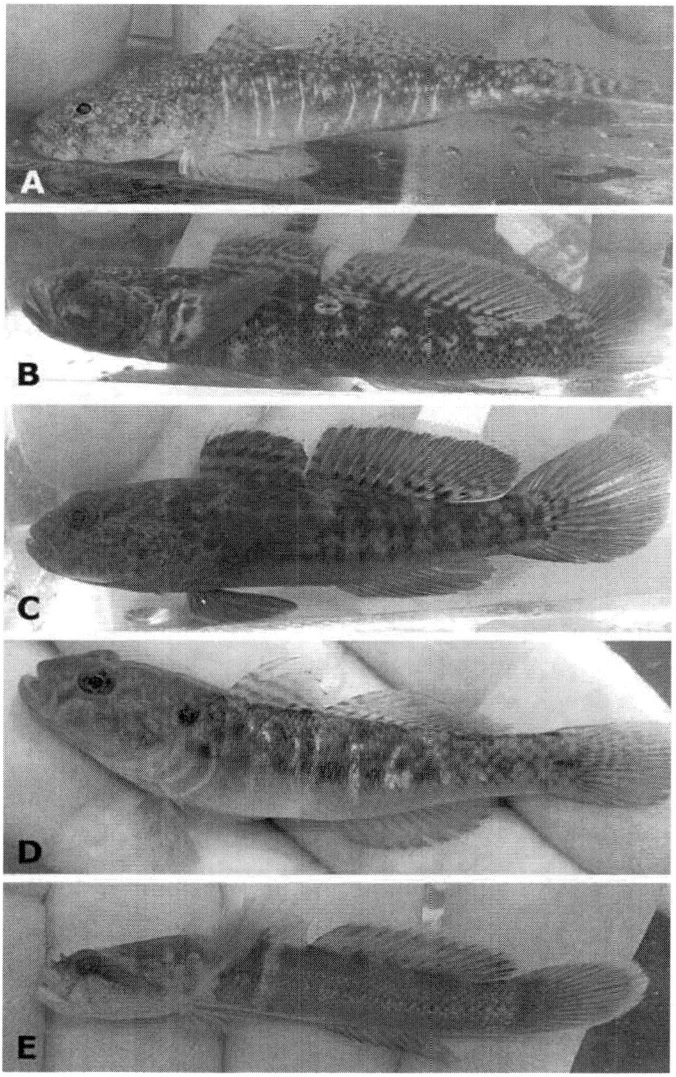

Fig. 2.3.2 A. *Psammogobius knysnaensis* (Nenga River mouth at Coffee Bay), **B.** *Caffrogobius caffer* (Tide Pools at Hole in Wall, South Africa), **C.** *Caffrogobius gilchristi* (Bomvu River mouth close to Coffee Bay, South Africa), **D.** *Caffrogobius nudiceps* (Kowie River mouth, South Africa), **E.** *Hetereleotris zonata* (Tide Pools at Hole in Wall, South Africa). Photos U. Schliewen.

Color image of this figure appears in the color plate section at the end of the book.

Ivory Coast and Congo (Harrison *et al.*, 2003a, 2007a). Relationships among Latin American and African sicydiine gobies are not yet analysed in detail.

Familie Gobiidae

Genus *Bathygobius* Bleeker, 1878

The circumtropical genus *Bathygobius* with approximately 28 species is one of the few genera in the region which are represented both on the Atlantic and on the Indian Ocean side (Miller and Smith, 1989; Miller and Stefanni, 2001). However, only two of three western African species *B. burtoni* (O'Shaughnessy, 1875) and *B. casamancus* (Rochebrune, 1880) are endemic to the region (Miller and Smith, 1989). The third species is *B. soporator* (Valenciennes, 1837). The two species, that inhabit brackish as well as marine waters, i.e. *B. soporator* and *B. casamancus,* occur from the Cape Verde islands over Senegal to Angola including the Gulf of Guinea islands (Brito and Miller, 2001; Miller and Smith, 1989; Harrison *et al.*, 2003a, 2007a). The strictly marine species *B. burtoni* is restricted to the Gulf of Guinea islands and the coastline between Ghana and Cameroon (Miller and Smith, 1989; Miller, 1990; Wirtz *et al.*, 2007). Miller and Smith (1989) compared the Western African species with 11 other species of *Bathygobius* using 46 morphological characters and concluded that B. *soporator* is only distantly related to the endemic species. Both, *B. casamancus* and *B. burtoni* were consistently placed in a monophyletic clade with the Indo-Pacific species *B. cocosensis* and *B. petrophilus* and the Western Atlantic *B. curacao.* All other seven species in the region are widespread Indo-Pacific marine species with their southwestern limits of distribution occurring in Natal subprovince (Hoese, 1986; Froese and Pauly, 2009 (SAIAB entries only); Winterbottom, 1976).

Genus *Caffrogobius* Smitt, 1900

Caffrogobius is endemic to southern Africa (except for *C. dubius* (Smith, 1959) – see below) and was recently revised (Goren, 1996). Three of the six species are predominantly marine inshore species and are restricted to the Agulhas Province: *C. caffer* (Günther, 1874) (Fig. 2.3.2b), *C. gilchristi* (Boulenger, 1898) (Fig. 2.3.2c), *C. natalensis* (Günther, 1874). The two species from deeper waters, *C. saldanha* (Barnard, 1927) and *C. agulhensis* (Barnard, 1927), as well as the estuarine species *C. nudiceps* (Valenciennes, 1837) (Fig. 2.3.2d) extend their distribution into the eastern Atlantic Benguela Province (Goren, 1996; Froese and Pauly, 2009 (SAIAB entries only)).

Caffrogobius dubius is known from the small holotype (27.7 mm SL) in bad condition, which was collected from the Seychelles (Smith, 1959). Originally described as *Coryphopterus* and not being unambigously identifiable by counts as *Caffrogobius*, this species is only referable to *Caffrogobius* based on the "general pattern of the genus *Caffrogobius*" despite the facts that "the type prevents an accurate count of the cephalic sensory papillae" (Goren, 1996). Therefore it is suggested here that only fresh material will allow solving the problem of the generic status of this species. Consequently, the genus *Caffrogobius* may be provisionally considered to be endemic to southern Africa, and nearly endemic to region considered here because of sporadic records of *Caffrogobius* from north of Coffee Bay to Delagoa Bay (southern Mozambique). Several authors consider *Caffrogobius* to be closely related to eastern Atlantic gobies and discussed potential affinities to *Gobius* and *Zebrus* (Miller, 1978a), to *Gammogobius*, *Zebrus* but also to *Coryogalops* (Van Tassell *et al.*, 1988), whereas Birdsong *et al.* (1988) and Gill and Bradley (1992) place the genus in the *Bathygobius* group (Birdsong *et al.*, 1988) including *Bathygobius*, *Caffrogobius*, *Glossogobius*, *Gorogobius*, *Hetereleotris*, *Lesueurigobius*, *Coryogalops* and *Nematogobius*. Within the *Bathygobius*-group, Goren (1996) suggests a closer relationship with the genera with transverse papillae patterns, i.e. with *Gorogobius*, *Hetereleotris*, *Coryogalops* and *Nematogobius*. Obviously, the biogeographic enigma of the origin of the genus *Caffrogobius* needs more detailed phylogenetic inferences with a full taxon sampling of all potential relatives.

Genus *Corcyrogobius* Miller, 1972

The small genus *Corcyrogobius* harbours two widely disjunctly distributed species, *C. lichtensteini* (Kolombatović, 1891) from the Mediterranean and C. *lubbocki* Miller, 1988 from the tropical eastern Atlantic. The latter is known only from three locations over rocky bottom or caves off Ghana, Annobon and from Sao Tome islands (Miller, 1988a; Wirtz *et al.*, 2007). The distribution pattern might be a sampling artefact as the preferred habitat of this species is difficult to sample without SCUBA gear (Herler *et al.*, 1999). Miller (1978b) and van Tassell *et al.* (1988) discussed a close relationship of the genera *Vanneaugobius*, *Wheelerigobius*, *Corcyrogobius* and *Odondebuenia*, mainly based on similarities in the head lateral line system.

Genus *Coryogalops* Smith, 1958

The nine currently recognized species of *Coryogalops* are restricted to the central and western Indian Ocean and the Red Sea (Goren, 1991;

Randall, 1994). Published literature suggests that two of these species, *C. bretti* Goren, 1991 and *C. william* (Smith, 1948) are almost endemic to Agulhas Province in South Africa (several records for *C. william* are from the Delagoa subregion of the Western Indo-Pacific ecoregion (Froese and Pauly, 2009 (SAIAB entries only). However, there is a record from Mauritius for *C. bretti*, which needs to be critically checked (Froese and Pauly, 2009 (SAIAB entries only), because the species is otherwise known only from a single specimen of the Kwelera estuary in South Africa. *Coryogalops* appears closely related to *Hetereleotris* (Goren, 1991; Hoese, 1986b). They are small marine gobies living predominantly in calm water of estuaries and lagoons.

Genus *Didogobius* Miller, 1966

The genus comprises six species of which three are known from the tropical eastern Atlantic, i.e. *D. kochi* van Tassell, 1988 and *D. wirtzi* Schliewen and Kovačić, 2008 from Cabo Verde (otherwise known only from the Canaries) and *D. amicuscaridis* Schliewen and Kovačić, 2008 from Sao Tome islands. *Didogobius* belongs to a species-complex together with *Chromogobius*, *Gammogobius* and an undescribed species from the Canary islands (Brito and Miller, 2001; Schliewen and Kovačić, 2008). The generic status of all members of that species complex remains to be critically tested as the type species of the genus, the mediterranean *D. bentuvii* Miller, 1966, is the most distinctive of the genus. Two sister taxa *D. wirtzi* and *D. amicuscaridis* are noteworthy for living in burrows of an axiid shrimp (Wirtz *et al.*, 2007).

Genus *Ebomegobius* Herre, 1946

With *Ebomegobius goodi* Herre, 1946 this enigmatic genus is monotypic and known only from a single specimen, that was collected from the lower Ebome stream, a tiny coastal stream close to Kribi, southern Cameroon. It shares substantial similarities with the Indo-Pacific genus *Silhouettea* Smith, 1959, however bad preservation status of the holotype prevents further inferences about relationships (Miller, 1988b).

Genus *Glossogobius* Gill, 1859

This species-rich Indo-Pacific genus with more than 20 species is urgently in the need of revision. It contains several purely freshwater species with localized distributions as well as widespread brackish water species. Depending on the generic status of the tropical "*Gobius*" *biocellatus*

Valenciennes, 1987 as either belonging to *Glossogobius* (as e.g. in Akihito *et al.*, 1984) or to *Psammogobius* (as e.g. in Larson *et al.*, 2008), there are two or three species in South Africa. In addition to the aforementioned species, there occurs the tropical and widespread *G. giuris* (Hamilton, 1822) with its southeastern limits in northern South Africa (Hoese, 1986; Skelton, 2001; Froese and Pauly, 2009 (SAIAB entries only)) and *G. callidus* (Smith, 1937). The latter is almost endemic to the Agulhas Province with its northernmost record in southern Mozambique (Hoese, 1986; Skelton, 2001; Froese and Pauly, 2009 (SAIAB entries only)). In different estuaries and river systems, this species exhibits extremely elevated levels of genetic differentiation (Engelbrecht and Mulder, 1999), indicating the presence of a *G. callidus*—species complex rather than a single species.

Genus *Gobius* Linnaeus, 1758

Gobius is a predominantly north-eastern Atlantic and Mediterranean genus with close but unresolved relationships to several other genera from the eastern Atlantic and southern Africa (see for example McKay and Miller, 1991). Compared to the species richness in the north-eastern Atlantic and Mediterranean, only six species are known from the tropical eastern Atlantic. Two reach their southernmost limit slightly south of the border between the temperate and tropical Atlantic (Cape Verde Peninsula, Senegal): *G. paganellus* Linnaeus, 1758 and *G. cruentatus* Gmelin, 1879; two are local endemics of the Cabo Verde islands off Senegal: *Gobius ateriformis* Brito and Miller, 2001 and *G. tetrophthalmus* Brito and Miller, 2001; and two poorly known species have colonized the tropical coast of continental Africa from Mauritania to Angola. *G. senegambiensis* Metzelaar, 1919 and *G. rubropunctatus* Delais, 1951) (Brito and Miller, 2001; Miller, 1990; Poll, 1959; Wirtz *et al.*, 2007, pers. obs.). Relationships of the four african species remain to be explored, but similarities with north-eastern Atlantic and Mediterranean species are evident for the Cape Verde islands species, i.e. of *G. ateriformis* Brito and Miller, 2001 with *G. ater* Bellotti, 1888, and of *G. tetrophthalmus* Brito and Miller, 2001 with *G. bucchichi* Steindachner, 1870 (Brito and Miller, 2001).

Genus *Gorogobius* Miller, 1978

Gorogobius is a tropical western african coast endemic with two marine species. *G. nigricinctus* (Delais, 1951) is comparatively widespread occuring from Senegal to Ghana and in the Gulf of Guinea islands (Delais, 1951; Miller, 1978b; Wirtz *et al.*, 2007). *Gorogobius stevcici* Kovačić and Schliewen,

2008 is known only from the type locality on the Sao Tome islands (Kovačić and Schliewen, 2008). Both species appear to live preferentially in rocky areas, sometimes in caves (Wirtz *et al.*, 2007). Miller (1978b) and van Tassell *et al.* (1988) discuss a close relationship of *Gorogobius* with the genera *Vanneaugobius*, *Wheelerigobius*, *Corcyrogobius* and *Odondebuenia* mainly based on similarities in the head lateral line system.

Genus *Hetereleotris* Bleeker, 1874

The genus *Hetereleotris* is treated here as having an endemic representative, because Hoese (1986b) points out, that *Hetereleotris zonata* (Fowler, 1934) (Fig. 2.3.2e) has a largely disjunct distribution and that specimens of the two disjunct areas, i.e. India/Pakistan and South Africa differ in several characters. He points out that additional material is necessary to answer the question, whether the disjunct populations represent discrete species or not. *Hetereleotris* comprises 17 species (if *Pascua* Randall, 2005 is considered to be a synonym of *Hetereleotris*) (Hoese and Larson, 2005). Since Hoese´s treatment new material apparently become available from other regions including the Comoros (Froese and Pauly, 2009 (SAIAB entries only)), but this question has not been treated yet.

Genus *Lesueurigobius* Whitley, 1950

Deepwater gobies of this genus with four species are endemic to muddy bottoms of the eastern Atlantic and Mediterranean from southern Norway down to the Namaqua ecoregion of the Benguela province (Miller, 1990). Two species are known from the area, i.e. *L. sanzi* (de Buen, 1918) and *L. koumansi* (Norman, 1935). The distribution of the first species is noteworthy, as it is recorded from an area from off Namibia, but otherwise known only in the north-eastern Atlantic and western Mediterranean (Miller, 1986, 1990; Olivar, 1989). Inter- and intrarelationships of the genus remain to be studied in detail.

Genus *Mauligobius* Miller, 1981

Mauligobius is represented by only two species, the type species *M. maderensis* (Valenciennes, 1837) and *M. nigri* (Günther, 1861), which are apparently closely related to the members of the genus *Gobius, Neogobius, Mesogobius, Proterorhinus* and *Zosterisessor* (Miller, 1984). Wirtz *et al.* (2008) synonymize it with *Gobius*, an idea which already had been discarded by Harrison *et al.* (2007a). Waiting for a detailed phylogenetic analysis of *Gobius* and related genera, *Mauligobius* is treated here as valid following Harrison *et al.* (2007a). The single species in the area, *M. nigri* is known with certainty only from the Cabo Verde islands (Brito and Miller, 2001)[1]. Records from Nigeria Cameroon and Bioko are doubtful (Miller 1984; Brito and Miller, 2001). Wirtz *et al.* (2008) consider *G. paganellus* Linnaeus, 1758 and *M. maderensis* to be sister taxa, but do not comment on *M. nigri*.

Genus *Microdesmus* Günther, 1864

Microdesmus are elongate gobiids ("wormfishes") from muddy inshore littoral or estuarine habitats of amphiatlantic and tropical eastern Pacific areas (Thacker, 2000). African species are rarely collected and are known from altogether ten specimens (Dawson, 1979, 1990). The three species in the area are *M. longipinnis* (Weymouth, 1910), a western Atlantic species that is known from a single specimen in Senegal, *M. africanus* Dawson, 1979, known from only two specimens of the Congo estuary and *M. aethiopicus* (Chabanaud, 1927) known from seven specimens from Cameroon, Bioko and the Congo estuary (Dawson, 1979, 1990; Harrison *et al.*, 2007b). Phylogenetic intrageneric relationships have been studied by Thacker (2000) using morphological characters. She identified *M. africanus* as the sister taxon to the eastern central Pacific *M. suttkusi* Gilbert, 1966 and western central Atlantic *M. carri* Gilbert, 1966, *M. longipinnis* as the sister taxon to the eastern Central Pacific *M. affinis* Meek and Hildebrand, 1928, and *M. aethiopicus* as sister to the eastern Central Pacific *M. retropinnis* Jordan and Gilbert, 1882 and the Carribean and Brazilian *M. bahianus* Dawson, 1973.

Genus *Nematogobius* Boulenger, 1910

One of the few endemic goby genera of the western african coast, *Nematogobius* encompasses two species, the inshore marine *N. brachynemus* Pfaff, 1933 and the euryhaline *N. maindroni* Boulenger, 1910. Both species are widespread along the western african coast between Senegal and Congo (*N. brachynemus*) or southern Angola (*N. maindroni*) (Harrison, 1990; Harrison *et al.*, 2003a, 2007a; Miller, 1990; Wirtz *et al.*, 2007). The character of paired mental barbels is shared with *Sufflogobius*, which formerly was classified under *Nematogobius* (Miller, 1990).

[1]ZSM material is incorrectly cited by Brito and Miller (2001) as being from "Ilha de Sao Tiago". The correct collection locality of lot ZSM 38667 is "Fogo island near São Filipe, Cabo Verde".

Genus *Porogobius* Bleeker, 1874

The monotypic genus is represented by *Porogobius schlegelii* (Günther, 1861), is found in brackish and rarely coastal freshwaters of the western African coast between Senegal and Congo (Harrison *et al.*, 2003a, 2007a). Closer relationships remain unstudied.

Genus *Psammogobius* Smith, 1935

The type species of this is genus is the common southern African estuarine sand-dweller *P. knysnaensis* Smith, 1935 (Fig. 2.3.2a) which occurs in Namaqualand, Agulhas bank and Natal ecoregions (Hoese, 1986; Froese and Pauly, 2009 (SAIAB entries only)). Depending on the generic status of the tropical "*Gobius*" *biocellatus* Valenciennes, 1987 sometimes a second species is classified by some authors as *Psammogobius* (see *Glossogobius* account in this chapter), which is recorded only sporadically at the north-easternmost limit of the region treated here (Hoese, 1986; Froese and Pauly, 2009 (SAIAB entries only).

Genus *Sufflogobius* Smith, 1956

The pelagic goby *S. bibarbatus* (von Bonde, 1923) is the only member of *Sufflogobius* and almost endemic to the cold upwelling waters of the Namaqua ecoregion in southern Africa (Hoese, 1986; Miller, 1990; Froese and Pauly, 2009 (SAIAB entries only)). Sporadic records of this species are listed in the Froese and Pauly, 2009 (SAIAB entries only) from the Agulhas bank. Although relationships are unstudied, the two name-bearing chin barbels are shared with the tropical eastern Atlantic genus *Nematogobius* (Harrison, 1990), which might indicate a close relationship. It is the only pelagic member of the area (Melo and Clus, 2005).

Genus *Thorogobius* Miller, 1969

Four species of *Thorogobius* species are known, of which *T. rofeni* Miller, 1988 and *T. angolensis* (Norman, 1935) are represented in the area (Miller, 1988a, 1990; Sauberer and Ahnelt, 2008; pers. obs.). *Thorogobius rofeni* is known only from very few specimens from the type locality (Gulf of Guinea) and southern Angola, whereas *T. angolensis* (Norman, 1935) appears to be more widespread from Senegal to Angola (based on records of grey literature available through Fishbase (SIAP, SCRP, EC & SAUP, 2002). However, gobies are often misidentified and it is not unlikely that both species are more widespread than published literature suggests.

Relationships to the two other species of the north-eastern Atlantic and Mediterranean, *T. ephippiatus* and *T. macrolepis*, art not yet studied in detail.

Genus *Wheelerigobius* Miller, 1981

Wheelerigobius is an endemic western tropical atlantic genus with two strictly marine species: *W. maltzani* (Steindachner, 1882), occurring from Senegal to Cameroon and from this Gulf of Guinea islands, and *W. wirtzi* Miller, 1988 known only from the type locality in Cameroon and from Bioko and Sao Tome islands (Afonso *et al.*, 1999; Miller, 1981, 1988a, 1990; Wirtz *et al.*, 2007; pers. obs.). Miller (1978b, 1981) and van Tassell *et al.* (1988) discuss a close relationship of *Wheelerigobius* with *Gorogobius, Vanneaugobius, Corcyrogobius, Odondebuenia,* but also with *Millerigobius* mainly based on similarities in the head lateral line system.

Genus *Yongeichthys* Whitley, 1932

One of the four species in the genus, *Y . thomasi* (Boulenger, 1916) is endemic to brackish waters along the West African coastline between Senegal and Congo (Harrison *et al.*, 2003a, 2007a). The other three species are of Indo-Pacific origin.

PATTERNS OF DISTRIBUTION AND ENDEMICITY

Altogether 87 gobioid species belonging to 31 genera are known from the area (see Table 2.3.1), the majority of species (n = 51) and genera (n = 9) being strictly endemic to the area. It makes no sense to analyze here incomplete presence/absence data of Table 2.3.1 without scant phylogenetic information in order to reconstruct biogeographic areas in the area based on gobies. Therefore, a subjective approach is applied to summarize patterns of common species occurrences as a first step: Non-endemics and endemics are tentatively grouped according to presence/absence of data in biogeographical provinces (Spalding *et al.*, 2007) as deduced from Table 2.3.1 (based on information provided with citations in the first part of this chapter). A full molecular phylogenetic hypothesis for north-eastern Atlantic, Mediterranean, Ponto-Caspian gobies and those from the tropical eastern Atlantic and Southern Africa is currently being prepared in the framework of world gobies by a team of collaborators including the author. This will serve as a novel basis to discuss character states of squamation, osteology, lateral line system and other characters described as relevant for gobioid phylogeny.

Table 2.3.1 Presence of marine, euryhaline and amphidromous gobioid species in the Marine Biogeographic Provinces and Ecoregions (Spalding *et al.*, 2007; compare with Fig. 2.3.1). Provinces/Ecoregions are in the West African Transition Province (16): Cape Verde (79) and Sahelian Upwelling (80); in the Gulf of Guinea Province (17): West (81), Upwelling (82), Central (83), Islands (84), South (85) and Angolan (86); in the Benguela Province: Namib (190) and Namaqua (191); in the Province (up to Coffee Bay, South Africa): Agulhas Bank (192) and Natal (193). Data are compiled on the basis of the literature and databases provided in this chapter, family assignation of species follows Thacker (2009).

	Marine Biogeographic Ecoregion No. (after Spalding *et al.*, 2007)											
	79	80	81	82	83	84	85	86	190	191	192	193
Species												
Eleotridae												
Dormitator lebretonis		+	+	+	+	+	+	+	+			
Eleotris annobonensis						+						
Eleotris daganensis		+	+	+	+	+	+	+	+			
Eleotris fusca												+
Eleotris melanosoma												+
Eleotris senegalensis		+	+	+	+	+	+					
Eleotris vittata			+	+	+	+	+					
Butidae												
Bostrychus africanus		+	+	+	+	+	+	+				
Butis koilomatodon												
Gobiidae												
Bathygobius albopunctatus											+	
Bathygobius burtoni				+	+	+						
Bathygobius casamancus	+	+	+	+	+	+	+	+				
Bathygobius cocosensis											+	
Bathygobius cotticeps												+
Bathygobius cyclopterus										+		
Bathygobius laddi										+		

Contd...

Contd...

	Marine Biogeographic Ecoregion No. (after Spalding *et al.*, 2007)											
	79	80	81	82	83	84	85	86	190	191	192	193
Bathygobius niger											+	
Bathygobius padangensis											+	+
Bathygobius soporator	+	+	+	+	+	+	+	+				
Caffrogobius agulhensis											+	+
Caffrogobius caffer											++	
Caffrogobius gilchristi											++	
Caffrogobius natalensis											++	
Caffrogobius nudiceps										+	++	
Caffrogobius saldanha										+	++	
Corcyrogobius lubbocki				+		+						
Coryogalops bretti											+	
Coryogalops william											+	+
Didogobius amicuscaridis						+						
Didogobius kochi	+											
Didogobius wirtzi	+											
Ebomegobius goodi					+							
Glossogobius biocellatus											+	
Glossogobius callidus											+	+
Glossogobius giuris												+
Gobius ateriformis	+											
Gobius cruentatus		+										
Gobius paganellus	?	+										
Gobius rubropunctatus		+		+		+		+				
Gobius senegambiensis		+				+		+				
Gobius tetrophthalmus	+											

Contd...

Contd...

	Marine Biogeographic Ecoregion No. (after Spalding *et al.*, 2007)											
	79	80	81	82	83	84	85	86	190	191	192	193
Gorogobius nigricinctus		+	+			+						
Gorogobius stevcici						+						
Heterelotris zonata											+	
Lesueurigobius koumansi			+	+	+		+	+	+			
Lesueurigobius sanzi		+							+			
Mauligobius nigri	+				?							
Microdesmus aethiopicus					+		+					
Microdesmus africanus							+					
Microdesmus longipinnis		+										
Nematogobius brachynemus		+	+	+	+	+	+					
Nematogobius maindroni		+	+	+	+	+	+	+				
Porogobius schlegelii	?	+	+	+	+	+	+					
Priolepis cinctus												
Psammogobius knysnaensis										+	+	
Sufflogobius bibarbatus										+	+	
Thorogobius angolensis		+	+	+	+	+	+	+				
Thorogobius rofeni					+			+		+		
Vanneaugobius canariensis	+											
Wheelerigobius maltzani		+		+	+	+						
Wheelerigobius wirtzi					+	+						
Yongeichthys thomasi		+	+	+	+	+	+					
Gobionellidae												
Awaous aeneofuscus												+
Awaous bustamantei						+						
Awaous lateristriga		+	+	+	+	+	+	+				

Contd...

Contd...

	Marine Biogeographic Ecoregion No. (after Spalding *et al.*, 2007)											
	79	80	81	82	83	84	85	86	190	191	192	193
Ctenogobius lepturus		+	+	+	+	+	+					
Gnatholepis anjerensis											+	
Gnatholepis c. cauerensis	+			+		+						
Gobioides africanus				+	+	+	+					
Gobioides sagitta		+	+	+	+	+	+					
Gobionellus occidentalis		+	+	+	+	+	+					
Mugilogobius mertoni												+
Oligolepis acutipennis												+
Oligolepis keiensis												+
Oxyurichthys microlepis												+
Oxyurichthys ophthalmonema												+
Pandaka silvana											+	
Parasicydium bandama				+	+		+					
Periophthalmus argentilineatus												+
Periophthalmus barbarus		+	+	+	+	+	+					
Periophthalmus kalolo												+
Redigobius bikolanus												+
Redigobius dewaali											+	+
Sicydium brevifile					+	+						
Sicydium bustamantei					?	+						
Sicydium crenilabrum				+	+		+					
Taenioides esquivel												+

Non-endemics

The 36 non-endemic species may be partitioned according to the distribution outside of the area into four different groups, which indicate existing or very recent migration between the focus area and marginal regions, or alternatively, that some non-endemic species are at the edge of their range:

- 26 tropical Western Indian Ocean or widespread Indo-Pacific species reach their southeastern limits in the Natal ecoregion of the Agulhas Province. If not otherwise mentioned the following records are based on Hoese, (1986a); Skelton, (2001) and Froese and Pauly, 2009 (SAIAB entries only): *Eleotris fusca, E. melanosoma; Bathygobius albopunctatus, B. cocosensis, B. cotticeps, B. cyclopterus, B. laddi, B. niger, B. padangensis, Glossogobius biocellatus, G. giuris, Heterelеotris zonata* (but see comments above), *Priolepis cinctus (pers. obs.), Awaous aeneofuscus, Gnatholepis anjerensis* (Randall and Greenfield, 2001), *G. cauerensis* (Brito and Miller, 2001; Miller, 1990; Randall and Greenfield, 2001, 2007; Rocha *et al.*, 2005; Wirtz *et al.*, 2007), *Mugilogobius mertoni, Oligolepis acutipennis, O. keiensis, Oxyurichthys microlepis, O. ophthalmonema, Pandaka silvana, Periophthalmus argentilineatus, Periophthalmus kalolo, Redigobius bikolanus, Taenioides esquivel.*
- Five species which are otherwise known from the north-eastern Atlantic, have been recorded in the Sahelian Upwelling Province and in the Benguela upwelling zone of the Namib ecoregion: *Didogobius kochi* (Cape Verde islands; otherwise known only from islands in the north-eastern Atlantic Brito and Miller, 2001), *Gobius paganellus* and *G. cruentatus* (Senegal: Miller, 1990), *Lesueurigobius sanzi* (off Namibia and from Sahelian Upwelling ecoregion (Miller, 1986, 1990; Olivar 1989) and *Vanneaugobius canariensis* (Cape Verde islands; otherwise known only from islands in the north-eastern Atlantic; Brito and Miller, 2001).
- Four species exhibit an amphiatlantic distribution or they are morphologically so similar to another species on the opposite side of the Atlantic that their separate taxonomic status needs confirmation: *Eleotris daganensis* and *E. annobonensis* (Pezold and Cage, 2002), *Microdesmus longipinnis* (Dawson, 1979, 1990; Harrison *et al.*, 2003b), *Bathygobius soporator* (Miller and Smith, 1989; Miller and Stefanni, 2001). The morphological similarity of these taxa indicates either a recent exchange between the eastern and western tropical

Atlantic, or, if not, a large degree of morphological stasis maintaining a close similarity between amphiatlantic species despite prolonged interruption of geneflow.

- One widespread Indo-Pacific brackish water species, *Butis koilomatodon,* has recently invaded brackish waters of the eastern Atlantic ocean—most likely by anthropogenic action. Records are from the Niger Delta (Nigeria), Ndian River Mangroves (Cameroon) and Guinea (Miller *et al.*, 1989; Harrison *et al.*, 2003c; Pezold *et al.*, 2007; Vreven *et al.*, 2007).

Endemics

The 51 endemic species in the area may be grouped as following:

- Eight species may be characterised as endemics to the Cape Region of southern Africa (Benguela and Agulhas Provinces), i.e. all members of the genus *Caffrogobius* except *for C. dubius), Sufflogobius bibarbatus* and *Psammogobius knysnaensis.* Interestingly, not only all species are endemic, but all genera, too (if *Psammogobius* is accepted as not including *"Gobius" biocellatus*). Among the widespread taxa, only *C. saldanha* and *C. agulhensis,* a closely related species pair within widespread *Caffrogobius,* are predominantly offshore marine, while all other species are inshore and estuarine species (Goren, 1996). *Sufflogobius bibarbatus* is largely restricted to the coldwater upwelling area of the Benguela province, i.e. the Namaqua ecoregion.
- Four species are nearly endemic to inshore waters of the southwestern Indian Ocean in South Africa and southern Mozambique (Agulhas Province and Delagoa ecoregion of Western Indian Ocean Province): *Coryogalops bretti* and *C. william, Glossogobius callidus* and *Redigobius dewaali.* In addition, if the southern *Hetereleotris zonata* finally turns out to be distinctive, this taxon might be added to this group. As distributional data on the small fishes of the Western and Central Indian Ocean are quite incomplete, it is not unlikely, that these four taxa are more widespread, and rather endemic to the greater area of the East African Coast as it is true for several Indo-Pacific species with a similarily restricted distribution, i.e. *Awaous aeneofuscus, Oligolepis acutipennis, O. keiensis, Pandaka silvana* and *Taenioides esquivel.*

Along the western African coastline a distinction appears possible according to geographic areas and ecology:

- Four marine species are endemics to the Cape Verde islands: *Didogobius wirtzi*, *Gobius ateriformis*, *G. tetrophthalmus*, and (possibly) *Mauligobius nigri* (but see comment above).
- 14 marine species are distributed along the western African shoreline between Cape Verde (Sahelina Upwelling ecoregion) and in the Gulf of Guinea Province including the islands: *Bathygobius burtoni, B. casamancus, Corcyrogobius lubbocki, Didogobius amicuscaridis, Gobius rubropunctatus, G. senegambiensis, Gorogobius nigricinctus, G. stevcici, Lesueurigobius koumansi, Nematogobius brachynemus, Thorogobius angolensis, T. rofeni, Wheelerigobius maltzani* and *W. wirtzi*. Five of these species appear to be restricted to the Central Gulf of Guinea Province, i.e. the islands, Central and Guinea Upwelling ecoregions. Whether this is a sampling artefact or a real pattern remains to be investigated.
- 15 are euryhaline ("brackish water") fishes appear to be restricted to estuarine habitats and mangroves *Bostrychus africanus, Dormitator lebretonis, Ebomegobius goodi, Eleotris senegalensis, Eleotris vittata, Periopthalmus barbarus, Nematogobius maindroni, Porogobius schlegelii, Yongeichthys thomasi, Gobioides africanus, Gobioides sagitta, Microdesmus aethiopicus, Microdesmus africanus, Ctenogobius lepturus, Gobionellus occidentalis.*
- Six species are amphidromous freshwater species as deduced from the occurrence of adults in freshwater habitats and from knowledge about related species, which are known to have marine larvae (i.e. Keith and Lord, 2011, Chapter 2.5 of this volume): *Awaous bustamantei, Awaous lateristriga, Parasicydium bandama, Sicydium brevifile, Sicydium crenilabrum, Sicydium bustamantei*. Noteworthy for this group is, that three of the six species (*A. bustamantei, S. brevifile, S. bustamantei*) exhibit a restricted distribution in the Gulf of Guinea Province and the islands—a pattern, that is similar to the marine species of this area.

The full phylogenetic hypothesis including node-age estimates will hopefully enable testing of biogeographic hypotheses, that have been put forward for the evolution of the ichthyofauna in the tropical Eastern Atlantic and Southern Africa (summarized in Floeter *et al.*, 2007). Overall gobioid diversity patterns in the area as delineated above provide several alternative hypotheses, which partially have already been discussed by several authors, e.g. Miller (1998), Miller and Smith (1989) or Pezold and Cage (2002). These include—among others—the role of the final closure of the Tethys ("Terminal Tethyan event"; Steininger and Rögl, 1984), the evolution of intraoceanic "soft barriers" as migration barriers for exchange of tropical species in the Pliocene and Pleistocene between the Indian and Atlantic Ocean in the vicariance speciation (Peeters *et al.*, 2004; Rocha *et al.*, 2005), the ancient or ongoing intercontinental exchange of coastal faunas between the New World origin and the western African coast (summarized in Floeter *et al.*, 2007; Wirtz *et al.*, 2007), as well as the role of Miocene changes of sea surface temperatures after the closure of the Isthmus of Panama and for extinction, recolonization and diversification of the tropical eastern Atlantic (Le Louff and von Cosel, 1998).

Acknowledgements

Detailed comments by the three friendly reviewers Ian Harrison, Marcelo Kovačić and Peter Miller helped to improve the first version of this review substantially—thank you! Ruth Kühbandner kindly prepared Fig. 2.3.1.

References

Afonso, P., F.M. Porteiro, R.S. Santos, J.P. Barreiros, J. Worms and P. Wirtz. 1999. Coastal marine fishes of Sao Tome Island (Gulf of Guinea). *Arquipelago. Life and Marine Sciences* 17: 65-92.

Akihito, Prince, M. Hayashi and T. Yoshino. 2004. Suborder Gobioidei. In: *The Fishes of Japanese Archipelago*, H. Masuda, K. Amaoka, C. Araga, and T Uyeno (eds.), Tokai University Press, Tokyo, pp. 236-239.

Birdsong, R.S., E.O. Murdy and F.L. Pezold. 1988. A study of the vertebral column and median fin osteology in gobioid fishes with comments on gobioid relationships. *Bulletin of Marine Science* 42: 174-214.

Brito, A. and P.J. Miller. 2001. Gobiid fishes from the Cape Verde Islands, including two new species of *Gobius* (Teleostei: Gobioidei). *Journal of Natural History* 35: 235-277.

Dawson, C.E. 1979. A new wormfish (Pisces: Microdesmidae) from the eastern tropical Atlantic. *Copeia* 1979: 203-205.

Dawson, C.E. 1990. Microdesmidae. In: *Checklist of the Fishes of the Eastern Tropical Atlantic.* J.C. Quero, J.C. Hureau, C. Karrer, A. Post, and L. Saldanha (eds.), JNICT, EUI and UNESCO, Lisbon, pp. 960-961.

Delais, M. 1951. Notes d'ichtyologie Ouest Africaine II. Les Gobiidae d'Afrique occidentale Francaise en collection au Laboratoire de biologie marine de l'I.F.A.N. a Goreé. *Bulletin de l'Institut Francais d'Afrique Noire Series A Sciences Naturelles* 13: 343-369.

Engelbrecht, G.D. and P.F.S. Mulder. 1999. Extremely high genetic differentiation between two populations of the river goby, *Glossogobius callidus* (Smith, 1937). *Water SA* 25: 85-90.

Froese, R. and D. Pauly. (eds.). 2009. FishBase. World Wide Web electronic publication. www.fishbase.org, version (01/2009)—only entries from South African Institute of Aquatic Biodiversity (SAIAB) were used for compilation of distribution data.

Floeter, S.R., L.A. Rocha, D.R. Robertson, J.-C. Joyeux, W.F. Smith-Vaniz, P. Wirtz, A.J. Edwards, J.P. Barreiros, C.E.L. Ferreira, J.L. Gasparini, A. Brito, J.M. Falcon, B.W. Bowen and G. Bernardi. 2008. Atlantic reef fish biogeography and evolution. *Journal of Biogeography* 35: 22-47.

Gill, H.S. and J.S. Bradley. 1992. Validation of the use of cephalic lateral-line papillae pattern for postulating relationships among gobioid genera. *Zoological Journal of the Linnean Society* 106: 97-114.

Goren, M. 1991. A new *Coryogalops* species (Pisces: Gobiidae) from South Africa. *J.L.B. Smith Institute of Ichthyology Special Publication* 1991: 1-7.

Goren, M. 1996. A review of the southern African gobiid fish genus *Caffrogobius* Smitt, 1900. *J.L.B. Smith Institute of Ichthyology Special Publication* 1996: 1-28.

Harrison, I.J. 1990. A revised synonymy of *Nematogobius maindroni* (Sauvage, 1880) (Teleostei, Gobiidae). *Cybium* 14: 27-35.

Harrison, I.J. 1993. The West African sicydiine fishes, with notes on the genus *Lentipes* (Teleostei: Gobiidae). *Ichthyological Exploration of Freshwaters* 4: 201-232.

Harrison, I.J., P.J. Miller and F. Pezold. 2003a. Gobiidae. In: *The Fresh and Brackish Water Fishes of West Africa,* Vol. 2, D. Paugy, C. Léveque, and G.G. Teugels (eds.), IRD Editions, Paris, pp. 625-666.

Harrison, I.J., P.J. Miller and F. Pezold. 2003b. Microdesmidae. In: *The Fresh and Brackish Water Fishes of West Africa,* Vol. 2, D. Paugy, C. Léveque, and G.G. Teugels, (eds.), IRD Editions, Paris, pp. 667-669.

Harrison, I.J., P.J. Miller and F. Pezold. 2003c. Eleotridae. In: *The Fresh and Brackish Water Fishes of West Africa,* Vol. 2, D. Paugy, C. Léveque, and G.G. Teugels, (eds.), IRD Editions, Paris, pp. 670-690.

Harrison, I.J., P.J. Miller and F. Pezold. 2007a. Gobiidae. In: *The Fresh and Brackish Water Fishes of Lower Guinea, West-Central Africa,* Vol. 2, M.L.J. Stiassny, G.G Teugels, and C.D. Hopkins (eds.), IRD Editions, Paris, pp. 491-544.

Harrison, I.J., P.J. Miller and F. Pezold. 2007b. Microdesmidae. In: *The Fresh and Brackish Water Fishes of Lower Guinea, West-Central Africa,* Vol. 2, M.L J. Stiassny, G.G. Teugels, and C.D. Hopkins (eds.), IRD Editions, Paris, pp. 545-547.

Herler, J., R.A. Patzner, H. Ahnelt and H. Hilgers. 1999. Habitat selection and ecology of two speleophilic gobiid fishes (Pisces: Gobiidae) from the western Mediterranean Sea. *Marine Ecology* 20(3): 49-62.

Hoese, D.F. 1986a. Families Gobiidae, Eleotridae and Kraemeriidae. In: *Smith's Sea Fishes*, M.M. Smith, and P.C. Heemstra (eds.), Springer Verlag, Berlin, Heidelberg, New York, pp. 774-811.

Hoese, D.F. 1986b. Descriptions of two new species of *Heterelectris* (Pisces: Gobiidae) from the western Indian Ocean, with discussion of related species. *J.L.B. Smith Institute of Ichthyology Special Publication* 1986: 1-25.

Hoese, D.F. and H.K. Larson. 2005. Description of two new species of *Heterelectris* (Gobiidae) from the South Pacific, with a revised key to species and synonymization of the genus *Pascua* with *Heterelectris*. *Zootaxa* 1096: 1-16.

Hoese, D. and R. Winterbottom. 1979. A new species of *Lioteres* (Pisces, Gobiidae) from KwaZulu, with a revised checklist of South African gobies and comments on the generic relationships and endemism of western Indian Ocean gobioids. *Life Sciences Occasional Papers. Royal Ontario Museum* 31: 1-13.

Jaafar, Z. and H.K. Larson. 2008. A new species of mudskipper, *Periophthalmus takita* (Teleostei: Gobiidae: Oxudercinae), from Australia, with a key to the genus. *Zoological Science (Tokyo)* 25: 946-952.

Keith, P. and C. Lord. 2011. Tropical freshwater gobies: Amphidromy as a life cycle. In: *The Biology of Gobies*, R.A. Patzner, J.L. Van Tassell, M. Kovačić and B.G. Kapoor (eds.), Science Publisher Inc., Enfield, pp. 243-277.

Kovačić, M. and U.K. Schliewen. 2008. A new species of *Gorogobius* (Perciformes: Gobiidae) from São Tomé Islands. *Zootaxa* 1686: 29-36.

Kovačić, M. and R.A. Patzner. 2011. North-Eastern Atlantic and Mediterranean Gobies. In: *The Biology of Gobies*, R.A. Patzner, J.L. Van Tassell, M. Kovačić and B.G. Kapoor (eds.), Science Publisher Inc., Enfield, pp. 177-206.

Larson, H.K. 2001. A revision of the gobiid fish genus *Mugilogobius* (Teleostei: Gobioidei), and its systematic placement. *Records of the Western Australian Museum (Supplement)* 2001: 1-233.

Larson, H.K. 2008. An annotated checklist of gobioid fishes from Singapore. *The Raffles Bulletin of Zoology* 56: 135-155.

Le Louff, P. and R. von Cosel. 1998. Biodiversity patterns of the marine benthic fauna on the Atlantic coast of tropical Africa in relation to hydroclimatic conditions and paleogeographic events. *Acta Oecologica* 19(3): 309-321.

McKay, S.I. and P.J. Miller. 1991. Isozyme criteria in the testing of phyletic relationships between species of *Gobius* and related eastern Atlantic-Mediterranean genera (Teleostei: Gobiidae). *Journal of Fish Biology* 39: 291-299.

Melo, Y.C. and F. Le Clus. 2005. Growth and reproduction of the pelagic goby *Sufflogobius bibarbatus* off the Orange River, southern Africa. *African Journal of Marine Science* 27: 265-273.

Miller, P.J. 1978a. The systematic position and origin of *Gobius ocheticus* Norman, 1927, from the Suez Canal. *Zoological Journal of the Linnean Society* 62: 39-58.

Miller, P.J. 1978b. The status of the West African fish *Gobius nigricinctus* with reference to New World autochthones and an Old World colour-analogue. *Zoological Journal of the Linnean Society* 64: 27-39.

Miller, P.J. 1981. The systematic position of a West African gobioid fish, *Eleotris maltzani* Steindachner. *Zoological Journal of the Linnean Society* 73: 273-286.

Miller, P.J. 1984. The gobiid fishes of temperate Macaronesia (eastern Atlantic). *Journal of Zoology (London)* 204: 363-412.

Miller, P.J. 1986. Gobiidae. In: *Fishes of the North-eastern Atlantic and the Mediterranean,* Volume III, P.J.P. Whitehead, M.-L. Bauchot, J.-C. Hureau, J. Nielsen, and E. Tortonese (eds.). United Nations Educational, Scientific and Cultural Organization, Paris, pp. 1019-1085.

Miller, P.J. 1988a. New species of *Corcyrogobius, Thorogobius* and *Wheelerigobius* from West Africa (Teleostei: Gobiidae). *Journal of Natural History* 22: 1245-1262.

Miller, P.J. 1988b. Studies on *Silhouettea* Smith 1959 and an account of *Ebomegobius* Herre 1946. *Senckenbergiana Biologica* 68: 241-273.

Miller, P.J. 1990. Gobiidae. In: *Checklist of the Fishes of the Eastern Tropical Atlantic,* J.C. Quero, J.C. Hureau, C. Karrer, A. Post, and L. Saldanha, (eds.), JNICT, EUI and UNESCO, Lisbon, pp. 925-959.

Miller, P.J. 1998. The West African species of *Eleotris* and their systematic affinities (Teleostei: Gobioidei). *Journal of Natural History* 32: 273-296.

Miller, P.J. and R.M. Smith 1989. The West African species of *Bathygobius* (Teleostei: Gobiidae) and their affinities. *Journal of Zoology (London)* 218: 277-318.

Miller, P.J. and S. Stefanni. 2001. The eastern Pacific species of *Bathygobius* (Perciformes: Gobiidae). *Revista Biologia Tropical* 49 (Supplement): 141-156

Miller, P.J., J. Wright and P. Wongrat. 1989. An Indo-Pacific goby (Teleostei: Gobioidei) from West Africa, with systematic notes on *Butis* and related eleotridine genera. *Journal of Natural History* 23(2): 311-324.

Murdy, E.O. 1989. A taxonomic revision and cladistic analysis of Oxudercine gobies (Gobiidae: Oxudercinae). *Records of the Australian Museum Suppl.* no. 11, 1-93.

Murdy, E.O. 1998. A review of the gobioid fish genus *Gobioides*. *Ichthyological Research* 45: 121-133.

Olivar, M.P. 1989. Spawning and development of *Lesueurigobius sanzoi* off northern Namibia. *South African Journal of Zoology* 24: 132-138.

Peeters F.J.C., R. Acheson, G.J.A. Brummer, W.P.M. de Ruijter, R.R. Schneider, G.M. Ganssen, E. Ufkes and D. Kroon. 2004. Vigorous exchange between the Indian and Atlantic Ocean at the end of the past five glacial periods. *Nature* 430 (7000): 661-665.

Pezold, F. 2004a. Phylogenetic analysis of the genus *Gobionellus* (Teleostei: Gobiidae). *Copeia* 2004: 260-280.

Pezold, F. 2004b. Redescriptions and synonymies of species of the American-West African genus *Gobionellus* (Teleostei, Gobiidae) with a key to species. *Copeia* 2004: 281-297.

Pezold, F. and B. Cage. 2002. A review of the spinycheek sleepers, genus *Eleotris* (Teleostei: Eleotridae), of the Western Hemisphere, with comparisons to the West African species. *Tulane Studies in Zoology and Botany* 31 (2): 19-63.

Pezold, F., T. Iwamoto and I.J. Harrison 2006. Multivariate analysis of sicydiines of São Tomé and Príncipe with redescription of *Sicydium brevifile* and *S. bustamantei* (Teleostei: Gobiidae) and a key to west African sicydiines. *The California Academy of Sciences Gulf of Guinea Expedition (2201). V. Proceedings of the California Academy of Sciences (Series 4)* 57 (34): 965-980.

Pezold, F., I.J. Harrison and P.J. Miller. 2007 Eleotridae. In: *The Fresh and Brackish Water Fishes of Lower Guinea, West-Central Africa,* Vol. 2, M. L. J. Stiassny, G. G. Teugels, and C.D. Hopkins (eds.), IRD Editions, Paris, pp. 472-490.

Poll, M. 1959. Teleosteens acanthopterygians. In: *Poissons V.-Téléostéens acanthoptérygiens (deuxième partie). Rés. Sci. Expéd. Océanogr. Belg. Eaux Cot. Afr. Atl. Sud.* 4 (3B): 1-417.

Poll, M. 1961. Contribution a l`étude des poissons `eau douce de Somalie apparenant aux genres *Gobius* et *Barbopsis. Bolletino dei Musei e degli Instituti Biologici dell`Università di Genova* 31 (183): 15-35.

Randall, J.E. 1994. A new genus and six new gobiid fishes (Perciformes: Gobiidae) from Arabian waters. *Fauna of Saudi Arabia* 14: 317-340.

Randall, J.E. and D.W. Greenfield 2001. A preliminary review of the Indo-Pacific gobiid fishes of the genus *Gnatholepis. Ichthyological Bulletin of the J.L.B. Smith Institute of Ichthyology* 2001: 1-17.

Randall, J.E. and D.W. Greenfield 2007. Redescription of *Gnatholepis cauerensis* (Bleeker, 1853), with discussion of the validity of the species. *Zoologische Mededelingen (Leiden)* 81: 303-308.

Rocha, L.A., D.R. Robertson, C.R. Rocha, J.L. Van Tassel, M.T. Craig and B.W. Bowen. 2005. Recent invasion of the tropical Atlantic by an Indo-Pacific coral reef fish. *Molecular Ecology* 14: 3921-3928.

Froese, R. and D. Pauly. (eds.). 2009. FishBase (South African Institute of Aquatic Biodiversity (SAIAB) entries only). Word Wide Web electronic publication. www.fishbase.org, version (01/2009).

Sauberer, M. and H. Ahnelt. 2008. First record of the rare deep-water gobiid *Thorogobius rofeni* (Gobiidae) from the Southeast Atlantic. *Cybium* 32: 277-278.

Schliewen, U.K. and M. Kovačić. 2008. *Didogobius amicuscaridis* spec. nov. and *D. wirtzi* spec. nov., two new species of symbiotic gobiid fish from São Tomé and Cape Verde islands (Perciformes, Gobiidae). *Spixiana* 32: 247-261.

SIAP, SCRP, EC and SAUP 2002. *Marine ecosystems of the Northwest African subregion.* SIAP, SCRP, EC and Sea around us project. CD-ROM data available through Fishbase.

Skelton, P. 2001. *A Complete Guide to the Freshwater Fishes of Southern Africa.* Struik Publishers, Cape Town, second edition.

Spalding, M.D., H.E. Fox, G.R. Allen, N. Davidson, Z.A. Ferdana, M. Finlayson, B.S. Halpern, M.A. Jorge, A. Lombana, S.A. Lourie, K.D. Martin, E. Mcmanus, J. Molnar, C. Recchia and J. Robertson. 2007. Marine ecoregions of the world: A bioregionalization of coastal and shelf areas. *BioScience* 57: 573-583.

Steininger, F.F. and F. Rögl. 2007. Paleography and palinspastic reconstruction of the Neogene of the Mediterranean and Paratethys. In: *The Geological Evolution of the Eastern Mediterranean*, J. E. Dixon and Robertson A.H.F. (eds.), Blackwell Scientific Publications, Oxford, pp. 659-668.

Thacker, C.E. 2000. Phylogeny of wormfishes (Teleostei: Gobioidei: Microdesmidae). *Copeia* 2000: 940-957.

Thacker, C.E. 2009. Phylogeny of Gobioidei and placement within Acanthomorpha, with a new classification and investigation of diversification and character evolution. *Copeia* 2009: 93-104.

van Tassell, J.L., P.J. Miller and A. Brito. 1988. A revision of *Vanneaugobius* (Teleostei: Gobiidae), with description of a new species. *Journal of Natural History* 22: 545-567.

Vreven, E.J., T. Musschoot, G. Boden and M.L.J. Stiassny. 2007. Introduced or alien species of Lower Guinea. In: *The Fresh and Brackish Water Fishes of Lower Guinea, West-Central Africa,* Vol. 1, M.L.J. Stiassny, G.G. Teugels, and C.D Hopkins (eds.), IRD Editions, Paris, pp. 58-101.

Watson, R.E. 1992. A review of the gobiid fish genus *Awaous* from insular streams of the Pacific Plate. *Ichthyological Exploration of Freshwaters* 3: 161-176.

Winterbottom, R. 1976. Notes on South African gobies possessing free upper pectoral fin rays (Pisces: Gobiidae). *J.L.B. Smith Institute of Ichthyology Special Publication* 1976: 1-11.

Wirtz, P., R. Fricke and M.J. Biscoito. 2008. The coastal fishes of Madeira Island—new records and an annotated checklist. *Zootaxa* 1715: 1-26.

Wirtz, P., C.E.L. Ferreira, S.R. Floeter, R. Fricke, J.L. Gasparini, T. Iwamoto, L. Rocha, C.L.S. Sampaio and U.K. Schliewen. 2007. Coastal Fishes of São Tomé and Príncipe islands, Gulf of Guinea (Eastern Atlantic Ocean)—an update. *Zootaxa* 1523: 1-48.

Zarske, A. 1993: Notizen zur Ichthyofauna der Insel São Tomé (Golf von Guinea) (Teleostei). *Faunistische Abhandlungen Staatliches Museum für Tierkunde Dresden* 19: 85-88.

CHAPTER

2.4

The Marine Temperate Gobioids of Southern Australia and the New Zealand Region

Helen K. Larson

INTRODUCTION

In contrast to the tropics, relatively few marine gobioid fishes live in the temperate regions of Australia and New Zealand and neighbouring islands. The marine fauna of this region is mostly confined within the Pacific Central-South Subtropical water mass, with New Zealand also being influenced by the southern Subtropical Convergence (Commonwealth of Australia 2006). In Australia, subtropical and temperate waters influence about half the continent's fauna, with considerable overlap and intermixing of tropical and temperate faunas on the west and east coasts (Wilson and Allen, 1987). For example, the seasonal southward-flowing Leeuwin Current of Western Australia brings tropical taxa to the southern coast of Western Australia, as exemplified by the coral reefs and associated gobioid fish fauna of the Houtman Abrolhos Archipelago at 29°S (Hutchins, 1994, 1997).

Author's address: 65 Reid Road, Wongaling Beach, Queensland 4852, Australia.
E-mail: helen.larson@nt.gov.au

The New Zealand region extends from Campbell Island in the Tasman Sea (52°S) north to the Norfolk and Kermadec island groups at 29°S and 30°S in the south-west Pacific. The main islands of New Zealand are influenced by two main water masses: subtropical in the north and subantarctic in the south (Francis, 1996). This region has seasonal shifts in currents as in southern Australia. New Zealand shares a considerable number of temperate fish species with Australia (Ayling and Cox, 1982). The southernmost islands of New Zealand do not support gobioid fishes (i.e. Snares, Auckland, Campbell Islands); gobioids are essentially a tropical-subtropical group.

Norfolk Island (which includes the smaller islands, Nepean and Philip, close to the main island) is politically part of Australia but sits atop the Norfolk Ridge north-west of New Zealand. The Kermadec group is scattered along the Kermadec Ridge to the north-east of New Zealand. Lord Howe Island (politically part of New South Wales, Australia) sits relatively close to the east coast of Australia, not far south of the isolated coral reefs of Middleton and Elizabeth.

The fishes of the southern temperate seas of Australia and New Zealand are highly endemic at genus and species levels (Wilson and Allen, 1987) and this endemism is reflected in the gobioid fauna. The Gobiidae is the third most speciose inshore marine fish family in temperate Australia (Wilson and Allen, 1987; Hoese and Larson, 2008), behind the more diverse Syngnathidae and Clinidae. The Australian gobioid fauna has recently been summarised in Hoese and Larson (2006b, 2008), while that of New Zealand appears in publications by McDowall (1978), Ayling and Cox (1982) and Francis (1996). The fish faunas of the isolated Lord Howe, Norfolk and Kermadec islands of the south-west Pacific, all of which share taxa with temperate Australia and New Zealand, have been discussed by Allen *et al.* (1976), Hoese *et al.* (1978) and Francis (1993).

SOUTHERN AUSTRALIA

Southern Australia is a warm-temperate to cool-temperate region, influenced by three main currents: the south-flowing Leeuwin along the west coast, the east-flowing Zeehan Current along the south coast and the south-flowing East Australian Current along the east coast. Seasonal shifts occur in these currents' intensity—for example during the summer, the Eastern Australian Current produces eddies of warm water which continue south, essentially bringing tropical waters as far south as Tasmania. More than 870 fish species are known from the south coast, the high diversity reflecting Australia's long isolation (Gomon *et al.*, 2008).

Australia has 95 genera and at least 380 species of gobies. Of these, 25 species in 15 genera occur in its southern waters, living over sand or mud in estuaries, the lower reaches of rivers, and in coastal bays. Four regional endemic genera are known: *Afurcagobius* (one species, *A. tamarensis*), *Arenigobius* (*A. bifrenatus* and *A. frenatus*), *Nesogobius* (10 species, with four described and the genus under review (Hoese and Larson, 2006a)) and *Tasmanogobius* (*T. lordi*). The southern Australian Gobiidae include six "tropical" genera at the extreme end of their range, such as *Callogobius* (*C. depressus* and *C. mucosus*), *Gobiopterus* (*G. semivestitus*), *Eviota* (*E. bimaculata*), *Mugilogobius* (*M. platynotus*), *Pseudogobius* (*P. olorum* and one undescribed) and *Redigobius* (*R. macrostoma*). Genera such as *Bathygobius* (*B. kreffti*) and *Favonigobius* (*F. lateralis*) are temperate to tropical in distribution. Three genera are introduced in temperate Australia: *Acanthogobius* (*A. flavimanus*), *Acentrogobius* (possibly *A. pflaumii*; generic review ongoing) and *Tridentiger* (*T. trigonocephalus*). All of these are native to the north-west Pacific and were introduced accidentally to Australia.

One endemic marine eleotrid genus is found in southern Australia, *Philypnodon* (*P. grandiceps* and *P. macrostomus*) . The other marine eleotrid genus *Thalasseleotris* (*P. adela*), is shared with New Zealand. Hoese and Roberts (2005) pointed out that: "Marine eleotrids are rare, with only three genera restricted to fully marine reef environments: the Indo-Pacific coral reef genus *Calumia* Smith, the temperate New Zealand genus *Grahamichthys* Whitley, and the temperate Australasian genus *Thalasseleotris* Hoese and Larson."

The fish fauna of much of the southern coast of Australia has recently been summarised in Gomon *et al.* (2008), but the complexity of the assemblages in this large region remains to be fully understood. The gobioid fauna is not large, with Victoria having the most species (27), due to its having eight of the 10 known species of the Australian temperate endemic genus *Nesogobius*. The "coolest" state, Tasmania, has 19 gobioid species (17 gobies and two gudgeons), but three of these actually belong to tropical genera: *Callogobius, Pseudogobius* and *Redigobius*. It has two temperate eleotrids, *Philypnodon* and *Thalasseleotris*.

Temperate Western Australia shares many species with southern and south-eastern Australia. It has several marine endemic fish species, none of which are gobioids. *Eviota bimaculata*, originally thought to be endemic to Western Australia (Hutchins, 1994), has since been found as far east as Ceduna in South Australia (Hutchins, 2001; Hoese and Larson, 2008). The southern waters of Western Australia hold only one marine eleotrid, *Thalasseleotris adela* and 21 gobiids. Western Australia is distinctive due to its very long coastline that includes tropical to temperate habitats and fish assemblages, with the south-flowing Leeuwin Current having considerable influence over fish distribution and abundance. For example, of the eight gobioid species recorded from Rottnest Island, just off Perth (32°S), four are tropical (three gobiids and one microdesmid) (Hutchins, 1994). Further north, the offshore Houtman Abrolhos Islands (29°S) of Western Australia host a mix of tropical (66%), warm-temperate (19%) and subtropical (13%) species, of which 23 are gobioids (Hutchins, 1994, 1997). Only two of these are warm-temperate species (*Callogobius mucosus* and *Eviota bimaculata*). The most extreme "south" for a Western Australian tropical goby is the extralimital record of the Indo-Pacific *Barbuligobius boelhkei* turning up in the Recherche Archipelago just west of the Great Australian Bight (Hutchins, 1994).

Lord Howe Island lies about 630 kilometers east of the coast of Australia at latitude 31°S, and contains the world's southernmost coral barrier reef and lagoon (Allen *et al.*, 1976). Its fish fauna includes temperate and tropical species, with the 26 species of gobies and two microdesmids recorded being of tropical Indo-Pacific origin (but there are no eleotrids known from the island) (Allen *et al.*, 1976; Francis, 1993). Norfolk Island is further isolated, being about 1375 kilometers east of Australia and 775 kilometers north-west of New Zealand, at latitude 29°S (Hoese *et al.*, 1978; Francis, 1993). It has only seven species of goby recorded from its rocky reefs and coral patch reefs (no microdesmids or marine eleotrids). Norfolk Island gobies belong to widespread coral reef genera: *Bathygobius, Callogobius, Eviota, Priolepis* and *Vanderhorstia* (Francis, 1993).

NEW ZEALAND

In New Zealand, marine teleosts include about 1065 species, belonging to 185 families (King *et al.*, 2009). Of the coastal fish species, about 25% are endemic, often with limited distributions. The greatest diversity of inshore species, such as gobioids, is to be found in the north-east of New Zealand. Further work may extend the distribution of some of the species with restricted distributions, as specimens from earlier collections are verified and further surveys carried out. For example, the goby *Priolepis psygmophilia*, thought to occur only in the Kermadecs, has been found at Lord Howe, Rapa and Easter islands (Hoese and Larson, in press).

The New Zealand region supports two genera of marine gobies, neither endemic: *Favonigobius* (*F. lateralis* and *F. lentiginosus*, both also occur in eastern Australia) and *Gobiopsis* (the endemic *G. atrata*). *Favonigobius* is a tropical-temperate genus, while all other species of *Gobiopsis* are tropical. One endemic eleotrid genus occurs in New Zealand, *Grahamichthys* (*G. radiata*), which is unusual in that it is monotypic, endemic to New Zealand and lives in loose schools (not typical behaviour for an eleotrid). New Zealand shares two eleotrid genera with Australia, *Thalasseleotris* (the endemic *T. iota*), and *Gobiomorphus* (*G. cotidianus*, *G. gobioides*, *G. hubbsi*, *G. huttoni*).

Thalasseleotris iota has a wide distribution in New Zealand coastal waters, from the Bay of Islands, North Island, to Stewart Island in the south. It has not been collected at the Chatham Islands, despite sampling at 17 rotenone stations at 0-21 m depth (Roberts *et al.*, 1991). Based on a number of shared derived characters, Hoese and Gill (1993) suggested a close relationship between *Grahamichthys* and *Thalasseleotris*.

Gobiomorphus has seven species in New Zealand, three being restricted to freshwater while three are diadromous estuarine species (*G. gobioides*, *G. hubbsi*, *G. huttoni*); however, *G. cotidianus* can be either. Australia's two *Gobiomorphus* species are both found only in freshwater. Stevens and Hicks (2009) have confirmed that the Australian and New Zealand species of *Gobiomorphus* are monophyletic, based on mitochondrial DNA and morphological evidence.

Several gobioids have been relatively recently reported from New Zealand: the gobiids *Acentrogobius pflaumii* (Francis *et al.*, 2003), *Arenigobius bifrenatus* (Willis *et al.*, 1999), *Gobiopterus semivestitus* (McDowall and David, 2008) and *Parioglossus marginalis* (a microdesmid; McDowall 2001). These are all from shallow estuarine or harbour environments in northern New Zealand. These recent records probably result from ship-assisted introductions or may be natural dispersals that had not been previously observed.

A few islands offshore from New Zealand support gobioid fishes. Francis (1996) discussed the distributions of *Eviota* sp. (not yet identified), *Gobiopsis atrata* and *Priolepis psygmophilia* in the New Zealand region. The fish fauna of the Chatham Islands, east of South Island, includes four eleotrids (*Gobiomorphus cotidianus, G. gobioides, G. huttoni* and *Grahamichthys radiata*) but no gobiids (Roberts, 1991). The *Eviota* is known so far only from the Kermadec Islands, at 30°S north of North Island, New Zealand, while *Gobiopsis atrata* and *Grahamichthys radiata* reach their southernmost limit at Stewart Island (just off the south coast of South Island, at 46° S) (Francis, 1996). These are the 'furthest south' gobioids in the entire Australia-New Zealand region.

References

Allen, G.R., D.F. Hoese, J.R. Paxton, J.E. Randall, B.C. Russell, W.A. Starck, F.H. Talbot and G.P. Whitley. 1976. Annotated checklist of the fishes of Lord Howe Island. *Records of the Australian Museum* 30: 365-454.

Ayling, T. and G.J. Cox. 1982. *Collins Guide to the Sea Fishes of New Zealand.* Collins, Auckland.

Commonwealth of Australia. 2006. A Guide to the Integrated Marine and Coastal Regionalisation of Australia, Version 4.0. Department of the Environment and Heritage, Canberra.

Francis, M.P. 1993. Checklist of the coastal fishes of Lord Howe, Norfolk and Kermadec Islands, southwest Pacific Ocean. *Pacific Science* 47: 136-170.

Francis, M.P. 1996. Geographic distribution of marine reef fishes in the New Zealand region. *New Zealand Journal of Marine and Freshwater Research* 30: 35-55.

Francis, M.P., C. Walsh, M.A. Morrison and C. Middleton. 2003. Invasion of the Asian goby, *Acentrogobius pflaumii*, into New Zealand, with new locality records of the introduced bridled goby, *Arenigobius bifrenatus. New Zealand Journal of Marine and Freshwater Research* 37: 105-112.

Gomon, M., D. Bray and R. Kuiter. 2008. *Fishes of Australia's Southern Coast.* Museum Victoria and New Holland Publishers, Chatswood.

Hoese, D.F. and H.K. Larson. 1987. New Australian fishes. Part II. A new genus and species of eleotridid (Gobioidei) from southern Australia, with a discussion of relationships. *Memoirs of the Museum of Victoria* 48: 43-50.

Hoese, D.F. and A.C. Gill. 1993. Phylogenetic relationships of eleotrid fishes (Perciformes: Gobioidei). *Bulletin of Marine Science* 52: 415-440.

Hoese, D.F. and C.D. Roberts. 2005. A new species of the eleotrid genus *Thalasseleotris* (Teleostei: Gobioidei) from New Zealand coastal waters. *Journal of the Royal Society of New Zealand* 35: 417-431.

Hoese, D.F. and H.K. Larson. 2006a. Description of two new species of *Nesogobius* (Pisces: Gobioidei: Gobiidae) from southern Australia. *Memoirs of Museum of Victoria* 63: 7-13.

Hoese, D.F. and H.K. Larson. 2006b. Gobiidae. Gobies. In: *Fishes*. D.F. Hoese, D.J. Bray, J.R. Paxton and G.R. Allen. In: *Zoological Catalogue of Australia, Volume 35, Parts 1-3*. P.L. Beesley and A. Wells (eds.). ABRS and CSIRO Publishing, Canberra, pp. 1612-1697.

Hoese, D.F. and H.K. Larson. 2008. Family Gobiidae. In: *The fishes of Australia's south coast.* M.F. Gomon, J.C.M. Glover and R.H. Kuiter (eds.). Museum Victoria, New Holland Publishers. pp. 781-809.

Hoese, D.F. and H.K. Larson. 2010. Description of two new species of the genus *Priolepis* from the Indo-Pacific, with redescription of *P. psygmophilia* and *P. profunda. Ichthyological Research* 57: 373-388.

Hoese, D.F., L.H. Hodgson and J.R. Paxton. 1978. Preliminary annotated checklist of the fishes of Norfolk Island. Unpublished report, Australian Museum. 78 pp.

Hutchins, B. 1994. A survey of the nearshore reef fish fauna of Western Australia's west and south coasts — the Leeuwin Province. *Records of the Western Australian Museum, Supplement* 46: 1-66.

Hutchins, B. 1997. Checklist of fishes of the Houtman Abrolhos Islands, Western Australia. In: *The Marine Flora and Fauna of the Houtman Abrolhos Islands, Western Australia. Proceedings of the Seventh International Marine Biological Workshop*, F.E. Wells (ed.). Western Australian Museum, Perth, pp. 239-253.

Hutchins, B. 2001. Checklist of the fishes of Western Australia. *Records of the Western Australian Museum, Supplement* 63: 9-50.

King, C.M., C.D. Roberts, B.D. Bell, R.E. Fordyce, R.S. Nicoll, T.H. Worthy, C.D. Paulin, R.A. Hitchmough, I.W. Keyes, A.N. Baker, A.L. Stewart, N. Hiller, R.M. McDowall, R.N. Holdaway, R.P. McPhee, W.W. Schwarzhans, A.J.D. Tennyson, S. Rust and I. Macadie. 2009. Phylum Chordata. Lancelets, fishes, amphibians, reptiles, birds, mammals. In: *The New Zealand Inventory of Biodiversity. Volume 1, Kingdom Animalia: Radiata, Lophotrochozoa, and Deuterostomia*, D.P. Gordon (ed.). Canterbury University Press, Christchurch, pp. 433-553.

McDowall, R.M. 1978. *New Zealand Freshwater Fishes. A Guide and Natural History*. Heinemann, Auckland.

McDowall, R.M. 2001. *Parioglossus* (Teleostei: Gobioidei: Microdesmidae) in New Zealand. *New Zealand Journal of Marine and Freshwater Research* 35: 165-172.

McDowall, R.M. and B.O. David. 2008. *Gobiopterus* in New Zealand (Teleostei: Gobiidae), with observations on sexual dimorphism. *New Zealand Journal of Marine and Freshwater Research* 42: 325-331.

Roberts, C.D. 1991. Fishes of the Chatham Islands, New Zealand: a trawl survey and summary of the ichthyofauna. *New Zealand Journal of Marine and Freshwater Research* 25: 1-19.

Roberts, C.D., C.D. Paulin and A.L. Stewart. 1991. Coast and lagoon fishes of Chatham Island: National Museum 1991 survey. Unpublished report, National Museum of New Zealand, Wellington. Held at Hector Library. 44 pp.

Stevens, M.I. and B.J. Hicks. 2009. Mitochondrial DNA reveals monophyly of New Zealand's *Gobiomorphus* (Teleostei: Eleotridae) amongst a morphological complex. *Evolutionary Ecological Research* 11: 109-123.

Willis, T.J., J.E.H. Saunders, D.L. Blackwood and J.E. Archer. 1999. First New Zealand record of the Australian bridled goby, *Arenigobius bifrenatus* (Pisces: Gobiidae). *New Zealand Journal of Marine and Freshwater Research* 33: 189-192.

Wilson, B.R. and G.R. Allen. 1987. Major components and distribution of marine fauna. In: *Fauna of Australia. Volume 1A*. Australian Government Publishing Service, Canberra, pp. 43-68.

CHAPTER

2.5

Tropical Freshwater Gobies: Amphidromy as a Life Cycle

Philippe Keith[1] and Clara Lord[2]

INTRODUCTION

In the Indo-Pacific areas, the Caribbean region, West Africa, Central America and North of South America, river systems are colonised by freshwater gobies with a life cycle adapted to the conditions in these distinctive habitats which are, particularly in islands, young oligotrophic rivers subject to extreme climatic and hydrological seasonal variation. These species spawn in freshwaters, the free embryos drift downstream to the sea where they undergo a planktonic phase, before returning to rivers to grow and reproduce (McDowall, 1997; Keith, 2003), hence they are called amphidromous (Fig. 2.5.1.) (McDowall, 1988, 1997, 2004). The practical details of their biological cycle and the parameters leading to such evolution in amphidromous gobies are poorly known, but our knowledge increases each year. These gobies contribute most to the diversity of fish communities in the Indo-Pacific and the Caribbean insular systems, and have the highest

Authors' address: Muséum national d'Histoire naturelle, UMR 7208, ichtyologie, CP 26, 57 rue Cuvier, 75231 Paris Cedex 05, France. E-mails: [1]keith@mnhn.fr, [2]claralord@mnhn.fr

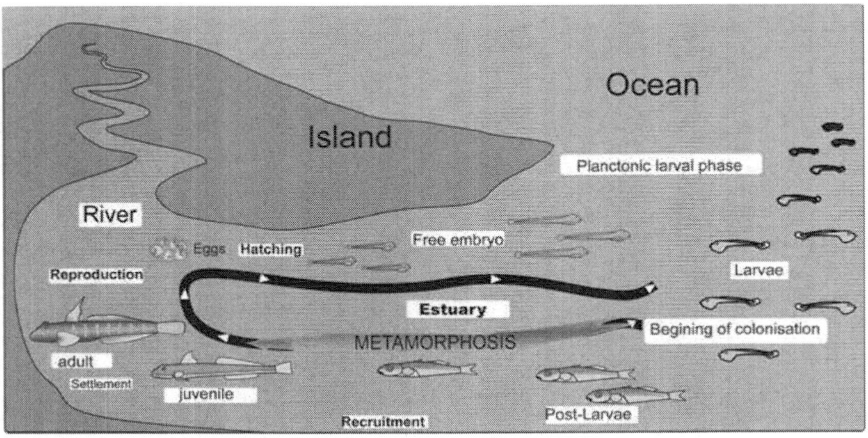

Fig. 2.5.1 Amphidromy in freshwater gobies. © P. Torres and P. Keith.

Color image of this figure appears in the color plate section at the end of the book.

levels of endemism (Keith, 2002a; Lim *et al.*, 2002; Marquet *et al.*, 2003; Keith, 2003; Lord and Keith, 2006; Lord and Keith, 2007).

At certain times of the year, the biomass of larvae migrating upstream is so great that they represent an important source of food for local human populations in certain archipelagos (Manacop, 1953; Delacroix, 1987; Bell, 1999; Keith *et al.*, 2008). However, harvesting of this food resource is highly unsustainable, on account of the complexity of the species' life cycle.

DISTRIBUTION AND BIOGEOGRAPHY

Amphidromous gobies belong mainly to the Sicydiinae sub-family, with the genera *Akihito* (2 species), *Lentipes* (12 species), *Cotylopus* (2 species), *Sicyopterus* (30 species), *Sicyopus* (15 species), *Stiphodon* (30 species), *Sicydium* (16 species) and *Parasicydium* (1 species), to the Gobionellinae sub-family with *Awaous* (14 species), *Stenogobius* (26 species), *Schismatogobius* (10 species) and *Rhinogobius* (50-70 species), and to the Gobiinae sub-family with *Glossogobius* (22 - 26 species). They represent nearly 200 species (Froese and Pauly, 2008).

All of these genera have a specific distribution. *Sicydium* is found throughout western and eastern Middle America, extending from Mexico to Peru along the Pacific coast, from Mexico to Venezuela along Atlantic coast, on Greater and Lesser Antilles islands of the Caribbean sea and in West Africa (Watson, 2000; Pezold *et al.*, 2006).

Sicyopterus, Stenogobius and *Glossogobius* are distributed in the Indo-Pacific area from the West Indian Ocean to the East Pacific Ocean (Watson *et al.*, 2000; Keith *et al.*, 2005a; Berrebi *et al.*, 2006). *Sicyopterus* and *Stenogobius* occur in the Indo-Pacific area throughout 18000 km from the west of the Indian Ocean (with Madagascar and the Comoros Islands) to the east of the Pacific Ocean (with the Austral Islands in French Polynesia and Hawai) (Keith *et al.*, 2005a). *Glossogobius* is absent in the eastern Pacific.

Stiphodon, Sicyopus, Lentipes and *Schismatogobius* are distributed from the East Indian Ocean to East Pacific Ocean (Watson *et al.*, 2001; Watson *et al.*, 2002; Keith and Marquet, 2007), *Cotylopus* is restricted to West Indian Ocean (Keith *et al.*, 2005b); *Parasicydium* is restricted to the West Africa (Pezold *et al.*, 2006), and *Akihito* (Watson *et al.*, 2007; Keith *et al.*, 2007) seems to be restricted to the west of the Pacific Ocean.

Of all amphidromous genera, *Awaous* is the most geographically extended as it is found throughout Central America, North of South America, on Greater and Lesser Antilles islands of the Caribbean sea, in West Africa and from the West of the Indian Ocean to the East of the Pacific Ocean (Watson, 1992).

Finally, the Asiatic freshwater goby genus *Rhinogobius* Gill, 1859, is widely distributed on some islands of the West Pacific including Japan, Taiwan, Hainan, Philippines, and also in continental Asia in Russia, Korea, China, Vietnam, Laos, Cambodia, and Thailand (Kawanabe and Mizuno, 1989; Chen and Kottelat, 2003). The life histories of *Rhinogobius* species indicate that the genus includes both amphidromous and landlocked species (Mizuno, 1960; Chen and Kottelat, 2003) and both temperate and tropical species.

Ninety percent of the tropical freshwater gobies are distributed in Indo-Pacific area, and only 10% occur in Atlantic and Caribbean regions. The higher rate of biodiversity in the Indo-Pacific area, seems to be due mainly to biogeography. The Indo-Pacific area contains thousands of marine and freshwater species which make this area the most diverse among marine habitats. Such a high diversity is not uniformly distributed but shows a gradient of species richness that extends from the highly diverse centre of the Indonesia-Philippines area to the surrounding archipelagos lying further east (Briggs, 1974). The reasons for the present-day distribution are complex and are the subject of fascinating studies, many of which are ongoing (Planes and Galzin, 1997).

Of several hypotheses that have been developed to explain biogeographic patterns in the Pacific Ocean (Rosen, 1988), three major theories can be distinguished: centre of origins, centre of accumulation, and centre of overlap (Palumbi, 1996; Planes and Galzin, 1997). These authors have proposed mechanisms that lead to a high diversity in the Indonesia-Philippines area. (a) The centre of origins theory suggests that the Indonesia-Philippines area is where speciation takes place and from where new species diffuse to peripheral habitats such as Pacific archipelagos (Ekman, 1953). (b) As opposing, it has been proposed that speciation is favoured in isolated peripheral archipelagos and new species would be transported to the Indonesia-Philippines area via prevailing currents (Ladd, 1960). In this case the Indonesia-Philippines area can be viewed as a centre of accumulation. Finally, (c) the third theory describes a centre of overlap where the high diversity of the Indonesia-Philippines area corresponds to a crossroad of several biogeographic provinces (Planes and Galzin, 1997). These biogeographic provinces could result from past geological events or present environmental conditions and overlap in the Indonesia-Philippines area where they create high species diversity. Many studies on corals, seagrasses, molluscs and fishes (Springer, 1982; Stoddart, 1992; Keith *et al.*, 2009) have investigated species distribution and support one or the other of the theories. However, up to now, it appears that there is no consensus on which model may be more generally suitable.

In the case of tropical freshwater gobies, it is necessary to consider all of these hypotheses to explain the existence of many endemic species and the broad or small distribution of the species. It is indeed probable that a complex system including marine currents, duration of the larval phase, paleohistory of the area (period of emergence of the islands, vicariousness and variation of the sea level, etc.), in addition to the particularities of the islands (altitude, substratum, age, velocity of river ...), are at the origin of the structure observed.

Finally, the most diverse group in tropical freshwater gobies is the subfamily Sicydiinae. Some studies have shown that there is some sister-group relationship between some Sicydiine genera. One of the most interesting is probably the relationship between *Sicyopterus* (in the Indo-Pacific area) and *Sicydium* (in the Caribbean and Atlantic areas). These two genera have a common ancestor, probably distributed in the past between the Pacific and Atlantic oceans.

REPRODUCTION AND EARLY LIFE STAGES

Nidification and Spawning

Few studies have investigated the reproductive biology of tropical freshwater Gobiidae, particularly in their natural habitat. Those species that have been studied live in various habitats, including middle reaches of rivers (Dotu and Mito, 1955; McDowall, 1965; Auty, 1978; Kim and Han, 1989; Daoulas *et al.*, 1993; Han *et al.*, 1998), streams with fast current (Delacroix, 1987; Kinzie, 1993*a*; Fitzsimons *et al.*, 1993; Ha and Kinzie, 1996; Takahashi *et al.*, 2001), lakes (Hidaka and Takahashi, 1987*a*; Hidaka and Takahashi, 1987*b*; Ito and Yanagisawa, 2000), estuaries or lower reaches (Hirashima and Tachihara, 2000; Takahashi, 2000; Tamada, 2000; Tamada, 2005a, b; Yamasaki and Tachihara, 2006, 2007 and others). Some amphidromous goby species are semelparous, others are annual spawners (Ego, 1956; Harrison *et al.*, 1991), while others spawn repeatedly (Torricelli *et al.*, 1985; Meadows and Meadows, 1988).

To date, there are few studies of the mating systems in amphidromous gobies. In the non-Sicydiine goby species studied, polygamy appears to be the commonest pattern, with males guarding nests containing eggs laid by several different females (Hidaka and Takahashi, 1987*a*; Daoulas *et al.*, 1993; Takahashi and Yanagisawa, 1999; Ito and Yanagisawa, 2000; Tamada, 2000). For Sicydiines gobies, aquarium based studies indicate that they are not polygamous (Kinzie, 1993*b*). They show no hermaphroditism, nor sex changes or other alternative sexual strategies, and this might also be the case for other freshwater gobioids. However, Watson (1991) reported two distinct phases in males of *Stenogobius caudimaculosus* Watson, 1991 from the Marquesas Islands. Though he added this was the only case of male dimorphism he had ever observed.

Clutch size is known for few amphidromous species (Mizuno, 1960; Timbol *et al.*, 1980, McDowall, 1994), such as *Sicyopterus lagocephalus* (Pallas, 1770), on the island of La Reunion, where the female fish produces 50 000 to 70 000 ova (Delacroix, 1987; Keith *et al.*, 1999), or [in Japan] *Sicyopterus japonicus* (Tanaka, 1909) which produces up to 225 000 ova (Dotu and Mito, 1955).

In general, the eggs are laid on the substratum, often on the underside of various supporting structures such as stones, rocks, crevices, but sometimes also on, or among, the vegetation. In *Lentipes concolor* (Gill, 1860) found in Hawaii, nests are situated preferentially under a large rock, away from the main river channel, close to the banks where they are safe from the full strength of the river current. The eggs are found from October to June, with a peak occurrence in February. The number of eggs in the ovaries varies from 12 000 to 14 000 (Kinzie, 1993*a*). Following the courtship display, the male fish leads the female to the spawning site (Fitzsimons *et al.*, 1993; Way *et al.*, 1998).

In Okinawa Island, southern Japan, *Stiphodon percnopterygionus* Watson and Chen, 1998, the spawning season was estimated as being from May to December. Standard length at maturity was approximately 20 mm in both sexes, and batch fecundity was approximately 1000-10 000 eggs per female (Yamasaki and Tachihara, 2006). The egg masses, guarded by the male, were laid on the undersurface of stones in freshwater. The pyriform eggs had long- and short-axis diameters of 0.54-0.58 mm and 0.49-0.50 mm, respectively. Newly hatched larvae (1.20-1.32 mm notochord length (NL)) were poorly developed, with large yolk sacs and unopened mouths. Three days after hatching (1.87-2.05 mm NL), eyes were fully pigmented and mouths were opened.

In *Awaous guamensis* (Valenciennes, 1837) from the Pacific area, the adult fish normally migrate downstream to the spawning grounds, which are situated in general in the first shallow rapids encountered upstream from the river mouth (Ego, 1956; Kido and Heacock, 1992). The females lay the eggs in the latter location, where they hatch within 24 to 70 hours (Kido and Heackock, 1992; Lindstrom and Brown, 1994). The females lay a vast number (up to 700000) of rather small eggs (0.23 cm in diameter) between August and December (Ego, 1956; Ha and Kinzie, 1996). The females usually leave the nest unattended, or are driven away by the male fish, immediately after releasing the eggs. The male fish care for and guard the eggs as in many other species of goby (Thresher, 1984; Gibson, 1970; Cole, 1982; Lindstrom and Brown, 1994). In *Awaous melanocephalus* (Bleeker, 1849), each nest contains 40000 - 220000 eggs guarded by males (Yamasaki and Tachihara, 2007).

In the rivers of Martinique (West Indies), Lim *et al.* (2002) showed that the female of *Sicydium punctatum* Perugia, 1896 produce 50000 to 60000 ova. The breeding season of *Sicydium plumieri* (Bloch, 1786) from Puerto Rico begins in early September and finishes in January with the end of the rainy season (Erdman, 1961; Lim *et al.*, 2002). Bell (1994) has shown that larvae were present virtually continuously, indicating that river gobies reproduce at all seasons.

Species of the genus *Rhinogobius* live in the river courses of Japan, Taiwan and Korea, and exhibit paternal care over nests that are typically found on the underside of rocks on the river bottom (Takahashi, 2000). The breeding season of the species *Rhinogobius* sp. DA (Dark type) is from mid-April to early July, and peaks in May. Half of the females spawn more than once in one breeding season and some guarding males receive eggs from several different females (Takahashi and Yanagisawa, 1999; Hirashima and Tachihara, 2000). Clutch sizes, in this Japanese species, ranging between 1043 and 4436, increased with increasing maternal standard length (Tamada, 2005b). Spawned egg volume ranged between 0.85 mm^3 and 1.38 mm^3. *Rhinogobius* sp. DA had a relatively larger egg and smaller clutch size compared to those of three other amphidromous *Rhinogobius* species inhabiting relatively larger rivers in Kii Peninsula. Interspecific egg size variations among the four species appear to result in differing larval survival rates during their migration to the sea (Tamada, 2005b).

Tamada (2001), studied three species of the *Rhinogobius* complex in a Japanese stream, and demonstrated that the distribution of their spawning grounds differed between species along the stream. The differences in spawning ground distribution and spawning season reduce interspecific competition for nest sites and enhance reproductive isolation. The difference in distribution of spawning grounds may be linked to other intraspecific differences in reproductive characteristics such as egg size. This is probably the result of a negative correlation between the survival of embryos during migration to the sea and the distance from the spawning ground to mouth of the river (Tamada, 2001). Hayashi (1972) noted for *Rhinogobius brunneus* (Temminck and Schlegel, 1845), that females were most abundant in the lower and middle reaches of streams while males are most abundant in the upper reaches of the stream.

Males of the stream goby *Rhinogobius* sp. DA (dark type) court females in deep pools in areas of streams with a fast current (Takahashi and Khoda, 2001) and care for their eggs under stones in shallow riffles. Although larger males won aggressive encounters with smaller male fish, large males did not monopolize courtship opportunities, and the frequency of successful courtships, after which males led the females to their nests, was not related to male body size (Takahashi *et al.*, 2001). Large males occupied larger nest stones, and the number of eggs received in the nest was correlated positively with nest size (area covered). Takahashi *et al.* (2001) suggested that male mating success of *Rhinogobius* sp. DA (dark type) is determined by male-male competition for large nests rather than competition for females.

Takahashi and Kohda (2001) demonstrated that neither male body size nor the length of the dorsal fin (sexually dimorphic traits) was related to female selectivity. Females favoured males that courted at locations where the water current was relatively fast (>1 l min^{-1}). Males in poor physical condition do not court in fast water currents. Hence, this mate choice pattern enables females to mate with males with the highest fitness.

Males of *Rhinogobius* sp. LD (Large Dark), selectively courted females of similar size prior to leading them to the nests, however females preferentially followed courting males that had larger body size than the female (Ito and Yanagisawa, 2000). Male-male competition for a female was relatively infrequent and the intensity was low. Developmental stages of eggs and egg numbers in one nest indicated that males receive 1-3 clutches during one breeding cycle. Males guarding multiple clutches frequently ate some of the eggs, but those guarding single clutches rarely consumed eggs. While in the nest, gravid females frequently cannibalized eggs laid by a previous female. This behaviour increases the area available for egg deposition and may reduce competition for oxygen among eggs (Ito and Yanagisawa, 2000).

The spawning behaviour, development of eggs and larvae of the Korean freshwater goby, *Rhinogobius brunneus* (Temminck and Schlegel, 1845) were studied by Han *et al.* (1998). The eggs were spawned as a monolayer, hanging from the underside of a small pebble, and guarded by one male. The eggs were elliptic, about 1.48 mm in length and 0.65 mm in breadth. Newly-hatched larvae were 3.1-3.3 mm in total length, and their mouth and anus were not open at this point. When the yolk sac and oil globules were nearly absorbed, the mouth and anus open, and the larvae begin to feed actively on rotifers.

The goby *Rhinogobius giurinus* (Rutter, 1897) examined in the Aizu River, Wakayama Prefecture, Japan, tended to have a larger clutch size, with smaller eggs, and matured at lower gonadosomatic index (GSI), compared with those in the three sympatric amphidromous species of *R. brunneus* complex (Tamada, 2005a). Higher water temperature and abundant food conditions may affect the optimal egg size in *R. giurinus*. Besides, higher larval survival rate during downstream migration may affect lower GSI at maturity in this goby that usually inhabits lower reaches of the river (Tamada, 2005a).

Tamada and Iwata (2005) have noted that for *Rhinogobius* sp CB (Cross Band type) both egg volume and clutch size increased with standard length or age of the maternal fish, while egg density in these clusters did not correlate with standard length. They noted that there was a significant positive correlation between egg size and the seventy two hours survival rate of unfed hatchlings. Intraspecific variation of egg volumes and clutch sizes in this species seems to be an adaptation for enhancing offspring survivorship during migration to the sea. Some females spawned a second time. Size of secondly spawned eggs sizes were smaller than firstly spawned eggs, although there was no difference in clutch between the two. Egg size variation between subsequent spawns may be an adjustment to the changes in seasonal environmental conditions (Tamada and Iwata, 2005).

Larval Life and Marine Phase

After hatching, free embryos of amphidromous gobiids are known to drift to the sea (or lake for some *Rhinogobius* species) mainly during the night (Dotu et Mito, 1955; Dotu, 1961; Iguchi and Mizuno, 1990, 1991; Bell and Brown 1994; Moriyama *et al.*, 1998; Valade, 2001; Keith, 2003; Keith *et al.*, 2008). This stage is described by Bell (1994) for *Sicydium*. The shift to salt water triggers the first transformations as the development continues (Manacop, 1953). Some experiments demonstrated that newly hatched larvae are better adapted physiologically to life in seawater, while prolonged exposure to freshwater retarded development and increased mortality markedly (Bell and Brown, 1994; Lindstrom and Brown, 1994; Lindstrom, 1998; Keith, 2003; Yokoi and Hosoya, 2005). Bell and Brown (1994) showed that there is an active salinity choice by the vesiculate larvae of *Sicydium punctatum* in Dominica (West Indies). Valade (2001) showed that *Sicyopterus lagocephalus* larvae die after 7 days if they have not reached the sea.

Free embryos of *Sicyopterus stimpsoni* (Gill, 1860) from Hawaii and of *S. lagocephalus* from La Reunion Island "swim" repeatedly upwards until they reach the water surface, then for a while cease "swimming" and sink and then move towards the surface again (Kinzie, 1993*a*; Keith *et al.*, 1999; Valade, 2001). This reaction facilitates their transport to the sea (Balon and Bruton, 1994). They appear in the marine environment when they have reached 1 to 4 mm in length (Han *et al.*, 1998; Keith *et al.*, 2008). According to Penaz (2001), the larval period begins with the capacity to catch and digest exogenous food. But this larval phase may be reached gradually given that the period of diet shift (endogenous/exogenous) is variable. The yolk was completely consumed in 3 to 7 days after hatching in *Rhinogobius* sp. (Hirashima and Tachihara, 2000). Moriyama *et al.* (1998) estimated that in low river flow, embryos from the upstream reaches, die of starvation before they reach the sea. As a result, Iguschi and Mizuno (1999) suggested that early starvation of larvae limits reproductive success of fish located far from the sea which would have a strong selection for fish reproduction in certain regions of rivers within a given distance from the sea. Tamada and Iwata (2005) have studied newly hatched *Rhinogobius* sp. CB (cross band type). Since larger and older individuals usually inhabit the upper reaches of rivers, larvae from larger females are more likely to suffer higher risks of starvation or predation during their longer migration to the sea.

Individuals can stay several months at sea before they get the signal to recruit into rivers. Indeed, within the same colonisation episode, the time spent at sea by *S. lagocephalus* individuals can vary greatly, from 133 to 256 days (Hoareau *et al.*, 2007). It varies from 90 to 169 days, for Hawaiian species (Radtke *et al.*, 1988; Radtke *et al.*, 1996), from 30 to 163 days for Tawainese species (Shen *et al.*, 1998) and from 173 to 253 days for *Sicyopterus japonicus* (Iida *et al.*, 2008). For the endemic *Cotylopus acutipinnis* (Guichenot, 1863), Hoareau (2005) has found 78 to 150 days. Post-larvae of the Hawaiian endemic goby *Lentipes concolor* settled after a mean length of larval life of 63-106 days at a mean size of 16.0 mm (14.1-17.9 mm) (Radtke *et al.*, 2001) and in Japan, *Stiphodon percnopterygionus* spend 78 to 146 days at sea (Yamasaki *et al.*, 2007). Bell *et al.* (1995) showed that the age at recruitment could vary slightly in the *Sicydium* species in Dominica (West Indies). In French Polynesia, the migration of post-larvae appears to be linked to body size (which is always approximately the same size for each species when they reach the estuary) (Keith, 2003).

This variability of larval duration at sea probably allows all these species to benefit from a broad window to find suitable climatic and ecological conditions for river colonisation. This marine phase is also a key element to the dispersion of species. It is important to note that these results about larval life duration are about twice or three times longer than that typically reported for tropical marine gobiids (Keith, 2003).

According to Murphy and Cowan (2007), the larva is competent when it is morphologically and physiologically capable of colonising freshwaters; the larva is able to respond to a signal triggering the processes leading to recruitment (i.e. beginning of the morphological transformation and migration towards the coast). From then on, they are referred to as post-larvae (Keith *et al.*, 2008) subject to the colonisation signal. Although the latter still has to be determined for all species, it is thought to be linked to the new moon period and to heavy rainfalls in La Réunion Island for *S. lagocephalus* (Delacroix, 1987; Keith *et al.*, 1999; Hoareau, 2005).

Where are larvae, when at sea? At this time nobody really knows. But Sorensen and Hobson (2005) used stable isotope analysis to evaluate the sources of nutrients used by amphidromous gobiid fishes (*Lentipes concolor*, *Sicyopterus stimpsoni*, *Awaous guamensis*) caught migrating into and living in Hakalau Stream, Hawaii. Although considerable variation amongst the stable isotope values of stream items was revealed during the 4 years of the study, the relationships between the fishes were relatively constant. Stable isotope values of recruiting gobies were consistently closer to those of both inshore plankton and freshwater adults than those of offshore plankton, suggesting that the larvae of these species derive much of their nutrition from inshore environments influenced by freshwater (Sorensen and Hobson, 2005). Small differences between the stable values of these species further suggested that their larvae come from different inshore locations. After entering freshwater, all species appear to swim rapidly upstream without feeding. Finally, once well upstream, adult *L. concolor* and *A. guamensis* appear to assume an omnivorous diet while adult *S. stimpsoni* rely upon autochthonous production within streams (Sorensen and Hobson, 2005).

RECRUITMENT AND METAMORPHOSIS

Recruitment into Rivers

Amphidromy seems to necessitate special migratory abilities in larvae (McDowall, 1988) to colonise rivers. After 30 to 260 days spent in the sea, depending on the species, they must locate a source of freshwater to colonise rivers as post-larvae. At this time, they have reached a length of 13 to 25 mm. In La Reunion Island, larvae are sometimes caught with planktonic nets in the sea (Keith *et al.*, 1999, 2006a). The various measurements of larvae made by Keith *et al.* (2006b) in French Polynesia and New Caledonia indicate that larvae occur at sea with a length that ranges from 13 to 16 mm in standard length (SL) for *Lentipes*, *Stenogobius*, *Awaous*, and *Eleotris*, and 23 to 25 mm for *Sicyopterus*. These measurements are similar to those made in Hawaii (Tate *et al.*, 1992). Some authors have made larval identification keys of amphidromous gobies of their area based on distinct physical characteristics and morphotypes (Tate *et al.*, 1992; Lindstrom, 1998; Lindstrom, 1999; Keith *et al.*, 2006b).

In Japan, Hayashi (1972) noted a post-larval length of 17 mm (SL) for *Rhinogobius brunneus*, Yuma *et al.* (2000) recorded a post-larval length of 16 to 26 mm (SL) for *Rhinogobius* sp. (orange form), and Iida *et al.* (2008), a larval length of 23.5 to 30 mm (SL) for *Sicyopterus japonicus*. In the Caribbean, the larvae of *Sicydium punctatum* reached 30 mm (Bell *et al.*, 1995; Lim *et al.*, 2002).

The recruitment of gobies is often associated with heavy rains (Delacroix, 1987; Keith, 2003). The freshwater flow seems to trigger their inland migration. Even intermittent rivers that flow only after major storms attract post-larvae (Delacroix and Champeau, 1992). Recruitment of post-larvae at the river mouth is possible all year round, but occurs most markedly in the Spring in Hawaii (Tate, 1995; Nishimoto and Kuamo'o, 1997) and during July-September in Japan for *Rhinogobius* sp. (orange form) (Yuma *et al.*, 2000). Keith *et al.* (2002a; 2006b) reported massive upstream movements of post-larvae of *S. lagocephalus* from December to February in Tahiti. For *Sicyopterus japonicus* the recruitment is initiated to a few months (Iida *et al.*, 2008).

The movement of these young fish from the ocean to the rivers is the key link in completing their amphidromous cycle. The maintenance of this system is crucial to the long term supply of resources in the freshwaters of insular systems (Keith, 2003; Lim *et al.*, 2002).

To access the rivers, the post-larvae choose to swim close to the shoreline, where the current is weaker and can even be reversed (i.e. a counter-current) with the tide or the swell. Some post-larvae travel with their spindle shaped body facing the current and climb from one rock to the next using their pelvic suction cups. Others can reach freshwaters following passages through cracks in the rocks, thus avoiding turbulence (Tate, 1997). Post-larvae first congregate near the shore. They will begin migration into the estuary at various moments during the day, depending on the species (Miller, 1984; Keith, 2003). Once in the river, the post-larvae then migrate upstream, towards the adult habitat: this is the colonisation phase.

In *Lentipes concolor,* the post-larvae enter the rivers mainly during daylight hours, with the waves of the incoming tide (Nishimoto and Kuamo'o, 1997). Migration peaks in the hours following sunrise, when post-larvae are bourne on the rising tide thereby reducing their energy expenditure. The same phenomenon has been reported for *Lentipes kaaea* Watson, Keith and Marquet, 2002, from New Caledonia (Watson *et al.*, 2002; Keith *et al.*, 2006b). When they enter freshwaters, *Lentipes* post-larvae swim immediately upstream. No amphidromous goby species (whether adults or young) appears to develop strict nocturnal activities, thus a morning initiated migration enables larvae to reach higher up into the river system.

Tate (1997) and Nishimoto and Kuamo'o (1997), reported that, in *Awaous guamensis,* the post-larvae were transported to the river mouth by waves and that they entered freshwaters either by day or by night, with a peak occurring early in the night. *A. guamensis* is a slower climber, and depends on tidal flooding to move upstream (Keith *et al.*, 2000). In *Stenogobius hawaiiensis* Watson, 1991, the post-larvae appear to move with the incoming tide during the day (Tate, 1997), or, at night, using cracks in the rocks. In comparison, in *Eleotris fusca* (Forster, 1801) (Eleotridae), the post-larvae have difficulties manoeuvring and swimming in turbulent waters, and this determines recruitment periods. They are transported mainly at night by the waves of the incoming tide. In the absence of a pelvic suction cup, this species move upstream by attaching itself to the substratum (Keith, 2003).

Rhinogobius sp. (orange form) from Japan, migrates between a lake and its tributaries (this is a landlocked form) (Kawanabe and Mizuno, 1989; Yuma *et al.*, 1998). The species spawns in rivers and from where the larvae drift down into the lake (Hidaka and Takahashi, 1987*b*). They live in the lake for one or two months, whereupon the juveniles massively ascend the river during July-September. The upstream migration begun after 09:00, was most active from 14:00 to 18:00, and ceased around 18:00 just before sunset (Yuma *et al.*, 2000).

Schooling

Schooling has an anti-predator and foraging function (Pitcher and Parrish, 1993). The journey undertaken by the post-larvae from the sea to the river, inevitably leads to encounters with predators. Major predators would include open water predators such as *Kuhlia* sp., and ambush predators such as *Eleotris* sp. (Kinzie, 1992; Kido, 1996*a*; Keith *et al.*, 2002a). As a protection, some amphidromous goby species have developed a strategy of migrating in schools. Schooling behaviour is present only in post-larvae, and rarely occurs in either juveniles or adults. In *Lentipes concolor*, the post-larvae do not swim in schools, while *Awaous guamensis* post-larvae swim in pairs, in single species schools, or in mixed species schools of 30 to 100 individuals (Tate, 1997).

In the Indo-Pacific region, *Sicyopterus lagocephalus* post-larvae travel alone or in small single or mixed schools of some 30 to 100 individuals, depending on the islands concerned: mass movements of "bichique" (a local name for the post-larvae) schools in La Reunion Island (Delacroix, 1987; Keith *et al.*, 2006a), smaller schools in Tahiti (Keith *et al.*, 2002a), and even smaller ones in New Caledonia (Keith, 2003; Marquet *et al.*, 2003; Lord and Keith, 2007). In Indonesia, large aggregations of *Stiphodon semoni* Weber, 1895, were observed undertaking a migration upstream from the estuary (Martin, 1999). In the Caribbean, the *Sicydium* species in the streams of Guadeloupe (West Indies) (Gillet, 1983; Hostache, 1992;

Fièvet and Le Guennec, 1998; Fièvet, 1999), Martinique (Lim *et al.*, 2002), Puerto Rico (Nieves, 1999), Costa Rica (Bussing, 1996), Dominica (Watson, 2000) and others are known to undertake mass migrations.

Metamorphosis

Soon after entering freshwaters, the post-larvae of sicydiines species undergo several changes in colour and fin shape (Manacop, 1953; Tomihama, 1972; Lau, 1973; Erdman, 1961, 1986; Keith *et al.*, 2008). The body of these post-larvae is mostly translucent and light coloured when they enter the estuary, except for the eyes and internal organs. The tail is mainly forked but gradually becomes truncated and the body acquires pigments as it metamorphoses (Nishimoto and Kuamo'o, 1997; Keith *et al.*, 2008). Pectoral and tail fins grow bigger, along with the pelvic disc needed for upstream movements. The mouth and the jaw structure are also modified, as well as the digestive system (Tomihama, 1972; Schoenfuss *et al.*, 1997). The post-larvae must switch from a planktonic feeding mode (copepods) to a benthic feeding mode (periphyton), when they "graze" on the substratum (Bielsa *et al.*, 2003). The rows of teeth will regularly be replaced so as to allow grazing such as in *Sicyopterus japonicus* (Mochizuki and Fukui, 1983).

In *Sicyopterus stimpsoni,* Schoenfuss *et al.* (1997) reported the observation of the metamorphosis. When the post-larvae of this species migrate from the ocean to rivers, they will remain in the estuary while undergoing a total restructuration of the cranium. Forty eight hours after entering freshwaters, both the length of the snout, and the height and width of the head increase considerably, whilst the total body length and the length of the head remain unchanged. The upper lip is enlarged and the mouth moves from a terminal to an almost ventral position. When the metamorphosis is complete, *S. stimpsoni* is able to climb over waterfalls by using alternately its pelvic suction cup and its lips; they are also able to graze on the diatoms growing on rock surfaces. This metamorphosis is correlated to the changes in diet and spatial distribution between larvae and juveniles. Yamasaki *et al.* (2007) have given a precise description of post-larvae of *Stiphodon percnopterygionus*, distinguishing pelagic and settled stages. A deposition of a metamorphosis check in the otolith was concomitant with a change in head morphology associated with a transition in feeding behaviour. Metamorphosis checks in otoliths provide further evidence of the usefulness of daily growth increments in reconstructing the early life history of diadromous fishes (Shen and Tzeng, 2002; Keith *et al.*, 2008).

Keith *et al.* (2008) characterised in La Réunion Island, the development of the metamorphosis of *Sicyopterus lagocephalus* and its post-larval stages during river recolonisation. The results show that cranium reorganisation and pectoral fins transformation are part of the first and major transformation, completed less than two weeks after freshwater colonisation. These transformations, mobilizing all the available calcium, seem to be critical for the survival in their new environment. Then, individuals progressively acquire pigmentation, complete scalation, and the fork at the end of the caudal fin disappears. The end of the recruitment phase would seem to take place 21 to 28 days after entering the river mouth, when the territorial behaviour appears.

Keith *et al.* (2008) also described the different post-larval stages for *Sicyopterus lagocephalus.* The Post-Larval Stage 0 (PL0) is characterised by the beginning of the morphological transformation needed to recruit; the sucker is already formed and lateral scales start to appear. This stage takes place in salt water and individuals acquire the physiological and morphological abilities for the starting of the metamorphosis and colonization of the adult environment. This stage is complete as post-larvae enter the estuary. The Post-Larval Stage 1 (PL1) starts as soon as larvae arrive in the estuary. It is characterised by the beginning of river colonisation and by changes in osmoregulation. The sucker is already formed; the individual is practically translucent; the mouth is terminal. Metamorphosis of the head has started. The caudal fin is forked. Fish in this stage remain at the river mouth less than 48 hours. The rake-like teeth used for grazing are not or just slightly visible. This stage is complete when the pectoral fin has finished transforming and when the teeth start to appear clearly. The Post-Larval Stage 2 (PL2) show vertical coloured stripes on its sides. The mouth is sub-inferior, the head metamorphosis is ongoing. Individuals in PL2 have been in freshwater for 2 to 14 days. The rake-like teeth used for grazing are in place. This stage is complete when the metamorphosis of the cranium is completed (Keith *et al.*, 2008); it is mainly found in the lower course of the river. After this stage, individuals will become juveniles (see "upstream migration").

DIET

In adult freshwater gobiids, there are other types of food nature. Some are carnivorous, others are omnivorous, psammivorous or herbivorous. Generally, the diet changes between larval and juvenile stages. The most spectacular changes are in sicydiines herbivorous species. For example, the mouth and the jaw structure of *S. lagocephalus* change over time from larval to juvenile stages, during metamorphosis. The upper lip is enlarged, the mouth moves from a terminal to an almost ventral position, and the corner of mouth angle shows a logarithmic evolution over time; the angle changed by more than 35° (Keith *et al.*, 2008). Metamorphosis of feeding structures from the offshore larval phase is vital for the survival of the fish once it enters freshwater. The appearance of premaxillary teeth adapted to scraping diatoms off the substrate, typical of the *Sicyopterus* genus, is progressive from stage PL1 through PL2. These changes are essential to allow new recruits to begin feeding on benthic algae (Schoenfuss *et al.*, 1997; Keith *et al.*, 2008).

Fitzsimons *et al.* (2003) have shown that *Sicyopterus stimpsoni* from Hawaii maintained 'gardens' of low-growing periphyton in swift water on the upper surfaces of large pebbles and boulders. These conspicuous patches of diatoms and short filaments of green and blue-green algae represent a food source and the area for the initiation of stereotypical social behaviour, including territoriality and courtship. Patches are located primarily in the main flow of a stream on the top and upstream sides of large hard substrates that provide an unobstructed view for territorial males. The association of food patches with the presence of the fish is sufficiently consistent that the patches alone can be used to assay the presence of *S. stimpsoni* (Fitzsimons *et al.*, 2003). Observations in the western Pacific indicate that gardening behaviour is not unique to the Hawaiian species of *Sicyopterus* and may even occur in other genera of stream gobies that feed on small benthic periphyton (Keith, 2003).

Gut contents of post-larvae, juvenile, and adult specimens of the Hawaiian gobiid fish *Sicyopterus stimpsoni* were examined by Julius *et al.* (2005) and Sherwood and Nishimoto (2005) to catalog the periphyton ingested by this species. It was dominated by diatom species and shifted from taxa representative of a marine, planktonic community in post-larval fish to a freshwater, benthic community in juvenile and adult fish. This change in diet corresponds with the migration of post-larvae to freshwater streams just prior to juvenile development in which rapid modification in mouth anatomy makes ingestion of planktonic species difficult (Julius *et al.*, 2005). Benthic diatoms found with juvenile and adult fish assemblages represented multiple genera that live in a narrow set of environmental conditions. These diatoms grow during a specific period in the development of the benthic community in Hawaiian streams. This suggests a highly specialized dietary behaviour that depends heavily on continually restarting the benthic periphyton successional pattern, which appears to be regulated by the hydrological cycles of streams on the island (Julius *et al.*, 2005).

In La Reunion Island, the biomass and the biochemical composition of periphyton were studied during one year in three streams. The results revealed that nutritional conditions for fish development, in particular for *Sicyopterus lagocephalus*, were most favourable during the dry season. Food availability was strongly affected by high flow; particularly during the exceptional hydrological events (cyclonic high flows) (Bielsa *et al.*, 2003). Inspite of unfavourable nutritional conditions caused by hydrological disturbances, fish continue their reproduction which coincides with the summer high flow period. The significant depletion of lipid reserves, which express deterioration in animal condition, is influenced by flooding period. At this time of gonad development involving an additional energy investment, fish were not able to recover a stable nutritional state after the complete achievement of the reproduction cycle. Except during this period, there is a weak participation of the reproductive physiological processes in the seasonal changes of somatic composition of this fish (Bielsa *et al.*, 2003).

UPSTREAM MIGRATION, DISTRIBUTION AND HABITAT

Upstream Migration

When the metamorphosis of the cranium of *Sicyopterus* species is completed, then post-larvae become juveniles and begin a real upstream migration. Keith *et al.* (2008) described two different stages for *Sicyopterus lagocephalus*. The Juvenile Stage 1 (J1) is characterised by an inferior mouth, the head metamorphosis is complete (the otoliths now have a metamorphosis check), and individuals can be found in the lower to medium courses of the river; they have been in freshwater for 14 to 21 days. This stage is complete when pigmentation is complete and scale growth restarts. The Juvenile Stage 2 (J2) is characterised by a pigmentation similar to that of non-mature adults and a caudal fin border nearly straight. Individuals in J2 are found in the lower and medium courses of the river and have been in freshwater for more than 21 days. This stage is characterised by the start of isomeric growth and is complete when this individual is mature with a colour showing sexual dimorphism. After this stage, individuals are adult.

The timing and direction to migrate upstream for post-larvae and juveniles, towards the adult habitat, and mechanisms required for the success of this migration remain mostly unknown. However, three studies have been undertaken on how gobies orientate over short distances. *Rhinogobiops nicholsii* (Bean, 1882), from the Pacific Ocean, is able to memorise the location of a refuge after a brief reconnaissance and is thus able to retreat to safety if a predator is detected (Markel, 1994). It takes only one high tide for the Caribbean goby, *Bathygobius soporator* (Valenciennes, 1837), to memorise the spatial location of neighbouring water pools and, at low tide, it can jump from one to the next when chased by a predator (Aronson, 1971). Finally, under experimental conditions, *Awaous guamensis*, *Sicyopterus stimpsoni* and *Lentipes concolor* locate quickly (in less than ten minutes) the freshwater current (1 l min^{-1} into the test pool) that will allow them to swim upstream, even if the flow is interrupted (Smith and Smith, 1998). The rapid acquisition of such directional preferences for upstream migration is crucial to these juveniles if they are to clear obstacles in intermittent flows (Keith, 2003).

Rheotaxis might not be the only phenomenon involved. Studying directional mechanisms is important if one is to understand the phenomenon involved. The existence of a solar or a magnetic compass may also be uncovered for the marine phases of migration. Leis *et al.* (1996) demonstrated that, in coral reefs, fish larvae could maintain their direction beyond the reefs. Kobayashi (1989) reports visual orientation in *Psilogobius* larvae towards the reef. These studies suggest that visual (non-rheotaxic) parameters can also be used during the oceanic larval phases (Smith and Smith, 1998).

The juveniles must migrate from the lower or medium reaches of river habitats to the habitats they will occupy as adults. Out of the four genera living highest up in river systems (*Lentipes, Akihito, Sicyopterus, Sicyopus*), *Lentipes* swims directly towards the upper reaches, while others take more time to complete their migration (Keith, 2003). In *Lentipes concolor,* for example, post-larvae remain in the estuary no longer than one day, and they migrate directly upstream at a speed of 90 m h^{-1}, until they reach the first obstacle (first waterfall). When this obstacle is cleared they reach a "protected area" where predators are less numerous (Tate, 1997). Yuma *et al.* (2000) noted for *Rhinogobius* sp. an upstream migration speed of 4.2 to 8.9 m min^{-1} depending on current velocities. Beyond the first major waterfall which acts as a barrier against predators; schooling and escape behaviours are replaced by agonistic interactions between juveniles competing for space and food (Keith, 2003).

In *Awaous guamensis,* many individuals spend over four weeks in the estuary. This was revealed due to a study on helminth parasites. Indeed, *Camallanus cottii*, an estuarine parasitic nematode, has been found in the stomach of the post-larvae of this species, and it takes four weeks for this parasite to become adult (Font and Tate, 1994).

Sicyopterus lagocephalus spend over 2-3 weeks in the lower part of the river in La Reunion Island (Keith *et al.*, 2008).

Waterfall Climbing Performance

In the family Gobiidae, the pelvic fins are joined together in a pelvic disc. In the subfamily Sicydiinae, the disk forms a kind of sucker fused to the abdomen along its entire length. Their strongly effective pelvic suction cup and well-developed pectoral fins allow them to access rapidly the upper reaches of the catchement area (Keith, 2003). To reach adult habitats, juveniles climb waterfalls up to 350 m high, over 10 000 times their body length. The demands of moving through such an extreme environment could constrain the range of viable locomotor mechanisms that these fish use. Previous qualitative observations indicated that *Lentipes concolor* and *Awaous guamensis* use 'powerbursts' of axial undulation to climb, whereas *Sicyopterus stimpsoni* `inches up' vertical surfaces by alternately attaching oral and pelvic suckers to the substrate (Schoenfuss and Blob, 2003). Bouts of powerburst climbing by *L. concolor* and *A. guamensis* typically begin in or near direct water flow and are initiated by a single, rapid adduction of the pectoral fins. Powerburst climbing bouts are rapid, but short in duration with few continuous locomotor cycles. Powerburst climbers use high amplitude undulations along the entire body, but minimum resultant velocities of these undulations are high. This suggests that these species may hybridize terrestrial propulsive mechanisms with aquatic mechanisms (Schoenfuss and Blob, 2003).

In contrast, climbing by inching in *S. stimpsoni* involves little axial undulation or fin movement. *Sicyopterus stimpsoni* typically exit the water outside of direct flow and seem to use terrestrial propulsive mechanisms. As the oral disc attaches to the substrate, it expands to almost twice its resting area, after which the posterior body is pulled upwards; once the pelvic disc attaches, the oral disc releases and the anterior body advances. The mouth is used as a secondary locomotor organ. The unusual behaviour and ontogenetic strategy of *S. stimpsoni* seem to be evolutionary novelties, rather than ancestral retentions, suggesting that the evolution of these features may have been closely correlated (Schoenfuss and Blob, 2003).

Potential effects of body size changes on locomotor performance were studied by Maie *et al.* (2007) in *Sicyopterus stimpsoni* from the Hawaiian Islands and *Sicydium punctatum* from Dominica, West Indies. They noted that the sucker area scaled isometrically with body length for both species, suggesting that climbing performance might decline as these fishes grow. Although the pectoral fin scaled differently in the two species, caudal-fin height showed strong positive allometry relative to body length in both species. The differing scaling patterns of sucker area and caudal-fin height relative to body length may reflect an increase in the importance of swimming over climbing in sicydiine gobies as they grow larger. Moreover, similarities in the scaling patterns of gobies from geographically distant islands suggest that species in this lineage may be subject to common ontogenetic constraints on performance (Maie *et al.*, 2007).

The importance of climbing for juvenile stream gobies is well established, but adult fish in upstream island habitats also face potential downstream displacement by periodic disturbances (Blob *et al.*, 2007). Thus, retention of climbing ability could be advantageous for adult stream gobies. Climbing performance might be expected to decline among adults, however, due to the tendency for mass-specific muscular power production to decrease with body size (Blob *et al.*, 2007).

To evaluate changes in waterfall-climbing ability with body size in Hawaiian stream gobies, Blob *et al.* (2007) compared climbing performance and kinematics between adults and juveniles from three species (*Awaous guamensis*, *Sicyopterus stimpsoni* and *Lentipes concolor*). For species in which juveniles climbed using 'powerbursts' of axial undulation, adult performance and kinematics showed marked changes: adult *A. guamensis* failed to climb, and adult *L. concolor* used multiple pectoral fin adductions to crutch up surfaces at slow speeds, rather than rapid powerbursts. Adult *S. stimpsoni*, like juveniles, still used oral and pelvic suckers to 'inch' up surfaces and climbed at speeds comparable to those of juveniles. However, unlike juveniles, adult *S. stimpsoni* also add pectoral fin crutching to every climbing cycle. Thus, although powerburst species appear to be particularly susceptible to size-related declines in waterfall-climbing performance, the addition of compensatory mechanisms prevents the loss of this novel function in some species (Blob *et al.*, 2007)

PREFERENTIAL HABITATS AND ZONATION

On the Opunohu River, in Moorea (French Polynesia), Keith *et al.* (2002a) showed that in *Sicyopterus lagocephalus* the numbers of adults increased while the numbers of juveniles decreased with distance from the river mouth. Yuma *et al.* (2000) showed that the number of upstream migrating juveniles of *Rhinogobius* sp. decreased progressively from the river mouth. They supposed that riffles in the mid- and upper reaches, having few slow current areas, probably presented barriers to upstream migration of the juvenile gobies, although both adults and subadults of the species are found in upper reaches beyond these obstacles. Balon and Bruton (1994), showed that the distribution of *S. lagocephalus* in Comoros Islands separated the smallest juveniles in the estuary and largest adults on the volcanic slopes. Finally, Keith *et al.* (2008), showed that post-larvae of this species in La Reunion island stayed in lower course of the river during three weeks before upstream migration.

In *Sicyopterus stimpsoni*, two forms access the rivers (Nishimoto and Fitzsimons, 1986). The first form develops coloured body markings used in signalling to defend a feeding territory; the second form does not develop such markings and individuals migrate further upstream when they are disturbed by the more colourful and aggressive form. The species' distribution along the river is determined by the aggression by coloured post-larvae and subsequent migration upstream of the least coloured individuals. When the least coloured post-larvae reach a site with a lower density of competitors, they stop migrating and develop coloured signals. At this time aggression becomes important in the distribution of the species, when adults and coloured juveniles displace non-coloured young (Nishimoto and Fitzsimons, 1986). Males of adjacent territories show no, or little, aggressive behaviour among themselves because the boundaries are exactly defined, however they will chase males passing through their territory (Nishimoto and Fitzsimons, 1986; Osugi *et al.*, 1998).

In the genus *Lentipes*, the adults are territorial (especially the males) (Fitzsimons and Nishimoto, 1990; Watson *et al.*, 2002). In *Lentipes concolor*, males establish a territory in the upper part of the river. These territories are always situated in areas that experience strong currents and good water clarity and vary in size from 0.1 to 4 m^2 (Fitzsimons and Nishimoto, 1990) and are larger in still waters (pools), and smaller in riffles. The size of this territory does not appear to be affected by the density of *Lentipes*, including the number of available females, and the number of competing males, or the size of the males (Fitzsimons and Nishimoto, 1990). Outside the mating season, the male attaches itself to one or more observation posts close to the centre of its territory to detect the presence of other males or drifting food. The same male may use the same territory for up to four and a half years (Fitzsimons and Nishimoto, 1990). The females move between the territories of different males and are courted when gravid. However, a number of females have been reported on the same territory for 32 months (Fitzsimons and Nishimoto, 1990).

Beyond the 'barrier' of the first waterfalls, most juveniles will develop dark colours and defend a temporary feeding and resting territory, forcing others of the same species to migrate upstream. In some species, the adult population is redistributed when adults swim downstream for reproduction, which creates space upstream for the juveniles (Ego, 1956).

As a rule, whatever the island studied, the first major waterfall is a crucial factor in species distribution along a river. This zonation has been described for the Arakitoach River in Palau (Fehlmann, 1960), the Asmafines River in Guam (Parham, 1995), the Pohnpei rivers (Maciolek and Ford, 1987), the rivers of Martinique island (Lim *et al.*, 2002) and in Puerto Rican streams (Holmquist *et al.*, 1998).

Keith *et al.* (2002a), Marquet *et al.* (2003) and Fossati *et al.* (2002) showed a specific longitudinal distribution in adult amphidromous gobies and decapod crustaceans inhabiting Polynesian and New Caledonian rivers. *Lentipes* are often the species present at the highest altitudes: in Hawaii, *L. concolor* is found at an altitude of up to 1000 m high, beyond a 130 m waterfall and multiple other falls adding up to 600 m of obstacles (Fitzsimons and Nishimoto, 1990; Nishimoto and Kuamo'o, 1991), while, in the Marquesas Islands, *Lentipes rubrofasciatus* Maugé, Marquet and Laboute, 1992, is found at an altitude of up to 500 m (Keith *et al.*, 2002a). In French Polynesia, New Caledonia, Mascareignes, Comoros, *Awaous* and *Stenogobius* are normally found below the lower reaches and the first waterfall (Keith *et al.*, 2002b, 2006a; Marquet *et al.*, 2003). A number of *Stiphodon* species do not reach beyond the major falls (5 to 10 m) (Keith *et al.*, 2002c), while other species, like *Sicyopus* swim a little further up; *Sicyopterus* clear major falls and may swim up to the upper reaches, where they mix with *Lentipes* (Keith *et al.*, 2002a). In the Caribbean, *Sicydium plumeri* (Bloch, 1786) may swim up to the upper reaches and can climb major waterfalls (Nieves, 1999; Lim *et al.*, 2002), and *Sicydium adelum* Bussing, 1996 is found between 10 and 800 m elevation in the streams of Costa Rica (Bussing, 1996).

Predatory species of the genus *Kuhlia* (Kuhliidae) abound in the lower reaches, below the first waterfall, but are absent beyond it; this is the case for *Kuhlia malo* (Val. *in* Cuvier et Valenciennes, 1831) in Polynesia (Keith *et al.*, 2002a), as well as for *K. marginata* (Cuvier, 1829) and *Kuhlia rupestris* (Lacepède, 1802) in New Caledonia (Keith, 2003; Marquet *et al.*, 2003). This limitation in the movements of predators of the genus *Kuhlia* affects the distribution of other organisms. In Guam, Parham (1995) showed that the type of habitat (riffle, pool, etc.) and the type of substratum determine species distribution and abundance. This is the case in Pohnpei and Palau (Nelson *et al.*, 1997), mainly for herbivorous species. The positive correlation between *Sicyopterus* density and the proportion of hard substratum is the result of the diet of these herbivorous species. These gobies feed by scraping diatoms or algae from hard surfaces in the river. Many of these species have teeth adapted for scraping (Kido, 1996a, b). Although the site with the highest density of gobies is the site with the highest proportion of hard substratum, the wide differences reported even for areas with large amounts of hard substratum show that other factors are to be taken into consideration (Nelson *et al.*, 1997).

The amount of plant cover on the banks of the river is another important factor where gobies' habitat is concerned. Nelson *et al.* (1997) showed that fish density is highest where plant cover is poorest, probably because primary productivity on rocks is higher in sunny locations, which provides more algae for the gobies to feed on. This might not be the only reason: it is possible that light plays a part in propagating the iridescent colours of displaying males (Keith, 2003). However, recent observations revealed that it is necessary to maintain also an important vegetational cover on the river banks: indeed it allows the water to stay cool and it provides sufficient water and nutrient supplies for omnivorous species. Whatever the cause, vegetation cover on river banks is a contributing factor in the distribution of gobies within streams. The local density of the grazing goby (*Sicyopterus japonicus*) and benthic algal production were investigated in the northern part of Okinawa Island, Japan (Abe *et al.*, 2003). The results indicate that the spatial heterogeneity in riparian shade controls the habitat use of the goby, with open-canopy areas with high algal renewal rates being preferred (Abe *et al.*, 2003, 2007).

Understanding the role of these two factors (density/plant cover, and density/hard substratum) is the first step towards ascertaining what the main habitat factors that are important for amphidromous gobies, and towards predicting the ecological impact of any environmental disturbance or modification. Working on fish populations in Sri Lanka, Moyle and Senanayake (1984) determined an "ecological key" for the species, based on current speed, substratum, depth and diet. Detailed ecological data on amphidromous gobies are required if one is to build and refine such "ecological keys", and use them with enough precision in management and conservation programs.

Conclusion

Tropical freshwater gobies have developed a specific life cycle. This group of fish is highly diverse, mainly in Indo-Pacific area. The mechanisms explaining species distribution and including larval production, retention, dispersal, and recruitment to freshwater are governed both by biological and physiological processes (Murphy and Cowan, 2007; Keith *et al.*, 2008).

Seasonal variables (e.g. rainfall, drought, floods, typhoons) have a major impact on the survival of populations: biological events such as reproduction, spawning, and the dispersal of larvae, are dependant on these events and are synchronised with them (Keith, 2003). On the islands, the impact of humans on aquatic habitats is highly significant, particularly on estuarine habitats which are crucial to amphidromous species. These have to undertake two migrations between freshwaters and the sea. The success of such a life cycle—i.e. on the one hand for larvae's downstream migration after hatching, and on the other hand for the post-larval and juvenile upstream colonisation after recruitment in freshwaters—depends on maintaining the mountain-ocean corridor open to allow movements between both habitats (McDowall, 1997; Fitzsimons *et al.*, 1997; Radtke *et al.*, 2001; Keith, 2003; Keith *et al.*, 2008). Anthropogenic alterations of the environment may have an impact on the performance of species, on their ability to find adequate food and to colonise the streams in altitude (Lord and Keith, 2006, 2007). Man-made developments on these streams can alter larval dispersion and therefore the recruitment success (Brasher, 2003; Houde, 1989). In other words, the lack or a reduced number of new recruits of amphidromous gobies might serve as an indicator of long-term health of estuaries (Keith *et al.*, 2008). The current state of knowledge on tropical freshwater gobies' life cycle is of direct relevance for management and conservation (Radtke *et al.*, 2001; Keith, 2003).

Acknowledgments

Many colleagues have contributed to provide data and discussion that have led to the realisation of this work. We want to thank : G.R. Allen (CI), J. Aoyama (ORI), P. Béarez (MNHN), K.N.I. Bell, P. Bosc (ARDA), E. Feunteun (MNHN), R. Galzin (EPHE), P. Gerbeaux (IUCN), M. Iida (ORI), M. Jégu (MNHN), A. Jenkins (WI), D. Kalfatak (EU), M. Kottelat, M. Kovacic (NHMR), H. Larson (NTM), P. Lim (ENSAT), G. Marquet, F. Meunier (MNHN), R.A. Patzner (Univ. Salzburg), F. Pezold (UL), S. Planes (EPHE), G. Ségura (ISE), K. Tsukamoto (ORI), S. Watanabe (ORI), P. Valade (ARDA), R. Vari (USMNH), E. Vigneux (HFC), R.E. Watson (UF), J. Williams (USNM).

The results obtained by the MNHN team were financially supported by the New Caledonian Government, the French Ichthyological Society, the French Ministery of Ecology, the French Foundation, the National Research Agency and the MNHN. We thank also the following for help and assistance during the various studies, trips and expeditions to Indian and Pacific Islands: the Environment Unit of Vanuatu, the Seychelles Environment Ministry, Conservation International from Madagascar and Pacific, the Provincial and Territorial authorities of New Caledonia and Wallis and Futuna, ARDA from La Réunion Island, and CNDRS from Comoros.

References

Abe, S., T. Yodo, N. Matsubara and K. Iguchi. 2007. Distribution of two sympatric amphidromous grazing fish *Plecoglossus altivelis* Temminck and Schlegel and *Sicyopterus japonicus* (Tanaka) along the course of a temperate river. *Hydrobiologia* 575: 415-422.

Abe, S., K. Iguchi, S. Ito, Y. Uchida, H. Ohnishi and K. Ohmori. 2003. Habitat use of the grazing goby (*sicyopterus japonicus*) in response to spatial heterogeneity in riparian shade. *Journal of Freshwater Ecology* 18: 161-168.

Aronson, L.R. 1971. Further studies on orientation and jumping behaviour in the gobiid fish, *Bathygobius soporator. Annals of New York Academy of Science* 188: 378-407.

Auty, H. 1978. Reproductive behaviour and early development of the empire fish *Hypseleotris compressus* (Eleotridae). *Australian Journal of Marine and Freshwater Research* 29: 585-597.

Balon, E.K. and M.N. Bruton. 1994. Fishes of the Tatinga River, Comoros, with comments on freshwater amphidromy in the goby *Sicyopterus lagocephalus. Ichthyological Exploration of Freshwaters* 5: 25-40.

Bell, K.N.I. 1994. *Life cycle, early life history, fisheries and recruitment dynamics of diadromous gobies of Dominica, W.I., emphasising* Sicydium punctatum *Perugia.* PhD Thesis, Memorial University of Newfoundland.

Bell, K.N.I. 1999. An overview of goby-fry fisheries. *Naga Manila* 22: 30-36.

Bell, K.N.I. and J.A. Brown. 1994. Active salinity choice and enhanced swimming endurance in 0-8-day-old larvae of diadromous gobies, including *Sicydium punctatum* (Pisces), in Dominica West Indies. *Marine Biology* 121: 409-417.

Bell, K.N.I., P. Pepin and J.A. Brown. 1995. Seasonal, inverse cycling of length and age at recruitment in the diadromous gobies *Sicydium punctatum* and *Sicydium antillarum* in Dominica, West Indies. *Canadian Journal of Fisheries and Aquatic Science* 52: 1535-1545.

Berrebi, P., T. Galewski and P. Keith. 2006. *Sicyopterus "lagocephalus"*, a unique widespread taxon confirmed by mtDNA sequences. *Molecular Phylogenetics and Evolution* 40: 903-904.

Bielsa, S., P. Francisco, S. Mastrorillo and J.P. Parent. 2003. Seasonal changes of periphytic nutritive quality for *Sicyopterus lagocephalus* (Pallas, 1770) (Gobiidae) in three streams of La Reunion Island. *Annales de Limnologie* 39: 115-127.

Blob, R.W., K.M. Wright, M. Becker, T. Maie, T.J. Iverson, M.L. Julius and H.L. Schoenfuss. 2007. Ontogenetic change in novel functions: waterfall climbing in adult Hawaiian gobiid fishes. *Journal of Zoology* 273: 200-209.

Brasher, A.M.D. 2003. Impact of human disturbances on biotic communities in Hawaiian streams. *Bioscience* 53: 1052-1060.

Briggs, J.C. 1974. *Marine Zoogeography*. McGraw Hill, New York.

Bussing, W.A. 1996. *Sicydium adelum*, new species of gobiid fish (Pisces: Gobiidae) from Atlantic slope streams of Costa Rica. *Revue de Biologie Tropicale* 44 (2B): 819-825.

Chen, I.-S. and M. Kottelat. 2003. Three new freshwater gobies of the genus *Rhinogobius* (Teleostei:Gobiidae) from Northeastern Laos. *The Raffles Bulletin of Zoology* 51: 87-95.

Cole, K.S. 1982. Male reproductive behaviour and spawning success in a temperate zone goby, *Coryphopterus nicholsi. Canadian Journal of Zoology* 60: 2309-2316.

Daoulas, C.H., A.N. Economou, Th. Psarras and R. Barbieri-Tseliki. 1993. Reproductive strategies and early development of three freshwater gobies. *Journal of Fish Biology* 42: 749-776.

Delacroix, P. 1987. *Etude des "bichiques", juvéniles des* Sicyopterus lagocephalus *(Pallas), poissons Gobiidae migrateurs des rivières de la Réunion (Océan Indien): exploitation, répartition, biologie de la reproduction et de la croissance.* PhD Thesis, University of La Reunion.

Delacroix, P. and A. Champeau. 1992. Ponte en eau douce de *Sicyopterus lagocephalus* (Pallas) poisson Gobiidae amphibionte des rivières de la Réunion. *Hydroécologie Appliquée* 4: 49-63.

Dotu, Y. 1961. The bionomics and life history of the gobioid fish, *Rhinogobius giurinus* (Rutter). *Bulletin of the Faculty of Fisheries, Nagasaki University* 10: 120-126.

Dotu, Y. and S. Mito. 1955. Life history of the gobioid fish, *Sicydium japonicum* Tanaka. *Science Bulletin of the Faculty of Agriculture, Kyushu University* 15: 213-221.

Ekman, S. 1953. *Zoogeography of the Sea.* Sidgwick and Jackson, London.

Ego, K. 1956. Life history of freshwater gobies. In: *Freshwater Games Fish Management Research,* Project No. 4-4-R. Department of Land and Natural Ressources, Honolulu, pp. 1-23.

Erdman, D.S. 1961. Notes on the biology of the gobiid fish *Sicydium plumieri* in Puerto Rico. *Bulletin of Marine Science* 11: 448-456.

Erdman, D.S. 1986. The green stream goby *Sicydium plumeri,* in Puerto Rico. *Tropical Fish Hobbyist* 2: 70-74.

Fehlmann, H.A. 1960. *Ecological distribution of fishes in a stream drainage in the Palau islands.* Ph.D. Thesis, Stanford University.

Fièvet, E. 1999. *Crevettes et poissons amphidromes des cours d'eau amenagés de Guadeloupe : exemple de relation dynamique aval-amont.* Ph.D. Thesis, Université Claude-Bernard, Lyon I.

Fièvet, E. and B. Le-Guennec. 1998. Migration de masse de *Sicydium* spp. (Gobiidae) dans les rivières de Guadeloupe: implications pour le schéma hydraulique des mini-centrales hydroelectriques au fil de l'eau. *Cybium* 22: 293-296.

Fitzsimons, J.M. and R.T. Nishimoto. 1990. Territories and site tenacity in males of the Hawaiian stream goby *Lentipes concolor* (Pisces : Gobiidae). *Ichthyological Exploration of Freshwaters* 1: 185-189.

Fitzsimons, J.M., R.T. Nishimoto and A.R. Yuen. 1993. Courtship and behavior in the native Hawaiian stream goby, *Sicyopterus stimpsoni*. *Ichthyological Exploration of Freshwaters* 4: 1-10.

Fitzsimons, J.M., H.L. Schoenfuss and T.C. Schoenfuss. 1997. Significance of unimpeded flows in limiting the transmission of parasites from exotics to Hawaiian stream fishes. *Micronesica* 30: 117-125.

Fitzsimons, J.M., M.G. McRae, H.L. Schoenfuss and R.T. Nishimoto. 2003. Gardening behavior in the amphidromous Hawaiian fish *Sicyopterus stimpsoni* (Osteicthyes: Gobiidae). *Ichthyological Exploration of Freshwaters* 14: 185-191.

Font, W.F. and D.C. Tate. 1994. Helminth parasites of native Hawaiian freshwater fishes: an exemple of extreme ecological isolation. *Journal of Parasitology* 80: 682-688.

Fossati, O., M. Mosseron and P. Keith. 2002. Atyidae (Crustacea: Decapoda) macro- and micro-distribution in the rivers of Nuku-Hiva island (French Polynesia), *Hydrobiologia* 472: 197-206.

Froese, R. and D. Pauly (eds). 2008. FishBase. World Wide Web electronic publication. www.fishbase.org, version (07/2008)

Gibson, R.N. 1970. Observations on the biology of the giant goby *Gobius cobitis* Pallas. *Journal of Fish Biology* 2: 281-288.

Gillet, C. 1983. Les peuplements de poissons et de crevettes des rivières de la Guadeloupe : Quelques données sur la biologie, la reproduction, la répartition des espèces. *Revue d'Hydrobiologie Tropicale* 16: 327-340.

Ha, P. and R.A. Kinzie III. 1996. Reproductive biology of *Awaous guamensis*, an amphidromous Hawaiian goby. *Environmental Biology of Fishes* 48: 383-396.

Han, K.H., Y.U. Kim and K.J. Choe. 1998. Spawning behavior and development of eggs and larvae of the Korea freshwater goby *Rhinogobius brunneus* (Gobiidae: Perciformes). *Journal of Korean Fisheries Society* 31: 114-120.

Harrison, J.T., R.A. Kinzie III, G.C. Smith, P.Y. Tamaru, D. Heacock, L. Honigman, A. Neuman, J.D. Parrish and M. Kido. 1991. *Baseline studies of the Hanalei River and other north shore rivers, Kauai*. Final Report. University of Hawaii Environmental Center, Honolulu.

Hayashi, M. 1972. A study on the freshwater goby, 'Yoshinobori' *Rhinogobius brunneus* Temminck et Schlegel in the Maedagawa River, Yokosuka City. *Sciences Report of Yokosuka City Museum* 19: 12-19.

Hidaka, T. and S. Takahashi. 1987a. Reproductive strategy and interspecific competition in the lake-living gobiid fish Isaza, *Chaenogobius isaza*. *Journal of Ethology* 5: 185-196.

Hidaka, T. and S. Takahashi. 1987b. Effects of temperature and day length on gonadal development of a goby, *Rhinogobius brunneus* (orange type). *Japanese Journal of Ichthyologie* 34: 361-367.

Hirashima, K. and K. Tachihara. 2000. Embryonic development and morphological changes in larvae and juveniles of two land-locked gobies, *Rhinogobius* spp. (Gobiidae), on Okinawa Island. *Japanese Journal of Ichthyology* 47: 29-41.

Hoareau, T. 2005. Dynamique structurale des populations de bichiques (*Sicyopterus lagocephalus*), Gobiidae amphidromes des rivières de La Réunion. PhD Thesis, University of La Reunion.

Hoareau, T., R. Lecomte-Finiger, H.-P. Grondin, C. Conand and P. Berrebi. 2007. Oceanic larval life of La Reunion 'bichiques', amphidromous gobiid post-larvae. *Marine Ecology Progress Series* 333: 303-308.

Holmquist, J.G., J.M. Schmidt-Gengenbach and B.B. Yoshioka. 1998. High dams and marine-freshwater linkages: effects on native and introduced fauna in the Caribbean. *Conservation Biology* 12: 621-630.

Hostache, G. 1992. *La vie dans les eaux douces de la Guadeloupe*. PNG, INRA, Pointe à Pitre.

Houde, E.D. 1989. Subtleties and episodes in the early life of fishes. *Journal of Fish Biology* 35 (Supplement A): 29-38.

Iida M., S. Watanabe, A. Shinoda and K. Tsukamoto, 2008. Recruitment of the amphidromous goby *Sicyopterus japonicus* to the estuary of the Ota river, Wakayama, Japan. *Environmental Biology of Fishes* (online March 2008).

Iguchi, K. and N. Mizuno. 1990. Diel changes of larval drift among amphidromous gobies in Japan, especially *Rhinogobius brunneus*. *Journal of Fish Biology* 37: 255-264.

Iguchi, K. and N. Mizuno. 1991. Mechanisms of embryonic drift in the amphidromous goby. *Rhinogobius brunneus*. *Environmental Biology of Fishes* 31: 295-300.

Ito, S. and Y. Yanagisawa. 2000. Mate choice and cannibalism in a natural population of a stream goby, *Rhinogobius* sp. *Ichthyological Research* 47: 51-58.

Julius, M.L., R.W. Blob and H.L. Schoenfuss. 2005. The survival of *Sicyopterus stimpsoni*, an endemic amphidromous Hawaiian gobiid fish, relies on the hydrological cycles of streams: evidence from changes in algal composition of diet through growth stages fish. *Aquatic Ecology* 39: 473-484.

Kawanabe, H. and N. Mizuno. 1989. *Freshwater Fishes of Japan*. Yama-to-Keikoku-sha, Tokyo.

Keith, P. 2002. Threatened fishes of the world: *Rhyacichthys guilberti* Dingerkus and Séret, 1992 (Rhyacichthyidae). *Environmental Biology of Fishes* 63: 40.

Keith, P. 2003. Biology and ecology of amphidromous Gobiidae in the Indo-Pacific and the Caribbean regions. *Journal of Fish Biology* 63: 831-847.

Keith, P. and L. Marion. 2002. Methodology for drawing up a Red List of threatened freshwater fish in France. *Aquatic Conservation* 12: 169-179.

Keith, P. and G. Marquet. 2007. *Stiphodon rubromaculatus*, a new species of freshwater goby from Futuna island (Teleostei: Gobioidei: Sicydiinae). *Cybium* 31: 45-49.

Keith, P., E. Vigneux and P. Bosc. 1999. *Atlas des poissons et crustacés d'eau douce de la Réunion*. Patrimoines naturels, vol. 39. Museum National d'Histoire Naturelle, Paris.

Keith, P., R.E. Watson and G. Marquet. 2000. Découverte d'*Awaous ocellaris* (Broussonet, 1782) (Perciformes, Gobiidae) en Nouvelle-Calédonie et au Vanuatu et conséquences biogéographiques. *Cybium* 24: 350-400.

Keith, P., E. Vigneux and G. Marquet. 2002a. *Atlas des poissons et des Crustacés d'eau douce de Polynésie française*. Patrimoines naturels, vol. 55. Museum National d'Histoire Naturelle, Paris.

Keith, P., R.E. Watson and G. Marquet. 2002b. *Stenogobius (insularigobius) yateiensis*, a new species of freshwater goby from New Caledonia (Teleostei: Gobioidei). *Bulletin Français de Pêche et de Pisciculture* 364: 187-196.

Keith, P., R.E. Watson and G. Marquet. 2002c. *Stiphodon julieni*, a new species of freshwater goby from Rapa, French Polynesia (Teleostei: Gobioidei). *Bulletin Français de Pêche et de Pisciculture* 364: 1-12.

Keith, P., G. Marquet and R.E. Watson. 2007. *Akihito futuna*, a new species of freshwater goby from the South Pacific (Teleostei: Gobioidei: Sicydiinae). *Cybium* 31: 471-476.

Keith, P., Lord C., Marquet G. and D. Kalfatak. 2009. Biodiversity and biogeography of amphidromous fishes from New Caledonia, a comparison to Vanuatu. *Mémoires du Muséum national d'Histoire naturelle.*

Keith P., T. Galewski, G. Cattaneo-Berrebi, T. Hoareau and P. Berrebi. 2005a. Ubiquity of *Sicyopterus lagocephalus* (Teleostei: Gobioidei) and phylogeography of the genus *Sicyopterus* in the Indo-Pacific area inferred from mitochondrial cytochrome *b* gene. *Molecular Phylogenetics and Evolution* 37: 721-732.

Keith, P., T. Hoareau and P. Bosc. 2005b. A new species of freshwater goby (Pisces: Teleostei: Gobioidei) from Mayotte island (Comoros) and comments about the genus *Cotylopus* endemic to Indian Ocean. *Journal of Natural History* 39: 1395-1405.

Keith, P., G. Marquet, P. Valade, P. Bosc and E. Vigneux. 2006a. *Atlas des poissons et crustacés d'eau douce des Comores, Mascareignes et Seychelles*. Patrimoines naturels, vol. 67, Muséum National d'Histoire Naturelle, Paris.

Keith, P., C. Lord and E. Vigneux. 2006b. In vivo observations on postlarval development of freshwater gobies and eleotrids from French Polynesia and New Caledonia. *Ichthyological Exploration of Freshwater* 17: 187-191.

Keith, P., T. Hoareau, C. Lord, O. Ah-Yane, G. Gimmoneau, T. Robinet and P. Valade. 2008. Characterisation of post-larval to juvenile stages, metamorphosis, and recruitment of an amphidromous goby, *Sicyopterus lagocephalus* (Pallas, 1767) (Teleostei: Gobiidae: Sicydiinae). *Marine and Freshwater Research* 59: 876-889.

Kido, M.H. 1996a. Morphological variation in feeding traits of native Hawaiian stream fishes. *Pacific Science* 50: 184-193.

Kido, M.H. 1996b. Diet and food selection in the endemic Hawaiian goby *Sicyopterus stimpsoni* (Pisces : Gobiidae). *Environmental Biology of Fishes* 45: 199-209.

Kido, M.H. and D.E. Heacock. 1992. The spawning ecology of o'opu-nakea (*Awaous stamineus*) in Wainiha River. In: *New Directions in Research Management and Conservation of Hawaiian Freshwater Stream, Biology and Fisheries Management,* W. Devick (ed.). Dept. of Aquatic Resources, Honolulu, pp. 142-157.

Kim, Y.U. and K.H. Han. 1989. Early life history of the marine animals. Egg development, larvae and juveniles of *Chaenogobius laevis* (Steindachner). *Bulletin of the Korean Fisheries Society* 22: 317-331.

Kinzie III, R.A. 1992. How unique are Hawaiian freshwater gobies? In: *New Directions in Research Management and Conservation of Hawaiian Freshwater Stream, Biology and Fisheries Management,* W. Devick (ed.). Dept. of Aquatic Resources, Honolulu, pp. 18-39.

Kinzie III, R.A. 1993a. Reproductive biology of an endemic, amphidromous goby *Lentipes concolor* in Hawaiian streams. *Environmental Biology of Fishes* 37: 257-268.

Kinzie III, R.A. 1993b. Evolution and life history patterns in freshwater gobies. *Micronesica* 30: 27-40.

Kobayashi, D.R. 1989. Fine-scale distribution of larval fishes: patterns and processes adjacent to coral reefs in Kaneohe Bay, Hawaii. *Marine Biology* 100: 285-293.

Ladd, H.S. 1960. Origin of the Pacific island molluscan fauna. *American Journal of Sciences* 258: 137-150.

Lau, E.T.K. 1973. *Dimorphism and speciation in the Hawaiian freshwater goby genus Lentipes*. M.S. Thesis, Department of Zoology. University of Hawaii, Honolulu, 31pp.

Leis, J.M., H.P.A. Sweatman and S.E. Reader. 1996. What the pelagic stages of coral reef fishes are doing out in blue water daytime field observations of larval behavioural capabilities. *Marine and Freshwater Research* 47: 401-411.

Lim, P., F. Meunier, P. Keith and P. Noël. 2002. *Atlas des poissons d'eau douce de la Martinique*. Patrimoines naturels, vol. 51. Muséum National d'Histoire Naturelle, Paris.

Lindstrom, D.P. 1998. *Reproduction, early development and larval transport dynamics of amphidromous Hawaiian Gobioids*. Ph.D. Thesis, University of Michigan, Ann Arbor.

Lindstrom, D.P. and C.L. Brown. 1994. Early development and biology of the amphidromous Hawaiian stream goby *Lentipes concolor*. In: *Systematics and Evolution of Indo-Pacific Fishes.* Proceedings of the Fourth Indo-Pacific Fish Conference. Faculty of Fisheries. Bangkok, Thaïland, pp. 397-409.

Lindstrom, D.P. 1999. Molecular species identification of newly hatched Hawaiian amphidromous gobioid larvae. *Marine Biotechnology* 1: 167-174.

Lord, C. and P. Keith. 2006. Threatened fishes of the world: *Protogobius attiti* (Watson and Pöllabauer, 1998) (Galaxiidae). *Environmental Biology of Fishes* 77: 101-102.

Lord, C., and P. Keith. 2007. Threatened fishes of the world: *Sicyopterus sarazini* Weber and De Beaufort (Gobiidae). *Environmental Biology of Fishes* (online 27/11/07).

Maciolek, J.A. and J.I. Ford. 1987. Macrofauna and environment of Nanpil-Kiepw River, Ponape, Easter Caroline Islands. *Bulletin of Marine Science* 4: 623-632.

Maie, T., H.L. Schoenfuss and R.W. Blob. 2007. Ontogenetic Scaling of Body Proportions in Waterfall-climbing Gobiid Fishes from Hawaii and Dominica: Implications for Locomotor Function. *Copeia* 2007: 755-764.

Manacop, P.R. 1953. The life history and habits of the goby *Sicyopterus extraneus* Herre (Anga) Gobiidae, with an account of the goby-fry fishery of Cagayan River, Oriental Misamis. *Philippine Journal of Fisheries* 2: 1-57.

Markel, R.W. 1994. An adaptive value of spatial learning and memory in the blackeye goby, *Coryphopterus nicholsi*. *Animal Behaviour* 47: 1462-1464.

Marquet, G., P. Keith and E. Vigneux. 2003. *Atlas des poissons et des Crustacés d'eau douce de Nouvelle-Calédonie*. Patrimoines naturels, vol. 56. Muséum National d'Histoire Naturelle, Paris.

Martin, K. 1999. Emergent aggregations in Semon's rock-climbing goby *Stiphodon semoni*. *Beagle* 15: 43-45.

McDowall, R.M. 1965. Studies on the biology of the red-finned bully *Gobiomorphus huttoni* (Ogliby) II. —Breeding and life history. *Transactions of the Royal Society of New Zealand* 5: 177-196.

McDowall, R.M. 1988. *Diadromy in Fishes: Migrations between Freshwater and Marine Environments*. Croom Helm, London.

McDowall, R.M. 1994. The Tarndale bully, *Gobiomorphus alpinus* Stokell (Pisces: Eleotridae) revisited and redescribed. *Journal of Royal Society of New Zealand* 24: 117-124.

McDowall, R.M. 1997. Is there such a thing as amphidromy? *Micronesica* 30: 3-14.

McDowall, R.M. 2004. Ancestry and amphidromy in island freshwater fish faunas. *Fish and Fisheries* 5: 75-85.

Meadows, R. and I. Meadows. 1988. Breeding the beautiful peacock gudgeon. *Tropical Fish Hobbyist* July 1988: 30-35.

Miller, P.J. 1984. The tokology of gobioid fishes. In: *Fish Reproduction: Strategies and Tactics*, G.W. Potts and R.J. Wooton (eds.). London Academic Press, pp. 119-152.

Mizuno, N. 1960. Study on a freshwater goby. *Rhinogobius similis* Gill, with a proposition on the relationships between landlocking and speciation of some freshwater gobies in Japan. *Memoirs Collection Sciences of University of Kyoto* B27: 97-115.

Mochizuki, K. and S. Fukui. 1983. Development and replacement of upper jaw teeth in gobiid fish, *Sicyopterus japonicus*. *Japan Journal of Ichthyology* 30: 27-36.

Moriyama, A., Y. Yanagisawa, N. Mizuno and K. Omori. 1998. Starvation of drifting goby larvae due to retention of free embryos in upstream reaches. *Environmental Biology of Fishes* 52: 321-329.

Moyle, P.B. and F.R. Senanayake. 1984. Resource partitioning among the fishes of rainforest streams in Sri Lanka. *Journal of Zoology London* 202: 195-223.

Murphy, C.A. and J.H. Cowan. 2007. Production, Marine Larval Retention or Dispersal, and Recruitment of Amphidromous Hawaiian Gobioids: Issues and Implications. *Bishop Museum Bulletin of Cultural and Environmental Studies* 3: 63-74.

Nelson, S.G., J.E. Parham, R.B. Tibbatts, F.A. Camacho, T. Leberer and B.D. Smith. 1997. Distribution and microhabitats of the amphidromous gobies in streams of Micronesia. *Micronesica* 30: 83-91.

Nieves, L.O. 1999. *Ecological study of the freshwater stream fishes of the upper Mameyes River (Bisley) in the Luquillo mountains of Puerto Rico*. Ph.D. Thesis, University of Michigan, Ann Arbor.

Nishimoto, R.T. and J.M. Fitzsimons. 1986. Courtship, territoriality, and coloration in the endemic Hawaiian freshwater goby, *Lentipes concolor*. In: *Indo-Pacific Fish Biology*. Proceedings of the 2nd International Conference on Indo-Pacific Fishes, Tokyo, pp. 811-817.

Nishimoto, R.T. and D.G.K. Kuamo'o. 1991. The occurrence and distribution of the native goby (*Lentipes concolor*) in Hawaiian Island streams with notes on the distribution of other native fish species. In: *Proceedings of the 1990 Symposium on Freshwater Stream Biology and Fisheries Management,* W. Devick (ed.). Dept. of Land and Natural Resources, Honolulu, pp. 77-95.

Nishimoto, R.T. and D.G.K. Kuamo'o. 1997. Recruitment of goby postlarvae into Hakalau stream, Hawaii island. *Micronesica* 30: 41-49.

Osugi, T., Y. Yanagisawa and N. Mizuno. 1998. Feeding of a benthic goby in a river where nektonic fishes are absent. *Environmental Biology of Fishes* 52: 331-343.

Parham, J.E. 1995. *Habitat use by an assemblage of tropical oceanic island streamfish*. MS Thesis in Biology, University of Guam.

Penaz, M. 2001. A general framework of fish ontogeny: a review of the ongoing debate. *Folia Zoologica* 50: 241-256.

Pezold, F., T. Iwamoto and I.J. Harrison. 2006. Multivariate analysis of sicydiines of São Tomé and Príncipe with redescription of *Sicydium brevifile* and *S. bustamantei* (Teleostei: Gobiidae) and a key to west African sicydiines. The California Academy of Sciences Gulf of Guinea Expedition. V. *Proceedings of California Academy of Science* (Ser. 4) 57 (no. 34): 965-980.

Pitcher, T.J. and J.K. Parrish. 1993. Functions of shoaling behavior in Teleosts. In: *Behavior of Teleost Fishes,* T.J. Pitcher (ed.). Chapman and Hall, London, pp. 363-439.

Palumbi, S.R. 1996. What can molecular genetics contribute to marine biogeography? An urchin's tale. *Journal of Experimental Marine Biology and Ecology* 203: 75-92.

Planes, S. and R. Galzin. 1997. New perspectives in biogeography of coral reef fish in the Pacific using phylogeography and population genetics approaches. *Vie Milieu* 4: 375-380.

Radtke, R.L., R.A. Kinzie III and S.D. Folsom. 1988. Age at recruitment of Hawaiian freshwater gobies. *Environmental Biology of Fishes* 23: 205-213.

Radtke, R.L., R.A. Kinzie III and D.J. Shafer. 1996. Life history transitions in the Hawaiian amphidromous goby, *Lentipes concolor*: otolith microstructure and microchemistry applications. Conference Paper. In: *76th Annual Meeting of the American Society of Icthyologists and Herpetologists*, New Orleans, pp. 13-19.

Radtke, R.L., R.A. Kinzie III and D.J. Shafer. 2001. Temporal and spatial variation in length of larval life and size at settlement of the Hawaiian amphidromous goby *Lentipes concolor*. *Journal of Fish Biology* 59: 928-938.

Rosen, B.R. 1988. Progress, problems and patterns in the biogeography of reef corals and other tropical marine organisms. *Helgen Meer* 42: 269-301.

Schoenfuss, H.L. and R.W. Blob. 2003. Kinematics of waterfall climbing in Hawaiian freshwater fishes (Gobiidae): vertical propulsion at the aquatic-terrestrial interface. *Journal of Zoology* 261: 191-205.

Schoenfuss, H.L., T.A. Blanchard and D.G. Kuamo'o. 1997. Metamorphosis in the cranium of postlarval *Sicyopterus stimpsoni*, an endemic Hawaiian stream goby. *Micronesica* 30: 93-104.

Shen, K.N. and W.N. Tzeng. 2002. Formation of a metamorphosis check in otoliths of the amphidromous goby *Sicyopterus japonicus*. *Marine Ecology Progress Series* 228: 205-211.

Shen, K.N., Y.C. Lee and W.N. Tzeng. 1998. Use of otolith microchemistry to investigate the life history pattern of gobies in a Taiwanese stream. *Zoological Studies* 37: 322-329.

Sherwood, A.R. and R.T. Nishimoto. 2005. Algal food preferences of the endemic Hawaiian stream goby *Sicyopterus stimpsoni* (Teleostei: Gobiidae). *Ichthyological Exploration of Freshwaters* 16: 97-106.

Springer, V.G. 1982. Pacific plate Biogeography with Special Reference to Shorefishes. *Smithsonian Contributions to Zoology* 367: 1-182

Smith, R.J. and M.J. Smith. 1998. Rapid acquisition of directional preferences by migratory juveniles of two amphidromous Hawaiian gobies, *Awaous guamensis* and *Sicyopterus stimpsoni*. *Environmental Biology of Fishes* 53: 275-292.

Sorensen, P.W. and K.A. Hobson. 2005. Stable isotope analysis of amphidromous Hawaiian gobies suggests their larvae spend a substantial period of time in freshwater river plumes. *Environmental Biology of Fishes* 74: 31-42.

Stoddart, D.R. 1992. Biogeography of the Tropical Pacific. *Pacific Science* 46: 276-293.

Takahashi, D. 2000. Conventional sex roles in an amphidromous *Rhinogobius* goby in which females exhibit nuptial coloration. *Ichthyological Research* 47: 303-306.

Takahashi, D. and Y. Yanagisawa. 1999. Breeding ecology of an amphidromous goby of the genus *Rhinogobius*. *Ichthyological Research* 46: 185-191.

Takahashi, D. and M. Kohda. 2001. Females of a stream goby choose mates that court in fast water currents. *Behaviour* 138: 937-946.

Takahashi, D., M. Kohda and Y. Yanagisawa. 2001. Male-male competition for large nests as a determinant of male mating success in a Japanese stream goby, *Rhinogobius* sp. DA. *Ichthyological Research* 48: 91-95.

Tamada, K. 2000. Distributions of the spawning grounds of four amphidromous gobies (*Rhinogobius* spp.) in the Aizu River, Wakayama Prefecture, Japan. *Japanese Journal of Ichthyology* 47: 55-59.

Tamada, K. 2001. Clutch size and egg size in three species of *Rhinogobius* complex dwelling in a single stream. *Japanese Journal of Ichthyology* 48: 49-52.

Tamada, K. 2005a. Egg and clutch sizes of a goby *Rhinogobius giurinus* in the Aizu River, Kii Peninsula, Japan. *Ichthyological Research* 52: 392-395.

Tamada, K. 2005b. Clutch and egg size in *Rhinogobius* sp. DA inhabiting small rivers. *Japanese Journal of Ichthyology* 52: 17-20.

Tamada, K. and K. Iwata. 2005. Intra-specific variations of egg size, clutch size and larval survival related to maternal size in amphidromous *Rhinogobius* goby. *Environmental Biology of Fishes* 73: 379-389.

Tate, D.C. 1995. *Behavioral and morphological development of immature Hawaiian freshwater fishes*. Ph.D. thesis, Louisiana State University.

Tate, D.C. 1997. The rôle of behavioral interactions of immature Hawaiian stream fishes (Pisces: Gobiodei) in population dispersal and distribution. *Micronesica* 30: 51-70.

Tate, D.C., J.M. Fitzsimons and R.P. Cody. 1992. Hawaiian freshwater fishes (Osteichthyes, Gobiodei): a field key to the species of larvae and postlarvae during recruitment into fresh water. *Occasional Papers of the Museum of Natural Science, Louisiana State University* 65: 1-10.

Thresher, R.E. 1984. *Reproduction of Reef Fishes*. T.F.H. Publications Inc., New York.

Timbol, A.S., A.J. Sutter and J.D. Parrish. 1980. *Distribution, relative abundance, and stream environment of* Lentipes concolor *(Gill 1860) and associated fauna in Hawaiian streams*. Water Ressources Research Center Coop. Rept. No. 6. University of Hawaii at Manoa, Honolulu.

Tomihama, M.T. 1972. *The biology of* Sicydium stimpsoni: *freshwater goby endemic to Hawaii*. B.S. (Honors) Thesis, Department of Zoology. University of Hawaii, Honolulu.

Torricelli, P.L., M. Lugli and M. Gandolfi. 1985. A quantitative analysis of the fanning activity in the male *Madogobius martensi* (Pisces: Gobiidae). *Behaviour* 92: 288-310.

Valade, P. 2001. *Etude de la biologie de reproduction et des premiers stades larvaires des cabots bouches ronde à l'île de la Réunion*. M.Sc. Thesis, University of La Reunion.

Watson, R.E. 1991. A provisional review of the genus *Stenogobius* with descriptions of a new subgenus and thirteen new species (Pisces: Teleostei: Gobiidae). *Records of the Western Australian Museum* 15: 627-710.

Watson, R.E. 1992. A review of the gobiid fish genus *Awaous* from insular streams of the Pacific plate. *Ichthyological Exploration of Freshwaters* 3: 161-176.

Watson, R.E. 2000. *Sicydium* from the Dominican Republic with description of a new species. *Stuttgarter Beiträge zur naturkunde* A(608): 1-31.

Watson, R.E., G. Marquet and C. Pöllabauer. 2000. New Caledonia Fish Species of the genus *Sicyopterus* (Teleostei:Gobiodei:Sicydiinae). *Aqua Journal of Ichthyology and Aquatic Biology* 4: 5-34.

Watson, R.E., P. Keith and G. Marquet. 2001. *Sicyopus (Smilosicyopus) chloe,* a new species of freshwater goby from New Caledonia (Teleostei:Gobioidei: Sicydiinae). *Cybium* 25: 41-52.

Watson R.E., P. Keith and G. Marquet. 2002. *Lentipes kaaea,* a new species of freshwater goby from New Caledonia (Teleostei:Gobioidei:Sicydiinae). *Bulletin Français de pêche et de Pisciculture* 364: 173-185.

Watson R.E., P. Keith and G. Marquet. 2007. *Akihito vanuatu,* a new genus and new species of freshwater goby from the South Pacific (Teleostei:Gobioidei: Sicydiinae).*Cybium* 31: 341-349.

Way, C.M., A.J. Burky, J.M. Harding, S. Hauand and W.K.L.C Puleloa. 1998. Reproductive biology of the endemic goby, *Lentipes concolor,* from Makamaka'ole Stream, Maui and Waikolu Stream, Moloka'i. *Environmental Biology of Fishes* 51: 53-65.

Yamasaki, N. and K. Tachihara. 2006. Reproductive biology and morphology of eggs and larvae of *Stiphodon percnopterygionus* (Gobiidae: Sicydiinae) collected from Okinawa Island. *Ichthyological Research* 53: 13-18.

Yamasaki, N. and K. Tachihara. 2007. Eggs and larvae of *Awaous melanocephalus* (Teleostei: Gobiidae). *Ichthyological Research* 54: 89-91.

Yamasaki, N., K. Maeda and K. Tachihara. 2007. Pelagic larval duration and morphology at recruitment of *Stiphodon percnopterygionus* (Gobiidae: Sicydiinae). *The Raffles Bulletin of Zoology suppl.* 14: 209-214.

Yokoi, K. and K. Hosoya. 2005. Larval salinity tolerance in the endangered goby *Rhinogobius* sp. BI (Gobiidae) from the Bonin Islands. *Japanese Journal of Ichthyology* 52: 31-34.

Yuma, M., K. Hosoya and Y. Nagata. 1998. Distribution of the freshwater fishes of Japan: an historical overview. *Environmental Biology of Fishes* 52: 97-124.

Yuma, M., A. Maruyama and B. Rusuwa. 2000. Behavior and distribution of upstream-migrating juvenile *Rhinogobius* sp. (the orange form). *Ichthyological Research* 47: 379-384.

CHAPTER

2.6

Diversity and Distribution of Freshwater Gobies from the Mediterranean, the Black and Caspian Seas

Jörg Freyhof

INTRODUCTION

Data on freshwater gobies from the Mediterranean, the Black and Caspian Seas have been summarized recently by Kottelat and Freyhof (2007). There are also detailed species accounts available in the books edited by Miller (2003) and Miller (2004). These books comprise some original views and much past data especially from relatively inaccessible eastern European and Russian sources, not only systematic and distributional but much general biology. Western and northern European freshwaters are rarely colonized by gobies and only one euryhaline species, *Pomatoschistus microps*, is widely known also from the North, Baltic and Atlantic Sea basins as well as from some areas in the western Mediterranean Sea. On the contrary, freshwater

Author's address: Leibniz-Institute of Freshwater Ecology and Inland Fisheries, Müggelseedamm 310, 12587 Berlin, Germany. E-mail: freyhof@igb-berlin.de gobies are common and often dominant in littoral habitats even far from the sea in Italy and the Eastern Mediterranean and especially in the basins of the Black and Caspian Seas. In these areas, there are about 50 goby species being restricted to freshwater habitats or regularly enter freshwaters. With two exceptions (*Padogobius*), these gobies can be grouped within so called sand gobies (*Economidichthys*, *Knipowitschia*, *Pomatoschistus*) and Paratethyan gobies (*Babka*, *Benthophiloides*, *Benthophilus*, *Caspiosoma*, *Mesogobius*, *Neogobius*, *Ponticola*, *Proterorhinus*). The monophyly of Paratethyan gobies was recently supported also by molecular evidence (Neilson and Stepien, 2009). Freshwater sand gobies are most diverse in the Adriatic and Aegean Sea basins from northern Italy south to Greece and West Anatolia. In this area, two euryhaline species are commonly found in estuaries and adjacent freshwaters (*Knipowitschia panizzae*, *Pomatoschistus canestrinii*). Both are related to a number of local endemics which are characterized by reduced patterns of head canals and squamation.

Paratethyan gobies are restricted to the tributaries of the Marmara, Black and Caspian Seas and none of them is ever recorded to permanently inhabit full marine waters. In this ecologically and morphologically diverse group of species, there are several real freshwater endemics (as *Ponticola kesslerii*) and several species which never enter pure freshwaters but are restricted to the brackish waters of the seas (as *Ponticola ratan*). Several species are euryhaline and are able to inhabit the sea as well as a wide range of freshwater habitats. Paratethyan gobies are a large radiation of real brackish- and freshwater gobies which has evolved in the today Black and Caspian Seas. They are connected to one of the key events forming the today diversity of Western Palearctic freshwater biota, the palaeographic processes dividing the Tethys into the Mediterranean and Paratethys Seas. While the Mediterranean fell dry during the late Miocene (Messinian), the Paratethys became more and more freshened during this time. During the Pliocene, the Paratethys was divided into the Black-, Azov-, Caspian and Aral Seas. These seas experienced massive water level fluctuations during the Pleistocene when they were disconnected from each others and from the Marmara and Mediterranean Seas during glacial phases, but connected by eight brief phases of high global sea levels corresponding with glacial melting phases and interglacial's. During the last glacial maximum, Paratethyan seas were fresh or slightly brackish lakes for at least 150,000 years and the sea level was about 150 m lower than today (see Aksu *et al.*, 1999 for overview).

Several species of Paratethyan gobies are actually expanding their range very fast following the massive habitat alternations and artificial river connections of the 20th century. Two of them were transported by ships from Europe to North America. The exact reasons for and modes of these massive upriver expansions of some Paratethyan gobies are still unknown. There are speculations, that there is partly an active upriver migration in rivers which have been transformed into shipping canals. But obviously, Paratethyan gobies are also able to use ships as vectors as there are several observations, that gobies are recorded suddenly far upriver or from places very distant from the closed distribution area. Recent examples are the records of *Babka gymnotrachelus* in the Danube close to Vienna, far upriver from the closed distribution area, *Neogobius melanostomus* from the lowermost Rhine (van Beek, 2006), which is a major harbour area and these fishes can only have been imported by ships or the first record of *N. fluviatilis* for the North sea basin from a harbour at the lower Rhine. Such initial introductions usually lead to a very fast downriver colonization of the major river and Paratethyan gobies as some *Neogobius*, *Ponticola* or *Proterorhinus* develop huge populations within one or two years and may spread several hundreds of km downriver in this time.

GOBIOID GENERA IN THE REGION

Family Odontobutidae

One species is introduced to Eastern Europe and is spreading to the west. *Micropercops swinhonis* is sometimes reported to have been introduced from East Asia to Eastern Europe, but there is no evidence that it has established somewhere.

Genus *Perccottus* Dybowski, 1877

The only species of this freshwater genus, *P. glenii*, is native to the Pacific basin of East Asia from Tugur drainage (Sea of Okhotsk) to northeastern Korea. It is an invasive species in Eastern Europe and Central Asia. It was first introduced in 1916 in St. Petersburg as an ornamental fish. In 1960s, it was also introduced to Kazakhstan, Uzbekistan, Turkmenia and Lake Baikal mixed with Chinese carps. This species is fast spreading and abundant in northern Russia, Belarus, Poland and Ukraine. There are also local records from Danube basin and an invasion of Central and Western Europe is expected. *Perccottus glenii* spreads actively via navigation canals and introduced with stocked fish.

Familie Gobiidae

Genus *Babka* Iljin, 1927

Babka is a monophyletic genus including the well known racer goby called *Neogobius gymnotrachelus* before (see under *Neogobius*). *Babka gymnotrachelus* is widespread all over the Marmara, Black and Caspian Sea basins and inhabits mostly freshwater and brackish waters with low salinity (< 2 %). This species belong to the group of invasive gobies and it is expanding its range since the late 1990s, when it invaded the rivers of northern Black Sea basin. In the Middle 1990th, it found its way to the Vistula River in the Baltic Sea basin in Poland, where it arrived from Dniepr via a navigation canal. *Babka gymnotrachelus* inhabits large rivers as well as medium sized and fast flowing streams.

Genus *Benthophiloides* Beling and Iljin, 1927

This monotypic genus is endemic to the Azov and Black Seas. *Benthophiloides brauneri* is very rarely observed and inhabit deep waters of estuaries, coastal lakes and lower part of rivers, at depths up to 15 m, in still, fresh and brackish water. It seems to be closely related to *Benthophilus* and *Caspiosoma.*

Genus *Benthophilus* Eichwald, 1831

There are some 20 species of *Benthophilus* and the genus was recently revised by Boldyrev and Bogutskaya (2004). *Benthophilus* are known from the Caspian, Azov and Black Seas. They do not enter fully marine conditions and are therefore absent from the Mediterranean Sea. Some are strictly marine (in littoral zone or in deep waters), others inhabit the estuaries and freshwaters. Two species (*B. durrelli, B. nudus)* are invasive in the large tributaries of the Caspian and Black Sea. All *Benthophilus* species inhabit marine habitats, lowland rivers, deltas and coastal lakes. Many species are euryhaline but some are restricted to brackish waters. In rivers, they are usually abundant in main stream. *Benthophilus* can dig themselves into fine sediments and usually inhabit muddy, silty or sandy bottoms with mollusc shells. They live only about 1 year, spawning in spring inside or under a mollusc shell. Adults die soon after spawning. All species feeds mainly on chironomid larvae, amphipods and molluscs, some are specialized on molluscs. Phylogenetic relationships of *Benthophilus* and the related genera *Benthophiloides* and *Caspiosoma* are still unresolved but Neilson and Stepien

(2009) suggest, that they are closely related to *Neogobius*, *Ponticola* and other Paratethyan gobies. *Benthophilus profundorum* a deep-water endemic from the Caspian Sea, is sometimes placed in the monotypic genus *Anatirostrum* but treated as a species of *Benthophilus* here.

Genus *Caspiosoma* Iljin, 1927

Caspiosoma caspium is the only species of this genus which is found in the lower reaches, deltas and limans of rivers draining to northern Black Sea, Sea of Azov and Caspian Sea. This small goby which grow to 50 mm SL is poorly known, but locally abundant, especially in marine habitats.

Genus *Economidichthys* Bianco, Bullock, Miller and Roubal 1987

With a size up to 25 mm SL, *E. trichonis*, which is endemic to Greek Lakes Trichonis and Lysimachia, is the smallest European freshwater fish species. There are two species in *Economidichthys*, both endemic to Western Greece. Both share a large peri-anal organ, distinguishing them from superficially similar *Knipowitschia*. Both are strict freshwater gobies with an extended pelagic phase.

Genus *Knipowitschia* Iljin, 1927

With 16 actually recognized species, *Knipowitschia* is one of the large genera of Western Palearctic gobies and all species are found in the Adriatic, Aegean, Black and Caspian Sea basins. Only one species, *K. iljini*, from the Caspian basin is strictly 'marine' inhabiting only the brackish water of that sea. There are additional four euryhaline species and 11 strict freshwater species, often with a very small distribution area in freshwater habitats adjacent to the habitat of an euryhaline species. Such local endemics are usually diagnosed by reduced head canals and/or reduced squamation and some seem to be very closely related to adjacent euryhaline species. While it might be speculated, that some of them are (as *K. ephesi*, *K. milleri*, *K. radovici*, *K. montenegrina* or *K. mrakovcici*) the result of allopatric speciation from an euryhaline ancestor identically or very similar to *K. panizzae* or *K. caucasica*, this seems to be unlikely for some other freshwater endemics. Due to the often very small distribution ranges, several *Knipowitschia* are of conservation concern. One species, *K. goerneri*, was not found again since 1983 in its only known habitat in a spring on Corfu Island (Greece) and a second species, *K. cameliae*, was not found since 1994 in its only habitat in Romania. Both species are for sure extirpated in their known habitats, but it cannot be excluded, that they exist still somewhere else. It can also not be excluded that they have gone extinct. *Hyrancogobius*, endemic to the Caspian basin is treated here as a synonym of *Knipowitschia*. Molecular data presented by Penzo *et al.* (1998) and Huyse *et al.* (2004) suggest, that at least one of the freshwater endemic species, *K. punctatissima*, might be related to *Pomatoschistus canestrinii* and this group is not closely related to *Knipowitschia* or to *Pomatoschistus* and might represent an own genus *Ninnigobius* Whitley, 1951 (see also Miller and Sanda, 2008). More data on the phylogeny of sand gobies are needed to resolve this problem.

Genus *Mesogobius* Bleeker, 1874

Only two large (up to 345 mm SL) predatory gobies (*M. batrachocephalus*, *M. nonultimus*) are actually placed in *Mesogobius*. The validity of *M. nigronotatus* from the Caspian Sea is questionable. They are found in inshore habitats, estuaries, brackish- and freshwater lagoons on sand or rock bottom; often very deep in summer (down to 100 m). While *M. batrachocephalus*, endemic to the Black Sea, locally enter pure freshwater, the Caspian species *M. nonultimus* is only found in brackish waters. Due to their large size, they have a certain commercial value in some areas and are popular for sport fishers.

Genus *Neogobius* Iljin, 1927

While *Neogobius* was long known as an artificial assemblage of Paratheyan gobies, only recently, Neilson and Stepien (2009) presented sufficient data to allow to split the Paratethyan gobies into natural, monophyletic groups which are treated as own genera as suggested by Neilson and Stepien (2009). Only four species remain in *Neogobius* (*N. fluviatilis*, *N. pallasi*, *N. caspicus* and *N. melanostomus*). All other species widely known as *Neogobius* are now placed in *Ponticola* or *Babka*. *Neogobius* is endemic to the Marmara, Black and Caspian Seas. *Neogobius caspicus* is restricted to the Caspian Sea and seems not to enter tributaries, while all others inhabit the sea as well as various kinds of freshwater bodies. No *Neogobius* species enter fully marine conditions and the genus is therefore absent from the Mediterranean Sea, but *N. fluviatilis* and *N. melanostomus* are known from the Marmara basin. *Neogobius fluviatilis* and *N. melanostomus* have reached the North and Baltic Sea basins through artificial waterways. *Neogobius melanostomus* has entered the Baltic Sea itself and is found to be abundant in the Sea in some areas. Beside the invasion of European rivers, this species is also invasive in the Great Lakes area in North America (Charlebois *et al.*, 1997). The invasion of *Neogobius* into new areas is often well documented. It is usually a very fast process downriver and a slow process upriver. Some species have colonized several hundreds km of large rivers within one or two years. Due to its large size of about 200 mm SL, *N. melanostomus* has a major commercial value in some areas, especially in the Azov Sea. It is usually salted, dried and consumed with beer.

Genus *Padogobius* Berg, 1932

There are actually two species in *Padogobius*, both distributed in Italy. While one of them, *P. nigricans*, is endemic to Central and Northern Italy, the second species, *P. bonelli*, is known from northern Italy and adjacent areas in Switzerland, Slovenia and Croatia. *P. bonelli* is also well known to be invasive in Italy and has massively expanded its range south since the last decades. Both species are pure freshwater inhabitants and known from a wide range of habitats. While all Paratethyan gobies seem to form one monophyletic lineage and all freshwater sand gobies are closely related to only two or three euryhaline species, both *Padogobius* species are not closely related to these groups. It is speculated, that both have independently originated from marine ancestors (see Penzo *et al.*, 1998) and molecular results by these authors also suggest, that *Padogobius* might be related to the marine rock goby *Gobius paganellus*.

Genus *Pomatoschistus* Gill, 1863

Sand gobies of the genus *Pomatoschistus* are widespread inhabitants of coastal water in Europe. There are only three species regularly found in freshwater habitats. From these, *P. canestrinii*, endemic to the Adriatic basin and *P. microps*, widespread in western and northern Europe are euryhaline inhabitants of coastal waters. While *P. canestrinii* is able to inhabit freshwaters permanently (see Kovačić, 2005 for details), *P. microps* seem to be an obligatory catadromous species. There is only one pure freshwater *Pomatoschistus* in Europe, *P. montenegrensis*, endemic to Lake Skadar basin in Montenegro and Albania. In fact, the position of both species within this genus is still a matter of discussion (see in *Knipowitschia*).

Genus *Ponticola* Iljin, 1927

Recently, Neilson and Stepien (2009) presented a molecular study suggesting, that *Ponticola* represents a monophyletic lineage within the radiation of

Paratethyan gobies. *Ponticola* is endemic to the Black and Caspian Seas and does not enter the Marmara or Mediterranean Seas. From the 13 species recognized, *P. ratan*, *P. platyrostris*, *P. bathybus* and *P. cephalargoides* are restricted to brackish water habitats and *P. eurycephalus* and *P. syrman* are usually found in the sea only, but enter estuaries in some areas. *Ponticola kessleri* and *P. gorlap* are invasive in tributaries of the Black and Caspian Seas following the massive river alternations in the 20th century. *Ponticola kessleri* has reached the North Sea basin through an artificial waterway connecting the Danube with the Rhine. This invasion is well-documented and within four years, the species colonized all suitable river habitats downstream of the invasion site. In *Ponticola*, there is a group of strict inhabitants of freshwater streams and rivers and some have a small distribution range as the *N. constructor* species complex, which include five species (*N. constructor*, *N. cyrius*, *N. rhodioni*, *N. rizensis* and *N. turani*), all endemic to small areas at or close to the Caucasus mountains. These species often inhabit cold mountain streams where they occur sympatric with trouts (*Salmo*).

Genus *Proterorhinus* Smitt, 1900

Four species of tubenose gobies are distributed in the Black and Caspian Seas and one of them; *P. semilunaris*, belong to the invasive Paratethyan gobies, which have massively expanded their range since the middle of the 20th century. It has also reached North America, where it is now invasive. One species, *P. marmoratus* seem to be a pure inhabitant of brackish waters while *P. semilunaris* and *P. tataricus* are found in freshwaters only (Stepien and Tumeo, 2006). All tubenoses are relatively small gobies which inhabit densely vegetated habitats or coarse stone fields as usually found on artificial river shores. The diversity of this genus is still not fully resolved and several undescribed species seem to exist.

Conclusion

Freshwater gobies from the Mediterranean, the Black and Caspian Seas are species rich and inhabit a diverse range of habitats. With the exception of *Padogobius*, they belong to two evolutionary lineages only. The first lineage is the large group of Paratethyan gobies, which is basically a freshwater and euryhaline radiation. It has its evolutionary roots in the Paratethys and its adaptation to brackish and freshwaters is likely to be related to the process of isolation of the Paratethys from the Tethys in the Middle Miocene (about 13 MY ago) and the subsequent freshening of the Paratethys in the latest Miocene (finally freshened about 6 MY ago) (Popov *et al.*, 2004). The second group are Mediterranean sand gobies which are related to sand gobies in the seas west and north of the Mediterranean. But in fact, sand gobies have their highest diversity in the Mediterranean and there are several endemics in the Black and Caspian Seas also. Therefore, Paratethyan origin of sand gobies with subsequent colonization of the Mediterranean is as speculative as their invasion of this area from the Atlantic. In contrast to Paratethyan gobies, the evolution of freshwater species might have been happened in several independent events in sand-gobies and there is some empirical evidence, that some species are the result of very recent allopatric speciation.

References

Aksu, A.E., R.N. Hiscott and D. Yasar. 1999. Oscillating quaternary water levels of the Marmara Sea and vigorous outflow into the Aegean Sea from the Marmar-Black Sea drainage corridor. *Marine Geology* 153: 275-302.

Boldyrev, V.S. and N.G. Bogutskaya. 2004. Revision of the tadpole-gobies of the genus *Benthophilus* (Teleostei: Gobiidae). *Ichthyological Exploration of Freshwaters* 18: 31-96.

Charlebois, P.M., J.E. Marsden, R.G. Goettel, R.K. Wolfe, D.J. Jude and S. Rudnika. 1997. The Round Goby, *Neogobius melanostomus* (Pallas), A Review of European and North American Literature. Illinois-Indiana Sea Grant Program and Illinois Natural History Survey, Illinois.

Huyse, T., J.V. Houdt and F.A.M. Volckaert. 2004. Paleoclimatic history and vicariant speciation in the "sand goby" group (Gobiidae, Teleostei). *Molecular Phylogenetics and Evolution* 32: 324-336.

Kottelat, M. and J. Freyhof. 2007. Handbook of European freshwater fishes. Kottelat, Cornol and Freyhof, Berlin.

Kovačić, M. 2005. Morphological variability of *Pomatoschistus canestrinii* (Gobiidae), with the reduction of squamation and head canals. *Cybium* 29: 373-379.

Miller, P.J. 2003. The Freshwater Fishes of Europe Vol. 8/I (Mugilidae, Atherinidae, Atherinopsidae, Blenniidae, Odontobutidae, Gobiidae 1). AULA-Verlag, Wiebelsheim.

Miller, P.J. 2004. The Freshwater Fishes of Europe Vol. 8/I (Gobiidae 2). AULA-Verlag, Wiebelsheim.

Miller, P.J. and R. Sanda. 2008. A new West Balkanian sand-goby (Teleostei: Gobiidae). *Journal of Fish Biology* 72: 259-270.

Neilson, M.E. and C.A. Stepien. 2009. Escape from the Ponto-Caspian: evolution and biogeography of an endemic goby species flock (Benthophilinae: Gobiidae: Teleostei). *Molecular Phylogenetics and Evolution* 52: 84-102.

Penzo, E., G. Gandolfi, L. Bargelloni and L. Colombo. 1998. Messinian salinity crisis and the origin of freshwater lifestyle in western Mediterranean gobies. *Molecular Biology and Evolution* 15: 1472-1480.

Popov, S.V., F. Rögl, A.Y. Rozanov, F.F. Steininger, I.G. Shcherba and M. Kovac. 2004. Lithological-Paleogeographic maps of Paratethys 10 maps Late Eocene to Pliocene. *Courier Forschungsinstitut Senckenberg* 250: 46 pages, maps 1-10.

Stepien, C.A. and M.A. Tumeo. 2006. Invasion genetics of Ponto-Caspian gobies in the Great Lakes: A "cryptic" species, absence of founder effects, and comparative risk analysis. *Biological Invasions* 8: 61-78.

van Beek, G.C.W. 2006. The round goby *Neogobius melanostomus* first record in the Netherlands. *Aquatic Invasions* 1: 42-43.

SECTION 3

General Biology

CHAPTER

3.1

Gobies as Predator and Prey

C. Dieter Zander

INTRODUCTION

Generally, gobies are carnivorous, often algae or other plants are ingested as secondary, rarely as primary prey. In some cases, these fish are aufwuchs feeders, which ingest plants as well as sessile animals like polychaetes or sponges, e.g. *Gobius vittatus* from the Gulf of Lion (Heymer and Zander, 1978). The kind of prey especially depends on two factors of the predator: size and way of life. Large species feed rather on large animals including fish, small species on small organisms. This rule can at best be demonstrated by ontogeny. Large gobies of more than 20 cm size are represented, e.g., by the Mediterranean *Gobius cobitis, Zosterisessor ophiocephalus*, the Ponto-caspian genera *Mesogobius* and *Neogobius,* or by species of the family Eleotridae. But the most gobies attain only sizes of 8-12 cm. Beside species which live several years, there are many gobies which spawn in the year after hatching and die thereafter; these species grow very fast to a size of 6 cm, some even to 9-10 cm (*Pomatoschistus minutus*), some to only 2 cm or less (*Pandaka pygmaea*). Secondly, the respective way of life defines also

Author's address: Biozentrum Grindel and Zoologisches Museum, University, Martin-Luther-King-Platz 3, 20146 Hamburg, Germany. E-mail: cedezet@zoologie.uni-hamburg.de the layer of foraging in the benthal, suprabenthal or pelagial. Whereas the largest gobies are bottom dwelling species, we find several smaller species which live in the suprabenthic (e.g. *Gobiusculus flavescens, Aphia minuta, Pomatoschistus quagga*), sometimes also in the pelagic layer (*Crystallogobius linearis)* and, therefore, feed mainly on planktonic organisms. Additionally, the suprabenthal is also exploited by benthic species, e.g. by *Pomatoschistus pictus* which regularly starts from the bottom into swarms of mysids in order to catch single specimen of these crustaceans.

Expressed as an ecological formula, the demand and acquisition of prey follows the energetic rule of MacArthur and Pianka (1966): **Eopt = Ei / Ts + Th**, where Eopt = optimal energy acquisition, Ei = energy content of prey, Ts = searching time, Th = handling time which includes persecution, overwhelming and feeding. An optimal concept, whch is also used by large gobies, is that of sit and wait for relatively large prey, because the predator gains much energy but invests a minimum of handling time. In contrast, plankton feeders, so called widely foragers, gain only low energy by a single tiny prey, but this is compensated by a minimum of searching time.

Regularly, the ontogeny of gobies begins after hatching with a pelagic phase during which the goby larvae feed on planktonic organisms. The duration of the pelagic phase varies according to the ontogeny patterns of the respective species. During their growth, they forage corresponding to their size at first for small invertebrate larvae and later on greater organisms, like adult Copepoda.

Regarding taxonomic units, the food spectrum is wide wherein the preferred group is clearly the Crustacea resulting from their great species richness and abundance in the aquatic environments, especially in marine habitats. In freshwater these are partly replaced by water dwelling insects or their larvae. The prey demand can be completed by Polychaeta, Mollusca and fish from the bottom or larvae of diverse groups from the pelagial.

The most important factors, which may influence and probably restrict the prey selection of these fish, are the quality and quantity of prey offer. In extreme habitats the potential food organisms can consist of only few species which occur in great abundance following Thienemann's (1939) biocenotic rule. Such environments like brackish waters, mud flats or mangrove swamps present high productivity and prey surplus. Gobies as generalists will then prefer those organisms which are due to their dominance caught easier and reduce searching time according to the energetic rule (MacArthur and Pianka, 1966). Thus, the predator seems to be a specialist dependant only to a dominant prey species. Other extreme habitats like caves comprise a small but diverse prey offer, which make the impression of an extreme generalist. If new environments like land or fresh water are conquered these may serve as an expansion to a habitat which is poor of competitors. Special ways of life like cleaning hosts from parasites consequently reduces the food spectrum on ectoparasites, but does not reduce the possibility to feed on other food items.

The productivity of an ecosystem can be mirrored by the fullness of the gut of a predator, expressed in the "fullness index" (Hureau, 1970): **Full = Wf / Wp** , where Wf = weight of food in the gut of a single predator, Wp = weight of the predator. The analysis of food spectra is focussed on three parameters which are set in relation: 1. frequency of occurrence (Hynes, 1950) which marks the importance of a food item for the predator population: **F = Ni / N** , where Ni = number of predators with a special food item / N = total number of predator specimens; 2. abundance which mark the numerical dominance of a food item: **Ab = Si / N**, where Si = number of a special food item; 3. weight which marks the affectivity and nutritive value of a food item: **W = Wi / N** , where Wi = weight of a special food item in a predator population. Several investigators use the volume "**V**" or volume of a food item "**Vi**" of prey as an alternative parameter to weight. The meaning of the respective parameters is different. Thus, the frequency may mirror the abundance of the offered food items, the abundance can express feeding activities (handling time), the weight or volume is the most important factor for the nutrition of a predator (Zander, 1982).

Several authors tried to combine two or three compartments with the aim to improve these analyses and to develop equations in order to characterise the significance of food items. In his nutrient coefficient, Hureau (1969) multiplied percentage of numbers with percentage of weight. The index of relative importance (Pinkas *et al.*, 1971) uses still the volume instead of weight: **IRI = % F (% Ab + % V)**. Simenstad (1979) replaced volume by weight. Because Zander (1982) assumes that frequency in the IRI is overvalued and volume is too inaccurate for small-sized fish, he developed the main food index: **MFI = √ [(% Ab + % F) % W / 2]**, where W = weight which is emphasised to use in small-sized fish like gobies as dry weight. The calculated values are classified as follows: > 75 = primary food, 51-75 = main food, 26-50 = secondary food, < 25 = insignificant food. The advantage lay in the presentation of the MFI by 3-axis graphs see below. Because volumes in IRI are only coarsely estimated, Nieder and Zander (1994) present a method of a more accurate evaluation which is especially suited for small-sized fish. Finally, the calculation of selectivity can find out the preferred food items of the predator by the following equation

(Berg, 1979): **Sel = log (number of ingested food / number of potentially available food)**. Numbers can be replaced in more accurate studies by weight. The calculation of Sel needs samples of free living communities which are only rarely done.

Because most species are of small size, gobies are a preferred prey for greater fish (also greater gobies), diving birds and mammals (seals and dolphins), but also for cuttlefish, jellyfish and anemones. Species which inhabit shallow waters are prey even for not diving birds like grebes and gulls. By this way, parasites can perform their life-cycles from intermediate (gobies) to final hosts (predator fish, birds and sea mammals). Gobies are also exploited by human fishery, which regards especially larger species in the Mediterranean and Black Sea, but also in tropic areas on tiny species like *Pandaka pygmaea,* which is one of the smallest vertebrate.

DIVERSE PATTERNS AND CHOICE OF GOBY PREY

The main factor that influences the predation of gobies is the habitat and its offer of suited prey organisms. We differ mainly between benthal, including soft and hard bottoms, and pelagial or suprabenthal, respectively. Therefore, we can appoint gobies to these feeding categories: algae/plants, meiobenthos/detritus, plankton, planktonic macrofauna (suprabenthos), macrobenthos or exclusively fish. Further differentiation concerns the relevant abiotic factors such as temperature, salinity and light.

Warm Temperate Climate, for example Mediterranean Sea

Whereas in cold temperate areas dominate gobies which live only a year plus a reproductive period, in warmer climate zones longer-living species dominate. In several researches, as much as 11 goby species were analysed off Banyuls-sur-Mer (Gulf of Lion, French Mediterranean coast) under the aspect of different sites, depths and peculiarities of the bottoms (Zander, 1982; Zander and Berg, 1984; Zander and Hagemann, 1989; Zander and Heymer, 1992). Among investigated *Gobius* spp., six are found on rocky substrates, whereas *G. geniporus* is the only which inhabit sand bottoms. Additionally, four short-living species are investigated which are found on or above sand bottoms.

The abundance analysis of prey regarding numbers clearly mirrors the way of life and the feeding habitat of the respective goby predator. It also regards behavioural aspects because every food item is singly ingested after a certain searching and handling time. An abundant meiofaunal component is copepods, which rather supply low energy contents: harpacticoids in seven benthic species, planktonic calanoids in the suprabenthic *P. quagga* (Fig. 3.1.1). Especially remarkable are algae found to be grazed by three *Gobius* spp., mysids to be caught in the suprabenthic layer by *Gobius xanthocephalus*, *Pomatoschistus bathi* and *P. quagga*, or sponges present in *G. vittatus*, which mark this species as an aufwuchs feeder. Also young fish are conspicuous prey in *P. bathi*, *Deltentosteus quadrimaculatus* and *Buenia jeffreysi* (probably *B. affinis*, Kovačic and Patzner, 2009) (Figs. 3.1.1, 3.1.2). Many benthic crustaceans are macrofaunal elements which provide a more effective energy supply such as polychaetes (found in all four *Gobius* spp.) and, in contrast to copepods, need longer searching times. *G. geniporus* is characterised by great quantities of sand dwelling gammarids and brachyuran decapods.

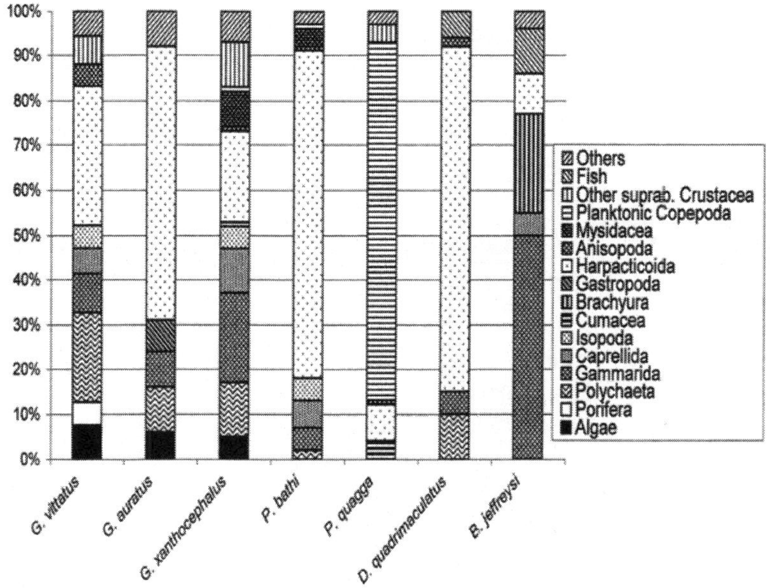

Fig. 3.1.1 Abundance of prey components of seven Mediterranean gobies from the Gulf of Lion. Names of genera: *B.* = *Buenia, D.* = *Deltentosteus, G.* = *Gobius, P.* = *Pomatoschistus*. Remarkable is the dominance of harpacticoids in three species, of gammarids in *B. jeffreysi* and of planktonic copepods in *P. quagga*. (Acc. to: Heymer and Zander, 1978; Zander, 1982; Zander and Berg, 1984; Zander and Hagemann, 1989; Zander and Heymer, 1992.)

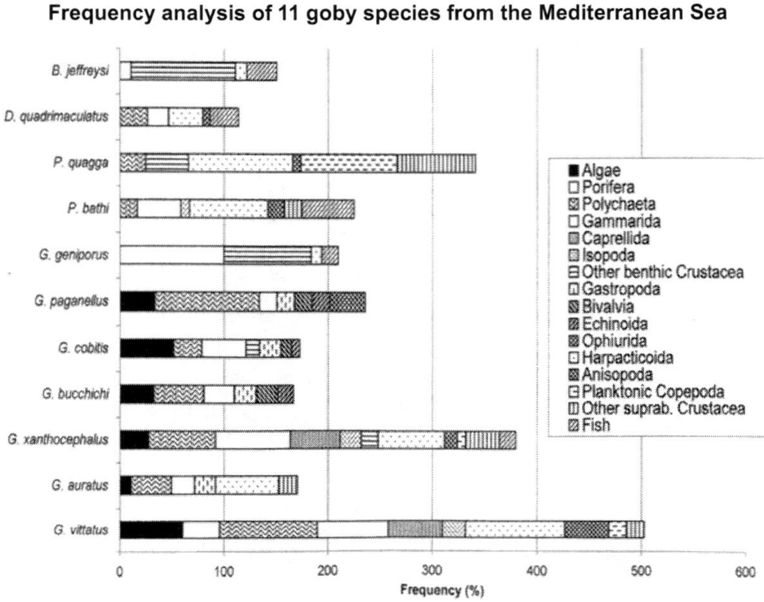

Fig. 3.1.2 Frequency of prey components of 11 Mediterranean gobies from the Gulf of Lion, presented in an arrangement of summed up values. High values of sums as in *G. vittatus* mean an increased searching of prey caused by its shortness. Low values as in *B. jeffreysi* mean the completion of a main food by only few components in low frequency. Names of genera as in Fig. 3.1.1. (Acc. to: Gibson, 1968; Heymer and Zander, 1978; Zander, 1982; Zander and Berg, 1984; Zander and Hagemann, 1989; Zander and Heymer, 1992.)

The frequency of prey occurrence which regards the food composition of a population reveals clear differences between the species (Fig. 3.1.2). Nevertheless, not only the kind of prey differs, but also their respective ratios. In total, 11 gobies are analysed including also three *Gobius* species, which were formerly investigated in the Banyuls region by Gibson (1968). *G. vittatus* and *G. xanthocephalus* prefer by more than 60% macrofauna especially Polychaeta, Gammarida and Harpacticoida, *G. paganellus* also Polychaeta, *G. geniporus* also Gammarida, *G. auratus, P. quagga* and *P. bathi* Harpacticoida, *B. jeffreysi* (vs. *B. affinis*) benthic Brachyura. Remarkably, 61% of *G. vittatus* ingest algae, which are, in lesser frequency, also prey for five other gobies. *P. quagga,* which forages in the suprabenthal, is characterised by high values of calanoids and harpacticoids. A remarkable part of *G. paganellus, G. cobitis* and *G. bucchichi* ingest molluscs and echinoderms (Gibson, 1968). A lesser part of *G. xanthocephalus, G. auratus* and *G. vittatus* additionally catch mysids or calanoids from the suprabenthic layer. To a lesser part, *G. xanthocephalus, G. geniporus, B. jeffreysi* (vs. *B. affinis*), *D. quadrimaculatus* and *P. bathi* feed on young fish from the same layer.

The frequency analysis can mark by its respective extent which quantitative rates of the prey spectrum that are supplied. The more components are present, the more this situation marks the gobies as generalists (Fig. 3.1.2). Six out of 10 components found in *G. vittatus* present wide ranges, especially polychaetes and harpacticoids. These high values may indicate a shortness of supply of other components which force the single specimen to take what is available (Heymer and Zander, 1978). In contrast, seven out of 11 components found in *G. xanthocephalus* have smaller ranges, which indicate a greater possibility for prey choice (Fig. 3.1.2) (Zander, 1982). In consequence, only four components are found in *G. geniporus* of which gammarids and decapods attain high values and, thus, reveal a sufficient offer and utilisation of these two components (Zander and Hagemann, 1989). These results are confirmed by values of the fullness index, which are high in sand and gravel bottoms, but low in rocky habitats (Table 3.1.1).

Table 3.1.1 Fullness index of 7 Mediterranean gobies from different habitats of the Gulf of Lion. Variable values mean results from different seasons. (Acc. to Zander 1982, Zander and Berg, 1984, Zander and Heymer, 1992.)

	Habitat						
Species	*Gravel*	*Sand*	*Muddy sand*	*Coarse sand*	*Rock/ sand*	*Rock walls*	*Coralligene*
Pomatoschistus bathi	2.4	2.2 - 2.8	1.7	1.8	2.0		
P. quagga	1.3						
Deltentosteus quadrimaculatus			2.2				
Buenia jeffreysi			1.8				
Gobius geniporus		2.2 - 2.5		1.1			
G. geniporus < 25 mm		3.7 - 4.6					
G. xanthocephalus	2.9				1.4 - 1.6		1.2
G. auratus						0.8	

The influence of different habitats on the predation of two *Gobius* spp. and three short living species from the Banyuls region is tested with help of the main-food-index (MFI, which includes and emphasises weights, see above) (Fig. 3.1.3) (Zander, 1982; Zander and Hagemann, 1989).

Pomatoschistus bathi was found on gravel and coarse sand bottoms in 7 m (A and A', Fig. 3.1.3), a muddy sand bottom in 10 m (B) and a rock sand ecotone in 14 m (C). Conspicuously, ratio of preyed young fish increases by depth whereas the importance of gammarids and harpacticoids decreases. There is also a clear difference of the two sand bottoms, because in one locality mysids and other suprabenthic crustaceans dominate (A') instead of harpacticoids and gammarids in the other locality (A). Planktonic calanoids dominate in the suprabenthic *P. quagga*, benthic crustaceans are main food for *G. geniporus*. For *D. quadrimaculatus*, compared with *P. bathi* in a muddy sand habitat (B), polychaetes are, beside young fish, of greater importance, but to a lesser extent harpacticoids. *G. xanthocephalus* from a rock sand habitat (C) presents lower values of components than in a coralligene habitat in a depth of 24 m (D). There, a broader range of components like algae and polychaetes attains great importance, whereas young fish as prey are absent there (Fig. 3.1.3). The increasing multiplicity and range extension of components by depth, as it is found in *G. xanthocephalus*, is due to increasing weights of macrofaunal prey and mean an evidence of a better prey supply.

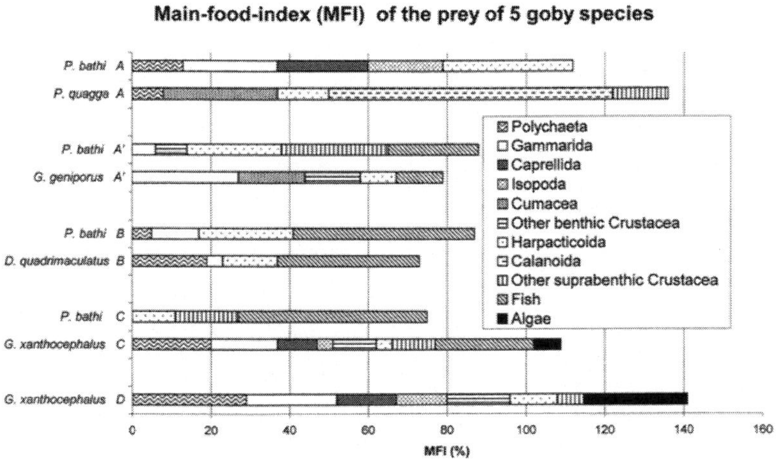

Fig. 3.1.3 Main-food-index of prey components of five Mediterranean gobies from the Gulf of Lion in an arrangement of summed up values. The values stress those components which are important for the nutrition. Capitals mean different habitats: A = gravel bottom, A' = pure sand bottom, B = muddy sand bottom, C = boulder/ sand bottom, D = coralligene. Names of genera as in Fig. 3.1.1. (Acc. to: Zander and Berg, 1984; Zander and Hagemann, 1989.)

The investigated gobies off Banyuls can, therefore, be classified as follows: One suprabenthic plankton feeder (*P. quagga*), two predator of sand macrofauna (*G. geniporus* and *B. jeffreysi* vs. *B. affinis*), two meiofauna feeder (*P. bathi* which is very small, and *D. quadrimaculatus*), one aufwuchs feeder (*G. vittatus*), four macrobenthos feeder on hard substrates ingest also shelled organisms like molluscs and echinoderms (*G. cobitis* which is of giant size, *G. bucchichi* and *G. paganellus* which live in shallow habitats, and *G. auratus* which inhabit deeper regions), and one macrofauna feeder from hard substrates which also feed on suprabenthic organisms (*G. xanthocephalus*). Conspicuously, seven species, four *Gobius* spp., catch young fish. Six *Gobius* species, *G. geniporus* excepted, ingest also algae and have to be characterised as omnivorous.

Cold Temperate Regions, for Example North Sea

The benthic and suprabenthic fish community of the rocks of Helgoland, North Sea, is poorer in species than that of the Mediterranean Sea, and presents an assemblage of several short living gobies. *Pomatoschistus microps, P. pictus, P. minutus* and *Gobiusculus flavescens* are found in diverse habitats or microhabitats, respectively. A longer living goby, *Gobius niger*, is not found in the entire North Sea which is caused still by unknown reasons, elsewhere it is distributed in the northern Atlantic, Mediterranean and the brackish Baltic Sea.

P. microps grows up to 6 cm and inhabits at Helgoland shallow waters on mixed grounds of sand and boulders and also rockpools, where it feeds mainly on epibenthic meiofauna like harpacticoids and small amphipods (Zander, 2005b) (Fig. 3.1.4). In deeper layers, *P. pictus* (sized up to 6 cm) is found in a similar environment where amphipods are the most abundant prey component followed by harpacticoids (Zander, 1979, 2005b). *G. flavescens* (sized up to 5.5 cm) is swimming above the rocks, eelgrass or algae, and consumes mainly copepods and other plankton (Berg, 1979; Zander, 2005b) (Fig. 3.1.4). *P. minutus and P. lozanoi* inhabit pure sand bottoms together with several flatfish species which exploit diverse benthic food components, respectively (Krüß, 1988). Because *P. lozanoi* is absent off Helgoland, the prey of these two gobies (sized up to 8 – 9 cm) are presented by results from the Belgian coast (Hamerlynck and Kattrijsse, 1994). *P. minutus* preys mostly on amphipods, and in spring additionally on planktonic copepods, whereas *P. lozanoi* exploits the benthos and ingests also copepods, as well as mysids and fish larvae, which play no role by numbers (Fig. 3.1.4), but dominate by weight. The part of Cumacea is greater in *P. lozanoi* than in *P. minutus,* which indicate a foraging activity at night when these crustaceans are active (Hamerlynck *et al.*, 1990).

The ratios of prey components of *G. flavescens* only differ insignificantly in a time span of 13 years (Fig. 3.1.4), the main food Calanoida attained 85% of abundance, only the other components vary to an minor extent. *P. pictus* present only slight alterations even after 14 years when the dominating Gammarida (abundance 80%) are to a minor part displaced by meiofaunal Harpacticoida (20%). If different habitats are compared, *P. pictus* ingested in the rock flat of Helgoland where isopods are very abundant 19% of this component but at the steep walls of the harbour only 2% (Zander, 1979, 2005b).

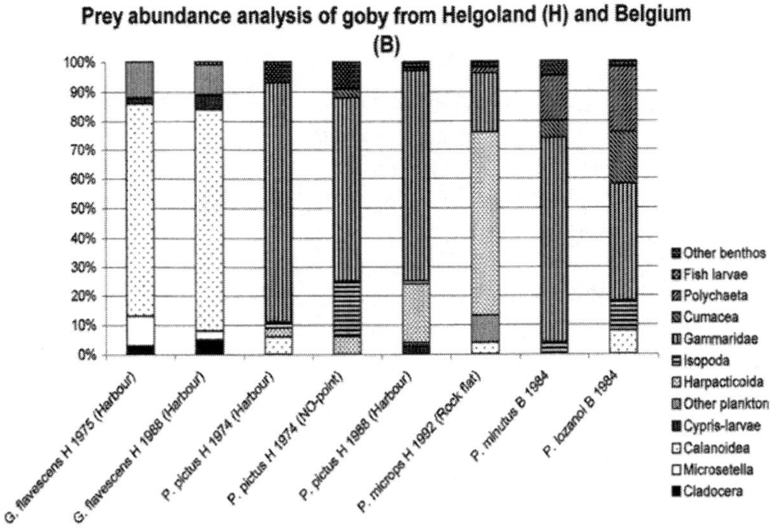

Fig. 3.1.4 Abundance of prey components of five gobies from the North Sea. Names of genera: *G.* = *Gobiusculus, P.* = *Pomatoschistus*. Remarkable is the dominance of harpacticoids in *G. flavescens,* of gammarids in *P. pictus* and *P. minutus*, and of harpacticoids in *P. microps*. (Acc. to: Zander 1979, 2005b; Hamerlynck and Kattrijsse, 1994.)

The results evidence that at comparable seasons the prey does not alter importantly the feeding habits of the goby predators, even after a greater time span. However, in temperate locations a variation of prey between seasons can generally occur (see below, p. 308). Only if different habitats of the same locality are compared, greater differences can be found which depend on the special supply of prey organisms, respectively.

Indo-pacific and other Tropical Regions

There are only few studies of the prey of gobies from the tropics. The extent investigations of Hiatt and Strasburg (1960) of reef fishes of the Marshall Islands, Central Pacific, comprised also the prey of six goby species, three of larger, three of smaller size (Table 3.1.2). The larger species live on sand bottoms or in tidal pools and exploit especially crustaceans of different sizes, the intertidal *Bathygobius fuscus* preys also on fish and not on small copepods. These larger species conspicuously ingest also algae. The smaller species live between the branches of corals where they exploit several crustacean groups, but only *Gobiodon rivulatus* preys on copepods. *G. erythrocephalus* profits from spawning of other fish between the ramose corals, because it feeds on their eggs.

Lophogobius cyprinoides from Florida is also an opportunistic feeder, which ingests diverse crustaceans, molluscs and barnacles, but the greatest part of their food comprises algae (Darcy, 1981).

Extreme Habitats and Ways of Life

Extreme habitats are characterised by a poorer species richness in comparison to bordering habitats. According to Thienemann's rule (1939) the number of species decreases if the strength of an extreme ecological factor increases, whereas density of these populations can increase until massive appearance. If few prey species dominate over others that occur in low densities, species diversity decreases. Species richness is only the number of species, whereas species diversity judges about the relationships of species of a community. It is measured by evenness "J" which is the relation of the Shannon-Wiener index "Hs" to the number of species regarded. J-values > 0.6 indicate homogeneity, > 0.8 great homogeneity (Zaret and Rand, 1971). Generalist feeders will exploit the prey according to their optimal availability which has to be the most abundant one. Corresponding to extreme habitats, also specialised or extreme ways of life can lead to unbalanced prey ingestion.

Brackish Waters

Brackish waters like the Baltic Sea present extreme habitats because these lie between marine and freshwater conditions. The threshold for the existence of marine and freshwater species is at 5-8 ‰ salinity which means a wide shift onto fresh water conditions, at this range only very tolerant species from both ecosystems may occur: species minimum (Remane, 1958). Therefore, prey of gobies is also of lesser component richness than in marine habitats. Evenness as a measure of diversity results in the Mediterranean Sea in 61% of the tests to homogeneity, in a brackish Baltic fjord only in

Table 3.1.2 Prey (% frequency) of gobies from the coral reefs of Marshall Islands, tropical Pacific (acc. to Hiatt and Strassburg, 1960).

Species	*Gnatholepis anjerrensis*	*Acentrogobius ornatus*	*Bathygobius fuscus*	*Gobiodon erythrospilus*	*Gobiodon rivulatus*	*Paragobiodon echinocephalus*
Habitat	*Sand bottoms*	*Burrows in sand bottoms*	*Intertidal reef, tide pools*	*Living heads of ramose corals*	*Living heads of ramose corals*	*Living heads of ramose* corals
N / size (mm)	5 / 32 - 40	13 / 31 - 65	14 / 32 - 66	5 / 28 - 31	11 / 21 - 33	8 / 12 - 19
Prey						
Copepoda	100	84			50	
Metanauplii	20	7			25	
Isopoda		69		×	25	×
Amphipoda		7				
Ostracoda	20	53				
Brachyura	20		30			
Natantia			23			×
Polychaeta		46	23	×		
Insecta			23			
Gastropoda	20	15				
Foraminifera		23				
Fish			30	eggs		
Algae	40	7	7			
Detritus	60	61			12	

12% (Zander and Hagemann, 1989; Zander and Westphal, 1992). Especially young specimens or very small species tend onto unbalanced relations. Prey components that occur in high abundance like *Gammarus* spp., calanoids, harpacticoids or sometimes chironomid and *Mytilus* larvae are consequently the most preferred prey of five goby species: *Gobius niger, Gobiusculus flavescens, Pomatoschistus microps, P. minutus* and *P. pictus* (Zander, 1979, 1994). Another characteristic pattern of brackish waters is a seasonal massive occurrence of, e.g. calanoid copepods in the plankton which leads to switching of goby predators to the most abundant and, therefore, best available food items (Zander, 1994, see resource partitioning p. 316).

These results from the SW Baltic, where salinity is about 10-20 ‰, differ from those of the eastern Baltic where salinity is about 6 ‰, which lies within the level of species minimum. The species richness is reduced again according to the food analysis of *Gobius niger*, *Pomatoschistus minutus* and *P. microps* which consume mainly chironomid larvae and pupae, gammarids, corophiids or zooplankton whereas other components are without importance (Thorman and Wiederholm, 1983, 1986). In the brackish Schlei-Fjord (SW Baltic), the species richness and diversity of the prey of *Pomatoschistus microps* and *P. minutus* increases from a locality with low salinity of 1 - 5 ‰, to a locality with 4 - 9 ‰ and a third with 11 - 19 ‰ (Fig. 3.1.5) (Zander and Westphal, 1992).

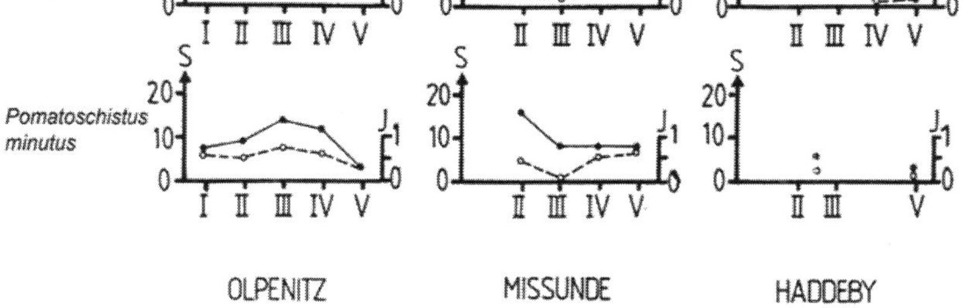

Fig. 3.1.5 Comparison of the number of prey components (S) and evenness (J) of the gobies *Pomatoschistus microps* and *P. minutus* from the brackish Schlei fjord, SW Baltic Sea. Roman ciphers: different seasons; salinity of different localities: Olpenitz = alpha-mesohaline, Missunde = beta-mesohaline (corresponds to species minimum), Haddeby = oligohaline brackish water. (From: Zander and Westphal, 1992.)

Another kind of brackish waters are lagoons where salinity are subject to strong alteration during seasons. In the Lagoon Maugio at the French Mediterranean coast the salinity varies between 10 and 37 ‰. *Gobius niger* consumes there predominantly fishes but also larger crustaceans and some algae, totally 16 components (Joyeux *et al.*, 1991). These food items represent the genuine lagoon organisms. In a Corsican lagoon with salinities between 9 and 50 ‰, *Pomatoschistus microps* feeds almost exclusively on harpacticoids, whereas in other lagoons *Gobius niger*, *G. paganellus*, *Zosterisessor ophiocephalus* and *Pomatoschistus marmoratus* include a broader food spectrum with a preference of diverse amphipods and chironomid larvae (Casabianca and Kiener, 1969).

The influence of salinity on the species richness is therefore mirrored by the food of predators, referred to a lesser level also by the group of gobies. They can thus confirm the rules of Thienemann (1939) for extreme habitats and of Remane (1958) especially for brackish waters.

Freshwater

Freshwater is an extreme environment for marine organisms. This may be also valid for gobies because only few species live permanently in such ecosystems. Freshwater needs additional energy for increased metabolism, especially osmoregulation. This stresses also fish which migrate from sea to freshwater, a behaviour presented also by several gobies, or vice versa.

The common goby, *Pomatoschistus microps*, lives in sea as well as brackish waters (see above) and migrates, if possible, into the limnetic sections of rivers, e.g. the Elbe estuary to Hamburg and 50 km further on until a barrier dam. The nutrition of common gobies was investigated in a freshwater mud flat near Hamburg, which, advantageously, can be compared with a marine mud flat in the German Wadden Sea (Hennig and Zander, 1981; Zander and Hartwig, 1982). The most conspicuous result is that gobies from fresh water forage not only for epibenthos but also for plankton and endobenthos, which are represented by cyclopid copepods or oligochaetes and the cladoceran *Ileocryptus*, respectively (Fig. 3.1.6). In contrast, marine common gobies exploit the mud flat in feeding only on epibenthos (harpacticoids, gammarids, isopods, polychaetes) which corresponds to the characteristic behaviour of this species (Fig. 3.1.7). The reason for this discrepancy is a shortness of harpacticoids and other epibenthic organisms in the freshwater locality, whereas endobenthic oligochaetes are by 75% abundant. This is well-documented by the selectivity index by which the rare harpacticoids are clearly preferred and the very abundant oligochaetes are clearly avoided (Fig. 3.1.6).

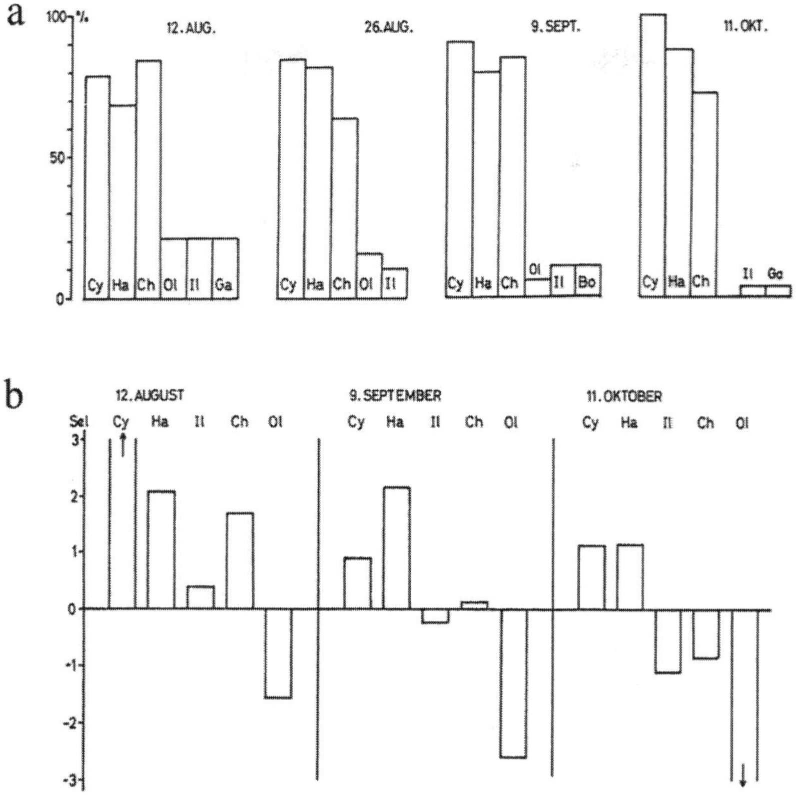

Fig. 3.1.6 Prey components of *Pomatoschistus microps* in a fresh water mud-flat of the Elbe river during summer and autumn 1976. a: Frequency of occurrence, b: selectivity index. Abbreviation of components: Ch = chironomid larvae, Cy = cyclopoids, Ga = *Gammarus*, Ha = Harpacticoids, Il = Ileocryptus, Ol = oligochaetes. (From: Hennig and Zander, 1981.)

For other gobies freshwaters are the main habitat, e.g. for species of *Knipowitschia* from the Mediterranean and Ponto-Caspian areas which attain a size of at most 40 mm. The prey of *K. caucasica* comprises small benthic crustaceans and chironomid larvae (Miller, 1986). Another Ponto-Caspian gobiid genus, *Neogobius*, comprises 16 species of which 8 prefer freshwater (e.g. *N. fluviatilis, N. kessleri* and *N. gymnotrachelus*) (Kovačic and Engin, 2008). They attain a size of more than 23 cm, therefore the main prey of benthic crustaceans is regularly completed by fish (Miller, 1986).

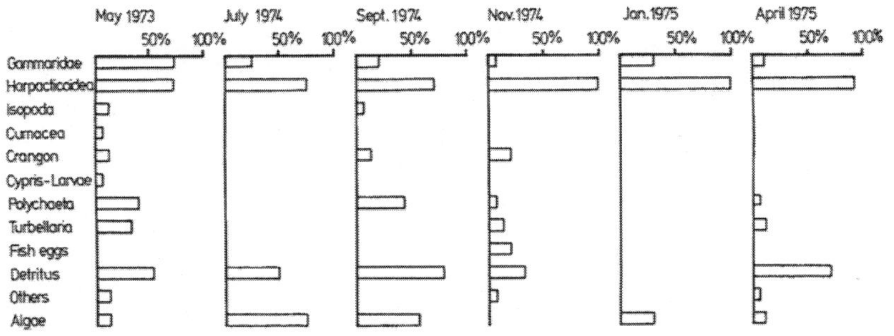

Fig. 3.1.7 Frequency of prey components of *Pomatoschistus microps* in a marine mud-flat in the German Wadden Sea. (From: Zander and Hartwig, 1982.)

Gobiomorus dormitor (Central America) or *Oxyeleotris marmorata* (SE Asia) are with a size of more than 60 cm the greatest gobies which live in freshwaters. They feed on all greater invertebrates and regularly also fish (Zaret and Paine, 1973; Larson and Murdy, 2001). On behalf of their large size, they play an important role for the local fishery.

In freshwater very small gobies as well as giant species are found. The latter seem to contradict two hypotheses: 1. that the searching for suited food in limnetic habitats is more difficult than in the sea, and 2. that osmoregulation needs as much energy that make such an extreme growth impossible. The reason lies in the situation of tropical habitats, e.g. estuaries, river deltas or eutrophic lakes, where a rich supply of potential prey is present. Thus, the only difference between feeding in marine and freshwater lies on the respective occurrence of prey which live in only one of the two milieus like, e.g. water insects and insect larvae or amphibians.

Cleaner

Elacatinus oceanops is a cleaner fish which lives in the Caribbean Sea and seek for ectoparasites (copopods, isopods, monogeneans and maybe ciliates) on skin and fins, in the mouth and in the gill chambers of greater fish like groupers and others (Eibl-Eibesfeldt, 1955; Chapter 4.3). The host fishes which are often predators of smaller fish respect the cleaner which is characterised by a conspicuous coloration: a dark blue horizontal band on an almost light body. This is a wide spread coloration for cleaner fish in tropical oceans for which the pilot fish, *Echineis naucratoides*, maybe the model. Among 15 *Elacatinus* species that occur in the Caribic, six species are cleaner of which four species present the cleaner colouration: *Elacatinus oceanops, E. evelynae, E. genie* and *E. prochilos* (Human, 1997). Facultative cleaners are *Elacatinus puncticulatus* in the tropic Pacific at the American coast and *E. nesiotes* at Galapagos but these do not present the cleaner colouration (Thompson *et al.*, 1979; Human and DeLoach, 2003).

Obligate cleaner fish seem to be specialised by an adaptation of mouth and teeth in order to be able to pick up tiny parasites from skin, fins and mouth of the hosts.

Terrestrial and Intertidal Ecosystems

For aquatic animals like fish it is a great step to discover terrestrial habitats. This step is successfully performed by several animal groups of which insects and vertebrates are the most spectacular ones. Terrestrial gobies have evolved from submerged living relatives and do not originate from that fish group which leads to the land vertebrates, the coelacanthids. The stimulus to conquer this exotic habitat is a rich food supply as a result of radiative evolution of insects and the development of other aquatic groups like crabs on land or in the intertidal zone. The terrestrial life needs fundamental adaptations of morphological and physiological patterns, especially for locomotion, breathing and sense organs as well as shelter of skin against desiccation (see Chapters 3.2 and 4.5).

Mudskippers like *Periophthalmus, Periophthalmodon, Boleophthalmus* and *Scartelaos* spp., occur mainly in tropic mangrove swamps on soft bottoms. The nests for reproduction until hatching of youngs are digged deeply into the mud bottom because the larvae need environments which lie below the water level (Magnus, 1981). Therefore, the prey of adults clearly differs from that of young specimens which are not adapted to a terrestrial way of life and, therefore, forage in the intertidal zone during high tide. Generally, *Periophthalmus* and *Periophthalmodon* spp. are carnivores, *Boleophthalmus* spp. herbivores and *Scartelaos* spp. omnivores, but principally all of them are omnivores, but differ in the importance of the respective kind of prey (Clayton, 1993). This is well-documented by gut lengths, which are shortest in carnivorous mudskipper, of middle extent in *Scartelaos* and longest in the herbivorous *Boleophthalmus*.

The carnivorous mud-skippers live at higher zones than the herbivorous. *Periophthalmus* and *Periophthalmodon* hunt for crabs by sudden tail-powered jumps (Khoo, 1966; Lim, 1971). *Boleophthalmus* uses two-thirds of its active time for grazing on the surface sediment with help of distinct sideward movements of the tail by skimming off diatoms and other algae. *Scartelaos* spp. feed by scraping on the top layer of the substrate and, additionally, bites into the substratum after slightly arching up (Clayton, 1993).

Periophthalmodon schlosseri, which is abundant in Asian estuaries, feeds mainly on crabs, especially *Uca*, but additionally bivalves and insects like ants. *Periophthalmus kalolo* from east Africa are active at low tide and hunt in the intertidal zone for invertebrates which are also active at low tide, e.g. amphibious fiddler crabs, *Uca* spp., and other crustaceans as well as polychaetes living on or near the surface (Magnus, 1981). In *Periophthalmus argentilineatus* from Singapore insects attain even 29% and are the second important prey after crabs (61%); their spectrum is completed by 9% gastropods. According to Kruitwagen *et al.* (2007), the metamorphosis of *P. argentilineatus* from east Africa begins with a size of 70 mm when contemporarily hunting during low tide begins. The part of crabs increases with size because only large mudskipper can handle and ingest greater specimen of fiddler and hermit crabs (Fig. 3.1.8).

Boleophthalmus boddarti and *Scartelaos viridis* from Singapore ingest algae and also harpacticoids, chromodorids and nematodes, but the last two invertebrates are only fed on in significant parts by *Scartelaos viridis* (Khoo, 1966). In other regions also polychaetes and fish eggs are found in the guts of *Boleophthalmus boddarti* (Mutsaddi and Bal, 1969). Contrarily, the young of *Boleophthalmus boddarti* feed mainly on crustaceans.

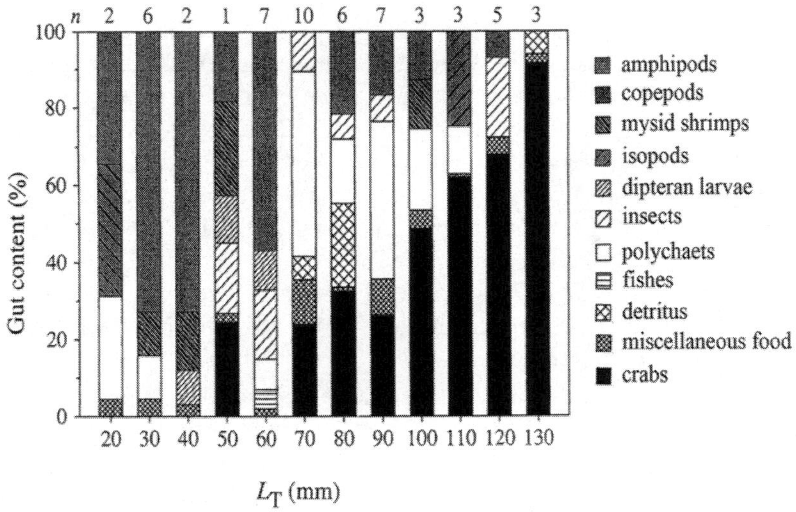

Fig. 3.1.8 Food items (% volume) ingested by mudskippers, *Periophthalmus argentilineatus*, from East Africa in correspondence to body size (LT). (From: Kruitwagen *et al.*, 2007.)

Another intertidal goby from the western American coast, *Ctenogobius sagittula*, inhabits mud or rock flats and is a mud siever (Todd, 1976). For food ingestion it leaves during low tides the home pool, takes a mouth full of mud and returns to its pool where it washes out the mud through its gill membrane and swallows the particles which are sieved by the gill rakers. Large specimens even scratch the film of food from the mud surface with their teeth.

A male and a female of the blind goby, *Typhlogobius californiensis*, live in the intertidal zone of South California together with a pair of shrimps, *Thalianassa californiensis*, in burrows, which are built by the latter (MacGinitie, 1939). Because the fish are not able to gain for food, they profit from the ventilation activity of the shrimps which create a water flux through the burrow enclosing several meiofauna, debris and detritus. These are ingested by the gobies, which are full dependent to their hosts because they have to die when the shrimps die.

GROWTH AND LIFE CYCLES

Generally, during growth of every organism an alteration of food is to be expected. Not only the width of the mouth is smaller in larval or young fish but these often live in other habitats than adults. Young fish forage for small prey organisms where these are offered in large quantities. Such a habitat is the free water with its rich plankton supply, mostly calanoid Copopoda. Gobies spawn on primary or hard bottoms, the larvae leave the parental care after hatching and live as member of the plankton. Because of their extreme small size they feed at first on copepod nauplii (Fonds, 1970).

The giant goby, *Gobius cobitis*, from the Mediterranean Sea and the north eastern Atlantic Ocean grows up to 30 cm and is, therefore, a suited object for studies on size caused changes of the food (Gibson, 1970). Whereas size increases the more algae and crabs are ingested; copepods, ostracods and small amphipods were mainly preyed on by young fish up to 80–90 mm. Conspicuously, algae are found as the only prey item in the largest investigated goby specimen of 190 mm.

Regarding small sized gobies, the first stages of many species, however, live in the free water but change toward the bottom at very different sizes (Zander and Hagemann, 1986, 1987, 1989). Young *Gobius xanthocephalus* swim above the bottom and mainly feed on suprabenthic mysids and benthic harpacticoids which are completed by planktonic calanoids and benthic gammarids. The adults (size up to 100 mm) prey exclusively on benthic organisms, especially gammarids, isopods, decapods and polychaetes (Fig. 3.1.9). The young of *G. geniporus,* which attains a size of 160 mm, live on the bottom where they prey on small benthic crustaceans like gammarids and harpacticoids, whereas the adults forage on larger crustaceans and fish (Fig. 3.1.9). Mazé (2004) found in *G. paganellus* (up to 120 mm) from northern Spain a correlation between predator size and some food items, which are negative for copepods and ostracods, but positive for leptostracans and decapods.

In *Deltentosteus quadrimaculatus* (up to 80 mm) the ratios of copepods decrease, the ratios of amphipods, cumaceans and decapods

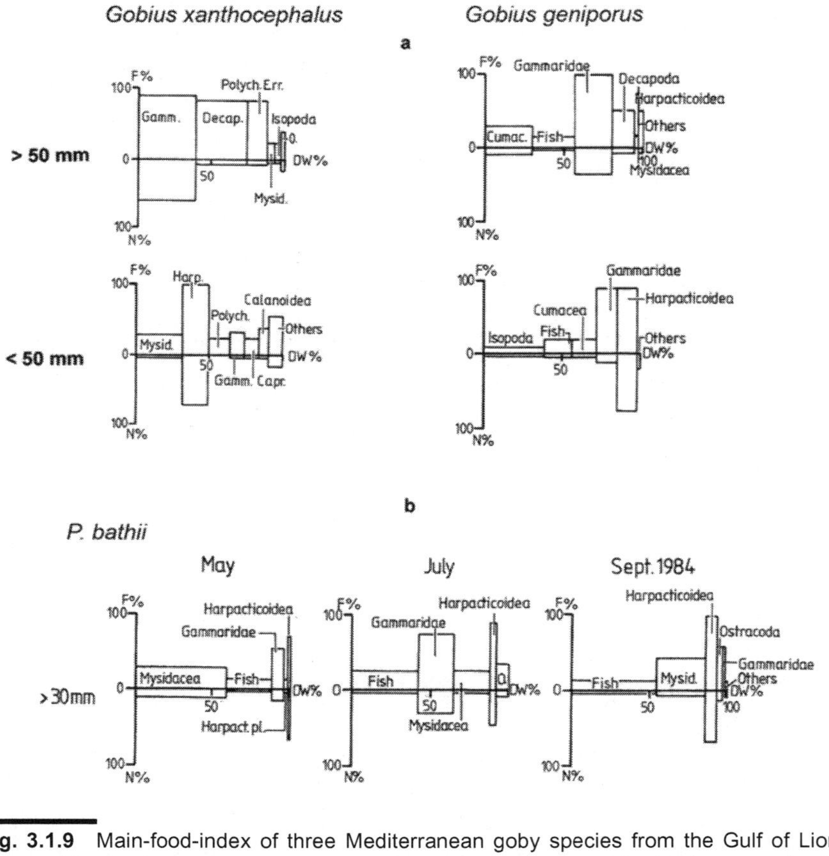

Fig. 3.1.9 Main-food-index of three Mediterranean goby species from the Gulf of Lion, arranged in 3-axis graphs where dry weight is arranged on the X-axis, frequency in the upper, abundance in the lower Y-axis. a: Adult and young *Gobius xanthocephalus* and subadult and young *G. geniporus*. b: Adult *Pomatoschistus bathi* investigated in different seasons. (From: Zander and Berg, 1984.)

increase with larger size (Villiers, 1980). Remarkably, the earliest larvae ingest planktonic calanoids, the larger young (about 25 mm) benthic harpacticoids. *Pomatoschistus bathi* (maximal size 35 mm), living together with *G. geniporus* on sand bottoms, prefers benthic prey (Gammarida and Harpacticoida) as well as fish, maybe own young and those of *G. geniporus*, and suprabenthic mysids. The young of *Gobius geniporus* ingest still smaller crustaceans like harpacticoids and ostracods (Fig. 3.1.9) (Zander and Berg, 1984).

Among short-living gobies from the Baltic the prey of the suprabenthic *Gobiusculus flavescens* (up to 50 mm) does not differ between young and adults, both prefer calanoids (except if plankton is rare, see below p. 308) (Zander and Hagemann, 1987). In contrast, adults of *Pomatoschistus minutus* (in the Baltic up to 75 mm) ingest mostly benthic components like gammarids, isopods and mussels. Subadults of 21-30 mm forage partially in the free water and partially on the bottom, what is confirmed by calanoids and mysids as well as by gammarids and harpacticoids in the guts. Specimen sized less than 20 mm prey exclusively on calanoids and mysids (Zander and Hagemann, 1987) (Fig. 3.1.10). Later analyses of this goby revealed that the transitional stage from free water to bottom can be extended until a size of 40 mm. The benthic *P. pictus* (up to 50 mm) exploits the bottom and the free water, which is indicated by the food spectrum consisting of gammarids and isopods, as well as mysids and fish larvae. The young rather feed on calanoids and mysids but also benthic isopods. Differences in the prey of juvenile and adult mud-skipper underlie metamorphosis to terrestrial life and are presented above (p. 308).

These studies clearly mirror the way of life and the environment of foraging of the predators. Beside harpacticoid copepods on the bottom mostly macrofaunal crustaceans or polychaetes and molluscs are supplied which need large mouths and, in the case of molluscs, suited teeth to crack possibly the shells.

SEASONAL CHANGES

Gobies as prey generalists adapt to the changing food offer of their habitat at different seasons. They follow the optimal foraging concept and especially regard the reduction of the searching time. As consequence, they may change also their characteristic habitat if the bordering areas or layers offer a better supply of prey organisms. Such changes are to expect in a greater extent in the temperate climate zone than in the subtropical and tropical zones with a more homogeneous climate.

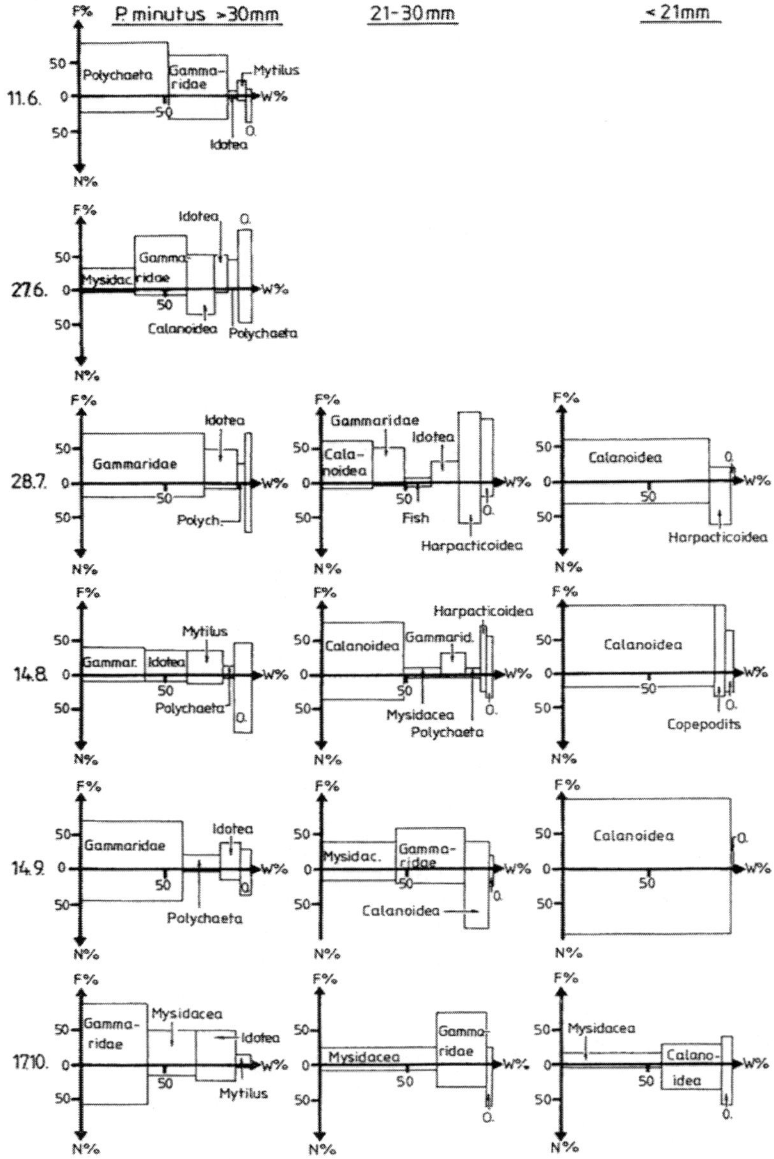

Fig. 3.1.10 Main-food-index of *Pomatoschistus minutus* from the SW Baltic Sea, arranged in 3-axis graphs where dry weight is arranged on the X-axis, frequency in the upper, abundance in the lower Y-axis. The diagram comprises different seasons and ontogenetic stages which present bottom living adults, subadults which switch between bottom and suprabenthal, and youngs and larvae which live in the free water. (From: Zander and Hagemann, 1987.)

Extended studies in the Mediterranean and Baltic Sea in spring, summer and autumn can underline these differences (Zander and Hagemann, 1987; Zander, 1994). Though gobies from cold temperate areas avoid temperatures lower than 3.5°C and migrate in winter from shallow into deeper areas, the prey choice, therefore, seems not generally to differ from that of other seasons.

However, in the Baltic the prey of gobies from shallow areas can change during warmer seasons. This situation is due to a surplus or a shortness of distinct prey organisms. Investigations on *Pomatoschistus minutus, P. pictus, P. microps, Gobiusculus flavescens* (these are short-lived species) and *Gobius niger* from Mecklenburg Bight revealed an extensive interdependence to plankton appearance (Zander, 1994, 2004) (Fig. 3.1.11).

The Mediterranean presents higher temperatures (minimum in winter 12°C) than the Baltic and shallower peaks of plankton, the higher of which occur in spring. Therefore, such switching as in the Baltic is not observed in the Mediterranean Sea regarding *G. xanthocephalus*, *G. geniporus* and *Pomatoschistus bathi* (Zander and Hagemann, 1989). The prey of the latter species consists of meiofauna, of suprabenthos (fish and mysids), and of some macrofauna which present only slight variance in the different seasons (Fig. 3.1.9).

There are only rare investigation on the winter food of gobies from temperate areas during this extreme season when the fish emigrate from the shallow habitat in deeper ones. Three studies on the foraging of the sand goby, *Pomatoschistus minutus*, are performed in east Scotland, North Sea (Healey, 1971), in the Belt Sea (Hesthagen, 1971) and in the Gdansk Bay (Morawski, 1978), both Baltic Sea. Whereas gobies from North Sea feed in winter mainly on amphipods but rarely on polychaetes, this group dominates together with copepods in specimen from the Belt Sea. Another effect of the latter population was the selection of mysids, though cumaceans are offered by eightfold numbers. In Gdansk Bay polychaetes, amphipods and mysids are important food items which dominance changes with depth of habitat and size of gobies (Morawski, 1978). The winter food of *P. microps* from the Belt Sea resembles that of *P. minutus* with the exception that polychaetes are of less importance (Hesthagen, 1971).

These results emphasise that gobies are able to adapt themselves to the respective prey supply and choose that items which are easiest to catch and to handle in order to attain optimal foraging.

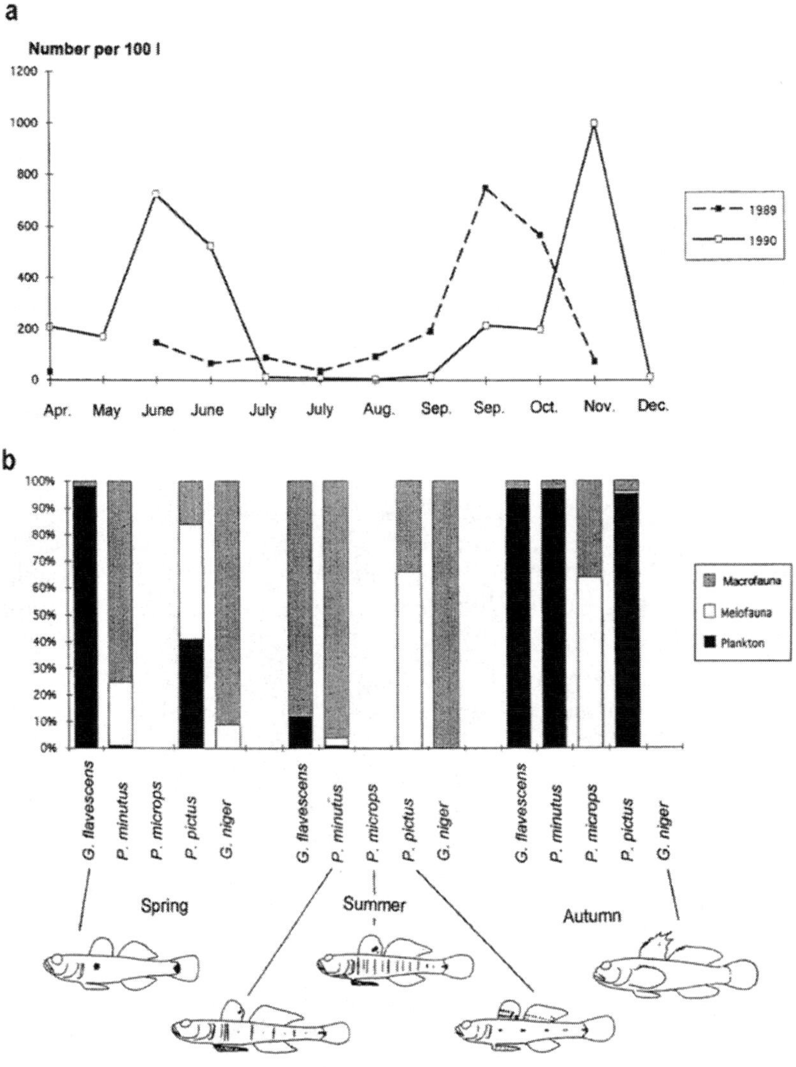

Fig. 3.1.11 Prey of five goby species from the SW Baltic Sea during three seasons in relation to the occurrence of planktonic copepods. a: Number of copepods in 100 l plankton samples during two years. b: Composition of the prey ingested by the gobies combined to the categories macrofauna, meiofauna and plankton. Special attention is paid to *Gobiusculus flavescens* in summer and *Pomatoschistus minutus* in autumn. *G. niger* = *Gobius niger*, *P.* = *Pomatoschistus*. (From: Zander, 2004.)

DAILY ACTIVITY

The feeding activity can very probable be combined with general activity, e.g. swimming in fish. Methods for studying this pattern can be the fullness (p. 293) of intestinal tracts and/or the gravimetric prey analysis during day and night. Hesthagen (1971) analysed the diurnal ratios of sand goby, *Pomatoschistus minutus,* in the Baltic during winter, which indicate an increased activity in night. Additionally, the ratio of empty stomachs is higher at day than at night. These results must be considered under the exceptional situation of studies during this season when gobies are emigrated from shallow into deeper waters. Contrarily, the feeding of the Mediterranean goby *Deltentosteus quadrimaculatus* begins at dawn and thereafter fullness decreases but increases later until dusk when feeding stops (Villier, 1982). Similar results were presented by del Norte-Campos and Temming (1994) of *Pomatoschistus minutus* and *P. microps* from the North Sea.

More detailed investigations were performed by Antholz *et al.* (1991) and Meyer-Antholz (1986) regarding the common goby, *Pomatoschistus microps* from a Baltic brackish fjord. Results of fullness present daily variations and decrease during night. Only young gobies feed on some benthos at night. Regarding the numbers of food items, calanoids prevail in spring in the guts of gobies which are caught during day, whereas in summer and autumn these are replaced by harpacticoids which ratio is not essentially lower in the night. In contrast, macrofauna like amphipods, isopods or polychaetes prevailed at night. This is explained by reduced digestion of the chitinuous or bristly integument of these organisms (Antholz *et al.*, 1991).

As laboratory studies reveal, the general and feeding activity of *P. microps* from a brackish water are in harmony and depend on temperature and light, to a lesser extent on salinity, but not on the length of days (Meyer-Antholz, 1986). Light intensity of less than 0.7 lux lowers moving activity as do temperature below 4°C and beyond 25°C. Salinity of more than 25‰ lead to hyperactivity, of more than 30 ‰ to death of specimen of this common goby population from brackish waters, whereas population from the marine North Sea are not effected by such values.

Peaks of food ingestion at dawn and dusk is a general phenomenon in gobies and also in most fishes as Hesthagen (1976, 1980) confirmed for *Gobius niger* and *Pomatoschistus pictus* from North Sea, or Casabianca and Kiener (1969) for *P. microps* from Mediterranean lagoons. An exception seems to be young gobies, which seek for food on the bottom also at night because their digestion ratio is in summer very high (Antholz *et al.*, 1991).

RESOURCE PARTITIONING, COMPETITION AND NICHE

Resource partitioning is the consequence of competition for the essential conditions of existence, space and food. An important part plays the respective ecological niche which width depends on physical, chemical and biotic (predators, parasites, competitors, prey) factors. In cases of a surplus of prey offer the strength of competition can decrease whereas, contemporarily, prey partitioning can increase.

Competition prevails when resources become short, a consequence for the respective predator is specialisation. However, generalists are able to exploit quantitatively the population alterations of several different prey organisms.

Prey partitioning is observed in the SW Baltic Sea in five related gobies. Whereas *Pomatoschistus minutus, P. microps, P. pictus* and *Gobius niger* are bottom dwellers, *Gobiusculus flavescens* lives suprabenthically and exploits mainly copepods from the plankton (Fig. 3.1.11) (Zander, 1994). *G. niger* and *P. minutus* prefer clearly macrobenthos like gammarids, isopods and polychaetes but in *P. microps* meiofauna like benthic copepods surpass macrofauna and are most abundant. During the seasons, the feeding behaviour can change according to the respective development of plankton which density is high in spring, very low in summer and highest in autumn. The consequence is that the plankton feeding *G. flavescens* is in summer forced to forage macrofauna on the bottom. However, in autumn the macrofauna feeding *P. minutus* switches over to the rich supply of plankton. *P. pictus* presents an intermediate strategy consuming in spring meiofauna, plankton and a smaller part of macrofauna, in the plankton-poor summer meio- and macrofauna, and in autumn mainly plankton (Fig. 3.1.11). Niche overlap is thus high in autumn but low (< 60%) in the other seasons.

Similar results are found by Thorman (1982) and Thorman and Wiederholm (1983) in gobies and other small fish at the Swedish west and east coasts. A wide niche overlap between *P. microps* and *P. minutus* is found only in early summer, but not between *G. niger* and these two species, mean values remain very low during a time-span of seven months. The *Pomatoschistus* species present also important overlap values with two stickleback species in summer only; *P. microps* and *Pungitius pungitius* are similar by 62% overlap regarding the yearly mean of adults. However, young *P. minutus* which prefer plankton even compete with *Gasterosteus aculeatus* by 91% in the mean of the year (Table 3.1.3) (Thorman and Wiederholm, 1983). Generally, it is obvious that ecological segregation of predators basing on foraging behaviours can be cancelled if supply of food is at its extreme ranges, short as in summer or extreme rich as in autumn (Zander, 1994). Gobies as generalists are able to switch to alternative prey items, as Edlund and Magnhagen (1981) found in laboratory experiments with *P. minutus* and *P. microps*. In these, *Corophium* (Amphipoda) and *Chironomus*-larvae (Insecta) are offered as prey organisms. When kept separately, at first *Chironomus* is preferred over *Corophium* which ratio increases with lower prey density, when kept together, *P. minutus* prefers *Chironomus* but *P. microps* feed almost entirely on *Corophium*.

Table 3.1.3 Mean food overlap of 6 small sized fish from the eastern Baltic, acc. to Thorman and Wiederholm (1983). Upper rows, adults, lower rows: young fish. Values of more than 0.6 mean great conformity.

	Pungitius pungitius	*Gasterosteus aculeatus*	*Pomatoschistus microps*	*Pomatoschistus minutus*	*Gobius niger*	*Phoxinus phoxinus*
P. pungitius	1	0.43	0.62	0.53	0.57	0.43
G. aculeatus	0.50	1	0.22	0.43	0.48	0.58
P. microps	0.35	0.52	1	0.37	0.39	0.06
P. minutus	0.46	0.91	0.42	1	0.44	0.40
G. niger	0.45	0.07	0.56	0.14	1	0.40

In the Mediterranean sand bottoms *Pomatoschistus bathi* present niche identity in different habitats (Zander, 1996). If this small species lives syntopically with juvenile and larger *Gobius geniporus* there is a clear niche overlap between *P. bathi* and juvenile but not with subadult *G. geniporus,* as is also between young and subadult *G. geniporus*. Thus, competition may rather occur between specimens of similar size albeit these belong to different species than between conspecifics of different sizes.

In the Tejo estuary (Portuguese Atlantic coast), *Pomatoschistus minutus* and *P. microps* are important members in the food web (Salgado *et al.*, 2005). Mysids are common prey of sand gobies, additionally shrimps and polychaetes. The common goby prefers polychaetes, whereas isopods, amphipods, bivalves and copepods are secondary food. With growth, the niche width of *P. minutus* increases. Due to different sizes of adults, the degree of interspecific overlap decreases with increasing body size.

Generalist predators like gobies can switch without difficulties to another kind of prey even these occur in other water levels (benthos vs. plankton). If these prey components are supplied in surplus and are easy to handle, competition with other species can be avoided.

PREY SELECTION

The selectivity index as developed by Berg (1979) (see p. 294) was applied by this author to the feeding behaviour of the two-spotted goby, *Gobiusculus flavescens*, from Helgoland, German Bight. By comparing the numbers of potentially available with ingested food of this small-sized goby from the suprabenthic layer, the most components of the plankton are avoided but the most abundant calanoid copepods are preferred such as also harpacticoid copepods, polychaete larvae and barnacle larvae (Fig. 3.1.12). Conspicuously, the selectivity index is low in the most abundant calanoids (59% of all available food components), whereas it is high in harpacticoids and barnacle larvae which occur only by 2 or 1% in the plankton.

An extended investigation on sand gobies, *Pomatoschistus minutus,* in the brackish Baltic regarded the selection of food items, as well as of selection of sizes of some food components was done by Zander (1990). This goby inhabits sand bottoms and is, therefore, also found in ecotones near hard bottoms. Whereas these gobies select Harpacticoida and sometimes Polychaeta and Oligochaeta in pure sand bottoms, they avoid mussels and snails. In the near of hard bottoms they prefer in most cases Isopoda and Gammarida but avoid mussels as well as polychaetes. Regarding sizes of single components, the most selected Gammarida were 4 - 5 mm, in autumn even 5 - 6 mm in size, which is less than the mean size of the available specimen (Fig. 13). In contrast, *Jaera* isopods of 2 - 3 mm size are selected and, thus, excel the mean size of the offered specimen of 1 - 2 mm (Fig. 5.1.13).

The above described gobies from the Mediterranean Sea have also clear preferences (see p. 295). In a coarse sand habitat, *Gobius xanthocephalus* prefers macrofaunal crustaceans like gammarids, caprellids and crabs, but avoid isopods and meiofaunal harpacticoids (Zander and Berg, 1984). In a deeper rock/sand ecotone also polychaetes are avoided. The short living *Pomatoschistus bathi* from the same bottoms prefer on the coarse sand only harpacticoids, but other components present such a rich surplus of available prey offer that these seem to be avoided. In a rock/sand ecotone also some macrofauna elements are preferred and only polychaetes and barnacle larvae are avoided which characterizes a more scarce prey offer than in the coarse sand (Fig. 3.1.14). Regarding sizes of harpacticoids, even the small *P. bathi* select specimen of larger size than present in the offer as does also *Gobius xanthocephalus* (Fig. 3.1.14). Regarding gammarids, *P. bathi* exploits specimens of about 2 mm which are of the same mean as supplied ones, whereas *Gobius xanthocephalus* prefers prey specimen of about 3 mm size.

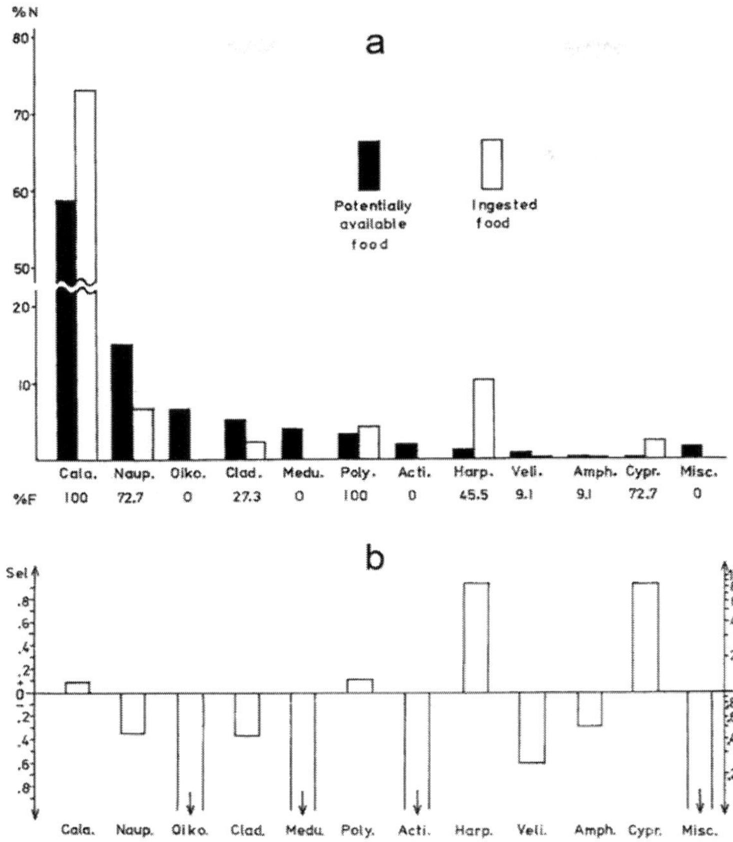

Fig. 3.1.12 a: Numerical percentages (%N) of food components available in a plankton sample and in stomachs of *Gobiusculus flavescens* from Helgooland, North Sea. Below X-axis: frequency of occurrence (%F). b: Feeding selectivity (Sel) of *G. flavescens* calculated from sample in a. Right scale is in according to Shorigin's index "K". Items: Cala. = Calanoida, Naup. = nauplii, Oiko = *Oikopleura*, Clad. = Cladocera, Medu = medusae, Poly. = Polychaeta, Acti. = actinotrocha larvae, Harp. = Harpacticoida, Veli. = veliger larvae, Amph. = Amphipoda, Cypr. = cypris larvae, Misc. = miscellaneous items. (From: Berg, 1979.)

The interpretation of the selectivity index regards only the ratio of ingested and potentially available food and says nothing about the importance of the respective food item. May be a rare component leads to an increased searching activity whereas others are, caused by high abundances, main food components. Shortness of preferred food can thus lead to high selectivity values.

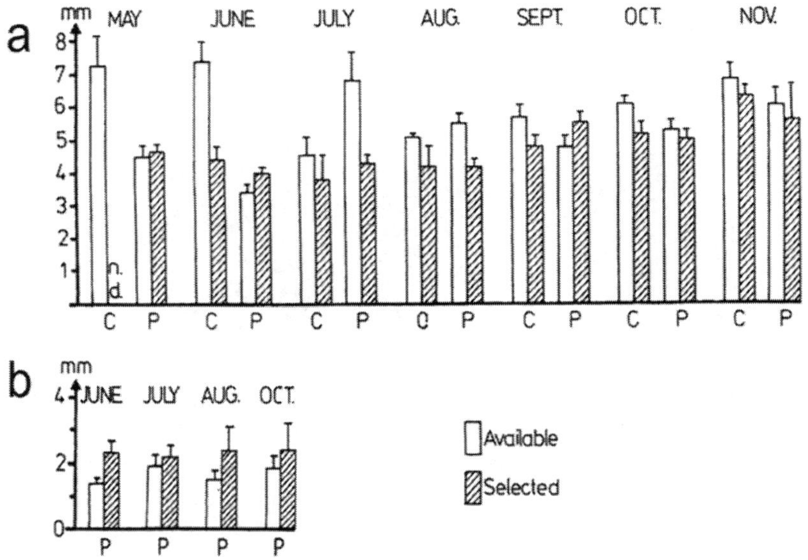

Fig. 3.1.13 Mean sizes of a: Gammarids, b: *Jaera* isopods available from food supply and selected as food by *Pomatoschistus minutus* from the SW Baltic Sea at different months. C = clay/sand ecotone, P = sand/rock ecotone at a pier. (From: Zander, 1990.)

DIGESTION AND DAILY CONSUMPTION

Digestion is primarily a physical and chemical process in order to supply energy for metabolism. Ratios of digestion indicate the amount of prey that is needed for growth, activity and reproduction of organisms. These ratios are secondarily influenced by many factors, such as temperature in ectotherms like fish, body size following the weight/volume rule that alter during ontogeny, or range of activities. Therefore, research on digestion is a contribution to the production ecology following the formula **P = A – R = C - FU** (where A = assimilated part of prey, C = consumption, FU = faeces/urea, P = production, R = respiration). The consumption from an ecosystem, C, therefore means the usable part of prey for digestion.

One of the most extended investigations of digestion ratios in gobies regard the common goby, *Pomatoschistus microps,* in the laboratory and in the natural environment these were done by Meyer-Antholz (1986). It is widely distributed in Europe and lives not only in marine but also in brackish ecosystems and migrates often into freshwaters. Regarding specimen from the brackish Baltic Sea, the greatest influence on digestion rates has temperature and quantity of prey, which both accelerate digestion

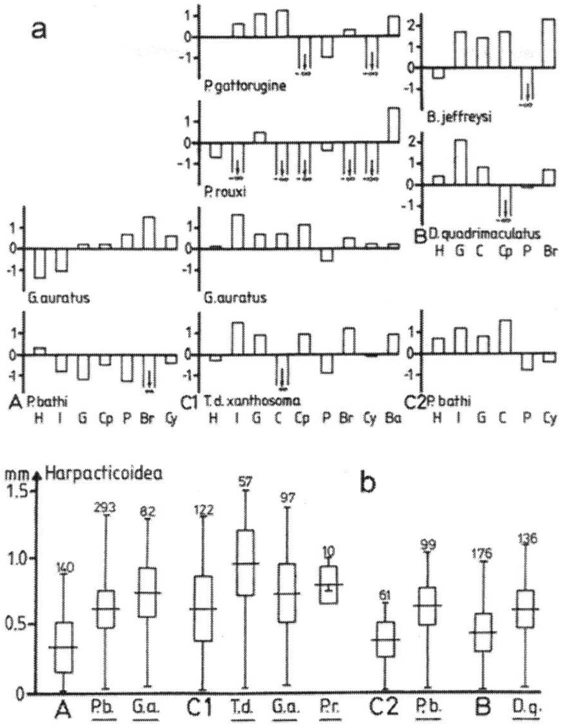

Fig. 3.1.14 Selectivity of four goby and three blennioid species from different bottoms from the Gulf of Lion. A = gravel bottom, B = muddy sand bottom, C1 = rock/sand bottom, C2 = sand bottom. a: Indices of prey components. *P. gattorugine* = *Parablennius gattorugine*; *G. auratus* is a misidentification of *Gobius xanthocephalus*. b: Size ranges of available (A, B, C1 and C2) and by gobies or blennioids ingested harpacticoids. P.b. = *Pomatoschistus bathi*, G.a. = *Gobius xanthocephalus*, D.q. = *Deltentosteus quadrimaculatus*, P.r. = *Parablennius rouxi*, T.d. = *Tripterygion delaisi xanthosoma*. (From: Zander and Berg, 1984)

with increasing values (Fig. 3.1.15). Also the kind of food organisms can play an important role because lipid-rich prey like crustaceans are faster digested than lipid-poor ones like oligochaetes and chironomid larvae. Additionally, greater prey organisms seem to be faster digested than smaller ones. The question whether size or weight of the predator is correlated with the speed of digestion remains not yet clear. Results from laboratory indicate that the young common gobies had in fact higher digestion rates than adults which are referred to growth, but in the natural environment higher temperatures have a more stimulating effect for adults than for young. A total evacuation is attained after 21 hours and follows at first a linear, after

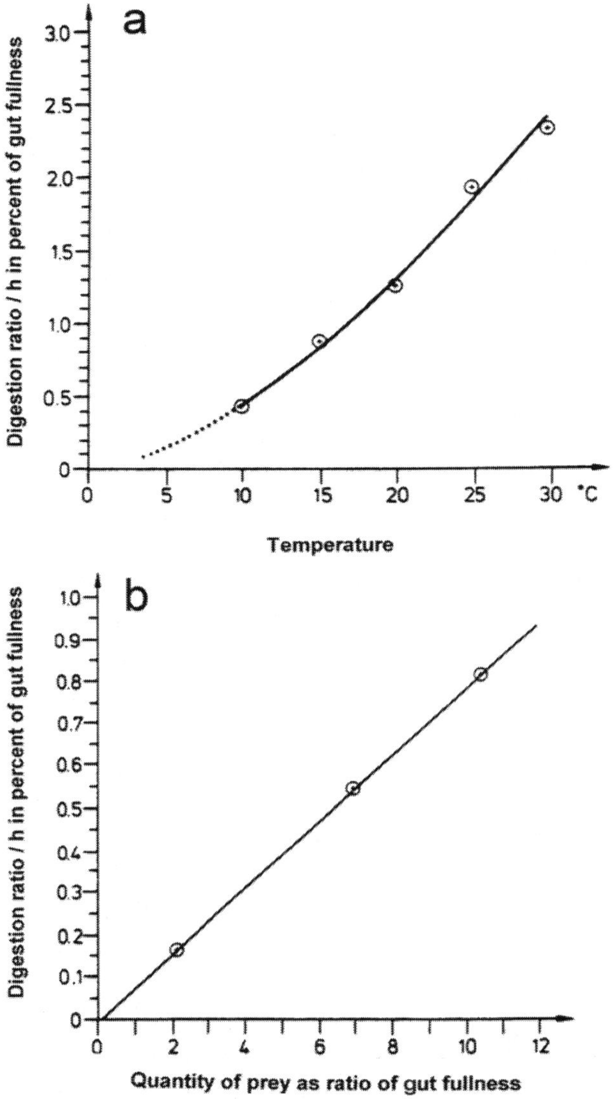

Fig. 3.1.15 Rates of digestion in the common goby, *Pomatoschistus microps*. Results from laboratory experiments on specimen from the brackish SW Baltic Sea. a: Influence of temperature. b: Influence of fullness of the guts under constant temperatures of 15°C. Significance is below 0.001. (From: Meyer-Antholz, 1986.)

12 hours a logarhythmic equation (Meyer-Antholz, 1986). The feeding activity comprises the whole day with peaks at dawn and dust and ceases at night (see above). Feeding activity is in accordance with locomotor activity which increases with temperature in a range from 4 to 25°C whereas low salinities are without influence.

Finally, this author gives a formula for the calculation of the daily prey ratio, which was calculated according to the digestion ratios of *P. microps*. The daily prey ratio "R" is calculated as 5.8% in adults, as 4.2% in young according to the fullness index of the stomach. The absolute prey demand per day correspond, therefore, to values of 4.8 and 1.4 mg (dry weight) per day, respectively. On the basis of these values Meyer-Antholz (1986) calculated a food conversion (P/C) of 0.25 in common gobies of different sizes and from different seasons. Anderson (1984) reports a daily consumption ratio of 8.6% for Baltic *P. microps* which is higher as that found by Meyer-Antholz (1986) and may be influenced by different temperatures in the respective investigations. Norte-Campos and Temming (1994) calculated a value of 27% of body weight in *P. microps* and, according to size, a mean of 13% in *P. minutus* from the German Wadden Sea which differs according to site groups of this predator. In west Sweden the daily consumption of *P. minutus* is calculated as values between 5 and 12% of body weight, regarding temperatures of 5°C or 15°C, respectively.

In mud-skippers, together with metamorphosis, prey change from diverse aquatic invertebrates to mainly terrestrial crustaceans. Contemporarily, also the chitinase and chitobiase activities in the guts increase which enables to disrupt and degrade exoskeletons of arthropods in order to gain additional source of nutrients (Kruitwagen *et al.*, 2007)

Investigations of consumption are together with growth studies and studies on standing stocks of great importance for production and energy flux in ecosystems. This complex will be regarded more extendedly below (food web).

THE ROLE OF GOBIES IN THE FOOD WEB

Because gobies are mainly of small size, their importance in the ecosystem is often underestimated. Four patterns are considered here in order to clear their role in the food web: 1. productivity of gobies, 2. predation impact of gobies on their main prey organisms, 3. predator impact on goby population, completed by human exploitation: fishery, 4. gobies as transmitter of parasites.

Productivity of Gobies

Production is the increase of biomass in time: P = dB/dt. High productivity is attained by rapid growth of individuals or/and population. Both patterns apply for gobies.

Prior conditions to gain values of productivity are investigations of biomass at least two times during a year, optimally at the start and the end of the vegetation time. Zander and Hagemann (1986) found in the SW Baltic Sea under regard of seven samples in the year with mean densities of 247 two-spotted gobies (*Gobiusculus flavescens*) and 107 sand gobies (*Pomatoschistus minutus*), maximally even 937 or 429 specimens above 1 m^2, respectively. Production was calculated by the weights which increased in the course of a year. The great density of gobies amount to production values of 2.7 (*G. flavescens*) or 2.6 g/m^2 and year (*P. minutus*). Whereas these results are found in the near of a mussel belt at the shore, *P. minutus* and *P. pictus* together attain in a 6 m deep habitat values of only 0.6 g/m^2 and year. At the Swedish west coast Thorman and Fladvad (1981) found values of maximally 0.75 g/m^2 and year in *G. flavescens* and *Pomatoschistus microps* each.

Predation Impact of Gobies (consumption ratios)

Predation impact means the part of supplied prey which is ingested by the goby predator. Under regard of Odum's (1963) model, in average only 10% of stored energy is transferred from one trophic level to the next trophic level, that means from herbivorous to carnivorous and higher carnivorous species and so on. According to Yang's (1982a) concept gobies may be characterised by a trophic level between 1.5 and 2.5, mostly 2.0, where level 1 represent herbivores, level 2 carnivores and level 3 higher carnivores. The lower end of this scale is represented by omnivorous gobies like *Gobius vittatus*, the upper end by partially piscivorous species like *Neogobius* spp. or *Gobiomorus dormitor*. If the quantity of ingested food (Ix) is correlated to the quantity of that item which is supplied by the respective environment (Sx), the consumption ratio or predation impact can be calculated: Cx = Ix / Sx.

Investigations on consumption ratio are performed in the SW Baltic together with the above mentioned productivity values. At three localities of the Schlei fjord with different salinities sand and common gobies (*Pomatoschistus minutus, P. microps*) most probably regulate the population of harpacticoids by extensive grazing (Schmidt-Moser and Westphal, 1981) In a shallow mussel belt two spotted gobies (*Gobiusculus flavescens*) consume 4.9, sand gobies 3.5 g (dry weight)/ m^2 and y, in a deeper bottom consume sand and painted gobies (*P. pictus*) together 4.2 g/m² and year (Zander and Hagemann, 1987; Zander, 1991). Daily consumption ratios can attain as much as 150 mg/m² in summer and less than 10 mg/m² in spring and autumn (Fig. 3.1.16) (Zander and Döring, 1989). The predation impact attains 0.7% if all potentially available food is regarded, but selectivity of gobies leads to higher values in specific items: *Idothea* spp. (Isopods) 6%, *Gammarus* spp. 12%, Harpacticoida even 78% (Zander and Hagemann, 1987). Very low predation impact is presented by mussels (0.04%) because only small specimen which settle after the planktonic period on the substrate can be handled by gobies. At Sweden's east coast energy input of common and sand gobies is calculated as 100 kJ/m² and year which changes in years with lesser goby density to lower values of 50 kJ/m² (Ehrenberg *et al.*, 2004).

Cage experiments with sand gobies in the German Wadden Sea result in high feeding pressure on annelids and a slight one on benthic copepods (Harpacticoida) (Reise, 1979) whereas in an eastern Scottish estuary no influence can be appointed to this goby (Jaquet and Raffaeli, 1989). Similar experiments involving *Gobionellus boleosoma* and *Ctenogobius shufeldti*

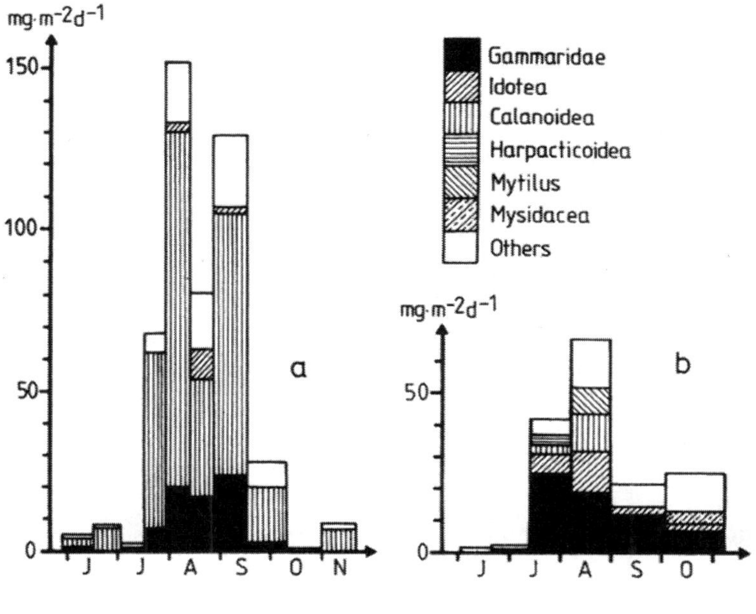

Fig. 3.1.16 Daily consumption rates of three gobies, *Gobiusculus flavescens, Pomatoschistus minutus* and *P. pictus*, at different months from the SW Baltic Sea. a: Results from a rock /sand ecotone in a depth of 3 m, b: Results from a sand/ clay ecotone in a depth of 6-7 m. (From: Zander and Döring, 1989.)

from a Caribean estuary result in significant decrease of nematodes and chironomid larvae in comparison with cages without gobies, but this is insignificant if polychaetes or copepods are regarded (Fitzhugh and Fleeger, 1985), though sizes of these predators correspond to that of sand gobies.

Also the Mediterranean gobies, regarded above (p. 295), present high predation impacts. *Gobius xanthocephalus* consumes on rocky substrates almost 10% of the mobile fauna of which gammarids attain 9%, polychaetes 17% and decapods 32% (Zander, 1996). *Gobius geniporus* and *Pomatoschistus bathi* from sand bottoms exploit together the supplied prey organism in a still more extended way: young fish 100%, Cumacea 83%, Gammarida 67%, Ostracoda 50%, Isopoda 16%, Harpacticoida 12%. The consumption ratio on the total, benthic as well as suprabenthic fauna, attains a value of 12%.

The yearly turn-over (production/biomass) of gobies can, therefore, be of great importance for the food webs. These ratios attain in gobies of the SW Baltic Sea as much as 2.7 – 3.9 (Zander and Hagemann, 1986). The various food items may play different roles according to the environmental and geographical patterns: these are to conclude from consumption ratios of more than 10% which are attained by several goby populations.

Gobies as Prey

A quantification of consumption ratios of higher and top predators is very difficult to investigate. According to Yang (1982b) the biomass of gobies in the whole North Sea is estimated in regard of *Lesueurigobius friesii, Gobius niger* and *Pomatoschistus minutus* by 6,000 metric tons each, all other gobies are rare and attain at most < 0.1 metric ton. Unfortunately, no production ratios of North Sea gobies are calculated. Though some studies are known which analyse the guts of goby predators, the size of their forage areas remains mostly unknown. An exceptional example is the hemi-enclosed brackish Lake Grevelingen in The Netherlands comprising an area of 108 km^2 (Doornbos, 1984; Doornbos and Twisk, 1987). By estimating a standing stock of sand and common gobies (*Pomatoschistus minutus, P. microps)* of 1.65 g/m^2, approximately 0.5 g are fed on by plaice and flounder, and 0.5 g by crested grebes and merganser. These predators cause a decrease of 60% of gobies from one to the next year though they use only 17% of goby production. Eels and cormorants are also predators of gobies but of lesser importance in this locality. In other localities, e.g. Sea of Azov, the prey of cormorants consisted of 68-78% *Neogobius melanostomus* (biomass), and as much as 20% of *Zosterisessor ophiocephalus* (Demtschenko *et al.*, 1999).

It was already mentioned (p. 295) that gobies are prey of gobies. This is a question of size relationships between prey and predator which depend on foraging behaviour and simply handling by the latter. In the North and Baltic Sea *Pomatoschistus minutus, P. pictus*, in the Mediterranean *P. bathi*, several *Gobius* spp., and other benthic small-sized gobies are predators of their own larvae and young. Obviously, suprabenthic species like *Gobiusculus flavescens* or *Pomatoschistus quagga* do not persuade fish larvae but rather pick up great quantities of planktonic copepods which are attained easier. As expected, large-sized gobies (20 cm and more) forage for invertebrates but also for smaller gobies. The Ponto-Caspian *Neogobius* spp. and *Mesogobius batrachocephalus* (35 cm) mainly prey on fish of which to a great part are smaller goby species or young (Miller, 1986). The same may be valid for still larger gobies like the sleeper *Gobiomorus dormitor* and *Oxyeleotris marmorata* (Zaret and Rand, 1973).

Quantitative results regarding gobies which are prey of other fishes are attained from investigations in North and Baltic Sea. In the SW Baltic Sea the weight of *Pomatoschistus minutus* preyed on by adult cods (*Gadus morhua*) attains 1%, as much as is also found in dab, flounder, plaice, rockling and whiting. But in young cod this goby comprises a ratio of 10%, in long rough dab (*Hippoglossoides platessoides*) even of 27% (Arntz, 1979). In young cods the ratios can alter in different localities between 6 and 34%, and in different seasons between 31% in winter and 0.002% in spring and summer, increasing in autumn onto 4% (Arntz, 1974). Three species of Gadidae from the Irish Sea feed regularly but in low weight ratios on gobies, especially on *Lesueurigobius friesii* (Armstrong, 1982). The ratios (weight) in cod, poor-cod (*Trisopterus minutus*) and pouting (*T. luscus*) vary mostly at a low level between 0 and 3%, but attain in subadult pouting 19% during spring and are therefore the second important prey following the decapod *Calocaris.* Also young poor-cod prey in summer and autumn on gobies at ratios of 7-8%.

Among higher vertebrates not only sea-birds but also mammals like seals and dolphins prey mainly on fish. In the German Wadden Sea, a common seal (*Phoca vitulina*) needs about 4-8 kg fish and shrimps per day, of which flatfish are the most important food item attaining 79% by weight or 43% by numbers; gobiid species attain a ratio of 41% per numbers and shrimps of 10% (Behrends, 1985). Young seals prefer prey of maximal 7 cm size, which correspond to gobies, whereas older ones prefer fish of 10-20 cm size which corresponds rather to flatfish. The Harbour purpoise (*Phocoena phocoena*) hunts for 14 fish species in the North Sea, especially flatfish, gadids, sandeels and gobies (Prochnow, 1998) and seven species in the Baltic sea of which cods attain 90% and gobies 1% by weight (Schulze, 1987).

Therefore, gobies play an important role at least in the food web, especially in the marine environment. Because of their mostly small size they are well-suited as prey for several predators of the next trophic level which can be according to Yang (1982a) the forth or fifth, that is the top level. The food web will be more complicated by cannibalism, which often occurs in carnivorous guilds like gobies.

Economics and Fishery

Gobies are though of small size also object of commercial fisheries, often just the small species and larval or postlarval stages are exploited by humans. These fisheries are focussed on regions with warm temperate and tropic climate which includes also the Mediterranean, Black and Caspian Sea. In cold temperate regions no fishery on gobies or goby fry is known, at most they occur in the by-catch of trawl fishery.

Fishery on goby fry is concentrated in several localities of the tropic region, especially in Philippines, Hawaii, Caribbean and Reunion, other areas present only unimportant activities. Mostly post-larvae of the genera *Sicyoptera* and *Sicydium* are caught during the emigration to the sea from freshwater where the adults spawn. Fishery is focussed on 2 days before and 9 days after full moon when the phase of recruitment occurs (Bell, 1999). Yields vary between 60,000 t/year in Luzon, 2 t in Hawaii, 0.9 t in Domenica (Caribbean) or meanly 40 t in Reunion. In other localities no data are attained. *Pandaka pygmaea*, with a size of 15 mm one of the smallest vertebrate, is part of Philippine fishery (Herre, 1953), but nowadays is seemingly extinct there. Remarkably, in all localities decreases the yield of fry-fishery.

In the western Mediterranean fishery the total yield in 1970-1980, mostly clupeids, attained 310000 t/year, gobies attained a part of 350 - 1100 t/year, which is exclusively from the region of Sardinia (Bas *et al.*, 1984). These fish play no role in the regions of Balearic Islands and Golf of Lion. In Italian waters trawling inside the 3 miles strip is forbidden but fishing by gears like Danish seine, small bottom trawl net or otter trawl are allowed during winter months. This fishery is directed to sardine fry and the transparent goby, *Aphia minuta*, but also sand gobies, *Pomatoschistus minutus*, and pilchards are caught (Auteri *et al.*, 1998; Ungaro *et al.*, 1994). At the Ligurian coasts total catches attain between 10000 and 25000 kg per year between 1991 and 1999, mostly *Aphia minuta* and sardine fry. Inshore bottom trawls get a yield of 1500 kg/year *Aphia minuta* until 1995 but decreases thereafter onto 500 kg/year. This fishery is also performed in the Adriatic Sea where the catches of vertical opening trawl nets comprise 37% sardines, 30% *Aphia minuta* and 13% pilchards regarding weights, otter trawls 28% *A. minuta*, 6% *Lesueurigobius sanzoi* and 2% *P. minutus*.

An important fishery on the great green goby (*Zosterisessor ophiocephalus*, maximal size 25 cm) is performed in the Venice Lagoon where a yield of 500 t/year is caught between 1972 and 1994 (Rossetto, 2000). This yield means the first place before sand smelts (*Atherina boyeri*) with 480 t and crabs. In 1997 sand smelts (170 t) surpass green gobies attaining only 110 t. A relative important fishery on gobies is performed in Black and Caspian Sea by the Romanian and Bulgarian governments. Main objects are the Black Sea herring (*Alosa caspia*) and sprat (*Sprattus sprattus*) which attained a yield of 3000-20000 t/year between 1961 and 1990. Gobies are caught especially in coastal lakes yielding 90-180 t/year between 1940 and 1960, 24 - 90 t between 1961 and 1990 in Bulgaria. Also in the entire Black Sea *Neogobius melanostomus* sized up to 22 cm and *Mesogobius batrachocephalus* which attains a size of 35 cm and is exclusively a fish predator are caught still to 30-40 or 10 t/year (Kissiov *et al.*, 2007). Fishery of Caspian Sea gobies is exploited especially by the governments of Iran, Uzbekistan and Kazakhstan.

Commercially used gobies are mostly 2 or 3 stage-carnivores, rarely omnivores. Therefore, they take the range between 3 and 4, or 2.5 in the graduation of Yang (1982a). This is valid if large benthic gobies as well as if post-larvae and fry from the free water are consumed. Consequently, humans which eat gobies are ranged between 3.5 and 4 on the scale of Yang which mean the third carnivore level or the role of a top predator in the food chain.

Gobies as Hosts and Transmitter of Parasites

Like other fish, also gobies are infected by several parasites: Protozoa, Aschelminthes, Nemathelminthes and Crustacea comprise the most important groups. The development of different larval stages which change between 1-4 different hosts or phases of free living larvae are especially characteristic for Digenea, Cestoda, Nematoda and Acanthocephala which present complicated developmental cycles ending in a final vertebrate host, also gobies (Fig. 3.1.17). But, because most of them are of small size they operate also as intermediate hosts and play an important role in transmitting parasites to final hosts, if they are prey of top predators: fish, birds, seals, whales or dolphins.

The ecological analysis of parasites is founded on the level concept of parasite populations and communities (Esch *et al.*, 1975; Holmes and Price, 1986, completed by Zander, 1998a). Thus, a population of parasite species in a single host is called infra population, all parasites in a single host means the infra community. The next level is called component population

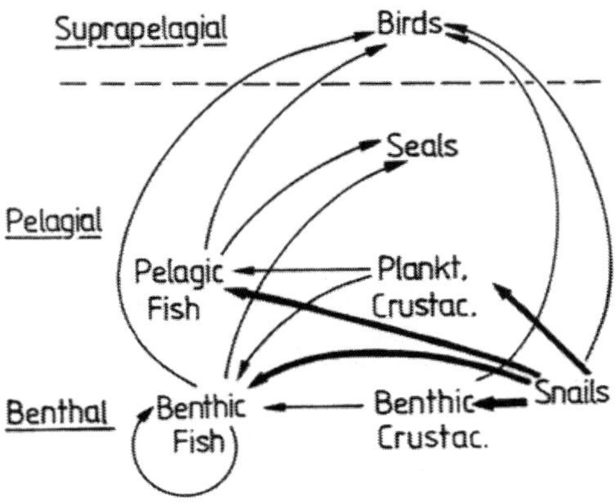

Fig. 3.1.17 Ways of infection by parasites in the marine food web. (From: Zander, 1993.)

which means a parasite population in a host population, and component community which regards a parasite community in a host population. The next higher level consider guild population and guild community where parasite population or parasite community is represented in a group of related and ecological similar hosts within a community. The last level, supra population and supra community, regard several host populations so far these comprise different stages of a parasite species, or all parasite species and stages in an ecosystem, respectively. The assemblage of several goby species are considered here as a guild. The mostly used measures are prevalence (= percent number of infected specimen in a host population), intensity (= mean number of parasite specimen in a host population) and abundance, which is the product of prevalence and intensity.

The common goby, *Pomatoschistus microps,* is the optimal second intermediate host of the digenean *Cryptocotyle concavum* in the brackish Baltic Sea. Its developmental cycle (Fig. 3.1.18) begins with the infection of snails by miracidia. They grow up to sporocysts and produce two new generations, the parasitic rediae (in many other Digenea replaced by 2. sporocysts) which generate free living cercariae. These infect actively the second host, a fish where they encyst mostly in the skin but in *P. microps* in the kidney and become metacercariae. If an infected fish is preyed on by a bird the parasites excyst and grow to adult in the guts. The settlement of a parasite stage in a host depends on environmental and immune response factors of both, parasite and host, which may vary among single specimen (Zander *et al.*, 1984; Zander, 2005a). Therefore, the infra population in common gobies can vary between an intensity of one and more than 2000

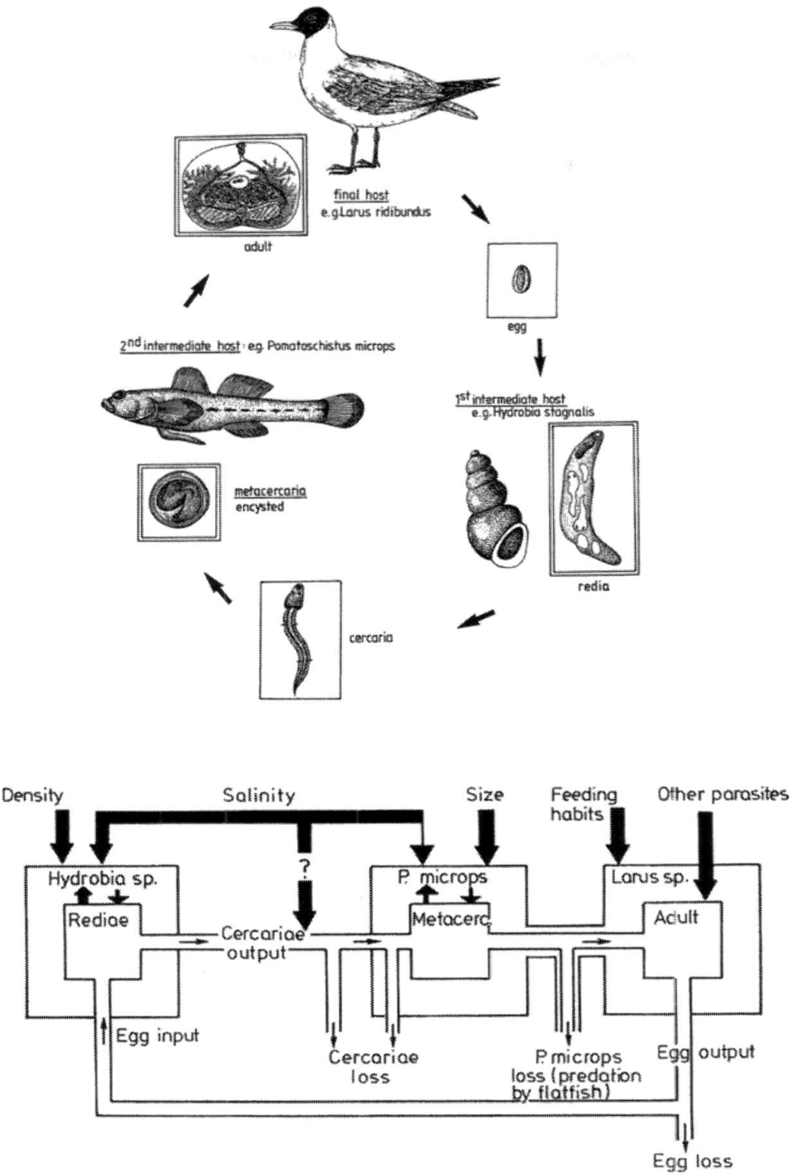

Fig. 3.1.18 Life cycle of the digenean *Cryptocotyle concavum.* Above: Free living and parasitic stages as well as the most frequent hosts. Below: Model of life cycle presenting host changes of parasites and influence of environmental factors on hosts and parasites. (From: Zander *et al.*, 1984.)

metacercariae (Fig. 3.1.19) (Zander, 1998b). By comparing the component populations of *C. concavum* in *P. microps* from three different localities, clear differences in mean intensities are obvious (Fig. 3.1.19), which may be caused by environmental factors, especially pollution. The guild population of this parasite also reveals clear differences between five goby hosts (Fig. 3.1.20) which can be referred to the development of different defence systems in each goby species (Zander, 1998b).

Considering parasite communities, the number of parasite species in fish, also in gobies, is very high. The parasites can, according to Esch *et al.* (1988), be divided into the following groups (Fig. 3.1.21): 1. autogenic

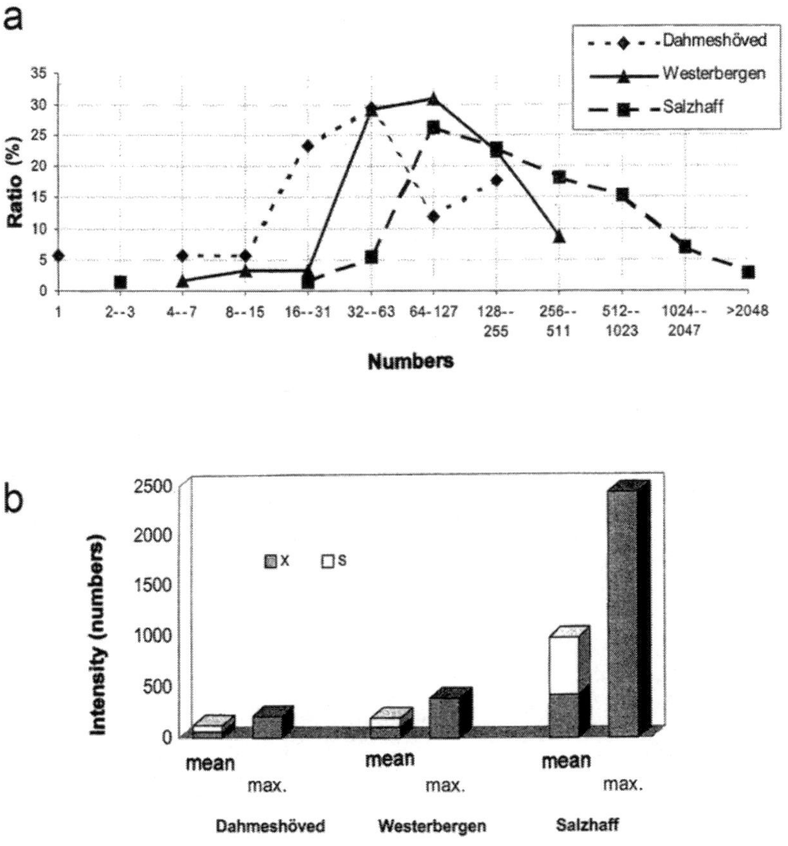

Fig. 3.1.19 Infection of kidneys of common gobies, *Pomatoschistus microps*, from different localities of the SW Baltic by metacercariae of the digenean *Cryptocotyle concavum*. a: Range of intensities along a geometrical scale. b: Mean and maximal values of intensities, x = mean, s = standard deviation. (From Zander, 1998b.)

species which remain in the same environment, 2. allogenic species which have final hosts in other environments, e.g. water living mammals migrating from locality to locality or birds living outside the water. The category of autogenic species can be subdivided in such parasites which infect fish as final and such which infect fish as intermediate hosts. The last group are transferred to top predators like great fish.

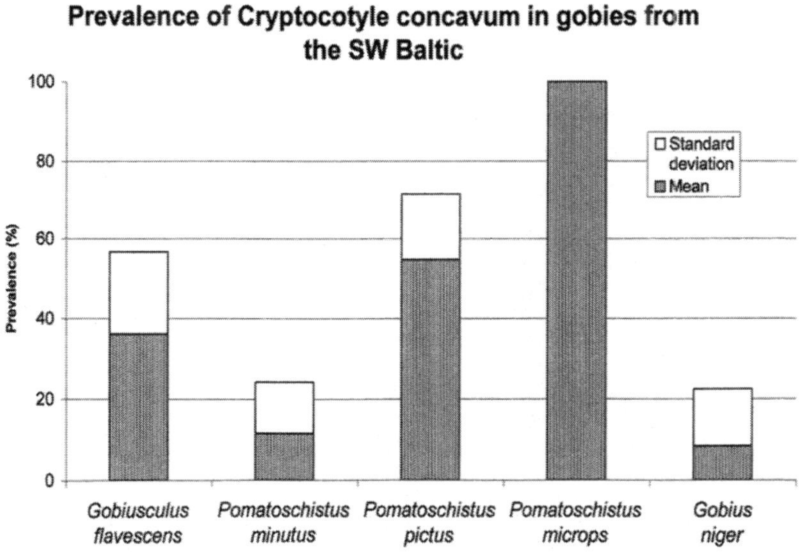

Fig. 3.1. 20 Prevalence of the digenean *Cryptocotyle concavum* in five goby hosts from the same locality in the SW Baltic. (Acc. to: Zander and Kesting, 1996; Zander, 1998b, 2005a.)

About infra communities some knowledge is gained from investigations of gobies from the Baltic (Zander *et al.*, 1999; Zander, 2004). Thus, as much as eight parasite species are found in a single specimen of *Pomatoschistus microps* in which the groups with 2-3 parasites per specimen dominate. In other gobies the groups with 1 or 2 parasites attain the greatest ratio (Fig. 3.1.22). The most research is done on component communities (parasites of a host population) in which the number of parasite species can surpass those of infra communities. As result, species number is not only different in the compared localities (component community) but also between host species (guild community) (Zander and Kesting, 1996; Zander *et al.*, 1999; Josten, 2004; Zander, 2005a). Populations of common gobies from 5 localities of North and Baltic Sea reveal great differences in abundances; e.g. the extremely abundant *Cryptocotyle concavum* is

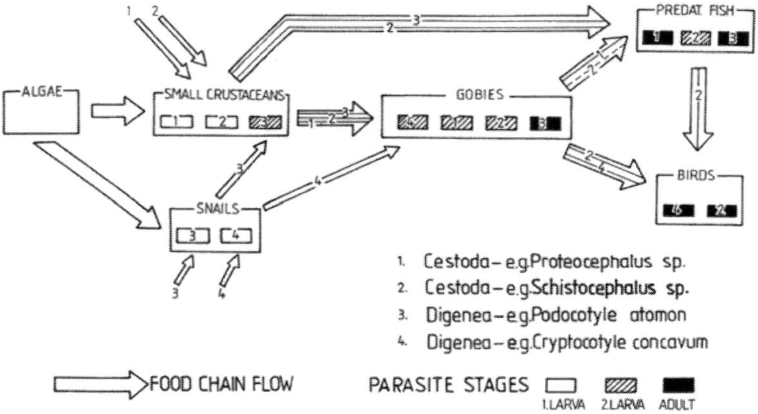

Fig. 3.1.21 The role of gobies as mediators of parasite transfer from 1. intermediate and 2. intermediate to final host. There are 4 pathways to differentiate, 1: autogenous parasites which final hosts are predatory fish; 2: allogenous parasites which intermediate and paratenic hosts are fish and final hosts are fish preying birds; 3: autogenous parasites which final hosts are among other fish gobies; 4: allogenous parasites which final hosts are fish preying birds. (From: Zander, 1998a.)

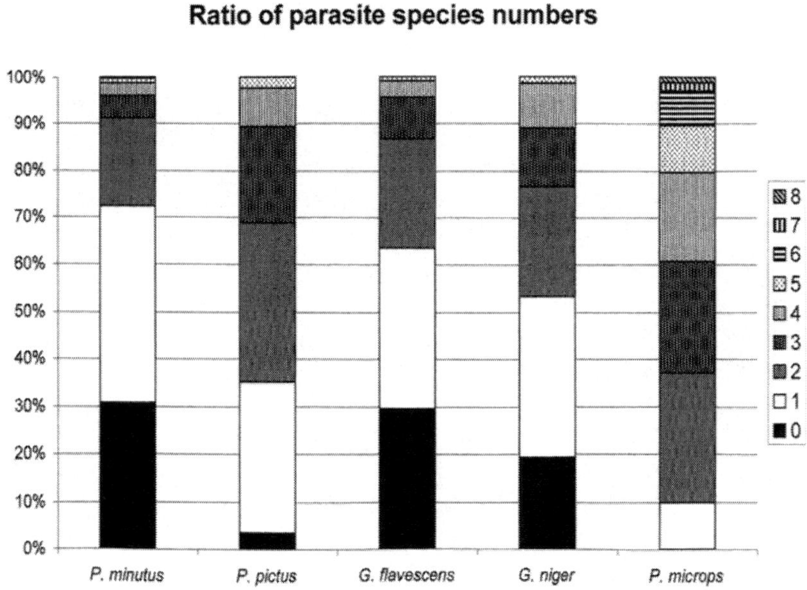

Fig. 3.1.22 Ratio of species richness in infra communities of five goby hosts. Numbers mean host specimen with the respective parasite numbers. (From: Zander, 2004.)

Table 3.1.4 Abundance of selected parasites in Pomatoschistus microps from 5 localities

	Helgoland	*Flensburg-fjord*	*Westerbergen*	*Dahmeshöved*	*Salzhaff*
Digenea					
Cryptocotyle concavum		28.1	227.7	75.7	101.8
C. lingua	0.2	0.1	0.3		0.2
Podocotyle atomon	0.1	0.5	0.7	0.3	0.01
Aphalloides timmi		0.02	3.3		6.3
Acanthostomum balthicum		0.04	0.2		0.01
Asymphylodora demeli		0.05	0.2		0.01
Brachyphallus crenatus		0.1	0.01		
Cestoda					
Ligula pawslowskii				0.01	0.01
Bothriocephalus scorpii		0.2		0.01	
Proteocephalus percae		0.01	0.03	0.01	0.01
Nematoda					
Hysterothylacium aduncum		0.01	0.01	0.8	0.01
Contracaecum sp.	0.05		0.04	0.4	0.01
Anisakis simplex		0.02		0.01	
Ascarophis arctica		0.03			0.01
Acanthocephala					
Echinorhynchus gadi	0.01	0.01		0.01	0.01

missing in the North Sea, as is also valid for the specialist of *P. microps*, *Aphalloides timmi*, which apparently occur only in the most polluted localities (Table 3.1.4). Autogenic parasites for which gobies are final hosts mostly prevail (Fig. 3.1.23). The next abundant group are allogenic parasites whereas the autogenic group for which gobies are intermediate hosts is only moderately present. But just the groups which are intermediate hosts play a central role in both, the food and parasite web. The single species of the goby guild play in this context different roles as consequence of microhabitat and infection: *P. microps* living in shallow waters can be attained as prey by gulls, deeper living gobies like *G. niger* can be caught rather by diving ducks or cormorants, suprabenthic *G. flavescens* also by terns. Therefore, the composition of bird parasite communities depends on the possibility of catching distinct fish, which may differ in quality and quantity of their parasite spectra.

The parasite communities in North Sea gobies are partly the same as in the Baltic (Zander, 2005b). Hamerlynck *et al.* (1989) found in *P. minutus* and *P. lozanoi* only 4 species, Zander (2005a) in *P. microps, P. pictus* and *G. flavescens* five species, but Groenewoldt *et al.* (1996) as much as 14 parasites in *P. minutus* of which nine species are adults. Parasites of Mediterranean gobies are rarely investigated. Sasal *et al.* (1997) reported only one acanthocephalan in *Gobius bucchichi* and *G. cruentatus*, four parasites in

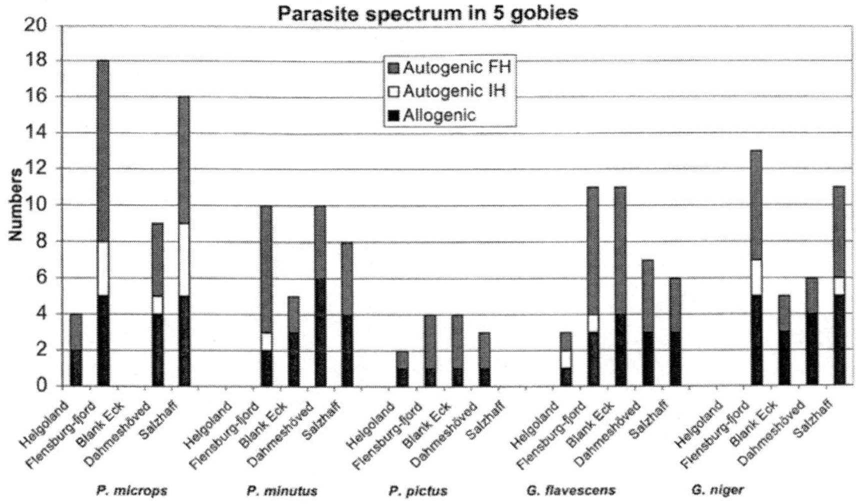

Fig. 3.1.23 Numbers of parasite species in five goby hosts combined into 3 categories of life cycles. (Acc. to: Zander and Kesting, 1996; Josten, 2004; Zander, 1998b, 2005b.)

G. niger. But Sasal *et al.* (1996) found differences in *G. bucchichi* from protected and unprotected environments regarding five parasite species. Because gobies sampled in protected areas become larger and older these can accumulate also more parasites.

Parasite communities of gobies revealed to be suited as indicator of eutrophication. The basic idea is that a surplus of nutrients promotes the frequency of herbivores like snails and small crustaceans, which are the transmitter of several first larvae of parasites (Zander, 1993, 2002). This increasing density has two consequences, 1. For snail and crustacean feeding birds the habitat becomes more attractive and, therefore, defecate eggs of parasites, and 2. also density of common parasites increases and, thus, is more effective in order to infect the next host in the developmental cycles, e.g. gobies. Rare parasites, often specialists, are finally suffered under the competition pressure of the increasing population of common species (Fig. 3.1.24).

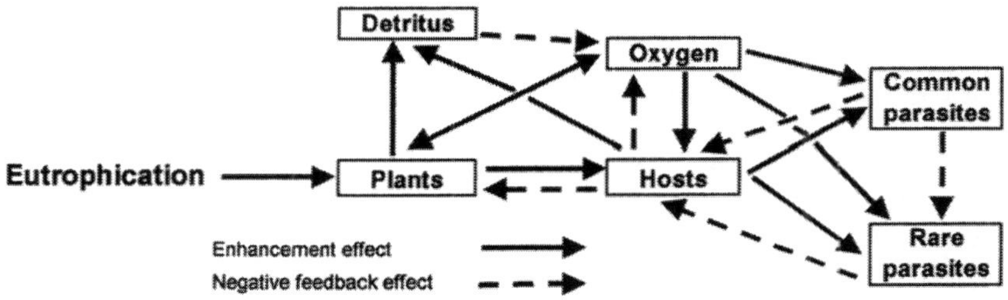

Fig. 3.1.24 Model of eutrophication effects on host and parasite populations comprising also negative feed-back reactions. The first consequence of eutrophication is increasing of plant (algal) density.

Acknowledgements

I wish to thank Marcelo Kovačić for many constructive criticisms.

References

Andersen, N.G. 1984. Depletion rates of gastrointestinal content in common goby (*Pomatoschistus microps*) (Kr.). Effects of temperature and fish size. *Dana* 3: 31-42.

Antholz, B., W. Meyer-Antholz and C.D. Zander. 1982. Feeding activities of two euryhaline small-sized fish in a western Baltic brackish fjord. *Helgoländer Meeresuntersuchungen* 45: 287-300.

Armstrong, M.J. 1982. The predator-prey relationships of Irish Sea poor-cod (*Trisopterus minutus* L.), pouting (*Trisopterus luscus* L.), and cod (*Gadus morhua* L.). *Journal de Conseil international de Exploration de Mer* 40: 135-152.

Arntz, W.E. 1974. Die Nahrung juveniler Dorsche (*Gadus morhua* L.) in der Kieler Bucht. *Ber. Deutsche Wissenschaftliche Kommission für Meeresforschung* 23: 97-120.

Arntz, W.E. 1979. Predation by demersal fish and its impact on the dynamics of macrobenthos. *The Belle W. Baruch Library in Marine Science* 11: 121-149.

Auteri, R., A. Albella, R. Baino, R. Righini, F. Serena, R. Silvestri and A. Volani. 1998. La pesca del Rossetto (*Aphia minuta*, Pisces, Gobiidae) in Toscana. *Biologie Marine Mediterrané* 5: 477-486.

Bas, C., E. MacPherson and F. Sarda. 1984. Fishes and fishermen. The exploitable trophic level. In: *Western Mediterranean*, R. Margaleff (ed.). Pergamon Press, Oxford, pp. 296-316.

Behrends, G. 1985. Zur Nahrungswahl von Seehunden (*Phoca vitulina* L.) im Wattenmeer Schleswig-Holsteins. *Zoologische Jagdwissenschaft* 31: 3-14

Bell, K.N.I. 1999. An overview of goby-fry fishery. *Naga* 22: 30-36.

Berg, J. 1979. Discussion of methods of investigating the food of fishes, with reference to a preliminary study of the food of *Gobiusculus flavescens* (Gobiidae). *Marine Biology* 50: 263-273.

Casabianca, M.L. and A. Kiener. 1969. Gobiidés des étangs Corses: Systématique, écologie, régime alimentaire et position dans les chaînes trophiques. *Vie et Milieu* A 20: 611-633.

Clayton, D.W. 1993. Mudskippers. *Oceonography and Marine Biology Annual Review* 37: 303-377.

Darcy, G.H. 1981. Food habits of the crested goby *Lophogobius cyprinoides*, in Two Dade Country, Florida, Waterways. *Bulletin of Marine Science* 31: 928-932.

Demtschenko, W.A., R.W. Pokussa, A.G. Semenow and L.S. Mitjaj. 1999. Artenzusammensetzung, Maße und Gewichte von Fischen in der Nahrung von Kormoranen auf den Inseln von Molotschnij Liman und in der Obitschnij Bucht des Asowschen Meeres. *Branta* 2: 172-178.

Doornbos, G. 1984. Piscivorous birds on the saline Lake Grevelingen; The Netherlands: abundance, prey selection and annual food consumption. *Netherland Journal of Sea Research* 18: 457-479.

Doornbos, G. and F. Twisk. 1987. Density, growth and annual food consumption of gobiid fish in the saline Lake Grevelingen, The Netherlands. *Netherland Journal of Sea Research* 21: 45-74.

Edlund, A.-M. and C. Magnhagen. 1981. Food segregation and consumption suppression in two coexisting fishes, *Pomatoschistus minutus* and *P. microps*; an experimental demonstration of consumption. *Oikos* 36: 23-27.

Ehrenberg, S.Z., S. Hansson and R. Elmgren. 2004. Sublittoral abundance and food consumption of Baltic gobies. *Jounal of Fish Biology* 67: 1083-1093.

Eibl-Eibesfeldt, I. 1955. Über Symbiosen, Parasitismus und andere besondere zwischenartliche Beziehungen tropischer Meeresfische. *Zeitschrift für Tierpsychologie* 12: 203-219.

Esch, G.W., J.W. Gibbons and J.E. Borouque. 1975. An analysis of relationship between stress and parasitism. *American Midland Naturalist* 93: 339-353.

Esch, G.W., C.R. Kennedy, A.O. Bush and J.M. Aho. 1988. Patterns in helminth communities in fresh water fish in Great Britain: alternative strategies for colonization. *Parasitology* 96: 519-532.

Fitzhugh, G.R. and J.W. Fleeger. 1985. Goby (Pisces: Gobiidae) interactions with meiofauna and small macrofauna. *Bulletin of Marine Science* 36: 436-444.

Fonds, M. 1970. Remarks on the rearing of gobies (*Pomatoschistus minutus* and *P. lozanoi*) for experimental purposes. *Helgoländer wissenschatliche Meeresuntersungen* 20: 620-628.

Gibson, R.N. 1968. The food and feeding relationships of littoral fish in the Banyuls region. *Vie et Milieu.* A 14: 447-456.

Gibson, R.N. 1970. Observations on the biology of the giant goby. *Gobius cobitis* Pallas. *Journal of Fish Biology* 2: 281-288.

Groenewoldt, S., R. Berghahn and C.D. Zander. 1996. Parasite communities of four fish species in the Wadden Sea and the role of fish discarded by the shrimp fisheries in parasite transmission. *Helgoländer Meeresuntersuchungen* 50: 69-85.

Hamerlynck, O. and A. Kattrijsse. 1994. The food of *Pomatoschistus minutus* (Pisces, Gobiidae) in Belgian coastal waters, and a comparison with the food of its potential competitor *P. lozanoi*. *Journal of Fish Biology* 44: 753-771.

Hamerlynck, O., A. Geets and P. van Damme. 1989. The parasites of two sympatric gobies *Pomatoschistus minutus* and *P. lozanoi* in the Belgian coastal waters. *Verhandelingen van het Symposium "Invertebraten van Belgie"* 1989: 27-30.

Hamerlynck, O., P. van der Vyver and C.R. Janssen. 1990. The trophic position of *Pomatoschistus lozanoi* (Piscas: Gobiidae) in the southern Bight. In: *Trophic Relationships in the Marine Environment*, M. Barnes and R.N. Gibson (eds.). Aberdeen University Press, pp. 183-190.

Healey, M.C. 1971. The distribution and abundance of sand gobies, *Pomatoschistus minutus*, in the Ythan estuary. *Jounal of Zoology, London* 163: 177-229.

Hennig, R. and C.D. Zander. 1981. Zur Biologie und Nahrung von Kleinfischen des Nord- und Ostseebereichs. III. Die Besiedlung eines Süßwasserwatts der Elbe durch euryhaline Fische. *Archiv für Hydrobiologie* Supplement 43: 487-505.

Herre, A.W.C.T. 1953. Checklist of Philippine fishes. *Research and Rapport of the U.S. Fish Wild Service* 20: 1-977.

Hesthagen, I. 1971. The winter food of the gobies from one of the deeper channels of the Belt Sea, with particular reference to the sand goby, *Pomatoschistus minutus* (Pallas). *Kieler Meeresforschung* 27: 28-35.

Hesthagen, I. 1976. Locomotory activity of the black goby, *Gobius niger* L. (Pisces, Gobiidae), under artificial light conditions, including a false dawn and dusk. *Sarsia* 62: 9-18.

Hesthagen, I. 1980. Locomotory activity in the painted goby, *Pomatoschistus pictus* (Malm) (Pisces), in relation to light intensity. *Sarsia* 65: 13-18.

Heymer, A. and C.D. Zander. 1978. Morphology and ecology of *Gobius vittatus* Vinciguerra, 1883, and its possible mimicry relationship to *Blennius rouxi* Cocco, 1833 in the Mediterranean. *Zeitschrift für Zoologische Systematik und Evolutionsforschung* 16: 132-143.

Hiatt, R.W. and D.W. Strasburg. 1960. Ecological relationships of the fish fauna on coral reefs of the Marshall Islands. *Ecological Monographs* 30: 67-127.

Holmes, J.C. and P.W. Price. 1986. Communities of parasites. In: *Community Ecology: Patterns and Processes*, D.J. Anderson and J. Kikkawa (eds.). Blackwell, Oxford, pp. 187-213.

Human, P. 1997. *Tauchführer. Karibik.* Jahr Verlag, Hamburg.

Human, P. and N. DeLoach. 2003. *Reef Fish Identification. Galapagos*, 2nd edition. New World Publ., Jacksonville, Florida.

Hureau, J.C. 1970. Biologie comparée des quelques poissons antarctiques (Nothotheneidae). *Bulletin de Institut Océanographie Monaco* 68: 1-250.

Hynes, H.B.N. 1950. The food of freshwater sticklebacks (*Gasterosteus aculeatus* and *Pygosteus pungitius*) with a review of methods used in studies of the food of fishes. *Journal of Animal Ecology* 19: 36-58.

Jaquet, N. and D. Raffaeli. 1989. The ecological importance of the sand goby *Pomatoschistus minutus* (Pallas). *Journal of Experimental Biology and Ecology* 128: 147-158.

Josten, N. 2004. Die Parasitengemeinschaft in der Flensburger Förde und die Populationsdynamik ihrer Wirte. Thesis, Univ Hamburg, 230 pp.

Joyeux, J.-C., J.-A. Thomasini and J.-L. Bouchereau. 1991. Le régime alimentaire de *Gobius niger* Linné, 1758 (Teleostei, Gobiidae) dans la lagune de Mauguio – France. *Annales des Sciences Naturelles, Zoologique* 12: 57-69.

Khoo, K.G. 1966. Studies of the periophthalmid fishes in Singapore. BSc thesis, Univ. of Singapore, 61 pp.

Kissiov, N., P. Kolarov, T. Dikov, S. Zlatanova, A. Boyadjiev and P. Petrov. 2007. Status and Assessment of Bulgarian Fish Resources. *www.wildlife org.*

Kovačic, M. and S. Engin. 2008. Two new species of *Neogobius* (Gobiidae) from northeastern Turkey. *Cybium* 32: 73-80.

Kovačic, M. and R.A. Patzner 2009. Reidentification of *Buenia* specimens (Gobiidae) from Banyuls-sur–Mer (France, western Mediterranean). *Cybium* 33: 175-176.

Kruitwagen, G., I. Nagelkerken, B.R. Lugendo, H.B. Pratap, X. Wendelaar and S.E. Bonga. 2007. Influence of morphology and amphibious lifestyle on the feeding ecology of the mudskipper *Periophthalmus argentilineatus*. *Journal of Fish Biology* 71: 39-52.

Krüß, A. 1988. Die benthische Fischfauna des Helgoländer Felssockels. Diploma, Univ. Karlsruhe, Germany, 172 pp.

Larson, H.K. and E.O. Murdy. 2001. Families Eleotridae, Gobiidae, In: *Species Identification Guide for Fishery Purpose. Bony Fishes Part 4. V. 6.* K.E. Carpenter and V.H. Niehm (eds.). FAO, Rome. pp. 3574-3603.

Lim, L.S. 1971. The feeding biology of *Periophthalmus vulgaris* and *Periophthalmodon schlosseri* (Pallas) Perciformes, Periophthalmidae. MSc thesis, Univ. of Malaya, 33 pp.

MacArthur, R.H. and E.P. Pianka. 1966. An optimal use of a patchy environment. *American Naturalist* 101: 377-385.

MacGinitie, G.E. 1939. The natural history of the blind goby, *Typhlogobius californiensis* Steindachner. *American Midland Naturalist* 21: 489-505.

Magnus, D.E.B. 1981. Bewegungsweisen, Nahrungserwerb und Fortpflanzungsverhalten des Schlammspringers *Periophthalmus kalolo*. *Publikation zu Wissenschaftlichen Filmen, Sektion Biologie, Ser. 14, Nr 7/D* 1282, 19 pp.

Mazé, R.A. 2004. Seasonal and ontogenetic diet shifts in an intertidal population of *Gobius paganellus* (Teleostei, Gobiidae) from the Cantabrian coast. *Vie et Milieu* 54: 1-6.

Meyer-Antholz, W. 1986. Untersuchungen über Verdauungsraten, Fressrhythmik und lokomotorische Aktivitäten der Strandgrundel, *Pomatoschistus microps* (Kroyer 1838) (Gobiidae, Pisces), und ihre Anwendung zur Bestimmung der Tagesration. Thesis, Univ. Hamburg, Germany, 152 pp.

Miller, P.J. 1986. Gobiidae. In: *Fishes of the North-eastern Atlantic and the Mediterranean 3*, P.J.P. Whitehead, M.-L. Bauchot, J.-C. Hureau, J. Nielsen and E.Tortonese (eds.). UNESCO, Paris, pp. 1019-1085.

Morawski, M.S. 1978. Winter feeding of the sand goby (*Pomatoschistus minutus* Pallas) in relation to the depth in Gdansk Bay. *Kieler Meeresforschung Sonderheft* 4: 122-127.

Mutsaddi, K.B. and D.N. Bal. 1959. Food and feeding of *Boleophthalmus dussumieri* (Cuv. and Val.). *Journal of the University of Bombay* 38: 42-55.

Nieder, J. and C.D. Zander. 1994. Nocturnal activity of a blenny *Lipophrys trigloides* (Pisces, Blenniidae) at the Spanish Mediterranean coast. *Miscellaneous. Zoology* 17: 189-197.

Norte-Campos, A.G.C. del and A. Temming. 1994. Daily activity, feeding and rations in gobies and brown shrimp in the northern Wadden Sea. *Marine Ecolology Progress Series* 115: 41-53.

Odum, E.P. 1963. Primary and secondary energy flow in relation to ecosystem structure. *Proceedings of the XVI International Congress of Zoology, Washington. D.C.*, pp. 336-338.

Pinkas, L., M.S. Oliphant and I.L.K. Iverson. 1971. Food habits of albacore, blue fin tuna, and bonito in Californian waters. *Fish Bulletin of California* 152: 1-105.

Prochnow, G. 1998. Der Schweinswal – Die einzige einheimische Walart an unseren Küsten. *Phoca* 1: 70-74.

Reise, K. 1979. Moderate predation on meiofauna by the macrobenthos of the Wadden Sea. *Helgoländer Wissenschaftliche Meeresuntersuchungen* 32: 453-465.

Remane, A. 1958. Ökologie des Brackwassers. In: *Die Biologie des Brackwassers*, A. Remane and C. Schlieper (eds.). *Schweizerbartsche Verlagsbuchhandlung, Stuttgart*, pp. 12: 1-216.

Rossetto, L. 2000. The management of fishery in the lagoon of Venice. *www. oregonstate.edu/dept/IIFET* – 10 pp.

Salgado, J.P., H.N. Cabral and M.J. Costa. 2005. Feeding ecology of the gobies *Pomatoschistus minutus* (Pallas, 1770) and *Pomatoschistus microps* (Kroyer, 1838) in the upper Tagus estuary, Portugal. *Scientia Marina* 68: 425-434.

Sasal, P., E. Falliex and S. Morand. 1996. Parasitism of *Gobius bucchichii* Steindachner, 1870 (Teleostei, Gobiidae) in protected and unprotected marine environments. *Journal of Wildlife Deseases* 32: 607-613.

Sasal, P., S. Morand and J.-F. Guegan. 1997. Determinants of parasite species richness in Mediterranean marine fishes. *Marine Ecology Progress Series* 149: 61-71.

Schmidt-Moser, R. and D. Westphal. 1981. Predation of *Pomatoschistus microps* (Kroyer) and *P. minutus* (Pallas) (Gobiidae, Pisces) on macro- and meiofauna in the brackish fjord Schlei. *Kieler Meeresuntersuchungen Sonderheft* 5: 471-478.

Schulze, G. 1987. Die Schweinswale. Ziemsen, Wittenberg Lutherstadt.

Simenstad, C.A. 1979. Fish food habits analysis. Princ. Invest. Rep. *Environmental Assessments of the Alaskan Continent and Shelf* 1979: 441-450.

Thienemann, A. 1939. Grundzüge einer allgemeinen Ökologie. *Archiv für Hydrobiologie* 35: 267-285.

Thompson, D.A., L.T. Findley and A.N. Kerstich. 1979. *Reef Fishes of the Sea of Cortez*. John Wiley and Sons, New York.

Thorman, S. 1982. Niche dynamic and resource partitioning in a fish guild inhabiting a shallow estuary on the Swedish west coast. *Oikos* 38: 32-39.

Thorman, S. and B. Fladvad. 1981. Growth and production of fish in Broälven estuary on the Swedish west coast. *Natural Swedish Environmental Protection Board* Report 1416.

Thorman, S. and A.-M. Wiederholm. 1983. Seasonal occurrence and food resource use of an assemblage of nearshore fish species in the Bothnian Sea, Sweden. *Marine Ecolology Progress Series* 10: 223-229.

Thorman, S. and A.-M. Wiederholm. 1986. Food, habitat and time niche in a coastal fish assemblage in a brackish water bay in the Bothnian Sea, Sweden. *Journal of Experimental Marine Biology and Ecology* 95: 67-86.

Todd, E.S. 1976. Terrestrial grazing by the eastern tropical Pacific goby *Gobionellus sagittula*. *Copeia* 1976: 374-377.

Ungano, N., N. Casavola, G. Martano and E. Rizzi. 1994. "Bianchetto" and "Rossetto" fry fisheries in the Manfredonia gulf: effort exerted and catch composition. *Oebalia* 20: 99-106.

Villiers, L. 1980. Changes in predation by the juvenile goby *Deltentosteus quadrimaculatus* (Teleostei, Gobiidae). *Netherland Journal of Sea Research* 14: 362-373.

Villiers, L. 1982. The feeding of juvenile goby *Deltentosteus quadrimaculatus* (Pisces, Gobiidae). *Sarsia* 67: 157-162.

Yang , J. 1982a. A tentative analysis of the trophic levels of North Sea fish. *Marine Ecology Progress Series* 7: 247-252.

Yang, J. 1982b. An estimate of the fish biomass in the North Sea. *Journée Conseil Internationale de Exploration de Mer* 40: 161-172.

Zander, C.D. 1979. On the biology and food of small-sized fish from the North and Baltic Sea areas. I. Investigations of *Pomatoschistus pictus* (Malm) (Gobiidae) from Helgoland. *Zoologischer Anzeiger* 202: 413-424.

Zander, C.D. 1982. Feeding ecology of littoral gobiid and blennioid fish of the Banyuls area (Mediterranean Sea). I. Main food and trophic dimensions of niche and ecotope. *Vie et Milieu* 32: 1-10.

Zander, C.D. 1990. Prey selection of the shallow water fish *Pomatoschistus minutus* (Gobiidae, Teleostei) in the SW Baltic. *Helgoländer Meeresuntersuchungen* 44: 147-157.

Zander, C.D. 1991. Die biologische Bedeutung der Lebensgemeinschaft "Miesmuschelgürtel" in der Ostsee. *Seevögel* 12, Sonderheft 1: 70-73.

Zander, C.D. 1993. The biological indication of parasite life cycles and communities from the Lübeck Bight, SW Baltic Sea. *Zeitschrift für angewandte Zoologie* 79: 377-389.

Zander, C.D. 1994. Einnischung von fünf Grundeln (Teleostei, Gobiidae) der Ostsee und ihre Deutung mit Hilfe der Präadaptationstheorie. *Journal of Zoological Systematic and Evolution Research* 32: 220-234.

Zander, C.D. 1996. The distribution and feeding ecology of small-size fish in the coastal Mediterranean Sea. In: *Biology and Ecology of Shallow Coastal Waters.* Eleftheriou, A., *et al.* (eds.). Olsen and Olsen, Fredenborg, Denmark, pp. 369-376.

Zander, C.D. 1998a. Ecology of host parasite relationships in the Baltic Sea. *Naturwissenschaften* 85: 426-436.

Zander, C.D. 1998b. Parasitengemeinschaften bei Grundeln (Gobiidae, Teleostei) der südwestlichen Ostsee. *Verhandlungen Gesellschaft für Ichthyologie,* Düsseldorf 1997, 1, pp. 241-252.

Zander, C.D. 2002. The influence of eutrophication on parasite communities in the Baltic Sea. *Proceedings of the 10th International Congress of Parasitology,* Vancouver 2002: 247-253.

Zander, C.D. 2004. Ecology meets genetics—niche occupation as a factor of evolution interpreted by Kosswig's concepts. *Mitteilungen aus dem Hamburger Zoologischen Museum und Institut* 101: 131-147.

Zander, C.D. 2005a. Four-year monitoring of parasite communities in gobiid fishes of the south-western Baltic. III. Parasite species diversity and applicability of monitoring. *Parasitology Research* 95: 136-144.

Zander, C.D. 2005b. Comparative studies on goby (Teleostei) parasite communities from the North and Baltic Sea. *Parasitology Research* 96: 62-68.

Zander, C.D. and J. Berg. 1984. Feeding ecology of littoral gobiid and blennioid fish of the Banyuls area (Mediterranean Sea). II. Prey selection and size preference. *Vie et Milieu* 34: 149-157.

Zander, C.D. and W. Döring. 1989. The role of gobies (Gobiidae, Teleostei) in the food web of shallow habitats of the Baltic Sea. *Proceedings of the 21. European Marine Biology Society, Gdánsk*, Sept. 1986: 499-508.

Zander, C.D. and T. Hagemann. 1986. Fluctuation of prey, abundance and biomass of gobies (Gobiidae, Pisces) in a shallow habitat of the western Baltic Sea. *Zoologischer Anzeiger* 216: 289-304.

Zander, C.D. and T. Hagemann. 1987. Predation impact and ecological efficiency of *Pomatoschistus* spp. (Gobiidae, Pisces) from a clay/sand ecotone of the western Baltic Sea. *Zoologischer Anzeiger* 218: 33-48.

Zander, C.D. and T. Hagemann. 1989. Feeding ecology of littoral gobiid and blennioid fishes of the Banyuls area (Mediterranean Sea). III. Seasonal variations. *Scientia Marina* 52: 441-449.

Zander, C.D. and E. Hartwig. 1982. On the biology and food of small-sized fish from the North and Baltic Sea areas. IV Investigations on an eulittoral mud flat at Sylt Island. *Helgoländer Meeresuntersuchungen* 35: 47-63.

Zander, C.D. and D. Westphal. 1991. Nahrungsbeziehungen von vier euryhalinen Kleinfischarten der Ostseeförde Schlei. *Zoologisches Jahrbuch für Systematik* 119: 339-364.

Zander, C.D. and A. Heymer. 1992. Feeding habits of *Gobius auratus* and other benthic small-sized fish from the French Mediterranean coast under regard of some alternating parameters. *Zoologischer Anzeiger* 228: 220-228.

Zander, C.D. and D. Westphal.1992. Kleinfischparasiten der Ostseeförde Schlei und ihre Einbindung in die Nahrungskette. *Seevögel* 12: 4-8.

Zander, C.D. and V. Kesting. 1996. The indicator properties of parasite communities of gobies (Teleostei, Gobiidae) from Kiel and Lübeck Bight, SW Baltic Sea. *Applied Parasitology* 37: 186-204.

Zander, C.D., L.W. Reimer and K. Barz. 1999. Parasite communities of the Salzhaff (Northwest Mecklenburg, Baltic Sea). I. Structure and dynamics of communities of littoral fish, especially small-sized fish. *Parasitology Research* 85: 356-372.

Zander, C.D., H.-G. Kollra, B. Antholz, W. Meyer and D. Westphal. 1984. Small-sized euryhaline fish as intermediate hosts of the digenetic trematode *Cryptocotyle concavum. Helgoländer wissenschaftliche Meeresuntersuchungen* 37: 433-443.

Zaret, T.M. and R.T. Paine. 1973. Species introduction in a tropical lake. *Science* 182: 449-455.

Zaret, T.M. and A.S. Rand. 1971. Competition in tropical stream fishes: support for the competitive exclusion principle. *Ecology* 52: 336-342.

CHAPTER

3.2

Morphological Adaptations to Special Environments of Gobies

C. Dieter Zander

INTRODUCTION

Gobies are generally small-sized fish which surpass only occasionally the size of 10 cm and become very seldom larger than 50 cm. They mostly live on the bottom (benthal), only some species swim over the substrate (suprabenthal) or in the free water (pelagial). Their body shape is terete, approximately cylindrically and tapered, or depress, dorso-ventrally flattened (Horn, 1999), but rarely compress. The pelvic (ventral) fins are situated at the throat where the pelvis is connected to the shoulder girdle (cleithrum). The pelvic fins can be fused together forming a disc with an anteriorly dermal fold, the frenum, which adapts this fin-complex to function as a sucker by which gobies can attach to the bottom and can withstand streams or flows. Whereas all members of the subfamily Eleotrinae are characterized by separated pelvic fins, the reduction of the pelvic disc in the Gobiinae must have developed independently in several lines. There are

Author's address: Biozentrum Grindel and Zoologisches Museum, University. Martin-Luther-King-Platz 3, 20146 Hamburg, Germany. E-mail: cedezet@zoologie.uni-hamburg.de various intermediate steps of this development, from reduction of frenum to deep incisions between the fins (Miller and El Tawil, 1974; Heymer and Zander, 1978). Reduced suckers can be found in gobies of deeper layers living of more than 10 m depth like the Mediterranean species *Gobius auratus, G. fallax, G. vittatus* and *G. xanthocephalus* (Heymer and Zander, 1978, 1992; Zander, 1992), *Lebetus guilleti, L. scorpioides* (Miller and El Tawil, 1974), *Odondobuenia balearicus* and *Vanneaugobius pruvoti*, all from coralligeneous habitats. Also the amphibious *Periophthalmus* spp. from mud flats reveal the same evolutionary trend (Fig. 3.2.1). Heymer and Zander (1978) observed that the separated pelvic fins of *Gobius vittatus* and *G. xanthocephalus* are used to support the body by turning forward-downward when resting on the ground (Fig. 3.2.2).

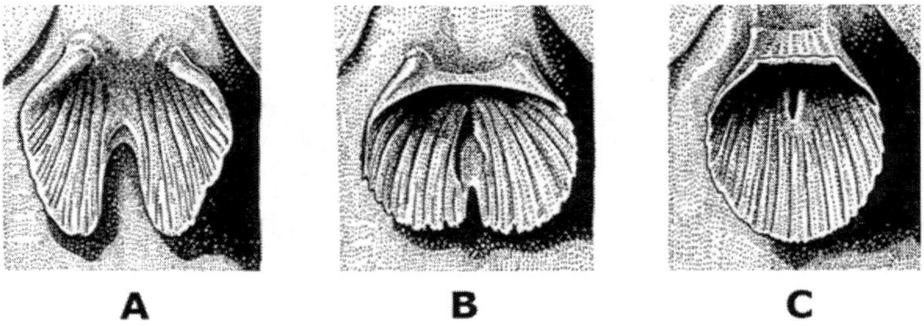

Fig. 3.2.1 Status of separation of pelvic fins in mudskippers. A *Periophthalmus argentilineatus,* B *Periophthalmodon schlosseri*, C *Periophthalmus chrysospilos*. From Sterba (1957).

Albeit the habit of gobies seems to be rather homogeneous their morphology can vary according to special ways of life. Here aspects of such alterations should be considered in four categories which follow to a greater part an idea of Miller (1979). Therefore, 4 environments are regarded: (1) adaptations to supra- and nectobenthic life in the free water; (2) adaptation to surf and intertidal zones; (3) adaptations to an amphibious epibenthic life on land; (4) adaptations to cryptobenthic life in caves and with symbionts. Fig. 3.2.3 regards these 4 categories and gives an overlook on the diversity of those habitats which are colonized by gobies.

ADAPTATIONS TO A WAY OF LIFE IN THE PELAGIAL AND SUPRABENTHOS

Preliminaries for the way of life in the free water are in teleosts (1) a well-developed swim bladder in order to reduce specific weights, and (2)

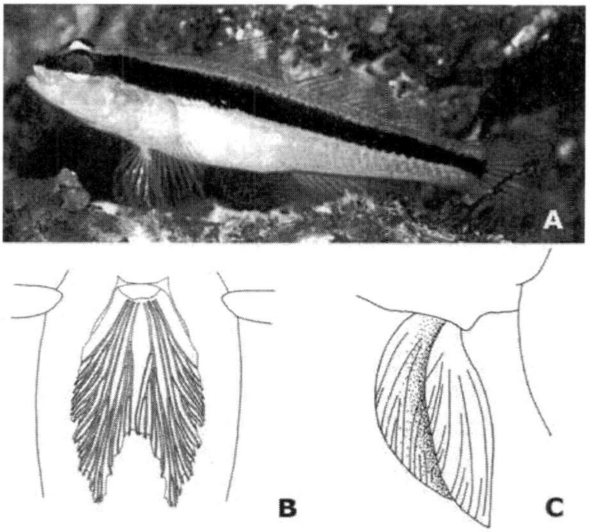

Fig. 3.2.2 *G. vittatus*, A habitus and view of pelvic fins B from ventral and C left side, showing their supporting function. From Heymer and Zander (1978).

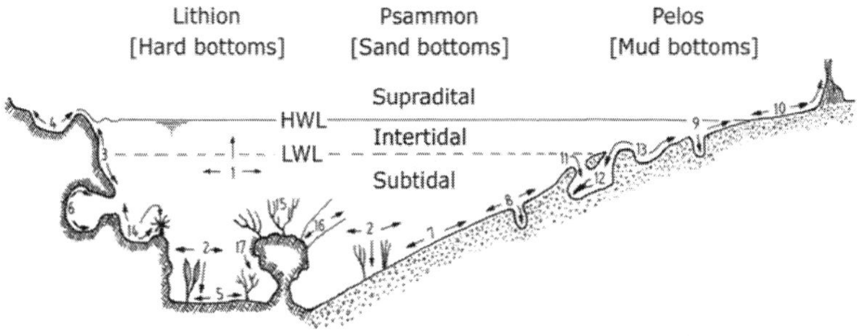

Fig. 3.2.3 Diverse marine habitats of gobies appointed to distinct species and their respective ways of life (numbers). **Pelagic.** 1 – *Crystallogobius linearis* (also Fig. 4). **Suprabenthic.** 2 – *Pomatoschistus quagga*, *Gobiusculus flavescens*. **Epibenthic.** 3 – intertidal, surf: *Gobius paganellus*, *G. cobitis*. 4 – intertidal, rock pools: *Pomatoschistus microps, Acentrogobius ornatus*, *Eviota meteori*. 5 – coralligen: *Gobius vittatus, G. xanthocephalus* (also Fig. 2). 6 – dim light habitats: *Speleogobius trigloides*. 7 – sand: *Pomatoschistus minutus*, *Gobius geniporus*. 8 – burrows: *Amblygobius* spp. 9 – intertidal, burrows: *Boleophthalmus*. 10 – supratidal, mangroves: *Periophthalmus* spp., *Periophthalmodon*. **Symbiotic,** in burrows of decapod crustaceans. 11 – facultative: *Lesueurigobius* spp., *Cryptocentrus* spp., *Clevelandia ios*. 12 – obligate: *Typhlogobius californiensis*. 13 – intertidal pools: *Gillichthys mirabilis*. On cnidarians 14 – on anemones: *Gobius bucchichii*. 15 – on ramous hard corals: *Gobiodon* spp., *Paragobiodon echinocephalus* (also Fig. 16). 16 – on flat corals: *Elacatinus* spp. 17 – on black and thorn corals: *Bryaninops* spp.

adaptations of the body shape to the water currents as at most streamlined as possible in order to avoid water resistance and turbulence. Regarding the body shape the gobies may be assorted to 3 types. The most frequent type is a predominantly bottom dweller which feeds, hides and spawn at the bottom but can make sometimes sojourns into the free water in order to catch some additional preys. These may be represented by *Gobius niger, Pomatoschistus microps, P. pictus, P. bathi, Acentrogobius ornatus* or *Bathygobius soporator*. Their body shape is terete as described above, slightly depress, the eyes are situated dorso-laterally in order to have also sight from the bottom into the free waters. They possess reduced swim-bladders which can be activated sometimes as was found in the sand goby *Pomatoschistus minutus* (Fig. 3.2.3: 7) in autumn when it is migrating in deeper water layers (Hesthagen and Koefoed, 1979). A second type which is very close bound to the bottom can be found in *Caspiosoma caspium, Benthophiloides brauneri* or, extremely, in *Benthophilus stellatus* which is shaped like an anglerfish (Fig. 3.2.4). These species from the Ponto-Caspian region show like other goby genera totally reduced swim-bladders, their larvae pass no planktonic phase but live after hatching like the adults on the bottom (Nikolski, 1959). These species have shifted their eyes dorsally and show also a trend for the reduction of scales. Thus, head and body of *Benthophilus stellatus* are covered with spinulose bony platelets (Pinchuk and Rahimov, 1979). On the other side, these species present also characteristic

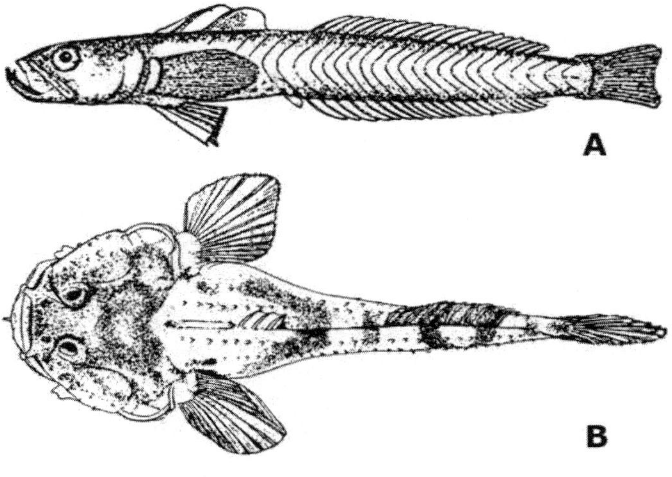

Fig. 3.2.4 Two contrastingly shaped gobies: A the pelagic *Cristallogobius linearis* and B the benthic and extremely flattened *Benthophilus stellaris*. From Miller (1986).

goby patterns like rounded caudal fins and ventral suckers. The third group regards suprabenthic species like *Gobiusculus flavescens* or *Pomatoschistus quagga* (Fig. 3.2.3: 2), and nektonic species like *Crystallogobius linearis* or *Pseudaphya ferreri* (Fig. 3.2.3: 1). The eyes of these species are situated laterally, they are fully scaled and possess functioning suckers, but their caudal fins are not rounded but slightly notched (Fig. 3.2.4) which provide for better manoeuvring. Due to their plankton feeding way of life the visceral skeleton is adapted by a prolonged lower jaw and an acute angle of 20-30° of the hyoid bone which contributes to a slender shaped head by reducing also water resistance (Mestermann and Zander, 1984, Fig. 3.2.5). In bottom dwelling gobies like *Pomatoschistus microps* the angle is between 40 and 50°. *P. quagga* additionally present small cteniae on the branchial arches which function as plankton filters (Mestermann and Zander, 1984).

Such patterns like presence of a swim-bladder or degree of the angle of the hyoid bone were used in the past as decisive tools for systematic of gobies. This view is not longer valid because the mentioned attributes are considered to be a result of parallel evolution and are, therefore, adaptations to different environments which comprise members from different genera as ecological equivalents.

ADAPTATIONS TO LIFE IN THE INTERTIDAL ZONE

The intertidal zone offers very different environments which are habitats for fish, also for gobies. We can discriminate the turbulent surf zone, the calm rock, sand and mud flats, and the estuaries influenced by strong currents. Because gobies like *Gobius paganellus* or *G. cobitis* (Fig. 3.2.3: 3) which live in the surf zone are found in the most sheltered sites (Gibson, 1972), only few morphological patterns can be considered as special adaptations. Whereas most fishes of the surf zone reduced or loosed their scales, gobies retain these organs by anchoring them especially firm in the skin (Gibson, 1969). Characteristic goby patterns such as body shape, sucker discs, reduction of swim bladder and lateral line of the body are advantageous pre-adaptations to this special habitat. Nevertheless, gobies do not display such adaptations like Gobiesocidae with very robust suckers (Chotkowski *et al.*, 1999) or Blenniidae with strong hooks on pectoral, ventral and anal fins (Zander, 1972) which can better withstand the forces of waves.

Whereas the inhabitants of steep rocks are exposed to the power of surf, fish of intertidal flats can survive at low tides by hiding between algae or remaining in rest pools. This is favoured by small size surpassing not more

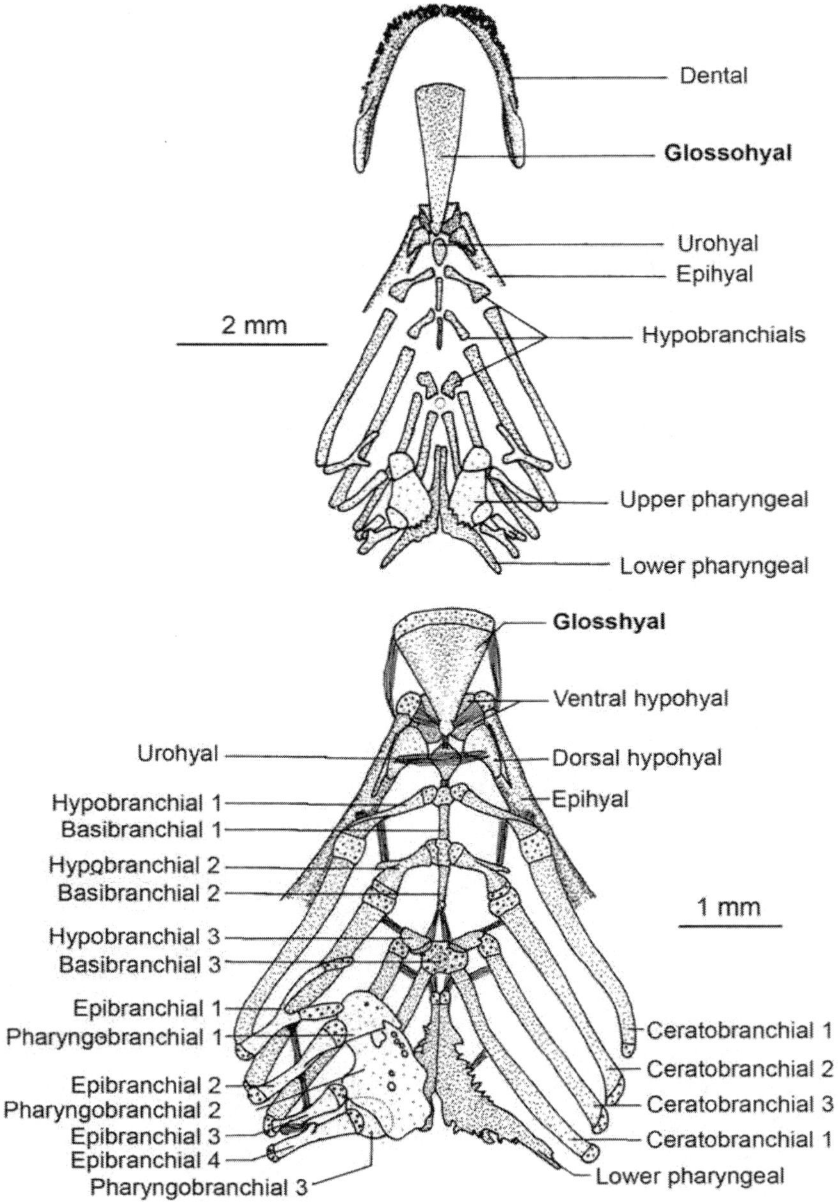

Fig. 3.2.5 Visceral head skeleton of *Pomatoschistus quagga* (left) and *P. microps.* Watch the general shape and the shape of the glosso-hyale bone. From Mestermann and Zander (1984).

than 12 cm and the ability to attach on small gravel or clam shells by the sucker. Examples are the common goby, *Pomatoschistus microps* (5 cm), in the sand flats of the north-eastern temperate zones, *Acentrogobius ornatus* (6 cm) and *Eviota meteori* (2.5 cm) (Fig. 3.2.3: 4) from tropical rock flats (Zander *et al.*, 1999). Beside morphological adaptations gobies from sand and rock flats developed physiological adaptations to the extremely changing temperature and salinity dependent to sun or rain. Heating of rest pools cause decreasing oxygen content so that some gobies had adapted to air breathing, like *Gillichthys mirabilis* (Fig. 3.2.3: 13) of the American Pacific coast. This goby gulps air from the water surface. The roof of its mouth cavity and, in lesser degree, the tongue is covered by a respiratory epithelium with dense blood capillaries where oxygen can be taken up (Todd and Ebeling, 1966). Similar organs present Australian gobies, *Cryptocentroides cristatus, Arenogobius bifrenatus, Chlamydogobius* sp., which activate the blood vessels in cases of hypoxia but one of these, *Mugilogobius palidus*, present blood vessels filled at all times (Todd and Ebeling, 1966; Gee and Gee, 1995). *Chlamydogobius eremius* lives in warm spring pools of the central Australian desert (Merrick and Schmieda, 1984). Also *Pseudapocryptes lanceolatus* and *Apocryptes bata* from mudflats of India are in cases of hypoxia in their pools or during short journeys on land air breathers. Their air breathing organs lie in the inner opercula where an artery splits into many capillaries which are sampled by a forth leading vein (Fig. 3.2.6) (Schöttle, 1931).

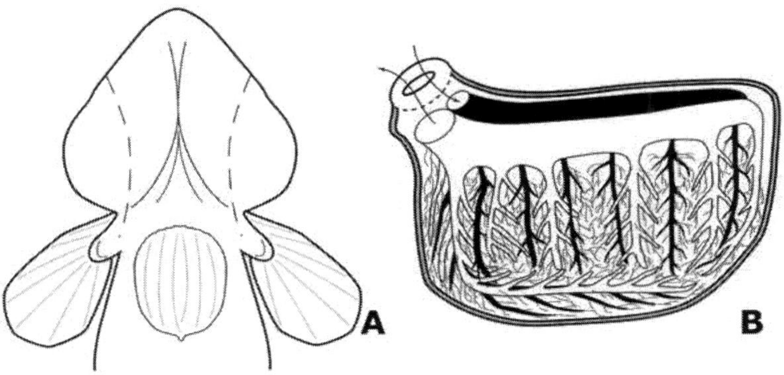

Fig. 3.2.6 *Pseudapocryptes lanceolatus*. A. ventral view with expanded opercular pouches (dashed lines). B. Internal surface of the right operculum with ramification of arteries (black) and veins (white). From Graham (1997).

ADAPTATIONS TO AN AMPHIBIOUS WAY OF LIFE

The transition of water living organisms to terrestrial life is a huge great step which demands drastic adaptations during evolution. Nevertheless, several taxonomic groups have successfully conquered an environment which was attractive as prey source and as shelter against water living predators. Preliminaries for an existence on land are: 1. suited extremities to move in a medium which has a far lower compactness than water where buoyancy allows swimming; 2. prevention of desiccation which concerns especially the external skin and the water respiration organs; 3. gills must be compensated by air breathing organs which had to be situated internally in the land living organisms; 4. alteration or reduction of sense organs which operate in water; e.g. eyes must be adapted to see in air where light rays pass a medium with another refraction patterns than water; sounds spread out in water four times faster than in air; lateral lines with current sense organs can only operate in water, not in air; 5. developmental stages, larvae and young are very sensitive against desiccation which requires special protection mechanisms.

Whereas *Boleophthalmus* spp. and *Periophthalmus chrysospilus* possess joint ventral fins, the suckers of other *Periophthalmus* and *Periophthalmodon* spp. present a trend of reduction (Fig. 3.2.1). This reduction is not a pattern of better adaptation to the way of living on land but may be rather dependant to the respective habitats. *P. chrysospilus* climbs on trees (Fig. 3.2.3: 10) where a sucker can be advantageous whereas these organs do not function on mud bottoms. *Boleophthalmus* spp. (Fig. 3.2.3: 9) are active during low water on the soft sediment but vanish by running aground when the water rises (Harms, 1929). Their skin is adapted to the boring way of life by an epidermis with vesicular cells and a thick cuticula which will be described below (Fig. 3.2.7).

Movements of the amphibious *Periophthalmus* and *Periophthalmodon* spp. on land are performed either by sliding with help of the pectoral fins or by leaps with the help of the tail, e.g. on flight, during courtship or when chasing competitors. They are not able to move as similar as the land vertebrates because the ventral fins are situated directly behind the pectoral fins and the pelvis is connected with the shoulder girdle; these morphological patterns exclude quadruped walking. Therefore, it is without importance whether the ventral fins are combined or secondarily separated because both patterns are found in this group. When walking, the prolonged pectoral fins are simultaneously swung forward, raised, moved backward and set down on the ventral fins (Harris, 1960; Magnus, 1978).

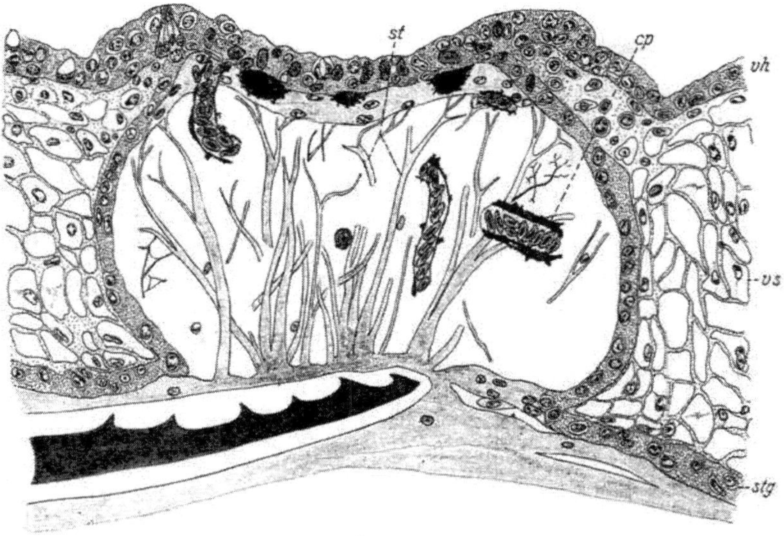

Fig. 3.2.7 Cut of the skin from the back of *Boleophthalmus boddaertii* showing a respiratory papilla. cp = capillary, st = supporting tissue, stg = germinative layer, vh = corneous layer, vs = vesicular cells. From Harms (1929).

These movements are then repeated several times. During leaping the tail is used as a robust third leg. The pectoral fins of mudskippers are the result of an evolution starting in submerged gobies. The developmental adaptation regards the shoulder girdle as well as the skeleton and muscles of the fin (Eggert, 1929): Thus, the cleithrum of *Boleophthalmus*, *Periophthalmus* and *Periophthalmodon* is broader than of submerged *Gobius* which make the onset of more muscles possible; the basal bones which link shoulder girdle and fin were prolonged, between the 2nd and 3rd basal is a hole by which muscles can pass (Fig. 3.2.8); the upper layers of abductor and adductor muscles were divided into two which cross one another; between the fin rays interradial muscles are developed which can in mudskippers be moved independently of the abductor muscles. Therefore, the mudskippers are able to rotate the pectoral fins by almost 90° so that these can be set plain on the ground (Eggert, 1929, Fig. 3.2.9).

A trend during the evolution of terrestrial vertebrates was a substitution of external water breathing organs like gills by internal ones, e.g. lungs, which are protected against desiccation. On land characteristic water breathing organs like gills would be unsuited for gas exchange because

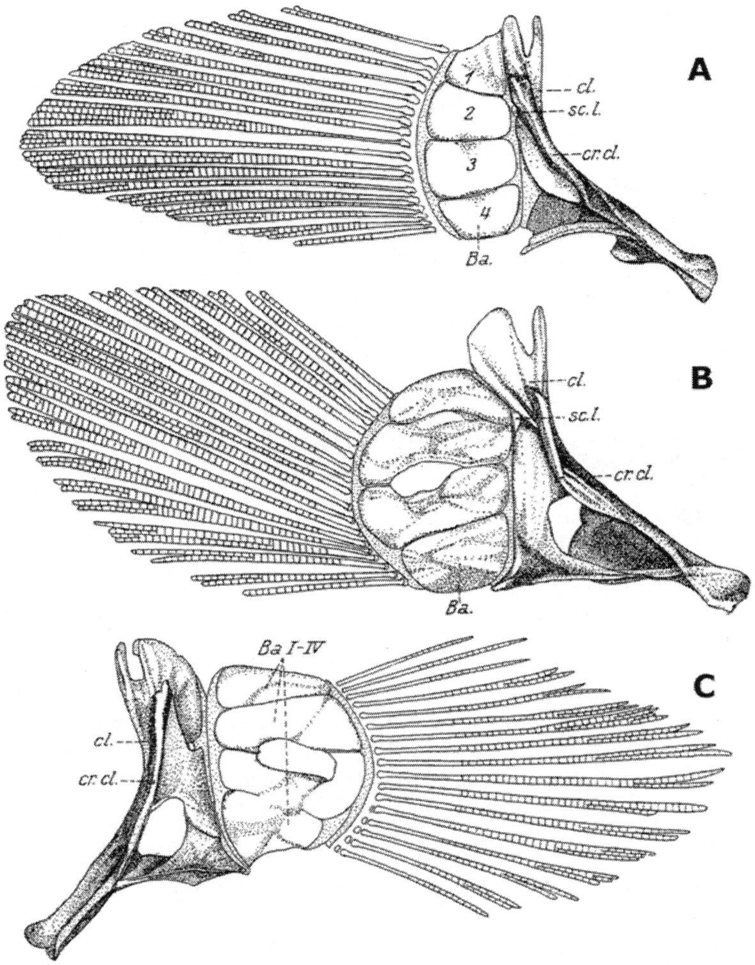

Fig. 3.2.8 Pectoral skeleton of A *Gobius* sp., B *Boleophthalmus boddaertii* and C *Periophthalmodon schlosseri*. Watch the differentiation of the shoulder girdle (cl) and of the basalia (Ba 1-4 or I-IV). cr.cl = crest of shoulder girdle, sc.l = aperture of scapula. From Eggert (1929).

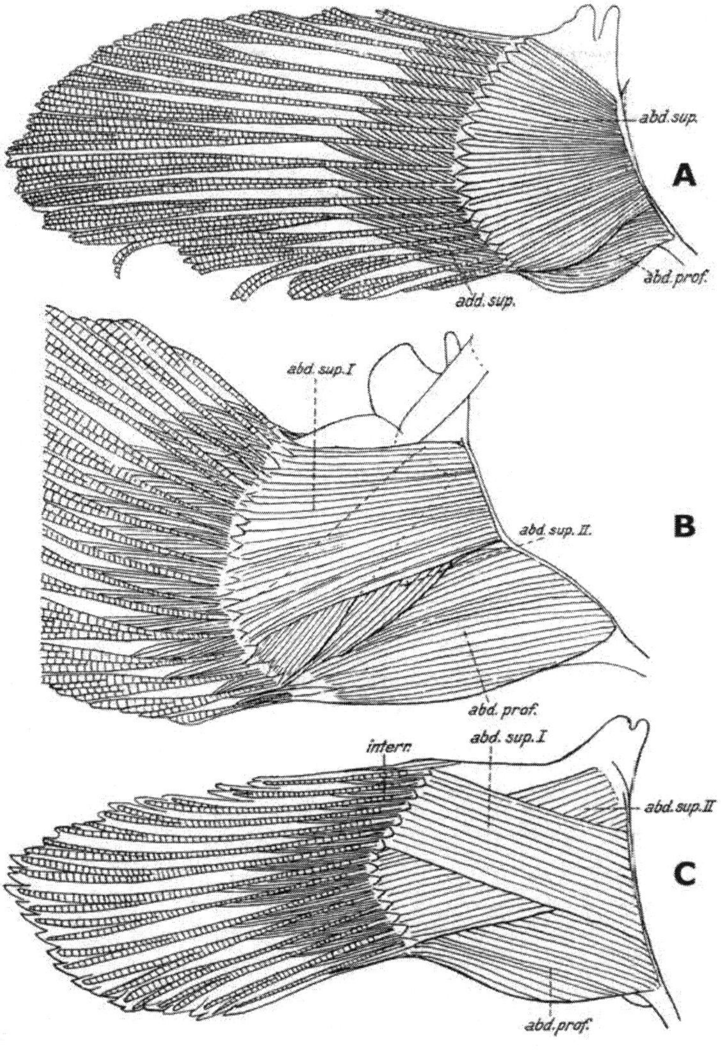

Fig. 3.2.9 Pelvic fin muscles of A *Gobius* sp., B *Boleophthalmus boddaertii* and C *Periophthalmodon schlosseri*. Watch the separation of the upper abductor muscles (abd. sup.). abd. prof.: lower abductor muscle, interr.: interadial muscles. From Eggert (1929).

these would collapse in air. The gills of mudskippers are, therefore, partially reduced, in *Periophthalmodon* even fused (Low *et al.*, 1988), and substituted by an extension of external skin blood vessels, by vascularisation of mouth and pharyngeal membranes, of gill and opercular chambers (Fig. 3.2.10) (Schöttle, 1931). The skin of *Boleophthalmus* which often are boring in the muddy sediment is supplied by a vesicular epidermis with chordoid cells and corneous papillae (Fig. 3.2.7). These are provided by arteries which extend into capillaries near the surface and then drawn off by veins (Fig. 3.2.11). *Periophthalmus* and *Periophthalmodon* spp. are lacking such papillae but dense packed capillaries lead to the skin surface of the head and the dorsal body (Fig. 3.2.11). The thickening of the corneous skin is in *Boleophthalmus* restricted to head and ventral body, it is spread in *Periophthalmus* and *Periophthalmodon* over the whole body. These are, therefore, better protected against desiccation than *Boleophthalmus*. The skin with capillaries must be kept moistly in order to ensure their function as air breathing organ (Graham, 1997). Therefore, the mudskippers live in habitats with high humidity like mangrove swamps; additionally, the fish present a behaviour of rolling on their longitudinal axis on the moist mud in order to save the air breathing function of the skin.

Further adaptations to a terrestrial life regard the sense organs. The eyes of *Boleophthalmus* and *Periophthalmus* are situated dorsally and are protruded (Fig. 3.2.10, 3.2.14), they can be moved independently of each other so that one can glimpse upward and the other downward. Regularly, the eyes are retracted in cups lying just below these in order to keep them wet (Graham, 1971). Like other bottom dwelling teleosts the eyes are protected externally by a secondary cornea, the spectacle. The lens is slightly flattened such as in other land vertebrates resulting in an index of

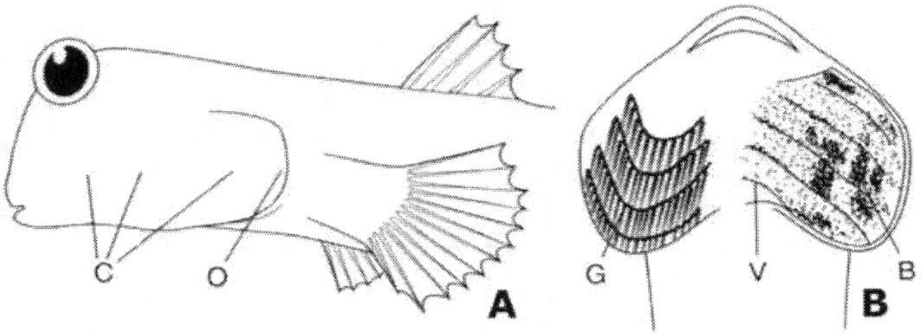

Fig. 3.2.10 *Periophthalmus* sp. A Side view presenting the expanded buccal, branchial and opercular chambers (C) and opercular opening (O). B Ventral view of opened branchial chamber with branchiostegal membrane (B) that wraps around the gills. Changed from Graham (1997).

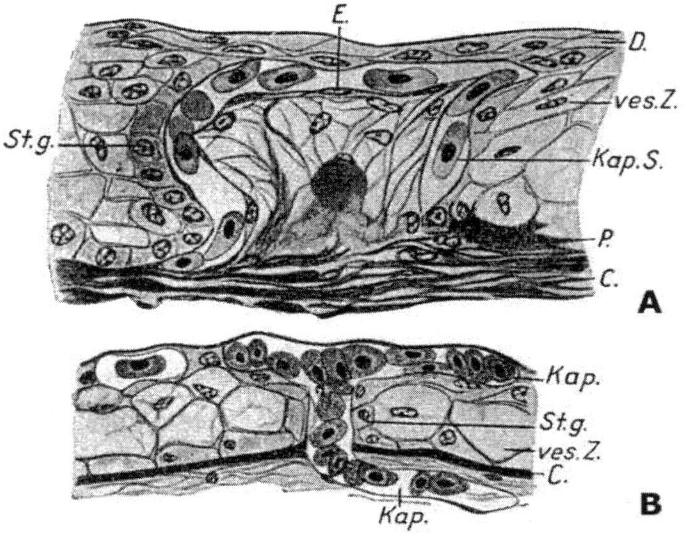

Fig. 3.2.11 Accessory dermal breathing organs in A *Boleophthalmus* sp., B *Periophthalmus* sp. Watch the capillaries in the outer epidermis of *Periophthalmus*. C = corium, D = outer layer of epidermis, E = endothel, Kap and KapS = capillaries, P = pigment, St.g = germinative layer, ves.c = vesicular cells. From Schöttle (1931).

1.14 and can be accommodated for near sight by moving it onto the iris by turning of the retractor muscles (Fig. 3.2.12) (Karsten, 1923; Harms, 1929). These two mechanisms enables the mudskipper to see acutely on land in the far as well as in the near.

The basal lateral sensory system of submerged living fish are in mudskippers reduced, the number of free neuromasts is increased at the ventral body, especially on the head. The hairs of several sensory cells are glued together to firm organs and emerge out of a pit above the skin (Fig. 3.2.13) (Harms, 1929). Taste buds are rare on the body but numbers are increased in the oral cavity. The nasal organ in mudskippers is well developed and is enlarged by an additional cave at the posterior end of this organ with many buds covered by layers of sensory cells. Hearing on land was assumed in mudskippers by Harms (1929). As special structures of the ear Harms (1929) found so called pillar cells which lie on the cristae of the outer ampulla and may be able to transport even slow vibrations to the sensory cells.

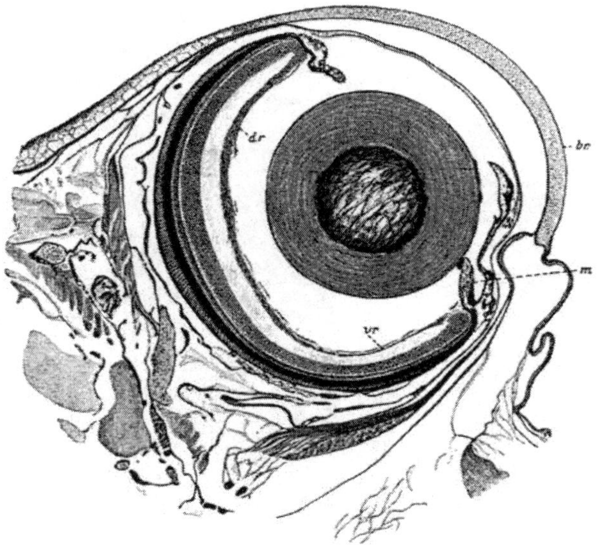

Fig. 3.2.12 Medial cut through the eye of *Periophthalmus argentilineatus*. Watch the slightly oval lens. br = "spectacle", dr = dorsal retina, m = lens muscle, vr = ventral retina. From Harms (1929).

The larvae of *Periophthalmus argentilineatus* are bread in deep holes which are filled with the ground water also at low-tides. They have a thin skin and functioning gills for breathing (Harms, 1929; Magnus, 1978). When required, in the case of oxygen deficits, the male brings bubbles of air from the surface to the breading chamber. At a size of 20 mm the young begin metamorphosis by which several organs develop characteristic patterns of mudskippers. The eyes are wandering to the upper head, the pectoral fins grow longer (Fig. 3.2.14). They still live mostly under water but make short sojourns on land. The next stage is the further improvement of land-living organs which allow life on land but flight is still into the water in contrast to adults which flee higher on the land (Harms, 1929; Magnus, 1978).

ADAPTATION TO CRYPTOBIOTIC AND SYMBIOTIC WAYS OF LIFE

Beside supra and epibenthic gobies there are several species which live in caves and burrows of very different sizes. In rocky environments these may range from small haptic holes to larger optic holes like crevices, cavities,

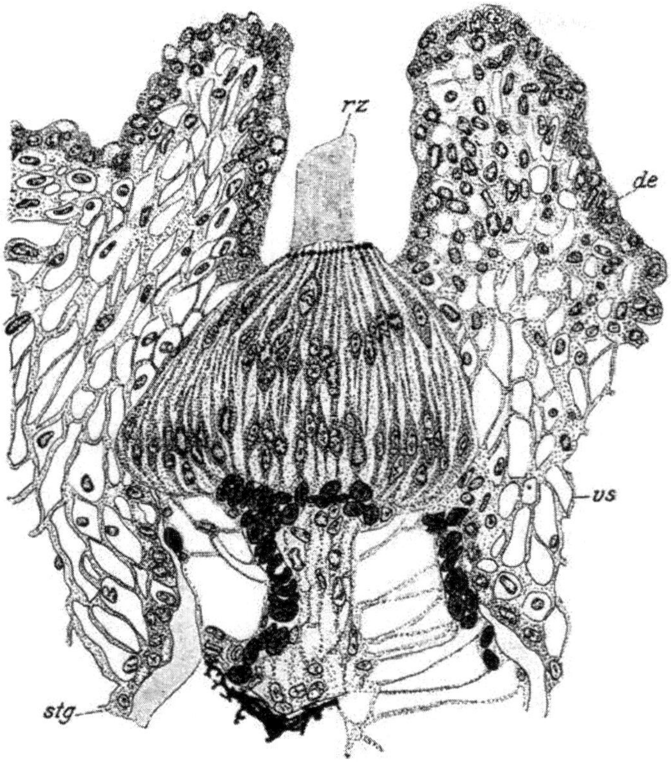

Fig. 3.2.13 Cut of a neuromast of *Boleophthalmus boddaertii*. de = outer layer of epidermis, rz = sensory hairs glued together, stg = germinative layer, vs = vesicular cells. From Harms (1929).

caves and grottos. *Trimma* spp. occur in caves and other dark habitats of the tropic Indo-pacific, whereas *Lebetus scorpioides* and *Speleogobius trigloides* (Fig. 3.2.3: 6) were found in similar habitats of the temperate NE Atlantic and the Mediterranean. These genuine cave dwellers present characteristic patterns like red colour and enlarged eyes which attain 7.5% of the total length in *S. trigloides* (Zander and Jelinek, 1976), a value which is also found in other small cave fishes (Zander, 1980; Zander and Heymer, 1976). But the most preferred holes of goby species lie in sand and mud bottoms as was already described for the retirement places of *Boleophthalmus* and the breeding places of *Periophthalmus*. Species of *Amblygobius* (Fig. 3.2.3: 8) burrow holes in sublittoral sediments. Also living organisms can be habitat for these fish like sponges, corals and other cnidarians, even greater anemones (Fig. 3.2.3: 14-17).

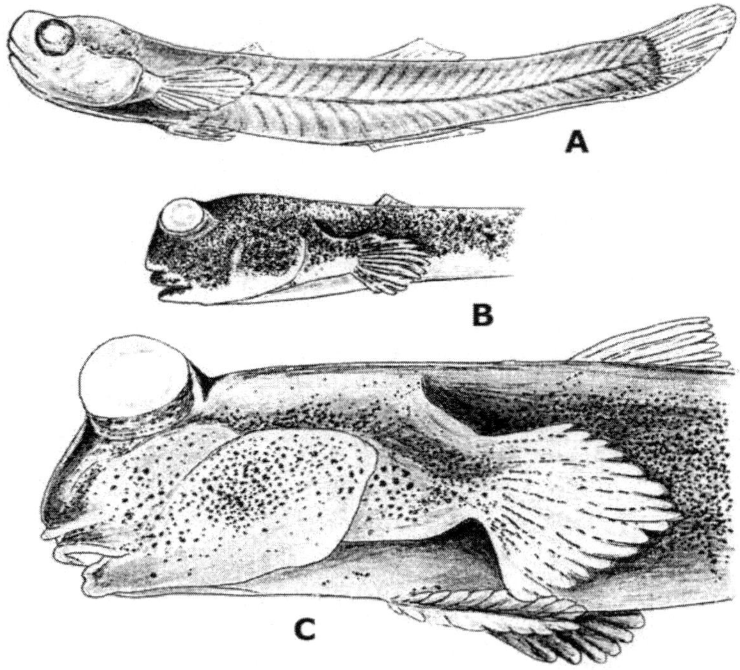

Fig. 3.2.14 Metamorphosis of *Periophthalmus argentilineatus* . A Larva of 9 mm size (above), B young with beginning protusion of eyes, C almost metamophosized specimen. From Harms (1929).

In diverse climate zones of the oceans the soft bottoms are regularly inhabited by burrowing decapod crustaceans which holes attract diverse commensals, also gobies. In the North Sea *Lesueurigobius friesei* (Fig. 3.2.3: 11) lives in holes of *Nephrops norvegicus*, in the Atlantic Ocean also other *Lesueurigobis* species. In the tropical Indopacific *Alpheus* shrimps with reduced eyes are dependent to the good view of their partner gobies (*Amblyeleotris*, *Cryptocentrus*, *Ctenogobiops* or *Vanderhorstia* spp.) which are contacted by the antennae. If the gobies detect potential enemies they cause at first the shrimps by special movements to disappear in the common burrow.

At the Californian coast an inversed situation is found in the joint burrows of the anomuran *Callianassa affinis* with full developed eyes and the blind *Typhlogobius californiensis* (Fig. 3.2.3: 12) which live together in pairs, respectively. The blind gobies are fully dependent to their hosts because these create a current flow which supports the necessary oxygen as well as small prey particles (MacGinitie, 1939). Therefore, the fish have to die if the shrimps die. Nevertheless, the water in the burrow is of low oxygen content which would not allow other fish to survive. This lack is compensated by additional breathing capillaries in the skin and operculum of the goby, similar to the adaptations described in *Gillichthys mirabilis* (above). The eyes of the blind goby are fully developed at hatching but growth ceased afterwards and the retina degenerates. In adults they lie so deep under the skin and is to such an extent overgrown by connective tissue (Fig. 3.2.15) that from outside only the pigment is seen (Ritter, 1893). MacGinitie (1939) assumed an evolutionary regression of related gobies beginning in *Clevelandia ios* with normal eyes, *Gillichthys mirabilis* with slightly reduced eyes to the blind *Typhlogobius californiensis*. Another blind goby, *Typhleleotris madagascarensis*, from freshwater caves of Madagascar is not yet investigated.

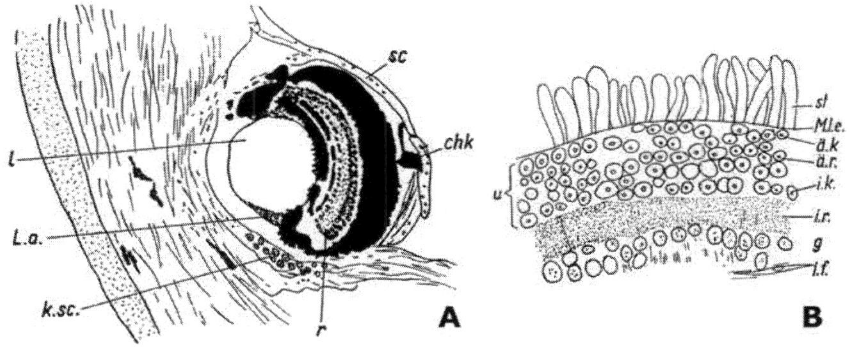

Fig. 3.2.15 *Typhlogobius californiensis*. a. medial cut of the eye, b. enlarged part of the degenerated retina. ä.k. = photoreceptor cell layer, a.r. = outer plexiform layer, chk = choroidean tissue, g = ganglion cell layer,, i.f. = fibres of optic nerve, i k. = bipolar cell layer, i.r. = inner plexiform layer, k.sc. = cartilaginous tissue, l = lens, l.a. = ligament, Mle = Lamina basalis, r = retina, sc = sclera, st = rods, u = degenerated part of retina. After Ritter (1893) from Schnakenbeck (1956).

Numerous fishes present symbiotic relationships to cnidarians like anemones, hard corals, gorgonians and black or thorny corals. Beside members of the family Pomacentridae especially Gobiidae are found between or on these polyps which normally sting visitors by their nematocysts. The Mediterranean *Gobius bucchichii* (Fig. 3.2.3: 14) flies if persuaded into the tentacles of the anemone *Anemonia sulcata*. *G. bucchichii* presents a

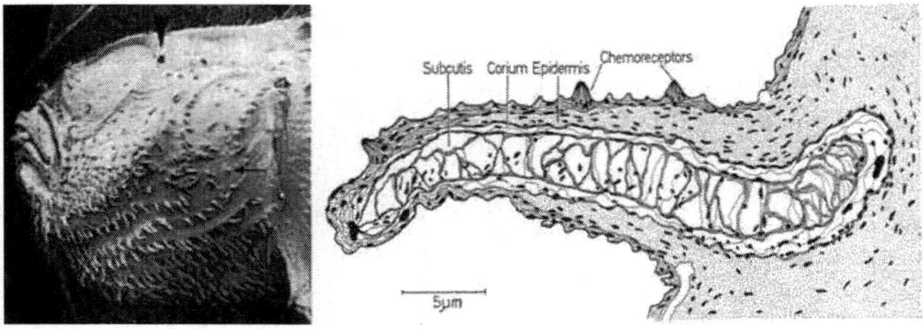

Fig. 3.2.16 *Paragobiodon echinocephalus.* REM-photo of the head with skin papillae (left) and drawing of a light microscope cut of a single papilla. From Zander (1979).

protection substance in the slime of its skin which development is inherited and is solely effective against *A. viridis* (= *A. sulcata*), not against other anemones (Abel, 1971). This does not correspond to the situation of the Indo-Pacific anemone fish *Amphiprion* which gains the protection substance against nematocysts with mean of acclimatization and cautious contacts to the anemone. Hard corals (Madreporaria) are visited regularly by several gobies which rest on the living polyps of plate corals like the cleaning gobies (*Elacatinus* spp., Fig. 3.2.3: 17) or hide and live between branching corals like *Gobiosoma* spp. or *Paragobiodon echinocephalus* without to be stung. There exist almost no investigations about the protection of the fish against the polyps. The Indo-Pacific *Paragobiodon echinocephalus* (Fig. 3.2.3: 15) which lives between the branches of *Acropora* corals shows on its head curious papillae. These are built by proliferations of corium and filled with vesicle cells of the subdermis (Zander, 1979) (Fig. 3.2.16) similar to those of *Boleophthalmus* and *Periophthalmus* spp. The papillae bear numerous mechano and chemo receptors and may have supporting and protection function, maybe also against nematocysts. Unknown is also the protection against nematocysts of *Bryaninops* spp. (Fig. 3.2.3: 16), one of the smallest gobies, which live on black and thorn corals.

Finally, the extreme tiny size of several goby species is to consider. These mature at a size of 7.5 - 30 mm (e.g. *Bryaninops* spp., *Eviota* spp., *Lebetus* spp., *Pandaka pygmaea, Pomatoschistus bathi, Trimmatom nanus*). The advantage of small body size lies in the possibility to feed on small food particles and the exploitation of spatially restricted habitats (Miller, 1979). The disadvantage is a greater susceptibility to predators. This is compensated by high fecundity and manifold reproduction within a year combined with early maturation (< 1 year) as well as cryptobenthic way of life which means the ability to hide in tiny crevices and holes (Miller, 1979).

CONCLUDING REMARKS

This compilation reveals that gobies can at least inhabit 17 marine habitats or microhabitats, completed by the conquer of several brackish and fresh-water environments. This wide choice of diverse habitats, combined with a small size, may have given rise for such an evolutionary success which render gobies to the species richest group of teleosts. Surely, it is to expect that several microhabitats may be added after further investigation in the future. The adaptations to special ways of life follow the general trends which are found also in other fish families. The patterns which are displayed are therefore results of parallel evolution which has developed independently to other fish but also, e.g. in the case of separation of the pelvic fins, independently within the gobies or even within genera (*Gobius* spp., *Periophthalmus* spp.). Other specialized patterns like air-breathing organs and skin-papillae are elements of the goby gene pool which can be activated by selection in the case that a conquest of new environments becomes necessary. The flattened lens of mudkippers is identical with those of the most land vertebrates. But also other ways for getting clear sight on land are evolved within teleost fish, e.g. in blennies (Zander, 1974). Nevertheless, there are some habitats which seem not to be colonized by gobies, e.g. the deep sea, haptic holes and, unexpectedly, the rocky supratidal zone—though the sucker organ seems to be an advantageous pre-adaptation. This result may be explained by competition with still more advantageous adapted blennies to this habitat (Zander, 1972).

References

Abel, E.F. 1961. Liaison facultative d'un poisson (*Gobius bucchichii* Steindachner) et d'une anémone (*Anemona sulcata* Penn.) en Méditerranée. *Vie et Milieu* 11: 517-531.

Chotkowski, M.A., D.G. Buth and K. Prochazka 1999. Systematics of intertidal fishes. In: *Intertidal Fishes—Life in Two Worlds*. M.H. Horn, K.L.M. Martin and M.A. Chotkowski (eds.) Academic Press, San Diego, pp. 297-331.

Eggert, B. 1929. Die Gobiidenflosse und ihre Anpassung an das Landleben. *Zeitschrift für wissenschaftliche Zoologie* 133: 411-440.

Gee, G.H. and P.A. Gee 1995. Aquatic surface respiration, buoyancy control and the evolution of air breathing in Gobies (Gobiidae: Pisces). *Journal of Experimental Biology* 198: 79-89.

Gibson, R.N. 1969. The biology and behaviour of littoral fish. *Oceanography and Marine Biology Annual Review* 7: 367-410.

Gibson, R.N. 1972. The vertical distribution and feeding relationships of intertidal fish on the Atlantic coast of France. *Journal of Animal Ecology* 41: 189-207.

Graham, J.W. 1971. Aerial vision in amphibious fish. *Fauna* 3: 14-23.

Graham, J.W. 1997. *Air-breathing Fishes. Evolution, Diversity and Adaptation.* Academic Press, San Diego.

Harms, J.W. 1929. Die Realisation von Genen und die consecutive Adaptation. *Zeitschrift für wissenschaftliche Zoologie* 133: 211-397. VII plates.

Harris, V.A. 1960. On the locomotion of the mudskipper *Periophthalmus koelreuteri* (Pallas): (Gobiidae). *Proeedings of the Zoological Society of London* 134: 107-135.

Hesthagen, I.H. and J.H. Koefoed. 1979. A note on the existence of a well-developed swim-bladder in the bottom-living *Pomatoschistus minutus*. *Journal of Fish Biology* 14: 224-228.

Heymer, A. and C.D. Zander. 1978. Morphology and ecology of *Gobius vittatus* Vinciguerra, 1883, and its possible mimicry relationship to *Blennius rouxi* Cocco, 1833 in the Mediterranean. *Zeitschrift für zoologische Systematik und Evololutionsforschung*16: 132-143.

Heymer, A. and C.D. Zander. 1992. Le statut de *Gobius auratus* et description de *Gobius xanthocephalus* n. sp. de la Méditerranée (Teleostei, Gobiidae). *Zoologisches Jahrbuch für Systematik* 119: 291-314.

Horn, M.H. 1999. Convergent evolution and community convergence: Research potential using intertidal fishes. In: *Intertidal Fishes—Life in Two Worlds*. M.H. Horn, K.L.M. Martin and M.A. Chotkowski (eds.), Academic Press, San Diego, pp. 356-372.

Karsten, H. 1923. Das Auge von *Periophthalmus koelreuteri*. *Jenaer Zeitschrift für Naturwissenschaften* 59: 115-154.

Low, W.P., D.J.W. Lane and Y.K. Ip. 1988. A comparative study of terrestrial adaptations of the gills of three mudskippers—*Periophthalmus chrysospilus* , *Boleopthalmus boddaerti*, and *Perophthalmodon schlosserei*. *Biological Bulletin* 175: 434-438.

MacGinitie, G.E. 1939. The natural history of the blind goby, *Typhlogobius californiensis* Steindachner. *American Midland Naturalist* 21: 489-505.

Magnus, D.E.B. 1978. Bewegungsweisen, Nahrungserwerb und Fortpflanzungsverhalten des Schlammspringers *Periophthalmus kalolo* (Freilandaufnahmen). *Publikationen zum Wissenschaftlichen Film, IWF Göttingen, Sektion. Biologie*, Serie 14 No. 7/D 1282.

Merrick, J.E. and G.E. Schmieda. 1984. *Australian Fresh Water Fishes: Biology and Management.* Griffin Press, Netley, South Australia.

Mestermann, K. and C.D. Zander. 1984. Vergleichende osteologische Untersuchungen an *Pomatoschistus*-Arten (Gobioidei, Pisces). *Zoologisches Jahrbuch für Anatatomie* 111: 501-542.

Miller, P.J. 1979. Adaptiveness and implications of small size in teleosts. *Symposium of the Zoological Society of London* 44: 263-306.

Miller, P.J. 1986. Gobiidae. In: *Fishes of the North-eastern Atlantic and the Mediterranean.* P.J.H. Whitehead, M.-L. Bauchot, J.-C. Hureau, J. Nielsen and E. Tortonese (eds.), UNESCO, Paris, pp. 1019-1085.

Miller, P.J. and M.Y. El Tawil. 1974. A multidisciplinary approach to a new species of *Gobius* (Teleostei: Gobiidae) from southern Cornwall. *Journal of Zoology, London* 174: 539-574.

Nikolski, G.W. 1959. *Spezielle Fischkunde.* Deutscher Verlag der Wissenschaften, Berlin.

Pinchuk, V.I. and D.B. Rahimov. 1979. *Benthophilus svetovidovi* Pinchuk and Rahimov n. sp. (Pisces, Gobiidae) from the Caspian Sea and a key of species of the genus *Benthophilus. Zoologik Zhurnal* 58: 515-519.

Ritter, E.W. 1893. On the eyes, the integumentary sense papillae, and the integument of the San Diego blind fish (*Typhlogobius californiensis* Steindachner). *Bulletin of the Harvard Museum of Comparative Zoo*logy 24: 51-102.

Schnakenbeck, W. 1956. Erste Klasse der Vertebrata—Pisces. In: *Handbuch der Zoologie, Founded by W. Kükenthal.* J-G. Helmcke, H. v. Lengerken. and D. Starck (eds.), W. de Gruyter and Co, Berlin, pp. 551-1115.

Schöttle, E. 1931. Morphologie und Physiologie der Atmung bei wasser-, schlamm- und landlebenden Gobiiformes. *Zeitschrift für wissenschaftliche Zoologie* 140: 1-114.

Sterba, G. 1959. *Süßwasserfische aus aller Welt.* Urania Verlag, Leipzig/Jena.

Todd, E.S. and A.W. Ebeling 1966. Aerial respiration in the longjaw mudsucker *Gillichthys mirabilis* (Teleostei: Gobiidae). *Biological Bulletin* 130: 265-288.

Zander, C.D. 1972. Beiträge zur Morphologie und Biologie von Blenniidae (Pisces) des Mittelmeeres. *Helgoländer wissenschaftliche Meeresuntersuchungen* 23: 193-231.

Zander, C.D. 1974. Beziehungen zwischen Körperbau und Lebensweise bei Blenniidae (Pisces) aus dem Roten Meer. III. Morphologie des Auges. *Marine Biology* 28: 61-71.

Zander, C.D. 1979. Morphologie von Hautstrukturen im Kopfbereich der Grundel *Paragobiodon echinocephalus* (Rüppell) (Gobiidae. Pisces). *Zoologischer Anzeiger* 202: 226-234.

Zander, C.D. 1980. Morphological and ecological investigations on sympatric *Lipophrys* species (Blenniidae, Pisces). *Helgoländer wissenschaftliche Meeresuntersuchungen* 34: 91-110.

Zander, C.D. 1992. Underwater census of epilithic fish along a depth gradient in the western Mediterranean Sea. *Raport de Commission Internationale Mer Méditerranée* 33 : 314-315.

Zander, C.D. and H. Jelinek 1976. Zur demersen Fischfauna im Bereich der Grotte von Banjole (Rovinj/YU) mit Beschreibung von *Speleogobius trigloides* n.

gen. n. sp. (Gobiidae, Perciformes). *Mitteilungen des Hamburger Zoologischen Museum und Institut* 73: 265-280.

Zander, C.D. and A. Heymer 1992. Feeding habits of *Gobius auratus* and other benthic small-sized fish from the French Mediterranean coast under regard of some alternating parameters. *Zoologischer Anzeiger* 228: 220-228.

Zander, C.D., J. Nieder and K. Martin 1999. Vertical distribution patterns. In: *Intertidal Fishes—Life in Two Worlds*. M. H. Horn, K.L.M. Martin and M.A. Chotkowski (eds.), Academic Press, San Diego, pp. 26-53.

CHAPTER

3.3

Morphological Organization and Variability of the Reproductive Apparatus in Gobies

Carlotta Mazzoldi[1], Robert A. Patzner[2] and Maria B. Rasotto[1]

INTRODUCTION

The goby family encompasses a great range of variation in ecology, behaviour and life histories with consequent variation in reproductive modalities. Indeed, despite a basic reproductive pattern shared by all gobies consisting of the laying of demersal eggs cared for by the male until hatching, differing life-spans (from semelparity to iteroparity), gender identities (from gonochoric to hermaphrodite), types of nest, levels of sperm competition (including alternative male mating tactics) and a variety of mating system (from monogamy to polygyny and promiscuity) have been described for this family. The goby reproductive apparatus, therefore, offers a unique setting for the study of both natural and sexual selection on morphological variability.

Authors' addresses: [1]Department of Biology, University of Padova, via U. Bassi 58/B, 35121 Padova, Italy. E-mails: mazzoldi@bio.unipd.it, rasotto@bio.unipd.it

[2]Department of Organismic Biology, University of Salzburg, Hellbrunnerstr. 34, 5020 Salzburg, Austria. E-mail: robert.patzner@sbg.ac.at

Teleost gonad morphology, at the anatomical, histological and fine structural levels, has long been studied with the aim of identifying annual reproductive cycles, the length of breeding seasons, the onset of reproductive maturity, spawning rhythms, fecundity and various other aspects of reproductive biology, or to be used in phylogenetic systematic analyses. Consequently some of the relationships between the reproductive modalities of the species and the morphology of the reproductive apparatus, such as the type of sexual pattern or the fecundity, have been widely investigated not only in gobies but in teleosts in general. However, in spite of the interest that sexual selection has received in recent decades, its influence on the development and performance of primary sexual traits still remains poorly known in fish.

Gobiid fishes, with their variability in life-cycle duration, mating system and sperm competition occurrence, provide excellent models to understand the influence of sexual selection pressures in shaping reproductive apparatus and gamete performance. For this reason this chapter will outline the morphology of the goby reproductive apparatus highlighting the influence of sexual selection on its variability, omitting discussion of the aspects related to sex differentiation and endocrinology, but will include some notes on pheromone production, and seasonal variation.

FEMALE REPRODUCTIVE APPARATUS

Ovary and Oviduct

The morphology of the female reproductive system, first investigated by Eggert (1931), is similar to that of other teleost fish. The ovaries are paired and fusiform in shape, anatomically located ventral to the kidney and suspended from the coelomic wall by a ligament, the mesovarium. Each ovary is enveloped by a thin layer of coelomic epithelium, consisting of one or two layers of smooth muscle. The two ovaries fuse posteriorly, forming a common oviduct which opens into the genital papilla. This is in contrast to other vertebrates in which the ovarian ducts arose from primordial renal duct (Hoar, 1969). In two *Gobiodon* species, *G. okinawae* and *G. oculolineatus* (Cole and Hoese, 2001; Cole, 2009), a pair of fully differentiated accessory structures are located laterally to the oviduct. These structures, whose function remains unknown, arise from extensive lobulation of the oviduct walls and are characterized by a low-to-tall columnar epithelium.

The ovary is a hollow organ. Numerous folds or lamellae, consisting of connective tissue stroma, nerves and blood vessels, protrude into the ovarian lumen (Eggert, 1931; De Vlaming, 1972; Cinquetti and Rinaldi, 1987). The stroma contains the developing follicle and oocyte complexes. A basement membrane separates the germinal epithelium, containing the oogonia and the primary oocytes, from the ovarian stroma (Grier, 2000). The ovary of the paedomorphic *Schindleria praematura*—one of the smallest vertebrates—differs in many aspects from that of other gobies (Thacker and Grier, 2005). It has a very small absolute size but is proportionally larger than that of larger gobies and contains only 25 to 30 oocytes. In contrast to other species, the oogonia in *S. praematura* are restricted to a germinal ridge, running ventrally along the length of the ovary. The stroma of the ovary is extremely reduced compared to that of other species, and there are no ovigerous lamellae: each ovary has a single continuous lumen (Thacker and Grier, 2005).

Ovaries differ not only in morphology but also in type (how they carry out their function) and goby females also present different types of ovary (Nagahama, 1983). Asynchronous ovaries, with oocytes at different stages of development, indicating the laying of several egg clutches (Fig 3.3.1A), have been described for temperate species with a long reproductive season (Miller, 1984; Cinquetti and Rinaldi, 1987; Mazzoldi, 1999; Caputo *et al.*, 2000; Mazzoldi and Rasotto, 2001; Kovačić, 2007). A group synchronous ovary, characterized by the presence of only few stages of egg development (oogonia, oocytes at the perinucleolus stage, and one stage of vitellogenic oocytes), has been observed in monogamous tropical species belonging to the genera *Amblygobius* and *Valenciennea* (Fig 3.3.1B), suggesting a protracted interval between the release of successive egg batches (Mazzoldi, 2001). In semelparous species, such as *Crystallogobius linearis*, a third type of ovary is found and only one stage of egg development can be observed (Caputo *et al.*, 2003).

The pattern of oocyte maturation in gobies is similar to that observed in other teleost fish (e.g. Bara, 1960, 1963) but comparison among species is often difficult because of the different terminology, as well as the number of oocyte maturation phases, adopted by different authors (Table 3.3.1). In many species, such as *Eucyclogobius newberryi* (Goldberg, 1977) and *Padogobius martensi* (Cinquetti and Rinaldi, 1987), the oocyte growth starts soon after the end of the spawning season whereas in others, such as *Gillichthys mirabilis* (De Vlaming, 1972), ovaries go through a quiescent period after the breeding season and only pre-vitellogenic oocytes are present in the stroma.

Table 3.3.1 Classification of ovary and oocyte stages by different authors. This table not only shows that a vast array of terms have been coined to describe different stages of oocyte development, it also shows that the terms are not directly interchangeable. Where one author used a single term, another may have used two terms to describe the same "stage". This lack of uniformity detracts somewhat from the value of studies if the data cannot be used by other authors for cross-study comparison and analysis.

Miller (1961)	*De Vlaming (1972)*	*Goldberg (1977)*	*Geevarghese and John (1983)*	*Hidaka and Takahashi (1987)*	*Cinquetti and Rinaldi (1987)*
	regressing ovary	regressing		regressing*	
immature	quiescent phase or phase of oogonial proliferation	quiescent	immature	yolkless stage	early oocyte
					early perinucleolus
developing virgin and recovering spent	phase of active vitellogenesis	vitellogenic	maturing virgin and recovered spent	beginning of yolk accumulation	late perinucleolus
			maturing	active yolk accumulation	primary yolk vesicle
early ripening					secondary yolk vesicle
late ripening	prespawning condition	pre-spawning		pre-spawning	primary yolk globule
ripe			ripe		secondary yolk globule
spent	post spawning condition	post-spawning	spent	post-spawning*	

* combined to one stage

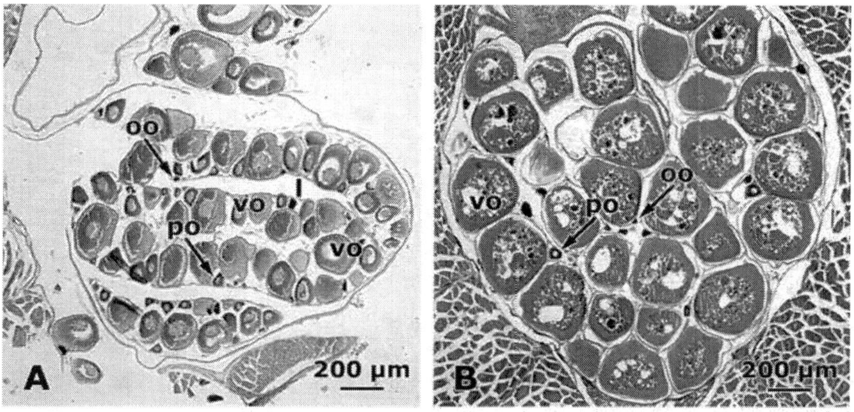

Fig. 3.3.1 Female reproductive apparatus: histological section of mature ovaries. **A**. *Pomatoschistus marmoratus*, asynchronous ovary showing different stages of vitellogenic oocytes, oocytes at perinucleolus stage and oogonia. **B**. *Amblygobius nocturnus*, group synchronous ovary showing only late vitellogenic oocytes, oocytes at early perinucleolus stage and oogonia. l= ovarian lumen; oo = oogonius; po = oocyte at perinucleolus stage; vo = vitellogenic oocyte.

Based on different life history traits, both "semelparous" species, as defined by Cole (1954), i.e. experiencing only one reproductive season, and "iteroparous" ones, i.e. reproducing more than once during their life span, have been found in gobies (Miller, 1984). Environmental conditions appear to influence the frequency of breeding seasons, the number of clutches, the number of eggs per clutch and the size of the eggs released by females in each season. In temperate areas—such as the Mediterranean—gobies have one reproductive season per year whereas in tropical ones two seasons may occur. However, the number and the duration of the breeding period may vary within species if its range of geographical distribution is wide, as documented in the mudskipper *Boleophthalmus dussumeri*. This species reproduces once a year along the western coast of India (Mutsadi and Bal, 1970) and on the Jodia coast (Soni and George, 1986) and twice a year on the Karachi coast, Pakistan (Hoda and Akhtar, 1985). Females may lay several egg batches during a reproductive season (Miller, 1984) but clutch number and size and egg size, despite generally being species specific, can vary within and/or among populations in response to several factors, such as female size, food availability, spawning season, or latitudinal distribution (e.g. Reavis, 1997; Mazzoldi *et al.*, 2002; Yamasaki and Takihara, 2006).

Eggs

Gobiid fishes lay their eggs in a dense patch of a single layer on the underside of a solid substrate, e.g. stones, rocks or shells (Miller, 1984), or on seagrass rhizomes (Scaggiante *et al.*, 1999). Egg deposition time appears to be related to body size, varying from less than an hour in smaller species to several hours in larger ones (Mazzoldi, 1999; Mazzoldi *et al.*, 2000; Mazzoldi and Rasotto, 2002) and eggs remain fertilizable for several hours (up to 40 hours in the grass goby, *Zosterisessor ophiocephalus*, Scaggiante *et al.*, 1999). Eggs are attached by filaments and hang downwards. Their shape varies inter-specifically, being elliptical, pyriform, cylindrical, close to spherical, etc. (Miller, 1984). The function of this variability has not been investigated. Different shapes may, however, be based on phylogenetic relationships as well as the ecological characteristics of the nest site. Indeed, eggs of different shapes may be differently packed in the egg patch, with a consequent variable water flow from fanning and possibly promoting sediment shed (Miller, 1984). Moreover, egg shape determines the surface/volume ratio, therefore influencing gas exchange. Hence it is rational to expect that variation in shape confers different performance to eggs, in terms of gas exchange and consequently developmental time, resistance to mechanical stress, and tendency to sediment deposition. Eggs may also be differently tightly packed within the layer regardless of size. Indeed in some species eggs are laid compactly, egg density is a function of egg size and egg number is a function of nest size (Mazzoldi, 1999; Mazzoldi *et al.*, 2002). On the contrary in the lagoon goby, *Knipowitschia panizzae*, spawning on anoxic muddy bottoms, nests are only partially covered by eggs that are loosely packed (Massironi *et al.*, 2005). As hypothesized for egg shape, such differences in deposition pattern could also be related to differences in nesting environment, with species spawning in anoxic environments limiting egg number and egg density for oxygen provisioning (Massironi *et al.*, 2005).

Gobiid eggs are known to vary in size both within and among species (Miller, 1984; Mazzoldi *et al.*, 2002; Tamada and Iwata, 2005). Egg diameter may vary from less than 0.4 mm to more than 2 mm for the major diameter (Katoh and Nishida, 1994; Mazzoldi *et al.*, 2002; Massironi *et al.*, 2005; Yamasaki and Takihara, 2007). Inter-specific variability in egg size could be related, as is egg shape, to phylogenetic relationships and nesting environment, however these ideas have not been investigated. Variability in egg size has been reported for the marbled goby *Pomatoschistus marmoratus* and the common goby *P. microps* (Bouchereau *et al.*, 1991; Mazzoldi *et al.*, 2002) at the intra-specific level, with egg size decreasing with the season and increasing with female size (Mazzoldi *et al.*, 2002). The influence of seasonal progression on egg size in temperate species is widely reported (Chambers, 1997). However, maternal size is the major factor affecting egg size in the marbled goby, and this finding is correlated with the high environmental variability of the lagoons and shallow waters where this species lives. The unpredictable variation in temperature and salinity fish experience in these waters could select for a wide reaction norm of offspring size, consequently promoting variability in egg size (Mazzoldi *et al.*, 2002). Correlation between egg and female size has been observed also in *Rhinogobius* sp., and in this case has been related to different spawning habitats. Larger females are known to spawn in the upper reaches of rivers where starvation can represent a consistent risk for larvae. Given the relationship between egg and larval size, the production of larger eggs insures higher survival to the offspring of larger females (Tamada and Iwata, 2005). A similar relationship between egg size and spawning habitat has been documented in the *Rhinogobius brunneus* complex (Katoh and Nishida, 1994).

The mature goby oocytes have a filament adhesion structure at the animal pole (Riehl, 1978; Miller, 1984; Giulianini and Ferrero, 2001; Giulianini *et al.*, 2001; Herler *et al.*, 2006; Kramer and Patzner, 2008) (Fig. 3.3.2A, B). Filaments are arranged around the micropylar pole providing the mode of attachment for these demersal eggs (Miller, 1984). Giulianini *et al.* (2001) have demonstrated that spermatozoa can pass through these nets of filaments to reach the micropyle and fertilize the egg, even after their attachment to the substrate. Kramer and Patzner (2008) have investigated and compared the mode of egg attachment of the adhesive filaments among four Caribbean *Coryphopterus* species. A similar pattern of egg attachment is observed in all these species. The attaching structure forms a circular, net-like pattern around the micropylar pole (Fig. 3.3.2C, D), which divides distally into single filaments approximately 2 μm in diameter (Fig. 3.3.2C, D). This net-like structure, which attaches the eggs to the substrate, is small in *C. thrix* and *C. dicrus*, and large and well-developed in *C. venezuelae* and *C. eidolon* (Kramer and Patzner, 2008).

The eggshell (zona radiata or chorion) varies in thickness, surface structure and adhesive filaments. Its main function is protecting the embryo from mechanical stress. Its thickness varies, on average, from 5 to 15 μm, which is common in teleosts. The thickness and the structure of the zona radiata often reflect adaptations to the reproductive ecology of the fish. Previous studies have come to the conclusion that fish whose eggs do not need protection from mechanical forces have a relatively thin zona radiata (Riehl and Greven, 1993; Riehl, 1995). The eggs of some gobies (e.g.

Coryphopterus) have a smooth surface and do not show any structural or specific pattern in addition to the adhesive filaments (Kramer and Patzner, 2008) (Fig. 3.3.2E), whereas in other species (e.g. Mediterranean *Gobius* species) the chorion is densely covered with microvilli (Herler *et al.*, 2006) (Fig. 3.3.2F); the reason for this is not known.

MALE REPRODUCTIVE APPARATUS

The reproductive apparatus of the male gobies typically consists of paired testes, a sperm transport system, i.e. the main testicular ducts and a sperm duct, and accessory structures (Fig. 3.3.3A; Miller, 1984, 1992). The apparatus opens at the tip of an erectile, conical urogenital papilla, tapered with a terminal slit, located between the anus and the anal fin (Egami, 1960; Arai, 1964; Miller, 1984).

Testes and Sperm Transport System

Testes are elongated organs, attached to the coelomic cavity by the mesorchia ligaments (Fig. 3.3.3B) and covered by tunica albuginea, formed by the thin coelomic wall and a layer of connective tissue rich in blood vessels, nerve terminations and smooth muscle cells (Nagahama, 1983). The germinal epithelium is organized in lobules and rests upon a basement membrane that separates the lobules from the interstitial compartment, which consists of a connective layer with myoid cells, blood vessels, amyelinic nerves, and Leydig cells, these last being the major source of steroid hormones in teleosts (Van den Hurk, 1973; Grier, 1981; Nagahama, 1983). This compartment is highly reduced in the paedomorphic species *Schindleria praematura* (Thacker and Grier, 2005). Each lobule presents a lumen and spermatogonia distributed along the entire lobule length (Miller, 1984; Mazzoldi, 1999; Caputo *et al.*, 2000, 2003; Cinquetti and Dramis, 2003; Thacker and Grier, 2005). Goby testes are therefore classified as of the lobular unrestricted type (Grier *et al.*, 1980; Grier, 1993; Parenti and Grier, 2004). An exception to this pattern is represented by the paedomorphic *S. praematura*, where spermatogonia are present only at the blind end of the lobules, near the testis wall. Earlier spermatogenic stages, such as spermatocytes, are concentrated in the distal part of the lobules, whilst later ones are concentrated near the main testicular duct and a true lobule lumen is lacking (Thacker and Grier, 2005). These features make the testis of this species similar to the lobular restricted type (Grier *et al.*, 1980; Grier, 1993; Parenti and Grier, 2004).

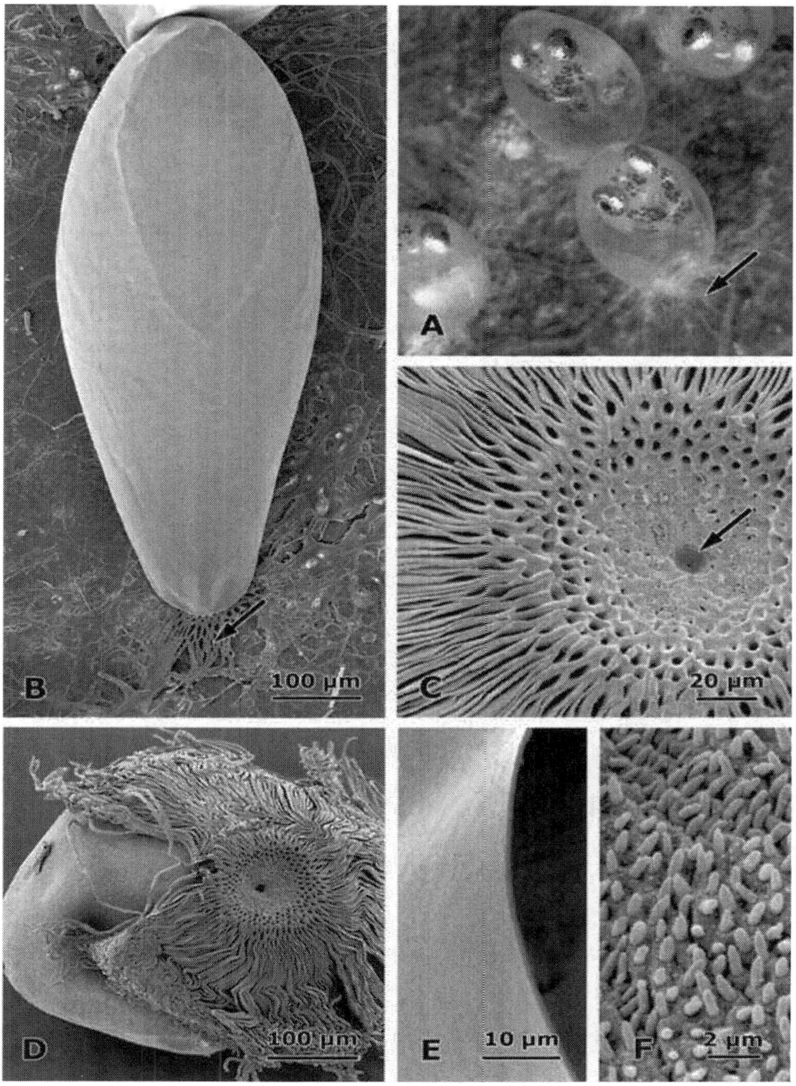

Fig. 3.3.2 Eggs. **A**. Deposited eggs of *Clamydogobius eremius*, arrow shows attaching filaments. **B**. Egg of *Coryphopterus venezuelae* with attaching structures (arrow) at the bottom. **C**. Egg of *Gobius auratus*, perforated plate at the basis of attaching filaments at the animal pole, surrounding the micropylar pit (arrow). **D**. Entire egg of *G. auratus* with attaching filaments at the base. **E**. Zona radiata of egg of *Coryphopterus dicrus*, left = outside of egg. **F**. Zona radiata of egg of *G. auratus*. SEM-images B, E by Kramer/ Patzner, C, D, F by Herler/Patzner.

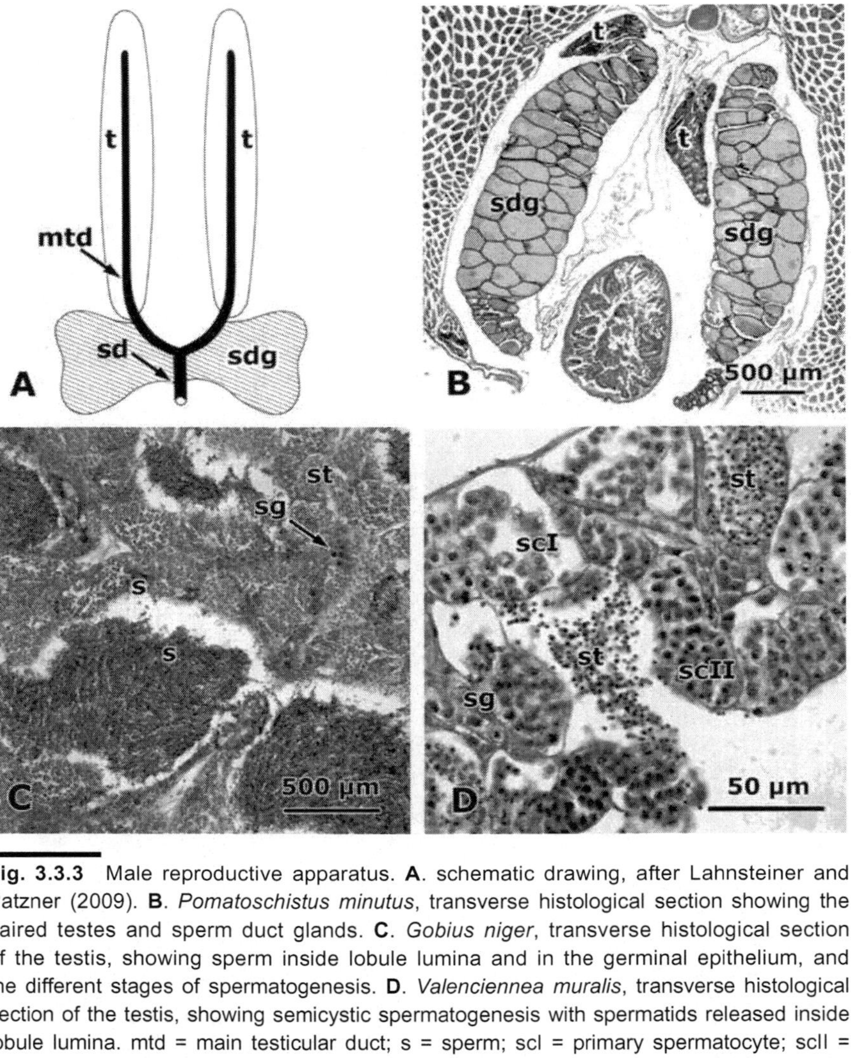

Fig. 3.3.3 Male reproductive apparatus. **A**. schematic drawing, after Lahnsteiner and Patzner (2009). **B**. *Pomatoschistus minutus*, transverse histological section showing the paired testes and sperm duct glands. **C**. *Gobius niger*, transverse histological section of the testis, showing sperm inside lobule lumina and in the germinal epithelium, and the different stages of spermatogenesis. **D**. *Valenciennea muralis*, transverse histological section of the testis, showing semicystic spermatogenesis with spermatids released inside lobule lumina. mtd = main testicular duct; s = sperm; scI = primary spermatocyte; scII = secondary spermatocyte; sd = sperm duct; sdg = sperm duct gland; sg = spermatogonium; st = spermatid; t = testis.

In most goby species, spermatogenesis occurs within cysts, formed by Sertoli cells, which open into the lobule lumen, releasing mature sperm (Fig. 3.3.3C; Patzner *et al.*, 1991; Caputo *et al.*, 2000; Cinquetti and Dramis, 2003; Mazzoldi *et al.*, 2005; Thacker and Grier, 2005). However, in some *Amblygobius*, *Cryptocentrus* and *Valenciennea* species, the spermatogenesis is semi-cystic (Mazzoldi, 2001; Mazzoldi *et al.*, 2005). In these gobies, spermatids are released into the lobule lumen, where spermiogenesis is completed and the sperm mature (Fig. 3.3.3D). It has been proposed that the semi-cystic spermatogenesis, described in other fish species, such as blennies (Lahnsteiner and Patzner, 1990), jawfish (Manni and Rasotto, 1997) and pipefish (Carcupino *et al.*, 1999), may be a mechanism favouring sperm economy since the asynchronous spermatid maturation along the lobule length greatly reduces the number of simultaneously mature sperm (Manni and Rasotto, 1997; Mazzoldi, 2001; Giacomello *et al.*, 2008). It is well known that spermatogenesis generates nontrivial costs (Dewsbury, 1989) and males have thus evolved mechanisms for the optimum allocation of finite numbers of sperm to maximize their reproductive success (for a review see Wedell *et al.*, 2002). A very low sperm production is expected to be particularly advantageous in those species where, in addition to a low female fecundity and a lack of sperm competition, males have high reproductive costs such as territory defence and parental care (Wedell *et al.*, 2002). The reproductive biology of goby species showing a semi-cystic spermatogenesis seems to support this hypothesis. Indeed these species are characterized by small body size, low fecundity and a monogamous mating system with no sperm competition (Mazzoldi, 2001; Mazzoldi *et al.*, 2005). In this scenario the number of sperm required for a successful fertilization is limited and males, whose reproductive costs also include nest building, territory defence, courtship, and parental care, might save energy by releasing small amounts of mature sperm at each spawning.

The sperm transport system of male gobies consists of the main testicular ducts and the sperm duct. The main testicular ducts run along the testes and collect the sperm from the lobules; posteriorly to the testes they fuse together, before reaching the urogenital papilla, in a sperm duct. The main testicular ducts and the sperm duct are linear in shape and the organization of their walls is usually similar, showing a thin external coelomic wall, a connective tissue layer with smooth muscle cells, and an inner secretory epithelium (Patzner *et al.*, 1991; Cinquetti, 1997).

Goby Male Accessory Structures

Mesorchial glands

In some goby species, most of the Leydig cells, instead of being spread along the interstitial tissue are aggregated to form a structure, called a mesorchial gland (MG), located dorsally in proximity to the mesorchium

(Eggert, 1931; Stanley *et al.*, 1965; Colombo and Burighel, 1974; Miller, 1984; Asahina *et al.*, 1985; Seiwald and Patzner, 1989). A distinct MG has been described in species belonging to the genus *Gobius*, such as *G. niger* and *G. paganellus* (Eggert, 1931; Vivien, 1939; Stanley *et al.*, 1965; Miller, 1984; Colombo and Burighel, 1974), in *Glossogobius olivaceus* (Asahina *et al.*, 1985) and in *Neogobius melanostomus* (Arbuckle *et al.*, 2005). The MG cells exhibit the typical features of steroid-producing cells (Colombo *et al.*, 1970; Colombo and Burighel, 1974; Seiwald and Patzner, 1989) and have been found to predominantly synthesize steroid metabolites (Colombo *et al.*, 1970, 1980; Asahina *et al.*, 1985; Arbuckle *et al.*, 2005). The MG of the black goby, *G. niger*, produces large quantities of 5β-reduced conjugated androgens, mainly steroid glucuronides such as etiocholanolone glucoronide (Colombo and Burighel, 1974; Colombo *et al.*, 1970, 1977). From an historical point of view the etiocholanolone glucoronide released by males with urine and ejaculate was the first sexual pheromone identified in fish (Stacey, 2003). Indeed, behavioural experiments, in which this conjugated steroid was pumped inside an empty nest, showed that it was able to attract ripe females to the nest, often eliciting egg deposition (Colombo *et al.*, 1980; Marconato, 1980). In the round goby, *N. melanostomus*, different conjugated steroids are known to increase ventilation rate in both males and females (Murphy *et al.*, 2001), and mixtures of male gonad steroids stimulate attraction responses in females (Moscicki *et al.*, unpublished cited in Jasra *et al.*, 2007). These data suggest that the MG may act as source of sexual pheromones in the round goby as well.

Sperm Duct Glands (Seminal Vesicles)

Goby males typically present a pair of accessory structures to the reproductive apparatus, called seminal vesicles or sperm duct glands (SDG; Figs. 3.2.3A, B). The SDGs are leaf-like paired structures growing out from the sperm duct and are organized in chambers, converging in separate collecting ducts that empty into the lateral sides of the sperm duct/s (Miller, 1984; Fishelson, 1991; Patzner *et al.*, 1991). The extension of the connecting portion between ducts and glands varies among species (Fishelson, 1991). The SDG wall consists, from outside to inside, of a thin coelomic wall, a layer of connective tissue, rich in smooth muscle cells and blood capillaries, a basal lamina and a single layered epithelium (Fishelson, 1991; Lahnsteiner *et al.*, 1992; Cinquetti, 1997; Scaggiante *et al.*, 1999). Similarly, the chamber walls present an innermost monolayered secretory epithelium that can exhibit transverse and/or longitudinal folds, a basal lamina, and a thin layer of connective tissue. Cells of the innermost epithelium have a nucleus in the basal or middle portion, the Golgi complexes in the mid-apical portion, and a well-developed rough endoplasmic reticulum (Lahnsteiner *et al.*, 1992; Cinquetti, 1997). The apical portion of the cells is characterized by the presence of vesicles (Lahnsteiner *et al.*, 1992; Cinquetti, 1997) involved in secretory activity through exocytosis and apocrine secretion (Lahnsteiner *et al.*, 1992). The secretion that fills the chamber lumina appears as an electron-dense fluid, characterized by positive staining to PAS, Alcian blue at pH 2.5 and mercury bromophenol blue, demonstrating that it consists of sialomucins and proteins (Young and Fox, 1937; Fishelson, 1991; Lahnsteiner *et al.*, 1992; Cinquetti, 1997; Scaggiante *et al.*, 1999; Mazzoldi, 1999; Rasotto and Mazzoldi, 2002; Mazzoldi *et al.*, 2005). Steroid producing cells have been found in the interstitial connective layer of the sperm duct glands of several species (Moiseyeva and Ponomareva, 1973; Asahina *et al.*, 1989; Fishelson, 1991; Lahnsteiner *et al.*, 1992; Cinquetti, 1997; Jasra *et al.*, 2007). The major products of these cells are conjugated steroids, hypothesized to act as pheromones (Asahina *et al.*, 1989; Cinquetti *et al.*, 1990; Jasra *et al.*, 2007).

First described by Eggert in 1931, SDGs were considered a synapomorphic trait for gobioid fishes for the following seventy years, as, according to Miller (1984, 1992) they are present in "virtually in all male gobiids that have been investigated". However it was recently shown that SDGs are lacking or extremely reduced in several tropical species (Mazzoldi, 2001; Mazzoldi *et al.*, 2005).

In temperate species, SDGs, as gonads, present a typical annual cycle. They are more developed during the breeding season, when chambers present wide lumina filled with secretion. In this period, the epithelium is distended; cells are flat and show high secretory activity (Cinquetti *et al.*, 1990; Lahnsteiner *et al.*, 1992). In the post-spawning period, SDGs appear highly reduced in size and the secretory activity stops. The chamber lumina are reduced, the thickness of the epithelium increases and groups of epithelial cells degenerate (Cinquetti *et al.*, 1990; Lahnsteiner *et al.*, 1992). In the inter-spawning period SDGs show their minimum sizes, with lumina filled with degenerating materials. Finally, in the pre-spawning period, SDGs grow at a remarkable rate, with new epithelial cells and the onset of secretory activity (Cinquetti *et al.*, 1990; Lahnsteiner *et al.*, 1992). This annual variation strongly indicates the involvement of SDGs in reproductive activities and several hypotheses were proposed speculating as to their function/s. Fishelson (1989) suggested that SDGs may be sperm storage organs. However, most hypotheses focus on the functions of their secretion, expected to be involved in nest compacting (Moskalkowa, 1967), the marking of oviposition sites (Giulianini *et al.*, 1989), enhancing sperm viability (Eggert, 1931; Weisel, 1948) or egg-sperm contact (Vivien, 1938), releasing antimicrobials for egg protection (Fishelson, 1991), or forming a layer of sperm on the substrate prior to egg deposition (Lahnsteiner *et al.*, 1992). This last hypothesis is based on the observation that male gobies often rub the urogenital papilla over the nest surface before egg deposition, suggesting the release of sticky ejaculates and an asynchronous release of gametes between males and females, a feature common in internal fertilizers but, among external ones, known to occur only in an eleotrid fish (Auty, 1978).

Some experimental studies undertaken over the last decade have demonstrated that, in fact, the main function of SDGs is the production of mucins allowing the release of ejaculate in the form of viscous, long lasting bands, called sperm trails (Marconato *et al.*, 1996; Ota *et al.*, 1996; Mazzoldi, 1999). Water samples were collected in the nests of different goby species both before and during egg deposition and acetate sheets placed on the inner surfaces of nests were coloured with specific stains for sperm and/or mucins. The results prove that males of several goby species lay sperm trails on the inner surface of the nest, even before egg deposition (Marconato *et al.*, 1996; Ota *et al.*, 1996; Mazzoldi, 1999). Moreover, the comparison of sperm trails released by males before and after the surgical removal of SDGs demonstrated that the mucous component of the trails was indeed produced by these accessory organs (Scaggiante *et al.*, 1999). SDG mucins are the major component of goby seminal fluid and, in this very peculiar type of ejaculate, sperm are inactive but, as a consequence of the slowly dissolving of mucus, they are released over prolonged periods of time (up to 20 hours in *Zosterisessor ophiocephalus*, Scaggiante *et al.*, 1999). It is known that sperm trail longevity, i.e. the time it takes to completely dissolve, depends on its mucin content (Scaggiante *et al.*, 1999; Mazzoldi *et al.* 2000; Rasotto and Mazzoldi, 2002). And thus, the role of the mucus in ensuring the long lasting presence of active sperm in the nest water shapes an important aspect of ejaculate performance.

In summary, goby males release sperm in trails and eggs do not need to be laid over sperm trails, given that sperm reaches them from the surrounding water (Scaggiante *et al.*, 1999; Giulianini *et al.*, 2001). This peculiar and asynchronous gamete release may be particularly rewarding for parental goby males. As egg deposition can last for several hours, males, thanks to the laying of sperm trails, do not need to spend all their spawning time close to the female and hence may defend the nest from predators and/or opportunistic males or attract additional females (Marconato *et al.*, 1996; Mazzoldi, 1999). Moreover, the release of sperm in the form of slowly dissolving trails reduces sperm dispersal, possibly allowing an economy in sperm production (Mazzoldi *et al.*, 2005). As sperm is costly (Dewsbury, 1989; for a review see Wedell *et al.*, 2002), the limitation of sperm dispersal may actively contribute to saving energy, and this aspect appears particularly relevant in species, such as gobies, in which parental males invest highly in egg care.

SDG secretion, from the grass goby, *Z. ophiocephalus,* was recently shown to exert antibacterial activity (Giacomello *et al.*, 2008). Indeed, the extract of SDG secretion inhibited the growth of two gram-positive, *Micrococcus luteus* and *Bacillus subtilis*, and of a gram-negative, *Tenacibaculum maritimum*, bacteria. This last species is the etiological agent of flexibacteriosis, a well known fish pathogen affecting both adult and juveniles (Toranzo *et al.*, 2005). The route of infection involved in flexibacteriosis is still uncertain but the microorganism has been isolated from sediments (Santos *et al.*, 1999). In this respect, it is noteworthy that *Z. ophiocephalus* males dig nests in mud sediment, underneath *Zostera* sp. rhizomes, and consequently the release of antimicrobial compounds from SDGs may play a crucial role in protecting both parental males and eggs from infections.

Considering the different functions demonstrated or strongly suggested for SDGs, the key role of their secretion in the peculiar fertilization dynamics of gobies suggests that this could be the main and ancestral function of these accessory organs, whereas the production of antimicrobial and/or pheromonal compounds might represent additional, secondary functions.

Sperm

Goby sperm morphology and performance are largely unknown. Sperm morphology has been investigated in less than 50 species (Mattei, 1970; Jamieson, 1991; Hara and Okiyama, 1998; Fishelson *et al.*, 1990; Zutshi and Murthy, 2001; Hara, 2004) and performance traits, such as longevity, velocity and viability, in only a few species (Scaggiante *et al.*, 1999; Locatello *et al.*, 2007; O. Svensson, unpublished data). The morphology of goby spermatozoa appears to be extremely diverse, showing most of the known teleostean traits (Hara, 2004). Sperm morphology has been suggested to be a useful tool for investigating phylogenetic relationships in this family (Hara, 2004). To this end, a classification of goby sperm into 6 different types has recently been proposed (Hara, 2004) (Fig. 3.3.4).

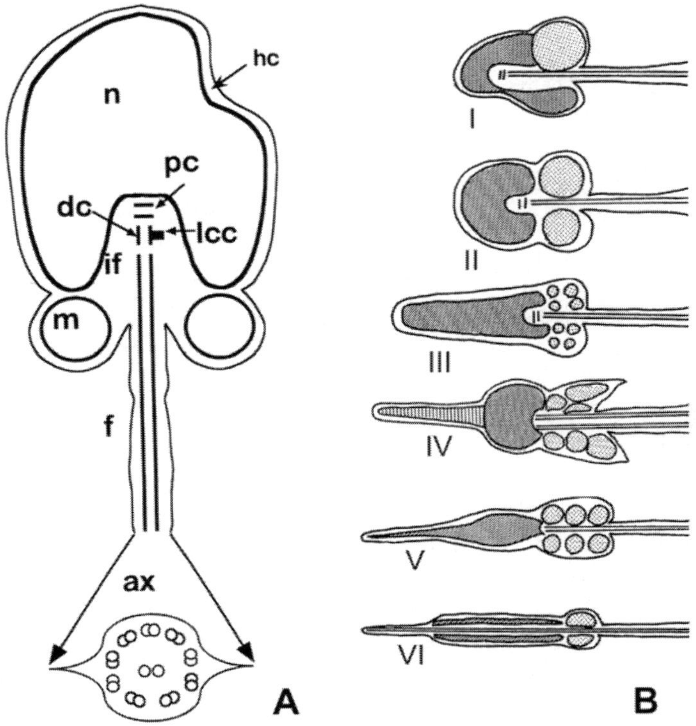

Fig. 3.3.4 Spermatozoa. **A**. fundamental structure; **B**. types I to VI. ax = axoneme; dc = distal centriole; f = flagellum with fin; hc = hiatus in chromatin (= nuclear notch); if = implantation fossa; lcc = lateral centriolar connection; m = mitochondrion; n = nucleus; pc = proximal centriole (after Hara, 2004).

On the basis of head (H), midpiece (M) and tail (T) characteristics, goby sperm can be defined as:

type I: H: the nucleus is ovoid to spherical in shape and shows distinct or indistinct nuclear notches (foldings) on the chromatin surface; M: one round unilateral mitochondrion is located opposite the chromatin material; and T: has a single flagellum. Examples: *Periophthalmus papilio* (Mattei, 1970), *Rhyacichthys aspro*, *Eleotris* spp., *Periophthalmus modestus* (Hara, 2004);

type II: H: the nucleus is round with no nuclear notches in the chromatin; M: numerous mitochondria are situated in one longitudinal layer around the base of the flagellum; and T: has a single flagellum. Examples: *Glossogobius giuris* (Zutschi and Murthy, 2001),

Sicyopterus japonicus, Odontamblyopus lacerpedii, Acanthogobius flavimanus (Hara, 2004);

type III: H: the nucleus is cone-shaped; M: numerous mitochondria are situated in several longitudinal layers around the base of the flagellum; and T: has a single flagellum. Examples: *Odontobutis* spp., *Gymnogobius* spp., *Luciogobius guttatus* (Hara, 2004);

type IV: H: the nucleus is round in shape and has a special longitudinal anterior accessory part; M: numerous irregular mitochondria are situated around the base of the flagellum; and T: has a double flagella. Example: *Leucopsarion petersi* (Hara, 2004);

type V: H: the nucleus is bell-shaped, gradually tapering off; M: numerous spherical mitochondria in several longitudinal layers lie around the base of the flagellum; and T: has a single flagellum. Example: *Ptereleotris hanae* (Hara, 2004);

type VI: H: the nucleus is elongated, cylindrical in shape and penetrated by the flagellum to well beyond the anterior apex; M: has one mitochondrion which encloses the base of the flagellum; and T: has a single flagellum. Examples: *Hypseleotris galii* (Jamieson, 1991), *Schindleria* sp. (Hara, 2004).

The little available information on the sperm performance of *Zosterisessor ophiocephalus* and *Pomatoschistus minutus* (Scaggiante *et al.*, 1999; O. Svensson and H. Elofsson, unpublished data), despite being collected with different methodologies, show a higher sperm longevity (from, on average, 83 min. in *Z. ophiocephalus* to more than 2 hours in *P. minutus*) than is found in other fish with external fertilization (Stockley *et al.*, 1997). Despite their limitations, these results make it very tempting to speculate that, in gobies, asynchronous gamete release and long-lasting egg deposition might have selected for the observed great sperm longevity. There is evidence that sperm quality may also vary within species. An intra-specific variability in other sperm traits related to performance, such as velocity, viability and ATP content, has been found in *Gobius niger* (see following section).

INTRA-SPECIFIC VARIABILITY IN MALE REPRODUCTIVE APPARATUS AND EJACULATE PERFORMANCE

In gobies, as noted in the previous section, testes and the structures associated with the male reproductive apparatus have roles that go beyond sperm production. Indeed, the production of seminal fluid, antimicrobials, and pheromones, as well as sperm storage, are important functions played by testes, MGs and SDGs (Colombo and Burighel, 1974; Colombo *et al.*, 1980; Fishelson, 1991; Van den Hurk and Resink, 1992; Marconato *et al.*, 1996; Mazzoldi, 1999; Giacomello *et al.*, 2008). Intra-specific differences in allocation to the different structures of the male reproductive apparatus and the consequent functional differences in the production of sperm, seminal fluid and pheromones have been described in some goby species where alternative male mating tactics occur (see Chapter 3.4 for alternative male mating tactics in gobies).

Males performing different mating tactics, due to different investment in female attraction, nest defence, parental care, and sperm competition, were expected and have been found to exhibit a strong dimorphism not only in secondary but also in primary sexual traits (Parker, 1990; Taborsky, 1994, 1998; Parker *et al.*, 1996; Stockley *et al.*, 1997; Petersen and Warner, 1998). However, the effort to characterize the intra-specific variability in the primary sex traits of species in relation to the adopted tactic has been primarily devoted to quantifying testis development and/or sperm production (Taborsky, 1994, 1998; Stockley *et al.*, 1997; Petersen and Warner, 1998). As observed in other fish (reviewed in Petersen and Warner, 1998; Taborsky, 2008), male gobies spawning in disfavoured roles (such as sneakers) show a much higher gonosomatic index, clearly indicating more developed testes, than territorial males who spend less energy on sperm production and more energy on defending the nest site and attracting mates (Mazzoldi, 1999; Scaggiante *et al.*, 1999; Rasotto and Mazzoldi, 2002). Other intra-specific differences, such as the size and the function of male accessory structures like SDGs and MGs, also occur in gobies. Indeed, in two species, the black goby *Gobius niger* and the grass goby *Zosterisessor ophiocephalus*, in which large territorial males monopolize females, build and/or defend a nest and provide egg parental care, while smaller and younger males attempt to parasitize parental male spawning (Mazzoldi, 1999; Mazzoldi *et al.*, 2000; Mazzoldi and Rasotto, 2002), territorial males exhibit larger SDGs and more abundant secretions (Fig. 3.3.5A) than sneaker males which adopt opportunistic spawning tactics and whose reduced SDGs mainly function as sperm storage organs (Fig. 3.3.5B; Scaggiante *et al.*, 1999; Mazzoldi *et al.*, 2000; Rasotto and Mazzoldi, 2002). A similar variability in SDGs, with those of parental males being well-developed and filled with mucins, and those of sneakers being smaller and used to store sperm rather than to produce mucins, also occurs in the bluebanded goby, *Lythrypnus dalli*, a species in which the presence of alternative male mating tactics has been hypothesized (Drilling and Grober, 2005).

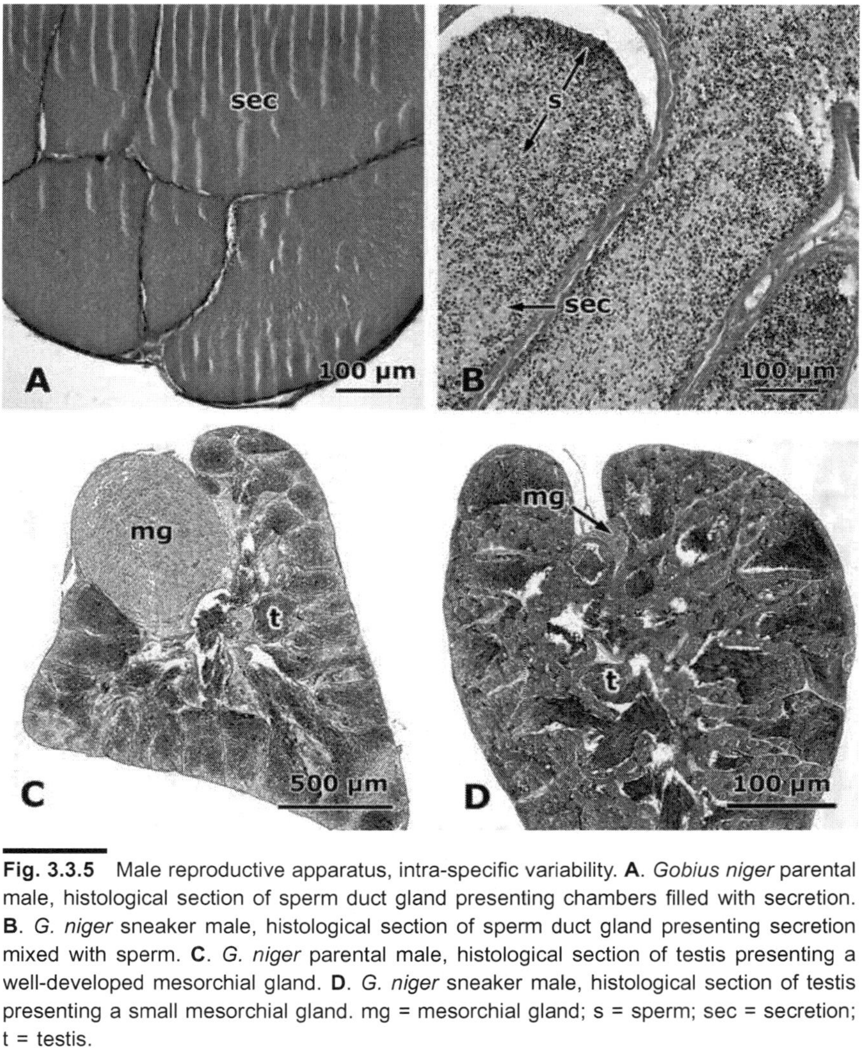

Fig. 3.3.5 Male reproductive apparatus, intra-specific variability. **A**. *Gobius niger* parental male, histological section of sperm duct gland presenting chambers filled with secretion. **B**. *G. niger* sneaker male, histological section of sperm duct gland presenting secretion mixed with sperm. **C**. *G. niger* parental male, histological section of testis presenting a well-developed mesorchial gland. **D**. *G. niger* sneaker male, histological section of testis presenting a small mesorchial gland. mg = mesorchial gland; s = sperm; sec = secretion; t = testis.

In the black goby and the grass goby, the intra-specific variability in SDG development is mirrored by differences in sperm trail performance. Those of territorial males, having relatively smaller testes and larger SDGs that produce more mucins, last longer but contain fewer sperm than those of sneaker males (Scaggiante *et al.*, 1999; Mazzoldi *et al.*, 2000; Rasotto and Mazzoldi, 2002). Sneaker males, with large testes and reduced SDGs storing sperm, produce sperm trails poorer in mucins, but with higher sperm density and a shorter duration than those of parental males (Mazzoldi *et al.*, 2000; Rasotto and Mazzoldi, 2002). These characteristics confer different qualities to ejaculates, with parental male trails guaranteeing a low but steady supply of sperm for hours, while sneaker males release a high and quickly diffusing number of sperm that outnumber those of parental males (on an average 5.2 : 1 in the grass goby and 10.4 : 1 in the black goby), even if only for a short time (Scaggiante *et al.*, 1999; Mazzoldi *et al.*, 2000; Rasotto and Mazzoldi, 2002).

In the black goby, the development of MGs is also influenced by the tactic the male adopts. Territorial males have significantly larger MGs than opportunistic sneaker males (Rasotto and Mazzoldi, 2002; Immler *et al.*, 2004) (3.2.5C, D). The pheromones produced by the MGs are released with both the urine and the ejaculate. Locatello *et al.* (2002) experimentally demonstrated that territorial males react to the ejaculate of other territorial males, with stereotypic aggressive behaviours, but not to the ejaculate of sneakers. Since sneaker male ejaculates appear to be pheromonally inconspicuous, it is tempting to speculate not only that the differential effects of territorial and sneaker ejaculates result from differential steroid glucuronides content, but also that altered pheromone function might be an important aspect of alternative male tactics in many species. Indeed sneaker males often resemble females, and the lack of chemical signals that could make them recognizable as potential competitors may minimize their chances of being detected by parental males, increasing their opportunities of parasitic spawning. Support for this hypothesis comes from experiments on change in mating tactic. One 2004 study demonstrated that sneakers are physiologically able to develop, in a few weeks, an MG, comparable in size to that of parental males, as consequence of transition from sneaker to parental tactic (Immler *et al.*, 2004).

The grass goby and the black goby have also provided excellent models to test theoretical predictions on ejaculate expenditure and sperm quality traits in response to sperm competition level. We have detailed information on their investment in testes and accessory structures and of the ejaculate performance of both these species. Both also exhibit high levels of sperm competition, with up to six sneakers congregating around the nest of territorial males and releasing their sperm when females spawn (Mazzoldi, 1999; Mazzoldi *et al.*, 2000; Mazzoldi and Rasotto, 2002). Since ejaculates are energetically costly to produce, males are expected and have been found to allocate sperm strategically, in terms of both number and quality, in response to varying levels of sperm competition (Parker, 1993, 1998; Parker *et al.*, 1996; Ball and Parker, 1996; Wedell *et al.*, 2002; Moore *et al.*, 2004).

Models dealing with sperm number allocation, in particular the so-called "intensity model" predict that an individual male that faces variable levels of sperm competition among successive spawns should release fewer sperm as the estimated number of competitors at a given spawning increases above two, because returns are diminishing for providing more sperm (Parker *et al.*, 1996; Parker, 1998). In other words, if a male can strategically allocate sperm among spawns, an increase in output will profit more when the intensity of sperm competition is low. The intensity model was first tested in the grass and black gobies, where, in accordance with theory, sneaker males of both species were found to release fewer sperm as the number of competitors increased (Pilastro *et al.*, 2002). Interestingly, territorial males did not modulate sperm expenditure but instead adjusted their effort in nest guarding according to the numbers of attending sneakers, by increasing the time spent patrolling the territory and attacking the competitors (Scaggiante *et al.*, 2005).

Theoretical models of sperm quality predict that the whole ejaculate expenditure, defined as the product of sperm number and any ejaculate property that is related to quality, covaries with the level of sperm competition (Ball and Parker, 1996). However, given that resources available to reproduction may be limited, trade-offs, between either different measures of sperm quality or sperm quality and number, are expected (Parker, 1993; Ball and Parker, 1996; Moore *et al.*, 2004). Gobies displaying alternative male mating tactics, with male ejaculates differing in both sperm number and seminal fluid content, are a perfect subject for investigating how sperm traits vary in response to different levels of sperm competition. An experimental study on the grass and black gobies showed that grass goby male sperm have a similar quality, whereas black goby sneaker sperm have faster swimming speeds, greater viability and a higher ATP content (about twice), than those of territorial males (Locatello *et al.*, 2007). This difference in the intra-specific variability of sperm quality between the two species has been attributed to differences in the effectiveness of territorial male mate guarding (Locatello *et al.*, 2007). The grass goby nest, a muddy chamber dug under sea grass rhizomes, has multiple entrances, enabling sneaker males to avoid a guarding territorial male and to remain inside the nest, releasing sperm in close proximity to the female (Mazzoldi *et al.*, 2000). In addition, inside the nest the degree of water mixing is low and, consequently, dilution of sperm is limited (Mazzoldi *et al.*, 2000). By contrast, black goby territorial male, which nest in rocky caves or under stones (Mazzoldi and Rasotto, 2002), may be better able to control the single nest entrance, forcing sneakers to release sperm at a greater distance from the females, which entails rapid sperm dilution (Mazzoldi, 1999). Indeed, absolute sperm longevity is higher and absolute sperm number is lower in the grass goby than in the black goby, as expected if the latter's sperm are dispersed more rapidly (Scaggiante *et al.*, 1999; Mazzoldi *et al.*, 2000; Rasotto and Mazzoldi, 2002). However, territorial males of both these goby species respond to the presence of sneakers only by increasing mate-guarding and not by adjusting their ejaculate effort (Scaggiante *et al.*, 2005), even though territorial black gobies are probably more effective in mate guarding than are territorial grass gobies. When shifting from sneaky to territorial tactics (Mazzoldi *et al.*, 2000; Rasotto and Mazzoldi, 2002; Immler *et al.*, 2004; Scaggiante *et al.*, 2004), *G. niger* males might successfully invest in traits favouring mate acquisition at the expense of both sperm number and quality, whereas *Z. ophiocephalus* territorial males, despite reducing their sperm number, might be constrained in maintaining sperm quality at the same level as sneakers by the low effectiveness of their mate guarding.

INTER-SPECIFIC VARIABILITY IN MALE REPRODUCTIVE APPARATUS

The influence of sperm competition in shaping testis size, with species experiencing higher levels of sperm competition exhibiting larger testes, has been demonstrated in several taxa (Birkhead and Møller, 1998) including fish (Stockley *et al.*, 1997; Petersen and Warner, 1998). Unfortunately information available on gobies is too limited to test if this pattern is repeated in this family. In contrast the inter-specific variability of SDGs has been the focus of several studies and the selective pressures acting on it are better known. Ever since the first inter-specific descriptions of SDG shape and development, their extreme variability was evident (Miller, 1984; Fishelson, 1991). In general, SDGs may be extended and leaf-like, or more elongated, with or without finger-like extensions, or partitioned in different regions (Miller, 1984; Fishelson, 1991). In the paedomorphic *Schindleria prematura* the SDGs appear reduced to a single chamber (Thacker and Grier, 2005), while in other paedomorphic species, such as *Aphia minuta* or *Crystallogobius linearis*, several chambers are present (Caputo *et al.*, 2000, 2003). In *Gobiodon okinawae*, *G. oculolineatus*, *G. quinquestrigatus* and *Paragobiodon echinocephalus* the occurrence of two pairs of SDGs has been observed (Cole, 1990; Cole and Hoese, 2001). These two pairs of glands both produce a secretion that positively reacts to PAS; however they present a different appearance, with one pair having fibrillar-granular secretion and columnar epithelium, while the other produces a colloidal secretion and has squamous-to-low-cuboidal epithelium (Cole, 1990; Cole and Hoese, 2001; Cole, 2009). The presence of two pairs of SDGs is, apparently, unusual among gobies. However, a partition in two regions of the SDG has been reported for *Pomatoschistus* spp. (Miller, 1984), although information on differences in secretory activity or epithelium appearance are lacking.

Gobiodon species live, mate and release eggs on coral surfaces, where toxic substances are present (Kuwamura *et al.*, 1993a; Nakashima *et al.*, 1996); consequently the mucosubstances produced by the both male and female SDGs could be involved in the protection of individuals and/or eggs from such toxic compounds. In *G. okinawae*, *G. oculolineatus*, females have also been found to present a well-developed pair of oviduct accessory structures, suggesting that those of males may serve function/s additional to the production of seminal fluid (Cole and Hoese, 2001; Cole, 2009).

SDGs have been found to secrete sialomucins and proteins in all species investigated thus far, but in some species sperm has been found to mix with the secretion in the chamber lumina (Fishelson, 1991; Caputo *et al.*, 2000, 2003). However, in these latter species, the possible presence of alternative male mating tactics and the mating tactic adopted by the analyzed males have not been assessed. As a result it is unclear whether these species represent an additional example of intra-specific variability in SDG function in relation to mating modalities or if, in some goby species, SDGs play the dual role of mucin production and sperm storage. The production of steroids in the SDG has been demonstrated in only a few species (Asahina *et al.*, 1989; Cinquetti, 1997; Jasra *et al.*, 2007) and deserves further investigation.

The widely reported variability in SDG development may be the result of different selective pressures. As SDG mucous secretion is involved in ejaculate structure, egg protection from pathogens and female attraction, the development of these organs may be affected by several factors, such as type of mating system, sperm competition level, nest environmental condition, mate availability, etc. At the intra-specific level, SDG development has been proved to be influenced by sperm competition, with males exhibiting larger SDGs releasing longer-lasting sperm trails and devoting more time to nest defence against opportunistic competitors (Mazzoldi, 1999; Scaggiante *et al.*, 1999; Mazzoldi *et al.*, 2000; Rasotto and Mazzoldi, 2002). However, if the release of longer lasting sperm trails favours sperm economy and permits an increase in nest guarding, SDG development might be expected to be influenced, at the inter-specific level, by factors such as, for instance, egg deposition time and degree of polygyny, as well as sperm competition level.

A comparative analysis of 12 goby species showed that the development of SDGs in parental males appears to be influenced by the mating system type and, in particular, by the degree of polygyny (Mazzoldi *et al.*, 2005). Indeed SDGs were absent or rudimentary in the monogamous species, while they were well-developed in the polygynous ones. In monogamous species, males usually maintain the pair bond for several consecutive spawning events, through which they fertilize just one egg batch at a time, and extra-pair spawning has never been observed (Reavis, 1997; Takegaki and Nakazono, 1999; Takegaki, 2000; Mazzoldi, 2001). In addition, the nest entrance is plugged during spawning (Takegaki and Nakazono 1999), thus isolating the mating pair from outside, hence drastically reducing not only predation risk but also sperm dispersal and the potential for sperm competition. Consequently, males of these species might not need either a large sperm supply or mucus, and in this respect it is noteworthy that they lack SDGs and show a semi-cystic type of spermatogenesis (Mazzoldi, 2001). In contrast, parental males of polygynous species fertilize eggs from several females, sometimes simultaneously (Mazzoldi *et al.*, 2000; Mazzoldi and Rasotto, 2002). Consequently, they may need a larger gamete supply and an increase in sperm retention at spawning can be particularly rewarding for them (Mazzoldi *et al.*, 2005).

These results should be considered indicative and not exhaustive of the possible influences of reproductive modalities on goby SDG size and secretion. Indeed the number of species considered in this study was limited (n=12), and sperm competition occurs in only a few of them (n=3) possibly explaining the observed lack of influence on SDG size. SDGs are also known to occur in some monogamous gobies, despite their size not being quantified (Cole, 1990; Cole and Hoese, 2001). Moreover, some of the examined species exhibited larger seminal vesicles than expected if only the type of mating system was affecting their development. This suggests that other factors, such as nest compacting and/or egg protection, might influence their size (Mazzoldi *et al.*, 2005). Indeed, in one of these species, *Zosterisessor ophiocephalus*, a goby nesting in mud (Mazzoldi *et al.*, 2000), SDG secretion was proven to contain antimicrobials with a specific effect against bacteria living in marine sediment (Giacomello *et al.*, 2008). It is also possible that males of this species use mucus to build the nest since its walls are particularly compact and viscous to the touch (C. Mazzoldi, pers. observation).

HERMAPHRODITIC GONADS

Around 88% of fish species are gonochorists, whilst the remaining 12%, comprising 27 out of 448 teleost families, are hermaphrodites: most of them protogynous and the others simultaneous hermaphrodites or protandrous (Patzner, 2008; for a review see de Mitcheson and Liu, 2008). In gobies, the most common sexual pattern is gonochorism. Protogynous hermaphroditism, including bi-directional sex change, also occurs (Patzner, 2008; reviewed by de Mitcheson and Liu, 2008). This last is difficult to prove as exemplified by *Lythrypnus dalli*, for which both simultaneous (Fishelson 1989; St. Mary, 1993) and protogynous hermaphroditism have been proposed (Reavis and Grober, 1999; Drilling and Grober, 2005). However, recent studies have demonstrated both female-to-male and male-to female sex change in known individuals (Rodgers *et al.*, 2005; Lorenzi *et al.*, 2006). In gobies, functional hermaphroditism has been reported for species of the genera *Coryphopterus* (Robertson and Justines, 1982; Cole and Robertson, 1988; Cole and Shapiro, 1990, 1992), *Eviota* (Cole, 1990), *Paragobiodon* (Cole, 1990, Kuwamura *et al.*, 1994), *Gobiodon* (Cole, 1990; Nakashima *et al.*, 1996; Munday *et al.*, 1998; Cole and Hoese, 2001), *Priolepis* (Cole, 1990; Sunobe and Nakazono, 1999), *Trimma* (Cole, 1990; Sunobe and Nakazono, 1993; Kobayashi *et al.*, 2005), *Lythrypnus* (Cole, 1988; St Mary, 1993, 1996, 1998, 2000), and in *Tigrigobius multifasciatus* (Robertson and Justines, 1982). This last species represents an exception. Within the genus *Tigrigobius* and the whole *Gobiosoma* group to which the genus belongs hermaphroditism has been documented only in *T. multifasciatus* (Cole, 2008), while for all the other studied species hermaphroditism is widespread in the genus.

Confirming sex change in gobiid fishes has been problematic because of the great variability in testicular and ovarian tissue presence and organization both before and during sex change (Cole, 1990). In *Coryphopterus* species, testicular tissue, in the form of small cysts of spermatocytes, appears within the ovarian stroma only during sex change and subsequently replaces all ovarian tissue. In contrast, in *Trimma unisquamis* presumptive testicular tissue is commonly present along the ovary wall during the female phase, hence long before sex change occurs and, in *Paragobiodon hipoliti*, the gonad comprises both testicular and ovarian tissues that are topographically separated and appear to be active simultaneously for some period of time before sex change is completed (Cole, 1990). Within the genus *Lythrypnus*, individuals, varying from pure female to pure male, present both gonadal tissues (St Mary, 1993, 2000). In addition, the testes of protogynous gobies do not always reflect the preceding ovarian morphology as occurs, for instance, in wrasses and parrotfish (de Mitcheson and Liu, 2008). Indeed, in these species the ovarian cavity disappears soon after sex change and sperm are not collected by peripheral sperm sinuses but typically exit the body via a newly formed anastomosing network of tubules lined with spermatogenic cysts (Cole, 1983, 1990; Cole and Robertson, 1988).

A morphological feature, associated with the gonads, that has proved a useful criterion for the diagnosis of hermaphroditism in this family is the presence of the 'precursive accessory gonadal structures' (pAGSs) (Cole 1988, 1990; Cole and Robertson, 1988; Cole and Shapiro, 1990). pAGSs have been detected in conjunction with the ovaries of functional females in protogynous species. During sex change these structures develop into a pair of secretory organs, lateral to the sperm duct and hence correspond to the SDG. Precursive SDG structures have also been described in immature females of functionally gonochoric species and all the gonochoric species presenting precursive SDG structures belong to phylogenetic groups in which functional hermaphroditism is reported, therefore suggesting hermaphroditism as the ancestral status of the group (Cole, 2008).

The observed differences in gonadal morphology among hermaphrodite gobies has been argued follow phylogenetic lines (Cole, 1990; St. Mary, 1998). In contrast, the occurrence of functional hermaphroditism in this family and the type of hermaphorditism exhibited, i.e., protogyny rather than bi-directional sex change, appear to be linked to the life history traits of the species and represent adaptive responses to the social-ecological context individuals experience (Munday *et al.*, 2006; Herler *et al.*, this volume).

CONCLUDING REMARKS

The experimental studies on various aspects of the functional morphology of goby reproductive apparatus were fundamental to understanding the modalities of fertilization dynamics and the influences of sexual selection in shaping testis size, accessory organ presence, size and function. These findings already serve as models; indeed similar patterns of testis and accessory structure variability, as well as fertilization mode, have been found in other demersal spawner species (Barni *et al.*, 2001; Neat *et al.*, 2003; Petersen *et al.*, 2004; Mazzoldi *et al.*, 2007; Giacomello *et al.*, 2008). By expanding the comparative studies to other goby species, we believe that the inter-specific and additional intra-specific differences that may be found in organization, size, and function of gonads and accessory structures will lead to a better understanding of the role of these organs in the evolution of modalities of reproduction and mating systems of teleosts in general. In addition, the differences in testis, sperm number and SDG allocation, characterizing species with alternative male mating tactics, might prove a useful tool to test the possible influence of environmental changes, such as variability in temperature and salinity, on the relative abundance of male phenotype and, consequently, their impact on species reproduction. The goby may, therefore, be expected to play a future role as model species in studies dealing with the effects of climate change.

References

Arai, R. 1964. Sex characters of Japanese gobioid fishes (1). *Bulletin of the National Science Museum (Tokyo)* 7: 295-306.

Arbuckle, W.J., A.J. Bélanger, L.D. Corkum, B.S. Zielinski, W. Li, S.-S. Yun, S. Bachynski and A.P. Scott. 2005. In vitro biosynthesis of novel 5β-reduced steroids by the testis of the round goby, *Neogobius melanostomus*. *General and Comparative Endocrinology* 140: 1-13.

Asahina, K., K. Suzuki, T. Hibiya and B. Tamaoki. 1985. Relationship between the structure and steroidogenic functions of the testes of the urohaze-goby (*Glossogobius olivaceus*). *General and Comparative Endocrinology* 57: 281-292.

Asahina, K., K. Suzuki, T. Hibiya and B. Tamaoki. 1989. Structure and steroidogenic enzymes of the seminal vesicles of the urohaze-goby (*Glossogobius olivaceus*). *General and Comparative Endocrinology* 74: 385-391.

Auty, E.H. 1978. Reproductive behavior and early development of the empire fish, *Hypseleotris compressus* (Eleotridae). *Australian Journal of Marine and Freshwater Research* 29: 585-597.

Ball, M.A. and G.A. Parker. 1996. Sperm competition games: external fertilization and 'adaptive' infertility. *Journal of Theoretical Biology* 180: 141-150.

Barni, A., C. Mazzoldi and M.B. Rasotto. 2001. Reproductive apparatus and male accessory structures in two batrachoid species (Teleostei, Batrachoididae). *Journal of Fish Biology* 58: 1557-1569.

Birkhead, T.R. and A.P. Møller. 1998. *Sperm Competition and Sexual Selection*. Academic Press, London.

Bouchereau, J.L., J.P. Quignard, J.C. Joyeux and J.A. Tomasini. 1991. Stratégies et tactiques de reproduction de *Pomatoschistus microps* (Krøyer, 1838) et de *Pomatoschistus minutus* (Pallas, 1770) (Pisces, Gobiidae) dans le Golfe du Lion (France). Nids, déterminismes de la sédentarité et de la migration. *Cybium* 15: 315-346.

Caputo, V., C. Candi, M. La Mesa and E. Arneri. 2000. Pattern of gonad maturation and the question of semelparity in the paedomorphic goby *Aphia minuta*. *Journal of Fish Biology* 58: 656-669.

Caputo, V., C. Candi, M. La Mesa and P.N. Cerioni. 2003. The reproductive biology of the crystal goby with a comparison to that of the transparent goby. *Journal of Fish Biology* 62: 375-385.

Carcupino, M., A. Baldacci, G. Corso, P. Franzoi, M. Pala and M. Mazzini. 1999. Testis structure and symplastic spermatid formation during spermatogenesis of pipefishes. *Journal of Fish Biology* 55: 344-353.

Chambers, R.C. 1997. Environmental influences on egg and propagule sizes in marine fishes. In: *Early Life History and Recruitment in Fish Population*, R.C. Chambers and L.E.A. Trippel (eds.), Chapman & Hall, London, pp. 63-102.

Cinquetti, R. 1997. Histochemical, enzyme histochemical and ultrastructural investigation on the sperm-duct glands of *Padogobius martensi* (Pisces, Gobiidae). *Journal of Fish Biology* 50: 978-991.

Cinquetti, R. and L. Rinaldi. 1987. Changes in gonadal histology of *Padogobius martensi* (Pisces: Gobiidae) during the reproductive cycle. *Bollettino di Zoologia* 54: 233-241.

Cinquetti, R. and L. Dramis 2003. Histological, histochemical, enzyme histochemical and ultrastructural investigations of the testis of *Padogobius martensi* between annual breeding seasons. *Journal of Fish Biology* 63: 1402-1428.

Cinquetti, R., L. Rinaldi, C. Azzoni and A. De Ascentiis. 1990. Cycling changes in the morphology of the sperm-duct glands of *Padogobius martensi* (Pisces, Gobiidae). *Bollettino di Zoologia* 57: 295-302.

Cole, L.C. 1954. The population consequences of life history phenomena. *Quarterly Review of Biology* 29: 103-137.

Cole, K.S. 1983. Protogynous hermaphroditism in a temperate zone territorial marine goby, *Coryphopterus nicholsi. Copeia* 1983(3): 809-812.

Cole, K.S. 1988. Predicting the potential for sex-change on the basis of ovarian structure in gobiid fishes. *Copeia* 1988(4): 1082-1086.

Cole, K.S. 1990. Patterns of gonad structure in hermaphroditic gobies (Teleostei: Gobiidae). *Environmental Biology of Fishes* 28: 125-142.

Cole, K.S. 2008. Transient ontogenetic expression of hermaphroditic gonad morphology within the *Gobiosoma* group of the Neotropical seven-spined gobies (Teleostei: Gobiidae). *Marine Biology* 154: 943-951.

Cole, K.S. 2009. Modifications of the reproductive complex and implications for the reproductive biology of *Gobiodon oculolineatus* (Teleostei: Gobiidae). *Environmental Biology of Fishes* 84: 261-273.

Cole, K.S. and D.R. Robertson. 1988. Protogyny in the Caribbean reef goby, *Coryphopterus personatus*: Gonad ontogeny and social influence on sex-change. *Bulletin of Marine Science* 42: 317-333.

Cole, K.S. and D.Y. Shapiro. 1990. Gonad structure and hermaphroditism in the gobiid genus *Coryphopterus* (Teleostei: Gobiidae). *Copeia* 1990(4): 996-1003.

Cole, K.S. and D.S. Hoese. 2001. Gonad morphology, colony demography and evidence for hermaphroditism in *Gobiodon okinawae* (Teleostei, Gobiidae). *Environmental Biology of Fishes* 61: 161-173.

Colombo, L. and P. Burighel. 1974. Fine structure of the testicular gland of the black goby, *Gobius jozo* L. *Cell and Tissue Research* 154: 39-49.

Colombo, L., C. Lupo di Prisco and G. Binder. 1970. Metabolism of pregnenolone-4-^{14}C by the testis of *Gobius paganellus* (Teleostei). *General and Comparative Endocrinology* 15: 404-419.

Colombo, L., P. Colombo Belvedere and A. Pilati. 1977. Biosynthesis of free and conjugated 5β-reduced androgens by the testis of the black goby, *Gobius jozo* L. *Bollettino di Zoologia* 44: 131-134.

Colombo, L., A. Marconato, P. Colombo Belvedere and C. Friso. 1980. Endocrinology of teleost reproduction: a testicular steroid pheromone in the black goby, *Gobius jozo* L. *Bollettino di Zoologia* 47: 355-364.

de Mitcheson, Y.S. and M. Liu. 2008. Functional hermaphroditism in teleosts. *Fish and Fisheries* 9: 1-43.

de Vlaming, V. 1972. Reproductive cycling in the estuarine gobiid fish, *Gillichthys mirabilis*. *Copeia* 1972(2): 278-291.

Dewsbury, D.A. 1982. Ejaculate cost and male choice. *The American Naturalist* 119: 601-610.

Drilling, C.C. and M.S. Grober. 2005. An initial description of alternative male reproductive phenotypes in the bluebanded goby, *Lythrypnus dalli* (Teleostei, Gobiidae). *Environmental Biology of Fishes* 72: 361-372.

Egami, N. 1960. Comparative morphology of the sex characters in several species of Japanese gobies with reference to the effect of sex-steroids on the character. *Journal of the Faculty of Science, University of Tokyo, Section IV, Zoology* 9: 67-100.

Eggert, B. 1931. Die Geschlechtsorgane del Gobiiformes und Blenniiformes. *Zeitschrift für wissenschaftliche Zoologie* 139: 249-517.

Fishelson, L. 1989. Bisexuality and pedogenesis in gobies (Teleostei: Gobiidae) and other fish, or: Why so many little fish in tropical areas. *Senckenbergiana Maritima* 20: 147-160.

Fishelson, L. 1991. Comparative cytology and morphology of seminal vesicles in male gobiid fish. *Japanese Journal of Ichthyology* 38: 17-30.

Fishelson, L., R.N. Gibson and Y. Delarea. 1990. Unusual cell organelles during spermiogenesis in two species of gobies (Gobiidae, Teleostei). *Cell and Tissue Research* 262: 397-400.

Giacomello, E., F.C. Neat and M.B. Rasotto. 2008. Mechanisms enabling sperm economy in blenniid fishes. *Behavioral Ecology and Sociobiology* 62: 671-680.

Giacomello, E., L. Marri, D. Marchini, C. Mazzoldi and M.B. Rasotto. 2008. Sperm duct gland secretion of the grass goby *Zosterisessor ophiocephalus* exhibits antimicrobial activity. *Journal of Fish Biology* 73: 1823-1828.

Giulianini, P.G. and E.A. Ferrero. 2001. Ultrastructural aspects of the ovarian follicle and egg envelope of the sea-grass goby *Zosterisessor ophiocephalus* (Osteichthyes, Gobiidae). *Italian Journal of Zoology* 68: 29-37.

Giulianini, P.G., D. Ota, M. Marchesan and E.A. Ferrero. 2001. Can goby spermatozoa pass through the filament adhesion apparatus of laid eggs? *Journal of Fish Biology* 58: 1750-1752.

Giulianini, P.G., M. Balestra, A. di Marcotullio, E.A. Ferrero, D. Ota and R.A. Patzner. 1989. Seminal vesicle secretion in *Zosterisessor ophiocephalus* (Pisces, Gobiidae): seasonal cycle and biochemical characterization. *Nova Thalassia* 10: 651-652.

Goldberg, S.E. 1977. Seasonal ovarian cycle of the tidwater goby, *Eucyclogobius newberryi*. *The Southwestern Naturalist* 22: 557-559.

Grier, H.J. 1981. Cellular organization of the testis and spermatogenesis in fishes. *American Zoologist* 21: 345-357.

Grier, H.J. 1993. Comparative organization of Sertoli cells including the Sertoli cells barrier. In: *The Sertoli Cells*, L.D. Russel and M.D. Griswold (eds.), Chace River Press, Clearwater, FL, pp. 703-739.

Grier, H.J. 2000. Ovarian germinal epithelium and folliculogenesis in the common snook *Centropomus undecimalis* (Teleostei: Centropomidae). *Journal of Morphology* 243: 265-281.

Grier, H.J., J.R. Linton, J.F. Leatherland and V.L. deVlaming. 1980. Structural evidence for two difference testicular types in teleost fishes. *American Journal of Anatomy* 159: 331-345.

Hara, M. 2004. Diversity of the sperm morphology in gobioid fishes. *Japanese Journal of Ichthyology* 51: 1-32.

Hara, M. and M. Okiyama. 1998. An ultrastructural review of the spermatozoa of Japanese fishes. *Bulletin of the Ocean Research Institute, University of Tokyo* 33: 1-138.

Herler, J., R.A. Patzner and H. Hilgers. 2006. Notes on the egg envelope and number of oocytes in two *Gobius* species (Teleostei: Gobiidae) from the northern Adriatic. *Zeitschrift für Fischkunde* 8: 1-7.

Hidaka, T. and S. Takahashi. 1987. Effects of temperature and daylength on gonadal development of a goby, *Rhinogobius brunneus* (orange type). *Japanese Journal of Ichthyology* 34: 361-367.

Hoar, W.S. 1969. Reproduction. In: *Fish Physiology, Vol. III*, W.S. Hoar and D.J. Randall (eds.), Academic Press, New York, pp. 1-72.

Hoda, S.M.S. and Y. Akhtar. 1985. Maturation and fecundity of the mudskipper *Boleophthalmus dussumeri* in the northern Arabian Sea. *Indian Journal of Fisheries* 32: 64-67.

Immler, S., C. Mazzoldi and M.B. Rasotto. 2004. From sneaker to parental male: change of reproductive traits in the black goby, *Gobius niger* (Teleostei, Gobiidae). *Journal of Experimental Zoology* 301A: 177-185.

Jamieson, B.G.M. 1991. *Fish Evolution and Systematics: Evidence from Spermatozoa.* Cambridge University Press, Cambridge.

Jasra, S.K., W.J. Arbuckle, L.D. Corkum, W. Li, A.P. Scott and B. Zielinski. 2007. The seminal vesicle synthesizes steroids in the round goby *Neogobius melanostomus*. *Comparative Biochemistry and Physiology, Part A* 148: 117-123.

Katoh, M. and M. Nishida. 1994. Biochemical and egg size evolution of freshwater fishes in the *Rhinogobius brunneus* complex (Pisces, Gobiidae) in Okinawa, Japan. *Biological Journal of the Linnean Society* 51: 325-335.

Kobayashi. Y., T. Sunobe, T. Kobayashi, Y. Nagahama and M. Nakamura. 2005. Gonadal structure of the serial-sex changing gobiid fish *Trimma okinawae*. *Development, Growth and Differentiation* 47: 7-13.

Kovačić, M. 2007. Reproductive biology of the striped goby, *Gobius vittatus* (Gobiidae) in the northern Adriatic Sea. *Scientia Marina* 71: 145-151.

Kramer, A. and R.A. Patzner. 2008. Morphological and developmental studies on the eggs of four *Coryphopterus* species (Actinopterygii: Gobiidae) in the Caribbean Sea. *Acta Ichthyologica et Piscatoria* 38: 97-102.

Kuwamura, T., Y. Nakashima and Y. Yogo. 1993. Size-assortative monogamy and parental care in a coral goby *Paragobiodon echinocephalus*. *Ethology* 95: 65-75.

Kuwamura, T., Y. Nakashima and Y. Yogo. 1994. Sex change in either directions by growth-rate advantage in the monogamous coral goby, *Parogobidon echinocephalus*. *Behavioural Ecology* 5: 434-438.

Lahnsteiner, F. and R.A. Patzner. 1990. Spermiogenesis and structure of mature spermatozoa in blenniid fishes (Pisces, Blenniidae). *Journal of Submicroscopic Cytology and Pathology* 22: 565-576.

Lahnsteiner, F. and R.A. Patzner. 2009. Male reproductive system: spermatic duct and accessory organs of the testis. In: *Reproductive Biology and Phylogeny of Fishes (Agnathans and Bony Fishes)*, B.G.M. Jamieson (ed.), Science Publishers, Enfield, pp. 143-186.

Lahnsteiner, F., M. Seiwald, R.A. Patzner and E.A. Ferrero. 1992. The seminal vesicles of the male grass goby *Zosterisessor ophiocephalus* (Teleostei, Gobiidae). *Zoomorphology* 111: 239-248.

Locatello, L., C. Mazzoldi and M.B. Rasotto. 2002. Ejaculate of sneaker males is pheromonally inconspicous in the black goby, *Gobius niger* (Teleostei, Gobiidae). *Journal of Experimental Zoology* 293: 601-605.

Locatello, L., A. Pilastro, R. Deana, A. Zarpelloni and M.B. Rasotto. 2007. Variation pattern of sperm quality traits in two gobies with alternative mating tactics. *Functional Ecology* 21: 975-981.

Lorenzi, V., R.L. Earley and M.S. Grober. 2006. Preventing behavioural interactions with a male facilitates sex change in female bluebanded gobies, *Lythrypnus dalli*. *Behavioral Ecology and Sociobiology* 59: 715-722.

Manni, L. and M.B. Rasotto. 1997. Ultrastructure and histochemistry of the testicular efferent duct system and spermiogenesis in *Opistognathus whitehurstii* (Teleostei, Trachinoidei). *Zoomorphology* 117: 93-102.

Marconato, A. 1980. Pheromones and sexual behaviour in *Gobius jozo* L. and *Carassius auratus* L. PhD thesis, University of Padova, Padova, Italy.

Marconato, A. and D.Y. Shapiro. 1996. Sperm allocation, sperm production and fertilization rates in the bucktooth parrotfish. *Animal Behaviour* 52: 971-980.

Marconato, A., M.B. Rasotto and C. Mazzoldi. 1996. On the mechanism of sperm release in three gobiid fishes (Teleostei: Gobiidae). *Environmental Biology of Fishes* 46: 321-327.

Massironi, M., M.B. Rasotto and C. Mazzoldi. 2005. A reliable indicator of female fecundity: the case of the yellow belly in *Knipowitschia panizzae* (Teleostei: Gobiidae). *Marine Biology* 147: 71-76.

Mattei, X. 1970. Spermiogenèse comparée des poissons. In: *Comparative Spermatology*, B. Baccetti (ed.), Academic Press, New York, pp. 57-69.

Mazzoldi, C. 1999. A comparative study on mating behaviour and fertilization dynamics in demersal spawner teleosts. PhD thesis, University of Padova, Padova, Italy.

Mazzoldi, C. 2001. Reproductive apparatus and mating system in two tropical goby species (Teleostei, Gobiidae). *Journal of Fish Biology* 59: 1686-1691.

Mazzoldi, C. and M.B. Rasotto. 2001. Extended breeding season in the marbled goby, *Pomatoschistus marmoratus* (Teleostei: Gobiidae), in the Venetian Lagoon. *Environmental Biology of Fishes* 61: 175-183.

Mazzoldi, C. and M.B. Rasotto. 2002. Alternative male mating tactics in *Gobius niger*. *Journal of Fish Biology* 61: 157-172.

Mazzoldi, C., C. Poltronieri and M.B. Rasotto. 2002. Egg size variability and mating system in the marbled goby, *Pomatoschistus marmoratus* (Pisces, Gobiidae). *Marine Ecology Progress Series* 233: 231-239.

Mazzoldi, C., C.W. Petersen and M.B. Rasotto. 2005. The influence of mating system on seminal vesicle variability among gobies (Teleostei, Gobiidae). *Journal of Zoological Systematics and Evolutionary Research* 43(4): 307-314.

Mazzoldi, C., V. Lorenzi and M.B. Rasotto. 2007. Variation of male reproductive apparatus in relation to fertilization modalities in the catfish families Auchenipteridae and Callichthyidae (Teleostei: Siluriformes). *Journal of Fish Biology* 70: 243-256.

Mazzoldi, C., M. Scaggiante, E. Ambrosin and M.B. Rasotto. 2000. Mating system and alternative male mating tactics in the grass goby *Zosterisessor ophiocephalus* (Teleostei: Gobidae). *Marine Biology* 13: 1041-1048.

Miller, P.J. 1961. Age, growth, and reproduction of the rock goby, *Gobius paganellus* L., in the Isle of Man. *Journal of the Marine Biological Association of the United Kingdom* 41: 737-769.

Miller, P.J. 1984. The tokology of gobioid fishes. In: *Fish Reproduction: Strategies and Tactics*, G.W. Potts and J.R. Wooton (eds.), Academic Press, London, pp. 119-153.

Miller, P.J. 1992. The sperm duct gland: a visceral synapomorphy for gobioid fishes. *Copeia* 1992: 253-256.

Moiseyeva, Y.B. and V.P. Ponomareva. 1973. Histophysiological characteristics of the testis and seminal vesicles of the round goby (*Gobius melanostomus* Pallas) in different stages of the sexual cycle. *Voprosy Ikhtyologii* 3: 368-381.

Mooi, R.D. 1990. Egg surface morphology of pseudochromoids (Perciformes: Percoidei), with comments on its phylogenetic implications. *Copeia* 1990(2): 455-475.

Moore, P.J., E.W. Harris, T.V. Montrose, D. Levin and A.J. Moore. 2004. Constraints on evolution and postcopulatory sexual selection: trade-offs among ejaculate characteristics. *Evolution* 58: 1773-1780.

Moskalkowa, K.I. 1967. Morfo-ekologicheskie oso bennosti razvitiya bychka-kruglyaka *Neogobius melanostomus* (Pall.). In: *Morfo-ekologicheskiĭ Analiz Razvitiya Ryb*, N.N. Disler (ed.). Nauka, Moskva, pp. 48-75.

Munday, P.L., J. Caley and G.P. Jones. 1998. Bi-directional sex change in a coral-dwelling goby. *Behavioral Ecology and Sociobiology* 43: 371-377.

Munday, P.L., Buston, P.M. and R.R. Warner. 2006. Diversity and flexibility of sex-change strategies in animals. *Trends in Ecology and Evolution* 21: 89-95.

Murphy, C.A., N.E. Stacey and L.D. Corkum. 2001. Putative steroidal pheromones in the round goby, *Neogobius melanostomus*: olfactory and behavioural responses. *Journal of Chemical Ecology* 27: 443-470.

Mutsadi, K.B. and D.V. Bal. 1970. Maturation and spawning of *Boleophthalmus dussumeri*. *Journal of the University of Bombay* 39: 58-76.

Nagahama, Y. 1983. The functional morphology of teleost gonads. In: *Fish Physiology, Vol. IX*, W.S. Hoar, D.J. Randall and E.M. Donaldson (eds.), Academic Press, New York, pp. 223-275.

Nakashima, Y., T. Kuwamura and Y. Yogo. 1996. Both-way sex change in monogamous coral gobies, *Gobiodon* spp. *Environmental Biology of Fishes* 46: 281-288.

Neat, F.C., L. Locatelo and M.B. Rasotto. 2003. Reproductive morphology in relation to alternative male reproductive tactics in *Scartella cristata*. *Journal of Fish Biology* 62: 1381-1391.

Ota, D., M. Marchesan and E.A. Ferrero. 1996. Sperm release behaviour and fertilisation in the grass goby. *Journal of Fish Biology* 49: 246-256.

Parenti, L.R. and H.J. Grier. 2004. Evolution and phylogeny of gonad morphology in bony fishes. *Integrative and Comparative Biology* 44: 333-348.

Parker, G.A. 1990. Sperm competition games: sneaks and extra-pair copulations. *Proceedings of the Royal Society of London. Series B: Biological Sciences* 242: 127-133.

Parker, G.A. 1993. Sperm competition games: sperm size and sperm number under adult control. *Proceedings of the Royal Society of London. Series B: Biological Sciences* 253: 245-254.

Parker, G.A. 1998. Sperm competition and the evolution of ejaculates: towards a theory base. In: *Sperm Competition and Sexual Selection*, T.R. Birkhead and A.P. Møller (eds.). Academic Press, London, pp. 1-54.

Parker, G.A., M.A. Ball, P. Stockley and M.J.G. Gage. 1996. Sperm competition games: individual assessment of sperm competition intensity by group spawners. *Proceedings of the Royal Society of London. Series B: Biological Sciences* 263: 1291-1297.

Patzner, R.A. 1984. The reproduction of *Blennius pavo* (Teleostei, Blennidae) II. Surface structure of the ripe egg. *Zoologischer Anzeiger* 213: 44-50.

Patzner, R.A. 2008. Reproductive strategies of fish. In: *Fish Reproduction: Cytology, Biology and Ecology*, M.J. Rocha, A. Arukwe and B.G. Kapoor (eds.), Science Publisher, Inc., Oxford, pp. 311-350.

Patzner, R.A., M. Seiwald, S. Angerer, E.A. Ferrero and P.G. Giulianini. 1991. Genital system and reproductive cycle of the male grass goby, *Zosterisessor ophiocephalus* (Teleostei, Gobiidae) in the Northern Adriatic Sea. *Zoologischer Anzeiger* 226: 205-219.

Petersen, C.W. and R.R. Warner. 1998. Sperm competition in fishes. In: *Sperm Competition and Sexual Selection*, T.R. Birkhead and A.P. Møller (eds.). Academic Press, London, pp. 435-463.

Petersen, C.W., C. Mazzoldi, Zarrella K.A. and R. Hale. 2004. Fertilization mode, sperm characteristics, mate choice and parental care patterns, in *Artedius* spp. (Cottidae). *Journal of Fish Biology* 67: 239-254.

Pilastro, A., M. Scaggiante and M.B. Rasotto. 2002. Individual adjustment of sperm expenditure accords with sperm competition theory. *Proceedings of the National Academy of Sciences of the United States of America* 99: 9913-9915.

Rasotto, M.B. and C. Mazzoldi. 2002. Male traits associated with alternative reproductive tactics in *Gobius niger*. *Journal of Fish Biology* 61: 173-184.

Reavis, R.H. 1997. The natural history of a monogamous coral-reef fish, *Valenciennea strigata* (Gobiidae): 2. behavior, mate fidelity and reproductive success. *Environmental Biology of Fishes* 49: 247-257.

Reavis, R.H. and M.S. Grober 1999. An integrative approach to sex change: social, behavioural and neurochemical changes in *Lythrypnus dalli* (Pisces). *Acta Ethology* 2: 51-60.

Riehl, R. 1978. Die Oocyten der Grundel *Pomatoschistus minutus*. II. Lichtmikroskopische Untersuchungen zur Kenntnis der Mikropyle. *Microscopica Acta* 80: 287-291.

Riehl, R. 1995. Die Eier und Eihüllen von Knochenfischen. In: *Fortpflanzungsbiologie der Aquarienfische*. H. Greven and R. Riehl (eds.), Birgit Schmettkamp Verlag, Bornheim pp. 11-16.

Riehl, R. and H. Greven. 1993. Fine structure of egg envelopes in some viviparous goodeid fishes with comments on the relation of envelope thinness to viviparity. *Canadian Journal of Zoology* 71: 91-97.

Robertson, D.R. and G. Justines. 1982. Protogynous hermaphroditism and gonochorism in four Caribbean reef gobies. *Environmental Biology of Fishes* 7: 137-142.

Rodgers, E.W., S. Drane and M.S. Grober. 2005. Sex reversal in pairs of *Lythrypnus dalli*: behavioral and morphological changes. *The Biological Bulletin* 208: 120-126.

Santos, Y., F. Pazos and J.L. Barja. 1999. *Flexibacter maritimus*, causal agent of flexibacteriosis in marine fish. In: *ICES Identification Leaflets for Diseases and Parasites of Fish and Shellfish*, No. 55, G. Olivier (ed.). International Council for the Exploration of the Sea, Copenhagen, pp. 1-6.

Scaggiante, M., C. Mazzoldi, C.W. Petersen and M.B. Rasotto. 1999. Sperm competition and mode of fertilization in the grass goby *Zosterisessor ophiocephalus* (Teleostei: Gobiidae). *Journal of Experimental Zoology* 283: 81-90.

Scaggiante, M., M.S. Grober, V. Lorenzi and M.B. Rasotto. 2004. Changes along the male reproductive axis in response to social context in a gonochoristic gobiid, *Zosterisessor ophiocephalus* (Teleostei, Gobiidae), with alternative mating tactics. *Hormones and Behavior* 46: 607-617.

Scaggiante, M., M.B. Rasotto, C. Romualdi and A. Pilastro. 2005. Territorial male gobies respond aggressively to sneakers but do not adjust their sperm expenditure. *Behavioral Ecology* 16: 1001-1007.

Seiwald, M. and R.A. Patzner. 1989. Histological, fine-structural and histochemical differences in the testicular glands of gobiid and blenniid fishes. *Journal of Fish Biology* 35: 631-640.

Soni, V.C. and B. George. 1986. Age determination and length–weight relationship in the mudskipper *Boleophthalmus dentatus*. *Indian Journal of Fisheries* 33: 234-321.

St. Mary, C.M. 1993. Novel sexual patterns in two simultaneously hermaphroditic gobies, *Lythrypnus dalli* and *Lythrypnus zebra*. *Copeia* 1993(4): 1062-1072.

St. Mary, C.M. 1996. Sex allocation in a simultaneous hermaphrodite, the zebra goby *Lythrypnus zebra*: insights gained through a comparison with its sympatric congener, *Lythrypnus dalli*. *Environmental Biology of Fishes* 45: 177-190.

St. Mary, C.M. 1998. Characteristic gonad structure in the gobiid genus *Lythrypnus* with comparisons to other hermaphroditic gobies. *Copeia* 1998(3): 720-724.

St Mary, C.M. 2000. Sex allocation in *Lythrypnus* (Gobiidae): variations on a hermaphroditic theme. *Environmental Biology of Fishes* 58: 321-333.

Stacey, N.E. 2003. Hormones, pheromones and reproductive behaviour. *Fish Physiology and Biochemistry* 28: 229-235.

Stanley, H.P., G. Chieffi and V. Botte. 1965. Histological and histochemical observations on the testis of *Gobius paganellus*. *Zeitschrift für Zellforschung und Mikroskopische Anatomie* 65: 350-362.

Stockley, P., M.J.G. Gage, G.A. Parker and A.P. Møller. 1997. Sperm competition in fishes: the evolution of testis size and ejaculate characteristics. *The American Naturalist* 149: 933-954.

Sunobe, T. and A. Nakazono. 1993. Sex change in both directions by alternation of social dominance in *Trimma okinawae*. *Ethology* 94: 339-345.

Taborsky, M. 1994. Sneakers, satellites, and helpers: parasitic and cooperative behavior in fish reproduction. *Advances in the Study of Behavior* 23: 1-100.

Taborsky, M. 1998. Title: Sperm competition in fish: 'bourgeois' males and parasitic spawning. *Trends in Ecology and Evolution* 13: 222-227.

Taborsky, M. 2008. Alternative reproductive tactics in fish. In: *Alternative Reproductive Tactics. An Integrative Approach*, R. Oliveira, M. Taborsky and H.J. Brokman (eds.), Cambridge University Press, pp. 251-299.

Takahashi, S. 1978. The adherent filaments of the eggs of Isaza (*Chaenogobius isaza* Tanaka, Pisces). *Zoological Magazine (Tokyo)* 87: 216-220.

Takegaki, T. 2000. Monogamous mating system and spawning cycle in the gobiid fish, *Amblygobius phalaena* (Gobiidae). *Environmental Biology of Fishes* 59: 61-67.

Takegaki, T. and A. Nakazono. 1999. Reproductive behavior and mate fidelity in the monogamous goby, *Valenciennea longipinnis*. *Ichthyological Research* 46: 115-123.

Tamada, K. and K. Iwata. 2005. Intra-specific variations of egg size, clutch size and larval survival related to maternal size in amphidromous *Rhinogobius* goby. *Environmental Biology of Fishes* 73: 379-389.

Thacker, C. and H. Grier. 2005. Unusual gonad structure in the paedomorphic teleost *Schindleria praematura* (Teleostei: Gobioidei): a comparison with other gobioid fishes. *Journal of Fish Biology* 66: 378-391.

Toranzo, A.E., B. Magarinos and J.L. Romalde. 2005. A review of the main bacterial fish diseases in mariculture systems. *Aquaculture* 246: 31-67.

Van den Hurk, R. 1973. The localization of steroidogenesis in the testes of oviparous and viviparous teleosts. *Proceedings of the Koninklijke Nederlandse Akademie van Wetenschappen Serie C* 76: 270-279.

Van den Hurk, R. and J.W. Resink. 1992. Male reproductive system as sex pheromone producer in teleost fish. *Journal of Experimental Zoology* 261: 204-213.

Vivien, J.H. 1938. Sue l'existence de glands annexes du tissu genital chez e male de *Gobius paganellus* L. *Comptes Rendus de l'Academie des Sciences. Paris* 206: 938-940.

Vivien, J.H. 1939. Etude préliminaire du tractus genital chez *Gobius paganellus* L. *Travaux de la station zoologique de Wimereux* 13: 713-721.

Wedell, N., M.J.G. Gage and G.A. Parker. 2002. Sperm competition, male prudence and sperm-limited females. *Trends in Ecology and Evolution* 17: 313-320.

Weisel, G.F. 1948 Relation of salinity to the activity of the spermatozoa of *Gillichthys*, a marine teleost. *Physiological Zoology* 21: 40-48.

Yamasaki, N. and K. Tachihara. 2006. Reproductive biology and morphology of eggs and larvae of *Stiphodon percnopterygionus* (Gobiidae: Sicydiinae) collected from Okinawa Island. *Ichthyological Reasearch* 53: 13-18.

Yamasaki, N. and K. Tachihara 2007. Eggs and larvae of *Awaous melanocephalus* (Teleostei: Gobiidae). *Ichthyological Reasearch* 54: 89-91.

Young, R.T. and D.L. Fox. 1937. The seminal vesicles of the goby, with preliminary chemical and physiological studies of the vesicular fluid. *Proceedings of the National Academy of Sciences of the United States of America* 23: 461-467.

Zutshi, B. and P.S. Murthy. 2001. Ultrastructural changes in testis of gobiid fish *Glossogobius giuris* (Ham) induced by fenthion. *Indian Journal of Experimental Biology* 39: 170-173.

CHAPTER

3.4

Early Development of Gobies

Rita Borges[1], Cláudia Faria[2], Fátima Gil[3] and Emanuel J. Gonçalves[4,*]

INTRODUCTION

Gobies are important constituents of coastal larval fish assemblages, reflecting the high density of goby species found in shallow rocky, coral or sandy bottoms of many nearshore habitats both in marine, estuarine and freshwater systems and on temperate and tropical regions. Nevertheless, the early life-history stages of gobies are relatively poorly-known, when considering that this is the second most species-rich family of vertebrates (only cyprinids have more species) and the most speciose of the marine families (Nelson, 2006). This lack of data prevents, at this point, any

Authors' addresses: [1]Centro de Ciências do Mar, Universidade do Algarve and Eco-Ethology Research Unit, Instituto Superior de Psicologia Aplicada, Rua Jardim do Tabaco 34, 1149-041 Lisboa, Portugal. E-mail: rborges@ispa.pt

[2]Centro de Investigação em Educação Faculdade de Ciências da Universidade de Lisboa and Eco-Ethology Research Unit, Instituto Superior de Psicologia Aplicada, Rua Jardim do Tabaco 34, 1149-041 Lisboa, Portugal. E-mail: cbfaria@fc.ul.pt

[3]Aquário Vasco da Gama, Rua Direita, Dafundo, 1495 Lisboa, Portugal. E-mail: fatimagil@mail.telepac.pt

[4]Eco-Ethology Research Unit, Instituto Superior de Psicologia Aplicada, Rua Jardim do Tabaco 34, 1149-041 Lisboa, Portugal. E-mail: emanuel@ispa.pt

[*]*Corresponding author*

meaningful comparison between the different families of the Gobioidei, and therefore this chapter will focus only on the Gobiidae.

General descriptions of early developmental stages (eggs, embryo and larvae) of gobies may be found in Bertolini *et al.* (1956), Russell (1976), Fahay (1983, 2007), Moser *et al.* (1984), Okiyama (1988), Moser (1996), Neira *et al.* (1998), Leis and Carson-Ewart (2000), Munk and Nielsen (2005) and Richards (2006).

Information on the early life-history of the Gobiidae is available for about 49% of the 210 genera and for 11% of the 1950 species recognized (Nelson, 2006) (Table 3.4.1). Considering the five subfamilies, Oxudercinae, Amblyopinae, Sicydiinae, Gobionellinae and Gobiinae, there is little information available for the first three. Embryonic descriptions exist for at least 74 species and larval descriptions for 106 species. The highest number of studied species belongs to the Gobiinae (with information for 127 species), followed by the Gobionellinae (72 species). The early stages of the other three subfamilies remain very poorly investigated, with information for only nine species of the subfamily Oxudercinae, eight Sycidiinae and only one Amblyopinae (Table 3.4.1).

The available information also exhibits an asymmetric geographic distribution with the Japanese waters, the north-eastern Atlantic and the Americas as the best-known regions. In addition, data on the early development of freshwater species is quite scarce.

Here we explore the existing information concerning the early life stages of the Gobiidae and discuss some of the general trends derived from the data. In spite of the limitations expressed above, we believe that the patterns extracted from the data realistically describe the general trends for the different groups and habitats, based on the current information.

MATERIAL AND METHODS

The information used in this review was retrieved from FishBase (Froese and Pauly, 2009) and completed with data from published sources. Each value of the dataset was confirmed using, whenever possible, the primary sources. A word of caution is however needed since some data could not be confirmed due to various reasons (e.g. lack of access to the original material). In those cases, we kept the recorded values as derived from a secondary source.

The basic classification of "environments" from FishBase was used and the species were divided in three groups: marine, freshwater and amphidromous. Due to the focus on the larval stage, some species were

Table 3.4.1 Compiled information of the early life-history characteristics of the Gobioidei and associated literature. Env. = environment: A – amphidromous, F – freshwater, M – marine; Region: T – temperate (including subtropical), TR – tropical; Develop. = duration of the embryonic period; Larval Descr. = existence of larval descriptions; Embr. Descr. = existence of embryonic descriptions (see text for details).

Subfamily	*Species*	*Env.*	*Region*	*Size at Hatching (mm)*	*Size at Settlement (mm)*	*PLD (days)*	*EGG L (mm)*	*EGG W (mm)*	*Develop. (days)*	*Larval Descr.*	*Embr. Descr.*	*References*
Amblyopinae	*Odontamblyopus rubicundus*	A	TR	3.4 (3.3-3.5)			1.30	0.70		+	+	1, 2, 3, 4, 5
Gobiinae	*Acentrogobius pflaumii*	M	T	2.7 (2.6-2.8)		28	1.54 (1.48-1.60)	0.58 (0.55-0.60)				4, 6
Gobiinae	*Acentrogobius viridipunctatus*	A	TR	3.1 (2.5-3.1)		13	2.29 (2.10-2.40)	0.62 (0.60-0.63)	6 (5-7)		+	4, 7, 8
Gobiinae	*Afurcagobius suppositus*	F	T	3.0	9.0					+		9, 10
Gobiinae	*Akko dionaea*	M	TR		9.9					+		11
Gobiinae	*Amblygobius phalaena*	M	TR			29 (21-46)						12
Gobiinae	*Amblygobius rainfordi*	M	TR			35 (23-41)						10, 12, 13, 14, 15
Gobiinae	*Aphia minuta*	M	T	4.0			1.16	0.63		+		7, 10, 16, 17, 18, 19, 20, 21, 22, 23, 24, 25, 26, 27, 28, 29
Gobiinae	*Arenigobius bifrenatus*	M	T	3.1	8.7 (7.4-10.0)					+		9, 10
Gobiinae	*Asterropteryx* semipunctata	M	TR	1.8 (1.6-2.0)			0.76 (0.67-0.84)	0.47 (0.41-0.52)	5 (4-5)		+	4, 30, 31, 32, 33
Gobiinae	*Bathygobius coalitus*	M	TR		9.0 (8.3-9.8)	41 (32-50)						34, 35
Gobiinae	*Bathygobius curacao*	M	TR		5.6					+		11, 36, 37

Contd...

Contd...

Subfamily	*Species*	*Env.*	*Region*	*Size at Hatching (mm)*	*Size at Settlement (mm)*	*PLD (days)*	*EGG L (mm)*	*EGG W (mm)*	*Develop. (days)*	*Larval Descr.*	*Embr. Descr.*	*References*
Gobiinae	*Bathygobius fuscus*	A	TR	2.3			1.80	0.35				4, 38
Gobiinae	*Bathygobius ostreicola*	A	TR				1.80	0.45				39
Gobiinae	*Bathygobius soporator*	M	TR	2.2 (1.9-2.2)	6.4 (5.3-7.0)	29 (26-31)	2.36	0.41	5 (4-7)	+	+	7, 11, 37, 39, 40, 41, 42, 43
Gobiinae	*Bollmannia communis*	M	TR							+		11
Gobiinae	*Bryaninops amplus*	M	TR		19.5 (16.0-23.0)							44
Gobiinae	*Buenia affinis*	M	T							+		10, 20
Gobiinae	*Buenia jeffreysii*	M	T							+		10,18, 24, 45, 46
Gobiinae	*Caffrogobius nudiceps*	M	T				1.80	0.97				7, 39, 47
Gobiinae	*Chromogobius britoi*	M	T			19 (16-22)						48
Gobiinae	*Coryphopterus alloides*	M	TR		9.2							49
Gobiinae	*Coryphopterus bol*	M	TR		11.2 (8.5-13.9)							49
Gobiinae	*Coryphopterus dicrus*	M	TR				0.60	0.40	3		+	10, 50
Gobiinae	*Coryphopterus eidolon*	M	TR	1.7	11.5		1.10	0.40	3		+	10, 49, 50
Gobiinae	*Coryphopterus glaucofraenum*	M	TR		7.3 (7.1-16.0)	27 (23-34)			4			49, 51
Gobiinae	*Coryphopterus hyalinus*	M	TR		8.9 (8.0-9.7)							49

Contd...

Contd...

Gobiinae	*Coryphopterus kuna*	M	TR			63						52
Gobiinae	*Coryphopterus personatus*	M	TR		9.5 (9.1-9.9)				6	+		49, 50, 53
Gobiinae	*Coryphopterus thrix*	M	TR	1.7	11.6		1.10	0.40	3	+	+	10, 49, 50
Gobiinae	*Coryphopterus venezuelae*	M	TR	1.8	13.5		1.10	0.40	3		+	10, 50
Gobiinae	*Crystallogobius linearis*	M	T				1.35 (1.20-1.50)	0.45 (0.40-0.50)				7, 10,17, 18, 19, 24, 26, 29, 39, 54
Gobiinae	*Economidichthys pygmaeus*	F	T	4.4 (4.4-4.7)		43	2.38	0.89			+	4, 55
Gobiinae	*Economidichthys trichonis*	F	T	2.2 (2.2-2.7)			0.64	0.58	1		+	55
Gobiinae	*Elacatinus evelynae*	M	TR	3.7	10.2 (9.9-10.5)	22 (30-40)	2.20 (1.10-3.30)	0.75 (0.50-1.00)	6	+	+	49, 56, 57
Gobiinae	*Elacatinus illecebrosus*	M	TR		10.1							49
Gobiinae	*Elacatinus oceanops*	M	TR	3.5 (3.0-4.0)	10.7	28 (18-40)	2.30		7 (6-7)		+	7, 49, 51, 58, 59, 60, 61
Gobiinae	*Elacatinus pallens*	M	TR		9.2							49
Gobiinae	*Elacatinus phthirophagus*	M	TR		8.8							49
Gobiinae	*Eviota abax*	M	TR	2.7			1.79 (1.18-2.50)	0.83 (0.60-1.10)	5		+	4, 30, 31, 62, 63, 64, 65
Gobiinae	*Eviota albolineata*	M	TR	1.6 (1.5-1.6)			0.65	0.41				4, 30, 31, 32, 66, 67
Gobiinae	*Eviota fasciola*	M	TR	2.0			0.90	0.50				30, 66, 67
Gobiinae	*Eviota lacrimae*	M	TR	2.3 (2.2-2.3)			1.26	0.49				30, 66
Gobiinae	*Eviota melasma*	M	TR	2.4 (2.3-2.4)	7.3 (7.0-7.5)	26	0.90	0.55				15, 30, 66, 67
Gobiinae	*Eviota prasina*	M	TR	2.6 (2.6-2.7)			1.40	0.58				30, 31, 32, 66, 67
Gobiinae	*Eviota prasites*	M	TR	2.8 (2.8-2.9)			1.65	0.60				30, 66, 67

Contd...

Contd...

Subfamily	*Species*	*Env.*	*Region*	*Size at Hatching (mm)*	*Size at Settlement (mm)*	*PLD (days)*	*EGG L (mm)*	*EGG W (mm)*	*Develop. (days)*	*Larval Descr.*	*Embr. Descr.*	*References*
Gobiinae	*Eviota queenslandica*	M	TR	2.1 (2.0-2.1)	7.3 (7.0-7.5)	24	1.05	0.47				15, 30, 66, 67
Gobiinae	*Eviota sigillata*	M	TR		7.3 (7.0-7.5)	25 (23-27)	0.75	0.55	14 (11-17)		+	15, 68
Gobiinae	*Eviota storthynx*	M	TR	2.0 (1.9-2.1)			1.25	0.40				4, 10, 30, 31, 62, 63, 69
Gobiinae	*Favonigobius gymnauchen*	A	TR	2.0		36	2.00	0.60				4, 6
Gobiinae	*Favonigobius lateralis*	M	T	2.0	12.3 (11.1-13.5)					+		9, 10
Gobiinae	*Glossogobius giuris*	A	TR				3.50		3		+	7, 70
Gobiinae	*Glossogobius olivaceus*	A	T	2.0		23				+	+	4, 31, 71
Gobiinae	*Gobiosoma bosc*	M	T	2.0	8.3 (7.0-12.0)		1.26 (1.15-1.37)	0.58 (0.57-0.59)	5	+	+	2, 7, 10, 11, 39, 72, 73, 74, 75, 76, 77, 78, 79, 80, 81
Gobiinae	*Gobiosoma ginsburgi*	M	T	2.0						+		2, 7, 10, 11, 72, 73, 79, 80, 81
Gobiinae	*Gobiosoma hildebrandi*	M	TR		9.7							49
Gobiinae	*Gobiosoma longipala*	M	TR	3.2	7.7 (7.0-12.6)					+		2, 11
Gobiinae	*Gobiosoma parri*	M	TR		14.8 (12-17.6)					+		2, 11, 79, 80, 82
Gobiinae	*Gobiosoma robustum*	M	TR		7.0		1.50	0.50		+		2, 7, 10, 11, 39, 83, 84, 85, 86

Contd...

Contd...

Gobiinae	*Gobiosoma spes*	M	TR		7.1 (6.7-7.6)							49
Gobiinae	*Gobius bucchichi*	M	T	2.2		19 (17-21)	1.28 (1.05-1.50)	0.54 (0.50-0.57)	6			45, 87, 88, 89
Gobiinae	*Gobius cobitis*	M	T	5.3 (5.1-5.5)	18.0 (13.0-19.0)	35 (35-44)	3.84 (3.40-5.45)	1.22 (1.10-1.54)	20 (15-24)	+	+	7,10, 20, 23, 26, 45, 90, 91, 92
Gobiinae	*Gobius cruentatus*	M	T	3.3 (3.2-3.3)			2.04 (1.90-2.10)	0.56 (0.50-0.60)	13		+	93
Gobiinae	*Gobius gasteveni*	M	T			20 (14-23)						48
Gobiinae	*Gobius niger*	M	T	3.1 (2.5-4.0)	9.0		2.14 (1.50-3.00)	0.64 (0.45-0.90)	9 (6-13)	+	+	4, 7,10,16,18, 19, 20, 24, 26, 29, 39, 94, 95, 96, 97, 98, 99, 100,101, 102
Gobiinae	*Gobius paganellus*	M	T	4.0 (3.6-4.8)	11.6 (9.5-17.0)	29 (22-51)	2.21 (1.60-3.00)	0.82 (0.70-0.90)	13 (9-20)	+	+	2, 7, 10, 16, 18, 22, 24, 48, 103, 104, 105, 106, 107, 108, 109, 110
Gobiinae	*Gobius xanthocephalus*	M	T	3.0 (2.9-3.3)	13.0	30 (24-39)				+		48, 111
Gobiinae	*Gobiusculus flavescens*	M	T	2.4 (2.2-2.6)			0.70 (0.60-0.80)	0.57	10	+	+	7, 10, 18, 19, 20, 24, 26, 29, 39, 97, 112, 113
Gobiinae	*Gobulus crescentalis*	M	T	2.3	11.3 (7.7-14.9)					+		10, 114
Gobiinae	*Istigobius campbelli*	M	TR	2.1 (1.9-2.3)		30	1.23 (1.18-1.27)	0.53 (0.50-0.55)		+		4, 115
Gobiinae	*Istigobius hoshinonis*	M	TR	2.1 (1.8-2.3)		30	1.39 (1.33-1.45)	0.53 (0.50-0.55)		+		4, 115
Gobiinae	*Istigobius ornatus*	M	TR	1.2 (1.1-1.3)			0.85	0.40	9 (8-10)	+	+	7, 116

Contd...

Contd...

Subfamily	*Species*	*Env.*	*Region*	*Size at Hatching (mm)*	*Size at Settlement (mm)*	*PLD (days)*	*EGG L (mm)*	*EGG W (mm)*	*Develop. (days)*	*Larval Descr.*	*Embr. Descr.*	*References*
Gobiinae	*Knipowitschia caucasica*	A	T	3.2 (3.0-3.5)			2.24 (1.5-2.6)	0.85 (0.70-0.90)	14	+		10, 30, 55, 117, 118
Gobiinae	*Knipowitschia longecaudata*	F	T				1.70	0.80				45
Gobiinae	*Lebetus guilleti*	M	T							+		10, 19, 24, 119
Gobiinae	*Lebetus scorpioides*	M	T							+		10, 24, 119, 120
Gobiinae	*Lesueurigobius friesii*	M	T				1.85 (1.80-1.90)	0.95 (0.90-1.00)				10, 121
Gobiinae	*Lesueurigobius sanzi*	M	T							+		122, 123
Gobiinae	*Lesueurigobius suerii*	M	T							+		10, 124
Gobiinae	*Lophogobius cyprinoides*	A	TR	1.6 (1.5-2.0)	9.0	26 (17-35)	1.30	0.40	5		+	7, 49, 125, 126
Gobiinae	*Lubricogobius exiguus*	M	TR	1.8 (1.7-1.8)			1.04	0.39		+	+	4, 127
Gobiinae	*Lythrypnus dalli*	M	T	2.1 (2.0-2.2)	11.3 (9.5-13.0)					+		10, 114, 128, 129
Gobiinae	*Lythrypnus nesiotes*	M	TR		7.7							49
Gobiinae	*Lythrypnus zebra*	M	T		11.0 (10.0-12.0)					+		10, 114
Gobiinae	*Mesogobius batrachocephalus*	M	T				4.61 (4.01-5.20)	2.38 (2.15-2.60)				7, 10, 45, 130
Gobiinae	*Microgobius carri*	M	TR		15.2 (12.8-17.6)							49
Gobiinae	*Microgobius signatus*	M	TR		14.0 (12.8-15.2)							49
Gobiinae	*Microgobius thalassinus*	M	T		12.5 (11.2-13.8)					+		2, 11, 73, 74, 79, 80, 131
Gobiinae	*Neogobius cephalargoides*	M	T				4.40	2.50				10, 45
Gobiinae	*Neogobius gymnotrachelus*	F	T				3.27	1.25 (0.14-1.64)				7, 10, 130, 132

Contd...

Contd...

Gobiinae	*Neogobius melanostomus*	A	T	5.4 (5.0-5.7)			3.60 (2.20-3.90)		18 (14-20)	+	+	10, 133, 134, 135, 136, 137, 138, 139, 140, 141, 142
Gobiinae	*Neogobius ratan*	M	T				3.73	1.91				7, 130
Gobiinae	*Neogobius syrman*	A	T				4.21	1.89				7, 130
Gobiinae	*Nes longus*	M	TR							+		11, 37, 143
Gobiinae	*Ophiogobius jenynsi*	M	T							+		144
Gobiinae	*Padogobius bonelli*	F	T				4.00					30, 145
Gobiinae	*Padogobius nigricans*	F	T							+	+	146
Gobiinae	*Papillogobius punctatus*	M	T	2.5	10.6 (10.0-11.1)					+		9, 10
Gobiinae	*Paragobiodon echinocephalus*	M	TR			36	1.10	0.50				10, 13, 14, 15, 30, 147
Gobiinae	*Paragobiodon lacunicolus*	M	TR			31	1.10	0.50			+	4, 10, 13, 14, 15, 30, 31, 32, 147, 148
Gobiinae	*Paragobiodon melanosomus*	M	TR			42 (39-47)						10, 13, 15
Gobiinae	*Paragobiodon modestus*	M	TR				1.10	0.50				30, 147
Gobiinae	*Paragobiodon xanthosomus*	M	TR				1.10	0.50				30, 147
Gobiinae	*Pleurosicya mossambica*	M	TR		16.5 (13.0-20.0)							44
Gobiinae	*Pomatoschistus marmoratus*	M	T	2.7 (2.3-3.5)			1.00 (0.80-1.00)	0.60				7, 10, 39, 149, 150
Gobiinae	*Pomatoschistus microps*	M	T	3.0 (2.5-4.0)			0.86 (0.70-1.00)	0.70 (0.65-0.80)	12 (10-14)	+	+	7, 10, 16, 18, 19, 20, 24, 26, 29, 95, 97, 151, 152, 153
Gobiinae	*Pomatoschistus minutus*	M	T	2.8	12.0		1.19 (0.90-1.40)	0.75 (0.70-0.80)	14	+	+	7, 10, 18, 19, 20, 24, 26, 29, 97, 151, 154, 155, 156, 157

Contd...

Contd...

Subfamily	*Species*	*Env.*	*Region*	*Size at Hatching (mm)*	*Size at Settlement (mm)*	*PLD (days)*	*EGG L (mm)*	*EGG W (mm)*	*Develop. (days)*	*Larval Descr.*	*Embr. Descr.*	*References*
Gobiinae	*Pomatoschistus norvegicus*	M	T							+		10, 18, 20, 24, 120
Gobiinae	*Pomatoschistus pictus*	M	T	2.9 (2.7-3.0)	10.0		0.78 (0.75-0.80)	0.64	14	+	+	7, 10, 18, 19, 24, 26, 29, 39, 95, 151, 158
Gobiinae	*Priolepis boreus*	M	T	2.3 (2.1-2.5)	7.0		1.33 (1.20-1.45)	0.53 (0.51-0.55)		+	+	7, 10, 31, 159, 160
Gobiinae	*Priolepis cinctus*	M	TR	2.0			0.94 (0.76-1.12)	0.47 (0.41-0.52)	4		+	160
Gobiinae	*Priolepis hipoliti*	M	TR	2.0						+		11, 37
Gobiinae	*Priolepis nocturna*	M	TR	1.9			0.83 (0.75-0.90)	0.51 (0.49-0.52)	5		+	161
Gobiinae	*Priolepis semidoliata*	M	TR	2.0			1.04 (0.94-1.14)	0.55 (0.50-0.59)		+	+	10, 31, 162, 160
Gobiinae	*Proterorhinus marmoratus*	A	T	5.0						+		163
Gobiinae	*Psilotris amblyrhynchus*	M	TR							+		11, 37, 143
Gobiinae	*Rhinogobiops nicholsii*	M	T	2.9 (2.8-3.0)	21.0 (15.0-29.0)		2.15 (2.10-2.20)	0.49 (0.48-0.50)		+		4, 7, 10, 30, 114, 164, 165, 166, 167, 168
Gobiinae	*Risor ruber*	M	TR		7.9							49
Gobiinae	*Silhouettea dotui*	M	T	1.9			2.00 (1.90-2.10)	0.40 (0.40-0.50)				4, 7, 10, 31, 39, 169
Gobiinae	*Sueviota lachneri*	F	TR	2.5			1.62 (1.52-1.72)	0.60 (0.56-0.67)	6	+	+	10, 170
Gobiinae	*Sufflogobius bibarbatus*	M	T							+		123, 171
Gobiinae	*Trimma grammistes*	M	TR	1.7			0.70 (0.65-0.75)	0.49 (0.46-0.54)	4	+	+	30, 62, 172, 173

Contd...

Contd...

Gobiinae	*Trimma nasa*	M	TR			34						174
Gobiinae	*Trimma okinawae*	M	TR	2.0 (2.0-2.1)			0.84 (0.78-0.89)	0.48 (0.45-0.51)	4	+	+	30, 62, 172
Gobiinae	*Valenciennea bella*	M	TR	1.7			1.10 (1.10-1.20)	0.48 (0.45-0.50)				4, 175
Gobiinae	*Valenciennea helsdingenii*	M	TR	1.7 (1.6-1.8)			0.87 (0.78-0.93)	0.44 (0.43-0.45)		+	+	4, 10, 30, 31, 32, 62, 175, 176
Gobiinae	*Valenciennea strigata*	M	TR					0.39 (0.35-0.43)				10, 177
Gobiinae	*Zosterisessor ophiocephalus*	M	T	3.4	5.9	13	2.9	0.85	8		+	10, 178, 179
Gobionellinae	*Acanthogobius flavimanus*	A	T	4.6 (4.1-4.9)	14.2 (12.0-18.0)		5.15 (3.10-5.80)	0.95	23	+	+	2, 4, 7, 10, 31, 32, 114, 167, 178, 180, 181, 182, 183, 184
Gobionellinae	*Acanthogobius lactipes*	A	T	3.7 (3.6-3.7)		38	2.85 (2.40-3.30)	0.55 (0.50-0.60)	7			4, 6
Gobionellinae	*Amblychaeturichthys hexanema*	M	T	4.2	21.0		2.90	0.80	11		+	7, 10, 39, 185, 186
Gobionellinae	*Astrabe lactisella*	M	T		13.8					+		2, 187
Gobionellinae	*Awaous guamensis*	A	TR	1.2 (1.0-1.4)	14.4 (13.9-14.8)	161 (150-169)		0.59 (0.23-0.40)	1.5 (1-3)	+	+	4, 188, 189, 190, 191, 192, 193, 194, 195, 196, 197, 198, 199
Gobionellinae	*Awaous melanocephalus*	A	TR	1.0 (0.9-1.1)		80	0.34 (0.33-0.35)	0.33 (0.32-0.34)		+		4, 195, 198, 200, 201
Gobionellinae	*Brachygobius xanthozonus*	F	TR						6 (5-7)		+	7, 202
Gobionellinae	*Chaenogobius castaneus*	A	T	7.8 (7.0-8.0)	20.0		4.10	1.30				2, 4, 7, 10, 31, 203, 204, 205
Gobionellinae	*Chaenogobius macrognathos*	M	T							+		206

Contd...

Contd...

Subfamily	*Species*	*Env.*	*Region*	*Size at Hatching (mm)*	*Size at Settlement (mm)*	*PLD (days)*	*EGG L (mm)*	*EGG W (mm)*	*Develop. (days)*	*Larval Descr.*	*Embr. Descr.*	*References*
Gobionellinae	*Chaenogobius scrobiculatus*	F	T	4.7 (4.7-4.9)			3.15 (3.10-3.20)	1.05 (1.00-1.10)				4, 7, 207
Gobionellinae	*Chasmichthys dolichognathus*	M	T	5.2			4.15	1.35		+		4, 7, 10, 31, 39, 208, 209
Gobionellinae	*Chasmichthys gulosus*	M	T	5.8 (5.6-6.0)			4.65	1.22				4, 7, 10, 11, 31, 39, 208
Gobionellinae	*Clariger cosmurus*	M	T	4.3								4, 210
Gobionellinae	*Clevelandia ios*	M	T	2.9 (2.0-3.8)	13.0 (13-17.2)	30	1.95 (1.30-2.60)	0.75 (0.70-0.80)	11 (10-12)	+	+	7, 10, 114, 164, 167, 168, 211, 212, 213, 214, 215, 216, 217
Gobionellinae	*Ctenogobius boleosoma*	A	TR	1.2	14.8			0.30	0.8	+	+	2, 7, 10, 11, 49, 72, 73, 74, 75, 77, 79, 80, 81, 218
Gobionellinae	*Ctenogobius saepepallens*	M	TR		10.9					+		2, 11, 37, 49, 79, 80
Gobionellinae	*Eucyclogobius newberryi*	M	T		14.9 (13.4-16.3)		5.50 (5.00-6.00)	2.00		+		10, 30, 114, 214, 219, 220, 221
Gobionellinae	*Eutaeniichthys gilli*	M	T	4.1	15.3		2.70	0.80	9		+	4, 7, 39, 222, 223
Gobionellinae	*Evorthodus lyricus*	A	TR	1.9	12.7		0.45	0.20	0.8 (0.7-0.8)	+	+	2, 11, 49, 74, 79, 80, 224, 225
Gobionellinae	*Favonigobius reichei*	A	TR	2.1			1.20	0.40	3		+	7, 39
Gobionellinae	*Gillichthys mirabilis*	M	T	3.5 (3.0-4.0)	12.8 (10.5-15.0)		3.10 (2.80-3.40)	1.10	11 (10-12)	+	+	7, 10, 114, 167, 226, 227, 228
Gobionellinae	*Gnatholepis thompsoni*	M	TR		9.0 (8.3-11.0)	73 (46-112)			2	+		2, 11, 14, 37, 49, 51, 79, 80, 229, 230
Gobionellinae	*Gobioides broussonnetii*	A	TR							+		2, 11

Contd...

Contd...

Gobionellinae	*Gobionellus oceanicus*	A	TR		16.0				+		2, 11, 37, 73, 79, 80, 218
Gobionellinae	*Gobiopterus lacustris*	F	TR			1.00					10, 231
Gobionellinae	*Gobiopterus semivestitus*	F	TR	2.2	12.8 (10.5-15.0)				+		9, 10, 32
Gobionellinae	*Gymnogobius heptacanthus*	M	T	4.6							4, 205
Gobionellinae	*Gymnogobius isaza*	F	T	3.1		1.55 (1.50-1.60)	0.60 (0.50-0.70)				4, 232
Gobionellinae	*Gymnogobius uchidai*	M	T	4.4	15.0	2.51	0.90		+		4, 7, 10, 31, 206, 233, 234
Gobionellinae	*Gymnogobius urotaenia*	A	T	4.9 (4.1-5.3)	25.0	3.00	1.00	14		+	4, 7,10, 31, 39, 235, 236, 237
Gobionellinae	*Hemigobius hoevenii*	A	TR			2.83					7, 238
Gobionellinae	*Ilypnus gilberti*	M	T	3.1	16.7 (13.4-20.0)	3.30			+		10, 114, 167, 214
Gobionellinae	*Lepidogobius lepidus*	M	T	3.1 (3.0-3.2)	26.5 (24.0-29.0)	1.53 (1.20-1.80)	0.90 (0.80-1.00)		+		10, 114, 164, 167, 168
Gobionellinae	*Lethops connectens*	M	T		25.0 (20.0-30.0)				+		10, 114
Gobionellinae	*Leucopsarion petersii*	A	T					13	+	+	9, 239, 240, 241
Gobionellinae	*Luciogobius elongatus*	M	T	2.8					+		2, 4, 241, 242
Gobionellinae	*Luciogobius grandis*	M	T	3.4					+		4, 241, 243
Gobionellinae	*Luciogobius guttatus*	A	T	3.5		2.50	0.70	10	+	+	4, 7, 39, 241, 244
Gobionellinae	*Luciogobius koma*	M	T	3.8 (3.6-3.9)							4, 245
Gobionellinae	*Luciogobius pallidus*	A	T						+		241, 246
Gobionellinae	*Luciogobius parvulus*	M	T	3.4							4, 247
Gobionellinae	*Luciogobius platycephalus*	M	T	3.7 (3.5-3.8)		2.17 (2.05-2.28)	0.72 (0.68-0.75)		+		248
Gobionellinae	*Luciogobius saikaiensis*	M	T	4.3 (4.1-4.5)		3.00	1.30	10		+	4, 7, 31, 39, 249

Contd...

Contd...

Subfamily	*Species*	*Env.*	*Region*	*Size at Hatching (mm)*	*Size at Settlement (mm)*	*PLD (days)*	*EGG L (mm)*	*EGG W (mm)*	*Develop. (days)*	*Larval Descr.*	*Embr. Descr.*	*References*
Gobionellinae	*Mistichthys luzonensis*	F	TR					0.38				7, 250
Gobionellinae	*Mugilogobius abei*	M	T	2.1 (2.0-2.2)	17.3 (15.0-19.5)	40	0.98	0.45	4	+	+	4,10, 31, 251
Gobionellinae	*Oligolepis acutipennis*	A	TR			33						4, 201
Gobionellinae	*Paedogobius kimurai*	M	TR							+		9, 239, 241
Gobionellinae	*Pandaka lidwilli*	M	TR	1.8			1.20	0.70	5		+	4, 7, 10, 31, 32, 39, 252
Gobionellinae	*Pseudogobius masago*	F	T	2.0			1.20	0.40	4		+	4, 7, 10, 31, 39, 253
Gobionellinae	*Pseudogobius olorum*	F	T	2.0	9.4 (8.2-10.5)					+		9, 10
Gobionellinae	*Pterogobius elapoides*	M	T				2.30	0.80				7, 39, 254
Gobionellinae	*Pterogobius zacalles*	M	T	5.1 (5.0-5.3)			4.31 (3.79-4.83)	0.78	27		+	4, 255
Gobionellinae	*Pterogobius zonoleucus*	M	T	3.5			2.48 (2.10-3.16)	0.69	16		+	4, 7, 31, 39, 256, 257
Gobionellinae	*Quietula y-cauda*	M	T	2.9 (2.8-3.0)	10.8 (9.5-12.0)		3.40			+		10, 114
Gobionellinae	*Redigobius bikolanus*	A	TR	1.9		49				+		4, 10, 31, 258
Gobionellinae	*Redigobius macrostoma*	A	TR	2.2	7.6 (6.7-8.5)					+		9, 10, 32
Gobionellinae	*Rhinogobius brunneus*	A	T	3.4 (3.1-3.8)	13.0	32	1.69 (1.48-1.90)	0.73 (0.65-0.80)		+		10, 31, 195, 196, 203, 259, 260
Gobionellinae	*Rhinogobius flumineus*	F	T				2.10					10, 31, 195, 203
Gobionellinae	*Rhinogobius gigas*	F	T		17.6							194, 261
Gobionellinae	*Rhinogobius giurinus*	A	T	2.6 (2.5-2.7)	7.0 (6.4-7.6)	41 (12-60)	2.40	0.45	4		+	4, 7, 10, 31, 262, 263, 264

Contd...

Contd...

Gobionellinae	*Rhinogobius similis*	F	T	4.0			2.50	0.63				4, 7, 39, 265
Gobionellinae	*Sagamia geneionema*	M	T	5.6 (5.0-6.1)			6.83 (6.25-7.40)	1.09 (1.00-1.18)	18			4, 266
Gobionellinae	*Stenogobius genivittatus*	A	TR		13.2 (12.4-13.9)	93						4, 201
Gobionellinae	*Stenogobius hawaiiensis*	A	TR			135 (119-151)			3 (2-3)	+		4, 188, 189, 190, 194, 197, 198, 267
Gobionellinae	*Synechogobius hasta*	A	T	6.0	12.0	16 (14-18)	5.50 (4.90-6.60)	1.10 (1.00-1.20)	15	+	+	4, 268, 269
Gobionellinae	*Tridentiger barbatus*	M	T	2.6 (2.5-2.7)			1.52 (1.46-1.60)	0.53 (0.50-0.58)	4		+	4, 7, 10, 31, 39, 270
Gobionellinae	*Tridentiger brevispinis*	A	T	2.7 (2.6-2.8)			0.88	0.57 (0.56-0.58)				4, 271
Gobionellinae	*Tridentiger kuroiwae*	A	T	2.8			1.21	0.67	7			4
Gobionellinae	*Tridentinger nudicervicus*	M	T	2.5 (2.0-3.0)			1.15 (1.06-1.25)	0.46 (0.42-0.50)	4		+	4, 7, 10, 31, 272, 273
Gobionellinae	*Tridentinger obscurus*	A	T	3.4 (3.3-3.5)			1.39	0.72			+	4, 31, 271
Gobionellinae	*Tridentiger trigonocephalus*	M	T	2.5 (2.4-2.7)	14.5 (14.0-15.0)		1.35	0.68	7	+		4, 31, 32, 114, 181
Gobionellinae	*Typhlogobius californiensis*	M	T	3.2 (3.0-3.3)	19.3 (11.7-26.9)		2.80 (2.70-2.90)	0.70 (0.68-0.72)	11 (10-12)	+	+	4, 7, 10, 114, 274
Oxudercinae	*Apocryptodon madurensis*	A	TR		20.0							7, 275
Oxudercinae	*Apocryptodon punctatus*	M	T				1.00	0.50		+		4, 31, 275
Oxudercinae	*Boleophthalmus boddarti*	A	TR				1.60					7, 8
Oxudercinae	*Boleophthalmus pectinirostris*	M	TR	3.0 (2.4-3.3)		39 (36-41)	1.70 (1.00-1.30)	0.68 (0.60-0.80)		+	+	4, 276, 277, 278, 279
Oxudercinae	*Periophthalmodon schlosseri*	A	TR	2.5 (2.2-2.8)			1.02 (0.83-1.43)	0.73 (0.56-1.13)		+	+	280

Contd...

Contd...

Subfamily	*Species*	*Env.*	*Region*	*Size at Hatching (mm)*	*Size at Settlement (mm)*	*PLD (days)*	*EGG L (mm)*	*EGG W (mm)*	*Develop. (days)*	*Larval Descr.*	*Embr. Descr.*	*References*
Oxudercinae	*Periophthalmus argentilineatus*	A	TR	2.2		29	0.93 (0.93-1.00)	0.59	7 (6-7)	+	+	4, 280, 281, 282
Oxudercinae	*Periophthalmus barbarus*	A	TR				0.76	0.61	26 (25-26)			7, 39, 283
Oxudercinae	*Periophthalmus modestus*	A	TR	2.5 (2.1-2.9)		43 (28-52)	1.00 (0.76-1.20)	0.69 (0.50-0.78)		+	+	4, 280, 283, 284, 285, 286
Oxudercinae	*Scartelaos histophorus*	A	TR	2.2			1.30	0.50		+		4, 287
Sicydiinae	*Cotylopus acutipinnis*	A	TR		22.5	114 (78-150)						288
Sicydiinae	*Lentipes concolor*	A	TR	1.8	16.0 (13.5-17.9)	90 (63-106)		0.45 (0.40-0.50)	3 (2-3)	+		4, 30, 190, 198, 289, 290, 291, 292, 293, 294
Sicydiinae	*Sicydium plumieri*	A	TR	1.5	30.6 (22.1-30.1)	87 (63-139)			1.1 (1-1.2)	+	+	36, 195, 196, 291, 295, 296
Sicydiinae	*Sicydium punctatum*	A	TR	1.8	23.0 (18.8-29.5)	83 (50-140)		0.50	1 (0.9-1.5)	+	+	181, 195, 196, 225, 291, 297, 298, 299, 300
Sicydiinae	*Sicyopterus japonicus*	A	T	1.5	31.3 (27.8-38.1)	179 (125-253)		0.48 (0.45-0.50)	2	+	+	4, 7, 182, 194, 201, 262, 288, 300, 301, 302
Sicydiinae	*Sicyopterus lagocephalus*	A	TR	1.1	29.3 (21.0-33.0)	214 (131-266)			1.5 (0.8-2.1)		+	7, 146, 194, 196, 225, 288, 293, 300, 303, 304, 305, 306
Sicydiinae	*Sicyopterus stimpsoni*	A	TR	1.8	27.9					+		196, 198, 289, 300, 307, 308, 309, 310
Sicydiinae	*Stiphodon percnopterygionus*	A	TR	1.3 (1.2-1.4)	15.4 (14.2-15.9)	99 (78-146)	0.56 (0.54-0.58)	0.50 (0.49-0.50)	0.8		+	190, 195, 311, 312

Contd...

Contd...

1 (Dotu, 1957a); 2 (Ruple, 1984); 3 (Leis and Carson-Ewart, 2000); 4 (Kon and Yoshino, 2002); 5 (Dotu and Takita, 1967); 6 (Uchida and Dotsu, 1980); 7 (Darcy, 1980); 8 (Jones, 1937); 9 (Neira *et al.*, 1998); 10 (Froese and Pauly, 2009); 11 (Yeung and Ruple, 2006); 12 (Bay *et al.*, 2006); 13 (Brothers *et al.*, 1983); 14 (Lester and Ruttenberg, 2005); 15 (Depczynski and Bellwood, 2006); 16 (Garcia and Moyano, 1990); 17 (Ehrenbaum, 1905-09); 18 (Lebour, 1919); 19 (Petersen, 1919); 20 (Padoa, 1956); 21 (Iglesias and Morales-Nin, 2001); 22 (Holt and Byrne, 1898); 23 (Lo Bianco, 1909); 24 (Russell, 1976); 25 (La Mesa *et al.*, 2005); 26 (Wheeler, 1969); 27 (Lopez *et al.*, 2000); 28 (Vodyanitzki and Kazanova, 1954); 29 (Munk and Nielsen, 2005); 30 (Privitera, 2001); 31 (Okiyama, 1988); 32 (http://larval-fishes.australianmuseum.net.au/); 33 (Dotu and Mito, 1963); 34 (Shafer, 2000); 35 (Shafer, 1998); 36 (Victor, 2006); 37 (Baldwin and Smith, 2003); 38 (Dotu, 1955d); 39 (Coleman, 1991); 40 (Peters, 1983); 41 (Tavolga, 1950); 42 (Tavolga, 1954); 43 (Breder, 1943); 44 (Leis *et al.*, 2002); 45 (Miller, 1986); 46 (Tâning, 1940); 47 (Gilchrist, 1916); 48 (Beldade *et al.*, 2007); 49 (Victor, 2009); 50 (Kramer and Patzner, 2008); 51 (Sponaugle and Cowen, 1994); 52 (Victor, 2007); 53 (Cole and Robertson, 1988); 54 (Fives, 1970); 55 (Daoulas *et al.*, 1993); 56 (Olivotto *et al.*, 2005); 57 (Taylor and Hellberg, 2003); 58 (Colin, 1975); 59 (Moe, 1975); 60 (Valenti, 1972); 61 (Feddern, 1967); 62 (Masuda *et al.*, 1984); 63 (Sunobe and Nakazono, 1987); 64 (Taru and Sunobe, 2002); 65 (Dotu *et al.*, 1965); 66 (Sunobe and Nakazono, 1995); 67 (Myers, 1999); 68 (Depczynski and Bellwood, 2005); 69 (Shinomiya *et al.*, 1981); 70 (Ishikawa and Nakamura, 1940); 71 (Suzuki *et al.*, 1988); 72 (Scotton *et al.*, 1973); 73 (Hildebrand and Cable, 1938); 74 (Foster and Fuiman, 1987); 75 (Kuntz, 1916); 76 (Breitburg, 1991); 77 (Fahay, 1983); 78 (Lippson and Moran, 1974); 79 (Fahay, 2007); 80 (Fritzsche, 1978); 81 (Able and Fahay, 1998); 82 (Acha, 1994); 83 (Breder, 1942); 84 (Springer and McErlean, 1961); 85 (Garcia *et al.*, 1988); 86 (Zavala *et al.*, 1988/1989); 87 (Macpherson and Raventos, 2006); 88 (Biagi *et al.*, 1998); 89 (Raventos and Macpherson, 2001); 90 (Spartà, 1950); 91 (Gil *et al.*, 1997); 92 (Gibson, 1970); 93 (Gil *et al.*, 2002); 94 (Ehrenbaum and Strodtmann, 1904); 95 (Petersen, 1917); 96 (Ballard, 1969); 97 (Petersen, 1892); 98 (Iglesias, 1979); 99 (Villegas Cuadros, 1981); 100 (Kinzer, 1960); 101 (Joyeux *et al.*, 1991); 102 (Spartà, 1934a); 103 (Spartà, 1934b); 104 (Hefford, 1910); 105 (Miller, 1961); 106 (Villegas, 1980); 107 (Sabatés, 1988); 108 (Borges *et al.*, 2003); 109 (Danois, 1913); 110 (Vivien, 1939); 111 (Monteiro *et al.*, 2008); 112 (Thys van den Audenaerde, 1988); 113 (Lie *et al.*, 1978); 114 (Watson, 1996); 115 (Shiobara and Suzuki, 1983); 116 (Vijayaraghavan, 1975); 117 (Economou *et al.*, 1994); 118 (Koblitskaya, 1961) ; 119 (Demir and Russel, 1971); 120 (Fage, 1918); 121 (Coleman, 1986); 122 (Olivar, 1989); 123 (Olivar and Furtuno, 1991); 124 (Aboussouan, 1964); 125 (Delmonte *et al.*, 1968); 126 (Darcy, 1978); 127 (Dotu and Fujita, 1963); 128 (Behrents, 1987); 129 (Wiley, 1976); 130 (Bil'ko, 1968); 131 (Richardson and Joseph, 1975); 132 (Grabowska, 2005); 133 (Logachev and Mordvinov, 1979); 134 (Sapota, 2006); 135 (Moiseyeva, 1983); 136 (Moskalkova, 1967); 137 (Moskalkova, 1978); 138 (Moskalkova, 1989); 139 (Moskalkova, 1996); 140 (Charlebois *et al.*, 1997); 141 (Skora *et al.*, 1999); 142 (http://www.caspianenvironment.org/biodb/eng/main.htm/); 143 (Smith and Baldwin, 1999); 144 (Herrera, 1984); 145 (Marconato *et al.*, 1989); 146 (D'Onofrio *et al.*, 1990); 147 (Lassig, 1976); 148 (Yamamoto, 1980); 149 (Mazzoldi *et al.*, 2002); 150 (Spartà, 1936); 151 (Lebour, 1920); 152 (Guitel, 1892); 153 (Nyman, 1953); 154 (Healey, 1971); 155 (Swedmark, 1958); 156 (Holt, 1890); 157 (Shann, 1910); 158 (Kunne, 1934); 159 (Shiogaki and Dotsu, 1974a); 160 (Sunobe and Nakazono, 1989); 161 (Wittenrich *et al.*, 2007); 162 (Sonoda and Imai, 1971); 163 (Leslie *et al.*, 2002); 164 (Matarese *et al.*, 1989); 165 (Wiley, 1973); 166 (Ebert and Turner, 1962); 167 (Wang, 1981); 168 (http://access.afsc.noaa.gov/ichthyo/index.cfm/); 169 (Dotu, 1958a); 170 (Sunobe and Shinomiya, 2002); 171 (O'Toole, 1978); 172 (Sunobe, 1995); 173 (Shiobara and Tanaka, 1994); 174 (Winterbottom and Southcott, 2008); 175 (Shiobara and Tanaka, 1995); 176 (Tanaka *et al.* 1982); 177 (Reavis, 1997); 178 (Privileggi *et al.*, 1997); 179 (Franco *et al.*, 2002); 180 (Dotu and Mito, 1955a); 181 (Breder and Rosen, 1966); 182 (Dotu and Mito, 1955b); 183 (Suzuki *et al.*, 1989); 184 (Miyazaki, 1940); 185 (Hsieh and Chiu, 2002); 186 (Dotu *et al.*, 1955); 187 (Dotsu and Shiogaki, 1971); 188 (Kinzie, 1988); 189 (Murphy and Cowen, 2007); 190 (Radtke *et al.*, 2001); 191 (Kido and Heacock, 1991); 192 (Ha and Kinzie, 1996); 193 (Lindstrom and Brown, 1994); 194 (Shen and Tzeng, 2008); 195 (McDowall, 2009); 196 (Keith, 2003); 197 (Radtke *et al.*, 1988); 198 (Lindstrom, 1999); 199 (Timbol et al, 1990); 200 (Yamasaki and Tachihara, 2007); 201 (Shen *et al.*, 1998); 202 (Roloff, 1935); 203 (Miller, 1984); 204 (Dotu, 1954); 205 (Dotsu, 1984); 206 (Koki *et al.*, 1999); 207 (Dotu, 1961a); 208 (Nakamura, 1936); 209 (Kim, 1975); 210 (Shiogaki and Dotsu, 1972b); 211 (Prasad, 1958a); 212 (Kent and Marliave, 1997); 213 (Dawson *et al.*, 2002); 214 (Brothers, 1975); 215 (MacGinitie, 1935); 216 (Prasad, 1958b); 217 (Carter, 1965); 218 (Wyanski and Targett, 2000); 219 (Swift *et al.*, 1989); 220 (Swenson, 1995); 221 (Swenson, 1997); 222 (Dotu, 1965); 223 (Dotu, 1955c); 224 (Wyanski and Targett, 1985); 225 (Bell, 1994); 226 (Barlow, 1961); 227 (Eigenmann, 1892); 228 (Weisel, 1947); 229 (Maddox, 1992); 230 (Shulman, 1998); 231 (Blanco, 1947); 232 (Kobayashi *et al.*, 1950); 233 (Dotu, 1958b);

Contd...

Contd...

234 (Dotu, 1957e); 235 (Dotu, 1955a); 236 (Hamada, 1968); 237 (Ishino, 1986); 238 (Szabados, 1937); 239 (Dotsu and Uchida, 1979); 240 (Arakawa *et al.*, 1999); 241 (Maeda *et al.*, 2008); 242 (Shiogaki and Dotsu, 1972a); 243 (Shiogaki *et al.*, 1974); 244 (Dotu, 1957b); 245 (Shiogaki and Dotsu, 1974b); 246 (Yoshida *et al.*, 2006); 247 (Shiogaki and Dotsu, 1971); 248 (Shiogaki and Dotsu, 1977); 249 (Dotu and Mito, 1958); 250 (Te Winkel, 1935); 251 (Kanabashira *et al.*, 1980); 252 (Dotu, 1957c); 253 (Dotu, 1958c); 254 (Dotu and Tsutsumi, 1959); 255 (Shiogaki, 1981); 256 (Dotu, 1955b); 257 (Dotu, 1956); 258 (Akihito *et al.*, 2000); 259 (Sakai and Yasuda, 1978); 260 (Han *et al.*, 1998); 261 (Shiao, 1998); 262 (Chang *et al.*, 2008); 263 (Tamada, 2005); 264 (Dotu, 1961b); 265 (Mizuno, 1960); 266 (Shiobara *et al.*, 1990); 267 (Radke *et al.*, 1988); 268 (Takita, 1975); 269 (Kuno and Takita, 1997); 270 (Dotu, 1957d); 271 (Kishi, 1978); 272 (Dotu, 1958d); 273 (Dotu, 1957f); 274 (MacGinitie, 1939); 275 (Dotu, 1961c); 276 (Chen and Ting, 1984); 277 (Zhang and Zhang, 1988); 278 (Zhang *et al.*, 1989); 279 (Dotsu and Suzuki, 1985); 280 (Tsukaho *et al.*, 2003); 281 (Brillet, 1976); 282 (Ishimatsu *et al.*, 2007); 283 (Kimura, 1958); 284 (Kobayashi *et al.*, 1972); 285 (Uchida, 1931); 286 (Hagiwara *et al.*, 1993); 287 (Saitoh, 1993); 288 (Hoareau *et al.*, 2007); 289 (Kinzie, 1993); 290 (Radtke and Kinzie, 1996); 291 (Bell *et al.*, 1995); 292 (Way *et al.*, 1998); 293 (Keith *et al.*, 2005); 294 (Lindstrom, 1994); 295 (Erdman, 1961); 296 (Erdman, 1986); 297 (Bell and Brown, 1995); 298 (Bell, 2007); 299 (Bell, 1999); 300 (Iida *et al.*, 2009); 301 (Shiao *et al.*, 1999); 302 (Iida *et al.*, 2008); 303 (Hoareau, 2005); 304 (Manacop, 1953); 305 (Hoareau *et al.*, 2004); 306 (Bielsa *et al.*, 2003); 307 (Fitzsimons *et al.*, 1993); 308 (Schoenfuss *et al.*, 1997); 309 (Nishimoto and Kuamoʻo, 1997); 310 (Tomihama, 1972); 311 (Yamasaki and Tachihara, 2006); 312 (Yamasaki *et al.*, 2007).

however reclassified. Therefore, a marine species that also occurs in estuaries was classified as marine and not as amphidromous. This group was restricted to species whose adults' and/or larvae move from freshwater to brackish and marine waters or the reverse. The FishBase classification regarding the geographical distribution of species was also used. Nevertheless, we also recognized in this case the need of reclassifying some species based on their geographical distribution. Species were classified as tropical whenever they occur mostly in tropical waters, even if their distribution limits extended to sub-tropical or temperate waters. Species classified as temperate and as sub-tropical in FishBase were grouped together as "temperate" given the difficulty in separating the two regions, due to the frequent overlapping in the geographical distribution of species included in these two categories and the lack of distributional details for many species.

The following life-history data were collected: egg size; embryonic developmental time; size at hatching; size at settlement; pelagic larval duration (PLD), which also includes the oceanic larval duration (OLD) of amphidromous species. The growth in the plankton, measured as the difference between size at settlement and size at hatching when both values were available for the same species, was also calculated. Additionally, the existence of embryonic and larval descriptions was also noted (Table 3.4.1).

In some cases, corrections were needed in order to allow for comparisons. For example, total length (TL) was used as a measure for larval body size. Whenever only the standard length (SL) or notochordal length (NL) information was available, TL was calculated using a linear regression analysis. Two independent regressions were computed: one for pre-flexion larvae, classified as newly-hatched length, and one for post-flexion larvae classified as length at metamorphosis. Data used for each regression analysis were derived, respectively, from 152 and 1538 gobiid larvae, respectively, from which we had both TL and SL or NL measurements (data collected by R. Borges from nine species). This procedure might introduce some errors in the extrapolation of SL or NL to TL sizes since different species or groups might have different TL/SL or NL/TL relationships. Nevertheless, it is a useful way of comparing the available data in the overall dataset due to the lack of standardization of body length measurements (see below).

In many situations, the available information for a given variable of a given species needed to be standardized, in order to allow comparisons of the different variables among species. In some cases, only one value was found in the literature for a given variable, whereas, for others, several studies performed in different regions present data for that variable (e.g.

size at hatching). The following procedure was adopted for these latter situations: the mean value for each variable was used, if given in the primary source; when only the range was available, the arithmetic average was computed; if different studies reported different measures, the overall mean for that variable was determined. Some caution must however be taken when considering average values and even more so when these values are computed from different studies from different geographical locations and environments, as temperature (among other factors) is known to influence early development characteristics and variability can occur even within the same species (Sponaugle and Cowen, 1994; Sponaugle, 2009). Nonetheless, the averaging of the existing values allows describing the general trends, for the different variables analysed, across groups and species.

The different traits were tested at the subfamily (taxonomic) level, the ecosystem level (marine, freshwater and amphidromous groups) and the regional level (tropical and temperate groups). At the subfamily level, most comparisons were restricted to the Gobiinae, Gobionellinae and Sicydiinae due to the very scarce information available for the other subfamilies. The environmental comparisons were performed grouping the ecosystem and region levels together, using the following categories: marine temperate; marine tropical; amphidromous temperate; amphidromous tropical; freshwater temperate; freshwater tropical. Available information for tropical freshwater and sometimes temperate freshwater species did not allow a meaningful statistical analysis of the data for most variables.

Comparisons were performed using one-way ANOVA with Student-Newman–Keuls tests for post hoc comparisons when homoscedascity assumptions were met, after being tested with the Levene's test. If needed, variables were log or square root transformed prior to the analysis. When variances were heterogeneous, the Kruskall–Wallis ANOVA was used and post hoc comparisons were performed with the Dunn's test. The *T*-student test and the Mann-Whitney U Test were used for comparisons between two samples. Pearson or Spearman correlations were also computed for several variables.

RESULTS

General Description of Early Life Stages

Embryonic Development

Gobiidae eggs are generally ovoid to ellipsoid, but some are round and others pear-shaped, and many contain filaments for attachment to a hard substrate (Ruple, 1984; Leis and Carson-Ewart, 2000). Their colour varies from clear and transparent to yellow, or darker, depending on yolk colouration (Hildebrand and Cable, 1938; Russell, 1976; Ruple, 1984). Eggs range in size (maximum axis) from as little as 0.34 mm (range: 0.33 – 0.35 mm) in *Awaous melanocephalus* to 6.83 mm (range: 6.25 - 7.40 mm) in *Sagamia geneionema* (Table 3.4.1).

The incubation period also varies widely, even within the same subfamily. For example, the Gobionellinae presented one of the shortest and the longest developmental periods (18 hours for *Evorthodus lyricus* and for *Ctenogobius boleosoma* and 27 days for *Pterogobius zacalles*) (Table 3.4.1).

A typical sequence of embryonic development is described for some of the Gobiidae species our group has been studying. The main events of the embryonic stage might be found in Ballard (1969) Gil *et al.* (1997, 2002) and Borges *et al.* (2003) (Fig. 3.4.1). In these species of the genus *Gobius* (*G. paganellus*, *G. cobitis* and *G. cruentatus*) occurring in the western Portuguese shore, the eggs are elongated and fusiform, with egg sizes (long axis) ranging in our study area from 1.60 to 3.74 mm (Gil *et al.*, 1997, 2002; Borges *et al.*, 2003). Embryonic development lasted from 9 to 24 days (depending on the species and temperature). For details on developmental sequences of other genera see Sonoda and Imai (1971), Takita (1975), Sunobe and Nakazono (1987, 1989, 1995), Sunobe (1995), Privileggi *et al.* (1997), Arakawa *et al.* (1999) and Sunobe and Shinomiya (2002).

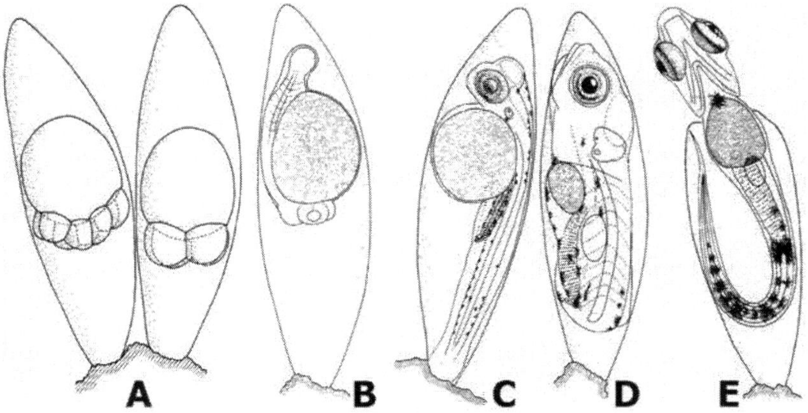

Fig. 3.4.1 Egg development of *Gobius cobitis* at different developmental stages showing the typical elongated egg type and the main events during ontogeny: (A) four and eight cell stages, day 0; (B) embryo with tail bud free of the yolk, day 4; (C) embryo as long as major axis, day 10; (D) embryo prior to hatching, day 21; (E) hatching process (from Gil *et al.*, 1997).

Newly Hatched Larvae

Larvae are planktonic with newly hatched individuals ranging in size from 1.0 mm (range: 0.9 – 1.1 mm) in *Awaous melanocephalus* to 7.8 mm (range: 7.0 – 8.0 mm) in *Chaenogobius castaneus* (Table 3.4.1). In general, they hatch with developed jaws, a functional mouth, partially to fully-pigmented eyes and a small yolk-sac. However, in some amphidromous Gobionellinae, such as *Awous melanocephalus* and *Rhinogobius brunneus*, and in amphidromous Sicydiinae, such as *Syciopterus lagocephalus*, hatching larvae are in a free embryo stage (sensu Balon, 1975) with a closed mouth and anus, unpigmented eyes and a large yolk (e.g. Han *et al.*, 1998; Keith, 2003; Keith *et al.*, 2005; Yamasaki and Tachiara, 2007).

A typical recently hatched larva may be found in the genus *Gobius* (Hefford, 1910; Lebour, 1919; Petersen, 1919; Spartá, 1934, 1950; Ballard, 1969; Iglesias, 1979; Gil *et al.*, 1997, 2002; Borges *et al.*, 2003; Monteiro *et al.*, 2008). The larvae show a slender and elongated body and almost no yolk (Fig. 3.4.2). They hatch in a relatively advanced stage of development with

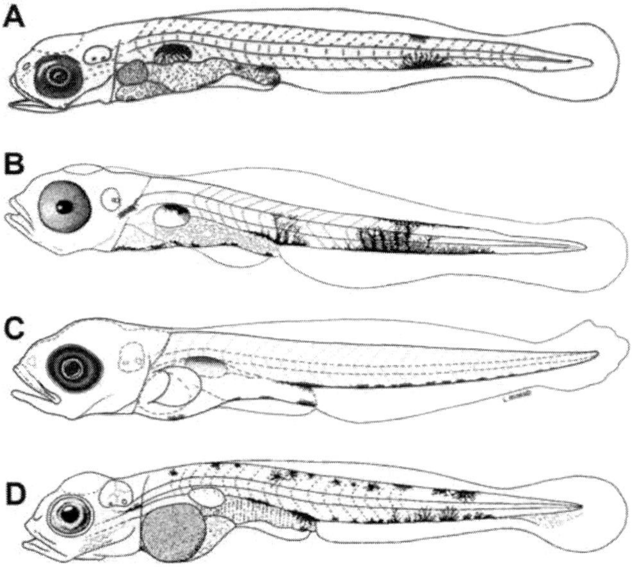

Fig. 3.4.2 Recently hatched larvae of the genus *Gobius*: (A) *Gobius paganellus* (3.9 mm) (from Borges *et al.*, 2003); (B) *G. cruentatus* (3.3 mm) (from Gil *et al.*, 2002); (C) *G. xantocephalus* (3.1 mm) (from Monteiro *et al.*, 2008); and (D) *G. cobitis* (5.5 mm) (from Gil *et al.*, 1997).

a functional mouth and anus, pigmented eyes and developed pectoral fins that allow active swimming and the initiation of exogenous feeding almost immediately after hatching. The liver is developed and the gas bladder is formed but not completely filled (the larvae were seen swimming towards the surface immediately after hatching presumable to fill the gas bladder; Borges *et al.*, 2003). The opercula are opened with the branchial arches already differentiated. They hatch at 2.2 to 5.50 mm, depending on species. The dorsal membrane of the gas bladder is fully pigmented. Ventrally there are some dispersed melanophores under the liver, the gut, and the anus and a series of post-anal melanophores. Dorsally, some melanophores may be present (Gil *et al.*, 1997, 2002; Borges *et al.*, 2003; Monteiro *et al.*, 2008).

Information for 99 species for which data for egg size (long axis) and size at hatching was compiled in our dataset (Table 3.4.1) revealed a strong positive linear correlation ($r^2 = 0.70$, $p < 0.05$) between these two traits (Fig. 3.4.3).

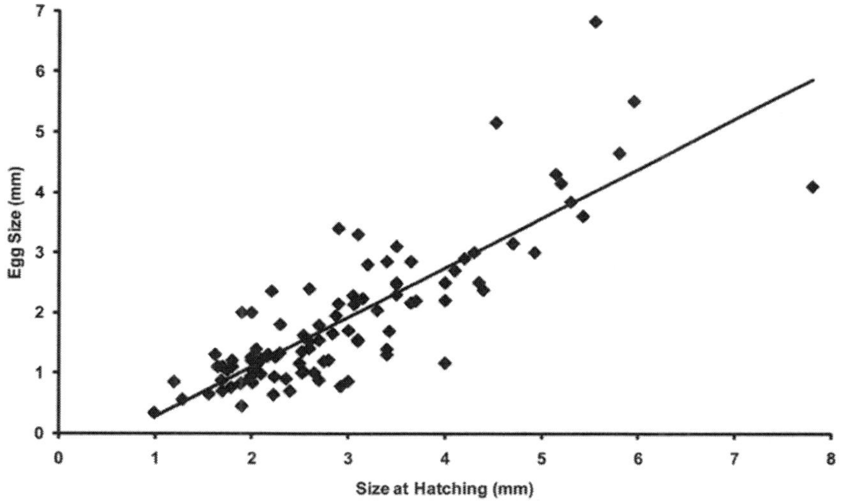

Fig. 3.4.3 Linear relationship between size at hatching and egg size (maximum axis) ($Y = 0.821\ X - 0.543$; $R^2 = 0.699$).

Larval Development

Gobiidae larval types are considerably variable but they can be generally easily distinguishable from those of other teleosts (Ruple, 1984). Pre-flexion stage larvae maintain, in general, the same body shape as the newly hatched larva, although allometric relationships of the different body measures may change with development (Monteiro *et al.*, 2008). The gut is either straight or gently curved, extending approximately to the midbody (Gobiinae) or relatively short and coiled (Amblyopinae) (Leis and Carson-Ewart, 2000). The head is in general of moderate length, slightly rounded and gently sloping and the eyes are essentially rounded and large (Ruple, 1984). The mouth tends to be horizontal and inferior, becoming rather oblique throughout development (Hildebrand and Cable, 1938).

The more developed stages can be easily recognized by their two separate dorsal fins and a prominent gas bladder which is present slightly anterior to midbody (it usually disappears in the juvenile stage) (Russell, 1976; Ruple, 1984; Leis and Carson-Ewart, 2000). However, some exceptions exist, such as the genus *Luciogobius* that lacks the first dorsal fin entirely (Shiogaki and Dotsu, 1972; Okiyama, 1988), and *Leucopsarion petersii* and *Paedogobius kimurai* that have the first dorsal fin totally unformed after the late post-flexion larval stage (Maeda *et al.*, 2008).

Head spines are lacking in all described larval gobiids. Median finfolds and pectoral fins are present at hatching or develop in early larvae. The caudal fin is the first to form differentiated rays. The second dorsal and anal fins are the next to develop, followed by the first dorsal and pectoral fins. The pelvic fins are the last to form, starting during the flexion or post-flexion stages (Ruple, 1984).

Melanistic pigments in gobiid larvae range from nearly absent, as is the case of some amphidromous species (Keith, 2003; Keith *et al.*, 2005; McDowall, 2009), to extensive (e.g. *Bathygobius curacao*, Richards, 2006; Victor, 2006), but more commonly are lightly distributed. Pigmentation on the gas bladder is a prominent feature of most known gobiids. There are however exceptions, such as *Paratrypauchen microcephalus* and *Taenioides cirratus*, which entirely lack pigmentation in the gas bladder during their early development (Ruple, 1984). The most pronounced pigmentation present in many gobiids is the one found along the ventral surface of the body, in the gut region and the anal fin base. Throughout development, pigmentation often appears on the caudal peduncle, along the dorsal surface of the body, on the otic vesicles, on the tip of the lower jaw, along the mid-lateral posterior portion of the body, and on various fins (Ruple, 1984; Leis and Carson-Ewart, 2000). Size at notochord flexion varies extensively between species, from 3.4 mm (range: 3.0 – 3.7 mm) in *Papillogobius punctatus* (Neira *et al.*, 1998) to 7.6 mm (range: 6.7 – 8.6 mm) in *Acanthogobius flavimanus* (Watson, 1996). Transformation from larvae to juveniles also occurs at very different sizes, with individuals settling from the plankton at a mean size of 13.4 mm (SD = 5.9; N = 90;

range = 5.6 mm in *Bathygobius curacao* to 31.3 mm [27.8 – 38.1 mm] in *Sicyopterus japonicus*) (Table 3.4.1).

The typical developmental sequence for the Gobiidae larval stage may be exemplified by the genus *Gobius*. This developmental sequence is consistent among all species for which there are larval descriptions in this genus: *G. cobitis* (Spartá, 1950; Gil *et al.*, 1997), *G. paganellus* (Hefford, 1910; Lebour, 1919; Spartá, 1934; Borges *et al.*, 2003), *G. niger* (Lebour, 1919; Petersen, 1919; Padoa, 1956; Russell, 1976) and *G. xantocephalus* (Monteiro *et al.*, 2008) (see Fig. 3.4.4). The pigmentation pattern changes throughout development with an increase in the number and intensity of pigment cells in the cephalic region. Ventrally, the row of melanophores extends from the throat to near the caudal peduncle with some variability between species. At metamorphosis the fishes became heavily pigmented. Notochord flexion is completed between 4.3 to 8.0 mm and larvae start to settle at a size of 9.0 mm in *G. niger* (Petersen, 1919; Padoa, 1956; Russell, 1976), 11.6 mm in *G. paganellus* (range = 9.5 – 17.0; Borges *et al.*, 2003), 17.0 – 18.0 mm in *G. cobitis* from reared larvae and juveniles collected in tidepools (range = 13.0 – 19.0mm, N = 64; Gil *et al.*, 1997) and, based on reared larvae, at a probable size of around 13.0 mm in *G. xanthocephalus* (Monteiro *et al.*, 2008). These values were determined from larvae raised in captivity, but also from field samples and otolith derived information (Beldade *et al.*, 2007; unpublished data).

For these species, the transformation from the larval to the juvenile stage was gradual (Gil *et al.*, 1997; Borges *et al.*, 2003). Initially, larvae started to touch the substratum returning immediately to the water column. Increasingly, they began to stay longer at the bottom until definitely acquiring a benthic mode of life. Metamorphosis is a gradual process with most fish showing a marked change in body form before the full acquisition of juvenile pigmentation (Borges *et al.*, 2003).

The pelagic larval duration (PLD) also varies extensively between species, with larvae spending a minimum of 13 days (*Zosterisessor ophiocephalus*) and a maximum of 214 days (range: 131 – 266, *Sicyopterus lagocephalus*) in the plankton (Table 3.4.1). These quite disparate values result from the equally contrasting life history characteristics adopted by the Gobiidae (see below). Data from otolith readings for species of the genus *Gobius* in our study area (W Portugal), show an average PLD of 23 days (*G. paganellus* = 25.5; *G. xantocephalus* = 23.7; *G. gasteveni* = 20), with a range from 14 to 39 days (Beldade *et al.*, 2007; unpublished data).

Considering the species for which there is published information on the size at hatching, size at settlement and PLD (N = 22), size at settlement

A

B

C

D

E

F

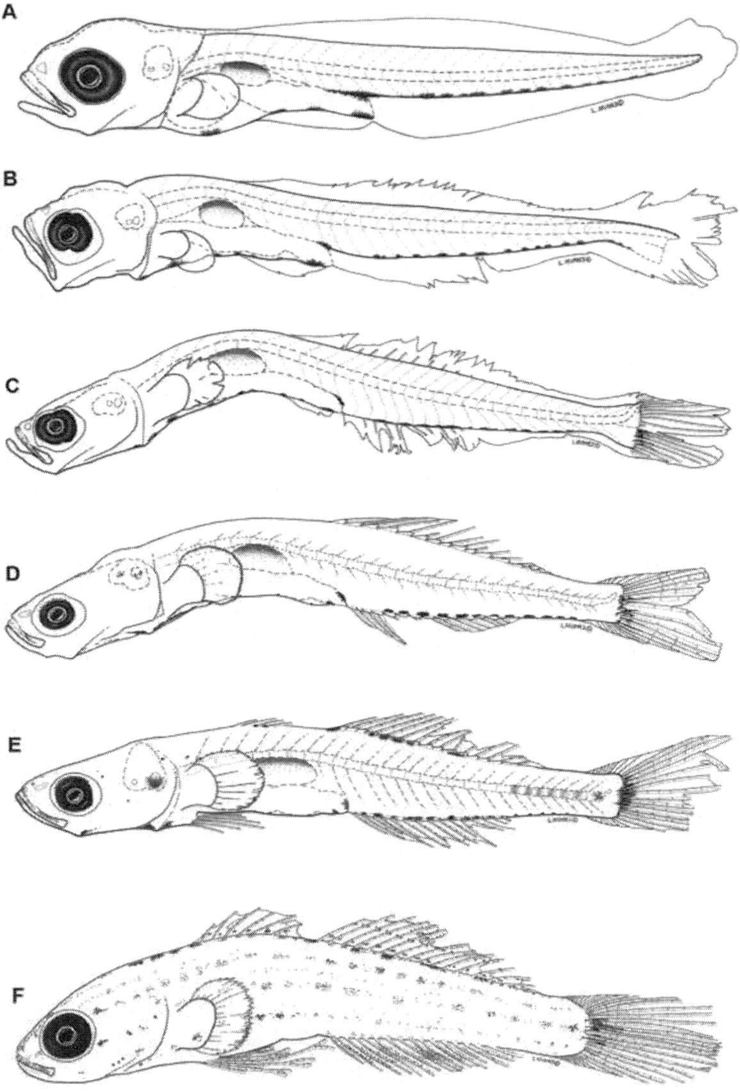

Fig. 3.4.4 Developmental sequence of *Gobius xanthocephalus* from reared larvae (A) and larvae collected in the plankton (B-F) (from Monteiro *et al.*, 2008). Recently hatched larvae : (A) 3.1 mm BL; Pre-flexion larva: (B) 3.95 mm BL; Post-flexion larva: (C) 5.6 mm BL; (D) 8.0 mm BL; (E) 11.1 mm BL; and (F) 15.0 mm BL.

shows a close relationship with the time spent in the pelagic environment (Spearman rank correlation, R = 0.78, p<0.001), regardless of the size at hatching, which showed a negative correlation with the PLD (Spearman rank correlation, R = -0.68, p < 0.001); size at settlement and size at hatching were negatively correlated (R = -0.42, p<0.05) (Fig. 3.4.5). Based on this information, the 'planktonic size' (= size at settlement – size at hatching) was determined, in order to derive the relationship between the amount of growth a larvae acquire while in the plankton, with the time spent there. There is a positive linear relationship between these two variables (PLD = 5.53 * planktonic size – 5.55; R^2 = 0.64, N = 22), although variation is larger for species that settle at a larger size (these are all amphidromous, see below). The number of species is however low and therefore it is not possible, at this stage, to derive a definite conclusion on the relative growth for larvae that spend a great amount of time in the plankton.

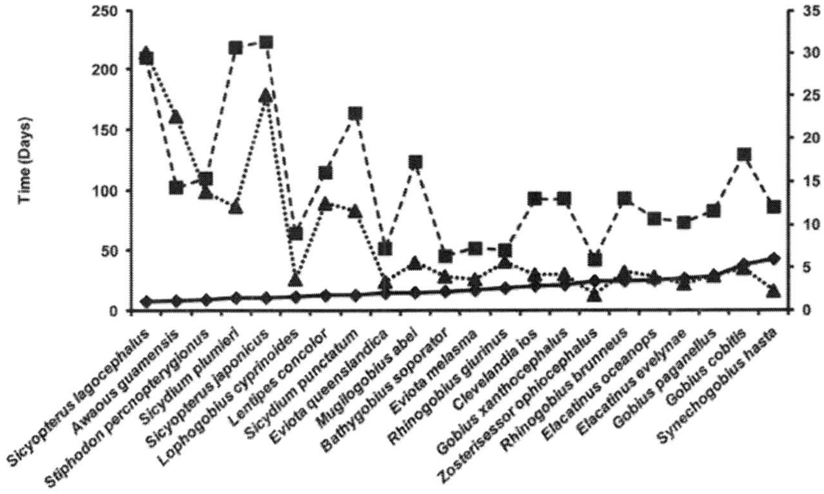

Fig. 3.4.5 Relationship between: Y1 (days) the pelagic larval duration (PLD) (triangles); and Y2 (mm) size at hatching (lozenges) and size at settlement (squares); for species with information for the three variables.

Taxonomic Comparison

We analysed the different available traits at the subfamily level for the variables for which we could compile enough information. The Gobionellinae exhibit larger eggs (Fig. 3.4.6a), longer embryonic developmental time (Fig. 3.4.6b) when compared to the other subfamilies

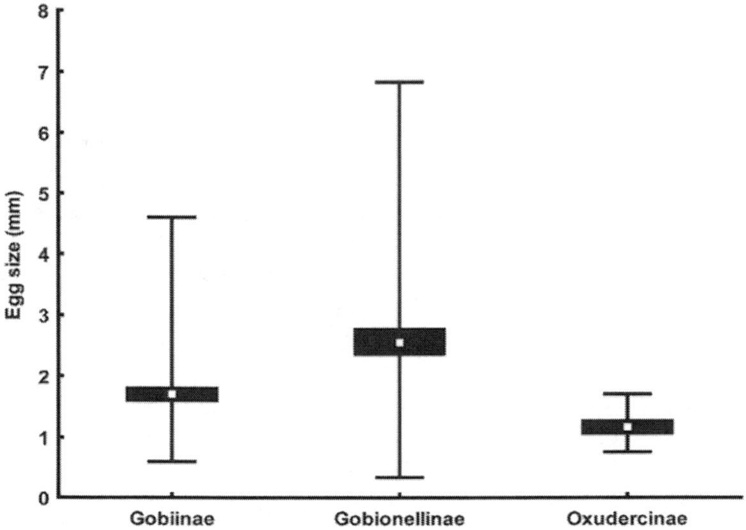

Fig. 3.4.6a Egg size (maximum axis) in the subfamilies Gobiinae , Gobionellinae and Oxudercinae (square = mean; black box = mean ± standard error; whiskers = range).

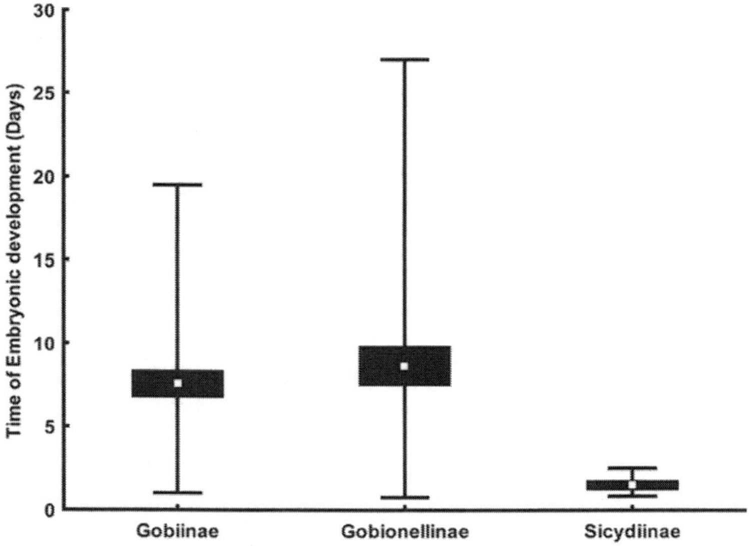

Fig. 3.4.6b Time of embryonic development in the subfamilies Gobiinae, Gobionellinae and Sicydiinae (square = mean; black box = mean ± standard error; whiskers = range).

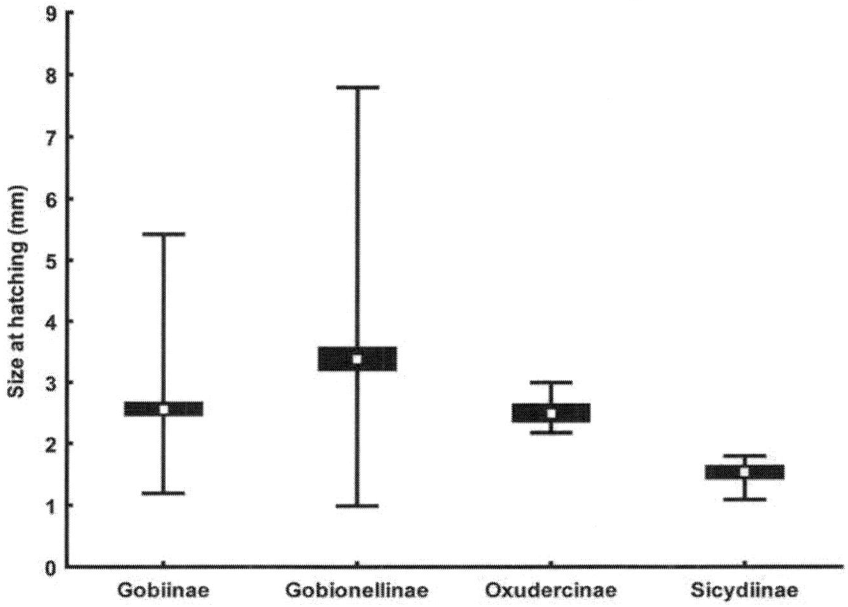

Fig. 3.4.6c Size at hatching in the subfamilies Gobiinae, Gobionellinae, Oxudercinae and Sicydiinae (square = mean; black box = mean ± standard error; whiskers = range).

(except the Oxudercinae, with information available for only 2 species), and larger newly-hatched larvae (Fig. 3.4.6c). On average, the Sicydiinae have smaller eggs (just one species with data, Table 3.4.1), the shortest embryonic developmental time (Fig. 3.4.6b) and the smallest newly-hatched larvae (Fig. 3.4.6c). On the other hand, they begin transformation at a larger size (Fig. 3.4.6d), presenting the longest PLD (Fig. 3.4.6e) and the largest planktonic size (mean = 23.2 mm; SD = 6.8; N = 7). Concerning other traits, the Gobiinae are smaller at metamorphosis, have a shorter PLD (Fig. 3.4.6d-e) and a smaller planktonic size (mean = 7.8 mm; SD = 3.3; N = 26). These results are statistically significant for most possible comparisons (considering only the subfamilies that had more than five species with information) (Table 3.4.2).

Ecological Comparison

Information on the species distribution and ecology was also compiled and species were pooled in temperate (including subtropical) and tropical regions and in marine, amphidromous and freshwater environments.

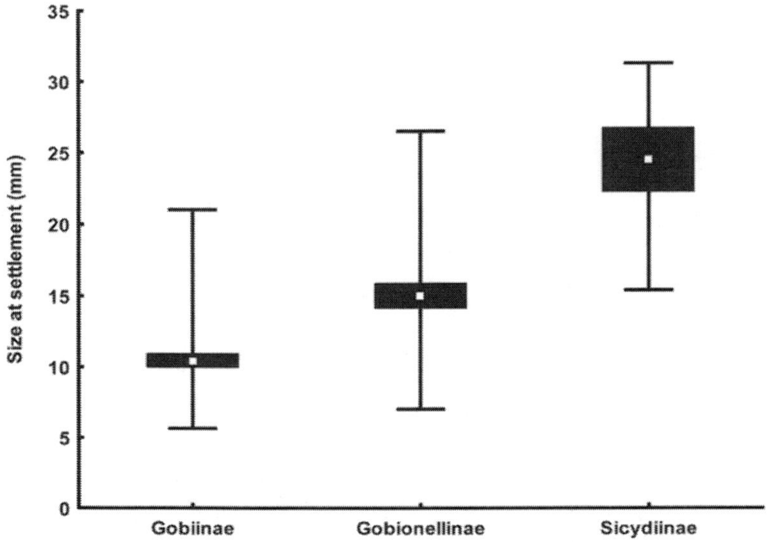

Fig. 3.4.6d Size at settlement in the subfamilies Gobiinae, Gobionellinae and Sicydiinae (square = mean; black box = mean ± standard error; whiskers = range).

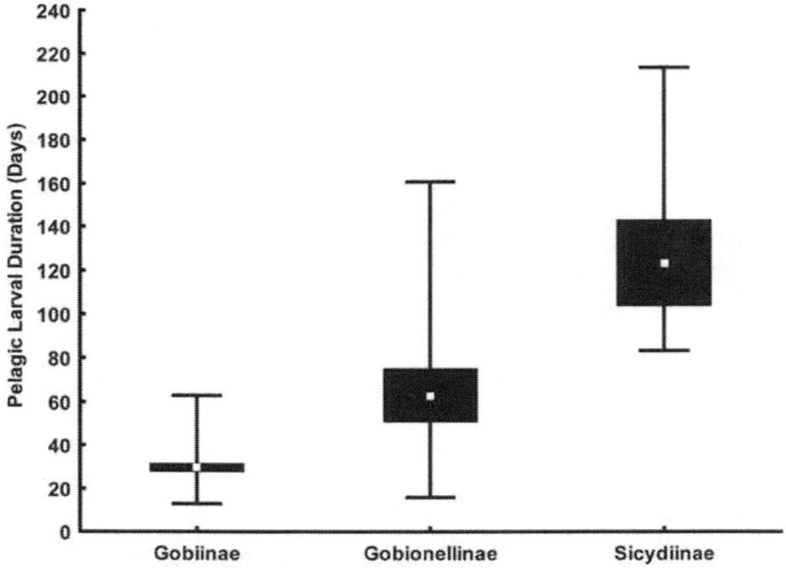

Fig. 3.4.6e Pelagic larval duration (PLD) in the subfamilies Gobiinae, Gobionellinae and Sicydiinae (square = mean; black box = mean ± standard error; whiskers = range).

Table 3.4.2 Taxonomical comparison of early life history traits among the subfamilies of the Gobiidae that presented enough data to allow statistical comparisons: (1) Gobiinae; (2) Gobionellinae; (3) Oxudercinae; (4) Sicydiinae. F = ANOVA; H = Kruskal-Wallis test; post-hoc comparisons obtained with the Newman-Keuls or Dunn's test, respectively. * = p < 0.05; ** = p < 0.01; *** = p < 0.001.

Variable	*Statistical comparison*	*N*	*Post-hoc comparisons*
Egg length	H = 17.58 ***	(1)72, (2)44, (3)8	1 × 2 *** 2 × 3 **
Embryonic developmental time	H = 13.06 **	(1)35, (2)30, (4)6	1 × 4 ** 2 × 4 **
Size at hatching	H = 28.78 ***	(1)65, (2)51, (3)5, (4)7	1 × 2 ** 1 × 4 * 2 × 4 ***
Size at settlement	F = 33.97 ***	(1)50, (2)31, (4)8	1 × 2 ** 1 × 4 *** 2 × 4 ***
Size in the plankton	F = 23.00 ***	(1)26, (2)23, (4)7	1 × 2 * 1 × 4 *** 2 × 4 ***
Pelagic larval duration (PLD)	H = 23.98 ***	(1)30, (2)13, (4)7	1 × 2 ** 1 × 4 ***

Tropical species exhibit smaller eggs (Fig. 3.4.7a), shorter embryonic developmental time (Fig. 3.4.7b) and smaller larvae at hatching (Fig. 3.4.7c), when compared with the temperate group, regardless of the environmental classification (marine, freshwater or amphidromous). Amphidromous species present different patterns on temperate and tropical regions. Tropical amphidromous have smaller eggs (Fig. 3.4.7a), shortest embryonic developmental time (Fig. 3.4.7b), the smallest larvae at hatching (Fig. 3.4.7c), the largest size at settlement (Fig. 3.4.7d), the largest planktonic size (mean = 16.7 mm, SD = 8.2, N = 11) and a much longer PLD (Fig. 3.4.7e). It is necessary to bear in mind that the amphidromous group contains species with different life-history trajectories in temperate and tropical regions (see Discussion). These differences are statistically significant for most possible comparisons (Table 3.4.3).

One result that should be highlighted is the difference in marine species between temperate and tropical regions since, contrary to what is many times assumed, temperate gobies presented a larger size at settlement but a shorter PLD than tropical gobies. If one compares size at hatching between temperate and tropical species, it becomes evident that tropical

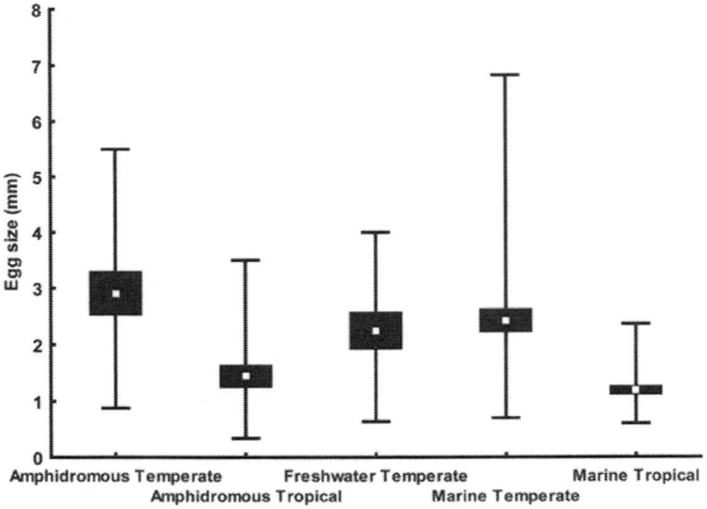

Fig. 3.4.7a Egg size (maximum axis) in the groups amphidromous temperate, amphidromous tropical, freshwater temperate, marine temperate and marine tropical (square = mean; black box = mean ± standard error; whiskers = range).

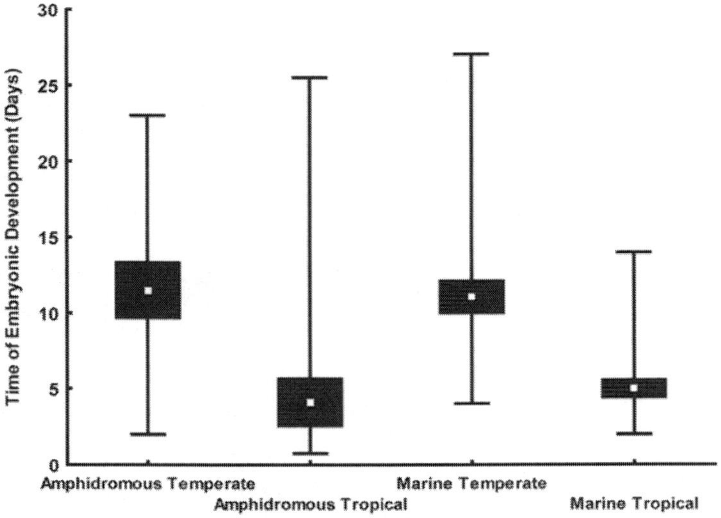

Fig. 3.4.7b Time of embryonic development in the groups amphidromous temperate, amphidromous tropical, marine temperate and marine tropical (square = mean; black box = mean ± standard error; whiskers = range).

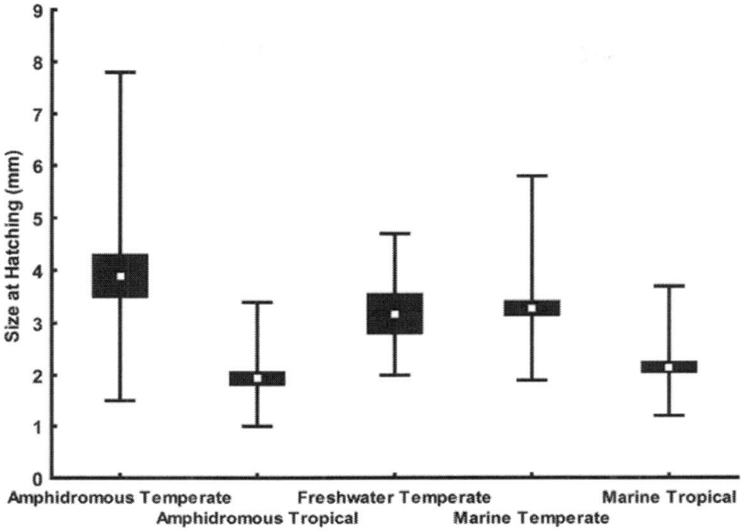

Fig. 3.4.7c Size at hatching in the groups amphidromous temperate, amphidromous tropical, freshwater temperate, marine temperate and marine tropical (square = mean; black box = mean ± standard error; whiskers = range).

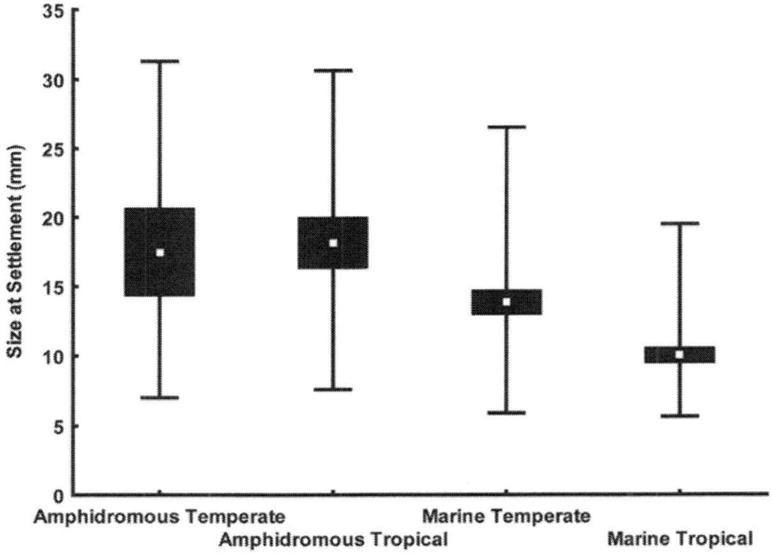

Fig. 3.4.7d Size at settlement in the groups amphidromous temperate, amphidromous tropical, marine temperate and marine tropical (square = mean; black box = mean ± standard error; whiskers = range).

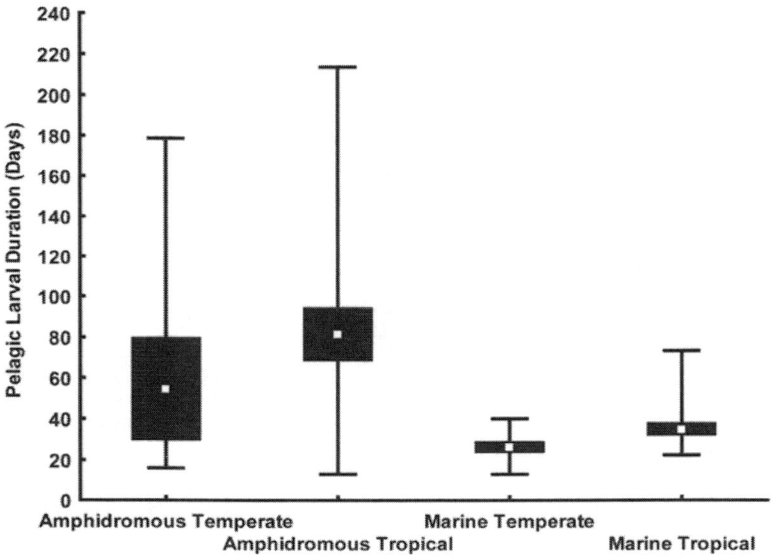

Fig. 3.4.7e Pelagic larval duration (PLD) in the groups amphidromous temperate, amphidromous tropical, marine temperate and marine tropical (square = mean; black box = mean ± standard error; whiskers = range).

species hatch significantly smaller than temperate species (mean temperate = 3.4 mm ± 1.2, N = 74; mean tropical = 2.1 ± 0.6, N = 55; t = 8.65, d.f. = 127, $p < 0.001$). In the case of the pelagic larval duration, temperate species spend less time in the plankton than tropical species, although this difference is not significant (mean temperate = 37.4 days ± 37.5, N = 17; mean tropical = 56.9 ± 44.1, N = 36; t= -1.58, n.s.). However, there are two factors that confound this result. On one side, tropical species include the amphidromous Sycidiinae which hatch very small and spend a great amount of time in the plankton. On the other hand, there are not enough species with data for both size at hatching and PLD to statistical disentangle this relationship when comparing both regions (see Discussion). Moreover, amphidromous species present a wide range of values for all this variables. When removing the amphidromous species from the analysis (comparing only marine species), the difference between size at hatching between the two regions is equally significant (mean temperate = 3.3 mm ± 1.0, N = 50; mean tropical = 2.2 ± 0.6, N = 31; t = 6.48, d.f. = 79, $p < 0.001$), and the difference in the PLD becomes less pronounced although it is still longer in the tropics (mean temperate = 26 days ± 8, N = 10; mean tropical = 35 days ± 13, N = 19; t = -1.88, p = n.s.).

Table 3.4.3 Taxonomical comparison of early life history traits among regions (temperate and tropical groups) and environments (marine, freshwater and amphidromous groups) that presented enough data to allow statistical comparisons: (1) amphidromous temperate; (2) amphidromous tropical; (3) freshwater temperate; (4) marine temperate; (5)marine tropical . F = ANOVA; H = Kruskal-Wallis test; post-hoc comparisons obtained with the Newman-Keuls or Dunn's test, respectively. * = $p < 0.05$; ** = $p < 0.01$; *** = $p < 0.001$

Variable	*Statistical comparison*	*N*	*Post-hoc comparisons*
Egg Length	H = 36.67 ***	(1)14, (2)18, (3)10, (4)46, (5)36	1 × 2 *
			1 × 5 ***
			2 × 4 *
			3 × 5 *
			4 × 5 ***
Embryonic developmental time	F = 9.54 ***	(1)11, (2)15, (4)24, (5)19	1 × 2 ***
			1 × 5 **
			2 × 4 **
			4 × 5 **
Size at hatching	F = 20.03 ***	(1)16, (2)22, (3)8, (4)50, (5)31	1 × 2 ***
			1 × 5 ***
			2 × 3 ***
			2 × 4 ***
			3 × 5 ***
			4 × 5 ***
Size at settlement	F = 10.75 ***	(1)7, (2)15, (4)31, (5)33	1 × 5 ***
			2 × 5 ***
			4 × 5 *
Size in the plankton	F = 4.35 **	(1)7, (2)11 , (4)26, (5)9	2 × 5 **
Pelagic larval duration (PLD)	H = 14.74 **	(1) 6, (2)17 , (4)10, (5)19	2 × 4 **
			2 × 5 *

DISCUSSION

This review attempts to systematize and revise the available life-history information for the family Gobiidae. Many egg and larval traits are highly variable within and between subfamilies and several factors contribute to this great variability, namely those related to the life history traits or the environment where these species thrive.

One of the main aspects emerging from our analysis is the uniqueness of the Sicydiinae studied so far. The larvae of Sicydiinae are clearly distinct from those of the other subfamilies: the embryonic developmental time is greatly reduced (may last only a few hours), the larvae hatch much smaller than all other gobies (just over 1mm in length), they settle at a larger size and spend up to 7-8 months in the plankton.

The particular early life traits of Sicydiinae gobies can be associated with their peculiar life cycle adapted to long dispersal periods, since these species are mainly distributed in oceanic islands (Iida *et al.*, 2009). In fact, sicydiines occur mostly in rivers of tropical islands and are amphidromous (McDowall, 2009). They hatch in rivers where the adults live and spawn and the larvae drift into the ocean where they develop and spend most of their pelagic phase, returning again to rivers just prior to settlement. Our analysis shows that amphidromous gobies have significantly longer larval phases when compared to marine gobies. Hoareau *et al.* (2007) found similar results when comparing the length of the larval phase of amphidromous species with that of marine demersal fishes and proposed that the highest habitat fragmentation was a plausible explanation for the observed differences.

These larvae usually hatch in a very early stage of development, often with the visual and digestive systems still non-functional and with considerable yolk reserves (McDowall, 2009). They hatch as an early free embryo (sensu Balon, 1975) which, together with physiological adaptations to the marine environment and some early behavioural patterns like positive phototaxis, may increase the chance of drifting downstream to the sea with reduced risk of starvation (Keith, 2003; McDowall, 2009).

The second important pattern revealed by the data is the difference found between temperate and tropical species. Within the amphidromous group, the larvae of the temperate gobies hatch bigger, spend less time in the plankton and settle smaller than the tropical ones. Once again, differences in the life history of these two groups must be considered: the kind of amphidromy is different given that, in temperate systems, it is not the larval stage but the adults that move between coastal areas, estuaries and rivers (as seems to be the case of *Acanthogobius flavimanus*, *Synechogobius hasta*, *Neogobius melanostomus* and *Chaenogobius castaneus*). In *Rhinogobius brunneus* it is the juvenile stage that moves to the sea, returning later to rivers. Only in *Rhinogobius giurinus*, a subtropical species, it is the larval stage that migrates, but in this case settlement occurs at a small size and the PLD is short (Chang *et al.*, 2008).

Within the marine species, differences were also found between tropical and temperate groups. Temperate species have bigger eggs and a longer embryonic development; the larvae hatch bigger and settle at a larger size when compared to tropical species. These results agree with those of

Privitera (2001) which found shorter incubation times, smaller eggs and smaller larvae at hatching in tropical gobies when compared to temperate ones. These differences were expected as temperature is known to influence metabolism and thus the developmental rates and early life history traits (Thresher, 1988). As noted by Privitera (2001), these comparisons were not calibrated for adult body size by, for example, comparing species with similar body sizes and/or using a correction factor to prevent possible allometric effects of body size on the derived trends, since temperate species tend to be bigger than tropical species.

Another relevant conclusion of this study is that, contrary to what is normally assumed, the pelagic larval duration (PLD) of tropical gobies was, if anything, longer than that of temperate species. This is in disagreement with some recent reviews (e.g. Bradbury *et al.*, 2008). These authors, found a significant increase in PLD with latitude using several groups of marine fish species. However, the majority of species analysed for temperate regions belong to families different from those included in the tropical group, both in terms of adult habitat use and in their early life-history traits. Although some caution must be taken given the reduced number of species analysed, the results presented herein call attention to the fact that the presumed differences between temperate and tropical systems may be smaller than what is normally assumed when the same taxonomic group is compared and the species' life history characteristics are considered.

Early Life History Data Requirements

One of the limitations of this type of reviews is the variety of procedures, methods and variables analysed in different studies and the lack of standardization of the datasets under study. In particular, data collected for taxonomic purposes are very distinct from those collected in ecological studies and it is sometimes difficult to bridge these datasets to allow an integration of both approaches. This is especially problematic since many early life-history traits are variable and dependent on environmental conditions such as temperature, location, season, maternal effects, etc. (Blaxter, 1969; Thresher, 1988; Bergenius *et al.*, 2005; Green and McCormick, 2005; Sponaugle *et al.*, 2002, 2006; Leis, 2006; Sponaugle and Grorud-Colvert, 2006; Cowen and Sponaugle, 2009; Sponaugle, 2009). Additionally, many times it would be very easy to include some measurements on a particular study, if common guidelines were available and broadly used. For example, in the dataset we collated, there were very few studies that included information on size/age at flexion, therefore preventing any meaningful comparison of this important life-history trait. On other occasions, egg size was not ascribed to a length or width measurement, in species in which the egg is not spherical. A greater attention to this type of details both by taxonomists, ecologists and fisheries biologists could greatly benefit our common knowledge on the early life-history of these fish.

Furthermore, when comparing ontogenetic events across species or between individuals developed under different conditions and temperature regimes, calibration is essential to guarantee an adequate interpretation of the patterns extracted from the data. In this sense, the Ontogenetic Index proposed by Fuiman (1994) (see also Fuiman *et al.*, 1998) should be widely adopted (an example of the usefulness of this index when applied to the development of gobies may be found in Monteiro *et al.*, 2008).

Guidelines for helping standardize methods and descriptions of larval fish have been described before (see Leis, 1993), but no wide dissemination of these procedures was accomplished. This has prevented a much needed quality increase both on the descriptions of fish larvae and on the associated ecological information, not withstand the difficulty *per se* of correctly identifying these early stages (Leis and Carson-Ewart, 2000). These guidelines include (see Leis, 1993 and references herein): i) showing how the larvae identity was established; ii) depositing voucher individuals in a Museum or similar institution with conditions to maintain larval collections and that keep them available for study; iii) providing comparisons with similar species and describe the unique characters to a certain species; iv) comparing reared larvae with field-caught individuals; v) discussing similarities with descriptions of related species from the literature; vi) including meristic and morphometric data in tabular or graphical form; vii) using the dynamic approach method (Leis and Rennis, 1983), by considering individual structures throughout development; viii) providing good quality illustrations; ix) staining larvae to better describe osteological structures such as spines; x) using clear terminology; xi) following the standard procedures published in Moser *et al.* (1984); xi) use as reference papers where these criteria have been met.

Further to this, we think that whenever possible, fish early life history information should contain the elements presented below.

Environmental data: (i) temperature; (ii) location; (iii) type of environment (marine, brackish, freshwater, etc.); (iv) season; (v) spawning area.

Rearing data: (i) temperature; (ii) rearing conditions (feeding, light, tanks, etc.).

Biological data of adults: (i) fish size; (ii) fecundity; (iii) number of eggs per batch; (iv) size of eggs per batch; (v) timing of spawning; (vi) spawning mode.

Early life measurements: (i) note on pre and post-fixation measurements (correct for shrinkage and indicate the preservative); (ii) use standard ontogenetic events, such as body length (use notochord length (NL) for pre-flexion larvae and standard length (SL) in post-flexion larvae; always note total length (TL) in intact specimens and present an SL/TL regression), size at hatching, size/age at flexion, size/age at fin development (specify for all fins and note the initiation or full completion of rays), size/age at settlement; (iii) measure changes with development on key traits, such as, pre-anal length, head length, eye diameter, body height and if possible provide allometric relationships (see Moser *et al.*, 1984; Monteiro *et al.*, 2008); (iv) egg shape and size (including both the long and short axis in non-spherical eggs).

Derived information: (i) pelagic larval duration (indicate the methods used: otolith readings on already settled or on recently settled juveniles; and estimations based on reared specimens); ii) otolith derived information (hatching date, growth rates, microchemistry interpretation of ecology and life-history data, etc.); iii) general pigmentation patterns and their dynamics throughout development, both from reared larvae and from field caught larvae.

Authors should also provide information on the sample size, range and coefficient of variation when computing average or modal values for all the variables under study.

The use of these guidelines could both increase the value of taxonomic studies to ecologists and ecological studies to taxonomists and would allow a more meaningful comparison than the one attempted in this review which, nevertheless, clearly revealed different early life strategies within the Gobiidae.

REFERENCES

Aboussouan, A. 1964. Contribution à l'étude des oeufs et larves pélagiques des poissons téléostéens dans le golfe de Marseille. *Revue Travaux de la Station Marine d'Endoume* 32: 87-171.

Able, K.W. and M.P. Fahay. 1998. *The First Year in the Life of Estuarine Fishes in the Middle Atlantic Bight*. Rutgers University Press, New Brunswick.

Acha, E.M. 1994. Development and occurrence of larvae of the goby, *Gobiosoma parri* (Ginsburg) (Gobiidae), in the estuary of the Rio de la Plata, Argentina. *Scientia Marina* 58: 337-343.

Akihito, I., K. Sakamoto, Y. Ikeda and A. Iwata. 2000. Gobioidei. In: *Fishes of Japan with keys to species*. 2nd Ed. T. Nakabo (Ed.). Tokai University Press, Tokyo, pp. 1139-310, 1606-28 (in Japanese).

Arakawa, T., Y. Kanno, N. Akiyama, T. Kitano, N. Nakatsuji and T. Nakatsuji. 1999. Stages of embryonic development of the ice goby (Shiro-uo), *Leucopsarion petersii. Zoological Science* 16: 761-773.

Baldwin, C.C. and D.G. Smith. 2003. Larval Gobiidae (Teleostei: Perciformes) of Carrie Bow Cay, Belize, Central America. *Bulletin of Marine Science* 72: 639-674.

Ballard, W.W. 1969. Normal embryonic stages of *Gobius niger jozo. Pubblicazioni della Stazione Zoologica di Napoli* 37: 1-17.

Balon, E.K. 1975. Terminology of intervals in fish development. *Journal of the Fisheries Research Board of Canada* 32: 1663-1670.

Barlow, G.W. 1961. Gobies of the genus *Gillichthys* with comments on the sensory canals as a taxonomic tool. *Copeia* 1961: 423-437.

Bay, L.K., K. Buechler, M. Gagliano and M.J. Caley. 2006. Intraspecific variation in the pelagic larval duration of tropical reef fishes. *Journal of Fish Biology* 68: 1206-1214.

Beldade, R., T. Pedro and E.J. Gonçalves. 2007. Pelagic larval duration of 10 temperate cryptobenthic fishes. *Journal of Fish Biology* 71: 376-382.

Bell, K.N.I. 1994. *Life Cycle, Early Life History, Fisheries and Recruitment Dynamics of Diadromous Gobies of Dominica, w.i., Emphasizing* Sicydium punctatum *Perugia.* PhD Thesis, Faculty of Science, Memorial University of Newfoundland, Canada.

Bell, K.N.I. 1999. An overview of goby-fry fisheries. *Naga, the ICLARM Quarterly* 22: 30-36.

Bell, K.N.I. 2007. Opportunities in stream drift: methods, goby larval types, temporal cycles, *in situ* mortality estimation, and conservation implications. In: *Biology of Hawaiian Streams and Estuaries.* N.L. Evenhuis & J.M. Fitzsimons (ed.). *Bishop Museum Bulletin in Cultural and Environmental Studies* 3: 35-61.

Bell, K.N.I. and J.A. Brown. 1995. Active salinity choice and enhanced swimming endurance in 0-d-old to 8-d-old larvae of diadromous gobies, including *Sicydium punctatum* (Pisces), in Dominica, West Indies. *Marine Biology* 121: 409-417.

Bell, K.N.I., P. Pepin and J.A. Brown. 1995. Seasonal, inverse cycling of length and age at recruitment in the diadromous gobies *Sicydium punctatum* and *Sicydium antillarum* in Dominica, West Indies. *Canadian Journal of Fisheries and Aquatic Sciences* 52: 1535-1545.

Behrents, K.C. 1987. The influence of shelter availability on recruitment and early juvenile survivorship of *Lythrypnus dalli* Gilbert (Pisces: Gobiidae). *Journal of Experimental Marine Biology and Ecology* 107: 45-59.

Bergenius, M.A.J., M.I. McCormick, M.G. Meekan and D.R. Robertson. 2005. Environmental influences on larval duration, growth and magnitude of settlement of a coral reef fish. *Marine Biology* 147: 291-300.

Bertolini, F., U. D'Ancona, E.P. Montalenti, S. Ranzi, L. Sanzo, A. Sparta, E. Tortonese and M. Vialli. 1956. *Uova, larve e stadi giovanili di Teleostei. Fauna e Flora del Golfo di Napoli,* Monograph 38: 1-1064.

Biagi, F., S. Gambaccini and M. Zazzetta. 1998. Settlement and recruitment in fishes: the role of coastal areas. *Italian Journal of Zoology* 65 (Suppl.): 269-274.

Bielsa, S., P. Francisco, S. Mastrorillo, and J.P. Parent. 2003. Seasonal changes of periphytic nutritive quality for *Sicyopterus lagocephalus* (Pallas, 1770) (Gobiidae) in three streams of Reunion Island. *Annales de Limnologie – International Journal of Limnology* 39: 115-127.

Bil'ko, V.P. 1968. Reproduction of Black Sea gobies in the Dnieper-Bug estuary. *Problems of Ichthyology* 8: 536-543.

Blanco, G.J. 1947. Artificial fertilization and embryology of *Mirogobius lacustris* Herre. *Philippine Journal of Science* 77: 83-87.

Blaxter, J.H.S. 1969. Development: eggs and larvae. In: *Fish Physiology, Volume III: Reproduction and Growth. Bioluminescence, Pigments, and Poisons*. W.S. Hoar and D.J. Randall (eds). Academic Press, New York, pp. 178-252.

Borges, R., C. Faria, M.F. Gil, E.J. Gonçalves and V.C. Almada. 2003. Embryonic and larval development of *Gobius paganellus* (Pisces: Gobiidae). *Journal of the Marine Biological Association of the United Kingdom* 83: 1151-1156.

Bradbury, I.R., B. Laurel, P.V.R. Snelgrove, P. Bentzen, S.E. Campana. 2008. Global patterns in marine dispersal estimates: the influence of geography, taxonomic category and life history. *Proceedings of the Royal Society B* 275: 1-7.

Breder, C.M., Jr. 1942. On the reproduction of *Gobiosoma robustum* Ginsburg. *Zoologica* 27: 61-65.

Breder, C.M., Jr. 1943. The eggs of *Bathygobius soporator* (Cuvier and Valenciennes) with a discussion of other non-spherical teleost eggs. *Bulletin Bingham Oceanographic Collection* 8: 1-49.

Breder, C.M. and D.E. Rosen. 1966. *Modes of Reproduction in Fishes*. T.F.H. Publications, Jersey City.

Breitburg, D.L. 1991. Settlement patterns and presettlement behavior of the naked goby, *Gobiosoma bosci*, a temperate oyster reef fish. *Marine Biology* 109: 213-221.

Brillet, C. 1976. Structure du terrier, reproduction et comportement des jeunes chez le poisson amphibie *Periophthalmus sobrinus* Eggert. *Revue d'Ecology: La Terre e La Vie* 30(3): 465-483.

Brothers, E.B. 1975. *The Comparative Ecology and Behavior of Three Sympatric California Gobies*. PhD Thesis, University of California, San Diego.

Brothers, E.B., D. McB. Williams and P.F. Sale. 1983. Length of larval life in twelve families of fishes at 'One Tree Lagoon', Great Barrier Reef, Australia. *Marine Biology* 76: 319-324.

Carter, W.R. 1965. Racial variations of the arrow goby, *Clevelandia ios* (Jordan and Gilbert, 1882), in Puget Sound and on the coast Ji-Washington State. M.S. Thesis, Univ. Wash. 91 pp.

Chang, M-Y., W-N. Tzeng, C-H. Wang and C-F. You. 2008. Differences in otolith elemental composition of the larval *Rhinogobius giurinus* (Perciformes, Gobiidae) among estuaries of Taiwan: implications for larval dispersal and connectance among metapopulations. *Zoological Studies* 47: 676-684.

Charlebois, P.M., J.E. Marsden, R.G. Goettel, R.K. Wolfe, D.J. Jude, and S. Rudnika. 1997. *The Round Goby,* Neogobius melanostomus *(Pallas), a Review of European and North American Literature*. Illinois-Indiana Sea Grant Program and Illinois Natural History Survey. INHS Special Publication No. 20, Illinois.

Chen, Y-D. and Y-Y. Ting. 1984. The preliminary report on inducing breeding experiments with mudskipper, *Boleophthalmus pectinirostris* (L.) by injection of HCG (in Chinese). *Bulletin of the Taiwan Fisheries Research Institute* 37: 133-144.

Cole, K.S. and D.R. Robertson. 1988. Protogyny in the Caribbean reef goby, *Coryphopterus personatus*: gonad ontogeny and social influences on sex change. *Bulletin of Marine Science* 42: 317-333.

Coleman, N. 1986. *Australian Sea Fishes South of 30°*. Double Day, Sydney, Australia.

Coleman, R.M. 1991. Measuring parental investment in nonspherical eggs. *Copeia* 1991: 1092-1098.

Colin, P.L. 1975. *Neon gobies*. T.F.H. Publishers, Neptune City.

Cowen, R.K. and S. Sponaugle. 2009. Larval dispersal and marine population connectivity. *Annual Revue of Marine Science* 1: 443-466.

Danois, E. 1913. Contribution a l'Étude systématique et biologique des Poissons de la Manche Ocidentale. *Annales de l'Institut Oceanographique*, Monaco 5: 1-214.

Daoulas, C., A.N. Economou, T. Psarras and R. Barbieri-Tseliki. 1993. Reproductive strategies and early development of three freshwater gobies. *Journal of Fish Biology* 42: 749-776.

Darcy, G.H. 1978. *Ecology of the Crested Goby,* Lophogobius Cyprinoides *(Pisces: Gobiidae), in Southern Florida.* M.Sc. Thesis, University of Miami.

Darcy, G.H. 1980. *Comparison of Ecological and Life History Information on Gobiid Fishes, with Emphasis on the South-Eastern United States*. NOAA Technical Memorandum NMFS-SEFC-15. U.S. Department of Commerce.

Dawson, M.N., K.D. Louie, M. Barlow, D.K. Jacobs and C.C. Swift. 2002. Comparative phylogeography of sympatric sister species, *Clevelandia ios* and *Eucyclogobius newberryi* (Teleostei, Gobiidae), across the California Transition Zone. *Molecular Ecology* 11: 1065-1075.

Delmonte, P.J., I. Rubinoff and R.W. Rubinoff. 1968. Laboratory rearing through metamorphosis of some Panamanian gobies. *Copeia* 1968: 411-412.

Demir, N. and F.S. Russel. 1971. On the postlarva of the goby *Lebetus*. *Journal of the Marine Biological Association of the United Kingdom* 51: 669-678.

Depczynski, M. and D.R. Bellwood. 2005. Shortest recorded vertebrate lifespan found in a coral reef fish. *Current Biology* 15: 288-289.

Depczynski, M. and D.R. Bellwood. 2006. Extremes, plasticity, and invariance in vertebrate life history traits: insights from coral reef fishes. *Ecology* 87: 3119-3127.

D'Onofrio, E., G. Gibertini, V. Margotta and S. Zerunian. 1990. The biology of *Gobius nigricans* (Osteichthyes, Gobiidae). 2. Aspects of embryonic development and morphology of the fry. *Rivista di Idrobiologia* 29: 185-199.

Dotsu, Y. 1984. The biology and induced spawning of the gobiid fish, *Chaenogobius heptacanthus*. *Bulletin of the Faculty of Fisheries, Nagasaki University* 55: 9-18.

Dotsu, Y. and M. Shiogaki. 1971. Larvae, juveniles, and young of the gobiid fish, *Astrabe lactisella*. *Japanese Journal of Ichthyology* 18: 182-186.

Dotsu, Y. and M. Suzuki. 1985. Studies on propagation and culture of mudskipper, *Boleophthalmus pectinirostris* -II. Distribution and rearing experiments of larvae. In: *Studies on Distribution and Propagation of Useful Organisms of Mudflats in the Ariake Sound*. Report of a Grants-in-Aid for Scientific Research from the Ministry of Education (No. 58480065) of Japan. pp. 14-24.

Dotsu, Y. and T. Uchida. 1979. Outline of life history of gobioid fishes exemplified by the ice-goby. *Aquabiology* 1: 9-16.

Dotu, Y. 1954. On the life history of a goby, *Chaenogobius castanea* O'Shaughnessy. *Japanese Journal of Ichthyology* 11: 133-138.

Dotu, Y. 1955a. The life history of a goby, *Chaenogobius urotaenia* (Hilgendorf). *Science Bulletin of the Faculty of Agriculture, Kyushu University* 15: 367-374.

Dotu, Y. 1955b. On the habits and egg-development of a goby, *Pterogobius zonolecus* Jordan et Snyder. *Science Bulletin of the Faculty of Agriculture, Kyushu University* 15: 483-487.

Dotu, Y. 1955c. On the life history of a gobioid fish, *Eutaeniichthys gilli* Jordan et Snyder. *Bulletin of the Biogeographical Society of Japan* 16-19: 338-344.

Dotu, Y. 1955d. Life history of a goby, *Gobius poecilichthys* Jordan et Snyder. *Science Bulletin of the Faculty of Agriculture, Kyushu University* 15: 77-85.

Dotu, Y. 1956. On the habits and egg-development of a goby, *Pterogobius zonoleuscus* Jordan et Snyder. *Science Bulletin of the Faculty of Agriculture, Kyushu University* 15: 483-487.

Dotu, Y. 1957a. On the bionomics and life history of the eel-like goby, *Odontamblyopus rubicundus* (Hamilton). *Science Bulletin of the Faculty of Agriculture, Kyushu University* 16: 101-110.

Dotu, Y. 1957b. The life history of the goby, *Luciogobius guttatus* Gill. *Science Bulletin of the Faculty of Agriculture, Kyushu University* 16: 93-100.

Dotu, Y. 1957c. The life history of the small, transparent goby, *Gobius lidwilli* McCulloch. *Science Bulletin of the Faculty of Agriculture, Kyushu University* 16: 85-92.

Dotu, Y. 1957d. The bionomics and life history of the goby, *Triaenopogon barbatus* (Gunther) in the innermost part of Ariake Sound. *Science Bulletin of the Faculty of Agriculture, Kyushu University* 16: 261-274.

Dotu, Y. 1957e. The bionomics and life history of a gobioid fish, *Peleatogobius uchidai* Takagi.] *Japanese Journal of Ichthyology* 5: 97-104 (in Japanese with English abstract).

Dotu, Y. 1957f. The bionomics and life history of two gobioid fishes, *Tridentiger undicervicus* Tomiyama and *Tridentiger trigonocephalus* (Gill) in the innermost part of Ariake Sound. *Science Bulletin of the Faculty of Agriculture, Kyushu* University 16: 343-358.

Dotu, Y. 1958a. The bionomics and life history of the gobioid fish, *Ctenogobius dotui* Takagi. *Science Bulletin of the Faculty of Agriculture, Kyushu University* 16: 427-432.

Dotu, Y. 1958b. The bionomics and life history of the gobioid fish, *Paleatogobius uchidae* Takagi. *Japanese Journal of Ichthyology* 6: 97-104.

Dotu, Y. 1958c. The life history of the gobioid fish, *Acentrogobius masago* (Tomiyama). *Science Bulletin of the Faculty of Agriculture, Kyushu University* 16: 359-370.

Dotu, Y. 1958d. The bionomics and life history of two gobioid fish, *Tridentiger undicervicus* Tomiyama and *Tridentiger trigonocephalus* (Gill) in the innermost part of Ariake Sound. *Bulletin of the Faculty of Agriculture, Kyushu University* 16: 343-358.

Dotu, Y. 1961a. The bionomics and life history of the gobioid fish, *Chaenogobius scrobiculatus* Takagi. *Bulletin of the Faculty of Fisheries, Nagasaki University* 10: 127-131.

Dotu, Y. 1961b. The bionomics and life history of the gobioid fish, *Rhinogobius giurinus* (Rutter). *Bulletin of the Faculty of Fisheries, Nagasaki University* 10: 120-125.

Dotu, Y. 1961c. The bionomics and life history of the gobioid fish, *Apocryptodon bleekeri* (Day). *Bulletin of the Faculty of Fisheries, Nagasaki University* 10: 133-139.

Dotu, Y. 1965. On the life history of a gobioid fish, *Eutaeniichthys gilli* Jordan et Snyder. *Bulletin of the Biogeographical Society of Japan* 16-19: 338-344.

Dotu, Y. and S. Fujita. 1963. The nesting behavior, egg development and larva of the gobiid fish, *Lubricogobius exiguus* Tanaka. *Bulletin of Japanese Society of Scientific Fisheries* 29: 969-975.

Dotu, Y. and S. Mito. 1955a. On the breeding-habits, larvae and young of a goby, *Acanthogobius falvimanus* (Temminck et Schlegel). *Japanese Journal of Ichthyology* 4: 153-161.

Dotu, Y. and S. Mito. 1955b. Life history of the gobioid fish, *Sicydium japonicum* Tanaka. *Scientific Bulletin of the Faculty of Agriculture, Kyushu University* 10: 120-126.

Dotu, Y. and S. Mito. 1958. The bionomics and life history of the gobioid fish, *Luciogobius saikaiensis* Dôtu. *Scientific Bulletin of the Faculty of Agriculture, Kyushu University* 16: 419-426.

Dotu, Y. and S. Mito. 1963. The bionomics and life history of the eleotrid fish *Asterropteryx semipunctatus* Rüpell. *Bulletin of the Faculty of Fisheries, Nagasaki University* 15: 10-16.

Dotu, Y. and T. Takita. 1967. Induced spawning by hormone operation, egg development and larva of blind gobioid fish, *Odontamblyopus rubicundus*. *Bulletin of the Faculty of Fisheries, Nagasaki University* 23: 135-144.

Dotu, Y. and T. Tsutsumi. 1959. The reproductive behavior in the gobioid fish, *Pterogobius elapoides* (Günther). *Bulletin of the Faculty of Fisheries, Nagasaki University* 8: 186-190.

Dotu, Y., S. Mito and M. Ueno. 1955. The life history of a goby, *Chaeturichthys hexanema* Bleeker. *Scientific Bulletin of the Faculty of Agriculture, Kyushu University* 15: 359-365.

Dotsu, Y., I. Arima and S. Mito. 1965. Biology of the eleotrid fishes, Eviota adax and Eviota zonura. Bulletin of the Faculty of Fisheries, Nagasaki University 18: 41-49.

Ebert, E.E. and C.H. Turner. 1962. The nesting behavior, eggs and larvae of the bluespot goby. *California Fish and Game* 48: 249-252.

Economou, A.N., C. Daoulas, T. Psarras and R. Barbieri-Tseliki. 1994. Further data on the reproduction and development of *Knipowitschia caucasica* (Gobiidae). *Journal of Fish Biology* 45: 360-362.

Ehrenbaum, E. 1905-1909. *Eier und Larven von Fischen*. Nordisches Plankton 1. Lipsius & Tischer, Kiel, Leipzig, pp. 1-414.

Ehrenbaum, E. and S. Strodtmann. 1904. Eier und Jungendformen der Ostseefische. *I. Bericht. Deutschen wissenschaftlichen Kommision für die internationale Meeresforschung. B. Biologischen Anstalt auf Helgoland* 1: 59-126.

Eigenmann, C.H. 1892. The fishes of San Diego, California. *Proceedings of the National Museum* 15: 123-178.

Erdman, D.S. 1961. Notes on the biology of the gobiid fish *Sicydium plumieri* in Puerto Rico. *Bulletin of Marine Science of the Gulf and Caribbean* 11: 448-456.

Erdman, D.S. 1986. The green stream goby *Sicydium plumeri*, in Puerto Rico. *Tropical Fish Hobbyist* 2: 70-74.

Fahay, M.P. 1983. Guide to the early stages of marine fishes occurring in the Western North Atlantic Ocean, Cape Hatteras to the southern Scotian Shelf. *Journal of Northwest Atlantic Fishery Science* 4: 296-301.

Fahay, M.P. 2007. *Early Stages of Fishes in the Western North Atlantic Ocean (Davis Strait, Southern Greenland and Flemish Cap to Cape Hatteras)*. Northwestern Atlantic Fisheries Organization, Dartmouth, Nova Scotia.

Fage, L. 1918. *Shore-fishes*. Report on the Danish Oceanographical Expeditions 1908-1910 to the Mediterranean and Adjacent Seas. 2 - Biology, A.3: 1-154.

Feddern, H.A. 1967. Larval development of the neon goby, *Elacatinus oceanops*, in Florida. *Bulletin of Marine Science* 17: 367-375.

Fitzsimons, J.M., R.T. Nishimoto and A.R. Yuen. 1993. Courtship and territorial behavior in the native Hawaiian stream goby, *Sicyopterus stimpsoni*. *Ichthyological Exploration of Freshwaters* 4: 1-10.

Fives, J.M. 1970. Investigations of the plankton of the west coast of Ireland – IV. Larval and post-larval stages of fishes taken from the plankton of the west coast in surveys during the year 1958-1966. *Proceedings of the Royal Irish Academy* 70: 15-93.

Foster, N.R. and L.A. Fuiman. 1987. Notes on the behavior and early life history of *Evorthodus lyricus*. *Bulletin of Marine Science* 41: 27-35.

Franco, A., S. Malavasi, F. Pranovi, P. Franzoi and P. Torricelli. 2002. Preliminary data on gonadal development and fecundity in the grass goby, *Zosterisessor ophiocephalus* (Pallas, 1811) from the Venice lagoon (northern Italy). *Acta Adriatica* 43: 43-48.

Fritzsche, R.A. 1978. *Development of Fishes of the Mid-AtlanticBight, an Atlas of Egg, Larval and Juvenile Stages. Vol. V: Chaetodontidae through Ophidiidae.* U.S. Department of Interior, Fish and Wildlife Service, Biology Service Program FWS/OBS-78/12.

Froese, R. and D. Pauly (eds.). 2009. FishBase. World Wide Web electronic publication. www.fishbase.org, version (07/2009).

Fuiman, L.A. 1994. The interplay of ontogeny and scaling in the interactions of fish larvae and their predators. *Journal of Fish Biology* 45: 55-79.

Fuiman, L.A., K.R. Poling and D.M. Higgs. 1998. Quantifying developmental progress for comparative studies of larval fishes. *Copeia* 1998: 602-611.

Garcia, A.M.A. and P.D. Moyano. 1990. Estados juveniles de la ictiofauna en los Caños de las salinas de la Bahia de Cadiz. Instituto de Ciencias Marinas de Andalucia. *Consejo Superior de Investigaciones Cientificas.*

Garcia, F.Z., C.F. Coto and M.d.D.M. Vargas. 1988. Desarrollo y distribucion larvaria de *Gobiosoma robustrum* Ginsburg (Pisces: Gobiidae). Laguna de terminos, Campeche. *Anales del Instituto de Ciencias del Mar y Limnologia Universidad Nacional Autonoma de Mexico* 15(1): 237-244.

Gibson, R.N. 1970. Observations on the biology of the giant goby, *Gobius cobitis* Pallas. *Journal of Fish Biology* 2: 281-288.

Gil, M.F., E.J. Gonçalves, C. Faria, V.C. Almada, C. Baptista and H. Carreiro. 1997. Embryonic and larval development of the giant goby *Gobius cobitis* (Pisces: Gobiidae). *Journal of Natural History* 31: 799-804.

Gil, F., R. Borges, C. Faria and E.J. Gonçalves. 2002. Early development of the red mouthed goby, *Gobius cruentatus* (Gobiidae). *Journal of the Marine Biological Association of the United Kingdom* 82: 161-163.

Gilchrist, J.D.F. 1916. Eggs and larvae of Cape fishes. *Marine Biological Report* (Union of South Africa) 3: 1-26.

Grabowska, J. 2005. Reproductive biology of racer goby *Neogobius gymnotrachelus* in the Wloclawski Reservoir (Vistula River, Poland). *Journal of Applied Ichthyology* 21: 296-299.

Green, B.S. and M.I. McCormick. 2005. Maternal and paternal effects determine size, growth and performance in larvae of a tropical reef fish. *Marine Ecology Progress Series* 289: 263-272.

Guitel, F. 1892. Observations sur les Moeurs du *Gobius minutus*. *Archives de Zoologie Experimentale et Generale* 10: 499-555.

Ha, P.Y. and R.A. Kinzie, III. 1996. Reproductive biology of *Awaous guamensis,* an amphidromous Hawaiian goby. *Environmental Biology of Fishes* 45: 383-396.

Hagiwara, K., N. Sakurai, Y. Kitajima, Y. Koshikawa and M. Machii 1993. Tideland ecology studies. 3. Development of seeding production of the mud-skipper *Periophthalmus modestus. Annual Report, Kajima Technical Research Institute* 41: 343-349.

Hamada, K. 1968. Development of a goby, *Chaenogobius urotaenia*, with special reference to the gill and chloride cell. *Bulletin of the Faculty of Fisheries, Hokkaido University* 19: 185-198.

Han, K.H., Y.U. Kim and K.J. Choe 1998. Spawning behavior and development of eggs and larvae of the Korea freshwater goby *Rhinogobius brunneus* (Gobiidae: Perciformes). *Journal of Korean Fisheries Society* 31: 114-120.

Healey, M.C. 1971. The distribution and abundance of sand gobies, *Gobius minutus*, in the Ythan estuary. *Journal of Zoology* 163: 177-229.

Hefford, A.E. 1910. Notes on the teleostean ova and larvae observed at Plymouth in spring and summer, 1909. *Journal of the Marine Biological Association of the United Kingdom* 9: 2-58.

Herrera, C.G. 1984. Description of post-embryonic stages of *Ophiogobius jenynsi* Hoese 1976 (Gobiidae, Blennioidei). *Revista de biologia marina*, Vina del Mar 20: 159-168.

Hildebrand, S.F. and L.E. Cable. 1938. Further notes on the development and life history of some teleosts at Beaufort, N.C. *Bulletin of the Bureau of Fisheries* 48: 505-642.

Hoareau, T. 2005. *Structural Dynamics of Bichiques (*s. lagocephalus*), Amphidromous Gobiids of la Reunion Rivers*. PhD Thesis, University of La Réunion.

Hoareau, T., R. Lecomte-Finiger, P. Bosc and P. Berrebi. 2004. Surprising larval life duration of *Sicyopterus* and *Cotylopus* gobiids in Mascarene islands. In: *Poster at the Third International Symposium on Fish Otolith Research and Application*, Townsville, Queensland, Australia, 11–16 July 2004.

Hoareau, T.B., R. Lecomte-Finiger, H.-R. Grondin, C. Conand and P. Berrebi. 2007. Oceanic larval life of La Réunion 'bichiques', amphidromous gobiid post-larvae. *Marine Ecology Progress Series* 333: 303-308.

Holt, E.W.L. 1890. On the ova of *Gobius*. *Annals and Magazine of Natural History* 6: 34-40.

Holt, E.W.L. and L.W. Byrne. 1898. Notes on the reproduction of teleostean fishes in the sout-western district. *Journal of the Marine Biological Association of the United Kingdom* 5: 333-340.

Hsieh, C.H. and T.S. Chiu. 2002. Summer spatial distribution of copepods and fish larvae in relation to hydrography in the northern Taiwan Strait. *Zoological Studies* 41: 85-98.

Iglesias, J. 1979. Eficiencias de crescimiento y fases embrionarias de *Gobius niger* L. 1758 (Gobiidae, Peces) sujeto a condiciones experimentales. *Boletin del Instituto Español de Oceanografia* 277: 162-172.

Iglesias, M. and B. Morales-Nin. 2001. Life cycle of the pelagic goby *Aphia minuta* (Pisces: Gobiidae). *Scientia Marina* 65: 183-192.

Iida, M., S. Watanabe, A. Shinoda and K. Tsukamoto. 2008. Recruitment of the amphidromous goby *Sicyopterus japonicus* to the estuary of the Ota River, Wakayama, Japan. *Environmental Biology of Fishes* 83: 331-341.

Iida, M., S. Watanabe and K. Tsukamoto. 2009. Life history characteristics of a Sicydiine goby in Japan, compared with its relatives and other amphidromous fishes. In: *Challenges for Diadromous Fishes in a Dynamic Global Environment.*

A.J. Haro, K.L. Smith, R.A. Rulifson, C.M. Moffitt, R.J. Klauda, M.J. Dadswell, R.A. Cunjak, J.E. Cooper, K.L. Beal, and T.S. Avery (eds.), Bethesda, Maryland. *American Fisheries Society, Symposium* 69: 355-373.

Ishikawa, M. and N. Nakamura. 1940. On the breeding habit and development of a goby, *Glossogobius brunneus* (Temminck and Schlegel). *Bulletin of Japanese Society of Science and Fisheries* 8: 365-366.

Ishimatsu, A., Y. Yoshida, N. Itoki, T. Takeda, H.J. Lee and J.B. Graham. 2007. Mudskippers brood their eggs in air but submerge them for hatching. *The Journal of Experimental Biology* 210: 3946-3954.

Ishino, K. 1986. Morphological and ecological studies on three types of *Caenogobius annularis* Gill. Special reference to speciation and adaptation. PhD Thesis, Hokkaido University, Japan.

Jones, S. 1937. Observations on the breeding habits and development of certain brackish water fishes of Adyar, Madras. *Proceedings of the Indian Academy of Science, Sector B* 5: 261-289.

Joyeux, J.-C., J.-L. Bouchereau and J.-A. Tomasini. 1991. Data about reproduction of *Gobius niger* (Pisces, Gobiidae) in brackish lagoon of Mauguio, France: Gonosomatic ratios, fecundities, spawning, eggs and larvae. *Vie et Milieu* 41: 97-106.

Kanabashira, Y., H. Sakai and F. Yasuda. 1980. Early development and reproductive behaviour of the Gobiid fish, *Mugiologobius abei. Japanese Journal of Ichthyology* 27: 191-198.

Keith, P. 2003. Biology and ecology of amphidromous Gobiidae of the Indo-Pacific and the Caribbean regions. *Journal of Fish Biology* 63: 831-847.

Keith, P., T. Galewski, G. Cattaneo-Berrebi, T. Hoareau and P. Berrebi. 2005. Ubiquity of *Sicyopterus lagocephalus* (Teleostei: Gobioidei) and phylogeography of the genus *Sicyopterus* in the Indo-Pacific area inferred from mitochondrial cytochrome b gene. *Molecular Phylogenetics and Evolution* 37: 721-732.

Kent, D.I. and J.B. Marliave. 1997. Early life history of the arrow goby, *Clevelandia ios* (Jordan & Gilbert), Gobiidae. *Micronesica* 30: 15-23.

Kido, M.H. and D.E. Heacock. 1991. The spawning ecology of ‘o’opu-nakea (*Awaous stamineus*) in Wainiha River and other selected north shore Kauai rivers. In: *New Directions in Research, Management and Conservation of Hawaiian Freshwater Stream Ecosystems: Proceedings of the 1990 Symposium on Freshwater Stream Biology and Fisheries Management*, W.S. Devick (ed.). Division of Aquatic Resources, Department of Land and Natural Resources, Honolulu, pp. 142-157.

Kim, Y.U. 1975. On the morphology of larval and young stages of *Chasmichthys dolichognathus* Hilgendorf. *Bulletin of the Korean Fisheries Society* 8: 225-233.

Kimura, M. 1958. Early development of the mud-skipper, *Periophthalmus cantonensis* (Osbeck). *Bulletin of the Japanese Society of Science and Fisheries* 23: 754-757.

Kinzer, J. 1960. Zur Ethologie und Biologie der Gobiiden, unter besonderer Berücksichtigung der Schwarzgrundel *Gobius jozo* L. *Zoologische Beiträge* 6: 207-290.

Kinzie, R.A. III. 1988. Habitat utilization by Hawaiian stream fishes with reference to community structure in oceanic island streams. *Environmental Biology of Fishes* 22: 179-192.

Kinzie, R.A. III. 1993. Reproductive biology of an endemic, amphidromous goby *Lentipes concolor* in Hawaiian streams. *Environmental Biology of Fishes* 37: 257-268.

Kishi, Y. 1978. Egg size difference among three populations of the goby, *Tridentiger obscurus*. *Japanese Journal of Ichthyology* 24: 278-284.

Kobayashi, S., Y. Yamanaka and R. Torii. 1950. Spawning habit of Chaenogobius isaza Tanaka (preliminary report). *Bulletin of the Japanese Society of Scientific Fisheries* 15: 808-812.

Kobayashi, T., Y. Dotsu and N. Miura. 1972. Egg development and rearing experiments of the larvae of the mud skipper, *Periophthalmus cantonensis*. *Bulletin of the Faculty of Fisheries, Nagasaki University* 33: 49-62.

Koblitskaya, A.F. 1961. New information on the biology of the bobyrg-goby *Pomatoshcistus caucasicus* (Kawarajsky) from the Volga delta. *Voprosi Ikhtiology* 2: 253-261.

Koki, K., K. Tetsu, S. Koichi and K. Hiroshi. 1999. Larvae and juveniles of two gobiids, *Chaenogobius uchidai* and *C. macrognathos*, collected from tidelands in the inner bay of Tokyo. *Bay Mer* 37: 59-68.

Kon, T. and T. Yoshino. 1992. Diversity and evolution of life histories of gobioid fishes from the viewpoint of heterochrony. *Marine and Freshwater Research* 53: 377-402.

Kramer, A. and R. Patzner. 2008. Morphological and developmental studies on the eggs of four *Coryphopterus* species (Actinopterygii: Gobiidae) in the Caribbean Sea. *Acta Ichthyologica et Piscatoria* 38: 97-102.

Kunne, C. 1934. Uber das Laichgeschaft der Bunten Grundel im Aquarium. *Blatter fur Aquarien- und Terrarienkunde* 45: 350-352.

Kuno, Y. and T. Takita. 1997. The growth, maturation and feeding habits of the gobiid fish *Acanthogobius hasta* distributed in Ariake Sound, Kyushu, Japan. *Fisheries Science* 63: 242–248.

Kuntz, A. 1916. Notes on the embryology and larval development of five species of teleostean fishes. *Bulletin of the United States Bureau Fisheries* 34: 407-429.

La Mesa, M., E. Arneri, V. Caputo and M. Iglesias. 2005. The transparent goby, *Aphia minuta*. Review of Biology and Fisheries of a Paedomorphic European Fish. *Reviews in Fish Biology and Fisheries* 15: 89-109.

Lassig, B. 1976. Field observations on the reproductive behavior of *Paragobiodon* spp. (Osteichthyes: Gobiidae) at Heron Island Great Barrier Reef. *Marine and Freshwater Behaviour and Physiology* 3: 283-293.

Lebour, M.V. 1919. The young of the gobiidae from the neighbourhood of Plymouth. *Journal of Marine Biological Association of the United Kingdom* 12: 48-80.

Lebour, M.V. 1920. The eggs of *Gobius minutus*, *pictus* and *microps*. *Journal of Marine Biological Association of the United Kingdom* 12: 253-260.

Leis, J.M. 1993. Minimum requirements for published descriptions of larval fish development. *Japanese Journal of Ichthyology* 40 (3): 393-395.

Leis, J.M. 2006. Are larvae of demersal fishes plankton or nekton? *Advances in Marine Biology* 51: 59-141.

Leis, J.M. and B.M. Carson-Ewart (eds.). 2000. *The Larvae of Indo-Pacific Coastal Fishes: an Identification Guide to Marine Fish Larvae.* Brill, Leiden.

Leis, J.M. and D.S. Rennis. 1983. *The Larvae of Indo-Pacific Coral Reef Fishes.* New South Wales University Press, Sydney and University of Hawaii Press, Honolulu.

Leis, J.M., B.M. Carson-Ewart and J. Webley. 2002. Settlement behaviour of coral-reef fish larvae at subsurface artificial-reef moorings. *Marine and Freshwater Research* 53: 319-327.

Leslie, J.K., C.A Timmins and R.G. Bonnell. 2002. Postembryonic development of the tubenose goby *Proterorhinus marmoratus* Pallas (Gobiidae) in the St. Clair River/Lake system, Ontario. *Archiv fur Hydrobiologie* 154: 341-352.

Lester, S.E. and B.I. Ruttenberg. 2005. The relationship between pelagic larval duration and range size in tropical reef fishes: a synthetic analysis. *Proceedings of the Royal Society of London, Series B* 272: 585-591.

Lie, U., D. Bjørstad and E. Oug. 1978. Eggs and larvae of fish from Lindåspollene. *Sarsia* 63: 163-167.

Lindstrom, D.P. 1994. Early larval development of the endemic Hawaiian stream goby, *Lentipes concolor. Pacific Science* 48: 204.

Lindstrom, D.P. 1999. Molecular species identification of newly hatched Hawaiian amphidromous gobioid larvae. *Marine Biotechnology* 1: 167-174.

Lindstrom, D.P. and C.L. Brown. 1994. Early development and biology of the amphidromous Hawaiian stream goby *Lentipes concolor*. In: *Systematics and Evolution of Indo-Pacific Fishes. Proceedings of the Fourth Indo-Pacific Fish Conference*, Bangkok, Thailand: Faculty of Fisheries, pp. 397-409.

Lippson, A.J. and R.L. Moran. 1974. Manual for identification of early developmental stages of fishes of the Potomac River Estuary. *Midland. Department of Natural Resources Power Plant Siting Program PPSP-MP* **13**, 282 pp.

Lo Bianco, S. 1909. Notize biologische riguardanti specialmente il periodo di maturita sessuale degli animali del Golfo di Napoli. *Mitteilungen aus der zoologischen Station zu Neapel* 19: 513-761.

Logachev, V.S. and Y.E. Mordvinov. 1979. Swimming speed and activity of larvae of round goby and some predatory crustaceans of the Black Sea. *Journal of Marine Biology* 5: 227-229.

Lopez Jaime, J.A., P. Escalante Blanque, R. Rodriguez Garcia and F. Zurita Escobar. 2000. Experimental culture of *Aphia minuta*. Study on the maintenance in captivity, nutritional requirements, ecological needs and possibility of controlled reproduction. *Cahiers Options Méditerranéennes* 47: 343-347.

MacGinitie, G.E. 1935. Ecological aspects of a California marine estuary. *The American. Midland Naturalist* 16: 629-765.

MacGinitie, G.E. 1939. The natural history of the blind goby, *Typhlogobius californiensis* Steindachner. *The American Midland Naturalist* 21: 489-505.

MacPherson, E. and N. Raventos. 2006. Relationship between pelagic larval duration and geographic distribution of Mediterranean littoral fishes. *Marine Ecology Progress Series* 327: 257-265.

Maddox, E. 1992. *Generic Identification and Patterns of External Pigment in Larval Gobies (Pisces: Gobiidae) from the Belize Barrier Reef.* M.Sc. Thesis, College of William and Mary, Virginia.

Maeda, K., N. Yamasaki, M. Kondo and K. Tachihara. 2008. Occurrence and morphology of larvae and juveniles of six *Luciogobius* species from Aritsu Beach, Okinawa Island. *Ichthyological Research* 55: 162-174.

Manacop, P.R. 1953. The life history and habits of the goby, *Sicyopterus extraneus* Herre (añga) Gobiidae with an account of the goby-fry fishery of Cagayan River, oriental Misamis (Province, Mindanao, Philippines). *Philippine Journal of Fisheries* 2: 1-60.

Marconato, A., A. Bisazza and G. Marin. 1989. Correlates of male reproductive success in *Padogobius martensi* (Gobiidae). *Journal of Fish Biology* 34: 889-899.

Masuda, H., K. Amaoka, C. Araga, T. Uyeno and T. Yoshino (eds.). 1984. *The Fishes of the Japanese Archipelago.* Tokyo, Tokai University Press.

Matarese, A.C., A.W. Kendall, D.M. Blood and M.V. Vinter. 1989. Laboratory guide to early life history stages of Northeast Pacific fishes. *NOAA Technical Reports NMFS* 80: 1-652.

Mazzoldi, C., C. Poltronieri and M.B. Rasotto. 2002. Egg size variability and mating system in the marbled goby *Pomatoschistus marmoratus* (Pisces: Gobiidae). *Marine Ecology Progress Series* 233: 231-239.

McDowall, R.M. 2009. Early hatch: a strategy for safe downstream larval transport in amphidromous gobies. *Reviews in Fish Biology and Fisheries* 19: 1-8.

Miller, P.J. 1961. Age, growth and reproduction of the rock goby, *Gobius paganellus* L. in the Isle of Man. *Journal of the Marine Biological Association of the United Kingdom* 41: 737-769.

Miller, P.J. 1984. The tokology of gobioid fishes. In: *Fish Reproduction: Strategies and Tactics, G.W.* Potts and R.J. Wootton (eds.). Academic Press, London, pp. 119-152.

Miller, P.J. 1986. Gobiidae. In: *Fishes of the North-eastern Atlantic and the Mediterranean.* P.J.P. Whitehead, M.-L. Bauchot, J.-C. Hureau, J. Nielsen and E. Tortonese (eds.) Volume 3. UNESCO, Paris. pp. 1019-1085.

Miyazaki, I. 1940. Studies on the Japanese common goby, *Acanthogobius flavimanus* (Temminck et Schlegel). *Bulletin of the Japanese Society of Scientific Fisheries* 9: 159-180.

Mizuno, N. 1960. Study on a freshwater goby, *Rhinogobius similis* Gill, with a proposition on the relationships between land-locking and speciation of some freshwater gobies in Japan. *Memoirs of the College of Science and Engineering, Kyoto University*, Ser. B. 27: 97-115.

Moe, M.A. Jr. 1975. Propagating the Atlantic neon goby. *Marine Aquarist* 6: 4-10.

Moiseyeva, Y.B. 1983. The development of the gonads of the round goby, *Neogobius melanostomus* (Gobiidae), during the embryonic period. *Journal of Ichthyology* 23: 64-74.

Monteiro, J., R. Borges, J. Robalo, V.C. Almada and E.J. Gonçalves. 2008. Larval development of *Gobius xanthocephalus* with genetic validation of larval identification. *Journal of Fish Biology* 73: 123-138.

Moser, H.G. (ed.). 1996. *The Early Stages of Fishes in the California Current Region.* California Cooperative Oceanic Fisheries Investigations (CalCOFI) Atlas No. 33. Allen Press, Inc., Lawrence, Kansas.

Moser, H.G., W.J. Richards, D.M. Cohen, M.P. Fahay, A.W. Kendall, Jr. and S.L. Richardson (eds.). 1984. *Ontogeny and Systematics of Fishes.* The American Society of Ichthyologists and Herpetologists, Special Publication N° 1.

Moskalkova, K.I. 1967. Ecological-morphological characteristics of round goby (*Neogobius melanostomus*) development. In: N.N. Disler (ed.). *Morphological analysis of fish* development. Nauka Press, Moskow. pp. 95-119.

Moskalkova, K.I. 1978. Development of Azov round goby *Neogobius melanostomus* (Pallas) in view of specifications of its existence in individual and historical development. In: N.N. Disler (ed.). *Ecological-morphological and ecological-physiological investigations of fish development.* Nauka Press, Moscow. pp.72-88.

Moskalkova, K.I. 1989. Anatomical histological and functional peculiarities of development of the intestine in the round goby, *Neogobius melanostomus*, a species with direct type of development. *Journal of Ichthyology* 29: 108-122.

Moskalkova, K.I. 1996. Ecological and morphophysiological prerequisites to range extention in the round goby *Neogobius melanostomus* under conditions of anthropogenic pollution. *Journal of Ichthyology* 36: 584-590.

Munk, P.N and J.G. Nielsen. 2005. *Eggs and Larvae of North Sea Fishes.* Biofolia, Denmark.

Murphy, C.A. and J.H., Jr. Cowen. 2007. Production, marine larval retention or dispersal, and recruitment of amphidromous Hawaiian gobioids: issues and implications. In: *Biology of Hawaiian Streams and Estuaries*, N.L. Evenhuis and J.M. Fitzsimons (eds.). Bishop Museum Bulletin in Cultural and Environmental Studies 3: 63-74.

Myers, R.F. 1999. *Micronesian Reef Fishes.* Coral Graphics, Barrigada, Guam.

Nakamura, S. 1936. Larvae and young of fishes in the vicinity of Kominata, II-VI. *Journal of the Imperial Fisheries Institute* (Japan) 31: 131-136.

Neira, F.M., A.G. Miskiewicz and T. Trnski. 1998. *Temperate Australian Fishes – Laboratory Guide for Larval Fish Identification.* Western Australia Press, Australia.

Nelson, J.S. 2006. *Fishes of the World.* 4th Edition, John Wiley & Sons, Inc., New York.

Nishimoto, R.T. and D.G.K. Kuamo'o. 1997. Recruitment of goby postlarvae into Hakalau Stream, Hawai'i Island. *Micronesica* 30: 41-49.

Nyman, K.J. 1953. Observations on the behaviour of *Gobius microps. Acta Societatis pro Fauna et Flora Fennica* 69: 1-11.

Okiyama, M. 1988. *An Atlas of the Early Stage Fishes in Japan*. Tokai University Press, Tokyo.

Olivar, M.P. 1989. Spawning and development of *Lesueurigobius sanzoi* off northern Namibia. *South African Journal of Zoology* 24: 132-138.

Olivar, M.P. and J.-M. Furtuno. 1991. Guide to ichthyoplankton of the southeast Atlantic (Benguela current region). *Scientia Marina* 55: 1-383.

Olivotto, I., A. Zenobi, A. Rollo, B. Migliarini, M. Avella and O. Carnevali. 2005. Breeding, rearing and feeding studies in the cleaner goby *Gobiosoma evelynae*. *Aquaculture* 250: 175-182.

O'Toole, M.J. 1978. Development, distribution and relative abundance of the larvae and early juveniles of the pelagic goby *Sufflogobius bibarbatus* (von Bonde) off South West. *Sea Fisheries Branch Investigational Report* 116. Department of Industries, RSA: Cape Town, RSA.

Padoa, E. 1956. Gobiidae. In: *Uova, larve e stadi giovanili di Teleostei. Fauna e Flora del* Golfo di Napoli, Bertolini *et al.* (ed.). Edizioni della Stazione Zoologica di Napoli Monograph 38: 648-678.

Peters, K.M. 1983. Larval and early juvenile development of the frillfin goby, *Bathygobius sporator* (Perciformes: Gobiidae). *Northeast Gulf Science* 6: 137-153.

Petersen, C.G.J. 1892. On the eggs and breeding of our Gobiid. *Reports of the Danish Biological Station* 2: 1-9.

Petersen, C.G.J. 1917. On the development of our common gobies (*Gobius*) form the egg to the adult stages. *Reports of the Danish Biological Station* 24: 5-16.

Petersen, C.G.J. 1919. Our gobies (Gobiidae). *Reports of the Danish Biological Station* 26: 45-66.

Prasad, R.R. 1958a. Reproduction in *Clevelandia ios* with an account of the embryonic and larval development. *Proceedings of the National Institute of Sciences of India*. Section B 25: 12-30.

Prasad, R.R. 1958b. Notes on the habitat and habits of *Clevelandia ios*. *Proceedings of the National Institute of Sciences of India*. Section B 24: 314-324.

Privileggi, N., D. Ota and E.A. Ferrero. 1997. Embryonic and larval development of the grass goby *Zosterisessor ophiocephalus* (Teleostei, Gobiidae). *Italian Journal of Zoology* 64: 201-207.

Privitera, L.A. 2001. Characteristics of egg and larval production in captive bluespotted gobies. *Journal of Fish Biology* 58: 1211-1220.

Radtke, R.L. and R.A. Kinzie. 1996. Evidence of a marine larval stage in endemic Hawaiian stream gobies from isolated high-elevation localities. *Transactions of the American Fisheries Society* 125: 613-621.

Radtke, R.L., R.A. Kinzie and S.D. Folsom. 1988. Age at recruitment of Hawaiian freshwater gobies. *Environmental Biology of Fishes* 23: 205-214.

Radtke, R.L., R.A. Kinzie III and D.J. Shafer 2001. Temporal and spatial variation in length of larval life and size at settlement of the Hawaiian amphidromous goby *Lentipes concolor*. *Journal of Fish Biology* 59: 928-938.

Raventos, N. and E. Macpherson. 2001. Planktonic larval duration and settlement marks on the otoliths of Mediterranean littoral fishes. *Marine Biology* 138: 1115-1120.

Reavis, R.H. 1997. The natural history of a monogamous coral-reef fish, *Valenciennea strigata* (Gobiidae): 2. behaviour, mate fidelity and reproductive success. *Environmental Biology of Fishes* 49: 247-257.

Richards, W.J. (ed.). 2006. *Early Stages of Atlantic Fishes. An Identification Guide for the Western Central North Atlantic.* Taylor & Francis, U.K.

Richardson, S.L., and E.B. Joseph. 1975. Occurrence of larvae of the green goby, *Microgobius thallasinus*, in the York River, Virginia. *Chesapeake Science* 16: 215-218.

Roloff, E. 1935. Die Pflege und Zucht von *Gobius xanthozona*. Wochenschrift für Aquarien und Terrarienkunde 32: 545-547; 562-564.

Ruple, D. 1984. Gobioidei: Development. In: *Ontogeny and Systematics of Fishes*, H.G. Moser, W.J. Richards, D.M. Cohen, M.P. Fahay, A.W. Kendall, Jr. and S.L. Richardson (eds.). The American Society of Ichthyologists and Herpetologists, Special Publication N° 1, pp. 582-587.

Russell, F.S. 1976. *The Eggs and Planktonic Stages of British Marine Fishes*. Academic Press, London, U.K.

Sabatés, A. 1988. *Sistematica Y distribucion espacio-temporal del ictioplancton en la costa catalana*. PhD Thesis. University of Barcelona, Barcelona.

Saitoh, K. 1993. *Early life history of the mudskipper,* Scartelaos histophorus *(Valenciennes) in Nakagusuku Bay, Okinawa Island, Japan*. B.Sc. Thesis, University of the Ryukyus, Okinawa, Japan.

Sakai, H. and F. Yasuda. 1978. Development of eggs and larvae of the freshwater goby, *Rhinogobius brunneus*. *Japanese Journal of Ichthyology* 25: 92-100.

Sapota, M.R. 2006. NOBANIS – Invasive Alien Species Fact Sheet – *Neogobius melanostomus*. – From: Online Database of the North European and Baltic Network on Invasive Alien Species – NOBANIS www.nobanis.org, Date of access July/2009.

Schoenfuss, H.L., T.A. Blanchard and D.G. Kuamo'o. 1997. Metamorphosis in the cranium of postlarval *Sicyopterus stimpsoni*, an endemic Hawaiian stream goby. *Micronesica* 30: 93-104.

Scotton, L.N., R.E. Smith, N.S. Smith, K.S. Price and D.P. de Sylva. 1973. *Pictorial Guide to Fish Larvae of Delaware Bay: With Information and Bibliographies Useful for the Study of Fish Larvae*. Delaware Bay Report Series. Vol. 7. College of Marine Studies, University of Delaware.

Shafer, D.J. 1998. *Early life history growth and settlement dynamics of a tropical reef fish (Gobiidae:* Bathygobius coalitus*)*. PhD Thesis, Department of Zoology, University of Hawai'i.

Shafer, D.J. 2000. Evaluation of periodic and aperiodic otolith structure and somatic-otolith scaling for use in retrospective life history analysis of a tropical marine goby, *Bathygobius coalitus*. *Marine Ecology Progress Series* 199: 217-229.

Shann, E.W. 1910. Some notes on the life-history and rate of growth in *Gobius minutus*. *Annals and Magazine of Natural History*, Ser. 5, 8: 217-239.

Shen, K.N. and W.N. Tzeng. 2008. Reproductive strategy and recruitment dynamics of amphidromous goby *Sicyopterus japonicus* as revealed by otolith microstructure. *Journal of Fish Biology* 73: 2497-2512.

Shen, K-N., Y.C. Lee and W.N. Tzeng. 1998. Use of otolith microchemistry to investigate the life history pattern of gobies in a Taiwanese stream. *Zoological Studies* 37: 322-329.

Shiao, J.C. 1998. *Early life history and fry resources of amphidromous gobies in Hsuikuluan River.* M.Sc. Thesis, National Tsing Hua University, Hsinchu, Taiwan.

Shiao, J.C., C.S.Tzeng, C.L. Leu and F.C. Chen. 1999. Enhancing the contrast and visibility of daily growth increments in fish otoliths etched by proteinase K buffer. *Journal of Fish Biology* 54: 302-309.

Shinomiya, A., K. Maeyama and S. Imai. 1981. Reproductive behaviour of the goby *Eviota storthynx. Memoirs of the Faculty of Fisheries Kagoshima University* 30: 237-246.

Shiobara, Y. and K. Suzuki. 1983. Life history of two gobioids, *Istigobius hoshinonsis* (Tanaka) and *I. campbelli* (Jordan et Snyder), under natural and rearing conditions. *Journal of the Faculty of Marine Science and Technology, Tokai University* 16: 193-205.

Shiobara, Y and Y. Tanaka. 1994. Development eggs and early larvae of the gobiid fish, *Trimma grammistes. Journal of the Faculty of Marine Science and Technology, Tokai University* 38: 257-264.

Shiobara, Y. and Y. Tanaka. 1995. Reproductive behavior, development of eggs and larvae of the gobiid fish, *Valenciennea bella*, reared in the aquarium. *Journal of the Faculty of Marine Science and Technology, Tokai University* 39: 207-220.

Shiobara, Y., K. Suzuki, and Y. Tanaka 1990. Life history of the hairychin goby, *Sagamia geneionema*, from the coast of Suruga Bay. *Journal of the Faculty of Marine Science and Technology, Tokai University* 31: 25-36.

Shiogaki, M. 1981. Life history of the gobiid fish *Pterogobius zacalles. Japanese Journal of Ichthyology* 28: 70-79.

Shiogaki, M., and Y. Dotsu. 1971. The life history of the gobiid fish *Expedio parvulus. Bulletin of the Faculty of Fisheries, Nagasaki University* 32: 17-25.

Shiogaki, M. and Y. Dotsu. 1972a. The life history of the gobiid fish, *Luciogobius elongatus. Bulletin of the Faculty of Fisheries, Nagasaki University* 34: 9-18.

Shiogaki, M. and Y. Dotsu. 1972b. The life history of the gobiid fish *Clariger cosmurus. Bulletin of the Faculty of Fisheries, Nagasaki University* 34: 19-27.

Shiogaki, M. and Y. Dotsu. 1974a. The life history of the gobiid fish, *Zonogobius boreus. Bulletin of the Faculty of Fisheries, Nagasaki University* 37: 1-8.

Shiogaki, M., and Y. Dotsu, 1974b. The life history of the gobiid fish *Inu koma. Bulletin of the Faculty of Fisheries, Nagasaki University* 38: 65-70.

Shiogaki, M. and Y. Dotsu. 1977. The life history of the gobiid fish, *Luciogobius platycephalus. Japanese Journal of Ichthyology* 24: 43-48.

Shiogaki, M., N. Miura and Y. Dotsu. 1974. The life history of the gobiid fish, *Luciogobius grandis*. *Bulletin of the Faculty of Fisheries, Nagasaki University* 38: 57-64.

Shulman, M.J. 1998. What can population genetics tell us about dispersal and biogeographic history of coral-reef fishes? *Australian Journal of Ecology* 23: 216-225.

Skora, K., S. Olenin and S. Gollasch. 1999. *Neogobius melanostomus* (Pallas, 1811). In: *Case Histories on Introduced Species: Their General Biology, Distribution, Range Expansion and Impact,* S. Gollasch, D. Michin, H. Rosenthal and M. Voight (eds.). Logos Verlag Berlin, pp. 69-73.

Smith, H.M. and C.C. Baldwin. 1999. *Psilotris amblyrhynchus*, a new seven-spined goby (Teleostei: Gobiidae) from Belize, with notes on settlement-stage larvae. *Proceedings of the Biological Society of Washington* 112(2): 433-442.

Sonoda, T. and S. Imai. 1971. On the spawning and embryonic development of a marine goby, *Zonogobius semidoliatus* (Valenciennes). *Memoirs of the Faculty of Fisheries, Kagoshima University* 20: 197-202.

Spartá, A. 1934a. Uova e larvae di Gobiidae. II. *Gobius jozo* L. *Memoria. R. Comitato Talassografico Italiano* 216: 8.

Spartá, A. 1934b. Uova e Larvae di Gobiidae. I - *Gobius paganellus*. *Report du Comitato Talassografico Italiano* 211: 1-11.

Spartá, A. 1936. Uova e larvae di Gobiidae. III. *Gobius ferrugineus* Kolomb. *Memoria. R. Comitato Talassografico Italiano* 218: 1-5.

Spartá, A. 1950. Uova e larvae di gobiidae. V. *Gobius capito*. *Bollettino di Pesca, Piscicoltura e Idrobiologia* 5: 99-104.

Sponaugle, S. and R.K. Cowen. 1994. Larval durations and recruitment patterns of two Caribbean gobies (Gobiidae): contrasting early life histories in demersal spawners. *Marine Biology* 120: 133-143.

Sponaugle, S. 2009. Daily otolith increments in the early stages of tropical fish. In: *Tropical Fish Otoliths: Information for Assessment, Management and Ecology.* B.S. Green, B.D. Mapstone, G. Carlos and G.A. Begg (eds.). Reviews: Methods and Technologies in Fish Biology and Fisheries 11: 93-132.

Sponaugle, S. and K. Grorud-Colvert. 2006. Environmental variability, early life history traits, and survival of new recruits of a coral reef fish. *Integrative and Comparative Biology* 46: 623-633.

Sponaugle, S., R.K. Cowen, A. Shanks, S.G. Morgan, J.M. Leis, J. Pineda, G.W. Boehlert, M.J. Kingsford, K.C. Lindeman, C. Grimes and J.L. Munro. 2002. Predicting self-recruitment in marine populations: biophysical correlates and mechanisms. *Bulletin of Marine Science* 70(1): 341-375.

Sponaugle, S., K. Grorud-Colvert and D. Pinkard. 2006. Temperature-mediated variation in early life history traits and recruitment success of the coral reef fish *Thalassoma bifasciatum* in the Florida Keys. *Marine Ecology Progress Series* 308: 1-15.

Springer, V.G. and A.J. McErlean. 1961. Spawning seasons and growth of the code goby, *Gobiosoma robustum* (Pisces: Gobiidae), in the Tampa Bay area. *Tulane Studies in Zoology* 9: 87-98.

Sunobe, T. 1995. Embryonic development and larvae of three gobiid fish, *Trimma okinawae*, *Trimma grammistes* and *Trimmatom* sp. *Japanese Journal of Ichthyology* 42: 11-16.

Sunobe, T. and A. Nakazono. 1987. Embryonic development and larvae of genus *Eviota* (Pisces: Gobiidae), *Eviota abax* and *E. storthynx*. *Journal of the Faculty of Agriculture, Kyushu University* 31: 287-295.

Sunobe, T. and A. Nakazono. 1989. Embryonic development and pré-larva of a gobiid fish *Priolepis naraharae*. *Japanese Journal of Ichthyology* 35: 484-487.

Sunobe, T. and A. Nakazono. 1995. Embryonic development and larvae of genus *Eviota* (Pisces: Gobiidae). II. Description of seven species. *Natural History Research* 3: 152-159.

Sunobe, T. and A. Shinomiya. 2002. Reproductive behavior, embryonic development and yolk-sac larva of *Sueviota lachneri* (Pisces: Gobiidae) in the laboratory. *Natural History Research* (Chiba) 7: 95-99.

Suzuki, N., N. Sakurai and T. Sugihara. 1988. Development of eggs, larvae and juveniles of the goby *Glossogobius olivaceus* reared in the laboratory. *Suisan Zoshoku* 35(4): 203-212.

Suzuki, N., N. Sakurai and T. Sugihara. 1989. Development of eggs, larvae and juveniles of the goby *Glossogobius olivaceus* reared in the laboratory. *Suisan Zoshoku* 35(4): 203-212.

Swedmark, M. 1958. Sur la variation géographique de *Gobius minutus* Pallas. I. Biologie et croissance. *Archives de Zoologie Experimentale et Generale* 95: 32-51.

Swenson, R.O. 1995. *The Reproductive Behavior and Ecology of the Tidewater Goby,* Eucyclogobius Newberryi *(Teleostei: Gobiidae).* Ph.D. Thesis, University of California, Berkeley, California.

Swenson, R.O. 1997. Sex-role reversal in the tide-water goby, *Eucyclogobius newberryi*. *Environmental Biology of Fishes* 50: 27-40.

Swift, C.C., J.L. Nelson, C. Maslow and T. Stein. 1989. Biology and distribution of the tidewater goby, *Eucyclogobius newberryi* (Pisces: Gobiidae) of California. *Natural History Museum of Los Angeles County, Contributions in Science* 404: 1-19.

Szabados, A. 1937. Die laichabgabe der Marmorgrundlen. *Wochenschrift fuer Aquarien und Terrarienkunde* 34: 367.

Takita, T. 1975. Spawning and embryonic development of the *Synechogobius hasta* in an aquarium, with description of larvae and juveniles. *Japanese Journal of Ichthyology* 22: 31-39.

Tamada, K. 2005. Egg and clutch sizes of a goby *Rhinogobius giurinus* in the Aizu River, Kii Peninsula, Japan. *Ichthyological Research* 52: 392-395.

Tanaka, Y., K. Suzuki and A. Doi. 1982. Reproduction and early life history of the gobioid fish, *Eleotriodes helsdingeni* Bleeker. *Journal of the Faculty of Marine Science and Technology, Tokai University* 15: 367-376.

Tåning, Å.V. 1940. Island-Faroe-Omraadets Kutlinger (Gobiidae). *Vidensk. Meddr dansk. Naturh. Foren* 104: 267-283.

Taru, M. and T. Sunobe. 2002. Reproductive ecology of the gobiid fish *Eviota abax* at Nobeoka, Japan, with notes on geographic variation. *Pacific Science* 56: 35-40.

Tavolga, W.N. 1950. Development of the gobiid fish, *Bathygobius soporator. Journal of Morphology* 87: 467-492.

Tavolga, W.N. 1954. Reproductive behavior in the gobiid fish *Bathygobius soporator. Bulletin of the American Museum of Natural History* 104: 427-460.

Taylor, M.S. and M.E. Hellberg. 2003. Genetic evidence for local retention of pelagic larvae in a Caribbean reef fish. *Science* 299: 107-109.

Te Winkel, L.E. 1935. A study of *Mistichthys luzonensis* with special reference to conditions related with reduced size. *Journal of Morphology* 58: 463-535.

Thresher, R.E. 1988. Latitudinal variation in egg sizes of tropical and sub-tropical North Atlantic shore fishes. *Environmental Biology of Fishes* 21: 17-25.

Thys van den Audenaerde, D.F.E. 1988. Natural distribution of tilapias and its consequences for the possible protection of genetic resources. In: *Tilapia Genetic Resources for Aquaculture*, R.S.V. Pullin (ed.). ICLARM Conference Proceedings 16, pp. 1-11.

Timbol, A.S., M. Kido and D. Heacock. 1990. *The Breeding Biology of the Hawaiian Endemic Freshwater Goby,* Awaous stamineus *('o'opu-nakea); Larvae Rearing in the Laboratory.* Honolulu, Department of Land and Natural Resources, Hawaii.

Tomihama, M.T. 1972. *The Biology of* Sicydium Stimpsoni*: a Freshwater Goby Endemic to Hawaii.* B.Sc. Thesis, Department of Zoology, University of Hawaii, Honolulu.

Tsukaho, Y., A. Ishimatsu, T. Takeda, K.K. Huat and K. Tachihara. 2003. The eggs and larvae of the giant mudskipper, *Periophthalmodon schlosseri*, collected from a mudflat in Penang, Malaysia. *Ichthyological Research* 50: 178-181.

Uchida, K. 1931. Spawning of the mudskippers, *Boleophthalmus pectinirostris* and *Periophthalmus cantonensis. Kagaku* 1: 226-227.

Uchida, T. and Y. Dotsu. 1980. Larvae and juveniles of three Japanese common gobiid fishes reared in vessels. *Bulletin of the Faculty of Fisheries, Nagasaki University* 19: 25-33.

Valenti, R.J. 1972. The embryology of the neon goby, *Gobiosoma oceanops. Copeia* 1972: 477-482

Victor, B. 2006. The late-stage larvae of Caribbean gobies, eleotrids, and microdesmids: identification guide and patterns of size and age at settlement. *Program book and abstracts, 2006 joint meeting of the American Society of Ichthyologists and Herpetologists July 12–17, 2006.* Allen Press, Lawrence, Kansas, pp. 128-129.

Victor, B. 2007. *Coryphopterus kuna*, a New Goby (Perciformes: Gobiidae: Gobiinae) from the Western Caribbean, with the identification of the late larval stage and an estimate of the pelagic larval duration. *Zootaxa* 1526: 51-61.

Victor, B. 2009. The photographic web guide to the larvae of coral reef fishes: the 21st century at http://www.coralreeffish.com (accessed in May 2009).

Vijayaraghavan, P. 1975. Studies on fish eggs and larvae from Indian waters. I. Development of eggs and larvae of *Acentrogobius ornatus* (Rüppell). *Indian Journal of Fisheries* 20: 523-532.

Villegas, M.L. 1980. Postlarvas de gobidos del mar Cantabrico. *Boletin del instituto Espanhol de Oceanografia* 6: 19-45.

Villegas Cuadros, M.L. 1981. Contribución al conocimiento de huevos y larvas de Gobidos. I.- Forma jozo de *Gobius niger* Linnaeus, 1758. *Boletin de Ciencias de lo Real instituto de Estudios Asturianos* 28.

Vivien, J.H. 1939. Étude préliminaire des annexes du tractus genital d'un Gobiidé: *Gobius paganellus* L. *Travaux de la Station Zoologique Wimereux* 13: 713-721.

Vodyanitzki, W.A. and I.I. Kazanova. 1954. Opredelitel pelaghiceski ikrino i licino rib Cernogo Morea (Key for identification of the pelagic eggs and fish larvae in the Black Sea). Trudy vses. nauchno-issled. *Inst. Morsk. Ryb. Khoz. Okeanogr.* 28: 240-327.

Wang, J.C. 1981. *Taxonomy of the Early Life Stages of Fishes: Fishes of the Sacramento-San Joaquin Estuary and Moss Landing Harbor-Elkhorn Slough, California.* Ecological Analysts, Inc., Concord, California.

Watson, W. 1996. Gobiidae: gobies. In: *The Early Stages of Fishes in the California Current Region.* California Cooperative Oceanic Fisheries Investigations (CalCOFI) Atlas No. 33, H.G. Moser (ed.). Allen Press, Inc., Lawrence, Kansas, pp. 1214-1245.

Way, C.M., A.J. Burky, J.M. Harding, S. Hau and W.K.L.C. Puleloa. 1998. Reproductive biology of the endemic goby, *Lentipes concolor*, from Makamak'ole Stream, Maui and Waikolu Stream, Moloka'i. *Environmental Biology of Fishes* 51: 53-65.

Weisel, G.F. 1947. Breeding behavior and early development of the mudsucker, and a gobiid fish of California. *Copeia* 1947: 77-85.

Wheeler, A. 1969. *The Fishes of the British Isles and North-West Europe.* MacMillan and Co., Ltd., London.

Wiley, J.W. 1973. Life history of the western North American goby *Coryphopterus nicholsii* (Bean). *Transactions of the San Diego Society of Natural History* 17: 187-208.

Wiley, J.W. 1976. Life histories and systematics of the western North American gobies *Lythrypnus dalli* (Gilbert) and *Lythrypnus zebra* (Gilbert). *Transactions of the San Diego Society of Natural History* 18: 169-184.

Winterbottom, R. and L. Southcott. 2008. Short lifespan and high mortality in the western Pacific coral reef goby *trimma nasa. Marine Ecology Progress Series* 366: 203-208.

Wittenrich, M.L., R.G. Turingan and R.L. Creswell. 2007. Spawning, early development and first feeding in the gobiid fish *Priolepis nocturna. Aquaculture* 270: 132-141.

Wyanski, D.M. and T.E. Targett. 1985. Juvenile development of the lyre goby, *Evorthodus lyricus* (Girard) (Pisces: Gobiidae), with a discussion of early life history. *Bulletin of Marine Science* 36: 115-123.

Wyanski, D.M. and T.E. Targett. 2000. Development of transformation larvae and juveniles of *Ctenogobius boleosoma*, *Ctenogobius shufeldti*, and *Gobionellus oceanicus* (Pisces: Gobiidae) from western North Atlantic estuaries, with notes on early life history. *Bulletin of Marine Science* 67: 709-728.

Yamamoto, T. 1980. Embryonic development in *Paragobiodon lacunicola* and the spawning sites of *Gobiodon* spp., and *Paragobiodon* spp. (Pisces : Gobiidae). *Biological Magazine of Okinawa* 18: 17-24.

Yamasaki, N. and K. Tachihara. 2006. Reproductive biology and morphology of eggs and larvae of *Stiphodon percnopterygionus* (Gobiidae Sicydiinae) collected from Okinawa Island. *Ichthyological Research* 53: 13-18.

Yamasaki, N. and K. Tachihara. 2007. Eggs and larvae of *Awaous melanocephalus* (Teleostei: Gobiidae). *Ichthyological Research* 54: 89-91.

Yamasaki, N., K. Maeda and K. Tachihara. 2007. Pelagic larval duration and morphology at recruitment of *Stiphodon percnopterygionus* (Gobiidae: Sicydiinae). *The Raffles Bulletin of Zoology* 14: 209-214.

Yeung, C. and D. Ruple. 2006. Gobiidae: Gobies. In: *Early Stages of Atlantic Fishes. An Identification Guide for the Western Central North Atlantic*, W.J. Richards (ed.). CRC Press, Taylor and Francis, Boca Raton, Florida, pp. 2029-2075.

Yoshida, T., Y. Dotsu, G. Fukagawa and K. Miyaki. 2006. Ecology, life history and rearing of the well goby, *Luciogobius* sp. O in Oomura Bay, Nagasaki Prefecture. *Transactions of the Nagasaki Biological Society* 61: 13-25.

Zavala, G.F., C.C. Flores and M.V.M. De Lourdes. 1988/1989. Early life history and larval distribution of *Gobiosoma robustum* Ginsburg (Pisces: Gobiidae), Terminos Lagoon, Campeche. *Anales del Instituto de Ciencias del Mar y Limnologia Universidad Nacional Autonoma de Mexico* 15(1): 237-244.

Zhang, Q. and J. Zhang. 1988. On the feeding, growth and survival of larval mudskipper (*Boleophthalmus pectinirostris*). *Journal of Fisheries of China* 12: 203-211. (in Chinese).

Zhang, Q-Y., W-S. Hong, Q-N Dai, J. Zhang, Y-Y. Cai and J-L. Hong. 1989. Studies on induced ovulation, embryonic development and larval rearing of the mudskipper (*Boleophthalmus pectinirostris*). *Aquaculture* 83: 375-385.

Electronic references

http://larval-fishes.australianmuseum.net.au/, accessed July 2009.

http://www.caspianenvironment.org/biodb/eng/main.htm/, accessed July 2009

http://access.afsc.noaa.gov/ichthyo/index.cfm/, accessed July 2009

SECTION 4

Ecology

CHAPTER

4.1

Planktonic and Paedomorphic Gobioids

Mario La Mesa

INTRODUCTION

The suborder Gobioidei (Teleostei, Perciformes) comprises about 270 nominal genera and more than 2000 species (Hoese, 1984), which are distributed worldwide except for Arctic and Antarctic areas. Gobioids are widespread in tropical and temperate seas from subtidal to more than 300 m deep waters, as well as in estuarine and freshwater environments (Nelson, 1994). The suborder includes relatively small fish ranging from 8 mm to about 50 cm SL, most of them being generally benthic dwellers as adult with planktonic larval stages. Some gobioid genera, however, exhibit a planktonic life mode throughout the whole life span. Among the diverse mode of life of gobioids, species living in the water column have been considered of particular zoological interest because of the persistence into adulthood of several larval anatomical characteristics (Miller, 1973), which are of adaptive significance for a planktotrophic life in midwater (Brunelli

Author's address: ISMAR-CNR, Istituto di Scienze Marine, Sede di Ancona, Largo Fiera della Pesca, 60125 Ancona, Italy. E-mail: m.lamesa@ismar.cnr.it and Atella, 1914; Brunelli, 1919). Generally, they are characterized by transparent body, scarcity of melanophores, persistence of the swimbladder and short and straight alimentary canal (Miller, 1973, 1989). There is also an evolutionary trend toward a reduction of the first dorsal fin, a change from ctenoid to cycloid scales and a more or less pronounced sexual dimorphism in dentition (Miller, 1973).

The retention of larval morphology when adult, or paedomorphosis, is achieved through heterochrony, a rather common process of adaptation and radiation among gobioids, consisting of a change in the timing and/or rates of processes underlying the ontogenetic formation of morphological traits compared to the ancestral species (Gould, 1977). Among different types of heterochrony (Reilly *et al.,* 1997), paedomorphosis is produced by a slower development rate or deceleration (neoteny sensu Gould, 1977), an earlier offset time of shape or hypomorphosis, sometime coupled with early maturity (progenesis sensu Gould, 1977), or a later onset time of shape or post-displacement (Kon and Yoshino, 2002a). It is suggested that paedomorphic ontogeny of descendants, expressed as progenesis or neoteny, is constrained by seasonally restricted maturity (Kon and Yoshino, 2002b).

Typical marine gobioid genera with paedomorphic traits when adult, such as transparent body, and planktonic life mode are the European Aphia Risso, Crystallogobius Gill and Pseudaphya Iljin, and the Indo-Pacific Paedogobius Iwata, Hosoya and Larson, Leucopsarion Hilgendorf and Schindleria Giltay. All these genera are monotypic, except for Schindleria, which includes three paedomorphic species, namely *S. brevipinguis*, *S. praematura* and *S. pietschmanni* (Johnson and Brothers, 1993; Watson and Walker, 2004). According to some authors (Miller, 1973, 1997; Dotsu and Uchida, 1979; Caputo *et al.,* 2000; Kon and Yoshino, 2002a, b; Kitano, 2002), *Paedogobius kimurai* and *Schindleria* spp. are progenetic, whereas *Aphia minuta, Crystallogobius linearis, Leucopsarion petersii* and *Pseudaphya ferreri* are neotenic. The progenetic gobioids *P. kimurai* and *Schindleria* spp. are distributed in warm waters and inhabit coral reefs and muddy bottoms, respectively (Johnson and Brothers, 1993; Iwata *et al.,* 2001; Kon and Yoshino, 2002a). The neotenic *L. petersii* and the three European paedomorphic gobioids are distributed in temperate and cold waters (Kon and Yoshino, 2002a), in shallow coastal areas over sandy and muddy bottoms (Miller, 1973, 1986; Matsui, 1986). Other than paedomorphosis achieved by heterochrony, all these species have in common similar life-history strategies, showing generally a short life span, a rapid achievement of sexual maturity (especially in progenetic species), a planktotrophic habit and, often, a sudden death of breeders soon after spawning.

In this chapter, the available data on the biology and ecology of paedomorphic and planktonic gobioids concerning taxonomy, habitat and distribution, size and body morphology, reproduction, early life stages, age, growth and mortality, prey and predators and paedomorphosis, ontogeny and phylogeny are critically reviewed, with the aim to provide an up-to-date and as comprehensive as possible review of these interesting fish species.

METHODOLOGY

Age-based heterochrony is a very common process of adaptation and radiation in the gobioids, involving several different genera distributed worldwide (Kon and Yoshino, 2002a). Nevertheless, I decided to consider only paedomorphic species, either neotenic or progenetic, which exhibit a pelagic lifestyle in marine waters over most or part of their life span. Hence, this chapter deals with species reported in Table 4.1.1.

Table 4.1.1 Nomenclature and taxonomy of the pelagic and paedomorphic gobioids investigated.

Taxa	*Common name*
Family: Gobiidae Subfamily: Gobiinae	
Aphia minuta (Risso, 1810)	transparent goby, chanquete, nonnat
Crystallogobius linearis (Düben, 1845)	crystal goby
Pseudaphya ferreri (De Buen and Fage, 1908)	Ferrer's goby
Family: Gobiidae Subfamily: Gobionellinae	
Leucopsarion petersii Hilgendorf, 1880	ice goby, shiro-uo
Paedogobius kimurai Iwata, Hosoya and Larson, 2001	wide-gape paedomorphic goby shirasukiba-haze
Family: Schindleriidae	
Schindleria brevipinguis Watson and Walker, 2004	stout infantfish
Schindleria pietschmanni Schindler, 1931	-
Schindleria praematura Schindler, 1930	Schindler's fish, shirasu-uo

The criterion of selection from the wide array of available literature, at least for some of the aforementioned species, was to consider all the material concerning several aspects of their biology and ecology, based almost exclusively on published data. Data from non peer-reviewed grey literature, consisting mainly of technical reports, are often confidential and need permission by the authors to be cited. However, most of them were published later in peer reviewed journals, becoming available for a wider range of people.

Several factors, such as commercial interest in small-scale fisheries, geographical distribution patterns and timing of first record and description determine a strong difference in the amount of data available for each of the species selected for this review. As a consequence, most of topics discussed below concerned only some of them. Finally, several papers on the Indo-Pacific paedomorphic gobioids are published only in Japanese, enabling me to acquire useful information solely from abstract and/or figure captions.

BIOLOGY

Habitat and Distribution

Among the European paedomorphic species, *P. ferreri* is distributed in the western basin of Mediterranean, including the Adriatic Sea, and occurs inshore over sand, whereas *A. minuta* and *C. linearis* are widespread in the north-eastern Atlantic, from the Strait of Gibraltar to the coasts of Norway, and in the whole Mediterranean basin (the former also in the Black Sea and Azov Sea) (Iljin, 1949; Miller, 1986) (Fig. 4.1.1). *A. minuta* is a neritic species, widely distributed in inshore and estuarine waters from the surface to 80-100 m depth over sand, mud or sea-weed (*Zostera marina* and *Cystoseira*) and sea-grass (*Posidonia oceanica*) (Bini, 1969; Tortonese, 1975). *C. linearis* is distributed from coastal to offshore waters up to 400 m depth on sandy-muddy bottoms (Miller, 1986).

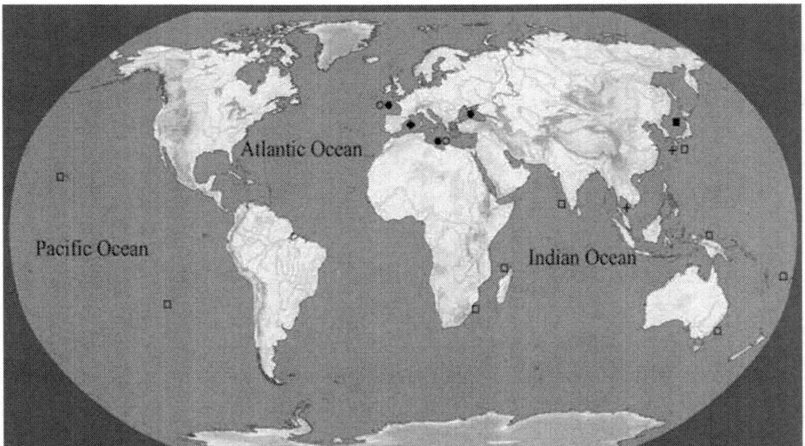

Fig. 4.1.1 Map of the world with the geographical distribution of marine paedomorphic and pelagic gobioids. (•) *A. minuta*, (○) *C. linearis*, (♦) *P. ferreri*, (■) *L. petersii*, (+) *P. kimurai* and (□) *Schindleria* spp.

All other species reported herein have an Indo-Pacific distribution (Fig. 4.1.1). *Schindleria* spp. are lagoon completer, namely are able to complete their pelagic phase in lagoons (Leis, 1994; Leis *et al.*, 2003) and are distributed broadly in the coral reefs of the Indo-Pacific region. In particular, *S. pietschmanni* is distributed around Hawaii and off Taiwan down to 60 m depth, whereas *S. praematura* is widespread in the western Pacific (Ryukyu Islands, the South China Sea and Papua-New Guinea) and southeast Pacific (Nazca and Sala y Gómez; including Easter Island) down to 130 m depth (Bruun, 1940; Ozawa and Matsui, 1979; Leis, 1982; Masuda *et al.*, 1984; Belyanina, 1989; Clarke, 1991; Myers, 1999; Young and Chiu, 2000; Chiu and Hsieh, 2001; Landaeta *et al.*, 2002; Mundy, 2005). Both species were also found in coral reefs off south-eastern India and south-eastern Africa (Jones and Kumaran, 1964; Sardou, 1974; Harris and Cyrus, 1996). Conversely, *S. brevipinguis* is currently known only from the Lizard Island-Carter reef area of the Great Barrier Reef in Australia and from Osprey Reef, nearby the Coral Sea (Watson and Walker, 2004).

P. kimurai is distributed in inner bays of Ryukyu Islands (Japan), Thailand and Australia (Neira *et al.*, 1998; Iwata *et al.*, 2001; Nakabo, 2002). *L. petersii* is widely distributed in the shallow coastal waters down to 10 m depth in Japan, except for Okinawa, China and the southern Korean Peninsula (Masuda *et al.*, 1984; Matsui, 1986; Randall and Lim, 2000; Kim *et al.*, 2005). It is an anadromous species, ascending rivers before the spawning season (spring) (McDowall, 1997; Arai, 2006).

Size and Body Morphology

All the species described in this chapter have in common several morphological traits, as a result of paedomorphosis and pelagic mode of life. The maximum size of adults is very small (Table 4.1.2), ranging from about 58 mm TL in *A. minuta* male (La Mesa *et al.*, 2005) to 9.1 mm TL in *S. brevipinguis* female, which represents almost certainly the world's smallest fish and smallest vertebrate (Watson and Walker, 2004). The body is generally transparent, with no or a few scattered chromatophores (Fig. 4.1.2). There is evidence of an evolutionary trend from ctenoid scales covering the body in *P. ferreri* (Miller, 1973) to cycloid scales in *A. minuta* (Miller, 1986), to a complete loss of scales in *C. linearis* (Miller, 1986), *L. petersii* (Dotsu and Uchida, 1979), *P. kimurai* (Iwata *et al.*, 2001) and *Schindleria* spp. (Johnson and Brothers, 1993; Watson and Walker, 2004). There is also a progressive reduction of the first dorsal fin, which is fully developed in *A. minuta* and *P. ferreri*, reduced to II rays in males of

Table 4.1.2 Maximum size reported for marine paedomorphic gobioids.

Species	*Size (mm)* ♀	♂	*Source*
A. minuta	53	58	La Mesa *et al.*, 2008
C. linearis	39	47	Miller, 1986
P. ferreri	35	-	Fage, 1910
L. petersii	50	46	Matsui, 1986
P. kimurai	16.5	16.9	Iwata *et al.*, 2001
S. brevipinguis	9.1	8.6	Watson and Walker, 2004
S. pietschmanni	15.1	-	Jones and Kumaran, 1964
S. praematura	20.1	-	Jones and Kumaran, 1964

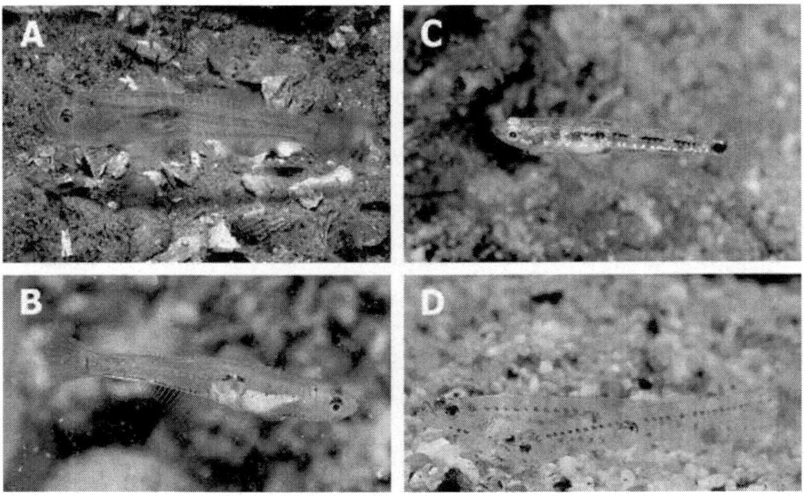

Fig. 4.1.2 Live pictures of some paedomorphic gobioids, showing the characteristic body morphology and colouration pattern. (**A**) *A. minuta*, (**B**) *C. linearis*, (**C**) *P. ferreri* and (**D**) *L. petersii* (photographs by kind permission of Dr. Mat Vestjens).

Color image of this figure appears in the color plate section at the end of the book.

C. linearis, and completely absent in females of *C. linearis*, *P. kimurai* and *Schindleria* spp. Similarly, pelvic fins are reduced in *L. petersii* and lacking in *P. kimurai* females and *Schindleria* spp. Yet, all these species retain into adulthood the swimbladder and a short and straight alimentary canal. Most of them exhibit a more or less pronounced sexual dimorphism in dentition, as enlarged caniniform teeth are developed in the jaws of adult males of *A. minuta*, *C. linearis* and *P. kimurai* (especially secondary males) (Fig. 4.1.3). Finally, as an adaptation to nektonic lifestyle in pelagic habitats, extremely

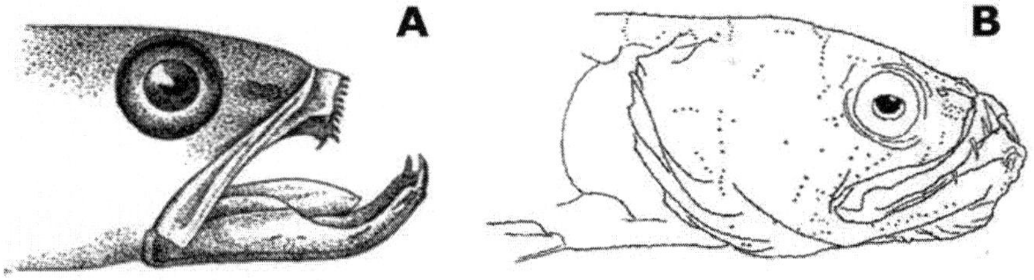

Fig. 4.1.3 Examples of sexual dimorphism in dentition, showing the enlarged caniniform teeth in (**a**) adult male of *C. linearis* and (**b**) secondary male of *P. kimurai* (modified from Collett, 1878 and Iwata *et al.*, 2001, respectively).

high numbers of sensory papillae (or free neuromasts) have been observed on body and caudal fin of *A. minuta*, *L. petersii* and *P. ferreri* (Mortara, 1918; Scattolin, 2003; Scattolin *et al.*, 2007).

Reproduction

The spawning period of *A. minuta* varies largely across its wide zoogeographical range. It spawns between June-August in Atlantic waters and May-July in the Black Sea (Fage, 1910, 1918; Clark, 1914; Allen, 1917; Borcea, 1933; Miller, 1986; Arruda *et al.*, 1993). A strong difference in the spawning period occurs among different localities of the Mediterranean Sea as well. *A. minuta* spawns in January-March off Murcia (south-eastern Spain) (Martínez-Baño *et al.*, 1993), and in December-April off Majorca (De Buen, 1940; Iglesias, 1995; Iglesias and Morales-Nin, 2001). However, on the basis of the hatchdates distribution derived by age estimates, some authors suggest an additional spawning period between June-November (Iglesias, 1995; Iglesias and Morales-Nin, 2001). Two main spawning periods (February-March and September) are also reported from the Ligurian and Tyrrhenian Sea (Baino *et al.*, 1996; Auteri *et al.*, 1998, 2000; Relini *et al.*, 1998). Finally, the breeding season of *A. minuta* lasts at least six months (i.e., from April to September) in the Adriatic Sea, with three main spawning peaks in spring, summer and early autumn (Froglia and Gramitto, 1989; La Mesa, 1999; Caputo *et al.*, 2000).

Several studies indicate *A. minuta* to be a semelparous species, with a short reproductive season followed by the death of all breeders (Collett, 1878; Fage, 1910; Mancini and Cavinato, 1969; Miller, 1973, 1989, 1997; Martínez-Baño *et al.*,1993; Iglesias *et al.*,1997). A recent histological analysis of gonads demonstrates that this species has an abbreviate iteroparous mode of reproduction, because a single individual can spawn at least twice, by means of different oocyte batches in which vitellogenesis occurs asynchronously. In particular, older females spawn in spring and summer and younger females spawn in summer and autumn, so that three different spawning events occur overall at the population level (Caputo *et al.*, 2000).

Also the range of absolute fecundity in *A. minuta* varies considerably between different sites, ranging between 900-2700 oocytes per female off Majorca (Iglesias and Morales-Nin, 2001), 1500-7000 oocytes in the Adriatic Sea (Froglia and Gramitto, 1989) and 2500-5000 oocytes in the Tyrrhenian Sea (Auteri *et al.*, 2000). Egg size in ripe females measures between 0.30-0.40 mm in the Adriatic Sea, 0.45-0.55 mm in the Ligurian Sea and 0.60 mm off Majorca (Froglia and Gramitto, 1989; Relini *et al.*, 1998; Caputo *et al.*, 2000; Iglesias and Morales-Nin, 2001). Gonadosomatic index (GSI) in ripe females is relatively high, ranging between 25 - 28 % of body weight (Iglesias, 1995; Arruda *et al.*, 1993; Relini *et al.*, 1998). Size at sexual maturity is between 28-30 mm and 42-45 mm TL in the central Mediterranean (Froglia and Gramitto, 1989; Relini *et al.*,1998; Auteri *et al.*, 2000; Caputo *et al.*, 2000) and 28-40 mm in the western Mediterranean (Iglesias and Morales-Nin, 2001).

C. linearis is generally considered a single spawner (i.e., semelparous sensu Cole, 1954), with a single breeding season between March-August and July-September in north Atlantic and Mediterranean waters, respectively (Collett, 1878; Padoa, 1956). A recent histological study on gametogenesis confirms that *C. linearis* is a truly semelparous species, as mature gonads are unable to provide gametes for a further spawning in both sexes (Caputo *et al.*, 2003). In the Adriatic Sea the spawning takes place in June-July, when most of males and females are mature. After spawning, males of *C. linearis* guard the eggs in the serpulid tubes (Raffaele, 1895). Size at sexual maturity of females is about 27.4 mm (TL_{50}), and mean size of ripe eggs ranges between 0.24-0.55 mm. Absolute and relative fecundity are relatively low, ranging between 209-696 eggs/females and 4.2-11.9 egg/g, respectively (Caputo *et al.*, 2003).

In comparison with the other two European paedomorphic gobioids, the reproductive traits of *P. ferreri* are poorly known. According to Fage (1910), the breeding season of this species would occur in June, on the basis of a ripe female of 35 mm caught off Cap d'Ail (Monaco). Yet, size at sexual maturity of *P. ferreri* is approximately 26-27 mm SL (Miller, 1973).

Across its wide geographical distribution in the Pacific Ocean, *S. praematura* spawns between October-December in the southern hemisphere (Tahiti, Samoa) (Bruun, 1940), in March near the equator (New Guinea) (Giltay, 1934) and between February-August in the northern hemisphere (Hawaii, Taiwan) (Schindler, 1932; Young and Chiu, 2000). In the Indian Ocean, the spawning period of *S. praematura* and *S. pietschmanni* takes place in March-May off South Africa (Harris and Cyrus, 1996), between October-December in the Mozambique Channel (Sardou, 1974) and respectively in December and April off India (Jones and Kumaran, 1964). A single mature female of *S. brevipinguis* (8.4 mm SL), bearing approximately 12 large and irregularly shaped eggs of 0.3-0.5 mm, was caught in Australian coral reefs in January (Watson and Walker, 2004). The absolute fecundity is very low in *S. praematura* as well, ranging between 25-110 eggs/female, and eggs are disproportionately large (up to 0.40 mm) compared to the ovary that contains them (Sardou, 1974; Landaeta *et al.*, 2002; Thacker and Grier, 2005). Interestingly, *S. praematura* exhibits an unusual gonad structure compared to most other species of gobioids. The stromal ovarian tissue and testicular interstitial tissue are largely reduced, and both the number and location of germinal cells are restricted (Thacker and Grier, 2005). In males, the restricted spermatogonial type (sensu Grier *et al.*,1980) observed in *S. praematura* is common only in atherinomorphs, whereas the restriction of germinal cells in the ovary is extremely rare in fish, being reported only from one other species (*Hippocampus erectus*).

L. petersii is an anadromous fish, whose mature adults migrate from the sea to freshwater in early spring (February) until April. After migration, adults spend several weeks into rivers to complete gonad maturation, attaining a GSI of 10-15 % in May, when they spawn (Matsui, 1986; Arai, 2006). In the southern rivers of Japan, the spawning season takes place before, i.e., in January (Matsui, 1986; Akiyama and Ogasawara, 1994; Akiyama *et al.*,1995). Under rearing condition, *L. petersii* spawns approximately nine months after hatching, when females attain a size of 5 cm TL (Arakawa *et al.*,1999; Kitano, 2002). Egg masses, consisting of 380-650 ripe eggs/female of approximately 0.6 - 0.9 mm, are laid down simultaneously in nests located under side of rocks and cared by males until hatching (Matsui, 1986; Nakatsuji *et al.*,1997; Arakawa *et al.*,1999; Harada *et al.*,2003). Downstream migration of early larvae of *L. petersii* in the northern Japan is relatively fast, as it occurs within two weeks from the end of May to mid-June (Tomiyama *et al.*, 1999).

Early Life Stages

Except for *P. ferreri*, the early life stages of the European paedomorphic species are relatively well-known. The eggs of *A. minuta* and *C. linearis* are demersal and very similar to those observed in *Gobius*, with a network of filaments used to adhering to hard substrata, such as stones, shells or serpulid tubes (Raffele, 1895; Grieg, 1896; Holt and Byrne, 1898; Lo Bianco, 1909; Clark, 1914; De Buen, 1931). The unfertilized eggs of *A. minuta* are slightly ovate with a wide perivitelline space, measuring about 0.8 mm (Iglesias, 1995), whereas those of *C. linearis* are club-shaped and approximately 1.5 × 0.5 mm (Raffele, 1895; Grieg, 1896).

The newly hatched larvae of *A. minuta* have been described only recently from specimens reared in aquarium (Lopez Jaime *et al.*, 2000). The yolk-sac larva, measuring about 3-4 mm, is strongly pigmented with melanophores on the head, on the lower jaw and on the ventral abdominal contour. Two rows of melanophores, more scattered dorsally, form dorsal and ventral contours. The overall pigmentation is largely reduced with ontogeny. The postlarvae from 7 to 20 mm have the head totally unpigmented, some stellate melanophores on the ventral side of the abdomen, a typical dorsal pigmentation on the swimbladder and a postanal ventral row of melanophores, extending to the posterior end of anal fin (Fig. 4.1.4a) (Ehrenbaum, 1905; Lebour, 1919; Petersen, 1919; Padoa, 1956; Russell, 1976). The second dorsal, the caudal and the anal fins are already developed at 7-8 mm of larval size. The first dorsal fin begins to develop at approximately 16-17 mm and it is not completely formed even at 20 mm length, when pectoral fin rays appear. The postlarvae of *A. minuta* have been recorded off Plymouth and in the North Sea from June to September, with a main peak in July and August (Ehrenbaum, 1905; Clark, 1914; Allen, 1917).

The newly hatched larva of *C. linearis* has not yet been described. The postlarva of 4-5 mm is strongly pigmented with melanophores on the head, in front of eyes, on the lower jaw and on the abdominal region. There are also five large evenly spaced melanophores along the dorsal contour and a continuous row from behind the pigmented swimbladder to the caudal fin. Most of pigmentation disappears with ontogeny also in *C. linearis*, in particular in the dorsal region (Fig. 4.1.4b, c, d). Second dorsal and anal fins are fully developed at 9 mm, whereas pelvic and pectoral fins are beginning to develop at about 25 mm. The first dorsal fin, with only two rays, is present exclusively in males and develops later on (Fage, 1918; Lebour,

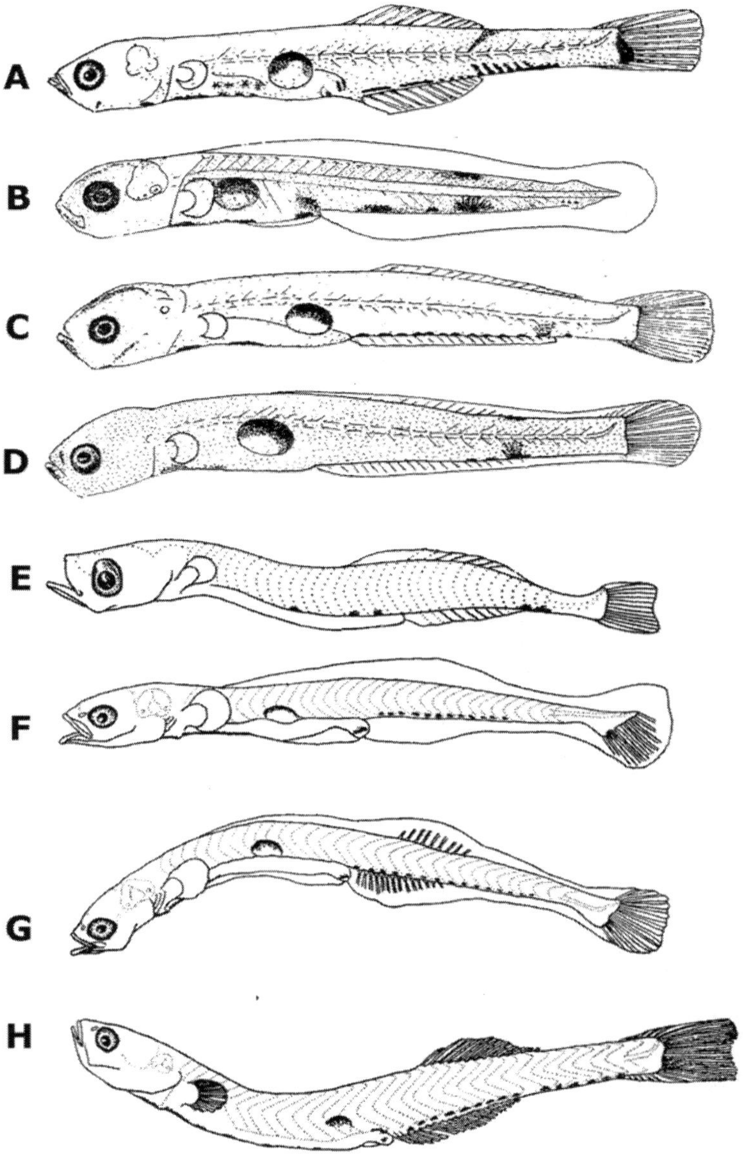

Fig. 4.1.4 Larval stages of paedomorphic gobioids. (**A**) *A. minuta* 10 mm; (**B**, **C**, **D**) *C. linearis* 6 mm, 9 mm and 11 mm; (**E**) *S. praematura* 3.6 mm, (**F**, **G**, **H**) *L. petersii* 6.1 mm, 7.4 mm and 15.1 mm (modified from Petersen, 1919; Ozawa and Matsui, 1979; Matsui, 1986).

1919; Petersen, 1919; Padoa, 1956; Russell, 1976). Eggs and postlarvae of *C. linearis* were recorded in June and between May-October off Plymouth, respectively (Allen, 1917; Clark, 1920), and postlarvae 3.8 - 3.9 mm TL were caught in June in the Cantabrian Sea (Villegas, 1980).

The early life stages of schindleriids have been reported exclusively for *S. praematura* and *S. pietschmanni*. The largest ovarian eggs lack oil droplets and are irregular in shape or oval, measuring 0.35-0.40 mm in *S. pietschmanni* and 0.30-0.65 in *S. praematura* (Watson *et al.*,1984). As seen previously in European species, the egg surface morphology in *S. praematura* is typical of gobies, being characterized by broad filaments surrounding the micropyle (Johnson and Brothers, 1993; Thacker and Grier, 2005).

Size at hatching is currently unknown for both species. However, *S. pietschmanni* at 1.9 mm has a rather large yolk-sac. Notochord flexion occurs between 2.7-3.5 mm in *S. pietschmanni* and before 4.3 mm in *S. praematura* (Watson *et al.*,1984; Watson, 2000). The larval pigmentation is very similar in both species, being characterized by one to four pairs of melanophores along the side of the gut, one to four melanophores along the ventral midline of the caudal fin and pigment on the posterior dorsal surface of the swimbladder (Fig. 4.1.4e). Except for the swimbladder, all the pigment is lost during the juvenile stage. Dorsal and anal fin rays are the first to develop, followed by the pectoral fin rays (Ozawa and Matsui, 1979; Sardou, 1974; Watson *et al.*,1984; Watson, 2000). Typically, the early life stages of *Schindleria* occur inside coral-reefs lagoons, where they complete all the larval development (Leis, 1982, 1994; Leis *et al.*, 2003).

The morphogenesis of early life stages of *L. petersii*, including the embryonic development from egg fertilization to larval hatching, has been recently described by several authors under rearing condition (Nakatsuji *et al.*,1997; Arakawa *et al.*,1999; Kitano, 2002; Saito *et al.*, 2001, 2002). The chorion of unfertilized eggs is round and close to the egg surface, with a wide perivitelline space and a tuft of filaments near the animal pole used for attachment to the underside of rocks. Hatching occurs between 10-18 days after egg fertilization, and the newly hatched larva as long as 4.3 mm is completely transparent (Matsui, 1986; Arakawa *et al.*,1999; Arai, 2006). Yolk absorption in early larvae occurs at 6-7 mm TL. Dorsal, anal and pectoral fins are fully developed at 10.0-12.5 mm TL, whereas pelvic fins are completed at 37-38 mm TL (Fig. 4.1.4f, g, h) (Matsui, 1986). Larval stages of *L. petersii* occur between June-August in Otsuchi Bay (north-eastern Japan) (Yamashita and Aoyama, 1984).

Larval stages of *P. kimurai* are still unknown. Conversely, the eggs are long and ellipsoid, 2.6-2.8 mm × 0.5 mm, with a series of filament used to keep them in a single mass (Iwata *et al.,* 2001).

Age, Growth and Mortality

Data on growth rate and longevity of the European paedomorphic gobioids are limited to *A. minuta* and *C. linearis*. Based on either otolith microincrement counts (Fig. 4.1.5a) or modal progression of length frequency distributions, a series of von Bertalanffy growth curves have been fitted for different populations of *A. minuta* from central and western Mediterranean (La Mesa *et al.*, 2005). Notwithstanding, the growth performance (Φ') of this species, which represents a useful tool for comparing growth curves fitted for populations of the same species living in different environments, is practically the same throughout the geographical range investigated (Table 4.1.3). The longevity (i.e., the maximum estimated age) of *A. minuta* reported from different sites is very similar from each other as well, ranging between 7-8 months in the western Mediterranean (Iglesias and Morales-Nin, 2001; Iglesias *et al.*, 1997) and 9 months in the central Mediterranean (Auteri *et al.*, 2000; La Mesa, 1999). It closely resembles the maximum age estimated for *C. linearis* by microincrement counts (Fig. 4.1.5b), namely 7 months in both sexes (La Mesa, 2001). Both species attain sexual maturity relatively early, as younger females of *A. minuta* and *C. linearis* are mature at 3 months and 30 mm TL and 4 months and 27 mm TL, respectively (Caputo *et al.*, 2000; La Mesa 2001). In agreement with several authors (Collett, 1878; Fage, 1910; Brunelli and Atella, 1914; De Buen, 1931; Miller, 1973), more recent studies confirm that the life span of these species is annual, with a strong mortality just after the spawning season, leading to a sudden disappearance of the whole annual cohort (La Mesa, 1999, 2001; Caputo *et al.*, 2000, 2003; La Mesa *et al.*, 2008). Such a mass mortality of breeders after spawning is probably triggered by apoptotic processes involving the intestinal epithelium (Caputo *et al.*, 2002), which cause a loss of functionality of the alimentary canal, as already suggested in *P. ferreri* too (Fage, 1910).

Compared to European species, *S. praematura* (and possibly of the other two schindleriids) attains sexual maturity at an earlier age and a short life span. Based on otolith increment counts (Fig. 4.1.5c), *S. praematura* females mature on average at 28-43 days old and exhibit an age range of 23-60 days (at 15-30 mm SL), suggesting a life span of no more than 2-3 months (Landaeta *et al.*,2002; Kon and Yoshino, 2002b). Also *P. kimurai*

Table 4.1.3 Von Bertalanffy growth parameters and growth performance index (Φ') of *A. minuta* from different localities of the Mediterranean Sea. 1: Iglesias *et al.*, 1997; 2: La Mesa, 1999; 3: Auteri *et al.*, 2000.

Sex	L_∞ *(mm)*	*k* *(year^{-1})*	t_0 *(year)*	*Φ'*	*Site*	*Source*
♂	61.01	1.68	–0.036	3.79	Balearic Sea	1
♂	74.26	1.41	0.074	3.89	Adriatic Sea	2
♀	49.58	2.76	0.019	3.84	Balearic Sea	1
♀	65.60	1.61	0.075	3.84	Adriatic Sea	2
♂ + ♀	53.69	2.23	–0.005	3.81	Balearic Sea	1
♂ + ♀	57.90	1.75	–	–	Tyrrhenian Sea	3
♂ + ♀	68.70	1.55	0.077	3.86	Adriatic Sea	2

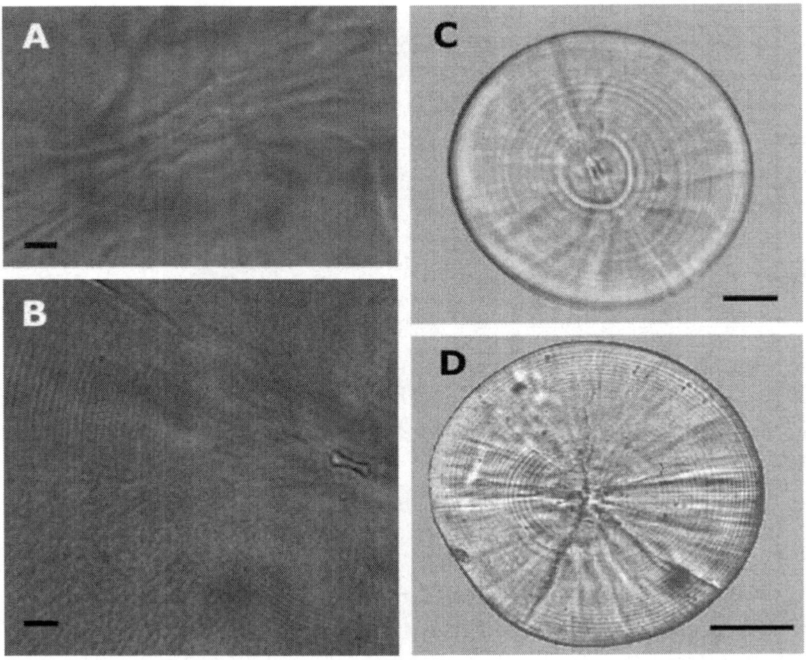

Fig. 4.1.5 Photomicrographs of sagittal otoliths of paedomorphic gobioids, with the elongate primordium which is characteristic of and unique to the Gobioidei. (**A**) *A. minuta*, (**B**) *C. linearis*, (**C**) *S. praematura* and (**D**) *P. kimurai* (modified from La Mesa 1999, 2001; Kon and Yoshino, 2002b).

has a similar short lifespan, being aged between 42-67 days (at 12-16 mm SL) (Fig. 4.1.5d) (Kon and Yoshino, 2002b). On the other hand, *L. petersii* reaches sexual maturity after 9 months of age (at 42-43 mm TL), when adults become dimorphic and ascend rivers to spawn ceasing to feed on, and die after spawning (Dotsu and Uchida, 1979; Matsui, 1986; Kitano, 2002).

Prey and Predators

The role of paedomorphic gobioids in the food web of neritic environments is still poorly known. Available data on feeding habits of most species can be considered almost exclusively anecdotal, based mainly on their pelagic mode of life. As such, all of them possibly rely on zooplanktonic prey.

Differently from the other paedomorphic species, the feeding habits of *A. minuta* have been studied to some extent. In general terms, it is a plankton feeder that relies on copepods, cirripede and mysid larvae (Hesthagen, 1971; Miller, 1986). In the Black Sea, for example, *A. minuta* feeds mostly on small copepods such as *Acartia clausi* and *Pseudocalanus elongatus*, sharing them with the Baltic sprat and the Azov anchovy (Chesalin *et al.*, 2004). In the northwestern Adriatic Sea, the diet composition of *A. minuta* consists mainly of the calanoid copepods *Acartia clausi* and *Temora longicornis*, followed in decreasing importance by other copepods and larvae of decapods, polychaetes and bivalves (La Mesa *et al.*, 2008). *A. minuta* adopts a generalistic feeding strategy, exhibiting a relatively broad niche width and an ontogenetic shift of food preferences (Fig. 4.1.6a), which probably contributes to mitigate the intraspecific competition for prey (La Mesa *et al.*, 2008).

In turns, *A. minuta* is known to be a common prey of several demersal fishes, such as *Merlangius merlangus*, *Mullus barbatus*, *Platichthys flesus*, *Scorpaena porcus*, *Solea* and *Trachinus* spp. in the Black Sea (Fortunatova, 1949), *Merluccius merluccius*, *Thunnus thynnus* and *Trachurus trachurus* in Spanish waters and *Serranus cabrilla* and *S. hepatus* in Greek waters (Iglesias, 1995; Labropoulou and Eleftheriou, 1997; Cartes *et al.*,2004). Similarly, *C. linearis* is one of the main prey of *Trisopterus minutus* off the west coast of the Isle of Man (Armstrong, 1982).

Ontogenetic and habitat-related changes in feeding habits of *L. petersii* are reported off the southwestern coasts of Japan (Matsui, 1986). Copepods represent by far the most important prey, being consumed in large amount by all the fish size groups. Amphipods become important prey for adult fish larger than 35 mm TL, which seldom rely on rotiferans and polychaete

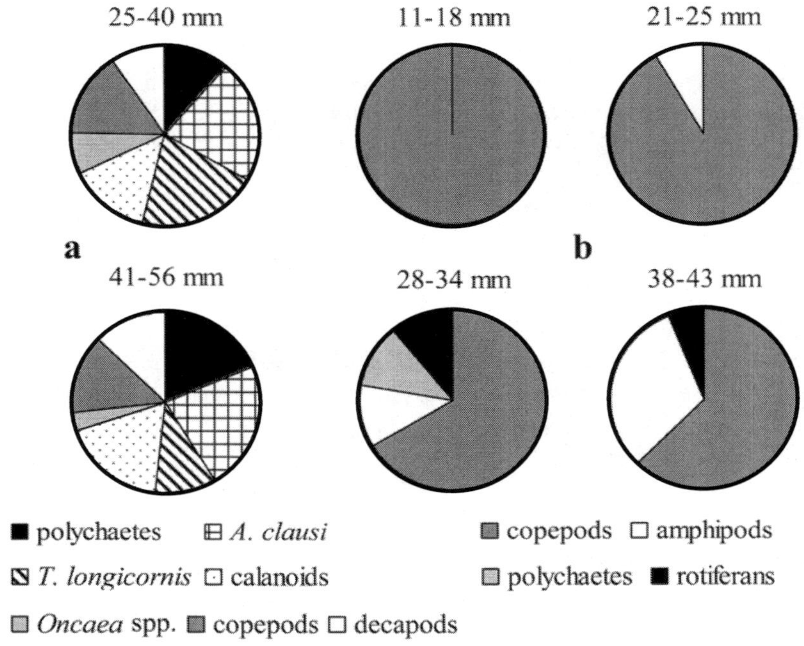

Fig. 4.1.6 Ontogenetic change in diet expressed as percentage of frequency of occurrence for (**A**) *A. minuta* and (**B**) *L. petersii*. Size groups are indicated above diagrams (data rearranged from Matsui, 1986 and La Mesa *et al.*, 2008).

larvae (Fig. 4.1.6b) (Matsui, 1986). Habitat-related changes in diet are strictly linked to the life cycle of *L. petersii*, since they rely on a wide prey spectrum when at sea. As adult start upstream migration for spawning, they feed only on copepods and amphipods in the brackish environment of the rivers mouth, ceasing completely to feed inside the river (Matsui, 1986).

Paedomorphosis, Ontogeny and Phylogeny

The heterochronic phenomenon of paedomorphosis, or the expression in later developmental stages of feature that characterize earlier stages ancestrally, is categorized in two major types in relation to life history strategies: progenesis (i.e., truncation of somatic development with early sexual maturity) and neoteny (i.e., slower somatic development or deceleration) (Gould, 1977; Alberch *et al.*, 1979). To assign the type of paedomorphosis, it is necessary to know the phylogeny of targeted taxa and determine the polarity of relevant developmental characters (Kon and Yoshino, 2002a). The phylogeny within gobioids is, however, difficult to estimate, since most of them have diverged simultaneously or during a very short time, as indicated by molecular evidence (Akihito *et al.*, 2000). However, the plesiomorphic ontogeny of gobioids, including eleotrids (*sensu* Hoese, 1984), would consist of several months planktonic-larval duration and attainment of sexual maturity at 6-12 months (Kon and Yoshino, 2002a). In comparison, the paedomorphic gobioids described herein exhibit both types of heterochrony (i.e., progenesis and neoteny), as well as an intermediate position (Fig. 4.1.7).

Some of the larval traits observed in the sexually mature individuals of *Schindleria* spp. and *P. kimurai*, including functional pronephros, transparent and scaleless body, lack of first dorsal fin and largely unossified skeleton, coupled with a very early attainment of sexual maturity (1-2 months of age), clearly indicate that these species are progenetic. On the opposite end, though retaining several larval traits as adult, *L. petersii* become sexually mature and complete pelvic fin development at nine months of age, showing a slower somatic developmental rate (i.e., neoteny) than that of plesiomorphic gobioids. Finally, the European paedomorphic gobioids *A. minuta* and *C. linearis* exhibit an intermediate tendency between progenesis and neoteny, being mature at 3-4 months of age and sharing several larval traits as adult.

From a taxonomic point of view, *A. minuta*, *C. linearis* and *P. ferreri* are assigned unequivocally to the family Gobiidae (Miller, 1986), whereas their phylogeny is still controversial. Sharing several paedomorphic traits, all of them were previously placed into the subfamily Gobiinae as a monophyletic group (tribe Aphyini) (Miller, 1973). However, on the basis of molecular data, a recent study strongly supports the polyphyletic origin of the three European paedomorphic species, indicating that the heterochronic process leading to the retention of larval traits in adults occurred independently in their ancestors (i.e., the shared larval traits are homoplasious) (Giovannotti *et al.*, 2007).

Since its original description, the systematic position and phylogeny of the genus *Schindleria* have been debated for a long time. Giltay (1934), for example, proposed that schindleriids are percomorphs, possibly related to the Blennioidei. Yet, further classifications have placed them close to the Ammodytoidei (Greenwood *et al.*, 1966). More recently, on the basis of osteological data, as well as of otolith primordium and gonad morphology, Johnson and Brothers (1993) conclude that *Schindleria* spp. are gobioids, although their precise relationships within the Gobioidei remain uncertain.

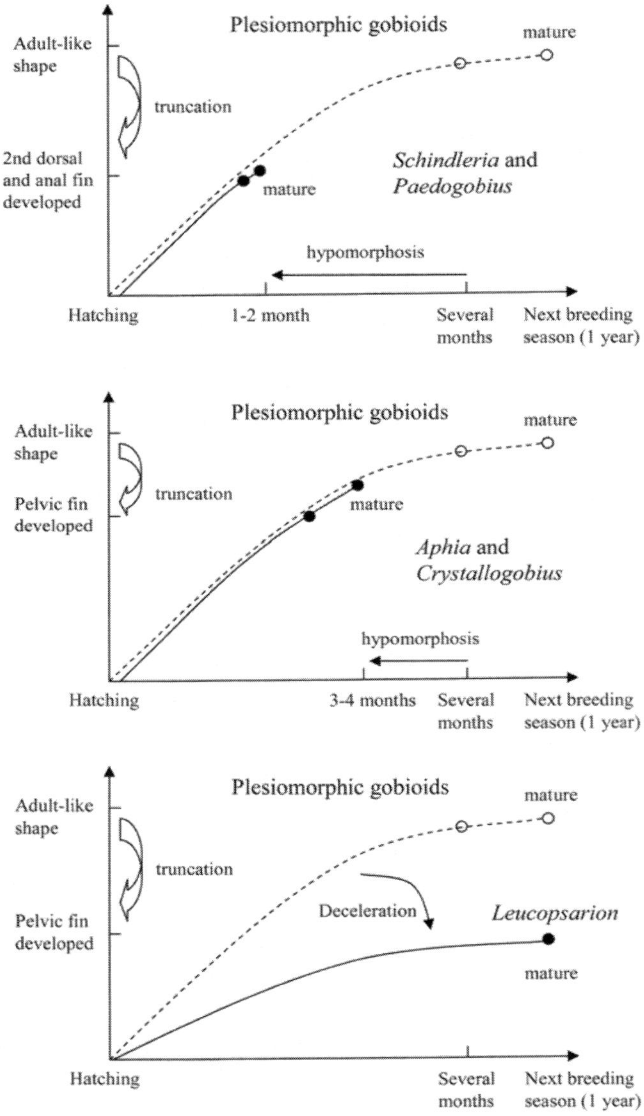

Fig. 4.1.7 Ontogenetic trajectories of paedomorphic species compared to plesiomorphic gobioids (partially redrawn from Kon and Yoshino, 2002a).

Finally, according to Iwata *et al.* (2001), *P. kimurai* belongs to the *Gobiopterus* group (sensu Birdsong *et al.*, 1988), in having similar vertebral count (10 + 15) and geographic distribution.

CONCLUSIONS

In an evolutionary perspective, heterochrony is considered to be one of the processes of adaptation and radiation more contributing to the high diversity of the Gobioidei (Kon and Yoshino, 2002a). In particular, paedomorphosis can be regarded as a heterochronic phenomenon that, through the retention of larval traits into adulthood, allows these species to maintain a small body size and a planktonic lifestyle. Body miniaturization, or the evolution of extremely small adult body size, has several important consequences for both organismal biology and phyletic diversification above the species level. The miniaturized phenotype is a complex combination of ancestral and derived traits, including reduction and structural simplification and morphological novelty (Hanken and Wake, 1993). Living in the water column for most (or entirely) of their life span, paedomorphic species are able to utilize efficiently zooplankton as prey resource, as well as to colonize trophic and spatial niches different from those typically occupied by other gobioid fishes. On the other hand, the pelagic lifestyle considerably increases the risk of predation also for adults, so that the more suitable adaptive response is represented by an early attainment of sexual maturity with a high reproductive effort, obtained through progenesis (Miller, 1989). Aiming to maximize fitness or lifetime reproductive output, optimal age or size at sexual maturity is determined by trade-offs between fecundity, growth rate and mortality (Roff, 1992).

Driven by their own physiological constraints, as well as by the local ecological and environmental features, the pelagic and paedomorphic gobioids have evolved life strategies which are similar from a general point of view, but very different if examined in detail. Paedomorphic species which live in tropical waters, such as *Schindleria* spp. and *P. kimurai*, exhibit a truncation of somatic development with a very early sexual maturity (i.e., through progenesis), giving rise to a rapid alternation of generations throughout the year. Tropical waters are typically characterized by environmental conditions, such as water temperature and zooplankton availability, which are rather stable all the year round. These factors probably sinergically interact, the stable warm water temperature increasing the rate of metabolic processes involved in gonadal maturation, the continuous food availability enhancing the larval survival of all generations throughout the year.

On the other hand, paedomorphic gobioids which live in temperate-cold waters, such as the European species and *L. petersii*, exhibit a gradual trend towards neoteny. To overcome the lower environmental predictability, they are developed two different reproductive strategies. *A. minuta* attains sexual maturity relatively early and has an abbreviate iteroparous mode of reproduction, with more than one spawning per reproductive season. As a result, the reproductive effort is spread over a relatively long time, balancing the detrimental effects of the fluctuation of environmental conditions on the larval survival. Conversely, *C. linearis* and *L. petersii*, which are truly semelparous spawner spending all their reproductive potential in a single event, evolved similar type of parental care in which males guard eggs until hatching, trading off the low fecundity and the single batch of eggs per female.

References

Akihito, P., A. Iwata, T. Kobayashi, K. Ikeo, T. Imanishi, H. Ono, Y. Umehara, C. Hamamatsu, K. Sugiyama, Y. Ikeda, K. Sakamoto, A. Fumihito, S. Ohno and T. Gojobori. 2000. Evolutionary aspects of gobioid fishes based upon a phylogenetic analysis of mitochondrial cytochrome b genes. *Gene* 259: 5-15.

Akiyama, N. and Y. Ogasawara. 1994. Reproductive behavior of the ice goby *Leucopsarion petersii* in captivity. *Suisanzoshoku* 42: 577-584. (In Japanese.)

Akiyama, N., T. Kitano, K. Hikichi and Y. Ogasawara. 1995. Relationship between nest-building activity of ice goby, *Leucopsarion petersii* and either sand particle size or stream velocity. *Suisanzoshoku* 43: 289-296. (In Japanese.)

Alberch, P., S.J. Gould, G.F. Oster and D.B. Wake. 1979. Size and shape in ontogeny and phylogeny. *Paleobiology* 5: 296-317.

Allen, E.J. 1917. Post-larval teleosteans collected near Plymouth. *Journal of the Marine Biological Association of the United Kingdom* 11: 207-250.

Arai, T. 2006. Comparison of habitat use during early life stage between ayu *Plecoglossus altivelis* and ice goby *Leucopsarion petersi* along the Sanriku coast of Japan, as determined from otolith microchemistry. *Fisheries Science* 72: 382-387.

Arakawa, T., Y. Kanno, N. Akiyama, T. Kitano, N. Nakatsuji and T. Nakatsuji. 1999. Stages of embryonic development of the ice goby (shiro-uo), *Leucopsarion petersii*. *Zoological Science* 16: 761-773.

Armstrong, M.J. 1982. The predator-prey relationships of Irish Sea poor-cod (*Trisopterus minutus* L.), pouting (*Trisopterus luscus* L.), and cod (*Gadus morhua* L.). *Journal du Conseil International pour l'Exploration de la Mer* 40: 135-152.

Arruda, L.M., J.N. Azevedo and A.I. Neto. 1993. Abundance, age-structure and growth, and reproduction of gobies (Pisces; Gobiidae) in the Ria de Aveiro Lagoon (Portugal). *Estuarine, Coastal and Shelf Science* 37: 509-523.

Auteri, R., R. Baino and A. Abella. 2000. Biology and population dynamic of the transparent goby: a locally important resource of artisanal fisheries. *Biologia Marina Mediterranea* 7: 144-157.

Auteri, R., A. Abella R. Baino, R. Righini, F. Serena, R. Silvestri and A. Voliani. 1998. La pesca del rossetto (*Aphia minuta*, Pisces, Gobidae) in Toscana. *Biologia Marina Mediterranea* 5: 477-486.

Baino, R., R. Auteri and L. Donati. 1996. Crescita e reclutamento alla pesca del rossetto. *Biologia Marina Mediterranea* 3: 525-526.

Belyanina, T.N. 1989. Ichthyoplankton in the regions of the Nazca and Sala y Gomes submarine ridges. *Journal of Ichthyology* 29: 84-90.

Bini, G. 1969. *Atlante dei Pesci delle Coste Italiane.* Mondo Sommerso, Roma.

Borcea, I. 1933. Révision systématique et distribution géographique des Gobidés de la Mer Noire et particulièrement des eaux Roumaines. *Annales Scientifiques de l'Université de Jassy* 19: 1-231.

Brunelli, C. 1919. Ricerche sull'anatomia e fisiologia comparata dei pesci. *Rivista di Biologia* 1: 400-404.

Brunelli, C. and F. Atella. 1914. Ricerche sugli adattamenti alla vita planctonica (i Gobidi planctonici). *Biologisches Zentralblatt* 34: 458-466.

Bruun, A.F. 1940. A study of a collection of the fish *Schindleria* from South Pacific waters. *Dana-Report Carlsberg Foundation* 21: 1-12.

Caputo, V., G. Candi, M. La Mesa and E. Arneri. 2000. Pattern of gonad maturation and the question of semelparity in the paedomorphic goby *Aphia minuta*. *Journal of Fish Biology* 58: 656-669.

Caputo, V., M. La Mesa, G. Candi and P. Nisi Cerioni. 2003. The reproductive biology of the crystal goby with a comparison to that of the transparent goby. *Journal of Fish Biology* 62: 375-385.

Caputo, V., G. Candi, E. Arneri, M. La Mesa, C. Cinti, M. Provinciali, P. Nisi Cerioni and A. Gregorini. 2002. Short lifespan and apoptosis in *Aphia minuta*. *Journal of Fish Biology* 60: 775-779.

Cartes, J.E., J. Rey, D. Lloris and L. Gil de Sola. 2004. Influence of environmental variables on the feeding and diet of European hake (*Merluccius merluccius*) on the Mediterranean Iberian coasts. *Journal of the Marine Biological Association of the United Kingdom* 84: 831-835.

Chesalin, M.N., G.V. Zuev and T.L. Chesalina. 2004. Transparent goby *Aphia minuta* (Gobiidae) in the pelagic zone of the Black Sea. *Journal of Ichthyology* 44: 587-591.

Chiu, T.S. and C.H. Hsieh. 2001. Comparison of ichthyoplankton assemblages with hydrographic condition across the northern margin of the Taiwan Strait as the monsoon interim. *Journal of the Fisheries Society of Taiwan* 28: 163-175.

Clark, R.S. 1914. General report of the larval and postlarval teleosteans in Plymouth waters. *Journal of the Marine Biological Association of the United Kingdom* 10: 327-394.

Clark, R.S. 1920. The pelagic young and early bottom stages of Teleosteans. *Journal of the Marine Biological Association of the United Kingdom* 12: 159-240.

Clarke, T. 1991. Larvae of nearshore fishes in oceanic waters near Oahu, Hawaii. *NOAA Technical Report NMFS* 101: 1-19.

Cole, L.C. 1954. The population consequences of life history phenomena. *The Quarterly Review of Biology* 29: 103-137.

Collett, R. 1878. On *Latrunculus* and *Crystallogobius*, two remarkable forms of gobioid fishes. *Proceedings of the Zoological Society of London* 46: 318-339.

De Buen, F. 1931. Notas a la familia Gobiidae. Observaciones sobre algunos generos y synopsis des especies ibéricas. *Notas y Resúmen Instituto Español de Oceanografia* 2: 1-76.

De Buen, F. 1940. Les Gobiides pélagiques ou vivants sur les fonds d'algues calcaires de l'Europe Occidentale. *Bulletin de l'Institut Océanographique de Monaco* 790: 1-16.

Dotsu, Y. and T. Uchida. 1979. Outline of life history of gobioid fishes exemplified by the Ice-Goby. *Aquabiology* 1: 9-16. (In Japanese.)

Ehrenbaum, E. 1905. *Eier und Larven von Fischen der Nordischen Planktons.* Teil I, 216 pp.

Fage, L. 1910. Recerches sur les stades pélagiques de quelques Téléostéens de la mer de Nice (parages de Monaco) et du Golfe du Lion. *Annales de l'Institut Océanographique de Monaco* 1: 1-53.

Fage, L. 1918. Shore Fishes. *Report on the Danish Oceanographical Expeditions 1908-1910 to the Mediterranean and Adjacent Seas* 2: 1-154.

Fortunatova, K.R. 1949. Biology and feeding of the scorpion fish. *Proceedings of the Sevastopol Biological Station* 7: 193-235.

Froese, R. and D. Pauly (eds.) 2008. *FishBase.* World Wide Web electronic publication. www.fishbase.org, version (10/2008).

Froglia, C. and M.E. Gramitto. 1989. La pesca del rossetto (*Aphia minuta*) nel medio Adriatico. *Nova Thalassia* 10: 447-455.

Giltay, L. 1934. Les larves de Schindler sont elles des Hemiramphidae? *Bulletin du Museum Royal d'Histoire Naturelle de Belgique* 10: 1-10.

Giovannotti, M., P. Nisi Cerioni, M. La Mesa and V. Caputo. 2007. Molecular phylogeny of the three paedomorphic Mediterranean gobies (Perciformes: Gobiidae). *Journal of Experimental Zoology* 308B: 722-729.

Gould, S.J. 1977. *Ontogeny and Phylogeny.* Harvard University Press, Harvard.

Greenwood, P.H., D.E. Rosen, S.H. Weitzman and G.W. Myers. 1966. Phyletic studies of teleostean fishes, with a provisional classification of living forms. *Bulletin of the American Museum of Natural History* 131: 339-456.

Grieg, J.A. 1896. Ichthyologiske notiser. *Bergens Museum Aarbok (1894-95)* 5: 1-20.

Grier, H.J., J.R. Linton, J.F. Leatherland and V.L. DeVlaming. 1980. Structural evidence for two different testicular types in teleost fishes. *American Journal of Anatomy* 159: 331-345.

Hanken, J. and D.B. Wake. 1993. Miniaturization of body size: organismal consequences and evolutionary significance. *Annual Review of Ecology and Systematics* 24: 501-519.

Harada, Y., S. Harada, I. Kinoshita, M. Tanaka and M. Tagawa. 2003. Thyroid gland development in a neotenic goby (ice goby, *Leucopsarion petersii*) and a common goby (ukigori, *Gymnogobius urotaenia*) during early life stages. *Zoological Science* 20: 883-888.

Harris, S.A. and D.P. Cyrus. 1996. Occurrence of Schindler's fishes, genus *Schindleria* (Teleostei: Gobioidei), at a small reef in the mouth of the Kosi Estuary, Kwazulu-Natal: a first record for southern Africa. *Bulletin of Marine Science* 59: 228-234.

Hesthagen, I.H. 1971. The winter food of the gobies from one of the deeper channels of the Kieler Bucht, with particular reference to the sand goby, *Pomatoschistus minutus* (Pallas). *Kieler MeeresforSchung* 27: 28-35.

Hoese, D.F. 1984. Gobioidei: relationships. In: *Ontogeny and Systematics in Fishes*, H.G. Moser (ed.), Allen Press, Lawrence, pp. 588-591.

Holt, E.W.L. and L.W. Byrne. 1898. Notes on the reproduction of the teleostean fishes in the south-western district. *Journal of the Marine Biological Association of the United Kingdom* 5: 333-340.

Iglesias, M. 1995. *Biología y pesca de Aphia minuta en la isla de Mallorca*. PhD Thesis, Universidad de las Islas Baleares, Palma de Majorca, 314 pp.

Iglesias, M. and B. Morales-Nin. 2001. Life cycle of the pelagic goby *Aphia minuta* (Pisces: Gobiidae). *Scientia Marina* 65: 183-192.

Iglesias, M., E.B. Brothers and B. Morales-Nin. 1997. Validation of daily deposition in otoliths. Age and growth determination of *Aphia minuta* (Pisces: Gobiidae) from the Northwest Mediterranean. *Marine Biology* 129: 279-287.

Iljin, B.S. 1949. Brief review of Black Sea gobies (Pisces, Gobiidae). *Byulletin Moskovskovo Obshchestva Ispytatelei Prirody, Otdel Biologicheskii* 54: 16-30.

Iwata, A., S. Hosoja and H.K. Larson. 2001. *Paedogobius kimurai*, a new genus and species of goby (Teleostei: Gobioidei: Gobiidae) from the west Pacific. *Records of Australian Museum* 53: 103-112.

Johnson, G.D. and E.B. Brothers. 1993. *Schindleria*: a paedomorphic goby (Teleostei: Gobioidei). *Bulletin of Marine Science* 52: 441-471.

Jones, S. and M. Kumaran. 1964. On the fishes of the genus *Schindleria* Giltay from the Indian Ocean. *Journal of the Marine Biological Association of India* 6: 257-264.

Kim, I.S., Y. Choi, C.L. Lee, Y.J. Lee, B.J. Kim and J.H. Kim. 2005. *Illustrated Book of Korean Fishes*. Kyo-Hak Pub Co. Seoul. (In Korean.)

Kitano, T. 2002. Morphogenesis of the ice goby, *Leucopsarion petersii*, under rearing condition. *Journal of the School of Marine Science and Technology Tokai University* 54: 37-48. (In Japanese.)

Kon, T. and T. Yoshino. 2002a. Diversity and evolution of life history of gobioid fishes from the viewpoint of heterochrony. *Marine and Freshwater Research* 53: 377-402.

Kon, T. and T. Yoshino. 2002b. Extremely early maturity found in Okinawan gobioid fishes. *Ichthyological Research* 49: 224-228.

La Mesa, M. 1999. Age and growth of *Aphia minuta* (Pisces, Gobiidae) from the central Adriatic Sea. *Scientia Marina* 63: 147-155.

La Mesa, M. 2001. Age and growth of *Crystallogobius linearis* (von Düben, 1845) (Teleostei, Gobiidae) from the Adriatic Sea. *Scientia Marina* 65: 375-381.

La Mesa, M., E. Arneri, V. Caputo and M. Iglesias. 2005. The transparent goby, *Aphia minuta*: review of biology and fisheries of a paedomorphic European fish. *Reviews in Fish Biology and Fisheries* 15: 89-109.

La Mesa, M., D. Borme, V. Tirelli, E. Di Poi, S. Legovini and S. Fonda Umani. 2008. Feeding ecology of the transparent goby *Aphia minuta* (Pisces, Gobiidae) in the northwestern Adriatic Sea. *Scientia Marina* 72: 99-108.

Labropoulou, M. and A. Eleftheriou. 1997. The foraging ecology of two pairs of congeneric demersal fish species: importance of morphological characteristics in prey selection. *Journal of Fish Biology* 50: 324-340.

Landaeta, M.F., R. Veas and L.R. Castro. 2002. First record of the paedomorphic goby *Schindleria praematura*, Easter Island, South Pacific. *Journal of Fish Biology* 61: 289-292.

Lebour, M.V. 1919. The young of the Gobiidae from the neighbourhood of Plymouth. *Journal of the Marine Biological Association of the United Kingdom* 12: 48-80.

Leis, J.M. 1982. Nearshore distributional gradients of larval fish (15 taxa) and planktonic crustaceans (6 taxa) in Hawaii. *Marine Biology* 72: 89-97.

Leis, J.M. 1994. Coral Sea atoll lagoons—closed nurseries for the larvae of a few species of coral-reef fishes. *Bulletin of Marine Science* 54: 206-227.

Leis, J.M., T. Trnski, V. Dufour, M. Harmelin-Vivien, J.-P. Renon and R. Galzin. 2003. Local completion of the pelagic larval stage of coastal fishes in coral-reef lagoons of the Society and Tuamotu Islands. *Coral Reefs* 22: 271-290.

Lo Bianco, S. 1909. Notizie biologiche riguardanti specialmente il periodo di maturità sessuale degli animali del golfo di Napoli. *Mitteilen Zoologische Station Neapel* 19: 513-761.

Lopez Jaime, J.A., P. Escalante Blanque, R. Rodriguez Garcia and F. Zurita Escobar. 2000. Experimental culture of *Aphia minuta*. Study on the maintenance in captivity, nutritional requirements, ecological needs and possibility of controlled reproduction. *Cahiers Options Méditerranée* 47: 343-347.

Mancini, L. and P.G. Cavinato. 1969. Osservazioni morfologiche e biometriche sul gobiide *Aphia minuta* nell'Adriatico centrale e alcune considerazioni in rapporto alla pesca. *Bollettino di Pesca, Piscicoltura e Idrobiologia* 24: 49-60.

Martínez-Baño, P., F. Vizuete and J. Mas. 1993. The fishery of the transparent goby *A. minuta* (Risso, 1810) on the fishing grounds off Murcia (South-eastern Spain). *Scientia Marina* 57: 199-205.

Masuda, H., K. Amaoka, C. Araga, T. Uyeno and T. Yoshino. 1984. *The Fishes of the Japanese Archipelago*. Vol. 1. Tokai University Press, Tokyo.

Matsui, S. 1986. Studies on the ecology and propagation of the ice goby, *Leucopsarion petersii* Hilgendorf. *Science Bulletin of the Faculty of Agriculture, Kyushu University* 40: 135-174.

McDowall, R.M. 1997. The evolution of diadromy in fishes (revisited) and its place in phylogenetic analysis. *Reviews in Fish Biology and Fisheries* 7: 443-462.

Miller, P.J. 1973. The species of *Pseudaphya* (Teleostei: Gobiidae) and the evolution of aphyiine gobies. *Journal of Fish Biology* 5: 353-365.

Miller, P.J. 1986. Gobiidae. In: *Fishes of the North-eastern Atlantic and the Mediterranean*, P.J.P. Whitehead, M.L. Bauchot, J.C. Hureau, J. Nielsen and E. Tortonese (eds.), UNESCO, Paris, pp. 1019-1085.

Miller, P.J. 1989. The tokology of gobioid fishes. In: *Fish Reproduction: Strategies and Tactics*, G.W. Potts and R.J. Wootton (eds.), Academic Press, New York, pp. 118-153.

Miller, P.J. 1997. Persistent postlarva — the case of progenetic gobies. *Journal of Fish Biology* 51: 412.

Mortara, S. 1918. La disposizione degli organi ciatiformi del genere *Aphya* e i suoi rapporti con quella del genere *Gobius*. *Memorie del Regio Comitato Talassografico Italiano* 65: 1-23.

Mundy, B.C. 2005. Checklist of the fishes of the Hawaiian Archipelago. *Bishop Museum Bulletins in Zoology* 6: 1-704.

Myers, R.F. 1999. *Micronesian Reef Fishes: A Comprehensive Guide to the Coral Reef Fishes of Micronesia*. 3rd revised and expanded edition. Coral Graphics, Barrigada, Guam.

Nakabo, T. 2002. *Fishes of Japan with Pictorial Keys to the Species*, English edition. Tokai University Press, Japan.

Nakatsuji, T., T. Kitano, N. Akiyama and N. Nakatsuji. 1997. Ice goby (shiro-uo), *Leucopsarion petersii*, may be a useful material for studying teleostean embryogenesis. *Zoological Science* 14: 443-448.

Neira, F.J., A.G. Miskiewicz and T. Trnski. 1998. *Larvae of Temperate Australian Fishes: Laboratory Guide for Larval Fish Identification*. University Western Australia Press, Nedlands.

Nelson, J.S. 1994. *Fishes of the World*, Wiley, NewYork, 3rd Edition.

Ozawa, T. and S. Matsui. 1979. First record of the schindlerid fish, *Schindleria praematura*, from southern Japan and the South China Sea. *Japanese Journal of Ichthyology* 25: 283-285.

Padoa, E. 1956. Triglidae, Peristediidae, Dactylopteridae, Gobiidae, Echeneidae, Jugulares, Gobiesocidae, Heterosomata, Pediculati. *Fauna e Flora del Golfo di Napoli* 38: 627-888.

Petersen, C.G.J. 1919. Our gobies (Gobiidae) from the egg to the adult stages. *Reports of the Danish Biological Station* 26: 47-66.

Raffaele, F. 1895. Uova di *Scomberesox*, di *Exocoetus* e di *Crystallogobius*. *Bollettino della Società dei Naturalisti in Napoli* 8: 127-130.

Randall, J.E. and K.K.P. Lim. 2000. A checklist of the fishes of the South China Sea. *Raffles Bulletin of Zoology Suppl.* 8: 569-667.

Reilly, S.M., E.O. Wiley and D.J. Meinhardt. 1997. An integrative approach to heterochrony: the distinction between interspecific and intraspecific phenomena. *Biological Journal of Linnean Society* 60: 118-143.

Relini, G., G. Palandri, M. Relini, F. Garibaldi, G. Torchia, C. Cima, E. Massaro, A. Piccone and M. Bellingeri. 1998. Pesca sperimentale del rossetto in Liguria. *Biologia Marina Mediterranea* 5: 487-502.

Roff, D.A. 1992. *The Evolution of Life Histories: Theory and Analysis.* Chapman and Hall, New York.

Russell, F.S. 1976. *The Eggs and Planktonic Stages of British Marine Fishes.* Academic Press, London.

Saito, T., S. Otani, T. Nagai, T. Nakatsuji, K. Arai and E. Yamaha. 2002. Germ cell lineage from a single blastomere at 8-cell stage in shiro-uo (ice goby). *Zoological Science* 19: 1027-1032.

Saito, T., M. Kamimoto, A. Miyake, E. Yamaha, T. Suzuki, N. Nakatsuji and T. Nakatsuji. 2001. Mesoderm formation by isolated and cultivated 8-cell stage blastomeres of the teleost, *Leucopsarion petersii* (shiro-uo). *International Journal of Developmental Biology* 45: 661-668.

Sardou, J. 1974. Contribution a la connaissance de la faune ichthyologique Malgache: découverte de poissons de la famille des Schindleriidae dans le canal de Mozambique, à Nosy-Bé et étude d'une collection de *Schindleria. Cahiers ORSTOM, Série Oceanographie* 12: 3-15.

Scattolin, G. 2003. *Das Seitenliniensystem cryptobentischer und pelagischer Meergrundeln (Teleostei: Gobiidae) - ein Vergleich.* M.Sc. Thesis, University of Vienna, pp. 1-146.

Scattolin, G., H. Ahnelt and M. Sauberer. 2007. The free neuromast pattern on the caudal fin of pelagic Gobiidae (Teleostei). *Annalen des Naturhistorischen Museums in Wien* 108B: 7-12.

Schindler, O. 1932. Sexually mature larval hemirhamphidae from the Hawaiian islands. *Bernice Pauahi Bishop Museum Bulletin* 97: 1-28.

Thacker, C. and H. Grier. 2005. Unusual gonad structure in the paedomorphic teleost *Schindleria praematura* (Teleostei: Gobioidei): a comparison with other gobioid fishes. *Journal of Fish Biology* 66: 378-391.

Tomiyama, T., H. Honda, A. Okata and M. Omori. 1999. Larval drift patterns of diadromous gobioids in the lower course of the Natori River, northern Japan. *Tohoku Journal of Agricultural Research* 50: 11-21.

Tortonese, B. 1975. *Osteichthyes. Pesci Ossei. Parte seconda, Fauna d'Italia XI*, Edizioni Calderini, Bologna.

Villegas M.L. 1980. Postlarvas de gobidos del mar Cantabrico. *Boletin del Instituto Español de Oceanografia* 6: 17-45.

Watson, W. 2000. Schindleriidae (Schindler's fishes). In: *The Larvae of Indo-Pacific Coastal Fishes*, J.M. Leis and B.M. Carson-Ewart (eds.), Brill, Leiden, pp. 633-636.

Watson, W. and H.J. Jr. Walker. 2004. The world's smallest vertebrate, *Schindleria brevipinguis*, a new paedomorphic species in the family Schindleriidae (Perciformes: Gobioidei). *Records of Australian Museum* 56: 139-142.

Watson, W., E.G. Stevens and A.C. Matarese. 1984. Schindleroidei: development and relationships. In: *Ontogeny and Systematics of Fishes*, H.G. Moser, W.J. Richards, D.M. Cohen, M.P. Fahay, A.W. Kendall and S.L. Richardson (eds.), American Society of Ichthyologists and Herpetologists, Special Publication 1, pp. 552-554.

Yamashita, Y. and T. Aoyama. 1984. Ichthyoplankton in Otsuchi Bay on northeastern Honshou with reference to the time-space segregation of their habitats. *Bulletin of the Japanese Society of Scientific Fisheries* 50: 189-198.

Young, S.S. and T.S. Chiu. 2000. New records of a paedomorphic fish, *Schindleria praematura* (Pisces: Schindleriidae), from waters of Taiwan. *Acta Zoologica Taiwanica* 11: 127-133.

CHAPTER

4.2

Gobies on Coral Reefs

Juergen Herler[1,*], Philip L. Munday[2] and Vanessa Hernaman[3]

INTRODUCTION

The family Gobiidae comprises at least 2000 species (Nelson, 2006), with many still undiscovered or undescribed. More than half of all goby species are found on coral reefs and thus they contribute greatly to the astonishing fish diversity of this environment. Most coral reef gobies are closely associated with the substratum, where they seek shelter from predators in the complex physical structure of reef habitat (Munday and Jones, 1998). As a consequence of their small size and cryptic life-style, careful investigations by trained experts are necessary to reveal their full diversity. Although there are few such researchers, new species are still being described (e.g. *Coryphopterus* (Victor, 2007), *Eviota* (Gill and Jewett, 2004),

Authors' addresses: [1]Department of Theoretical Biology, Morphology Section, Faculty of Life Sciences, University of Vienna, Althanstrasse 14, 1090 Vienna, Austria.
E-mail: juergen.herler@univie.ac.at

[2]ARC Centre of Excellence for Coral Reef Studies / School of Marine and Tropical Biology, James Cook University, Townsville, Queensland 4811, Australia.

E-mail: philip.munday@jcu.edu.au

[3]Centre for Marine Studies, University of Queensland, St Lucia, Queensland 4072, Australia.
E-mail: vanessa.hernaman@gmail.com

[*]*Corresponding author*

Gobiodon (Harold and Winterbottom, 1995; Harold and Winterbottom, 1999; Winterbottom and Harold, 2005), *Priolepis* (Nogawa and Endo, 2007) and *Trimma* (e.g. Winterbottom, 2004, 2006; Hagiwara and Winterbottom, 2007; Suzuki and Senou, 2007; Winterbottom and Zur, 2007). Despite the diversity of coral reef gobies—the genus *Trimma*, for example, may include nearly 100 species (Winterbottom and Southcott, 2008)—most species are easily overlooked and are commonly excluded from sampling or ecological surveys. However, a recent upsurge of interest in the ecological role of small reef fishes (e.g. Munday and Jones, 1998; Depczynski and Bellwood, 2003) and the use of several coral-reef goby genera as model groups in ecology and evolution (e.g. Forrester and Steele, 2000; Munday *et al.*, 2001; Taylor and Hellberg, 2003; Munday, 2004b) has greatly increased our understanding of the biology and ecology of this diverse and interesting group of coral reef fishes.

Gobies are not only diverse in terms of species richness, but also with regard to their life-styles and life-histories. Gobies occupy all major habitats around coral reefs, including sand and rubble, seagrass meadows, coral reef matrix, live corals and numerous other living organisms (Greenfield and Johnson, 1999; Nakamura and Sano, 2004; Syms and Jones, 2004). Size diminution has played an important role in the evolution of coral reef gobies and has enabled them to utilize many restricted or specialized microhabitats (Munday and Jones, 1998; Herler, 2007). The most habitat-specialized gobies are associated with invertebrates such as sponges, soft coral and hard coral (Larson, 1985, 1990; Munday *et al.*, 1997; Herler and Hilgers, 2005). Others do not use invertebrates as habitat, but use the structures created by them, such as the burrows of shrimps. Mutualistic relationships between shrimps and gobies are found in about 15 genera and most are obligate associates of burrowing alpheids.

Coral reef gobies hold two world records: the smallest size and the shortest lifespan among marine vertebrates (Winterbottom, 1990; Depczynski and Bellwood, 2005; Winterbottom and Southcott, 2008). Most reef gobies are short-lived species. Many are annuals (Kritzer, 2002; Hernaman and Munday, 2005a) and some have generation times of just a few months (Depczynski and Bellwood, 2005; Depczynski *et al.*, 2007; Winterbottom and Southcott, 2008). They are also among the most abundant families of fishes on coral reefs (Ackerman and Bellwood, 2000). High abundance combined with short lifespans means that they account for a very large proportion of the total growth production of reef fishes (Depczynski *et al.*, 2007). Consequently, these tiny reef fish contribute significantly to energy transfer throughout coral reef systems. Minute body size and short lifespan are also associated with a range of morphological adaptations and reductions in the skeleton, fins, squamation and lateral line system in the smallest species (Winterbottom, 1990; Herler, 2007; Winterbottom and Zur, 2007).

High predation risk and a short lifespan are associated with early maturation in many small coral reef gobies. Early maturation is advantageous because it results in a higher probability of survival to first reproduction and shorter generation times (Stearns, 1992). A disadvantage is that maturation at a small body size limits batch fecundity and, potentially, spawning opportunities (i.e., in social systems where larger size confers greater success in competition for mates or spawning sites). Perhaps as a result of these limitations, many coral reef gobies invest in somatic growth throughout life including during reproduction (Hernaman and Munday, 2005a), which contrasts with many small, long-lived coral reef fishes where growth ceases soon after sexual maturation (e.g. Choat and Robertson, 2002).

Coral reef gobies also exhibit a diversity of reproductive behaviours and patterns of social organisation. Most, if not all, coral reef gobies are substrate spawners with parental egg care, but mating strategies vary from monogamy to polygyny (e.g. Kuwamura *et al.*, 1993; Sunobe and Nakazono, 1999a, 1999b; Takegaki and Nakazono, 1999; Hernaman and Munday, 2005b). Some species form complex social hierarchies where individuals queue for breeding rights, and growth rates are controlled by social interactions (Wong *et al.*, 2007, 2008a, 2008b). Many species also exhibit remarkable sexual flexibility and are able to change sex, in some cases more than once (Nakashima *et al.*, 1996; Munday *et al.*, 1998; Manabe *et al.*, 2007a).

In this chapter we examine the ecology, life-histories and reproductive behaviour of coral reef gobies. We also explore morphological adaptations that are associated with the unique life-styles exhibited by many coral reef gobies, and the extreme miniaturization of certain groups of species. Although frequently overlooked in studies of reef fish assemblages due to their small size and cryptic behaviour, we show that gobies play an important role in coral reef ecoystems, and that they have proven to be a useful model group for studying a broad range of topics in ecology and evolution, including: species diversification, ecological and morphological adaptation, the evolution of animal life history traits, mechanisms of community structure and population regulation, and the evolution of animal behaviour.

ECOLOGY

Habitat Associations

Gobies inhabit most microhabitats associated with coral reefs (Fig. 4.2.1), and at the same time are among the most habitat-specialized reef fish groups. This apparent contradiction occurs because at the family level, gobies occupy nearly every habitat type present on coral reefs, yet many species are habitat-specialist with precise microhabitat requirements (Munday *et al.*, 1997; Greenfield and Johnson, 1999; Depczynski and Bellwood, 2004; Herler, 2007). Most coral reef gobies are epibenthic or cryptobenthic (Miller, 1996); there are very few nektonic species. Coral reefs are physically complex structures that offer many options for cryptobenthic life-styles (Fig. 4.2.2). The small body size of gobies enables them to exploit habitats and food resources that are not available to larger species. The opportunity to utilize a wide range of microhabitats at small spatial scales appears to have contributed to the extraordinary diversity of species within some genera, such as *Trimma*, *Eviota* and *Gobiodon* (Munday and Jones, 1998; Munday *et al.*, 2004).

The most habitat-specialized gobies are found in obligate association with living hosts, including sponges (Larson, 1990; Rocha *et al.*, 2000; Colin, 2002; Taylor and Hellberg, 2005), hydrozoans (Larson, 1985; Herler and Hilgers, 2005), octocorals (Larson, 1985, 1990), scleractinian corals (Larson, 1985; Munday *et al.*, 1997; Herler, 2007), antipatharians (Munday *et al.*, 2002; Herler, 2007), ascidians (Larson, 1990) and even plants, such as seagrasses (Larson, 1990). The best studied of these habitat specialists are the coral-dwelling gobies from the genus *Gobiodon*, which live among the branches of scleractinian corals from the family Acroporidae. These coral-dwelling gobies have been used as a model group for understanding the evolution and ecology of habitat specialization (e.g. Munday *et al.*, 1997, 2001; Munday *et al.*, 2004), and the consequences of habitat specialization in the face of increasing anthropogenic stresses to coral reefs (Munday, 2004b; Feary, 2007; Schiemer *et al.*, 2009). Although all species of coral-dwelling gobies can be considered habitat-specialists in the broad sense, there is a vast difference in the degree of specialization among species. Some species of *Gobiodon* are associated with just a single coral species throughout their entire geographic range (Munday, 2002; Munday *et al.*, 2004), whereas other species inhabit up to 10 - 20 different species of coral and change preferences among locations depending on the availability of different coral species (Munday *et al.*, 1997; Munday, 2002; Dirnwöber and Herler, 2007).

Fig. 4.2.1 Ten Indo-Pacific gobiid species as examples of the various life-styles of coral reefs gobies. A: *Amblygobius albimaculatus* sifting sand in a seagrass meadow. B: a pair of *Valenciennea puellaris* at the entrance of their self-dug burrow. C: *Amblyeleotris sungami* which lives in a mutualistic relationship with a species of *Alpheus* snapping shrimp. D: *Gnatholepis anjerensis* on the edge between coral rock and sand bottom. E: *Asterropteryx semipunctata* between algae on coral rubble. F: *Trimma avidori* swimming upside down in a small coral rock cave. G: *Eviota zebrina* inhabits live corals and coral rock. H: *Bryaninops natans* is an obligate associate of *Acropora* corals. I: *Paragobiodon echinocephalus* in the pocilloporid coral *Stylophora pistillata*. J: *Gobiodon histrio* among branches of *Acropora gemmifera*. Photographs by J. Herler.

Color image of this figure appears in the color plate section at the end of the book.

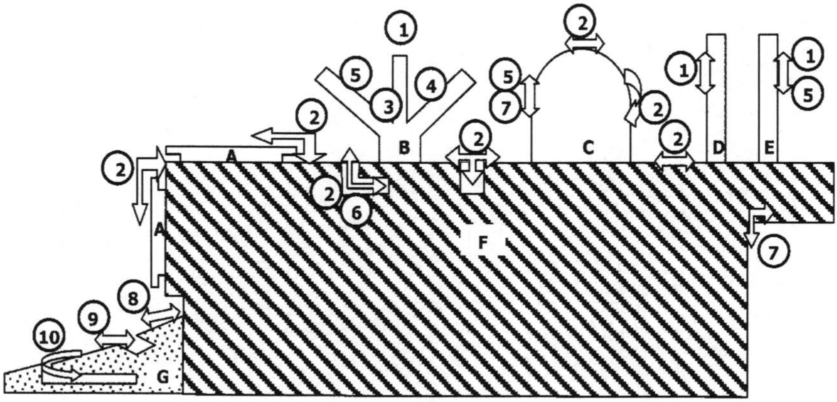

Fig. 4.2.2 Schematic presentation of coral, coral-rock and soft-sediment microhabitats used by coral reef gobies (modified from Herler, 2007). Grey structures represent living coral colonies, cross-hatched areas represent coral rock and stippled areas represent sand. Arrows indicate major movement and escape directions. Microhabitats are indicated by letters: A-C: different growth forms of scleractinian corals; A: plate-like and encrusting corals, B: branching corals (*Acropora* spp., *Stylophora pistillata*), C: massive corals; D: hydrozoan coral *Millepora dichotoma*; E: antipatharian and soft corals; F: coral rock; G: sand. Representative examples of gobiid genera are indicated by numbers: 1: *Bryaninops* spp., 2: *Eviota* spp.; 3: *Gobiodon* spp.; 4: *Paragobiodon* spp.; 5: *Pleurosicya* spp.; 6: *Priolepis* spp.; 7: *Trimma* spp.; 8: *Asterropteryx* spp., *Coryphopterus* spp., *Fusigobius* spp., *Gnatholepis* spp., *Istigobius* spp.; 9: *Amblygobius* spp., *Valenciennea* spp.; 10: *Amblyeleotris* spp., *Cryptocentrus* spp., *Ctenogobiops* spp., *Vanderhorstia* spp.

The intimate association of some gobies with coral habitat has implications for their distributions and abundances. Coral hosts appear to be a limiting resource for some species of *Gobiodon*, and consequently their abundances within and among reefs are often closely correlated with the abundance of their preferred coral species (Kuwamura *et al.*, 1994; Munday *et al.*, 1997; Munday, 2000). Furthermore, coral loss from bleaching or other disturbances usually causes dramatic declines in the abundances of coral-dwelling gobies (Munday *et al.*, 1997; Munday, 2004b; Feary *et al.*, 2007). The specialist species that inhabit just one or two species of coral usually fare worse when coral cover declines than do the generalist species (Munday, 2004b). The reason for this appears to be that the highly specialist species are unable to use alternative corals when the abundance of their preferred hosts decline, whereas the more generalist species shift to a range of alternative coral hosts even when their preferred habitats become less abundant.

Habitat specialization and the capacity to use alternative hosts is linked to the ecological performance of gobies within their habitat. Caley and Munday (2003) demonstrated that there was a trade-off between habitat specialization and growth rate among *Gobiodon* species. Two specialist species, *G. histrio* and *G. brochus*, grew much faster on their preferred host coral compared to a less preferred coral species. In contrast, the growth of two generalist species, *G. quinquestrigatus* and *G. oculolineatus*, differed little between the two coral species. Importantly, the growth of the specialists on their preferred coral host was significantly higher than that of the generalists on the same coral host, demonstrating that habitat specialists have superior performance on the preferred habitats. This illustrates that habitat specialists out-perform habitat generalists on their preferred habitat, but they are unable to maintain their performance between habitats. In contrast, generalists perform equally well on most habitat types. This trade-off between habitat specialization and performance helps explain why both habitat-specialist and generalist species are able to persist in animal communities. Furthermore, it helps explain why specialist species decline rapidly when their preferred habitats decline in abundance (Fig. 4.2.3).

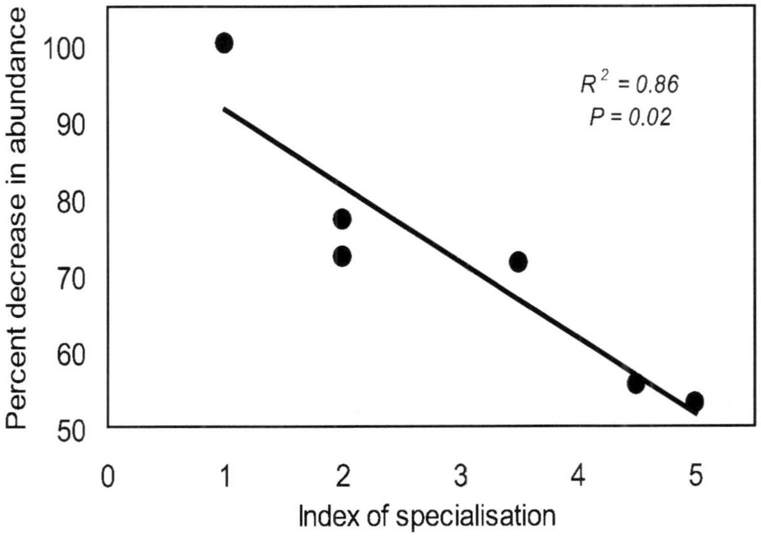

Fig. 4.2.3 Proportional decline in population size of six coral-dwelling goby species (genus *Gobiodon*) following severe loss of live coral habitat (modified from Munday, 2004b). Specialist species declined proportionally more than generalist species. The specialization index for each goby species was calculated by summing the total number of coral species that were used by each goby species either in proportion with availability (score of 0.5) or more frequently than expected from availability (score of 1.0). Habitat specialization data from Munday (2000).

Habitat specialization has also led to a range of other adaptive specializations in coral reef gobies. For example, coral-dwelling gobies exhibit remarkable capacity for hypoxia tolerance and air-breathing (Nilsson *et al.*, 2004, 2007). Oxygen levels among the branches of coral colonies can decline dramatically overnight due to the respiration of the coral host (Nilsson *et al.*, 2004). All species of coral-dwelling gobies are able to tolerate very low oxygen levels, presumably because such conditions are regularly encountered within their preferred habitats. Furthermore, species of *Gobiodon* associated with shallow-water corals that become air exposed during low tides have evolved a greater capacity for air-breathing than species associated with deeper-water corals that never become exposed to air (Nilsson *et al.*, 2007). Finally, the loss of scales that appears to be associated with the capacity for air-breathing across the skin (Nilsson *et al.*, 2007) is also associated with the presence of toxin glands in coral-dwelling gobies. In addition to their potential role in predator deterrence (Lassig, 1981; Schubert *et al.*, 2003), these toxins help protect the otherwise unprotected skin surface from attack by external parasites (Munday *et al.*, 2003). Therefore, it appears the specialized habitat associations of coral-dwelling gobies have led to the evolution of morphological adaptations to their coral habitat.

Many other coral reef gobies are much less specialized in their habitat choices than the coral-dwelling gobies discussed above, but nevertheless they still have habitat preferences. These preferences are seen in the distribution of species among reef environments (Greenfield and Johnson, 1999; Depczynski and Bellwood, 2004; Herler, 2007; Hernaman and Probert, 2008). Some species occur on shallow reef flats, others in caves on steep reef walls, or at the sand/reef interface at the base of reef slopes. Caves and patches of sand or rubble are particularly favoured habitat by many goby species (Depczynski and Bellwood, 2004). A few goby species, such as *Eviota queenslandica* and *Asterropteryx semipunctata*, are broad habitat generalists that are highly abundant in a range of reef environments (Depczynski and Bellwood, 2004; Hernaman and Probert, 2008). In contrast to the specialist coral-dwelling gobies, the abundances of gobies using these reef habitats is more closely linked to variation in patterns of settlement and post-settlement mortality (Forrester, 1995; Steele and Forrester, 2002; Hernaman and Probert, 2008) than it is to fluctuations in habitat availability.

A considerable number of gobies inhabit soft sediments and rubble beds between reefs, rather than the coral reef itself. Many of these species are cryptically coloured to match the sand and mud backgrounds where they permanently reside. A diverse group of species live in burrows constructed by alpheid shrimps (Polunin and Lubbock, 1977; Karplus, 1987). Others, such as species of *Amblygobius* and *Valenciennea*, dig their own burrows or hide in crevices and rubble near the reef's edge. Once again, habitat partitioning is observed among these species, with sediment type, depth, and distance from the reef being key environmental variables distinguishing the local-scale distributions of different species (Syms and Jones, 2004; Hernaman and Probert, 2008).

Diets and Feeding

In general, gobies tend to have broad diets and are opportunistic feeders. This applies equally to gobies on coral reefs (Depczynski and Bellwood, 2003; Hernaman *et al.*, 2009), although there are also clear examples of dietary specialization. For example, some species of *Gobiodon* supplement their diet by feeding on the tissue of their coral hosts (Cole *et al.*, 2008; Riedlecker and Herler, 2008) and differences in nutritional value among coral species could potentially explain significant differences in the growth rate and body condition of gobies inhabiting different coral species (Petrie, 2008). Other coral reef gobies have important ecological functions, such as operating as cleaner fish (Taylor and Van Tassell, 2002).

Most coral reef gobies hunt for prey close to the substratum. A few feed in the water column, but always seek shelter in near-by burrows (e.g. ptereleotrine gobies) or the reef matrix (e.g. several *Trimma* species) when threatened (Miller, 1996; Winterbottom and Southcott, 2008). Many of the sand-dwelling species are winnowers, spending much of the day sifting sand through gill rakers to capture small invertebrates (Hernaman *et al.*, 2009). Many cryptic reef gobies feed on meiofauna and small invertebrates that are inaccessible for larger fish and this contributes significantly to energy transfer to higher trophic levels (Depczynski *et al.*, 2007).

In one of the few detailed assessments of the diets of coral reef gobies, Depczynski and Bellwood (2003) showed that many reef gobies are detritivores in addition to carnivores. Most of the smallest goby species (e.g. *Pleurosicya*, *Trimma*, *Eviota* spp.) that they and other authors (Saeki *et al.*, 2005) examined preyed mostly on copepods and other small invertebrates. The intermediate-sized species (e.g. *Istigobius, Asterropteryx, Bathygobius* spp.), however, fed mostly on detrital matter. Three species of *Amblygobius* (*A. rainfordi*, *A. bynoensis* and *A. phalaena*) are herbivores (Depczynski and Bellwood, 2003; Hernaman *et al.*, 2009), feeding mostly on filamentous algae. These results indicate that the diets of coral reef gobies may be more diverse than often assumed.

A study into the feeding habits of five coral reef gobies inhabiting a reef flat at Orpheus Island on the Great Barrier Reef revealed a diverse range of strategies in feeding behaviours (Hernaman *et al.*, 2009). The small cryptobenthic *Asterropteryx semipunctata* was the most generalized feeder, feeding from all available sources (sand patches, detrital algal mats covering rock platforms, algal plants, and the water column), but spending relatively little time on feeding (i.e., mean bite rate of 12.2 bites per ten-minute observation period, and feeding recorded in only 31% of the observation period). At the other extreme, *Valenciennea muralis* was the most specialized feeder, feeding predominantly (96% of all bites) by sifting microinvertebrates from sand. Continuous foraging was commonly observed and feeding occupied a large proportion of the activity budget, with a mean feeding rate of 153.9 bites per ten-minute period, and feeding observed in 82% of the observation period. Whilst invertebrates represent a high quality food resource, high feeding rates might be necessary because of the patchy distribution of microinvertebrates within the sediments, and the high energetic costs and relative inefficiencies associated with the sieving action. Considerable amounts of sedimentary material were ingested during the sieving process (e.g. 24-27% by weight and volume in the gut of *V. muralis*), and high feeding rates may be required to overcome the ingestion of such large amounts of indigestible material.

Predation and Competition

Predation risk is exceptionally high for small fishes on coral reefs (Munday and Jones, 1998) and this likely explains the cryptic life-style of most coral reef gobies. Due to their small size, gobies are preyed upon by a broad suite of predators of all sizes. Even other small fishes that are themselves prey to piscivorous fishes, feed on small gobies that live among the reef matrix. The high risk of predation strongly influences the lifespans and life-history traits of coral reef gobies. High instantaneous mortality rates mean that most coral reef gobies live for less than 1 year on average (Reavis, 1997a; Kritzer, 2002; Hernaman and Munday, 2005a) and some of the smallest species have lifespans of just a few months (Depczynski and Bellwood, 2005, 2006; Winterbottom and Southcott, 2008).

Due to their small size and abundance, gobies have been used by some ecologists to test hypotheses relating to predation and competition in reef fish communities. In an elegant series of experiments, Graham Forrester and Mark Steele demonstrated the importance of shelter availability and predation by larger fishes in controlling the densities of gobies from the genera *Coryphopterus*, *Gnatholepis* and *Lythrypnus* (Forrester, 1995; Forrester and Steele, 2000, 2004). At the same time, they revealed differences among species in the relative importance of predation and competition in structuring adult populations, and differences in the manner in which mortality was related to population density. In one experiment, newly settled blackeyed gobies *Coryphopterus nicholsii* suffered intense density-dependent predation that dramatically modified patterns of abundance established at settlement (Steele and Forrester, 2002). In contrast, bridled gobies *Coryphopterus glaucofraenum* experienced moderate density-independent predation that only partially modified patterns of abundance established at settlement. Finally, goldspot gobies *Gnatholepis thompsoni* experienced low density-independent predation that did not affect patterns of abundance established at settlement. In another experiment, predation rates on *Coryphopterus glaucofraenum* were density dependent when shelter sites were uncommon, but density-independent when shelter sites were abundant (Forrester and Steele, 2004). These experiments have contributed substantially to our understanding of population regulation in coral reef fishes by demonstrating that the strength of density-dependent mortality can vary greatly among species and locations, and may be associated with the availability of shelter from predators. Competition did not appear to have a significant effect on mortality of these species, but clearly affected growth, which was significantly less at high densities (Forrester *et al.*, 2006).

Predation is greatly reduced in structurally complex habitats, such as among the close-knit branches of coral colonies, and it is here that competition for space becomes the key ecological process. Munday *et al.* (2001) showed that living space is a limited resource for some coral-dwelling gobies and that species compete for access to colonies of preferred coral species, such as *Acropora nasuta*. In lagoons and sheltered reefs, the superior competitors, *Gobiodon histrio* and *G. erythrospilus*, dominate the use of *A. nasuta* and force other species to inhabit alternative coral species (Munday *et al.*, 2001). The ecologically similar *G. histrio* and *G. erythrospilus* also compete for living space, with a lottery for vacant space determined by recruitment dynamics determining the relative abundance of the two species (Munday, 2004a). Competition for space also affects the social and spatial organisation of *G. histrio* and *G. erythrospilus*. Adults of both species form breeding pairs on large coral colonies, and nearly all colonies above a threshold size are occupied by a breeding pair (Hobbs and Munday, 2004). Juveniles are forced to live singly on small corals, but will readily move to form a pair if an adult loses its partner (Hobbs *et al.*, 2004; Wall and Herler, 2008).

Competition for space can have significant consequences for the life histories of coral-dwelling gobies. On the Great Barrier Reef, *G. histrio* and *G. brochus* coexist on the same reefs, but *G. brochus* is rarely able to inhabit colonies of *A. nasuta* because of the dominance of *G. histrio* in this coral. *Gobiodon brochus* is usually found inhabiting colonies of the coral *A. loripes*. However, *G. brochus* individuals inhabiting *A. loripes* grow more slowly and reach a much smaller size compared with individuals inhabiting *A. nasuta* (Munday, 2001). Therefore, competition for space with the dominant competitor, *G. histrio*, affects the maximum size, and ultimately the lifetime fecundity of *G. brochus* (Fig. 4.2.4).

MORPHOLOGY

Morphological Adaptations

Coral reef gobies occupy a wide range of microhabitats, including soft sediments, hard substrates, caves and crevices, and live corals. Accordingly, they exhibit a range of morphological adaptations to these microhabitats, the most obvious being different body shape (Herler, 2007). Most reef gobies, in particular the larger genera that live closely attached to the bottom (e.g. *Istigobius*, *Bathygobius*, *Valenciennea* and *Amblygobius*), have a rather fusiform body shape (Fig. 4.2.1). Other forms, such as the genera *Ptereleotris* and *Nemateleotris*, hover above the bottom and have elongated bodies. In contrast, the forms that live within 3-dimensional structures tend to be compressed (laterally flattened) and those living on surfaces of live corals tend to be depressed (dorso-ventrally flattened) in body shape (Fig. 4.2.5). The best example of highly compressed body shapes are species of the genus *Gobiodon*, some of which are surprisingly large when considering the restricted interbranch spaces offered by their preferred *Acropora* corals (Herler and Hilgers, 2005). One of the largest species *G. histrio* has a highly compressed body, which enables it to move among the coral branches, even when full adult size is attained. In contrast, the smaller *G. rivulatus* has a less compressed body shape, but can still move among the coral branches because of its smaller total body size. The relationship between body shape and habitat use in *Gobiodon* is further illustrated in *G. citrinus*, which inhabit open-branching staghorn corals. *Gobiodon citrinus* grows to a much larger size than *G. histrio*, but has a less compressed body, presumably because its shape is not constrained by coral interbranch space (J. Herler, unpublished data).

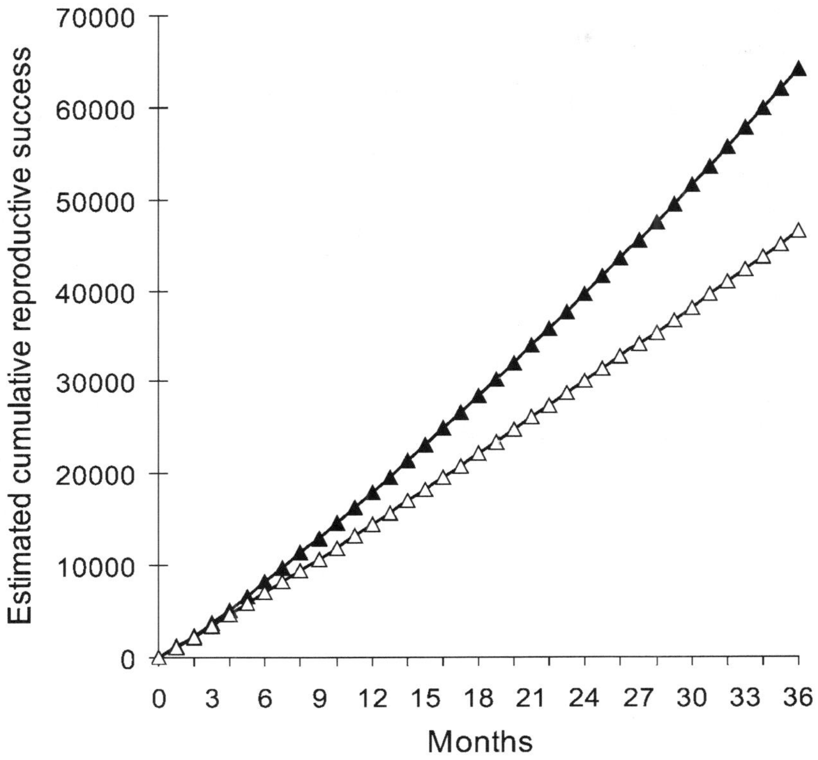

Fig. 4.2.4 Estimated cumulative reproductive success of *Gobiodon brochus* inhabiting either *Acropora nasuta* (filled triangles) or *Acropora loripes* (unfilled triangles). Calculations were based on: 1) age-based growth rates of *G. brochus* inhabiting *A. nasuta* and *A. loripes* reported by Munday (2001), 2) maximum body size of 49 mm for *G. brochus* (P.L. Munday, unpublished data) and 3) the size-fecundity relationships for coral-dwelling gobies reported by Kuwamura *et al.* (1993).

In contrast to species that live in branching corals, gobies that inhabit open surfaces, such as massive and plate-like corals or sea-whips, have depressed bodies to attain close physical contact with the substrate (Fig. 4.2.5). This is exemplified by species such as *Bryaninops yongei*, *B. ridens* and *Pleurosicya micheli* that inhabit seawhips, fire corals and massive scleractinians, respectively, and which are more depressed than congeneric species that inhabit branching corals (Herler, 2007; see also Fig. 4.2.5). Gobies associated with coral rock or caves have a more generalized, typically cylindrical, body form (Herler, 2007).

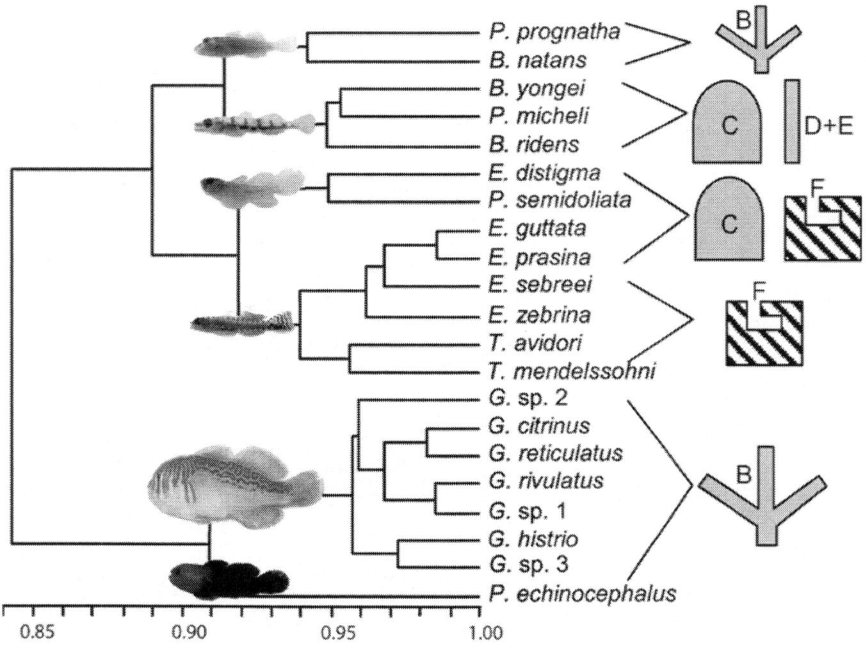

Fig. 4.2.5 Cluster analysis (Bray Curtis – similarity) of nine proportional morphometric variables (all expressed as percent of standard length) of 21 reef gobies from the Gulf of Aqaba, northern Red Sea (modified from Herler, 2007). The most frequented microhabitats are indicated (B = branching corals, C = massive corals, D = *Millepora dichotoma*, E = *Cirripathes* sp., F = coral rock). Fish genera abbreviations: B = *Bryaninops*, E = *Eviota*, G = *Gobiodon*, P = *Paragobiodon* (*echinocephalus*), *Pleurosicya* (*prognatha* and *micheli*), *Priolepis* (*semidoliata*), and T = *Trimma* respectively.

Color image of this figure appears in the color plate section at the end of the book.

In addition to overall body form, the morphology of fins is often associated with the preferred microhabitat. A common feature of gobies is united pelvic fins, forming a pelvic disc. Advanced developmental states of the pelvic disc are present in *Bryaninops*, *Pleurosicya*, *Gobiodon* and *Paragobiodon*, all of which inhabit live coral. In these species the pelvic fins form cup-shaped sucking discs. The two former genera comprise the most advanced state by having a lobed anterior frenum that is folded anteriorly to form a pocket-like structure (Larson, 1985, 1990; Herler, 2007). In contrast, coral-rock dwelling gobies have fins that may be united but are neither disc- or cup-shaped, nor with an anterior frenum (e.g. *Trimma* and *Priolepis*), or have almost separate fins (e.g. *Eviota*) (Herler, 2007). This suggests that a sucking disc is most useful on a smooth substrate, such as the surface of live corals, whereas, on rock or other rough substrate, a sucking function may not be possible. The pectoral fin might also play a role for locomotion within coral branches. Some species of *Gobiodon* have a wide pectoral fin base, which supports strong pectoral fin propulsion (J. Herler, unpublished data).

Apart from body form adaptation that may optimize locomotion and minimize predation risk, another adaptative feature that has likely evolved in response to the high predation is toxic skin mucous in the genus *Gobiodon* (Schubert *et al.*, 2003). Chemical alarm cues, which have been discovered in the small reef goby *Asterropteryx semipunctata* (McCormick and Larson, 2007), could also help reduce the risk of predation in coral reef gobies.

Morphological and Functional Consequences of Size Diminution

Miniaturization is the evolution of extremely small body size within a phylogenetic lineage (Hanken and Wake (1993). Gobies in general are a group of small fishes, with the majority being less than 10 cm (Larson and Murdy, 2001), and therefore, were good candidates for miniaturization. On coral reefs we find a unique diversity of small goby species that often do not exceed 3 cm maximum adult size. Spatial restriction of habitats has promoted miniaturization in some animal groups (Hanken and Wake, 1993). Such habitats are particularly diverse in coral reefs and may explain the diversity of small gobies in this ecosystem. Miniature gobies mostly occupy corals and coral-rock crevices, where they presumably gain better shelter from predators.

While the reduction of size was probably important for access to small shelter sites where predation risk is minimal, diminution has also had important consequences for the anatomy, ecology, physiology, and life histories of coral reef gobies (Miller, 1996; Munday and Jones, 1998). In reef gobies, size diminution and short lifespans may have been important promotors for reductions of squamation, locomotory body parts, and sensory systems. Scales are usually reduced in the smaller and/or more cryptic genera such as *Bryaninops*, *Eviota*, *Trimma* and *Trimmatom* (Winterbottom, 1990; Herler and Hilgers, 2005). At the extreme, scales are completely absent in the tiny *Trimmatom sagma*, *T. officius* and *T. nanus* (Winterbottom, 1990). Scales are also absent in the coral-dwelling *Gobiodon* (Herler and Hilgers, 2005; Harold *et al.*, 2008), but here squamation is replaced by a thick skin mucus layer that contains toxic substances and may prevent predation or parasite infestation (Munday *et al.*, 2003; Schubert *et al.*, 2003).

Reductions of fins in reef gobies involves the loss of fin-ray branching, or loss of entire rays in the median and pectoral fins (e.g. Lachner and Karnella, 1980; Larson, 1985, 1990; Winterbottom, 1990). Also the pelvic fin is often reduced: the fifth ray may be very short or almost absent (e.g. several species of *Eviota* and *Trimmatom*) (Lachner and Karnella, 1980; Winterbottom, 1990), resulting in the loss of pelvic sucking function (Herler, 2007). The supporting skeletal elements, such as the pterygiophores of median fins or the radials and scapula that support the pectoral fin, are often reduced or lack ossification in these tiny gobies (Winterbottom, 1990). These reductions were likely promoted by a cryptic life-style in spatially restricted habitats, where swimming may not be necessary, or possible. Similarly, a cryptic life-style may have reduced the importance of the lateral line system and promoted reductions in its development. In the genera *Priolepis*, *Trimma* and *Trimmatom* for example, head sensory canals have been lost and even the usually numerous rows of superficial head neuromasts may be reduced in number (Winterbottom, 1990, 2006).

Other important morphological changes associated with size reduction are caused by functional constraints of organs that do not proportionately follow overall body diminution. For example, relative eye size is greater in minute fish (Herler, 2007) due to physical constraints of eye development (Kirschfeld, 1976). This leads to disproportionate size development of head skeleton regions and can even affect trophic morphological structures. For example, Riedlecker and Herler (2008) found that eye size influences jaw suspensorium shape in different species of *Gobiodon*. Deeper-water species with large eyes had smaller suspensoria and a lower suspensorium/jaw-articulation angle than did shallow water species with smaller eyes. This may have important consequences for diet and feeding function.

Another important constraint set by diminution of body size is the limitation of bestowal and fecundity in small gobies. Although goby eggs are relatively small (Miller, 1979; Miller, 1996), life-time fecundity may be as little as several hundred eggs per female in the smallest species of *Eviota* (Depczynski and Bellwood, 2005). This has important consequences for population dynamics because any alteration of mortality rates will impact population sizes immediately due to their short adult lifespan and reproductive period. Behavioral adaptations that may partly compensate for low fecundity levels include egg guarding and egg fanning to increase oxygen supply (Depczynski and Bellwood, 2005).

LIFE HISTORY

Many small species of coral reef fishes are surprisingly long-lived, e.g. 18-32 years for several damselfishes (Worthington *et al.*, 1995; Meekan *et al.*, 2001) and 25-30 years for some small surgeonfishes (Choat and Axe, 1996; Hart and Russ, 1996; Choat and Robertson, 2002). These long-lived species tend to attain an asymptotic size very quickly (within 2-6 years) and thus spend much of their lifespan (often > 75%) at that size. In contrast, the life history patterns of coral reef gobies are typified by short lives and continued investment in growth for all or much of the lifespan. High mortality rates and short adult lifespans are important selection pressures influencing the timing of maturation and reproductive output in coral reef gobies.

Mortality and Longevity

Most coral reef gobies are short-lived, with maximum lifespans in the order of 2-4 years for the coral-dwelling species (Kuwamura *et al.*, 1996; Munday, 2001), 1-2 years for the cryptobenthic and burrow-dwelling species (Yanagisawa, 1982; Forrester, 1995; Reavis and Barlow, 1998; Kritzer, 2002; Hernaman and Munday, 2005a), and just 2-3 months for the smallest reef-dwelling species (Depczynski and Bellwood, 2006; Winterbottom and Southcott, 2008). *Eviota sigillata* reaches a maximum total length (TL) of 20 mm and has a maximum lifespan of just 8 weeks, making it the shortest recorded lifespan of any vertebrate (Depczynski and Bellwood, 2005). The similarly-sized *Trimma nasa* is also extremely short-lived, with a maximum lifespan of 12 weeks (Winterbottom and Southcott, 2008). These species experience extremely high daily mortality rates of 5-8%. Larger species (50-120 mm TL) that live in burrows (*Valenciennea muralis*, *Amblygobius bynoensis*, and *Amblygobius phalaena*) or cryptically amongst algal-covered rubble (*Asterropteryx semipunctata*) and rocks (*Istigobius goldmanni*) have slightly lower daily mortality rates (~ 1%), but are still relatively short-lived with maximum lifespans of 11-16 months (Hernaman and Munday, 2005a). Coral reef gobies are preyed upon by a wide range of reef fishes, including scorpaenids, muraenids, pinguipedids, synodontids and serranids (Luckhurst and Luckhurst, 1977; Sweatman, 1984; Parrish *et al.*, 1986; Norris and Parrish, 1988; Forrester, 1995). Predation probably accounts for much of the high mortality rates experienced by the cryptobenthic and burrow-dwelling species that must venture from their shelter site to feed, or, for females of the many polygynous species, to visit a nesting site to spawn. In contrast, coral-dwelling gobies do not need to leave their coral host to feed or mate, and the longer lifespans of these species may be related to the increased protection from predation afforded by the branches of the host coral colony.

Life History Patterns and the Timing of Maturation

The most extreme life history is exhibited by the coral reef pygmy goby *Eviota sigillata*, which settles from the plankton after 3 weeks, matures within 1-2 weeks of settlement, and has a maximum adult lifespan of less than 4 weeks. A similar life history is exhibited by the equally small *Trimma nasa*, which has a maximum of 7 weeks following settlement in which to grow, mature and reproduce. Such shortened lifespans heavily affect the benthic part of the life cycle by severely reducing the time available for maturation and reproductive events. They do not, however, appear to affect the pelagic part of the life cycle, with pelagic larval duration (PLD; the time between hatching and settlement to the adult habitat) being within the range exhibited by many goby species from a variety of habitats (Table 4.2.1). For *Trimma nasa* and species of *Eviota*, the PLD is 39-42% of the maximum lifespan. This compares to 6-11% for cryptobenthic and burrow-dwelling gobies (Table 4.2.1). In contrast, the PLDs of most reef fish species typically comprise < 1% of their expected lifespan (Depczynski and Bellwood, 2006). The observation that the PLD of extremely short-lived species is similar to longer-lived gobies indicates that there may be an optimal size/age at which fish larvae can successfully settle onto the reef habitat or that morphological and physiological development of larval gobies is constrained to a certain timeframe.

Life history theory predicts that high adult mortality will select for early maturation, which results in a higher probability of survival to first reproduction, shorter generation times, and a higher instantaneous rate of natural increase (Begon and Mortimer, 1981; Stearns, 1992). Hernaman and Munday (2005b) examined the timing of maturation in five goby species that experienced annual mortality rates of > 97%. Males of all five species and females of three species matured at a smaller size (relative to their mean maximum size) than is typically found for fishes, indicating selection for relatively early maturation. Mean size at maturation ranged from 50-60% of maximum size for females, and 41-52% for males (depending on species). Thus, males of all five species not only matured at a considerably smaller relative size than expected from life history patterns for fishes (65%; Charnov, 1993), but also at or below the lower threshold for organisms in general (50-80%; Charnov, 1993). A similar result was

Table 4.2.1 Mean size and age at settlement for coral reef gobies from a range of habitat types (SL, standard length; TL, total length; -- data unavailable)

Habitat	*Species*	*Mean PLD (days)*	*% of max. age represented by PLD (%)*	*Mean size at settlement (mm)*	*Source*
Cryptobenthic /epibenthic	*Eviota queenslandica*	24.2	24	7-7.5 TL	Depczynski and Bellwood (2006)
	Eviota sigillata	24.7	42	7-7.5 TL	Depczynski and Bellwood (2006)
	Istigobius goldmanni	25.7	7	8.1 TL	Hernaman (2003)
	Eviota melasma	26.2	27	7-7.5 TL	Depczynski and Bellwood (2006)
	Coryphopterus glaucofraenum	26.8	--	6.4 SL	Sponaugle and Cowen (1994)
	Asterropteryx semipunctata	27.5	6	7.6 TL	Hernaman (2003)
	Trimma nasa	33.9	39	--	Winterbottom and Southcott (2008)
Burrow-dwellers	*Amblygobius bynoensis*	28.2	8	10.1 TL	Hernaman (2003)
	Amblygobius phalaena	32.8	8	11.8 TL	Hernaman (2003)
	Valenciennea muralis	39.5	11	11.3 TL	Hernaman (2003)
	Amblygobius rainfordi	40.3	--	14.4 SL	Brothers *et al.* (1983)
Coral-dwellers	*Gobiodon erythrospilus*	17.6	<2	--	Munday and Hernaman, unpublished data
	Gobiodon histrio	21.2	<2	--	Munday and Hernaman, unpublished data
	Paragobiodon xanthosomus	23.0	--	--	Munday and Hernaman, unpublished data

found for females of the short-lived *Eviota queenslandica* and *E. melasma*, which on average matured at 51% and 44% of their mean maximum size respectively (Depczynski and Bellwood, 2006).

Although high adult mortality appears to have favoured early maturation in some species, not all gobies mature at such small relative sizes. Female *Asterropteryx semipunctata*, *Istigobius goldmanni*, and even the tiny *Eviota sigillata*, mature at around 63-65% of their mean maximum size (Hernaman and Munday, 2005b; Depczynski and Bellwood, 2006). The relatively larger size at maturation of these species may be associated with physiological constraints of a very small body size (*E. sigillata*) and the competitive benefits associated with a larger body size in polygynous species (*A. semipunctata* and *I. goldmanni*). The size at maturation (11 mm) of *Eviota sigillata* is very close to the physiological threshold at which maturation can occur (~10 mm; an exception may be *Trimmatom nanus*, which matures at 8 mm (Winterbottom, 1990)). The mean size at maturity for female *A. semipunctata* (27.7 mm TL) and *I. goldmanni* (30.7 mm TL) is 18-21 mm larger than the minimum size threshold of 10 mm. However, these species are much longer lived than *Eviota* and maturation may be delayed to obtain reproductive and competitive benefits associated with a larger body size. Both species are polygynous and tend to be found in high densities (Hernaman and Munday, 2007; Hernaman and Probert, 2008). The benefits associated with a large body size, in terms of increased fecundity and enhanced success in competition for shelter and spawning sites, may select against very early maturation in these species.

Seasonal influences on growth rates can also modify the timing of maturation. Hernaman and Munday (2005a) detected seasonal differences in growth rates for adults and juveniles of four goby species from three genera (*Valenciennea*, *Amblygobius* and *Istigobius*). Summer-growing individuals had significantly higher growth rates than winter-growing individuals, which had a considerable impact on the age at first reproduction by influencing the time required to attain a size suitable for maturation. Winter- and summer-growing *V. muralis* and *A. bynoensis* females matured at the same mean size, but due to seasonal differences in growth rates, summer fish attained that size one month quicker and thus matured at a significantly younger age than winter fish. This has important implications for short-lived coral reef species with seasonal reproduction, and suggests that the season of settlement may have a significant impact on the timing of maturation, the probability of survival to maturation, and reproductive success.

Patterns of Growth

Coral reef gobies typically invest in somatic growth throughout much or all of their lifespan, spending relatively little or no time at asymptotic size (e.g. Hernaman and Munday, 2005a; see also Fig. 4.2.6). Rapid near-linear growth has also been recorded in some species, including *Istigobius decoratus* (Kritzer, 2002), *I. goldmanni* (Hernaman and Munday, 2005a), *Trimma nasa* (Winterbottom and Southcott, 2008), and three species of *Eviota* (Depczynski and Bellwood, 2006). Similarly, the intertidal Hawaiian goby, *Bathygobius coalitus*, exhibited a linear pattern of growth for at least the first nine months of life, with no difference in growth rate between immature and mature gobies (Shafer, 1998).

The growth pattern of most coral reef gobies differs considerably to that exhibited by other relatively small, but long-lived coral reef fishes, such as certain acanthurids (e.g. Choat and Axe, 1996; Hart and Russ, 1996) and pomacentrids (e.g. Meekan *et al.*, 2001). These long-lived species often have a growth pattern characterized by rapid growth to asymptotic size and sexual maturity, followed by a relatively long period of reproductive activity but little additional somatic growth (Choat and Robertson, 2002). This growth pattern may reflect a physiological trade-off between growth and reproduction. Trade-offs play a central role in life history theory, and a physiological trade-off between growth and reproduction is expected where these two processes compete directly for a limited amount of energy (Begon and Mortimer, 1981; Partridge and Harvey, 1988; Stearns, 1992). Coral reef gobies are apparently able to acquire sufficient energy to satisfy demands for both growth and reproduction, with no apparent trade-off between these traits. However, this capacity to continue somatic growth throughout life may come at a cost of reduced longevity, possible due to increased exposure to predators while feeding to maintain sufficient energy to support rapid growth and high reproductive effort.

SOCIAL ORGANISATION AND REPRODUCTIVE STRATEGIES

In common with other gobies, coral reef species are demersal spawners with paternal egg care (Thresher, 1984). Resource defense polygyny, where males defend a nesting site that is visited by females, appears to be a common form of social organisation for coral reef gobies (e.g. Sunobe and Nakazono, 1990; Black *et al.*, 2005; Manabe *et al.*, 2007b; Manabe *et al.*, 2009) especially among species that live at high density populations (Hernaman and

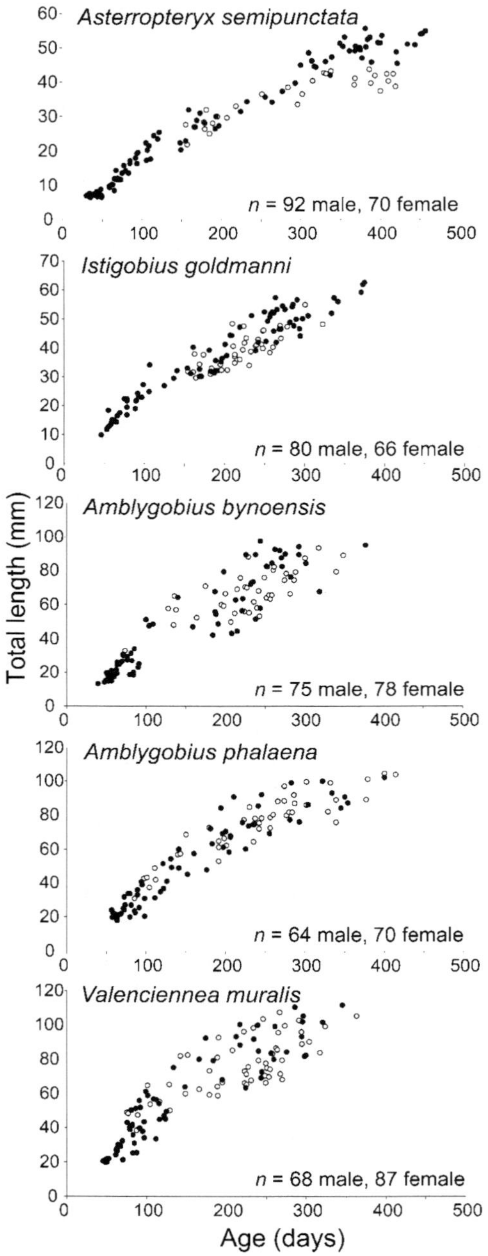

Fig. 4.2.6 Growth trajectories for five species of coral reef goby (males, filled circles; females, open circles. Unsexed juveniles used in both male and female data sets). Modified from Hernaman and Munday (2005a).

Munday, 2007). In contrast, pair-forming is common among species that live in low densities on sediments and rubble (e.g. Reavis, 1997b; Takegaki and Nakazono, 1999; Takegaki, 2000; Hernaman and Munday, 2007) and is the usual social structure of shrimp-associated gobies (Yanagisawa, 1982). These pair bonds are typically long-lasting and monogamous (Whiteman and Cote, 2004). Monogamy also occurs among obligate coral-dwelling gobies, but here it may be associated with social groups ranging in size from pairs up to 10 or more individuals (e.g. Kuwamura *et al.*, 1994; Munday *et al.*, 1998; Munday *et al.*, 2002; Thompson *et al.*, 2007). The presence of monogamy in large social groups increases the opportunities for conflict among individuals and has led to the evolution of complex behaviours to maintain social order within these goby societies (Wong *et al.*, 2007).

The most complex social structures are exhibited by group-forming species such as *Paragobiodon xanthosomus*. This species forms groups of up to 10 or more individuals on a single coral head, but only the two largest individuals are reproductively mature and they form a size-matched breeding pair. The presence of the dominant female keeps all the subordinate fish from becoming reproductively mature (Wong *et al.*, 2008a). Subordinates queue to reach the breeding positions, with the stability of the queue maintained by a strict size-based hierarchy. Each individual in the queue has a body-size ratio of approximately 90-95% of its immediate dominant. Any individual that tries to advance its position in the queue by increasing its size will be evicted from the group when the body-size ratio between itself and its immediate dominant exceeds the threshold for conflict between the two ranks (Wong *et al.*, 2007). Remarkably, individuals control their own growth rate so that they do not exceed this critical size threshold, and they can do this by regulating their dietary intake (Wong *et al.*, 2008b). However, if the dominant dies or is removed the subordinate will rapidly accelerate its growth until it reaches the size threshold with the next higher individual in the queue (Wong *et al.*, 2007). Similar control of individual growth rates has been observed in the pair-forming *G. histrio*, where males and females will adjust their growth rates to maintain a size-matched breeding pair (Munday *et al.*, 2006b).

Group size of *P. xanthosomus*, and some other coral-dwelling gobies, is influenced by the size of the coral host, with larger groups inhabiting larger coral colonies (Thompson *et al.*, 2007). Other species are strictly pair-forming regardless of coral colony size and the resident pair defend their host coral from any intruders (Munday *et al.*, 1998; Hobbs and Munday, 2004; Thompson *et al.*, 2007). Juveniles usually inhabit small coral colonies that are not big enough for breeding pairs. These juveniles will readily move to establish a pair with a single adult on a larger coral if there is an opportunity to do so. Juveniles delay maturation until they are able to form a pair, and then they rapidly mature into the opposite sex of the single adult (Kuwamura *et al.*, 1996; Hobbs *et al.*, 2004). Presumably delaying maturation allows juvenile gobies to invest additional energy in growth and also ensures they can mature into the correct sex to match available breeding opportunities (Hobbs *et al.*, 2004).

Alternative male mating tactics have been identified in several species, with small males either attempting to sneak spawn or intercept females before they reach the nesting sites of larger males (Sunobe and Nakazono, 1999a; Drilling and Grober, 2005; Manabe *et al.*, 2009). Perhaps the most intriguing reproductive characteristic of coral reef gobies, however, is the capacity of many species to change sex, in some cases more than once (e.g. Sunobe and Nakazono, 1993; Kuwamura *et al.*, 1994; Nakashima *et al.*, 1996; Munday *et al.*, 1998). Although many species are gonochoristic, many others are hermaphroditic, with the presence of gonochorism and hermaphroditism tending to be conserved within genera (Cole, 1990). At least 14 genera of gobies are confirmed to have hermaphroditic species, or are expected to be hermaphroditic based on behaviour or gonadal characteristics (de Mitcheson and Liu, 2008). Most of these species are found on or near coral reefs. In contrast, only a handful of temperate marine gobies are known to be hermaphroditic (e.g. Cole, 1983, 1990; St. Mary, 1996, 2000). Sex change is not known from freshwater species (de Mitcheson and Liu, 2008).

Sex change from female to male (protogyny) is exhibited by a number of species with a polygynous mating system (e.g. Cole and Robertson, 1988; Cole and Shapiro, 1992; Sunobe and Nakazono, 1999b). These species typically have a female biased sex ratio and the largest individual in a social group is male. Protogynous sex change can be favoured in a polygynous mating system because male reproductive success often increases more steeply with size or age than does the female reproductive success in this mating system. Therefore, an individual can maximize its lifetime reproductive success by first reproducing as a female and then changing sex to male when large enough to defend a nest site that is visited by multiple females (Munday *et al.*, 2006a). Intriguingly, there are many goby species with polygynous mating systems that do not change sex. It is unknown if phylogenetic constraints have prevented the evolution of sex change in these species, or there are some ecological characteristics that favour gonochorism over hermaphroditism in the species (Cole, 1990; de Mitcheson and Liu, 2008).

An increasing number of coral reef gobies are now recognized as being capable of bi-directional sex change (Munday *et al.*, 2010). In these species both males and females can change sex, and individuals may change sex more than once. Bi-directional sex change occurs in two different circumstances. In some polygynous gobies, such as *Trimma okinawae*, females change sex to males when there is the opportunity to defend a harem of females (Sunobe and Nakazono, 1990). Males may change sex back to female if they become single and the only breeding opportunity immediately available to them is to join another social group. They will change sex and reproduce as a female if they are smaller than the resident male in the new social group (Manabe *et al.*, 2007a, b). The advantage to sex change by males may lie in the fact that these small gobies experience high and unpredictable rates of mortality. On average a single male may have higher reproductive success by changing sex and reproducing as a female in a new harem than if he remains male and waits for a new harem to form (Munday *et al.*, 2006a).

Bi-directional sex change is also common among coral-dwelling gobies that form monogamous pairs (Kuwamura *et al.*, 1994; Nakashima *et al.*, 1996; Munday *et al.*, 1998). Here the advantage of sex change appears to be that there are very few single adults in the population and the ability to change sex in either direction reduces the risk of searching for a new mate, because a single adult could form a new breeding pair with any other single adult encountered (Nakashima *et al.*, 1996; Munday *et al.*, 2010). The common threads that link bi-directional sex change in polygynous and pair-forming species are the unpredictability of alternative mating opportunities for single fish and the high risk of predation in searching for new mates due to the small body size of gobies.

CONSERVATION

Coral reefs are under increasing stress from a variety of threats, including terrestrial pollution and runoff, overfishing, and climate change (Hughes *et al.*, 2003). Coral cover has declined on reefs in many parts of the world and in some locations a substantial proportion of reefs are already considered to be seriously degraded (Wilkinson, 2008). The relatively high abundance and short generation time of most coral reef gobies are characteristics that usually confer resilience to disturbance (Dulvy *et al.*, 2003). Many goby species are also geographically widespread, which greatly reduces their risk of extinction. The threat to goby species from degrading reef habitat will, therefore, mostly depend on their patterns of habitat use.

Species that utilize sand patches, rubble, or dead coral matrix may increase in abundance as these habitats become ubiquitous on degraded reefs. In contrast, populations of coral-dwelling species rapidly decline in abundance following coral loss (Munday *et al.*, 1997; Munday, 2004b; Pratchett *et al.*, 2008). Continued degradation of coral reef habitat will therefore cause shifts to goby community structure, towards communities with fewer coral specialists and more non-coral generalists. In an example of this shift in structure, Bellwood *et al.* (2006) showed that sand and rubble-dwelling gobies, such as *Eviota queenslandica* and several *Istigobius* species, increased in abundance following a major bleaching event, whereas coral-dwelling *Gobiodon* and *Paragobiodon* species disappeared or became uncommon.

Although some goby species will be favoured by the continued degradation of coral reef habitat, the net effect on the diversity of coral reef gobies is expected to be negative because a considerable number of species rely on living corals (e.g. species of *Gobiodon*, *Paragobiodon*, *Bryaninops*, *Pleurosicya* and some *Eviota* species) or other healthy invertebrate assemblages (e.g some species of *Pleurosicya*, *Luposicya* and *Gobiosoma*) for their survival. As healthy coral and benthic invertebrate communities decline, so too will the abundance of the many goby species that use these organisms as habitat. Furthermore, the species of *Acropora* and *Pocillopora* corals preferred by many coral-dwelling gobies are also the ones that are most susceptible to thermally-induced coral bleaching (Pratchett *et al.*, 2008). Therefore, the effects of global warming might be particularly severe for coral-dwelling gobies.

Even among coral-dwelling species, it appears that specialist species suffer more than generalist species from coral declines. Munday (2004b) found that populations of *Gobiodon* species that inhabit just one or two species of coral declined in abundance proportionally more than *Gobiodon* species that inhabit many coral species (Fig. 4.2.3). In addition, the specialist species had smaller population sizes before the disturbance (Munday, 2004b). Therefore, the specialist coral-dwellers were especially prone to local extinction because they had both smaller populations before the disturbance and larger percentage declines as a result of the disturbance. Although most coral-dwelling gobies have large geographic ranges (Munday *et al.*, 1999; Herler *et al.*, 2009), and are thus unlikely to be at risk of global extinction due to the effects of sporadic local extinctions, there are a few species of *Gobiodon* that appear to have very small geographic ranges (Munday *et al.*, 1999; Munday, 2004b; Herler and Hilgers, 2005; Herler *et al.*, 2009). The combination of specialization on live coral habitat, relatively small population sizes, and small geographic range, places these species at a greater risk of extinction than most other coral reef fishes (Munday, 2004b). Consequently, some of these small-range, coral-dwelling gobies may be among the most threatened of coral reef fishes.

CONCLUSIONS

The Gobiidae is the most species-rich family of coral reef fishes, yet many species still remain to be discovered or described. The small body size of most reef gobies makes them prone to predation and requires them to live cryptically. They occupy almost every type of microhabitat associated with coral reefs, including soft sediments, coral rock and rubble, caves and crevices, and even the burrows of shrimps. Many species use other reef organisms as habitat, especially scleractinian corals, to which some have evolved species-specific relationships. The different degrees of microhabitat specialization exhibited by coral reef gobies have implications for their distribution and abundance within and among reefs. The close association with coral habitats has also resulted in numerous morphological and behavioural adaptations, some of which are unique among reef fishes. The evolution of bi-directional sex change, cooperative growth regulation, hypoxia tolerance, skin toxins and various morphological characters is closely linked to microhabitat specialization, while morphological reductions characterize many of the smaller reef goby genera.

Evolutionary size diminution increases the risk of predation and this important source of mortality appears to have been a key factor shaping the life histories of coral reef gobies. High mortality rates mean that most reef gobies are short-lived. Many species live for a maximum of 1 - 2 years, while the smallest species may live just 1 - 2 months. Unlike most other coral reef fishes, gobies tend to exhibit continuous growth throughout life. At the same time they tend to mature early and exhibit high reproductive effort. Fast growth, early maturation and high reproductive effort are all characteristic of species with high mortality rates. Some coral-dwelling species are longer lived, presumably because they gain protection from predators by living among the close-knit branches of coral colonies. The use of coral habitat shelters individuals from predators, but it increases the intensity of competition for shelter space. In these species, intra- and interspecific competition for shelter space influences patterns of habitat use, local distributions and abundances, growth rates and survival. Habitat specialization and associated ecological adaptations also appear to have promoted species diversification in these coral-dwelling gobies.

Their high abundance and fast population turn-over mean that gobies play an important role in energy transfer through coral reef communities. The many planktivorous, micro-carnivorous and detritivorous species provide important linkages to higher trophic levels. In turn, many reef gobies—in particular those that are associated with live coral or other benthic invertebrates—depend on healthy reefs for their continued persistence. With coral reefs in decline due to various local and global anthropogenic stresses, we expect to observe shifts in the community structure of coral reef gobies over coming decades. Many of the highly specialist coral-dwelling species will decline in abundance, with a few species facing an elevated risk of extinction. In contrast, some generalist, rock, sand and rubble dwelling species may become more ubiquitous.

In this chapter we have demonstrated that, despite often being neglected in studies of reef fish assemblages, coral reef gobies are not only an ecologically important component of reef systems but are also good research models. Their exceptional species richness and widespread distribution offer excellent opportunities for investigating mechanisms of species diversification and maintenance of species diversity. Their small size and robust nature makes them well-suited to experimental manipulation, with recent experiments using gobies as model species making significant contributions to our understanding of how reef fish communities are structured and their populations regulated. As the smallest and shortest-living vertebrates they are critically important for understanding the evolution of animal life-history traits and for explaining scaling relationships of life history traits among species.

Acknowledgements

We are grateful to Robert Patzner for inviting us to write this chapter and to Marian Wong for reviewing our contribution. Helen Larson and Rick Winterbottom provided helpful comments on the draft version.

References

Ackerman, J.L. and D.R. Bellwood. 2000. Reef fish assemblages: a re-evaluation using enclosed rotenone stations. *Marine Ecology Progress Series* 206: 227-237.

Begon, M. and M. Mortimer. 1981. *Population Ecology. A Unified Study of Animals and Plants.* First, Blackwell Scientific Publications, London.

Bellwood, D.R., A.S. Hoey, J.L. Ackerman and M. Depczynski. 2006. Coral bleaching, reef fish community phase shifts and the resilience of coral reefs. *Global Change Biology* 12: 1587-1594.

Black, M.P., B. Moore, A.V. Canario, D. Ford, R.H. Reavis and M.S. Grober. 2005. Reproduction in context: field testing a laboratory model of socially controlled sex change in *Lythrypnus dalli* (Gilbert). *Journal of Experimental Marine Biology and Ecology* 318: 127-143.

Caley, M.J. and P.L. Munday. 2003. Growth trades off with habitat specialization. *Proceedings of the Royal Society of London Series B-Biological Sciences* 270: 175-177.

Charnov, E.L. 1993. *Life History Invariants: Some Explorations of Symmetry in Evolutionary Ecology.* Oxford University Press, Oxford.

Choat, J.H. and L.M. Axe. 1996. Growth and longevity in acanthurid fishes; an analysis of otolith increments. *Marine Ecology Progress Series* 134: 15-26.

Choat, J.H. and D.R. Robertson. 2002. Age-based Studies. In: *Coral Reef Fishes: Dynamics and Diversity in a Complex Ecosystem*, P. Sale (ed.), Academic Press, San Diego, pp. 57-80.

Cole, A.J., M.S. Pratchett and G.P. Jones. 2008. Diversity and functional importance of coral-feeding fishes on tropical coral reefs. *Fish and Fisheries* 9: 286-307.

Cole, K.S. 1983. Protogynous hermaphroditism in a temperate zone territorial marine goby, *Coryphopterus nicholsi*. *Copeia*: 1983: 809-812.

Cole, K.S. 1990. Patterns of gonad structure in hermaphroditic gobies (Teleostei, Gobiidae). *Environmental Biology of Fishes* 28: 125-142.

Cole, K.S. and D.R. Robertson. 1988. Protogyny in the caribbean reef goby, *Coryphopterus personatus*: gonad ontogeny and social influences on sex change. *Bulletin of Marine Science* 42: 317-333.

Cole, K.S. and D.Y. Shapiro. 1992. Gonadal structure and population characteristics of the protogynous goby *Coryphopterus glaucofraenum*. *Marine Biology* 113: 1-9.

Colin, P.L. 2002. A new species of sponge-dwelling *Elacatinus* (Pisces: Gobiidae) from the western Caribbean. *Zootaxa* 106: 1-7.

de Mitcheson, Y.S. and M. Liu. 2008. Functional hermaphroditism in teleosts. *Fish and Fisheries* 9: 1-43.

Depczynski, M. and D.R. Bellwood. 2003. The role of cryptobenthic reef fishes in coral reef trophodynamics. *Marine Ecology Progress Series* 256: 183-191.

Depczynski, M. and D.R. Bellwood. 2004. Microhabitat utilisation patterns in cyptobenthic coral reef fish communities. *Marine Biology* 145: 455-463.

Depczynski, M. and D.R. Bellwood. 2005. Shortest recorded vertebrate lifespan found in a coral reef fish. *Current Biology* 15: R288-R289.

Depczynski, M. and D.R. Bellwood. 2006. Extremes, plasticity, and invariance in vertebrate life history traits: Insights from coral reef fishes. *Ecology* 87: 3119-3127.

Depczynski, M., C.J. Fulton, M.J. Marnane and D.R. Bellwood. 2007. Life history patterns shape energy allocation among fishes on coral reefs. *Oecologia* 153: 111-120.

Dirnwöber, M. and J. Herler. 2007. Microhabitat specialization and ecological consequences for coral gobies of the genus *Gobiodon* in the Gulf of Aqaba, northern Red Sea. *Marine Ecology Progress Series* 342: 265-275.

Drilling, C.C. and M.S. Grober. 2005. An initial description of alternative male reproductive phenotypes in the bluebanded goby, *Lythrypnus dalli* (Teleostei, Gobiidae). *Environmental Biology of Fishes* 72: 361-372.

Dulvy, N.K., Y. Sadovy and J.D. Reynolds. 2003. Extiniction vulnerability in marine populations. *Fish and Fisheries* 4: 25-64.

Feary, D.A. 2007. The influence of resource specialization on the response of reef fish to coral disturbance. *Marine Biology* 153: 153-161.

Feary, D.A., G.R. Almany, G.P. Jones and M.I. McCormick. 2007. Coral degradation and the structure of tropical reef fish communities. *Marine Ecology Progress Series* 333: 243-248.

Forrester, G.E. 1995. Strong density-dependent survival and recruitment regulate the abundance of a coral reef fish. *Oecologia* 103: 275-282.

Forrester, G.E. and M.A. Steele. 2000. Variation in the presence and cause of density-dependent mortality in three species of reef fishes. *Ecology* 81: 2416-2427.

Forrester, G.E. and M.A. Steele. 2004. Predators, prey refuges, and the spatial scaling of density-dependent prey mortality. *Ecology* 85: 1332-1342.

Forrester, G.E., B. Evans, M.A. Steele and R.R. Vance. 2006. Assessing the magnitude of intra- and interspecific competition in two coral-reef fishes. *Oecologia* 148: 632-640.

Gill, A.C. and S.L. Jewett. 2004. *Eviota hoesei* and *E. readerae*, new species of fish from the Southwest Pacific, with comments on the identity of *E. corneliae* Fricke (Perciformes : Gobiidae). *Records of the Australian Museum* 56: 235-240.

Greenfield, D.W. and R.K. Johnson. 1999. Assemblage structure and habitat associations of Western Caribbean gobies (Teleostei: Gobiidae). *Copeia* 1999: 251-266.

Hagiwara, K. and R. Winterbottom. 2007. Two new species of *Trimma* (Gobiidae) from the Western Pacific. *Bulletin of the National Museum of Natural Sciences, Series A Suppl.* 1: 163-174.

Hanken, J. and D.B. Wake. 1993. Miniaturization of body size: organismal consequences and evolutionary significance. *Annual Review of Ecology and Systematics* 24: 501-519.

Harold, A.S. and R. Winterbottom. 1995. *Gobiodon acicularis*, a new species of gobioid fish (Teleostei: Gobiidae) from Belau, Micronesia. *Proceedings of the Biological Society of Washington* 108: 687-694.

Harold, A.S. and R. Winterbottom. 1999. *Gobiodon brochus*: A new species of gobiid fish (Teleostei: Gobioidei) from the Western South Pacific, with a description of its unique jaw morphology. *Copeia* 1999: 49-57.

Harold, A.S., R. Winterbottom, P.L. Munday and R.W. Chapman. 2008. Phylogenetic relationships of Indo-Pacific coral gobies of the genus *Gobiodon* (Teleostei: Gobiidae), based on morphological and molecular data. *Bulletin of Marine Science* 82: 119-136.

Hart, A.M. and G.R. Russ. 1996. Response of herbivorous fishes to crown-of-thorns starfish *Acanthaster planci* outbreaks. III. Age, growth, mortality and maturity indices of *Acanthurus nigrofuscus*. *Marine Ecology Progress Series* 136: 25-35.

Herler, J. 2007. Microhabitats and ecomorphology of coral- and coral rock-associated gobiid fish (Teleostei: Gobiidae) in the northern Red Sea. *Marine Ecology — An Evolutionary Perspective* 28: 82-94.

Herler, J. and H. Hilgers. 2005. A synopsis of coral and coral-rock associated gobies (Pisces: Gobiidae) in the Gulf of Aqaba, northern Red Sea. *Aqua, Journal of Ichthyology and Aquatic Biology* 10: 103-132.

Herler, J., S. Koblmueller and C. Sturmbauer. 2009. Phylogenetic relationships of coral-associated gobies (Teleostei, Gobiidae) from the Red Sea based on mitochondrial DNA data. *Marine Biology* 156: 725-739.

Hernaman, V. and P.L. Munday. 2005a. Life-history characteristics of coral reef gobies. I. Growth and life-span. *Marine Ecology Progress Series* 290: 207-221.

Hernaman, V. and P.L. Munday. 2005b. Life-history characteristics of coral reef gobies. II. Mortality rate, mating system and timing of maturation. *Marine Ecology Progress Series* 290: 223-237.

Hernaman, V. and P.L. Munday. 2007. Evolution of mating systems in coral reef gobies and constraints on mating system plasticity. *Coral Reefs* 26: 585-595.

Hernaman, V. and P.K. Probert. 2008. Spatial and temporal patterns of abundance of coral reef gobies (Teleostei: Gobiidae). *Journal of Fish Biology* 72: 1589-1606.

Hernaman, V., P.K. Probert and W.D. Robbins. 2009. Trophic ecology of coral reef gobies: interspecific, ontogenetic, and seasonal comparison of diet and feeding intensity. *Marine Biology* 156: 317-330.

Hobbs, J.-P.A. and P.L. Munday. 2004. Intraspecific competition controls spatial distribution and social organisation of the coral-dwelling goby *Gobiodon histrio*. *Marine Ecology Progress Series* 278: 253-259.

Hobbs, J.-P.A., P.L. Munday and G.P. Jones. 2004. Social induction of maturation and sex determination in a coral reef fish. *Proceedings of the Royal Society of London Series B-Biological Sciences* 271: 2109-2114.

Hughes, T.P., A.H. Baird, D.R. Bellwood, M. Card, S.R. Connolly, C. Folke, R. Grosberg, O. Hoegh-Guldberg, J.B.C. Jackson, J.A. Kleypas, J.M. Lough, P. Marshall, M. Nystörm, S.R. Palumbi, J.M. Pandolfi, B. Rosen and J. Roughgarden. 2003. Climate change, human impacts, and the resilience of coral reefs. *Science* 301: 929-933.

Karplus, I. 1987. The association between gobiid fishes and burrowing alpheid shrimps. *Oceanography and Marine Biology* 25: 507-562.

Kirschfeld, K. 1976. The resolution of lens and compound eyes. In: *Neural Principles in Vision*, F. Zettler and R. Weiler (eds.), Springer, Berlin, pp. 354-360.

Kritzer, J.P. 2002. Stock structure, mortality and growth of the decorated goby, *Istigobius decoratus* (Gobiidae), at Lizard Island, Great Barrier Reef. *Environmental Biology of Fishes* 63: 211-216.

Kuwamura, T., Y. Yogo and Y. Nakashima. 1993. Size-assortative monogamy and paternal egg care in a coral goby *Paragobiodon echinocephalus*. *Ethology* 95: 65-75.

Kuwamura, T., Y. Yogo and Y. Nakashima. 1994. Population dynamics of goby *Paragobiodon echinocephalus* and host coral *Stylophora pistillata*. *Marine Ecology Progress Series* 103: 17-23.

Kuwamura, T., Y. Nakashima and Y. Yogo. 1996. Plasticity in size and age at maturity in a monogamous fish: effect of host coral size and frequency dependence. *Behavioral Ecology and Sociobiology* 38: 365-370.

Lachner, E.A. and S.J. Karnella. 1980. Fishes of the Indo-Pacific genus *Eviota* with descriptions of eight new species (Teleostei: Gobiidae). *Smithsonian Contributions to Zoology* 315: 1-127.

Larson, H.K. 1985. A revision of the gobiid genus *Bryaninops* (Pisces), with a description of six new species. *The Beagle* 2: 57-93.

Larson, H.K. 1990. A revision of the commensal gobiid fish genera *Pleurosicya* and *Luposicya* (Gobiidae), with a description of eight new species of *Pleurosicya* and discussion of related genera. *The Beagle* 7: 1-53.

Larson, H.K. and E.O. Murdy. 2001. Gobiidae. In: *FAO Species Identification Guide for Fisheries Purposes. The Living Marine Resources of the Western Central Pacific. Volume 6. Bony Fishes Part 4 (Labridae to Latimeriidae), Estuarine Crocodiles, Sea Turtles, Sea Snakes and Marine Mammals*, K.E. Carpenter and V.H. Niem (eds.), FAO, Rome, pp. 3578-3603.

Lassig, B.R. 1981. Significance of the epidermal ichthyotoxic secretion of coral-dwelling gobies. *Toxicon* 19: 729-735.

Luckhurst, B.E. and K. Luckhurst. 1977. Recruitment patterns of coral reef fishes on the fringing reefs of Curacao, Netherlands Antilles. *Canadian Journal of Zoology* 55: 681-689.

Manabe, H., M. Ishimura, A. Shinomiya and T. Sunobe. 2007a. Field evidence for bi-directional sex change in the polygynous gobiid fish *Trimma okinawae*. *Journal of Fish Biology* 70: 600-609.

Manabe, H., M. Ishimura, A. Shinomiya and T. Sunobe. 2007b. Inter-group movement of females of the polygynous gobiid fish *Trimma okinawae* in relation to timing of protogynous sex change. *Journal of Ethology* 25: 133-137.

Manabe, H., K. Hagiwara, A. Yonemori, K. Fujiwara and A. Shinomiya. 2009. Semi-lunar spawning cycle and mating tactics in the marine goby *Asterropteryx semipunctata*. *Ichthyological Research* 56: 92-95.

McCormick, M.I. and J.K. Larson. 2007. Field verification of the use of chemical alarm cues in a coral reef fish. *Coral Reefs* 26: 571-576.

Meekan, M.G., J.L. Ackerman and G.M. Wellington. 2001. Demography and age structures of coral reef damselfishes in the tropical eastern Pacific Ocean. *Marine Ecology Progress Series* 212: 223-232.

Miller, P.J. 1979. Adaptiveness and implications of small size in teleosts. Symposia of the Zoological Society of London. pp. 263-306.

Miller, P.J. 1996. The functional ecology of small fish: some opportunities and consequences. In: *Miniature Vertebrates: The Implication of Small Body Size*, P.J. Miller (ed.), The Zoological Society of London, Oxford, pp. 175-199.

Munday, P.L. 2000. Interactions between habitat use and patterns of abundance in coral-dwelling fishes of the genus *Gobiodon*. *Environmental Biology of Fishes* 58: 355-369.

Munday, P.L. 2001. Fitness consequences of habitat use and competition among coral-dwelling fishes. *Oecologia* 128: 585-593.

Munday, P.L. 2002. Does habitat availability determine geographical-scale abundance of coral-dwelling fishes? *Coral Reefs* 21: 105-116.

Munday, P.L. 2004a. Competitive coexistence of coral-dwelling fishes: the lottery hypothesis revisited. *Ecology* 85: 623-628.

Munday, P.L. 2004b. Habitat loss, resource specialization, and extinction on coral reefs. *Global Change Biology* 10: 1642-1647.

Munday, P.L. and G.P. Jones. 1998. The ecological implications of small body size among coral-reef fishes. *Oceanography and Marine Biology: An Annual Review* 36: 373-411.

Munday, P.L., G.P. Jones and M.J. Caley. 1997. Habitat specialisation and the distribution and abundance of coral-dwelling gobies. *Marine Ecology Progress Series* 152: 227-239.

Munday, P.L., M.J. Caley and G.P. Jones. 1998. Bi-directional sex change in a coral-dwelling goby. *Behavioral Ecology and Sociobiology* 43: 371-377.

Munday, P.L., A.S. Harold and R. Winterbottom. 1999. Guide to coral-dwelling gobies, genus *Gobiodon* (Gobiidae), from Papua New Guinea and the Great Barrier Reef. *Revue Francaise d'Aquariologie* 26: 53-58.

Munday, P.L., G.P. Jones and M.J. Caley. 2001. Interspecific competition and coexistence in a guild of coral-dwelling fishes. *Ecology* 82: 2177-2189.

Munday, P.L., L. van Herwerden and C.L. Dudgeon. 2004. Evidence for sympatric speciation by host shift in the sea. *Current Biology* 14: 1498-1504.

Munday, P.L., P.M. Buston and R.R. Warner. 2006a. Diversity and flexibility of sex-change strategies in animals. *Trends in Ecology and Evolution* 21: 89-95.

Munday, P.L., A.M. Cardoni and C. Syms. 2006b. Cooperative growth regulation in coral-dwelling fishes. *Biology Letters* 2: 355-358.

Munday, P.L., S.J. Pierce, G.P. Jones and H.K. Larson. 2002. Habitat use, social organization and reproductive biology of the seawhip goby, *Bryaninops yongei*. *Marine and Freshwater Research* 53: 769-775.

Munday, P.L., M. Schubert, J.A. Baggio, G.P. Jones, M.J. Caley and A.S. Grutter. 2003. Skin toxins and external parasitism of coral-dwelling gobies. *Journal of Fish Biology* 62: 976-981.

Munday, P.L., T. Kuwamura and F.J. Kroon. 2010. Bi-directional sex change in marine fishes. In: *Reproduction and Sexuality in Marine Fishes: Patterns and Processes,* K.S. Cole (ed.). University of California Press, Berkeley, pp. 241-271.

Nakamura, Y. and M. Sano. 2004. Overlaps in habitat use of fishes between a seagrass bed and adjacent coral and sand areas at Amitori Bay, Iriomote Island, Japan: Importance of the seagrass bed as juvenile habitat. *Fisheries Science* 70: 788-803.

Nakashima, Y., T. Kuwamura and Y. Yogo. 1996. Both-ways sex change in monogamous coral gobies, *Gobiodon* spp. *Environmental Biology of Fishes* 46: 281-288.

Nelson, J.S. 2006. *Fishes of the World.* John Wiley and Sons, Inc., Hoboken. 4th edition

Nilsson, G.E., J.-P.A. Hobbs, P.L. Munday and S. Östlund-Nilsson. 2004. Coward or braveheart: extreme habitat fidelity through hypoxia tolerance in a coral-dwelling goby. *Journal of Experimental Biology* 207: 33-39.

Nilsson, G.E., J.P.A. Hobbs, S. Ostlund-Nilsson and P.L. Munday. 2007. Hypoxia tolerance and air-breathing ability correlate with habitat preference in coral-dwelling fishes. *Coral Reefs* 26: 241-248.

Nogawa, Y. and H. Endo. 2007. A new species of the genus *Priolepis* (Perciformes: Gobiidae) from Tosa Bay, Japan. *Bulletin of the National Museum of Natural Sciences, Series A Suppl.* 1: 153-161.

Norris, J.E. and J.D. Parrish. 1988. Predator-prey relationships among fishes in pristine coral reef communities. Proceedings of the 6th International Coral Reef Symposium. Townsville, Australia, pp. 107-113.

Parrish, J.D., J.E. Norris, M.W. Callahan, J.K. Callahan, E.J. Magarifuji and R.E. Schroeder. 1986. Piscivory in a coral reef fish community. In: *Contemporary Studies on Fish Feeding,* C.A. Simenstad and G.M. Cailliet (eds.), Dr. W. Junk Publishers, Dordrecht, pp. 285-297.

Partridge, L. and P.H. Harvey. 1988. The ecological context of life history evolution. *Science* 241: 1449-1455.

Petrie, M. 2008. *Relationships between coral characteristics and the ecological performance of coral associated fishes.* Honours thesis, James Cook University. Townsville.

Polunin, N.V.C. and R. Lubbock. 1977. Prawn-associated gobies (Teleostei: Gobiidae) from the Seychelles, western Indian Ocean: systematics and ecology. *Journal of Zoology (London)* 183 (pt 1): 63-101.

Pratchett, M.S., P.L. Munday, S.K. Wilson, N.A.J. Graham, J.E. Cinner, D.R. Bellwood, G.P. Jones, N.V.C. Polunin and T.R. Mcclanahan. 2008. Effects of climate-induced coral bleaching on coral-reef fishes — ecological and economic consequences. *Oceanography and Marine Biology: An Annual Review* 46: 251-296.

Reavis, R.H. 1997a. The natural history of a monogamous coral-reef fish, *Valenciennea strigata* (Gobiidae): 1. abundance, growth, survival and predation. *Environmental Biology of Fishes* 49: 239-246.

Reavis, R.H. 1997b. The natural history of a monogamous coral-reef fish, *Valenciennea strigata* (Gobiidae): 2. behavior, mate fidelity and reproductive success. *Environmental Biology of Fishes* 49: 247-257.

Reavis, R.H. and G.W. Barlow. 1998. Why is the coral-reef fish *Valenciennea strigata* (Gobiidae) monogamous? *Behavioral Ecology and Sociobiology* 43: 229-237.

Riedlecker, E. and J. Herler. 2008. Trophic morphology of the coral-associated genus *Gobiodon* (Teleostei: Gobiidae) from the Red Sea. *Journal of Zoological Systematics and Evolutionary Research* 47: 160-170.

Rocha, L.A., I.L. Rosa and B.M. Feitoza. 2000. Sponge-dwelling fishes of northeastern Brazil. *Environmental Biology of Fishes* 59: 453-458.

Saeki, T., Y. Sakai, H. Hashimoto and K. Gushima. 2005. Foraging behavior and diet composition of *Trimma caudomaculata* and *Trimma caesiura* (Gobiidae) on coral reefs of Okinawa, Japan. *Ichthyological Research* 52: 302-305.

Schiemer, L., S. Niedermüller and J. Herler. 2009. The influence of colony size and coral health on the occupation of coral-associated gobies (Pisces: Gobiidae). *Coral Reefs* 28: 137-142.

Schubert, M., P.L. Munday, M.J. Caley, G.P. Jones and L.E. Llewellyn. 2003. The toxicity of skin secretions from coral-dwelling gobies and their potential role as a predator deterrent. *Environmental Biology of Fishes* 67: 359-367.

Shafer, D.J. 1998. *Early life history, growth and settlement dynamics of a tropical reef fish (Gobiidae: Bathygobius coalitus).* Doctoral thesis, University of Hawaii, Manoa.

St. Mary, C.M. 1996. Sex allocation in a simultaneous hermaphrodite, the zebra goby *Lythrypnus zebra*: insights gained through a comparison with its sympatric congener, *Lythrypnus dalli. Environmental Biology of Fishes* 45: 177-190.

St. Mary, C.M. 2000. Sex allocation in *Lythrypnus* (Gobiidae): variations on a hermaphroditic theme. *Environmental Biology of Fishes* 58: 321-333.

Stearns, S.C. 1992. *The Evolution of Life Histories.* Oxford University Press, Oxford.

Steele, M.A. and G.E. Forrester. 2002. Early post-settlement predation on three reef fishes: effects on spatial patterns of recruitment. *Ecology* 83: 1076-1091.

Sunobe, T. and A. Nakazono. 1990. Polygynous mating system of *Trimma okinawae* (Pisces, Gobiidae) at Kagoshima, Japan with a note on sex change. *Ethology* 84: 133-143.

Sunobe, T. and A. Nakazono. 1993. Sex change in both directions by alteration of social dominance in *Trimma okinawae* (Pisces, Gobiidae). *Ethology* 94: 339-345.

Sunobe, T. and A. Nakazono. 1999a. Alternative mating tactics in the gobiid fish, *Eviota prasina. Ichthyological Research* 46: 212-215.

Sunobe, T. and A. Nakazono. 1999b. Mating system and hermaphroditism in the gobiid fish, *Priolepis cincta*, at Kagoshima, Japan. *Ichthyological Research* 46: 103-105.

Suzuki, T. and H. Senou. 2007. Two new species of the gobiid fish genus *Trimma* (Perciformes: Gobioidei) from southern Japan. *Bulletin of the National Museum of Natural Sciences, Series A Suppl.* 1: 175-184.

Sweatman, H.P.A. 1984. A field study of the predatory behaviour and feeding rate of a piscivorous coral reef fish, the lizardfish *Synodus englemani*. *Copeia* 1984: 187-194.

Syms, C. and G.P. Jones. 2004. Habitat structure, disturbance and the composition of sand-dwelling goby assemblages in a coral reef lagoon. *Marine Ecology Progress Series* 268: 221-230.

Takegaki, T. 2000. Monogamous mating system and spawning cycle in the gobiid fish, *Amblygobius phalaena* (Gobiidae). *Environmental Biology of Fishes* 59: 61-67.

Takegaki, T. and A. Nakazono. 1999. Reproductive behavior and mate fidelity in the monogamous goby, *Valenciennea longipinnis*. *Ichthyological Research* 46: 115-123.

Taylor, M.S. and M.E. Hellberg. 2003. Genetic evidence for local retention of pelagic larvae in a Caribbean reef fish. *Science* 299: 107-109.

Taylor, M.S. and J.L. Van Tassell. 2002. Observations on microhabitat utilization by three widely distributed neotropical gobies of the genus *Elacatinus*. *Copeia* 2002: 1134-1136.

Taylor, M.S. and M.E. Hellberg. 2005. Marine radiations at small geographic scales: speciation in neotropical reef gobies (*Elacatinus*). *Evolution* 59: 374-385.

Thompson, V.J., P.L. Munday and G.P. Jones. 2007. Habitat patch size and mating system as determinants of social group size in coral-dwelling fishes. *Coral Reefs* 26: 165-174.

Thresher, R.E. 1984. *Reproduction in Reef Fishes*. T.H.F. Publications, Hong Kong.

Victor, B.C. 2007. *Coryphopterus kuna*, a new goby (Perciformes: Gobiidae: Gobiinae) from the western Caribbean, with the identification of the late larval stage and an estimate of the pelagic larval duration. *Zootaxa*: 51-61.

Wall, M. and J. Herler. 2008. Post-settlement movement patterns and homing in a coral-associated fish. *Behavioral Ecology* 20: 87-95.

Whiteman, E.A. and I.M. Cote. 2004. Monogamy in marine fishes. *Biological Reviews* 79: 351-375.

Wilkinson, C. 2008. *Status of Coral Reefs of the World: 2008*. Global Coral Reef Monitoring Network and Reef and Rainforest Research Centre, Townsville, Australia.

Winterbottom, R. 1990. The *Trimmatom nanus* species complex (Actinopterygii, Gobiidae) — phylogeny and progenetic heterochrony. *Systematic Zoology* 39: 253-265.

Winterbottom, R. 2004. Three new species of *Trimma* (Pisces: Gobiidae) from the central, western and south Pacific. *Aqua, Journal of Ichthyology and Aquatic Biology* 9: 7-16.

Winterbottom, R. 2006. Two new species of the gobiid fish *Trimma* from the coral reefs of the western Pacific Ocean (Pisces; Perciformes; Gobioidei). *Zootaxa*: 55-68.

Winterbottom, R. and A.S. Harold. 2005. *Gobiodon prolixus*, a new species of gobiid fish (Teleostei: Perciformes: Gobiidae) from the Indo-West Pacific. *Proceedings of the Biological Society of Washington* 118: 582-589.

Winterbottom, R. and M. Zur. 2007. Three new species of the genus *Trimma* (Percomorpha: Gobiidae) from Palau, Western Pacific (Percomorpha: Gobiidae). *Aqua, International Journal of Ichthyology* 13: 13-24.

Winterbottom, R. and L. Southcott. 2008. Short lifespan and high mortality in the western Pacific coral reef goby *Trimma nasa*. *Marine Ecology Progress Series* 366: 203-208.

Wong, M.Y.L., P.M. Buston, P.L. Munday and G.P. Jones. 2007. The threat of punishment enforces peaceful cooperation and stabilizes queues in a coral-reef fish. *Proceedings of the Royal Society of London Series B-Biological Sciences* 274: 1093-1099.

Wong, M.Y.L., P.L. Munday, P.M. Buston and G.P. Jones. 2008a. Fasting or feasting in a fish social hierarchy. *Current Biology* 18: R372-R373.

Wong, M.Y.L., P.L. Munday, P.M. Buston and G.P. Jones. 2008b. Monogamy when there is potential for polygyny: tests of multiple hypotheses in a group-living fish. *Behavioral Ecology* 19: 353-361.

Worthington, D.G., P.J. Doherty and A.J. Fowler. 1995. Variation in the relationship between otolith weight and age: implications for the estimation of age of two tropical damselfish (*Pomacentrus moluccensis* and *P. wardi*). *Canadian Journal of Fisheries and Aquatic Sciences* 52: 233-242.

Yanagisawa, Y. 1982. Social behaviour and mating system of the gobiid fish *Amblyeleotris japonica*. *Japanese Journal of Ichthyology* 28: 401-422.

CHAPTER

4.3

Gobies as Cleaners

Isabelle M. Côté[1] and Marta C. Soares[2]

INTRODUCTION

In 1928, the renowned American ocean explorer and naturalist Charles William Beebe published his 12th book, *Beneath Tropic Seas: A Record of Diving among the Coral Reefs of Haiti.* The best-selling book relates in eclectic detail the first American expedition to explore the underwater world of the Caribbean Sea. Equipped with helmet, hose and pump, Beebe observed a wide variety of brilliantly coloured fish, corals and sponges. His observations, made in passing, of small fish picking at the mouth of parrotfish provide the earliest written record of cleaning behaviour. Cleaning interactions, in which a small fish or shrimp removes ectoparasites and other materials from the body surface of larger-bodied clients, have since proven to be widespread in the marine realm and have been the focus of intense study by behavioural and evolutionary ecologists over the past two decades (Côté, 2000). A number of cleanerfish species are found in the

Authors' addresses: [1]Department of Biological Sciences, Simon Fraser University, Burnaby BC, V5A 1S6, Canada. E-mail: imcote@sfu.ca

[2]Eco-Ethology Research Unit, Instituto Superior de Psicologia Aplicada, Rua Jardim do Tabaco 34, 1149-041 Lisboa, Portugal and Department of Zoology, University of Neuchâtel, Emile-Argand 11, 2009 Neuchâtel, Switzerland. E-mail: msoares@ispa.pt

Caribbean, but the object of William Beebe's attention were young wrasses. This is surprising because gobies, and not wrasses, are the most ubiquitous and only group of obligate cleanerfish in the region.

In this chapter, we focus on gobies as cleanerfish. The chapter is divided into three parts. First, we begin by examining the distribution of cleaning behaviour in the family Gobiidae, make inferences about the number of independent evolutionary occurrences of this behaviour, and describe the evolutionary correlates of cleaning in gobies. Second, we review current knowledge of the costs and benefits of cleaning by gobies and how the use of market and game theory has increased our understanding of how cleaning gobies interact with their fish clients. Finally, we examine the role cleaning gobies play in the fish communities in which they exist. Throughout, we draw parallels with cleaner wrasses—especially *Labroides dimidiatus*, the most ubiquitous cleaners in the Indo-Pacific and a model for the development of many ideas about the evolution of cleaning behaviour.

EVOLUTION OF CLEANING IN GOBIES

Who Cleans in The Goby Family?

Fourteen species of gobies are currently known to clean (Table 4.3.1). All cleaning gobies are restricted to the genus *Elacatinus* Jordan (formerly *Gobiosoma*), one of the largest neotropical genera of gobies. Within this genus, the number of cleaning species is distributed unevenly between the two recognized subgenera *Elacatinus* (*sensu stricto*) and *Trigrigobius* Fowler (Eschmeyer, 1998). The subgenus *Elacatinus* comprises 18 species, of which 10 clean. All but one (*E. puncticulatus*) are found in the western Atlantic. The subgenus *Tigrigobius* is composed of 13 species, of which only four clean. All are found in the eastern Pacific.

Evolutionary Correlates of Cleaning in Gobies

Taylor and Hellberg (2005) recently constructed a molecular phylogeny of 31 taxa in the *Elacatinus* group which provides a framework for exploring the evolutionary history and correlates of cleaning in this group. The overall phylogeny suggests the existence of three major clades: *Elacatinus* (*sensu stricto*) (13 Atlantic and 1 Pacific species), *Tigrigobius* (3 Atlantic species, none of which cleans) and *Risor* (5 Atlantic and 5 Pacific species) (Fig. 4.3.1). In the two clades with cleaning species, there is evidence of associations between cleaning, habitat, morphology and colour.

Table 4.3.1 Cleaning gobies *Elacatinus* spp. reported as cleaners in the wild (partially adapted from Côté, 2000). Common names, distribution and maximum length obtained from Fishbase (www.fishbase.org). TL = Total length, SL = Standard length. References pertain to information on reliance of species on cleaning as foraging mode. ? indicates uncertainty in assessing information on reliance on cleaning.

Subgenus	*Species*	*Authority*	*Common name*	*Distribution*	*Maximum size (cm)*	*Reliance on cleaning*	*Mouth position*	*Reference*
Elacatinus	*E. evelynae*	(Böhlke and Robins, 1968)	Sharknose goby	Atlantic	4	Obligate	Subterminal	(Losey, 1974) (Colin, 1975)
	E. figaro	Sazima, Moura and Rosa, 1997	Barber goby	Atlantic	4.5	Facultative?	Terminal	(Sazima, 1997) (Rocha *et al.*, 2000)
	E. genie	(Böhlke and Robins, 1968)	Cleaner goby	Atlantic	4.5	Obligate	Subterminal	(Colin, 1975)
	E. illecebrosus	(Böhlke and Robins, 1968)	Barsnout goby	Atlantic	4	Obligate	Subterminal	(Colin, 1975)
	E. oceanops	Jordan, 1904	Neon goby	Atlantic	5	Obligate	Subterminal	(Eibl-Eibesfeldt, 1955) (Colin, 1975)
	E. phthirophagus	Sazima, Carvalho-Filho and Sazima, 2008		Atlantic	3.2	Obligate?	Subterminal	(Sazima *et al.*, 2008)
	E. pridisi	Guimarães, Gasparini and Rocha, 2004	Trindade cleaner goby	Atlantic	2.8	Obligate	Subterminal	(Guimaraes *et al.*,2004)
	E. prochilos	(Böhlke and Robins, 1968)	Broadstripe goby	Atlantic	4	Obligate	Terminal	(Darcy *et al.*, 1974) (Colin, 1975) (Arnal and Côté, 2000)

Contd...

Contd...

	E. randalli	(Böhlke and Robbins, 1968)	Yellownose goby	Atlantic	4.6 TL	Obligate?	Terminal	(Colin, 1975)
	E. puncticulatus	(Ginsburg, 1938)		Pacific	4.4 SL	Facultative?	Terminal	(Michael, 1993), in (Sazima and Moura, 2000)
Tigrigobius	*E. digueti*	(Pellegrin, 1901)	Banded cleaner goby	Pacific	3.2 SL	Facultative	Terminal	(Hobson, 1968) (Colin, 1975; Allen and Robertson, 1994)
	E. inornatus	Bussing, 1990	Inornate goby	Pacific	3.3 SL	Facultative	Terminal	(Allen and Robertson, 1994)
	E. limbaughi	Hoese and Reader, 1991	Widebanded cleaning goby	Pacific	-	Facultative	Terminal	(Allen and Robertson, 1994)
	E. nesiotes	Bussing, 1990	-	Pacific	2.3 SL	Facultative?	Terminal	(Grove and Lavenberg, 1997)

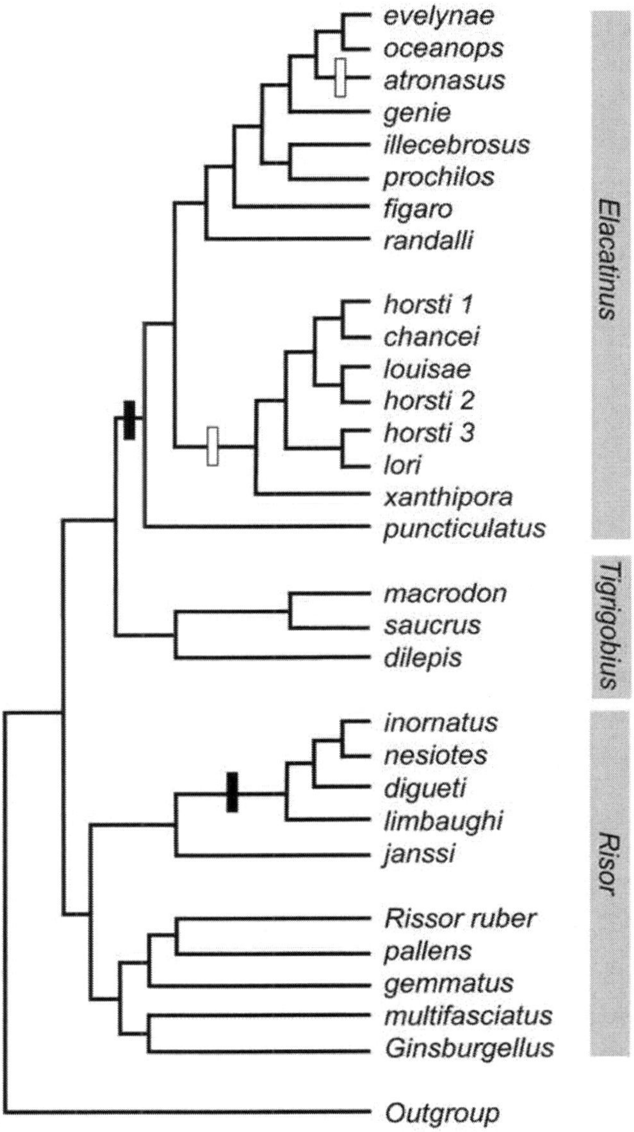

Fig. 4.3.1 Maximum likelihood phylogeny of *Elacatinus* gobies based on three independent molecular markers. Black bars represent possible evolutionary origins of cleaning behaviour. White bars represent possible losses of cleaning behaviour. Redrawn and modified from Taylor and Hellberg (2005).

The most parsimonious scenario to explain the distribution of cleaning in gobies is that this behaviour has evolved independently twice in this group (see also Rüber *et al.*, 2003). One evolutionary transition probably occurred following the divergence of the *Elacatinus* and *Tigrigobius* clades (Fig. 4.3.1), explaining the occurrence of cleaning in the Pacific species *E. puncticulatus* and the Atlantic species *E. evelynae*, *E. oceanops*, *E. genie*, *E. illecebrosus*, *E. prochilos*, *E. figaro* and *E. randalli*. Cleaning behaviour was then lost in *E. atronasus*, perhaps with the adoption of a schooling, non-benthic habit (Taylor and Akins, 2007), and in the *horsti* group (Rüber *et al.*, 2003). It is likely that cleaning in the two newly described Brazilian species *E. phthirophagus* (Sazima *et al.*, 2008) and *E. pridisi* (Guimaraes *et al.*, 2004) also arose in the same event. A second origin of cleaning must have occurred some time after the separation of the Atlantic and Pacific species of the *Risor* clade (Fig. 4.3.1), which explains the presence of cleaning in four of the five species in this group. The fifth, *Elacatinus janssi*, does not appear to clean (although little is known about this species).

In the *Elacatinus* clade, the absence of cleaning appears to be associated with a sponge-dwelling habit or, conversely, the presence of cleaning is associated with living on substrata other than sponge (Rüber *et al.*, 2003; Taylor and Hellberg, 2005). This alternative substratum is often, but not always, live coral (e.g., Sazima *et al.*, 2000; Whiteman and Côté, 2002; Sazima *et al.*, 2008). Interestingly, this habitat–feeding mode link also occurs intra-specifically. For example, in Barbados, the broadstripe cleaning goby *E. prochilos* occurs on both sponge and live coral (Whiteman and Côté, 2004a), and its foraging mode and social systems differ between the two substrata. Coral-dwelling *E. prochilos* are active cleaners (Arnal and Côté, 2000; Whiteman and Côté, 2002) that live in singly or in small groups (Whiteman and Côté, 2002) while sponge-dwelling *E. prochilos* occur in large, dominance-structured groups and feed predominantly on polycheate worms that burrow within sponge tissues (Whiteman and Côté 2004b). Variable habitat use has been noted in other species, such as *E. evelynae* (White *et al.*, 2007) and *E. figaro* (Rocha *et al.*, 2000), and in both cases, sponge-dwelling individuals did not clean. Note that there appears to be geographic variation in habitat associations: *E. evelynae* in Barbados occur virtually exclusively on live coral (Whiteman and Côté, 2002) while in the US Virgin Islands, they are found on coral as well as in basket sponges (White *et al.*, 2007).

Habitat associations are also apparent in the *Risor* clade (Fig. 4.3.1). Groups of non-cleaning species in this clade are specifically associated with chiton burrows, sea urchin spines, live coral or sponges (Taylor and

Hellberg, 2005). However, the group in which cleaning has arisen lacks strong habitat specialization. In both the *Elacatinus* and the *Risor* clades, it therefore appears that specialization in feeding mode (i.e., cleaning) has emancipated gobies from habitat specialization, although both types of specialization effectively lead to ecological separation that may have promoted the early stages of diversification in this group (Streelman and Danley, 2003).

Foraging specialization is often accompanied by morphological adaptations. It has been long thought that a 'picker-type', terminally located mouth may pre-adapt fish for cleaning (Hobson, 1971, 1976; McCourt and Thomson, 1984). Most *Elacatinus* (*sensu lato*) gobies do indeed have a terminal mouth; however, this ancestral morphology has been replaced in four species in the *Elacatinus* clade—*E. evelynae*, *E. oceanops*, *E. genie* and *E. illecebrosus*—by a subterminal, inferior mouth. Molecular evidence suggests a polyphyletic origin for this new mouth position (Taylor and Hellberg, 2005). In hindsight, the initial notion that terminal mouths would be well-suited to removing ectoparasites was a sensible one, but only for species that clean in mid-water, such as the damselfishes and wrasses observed by Hobson (1971, 1976) and McCourt and Thomson (1984). For species with a benthic habit, such as gobies which lie on their clients as they would on the substratum, a ventrally located mouth would facilitate cleaning. If so, the question becomes why has an inferior mouth evolved in some but not all cleaning gobies? The retention of a terminal mouth could be linked to flexibility in foraging mode. This could explain mouth morphology in cleaning gobies of the *Risor* clade, which have been described as facultative cleaners (Humann, 1993; Allen and Robertson, 1994), and in *E. prochilos* and *E. figaro*, which are both known to exhibit different foraging modes when occupying different substrata (Rocha *et al.*, 2000; Whiteman and Côté, 2002). However, *E. evelynae* also shows habitat-dependent foraging modes (White *et al.*, 2007) and yet has a subterminal mouth. Some insight into the evolutionary link between feeding morphology and cleaning behaviour could be gained by examining other cleaning taxa (e.g., the family Labridae).

The final potential correlate of cleaning behaviour is colour pattern. The presence of a cleaner signaling coloration was posited long ago by researchers who were struck by the similarity in colour patterns of cleaning species from various parts of the world (Eibl-Eibesfeldt, 1955; Wickler, 1963; Potts, 1973a). Côté (2000) provided the first test of this hypothesis and showed that cleaning gobies have a longer lateral stripe, relative to body length, than their non-cleaning relatives. By contrast, among more varied taxa of Caribbean facultative cleanerfish, there was no evidence that cleaners were more likely to be striped but they tended to be more brightly coloured than non-cleaning congeners. Although this analysis was based on taxonomy, the conclusions were later supported by Arnal *et al.* (2006) who, using a well-resolved molecular phylogeny, found that cleaning behaviour in wrasses is associated with the presence of a dark lateral stripe. In addition, in field experiments using wooden models of cleaner wrasses with and without stripes, the former attracted many more fish clients (Stummer *et al.*, 2004). In the *Elacatinus* clade, there appears to be an association between a subterminal mouth position and blue lateral stripe (Taylor and Hellberg, 2005). Incidentally, the spectral reflectance of the blue stripe in gobies appears to be similar to that of the ubiquitous Indo-Pacific cleaner wrasse *Labroides dimidiatus* (Taylor and Hellberg, 2005), which is distinct from the blue hue of virtually all other coral reef fishes (except for the cleanerfish mimic *Plagiotremus rhynorhinchos*, Cheney *et al.*, 2008). In the *Risor* clade, however, the four cleaning species, as well as their non-cleaning clade-mates are variously banded or spotted rather than stripped. Taken together, these studies suggest that a cleaner 'guild coloration' exists but it may not be universal. Much work remains to understand the exceptions to the stripe rule.

CLEANING BEHAVIOUR OF CLEANING GOBIES

Generic Cleaning Sequence

The behavioural patterns of cleaning gobies and their clients have been described in detail (Limbaugh, 1961; Losey, 1972; Colin, 1975), and these patterns bear similarities with those exhibited by cleaner wrasses (*Labroides* spp.), the well-studied counterparts of cleaning gobies on Indo-Pacific reefs (e.g., Potts, 1973b). Interactions between cleaning gobies and their fish clients, like those of cleaner wrasses, usually occur at cleaning stations—relatively fixed sites on reefs which are often, although not always, located on top or near the base of large coral heads or other prominent features (IMC and MCS, personal observations). Clients approach cleaning stations and adopt a species-specific posture which may entail an immobile head-stand or tail-stand position with all fins spread and the opercula flared (Losey, 1971; Côté *et al.*, 1998). These poses may be accompanied by dramatic colour changes (Feder, 1966) and have been shown to increase the likelihood of being cleaned (Côté *et al.*, 1998). Cleaning gobies may then swim onto and 'inspect' the client, before starting to feed on the client's body, in its mouth or gills, or returning to the station without feeding. Cleaning interactions may also be terminated by clients by twitching to signal their intention to depart and allow the cleaner to return to the station (Feder, 1966).

There are, however, marked differences in behaviour between cleaning gobies and cleaner wrasses. Cleaner wrasses often begin cleaning interactions by performing a 'dance'—a conspicuous, vertical zig-zag swimming pattern which appears to attract clients (Potts, 1973b; Gorlick *et al.*, 1978; Grutter, 2004). Cleaning gobies do not perform such advertising dances (IMC, MCS, personal observations). In addition, during inspections, cleaner wrasses often deliberately stroke the body surface of their clients with their pectoral fins—a behaviour termed 'tactile stimulation' (Bshary and Würth, 2001). The delivery of tactile stimulation appears to influence the decision of clients to remain at cleaning stations, both before cleaning and after cleaners deliver bites that appear painful to clients (Bshary and Würth, 2001). Its preferential use with potentially dangerous clients (e.g., hungry piscivores) suggests that it is also a pre-conflict management strategy (Grutter, 2004). This conspicuous behaviour is absent in cleaning gobies (Soares *et al.*, 2008a). We return to the significance of these behavioural differences later (see section on Insights from Biological Market Theory).

Clients of Cleaning Gobies

Cleaning gobies interact with a large variety of client species (Table 4.3.2). For example, Soares *et al.* (2007) noted a total of 46 different client species serviced by *E. evelynae* and *E. prochilos* on two Barbadian reefs (Table 4.3.2). In this instance, comparisons with cleaner wrasses are not useful, given the much greater diversity of potential client species in the Indo-Pacific. However, gobies do not hold a cleaning monopoly in the Caribbean. Other species of fish clean facultatively, particularly as juveniles, and several species of obligate cleaner shrimp also provide services. Few studies have compared the cleaning activity and client diversity between cleaning gobies and other cleaners within the same location but those that have repeatedly found that gobies attend a higher diversity and greater number of clients (Darcy *et al.*, 1974; Losey, 1974; Wicksten, 1998; Francini-Filho and Sazima, 2008; Table 4.3.2).

It is perhaps the interactions between cleaning gobies and piscivorous species that hold the greatest interest for biologists and lay people alike (Fig. 4.3.2A). The ability of cleaning gobies (and cleaner wrasses in the Indo-Pacific) to enter the mouth of piscivorous clients during inspections, and to re-emerge unscathed, was interpreted early on as an example of altruism

Table 4.3.2 Summary of client diversity of cleaning gobies (*Elacatinus* spp.) and other Caribbean cleaners.

Location	*Cleaning species*	*Number of clients species*	*Number of client families*	*Total observation time*	*Source*
Cleaning gobies					
Barbados	*E. evelynae* and *E. prochilos*	46	21	52 h	(Soares *et al.*, 2007)
Barbados	*E. evelynae* and *E. prochilos*	38	17	83 h	(Soares *et al.*, 2008c)
Barbados	*E. evelynae* and *E. prochilos*	25	14	25 h	(Arnal *et al.*, 2000)
St Croix (US Virgin Islands)	*E. evelynae* and *E. prochilos*	?	6	?	(Darcy *et al.*, 1974)
Barbados	*E. evelynae*	34	16	114 h	(Whiteman and Côté, 2002)
St Croix (US Virgin Islands)	*E. evelynae*	?	4	6.8 h	(White *et al.*, 2007)
St Croix (US Virgin Islands)	*E. evelynae*	25	14	?	(Johnson and Ruben, 1988)
Barbados	*E. prochilos*	28	14	78 h	(Whiteman and Côté, 2002)
Brazil	*E. figaro*	34	16	53 h	(Sazima *et al.*, 2000)
Brazil	*E. randalli*	22	13	~14 h	(Francini-Filho and Sazima, 2008)
Panama	*E. illecebrosus*	9	3	?	(Bohlke and McCosker, 1973)
Bonaire	*Elacatinus* spp.	43	19	?	(Wicksten, 1998)
Other cleanerfish					
Bonaire	*Thalassoma bifasciatum*	16	10	?	(Wicksten, 1998)
St Croix (US Virgin Islands)	*Thalassoma bifasciatum*	10	4	?	(Johnson and Ruben, 1988)
St Croix (US Virgin Islands)	*Thalassoma bifasciatum*	?	4	?	(Darcy *et al.*, 1974)

Contd...

Contd...

Brazil	*Thalassoma noronhanum*	19	11	~ 14 h	(Francini-Filho and Sazima, 2008)
Brazil	*Thalassoma noronhanum*	19	9	38 h	(Francihi-Filho *et al.*, 2000)
Bonaire	*Bodianus rufus*	12	9	?	(Wicksten, 1998)
St Croix (US Virgin Islands)	*Bodianus rufus*	15	9	?	(Johnson and Ruben, 1988)
Bonaire	*Pomacanthus paru*	2	2	?	(Wicksten, 1998)
Brazil	*Pomacanthus paru*	31	19	19 h	(Sazima *et al.*, 1999)
Cleaner shrimp					
Bonaire	*Periclimenes pedersoni*	28	12	?	(Wicksten, 1998)
St Croix (US Virgin Islands)	*Periclimenes pedersoni*	10	5	?	(Johnson and Ruben, 1988)
Bemuda	*Periclimenes anthophilus*	10	7	30 h	(Sargent and Wagenbach, 1975)
Bonaire	*Stenopus hispidus*	?	4 (3)	?	(Wicksten, 1998)

Fig. 4.3.2 A cleaning goby *Elacatinus* sp. inspecting the buccal area of a tiger grouper *Mycteroperca tigris* on a Little Cayman reef. Photo: Isabelle Côté. B Gnathiid isopod larva—the most ubiquitous ectoparasite found in stomach contents of cleaning gobies. Photo courtesy of Paul Sikkel.

Color image of this figure appears in the color plate section at the end of the book.

on the part of clients which could easily eat the cleaner (e.g., Trivers, 1971). An untested alternative explanation is that cleaning gobies are unpalatable (Colin, 1975). Nevertheless, cleaning gobies do inspect potentially dangerous clients regularly (Darcy *et al.*, 1974; Johnson and Ruben, 1988; Wicksten, 1998; Arnal *et al.*, 2000; Sazima and Moura, 2000; Sazima *et al.*, 2000; Soares *et al.*, 2007), and more often than cleaner shrimp or other non-obligate cleanerfish on Caribbean reefs (Losey, 1974; Wicksten, 1998). Wicksten (1998), for example, reported that 63% of groupers (Serranidae) seen at cleaning stations in Bonaire were cleaned by *Elacatinus* spp. Gobies are also the only Caribbean cleaners to have been observed interacting with elasmobranchs: whitetip reef sharks *Triaenodon obesus* and nurse sharks *Ginglymostoma cirratum* are cleaned by *E. puncticulatus*, Caribbean reef sharks *Carcharhinus perezi* by *E. phthirophagus* (Michael, 1993; Sazima *et al.*, 2008), large-eyed stingrays *Dasyatis macrophthalma* by *E. figaro* and Southern stingrays *D. americana* by *E. puncticulatus* (Sazima and Moura, 2000).

Cleaning gobies show distinct preferences towards some client species over others. They express these biases by cleaning a greater proportion of all visiting individuals of some species (Soares *et al.*, 2007), spending more time inspecting these species (Arnal *et al.*, 2000) and by attending to preferred visitors most quickly (Soares *et al.*, 2007). Ectoparasite load is a relatively good predictor of whether a visitor will be inspected or not (Soares *et al.*, 2007). However, swiftness of service is best explained by client trophic level: piscivorous clients are not made to wait (Soares *et al.*, 2007). This preference appears to be explained by the negative effect of the presence of piscivores on the abundance of other clients at cleaning stations. Hence, the quicker a piscivore is cleaned, the sooner it departs and other clients of cleaning gobies can return.

The Adaptive Significance of Cleaning

A function for cleaning was first suggested by the discovery of parasites in the gut of cleaner wrasses by Randall (1955, 1958). Soon after, a cleaner-removal experiment by Limbaugh (1961) from two Bahamian reefs resulted in dramatic immigration of fish from cleaning goby-free areas as well as a marked increase in disease in those individuals that remained. This apparently crucial role played by cleaning gobies in maintaining the health of their clients, combined with evidence that cleaners actually feed on ectoparasites, implied that cleaning was a co-operative interaction occurring between cleaner and client. Cleaning symbioses were viewed from then on as textbook examples of mutualistic interactions (Trivers, 1971; McFarland, 1985; Begon *et al.*, 1990; Thompson, 1994).

It has taken a surprisingly long time for researchers to verify this long-held perception. However, it is now agreed that the interactions between cleaner wrasses and their clients are largely mutualistic, on the basis of the significant decrease in client parasite loads effected by cleaner wrasses (e.g., Grutter, 1999) and the large effects of on client fish density and community structure resulting from cleaner removal (Bshary, 2003; Grutter *et al.*, 2003).

By contrast, the evidence has not been as strongly forthcoming for cleaning gobies.

Cleaning gobies do remove ectoparasites from the fish they inspect (Fig. 4.3.3). The identifiable diet of coral-dwelling *Elacatinus prochilos*, for example, is composed largely of larval gnathiid isopods (Whiteman and Côté, 2002; Fig. 4.3.2B), the commonest ectoparasites found in the stomach contents of cleanerfish worldwide. Client fish tend to visit cleaning stations more in the morning (Côté and Molloy, 2003; Sikkel *et al.*, 2004), which coincides with periods of peak ectoparasite infestation (Chambers and Sikkel, 2002; Sikkel *et al.*, 2006). There is also evidence that reef fish with access to cleaning gobies (*E. evelynae* and *E. prochilos*) have lower parasites loads. For example, Barbadian longfin damselfish *Stegastes diencaeus* that visit these cleaning gobies often, by virtue of having a cleaning station within their territories, have significantly fewer ectoparasites than damselfish that live further away from cleaners and hence visit them less frequently (Cheney and Côté, 2001). This difference is not simply due to differences in damselfish quality because a similar effect was observed at the end of a 6-month experiment in which half of the damselfish 'lost' the cleaning gobies that were originally in their territories (Cheney and Côté, 2003a). However, in both cases, the difference in ectoparasite load was small—approximately 1 gnathiid per damselfish—and such a small reduction did not result in a detectable decline in damselfish breeding output in Barbados (Cheney and Côté, 2003a).

The apparently small benefit of being cleaned by gobies may be biased by the fact that these studies have all been conducted in Barbados, where ectoparasite abundance is generally very low. In other parts of the Caribbean (e.g., in Puerto Rico) where there are more ectoparasites, the benefit of being cleaned by gobies can be twice as high (Cheney and Côté, 2005), but this still amounts to only two 1-mm long gnathiids per fish (Fig. 4.3.2B). Nevertheless, if gnathiids are an important source of skin irritation or vectors for blood diseases (e.g., Davies *et al.*, 1994), then the removal of a few parasites may have an appreciable impact on host fitness and health.

Geographic variation in benefit of being cleaned by gobies should be accompanied by variation in willingness of clients to incur costs of seeking cleaners. The costs of being cleaned, for clients, include travelling time to cleaning stations, risk of predation during travel, and resources lost from temporarily abandoned territories. Most of these are linked to travel distance (Cheney and Côté, 2001). Thus, in areas where ectoparasites are abundant and the scope of a benefit of being cleaned is high, clients should be more willing to travel long distances to be cleaned. Conversely, when the

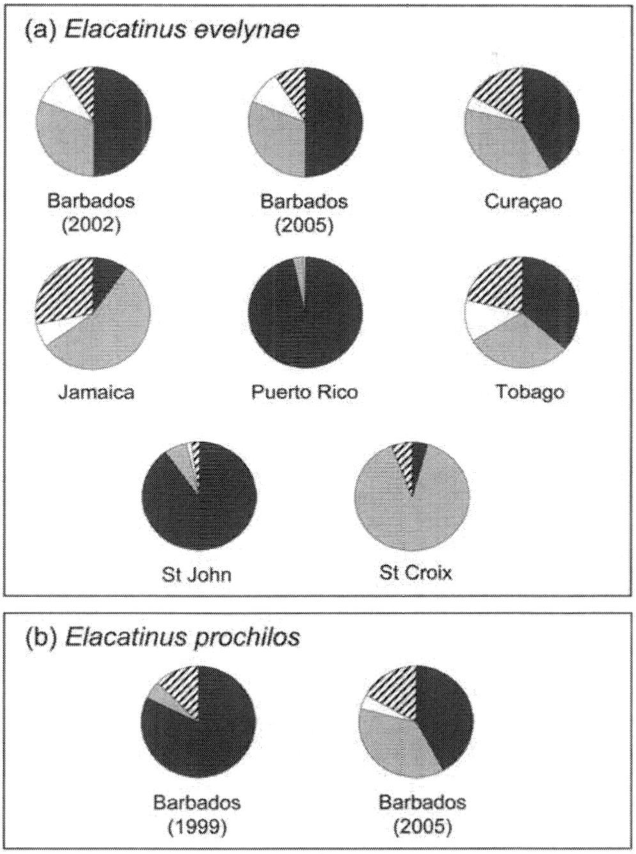

Fig. 4.3.3 Diet of cleaning gobies at various locations across the Caribbean. (a) *Elacatinus evelynae* and (b) *E. prochilos*. Dietary items shown are ectoparasites (black), fish scales (grey), fish mucus (white), and other items (hatched), which include non-parasitic, client-gleaned crustaceans as well as benthic invertebrates. Proportions are expressed as % of identified stomach contents. Sample sizes vary from 12 to 117 fish. Sources: *E. evelynae* and *E. prochilos* for Barbados (2005): M.C. Soares, unpublished data; *E. evelynae* for St Croix: (White *et al.*, 2007); *E. prochilos* for Barbados (1999): (Arnal and Côté, 2000); all others: (Cheney and Côté, 2005).

potential benefit of being cleaned is low, clients should only visit cleaners that are nearby. In this way, clients can attempt to tip the balance sheet, however slightly, in favour of a mutualistic relationship.

This careful accounting by clients can be thrown into disarray by cheating cleaners. It has long been known that cleanerfish remove not only parasites but mucus and scales from their clients (e.g., Gorlick, 1980). Cleaning gobies are no exception (Whiteman and Côté, 2002; Cheney and Côté, 2005) (Fig. 4.3.3). Dishonesty by gobies is clearly evidenced by the content of their stomachs, but also by the behaviour of their clients which jolt or shudder in response to cheating bites (Soares *et al.*, 2008b). Cleaning gobies take more scales and mucus at locations where ectoparasites are rare (Bansemer *et al.*, 2002; Cheney and Côté, 2005), which makes it sometimes difficult for cleaning relationships to remain mutualistic from the client's point of view. However, because ectoparasite abundance varies temporally, the outcome of cleaning at any locale will vary periodically from positive to negative, preventing the long-term breakdown of these interactions.

The issue of cheating is worth further examination because how clients deal with dishonest cleaners differs strikingly between cleaning gobies and cleaner wrasses. The clients of cleaner wrasses fall into two broad categories: those that live in small home ranges and have no choice of which cleaning station they visit, and those with larger home ranges which could take their custom to a number of different cleaners. Cleaner wrasses generally give priority of access to the latter class (Bshary and Schaffer, 2002; Bshary and Grutter, 2002a). These different classes of clients also retaliate to cheating by cleaner wrasses in different ways. Resident fish, which lack choice options, aggressively chase cheating cleaners after a bite, whereas transient fish with choice options simply swim away and withhold further visits for a while (Bshary and Grutter, 2002a). These retaliatory tactics result in a greater likelihood of cleaner honesty in subsequent interactions (Bshary and Grutter, 2002a, 2005). By contrast, the clients of cleaning gobies never chase cheating cleaners and do not take longer to return to a cleaner that had previously given a poorer service than to one that was honest (Soares *et al.*, 2008a). Similarities and differences between these two cleanerfish systems, and some of the general intricacies of cleaning mutualisms, become more clearly understood when placed in the framework of biological market theory.

Insights from Biological Market Theory

Ronald Noë and colleagues (Noë, 1990; Noë *et al.*, 1991; Noë and Hammerstein, 1994) first suggested that cooperative interactions among animals resemble various forms of human trading and proposed human market theory as a general framework for cooperation. In biological markets, just like in their human counterparts, two classes of traders exchange commodities to their mutual benefit. Such markets exhibit three defining characteristics: (1) competition within trader classes that is rarely overtly agonistic, but that occurs instead by outbidding, (2) preference for partners offering the highest value, and (3) conflicts over the exchange value of commodities (Noë and Hammerstein, 1994; Noë, 2001, 2006).

The parallels between biological markets and cleaning interactions were first noted by Bshary (2001). In cleaning, there are two classes of traders, each providing a reciprocal service: one removes parasites (the cleaners) and the other (the clients) offers a meal (in the form of parasites). Both classes advertise their services: the cleaners (at least in wrasses) by dancing and the clients by posing. Both classes prefer partners that give them the highest benefit, hence cleaners prefer clients that offer more food (e.g., larger clients) and clients prefer honest cleaners. However, conflicts may arise if clients only want their parasites removed but cleaners remove other items instead. Because interactions between individual cleaners and clients are repeated multiple times, partner choice becomes a powerful weapon for clients to enforce honesty in cleaners. Manipulative behaviours may also be expected to arise in dishonest traders to appease wronged parties and promote continued interactions. Bshary and colleagues have over the years produced a remarkable body of work, elucidating the cleaner wrasse market with evidence from both the field and from cleverly designed laboratory experiments (Bshary and Schaffer, 2002; Bshary and Grutter, 2002a, 2002b, 2003, 2005).

While cleaner wrasses and their clients fit perfectly the expectations of biological markets, this is not so much the case for cleaning gobies. As described previously, cleaning gobies do not advertise their services, they do not provide tactile stimulation to their clients, and while gobies do cheat, their clients do not appear to exercise strong control over cleaner dishonesty through partner choice. Why? The answer may lie in the low level of conflict between cleaning gobies and their clients (Soares *et al.*, 2008a). When given a choice between fish ectoparasites and mucus, cleaner wrasses almost invariably prefer mucus (Grutter and Bshary, 2003, 2004), giving rise to a conflict because mucus loss is costly for clients. In cleaning gobies, this conflict of interests appears to be much less important. Circumstantial evidence suggests cleaning gobies may primarily focus on removing client ectoparasites rather than mucus (Arnal *et al.*, 2001). In addition, cleaning gobies show a clear preference for clients with more ectoparasites (Soares *et al.*, 2007). This does not mean that cleaning gobies never cheat but simply that they may less readily do so than cleaner wrasses.

A reduced conflict of interest between cleaners and clients has several repercussions. First, there is little need to advertise cleaning services because variation among cleaners in service quality (i.e., how honestly they clean) is limited. Second, there is little need for clients to use tactics to control cheating by cleaners. Thus when cleaning gobies do cheat, their clients simply interrupt the interaction and swim away instead of punishing the cleaner by chasing or by withholding further visits (Soares *et al.*, 2008a). Finally, there is also little need for cleaners to attempt to manipulate the behaviour of their clients after cheating. If ectoparasites are the primary target of cleaning gobies, then gobies will only cheat once client ectoparasite loads have been depleted (Soares *et al.*, 2008b). There is therefore little incentive for cleaning gobies to encourage clients, via tactile stimulation, to remain at their cleaning station.

The application to biological market thinking to cleaning has undoubtedly deepened our understanding of these interactions. The fact that some cleanerfish species fit the model well and others, such as cleaning gobies, do not has nevertheless increased our appreciation of the diversity of cleaning systems.

ECOLOGICAL ROLE OF CLEANING GOBIES

The ecological role of some cleanerfish can go far beyond their function as ectoparasite consumers. This has been clearly shown to be the case for the cleaner wrasse *L. dimidiatus*. On reefs with few or no cleaner wrasses, either because of natural downturns in population size or experimental removals, reef fish species diversity can decrease by up to 50%, and abundance by ~75% over periods of 4 to 18 months (Bshary, 2003; Grutter *et al.*, 2003). Natural and experimental additions of cleaners to reefs have the reverse effect, which is noticeable in only a few weeks instead of a few months (Bshary, 2003). The effects are particularly marked for client species that move among reefs, but less so for resident clients, thus leading to changes in fish community composition. Surprisingly, the presence (or absence) of cleaners can also affect the distribution of non-client resident species (Bshary, 2003). The reasons for this latter effect are not clear, but Bshary (2003) suggested that if the higher species diversity of reefs with cleaners results in higher overall fish densities, this could lead to greater relative safety or lower ectoparasite availability (through dilution effects) for all fish, including those that rarely interact with cleaners.

In light of the relatively smaller apparent benefit for clients of being cleaned by gobies, one might surmise that gobies that clean play a more limited ecological role in Caribbean coral reef communities. In this respect, it is unfortunate that Limbaugh's (1961) early Caribbean cleaner removal experiment—which, admittedly, was carried out on only two reefs, included removal of all cleaners (not just gobies), and lacked a control—has never been replicated. There are nevertheless a few recent studies that shed some light on the influence of cleaning gobies on client distribution and abundance.

On a very small scale, if being cleaned is important, then proximity to cleaning stations should be an asset to fish that are limited in movement, such as territorial species. However, there is no indication of aggregations of territorial adult damselfish around cleaning stations operated by cleaning gobies (Whiteman *et al.*, 2002). An effect of cleaning gobies could be masked if young damselfish settling on a reef do indeed preferentially take up residence near cleaning stations, but damselfish longevity exceeds the tenure of gobies at a given site. Many adult damselfish are therefore near where cleaning stations used to be, but the cost of relocating closer to a current cleaning station may be too high given the limited benefit. However, this does not seem to be a plausible scenario because even when the cost of relocating is experimentally reduced (i.e. by creating a vacancy next to a cleaning station), damselfish still do not seek territories near cleaners (Cheney and Côté, 2003b).

At a larger, inter-reef scale, cleaning gobies appear, at first glance, to be more influential. As the density of coral-dwelling cleaning gobies increases, so does the total density of clients (Cheney and Côté, 2003c). This relationship does not seem to reflect a general habitat quality effect since there is no association between the abundances of sponge-dwelling gobies—which do not clean but feed instead on sponge-dwelling invertebrates ((Whiteman and Côté, 2002; White *et al.*, 2007)—and of fish clients (Cheney and Côté, 2003c). But does cleaning goby abundance drive client abundance, or vice-versa? The small-scale observations suggest that client abundance is probably the driver, i.e. cleaning gobies recruit to client-rich areas rather than the other way around. However, only replicated experiments in which either gobies or their clients are removed will provide a definite answer to this question.

Taken together, the evidence to date suggests that cleaning gobies do not affect client distribution or abundance greatly, certainly not to the extent demonstrated for cleaner wrasses. Immigration of clients away from cleaner-free reefs probably largely depends on client ectoparasite load: the greater the benefit of being cleaned, the more likely immigration becomes, and the greater the effect of cleaners on the distribution of their clients.

CONCLUSIONS AND FUTURE DIRECTIONS

Cleaning in gobies is definitely a more muted affair than it is in Indo-Pacific cleaner wrasses. The service is less conspicuous, by virtue of the smaller size of gobies and the lack of advertisement dancing, the transaction more straightforward, owing to the absence of tactics to counter cheating, and the outcome in terms of parasite removal only marginally beneficial, at least from the client's point of view. There has even been a suggestion that adopting a cleaning lifestyle is making the best of a bad situation, based on the observation that the growth rates of juvenile and survival rates of adult *E. evelynae* that live on sponges (where they do not clean) are higher than those of *E. evelynae* that live on coral (where they do clean) (White *et al.*, 2007). This is clearly not the case everywhere or for every cleaning goby species with dichotomous habits (e.g., Whiteman and Côté, 2004a), but such observations highlight the fact that cleaning behaviour, particularly in gobies, is far from well understood, either evolutionarily or ecologically.

Key questions remain to be answered. One concerns the outcome of cleaning. How substantial is the benefit of being cleaned by cleaning gobies? The studies that have focused on ectoparasite numbers suggest a very small effect, but other metrics might be important. The issue of ectoparasites as vectors for diseases was raised earlier. In addition, there is evidence that cortisol levels—a marker of physiological stress—are lower in clients interacting with cleanerfish (Bshary *et al.*, 2007). Thus, as a result of being cleaned, the general health of clients may therefore be improved in ways that are poorly reflected by simple parasite counts.

Another important question relates to the ultimate cause of the discrepancies observed between cleaner wrasses and cleaning gobies. We outlined earlier how a preference by cleaning gobies for ectoparasites over mucus or other client-gleaned items could potentially explain these large interspecific differences, but the evidence to date is only circumstantial. Experiments of the type carried out with cleaner wrasses, in which cleaners are simultaneously offered different food items (e.g., Grutter and Bshary, 2003, 2004) or allowed to forage on an anaesthetised client that cannot retaliate to cheating (Bshary and Grutter, 2002a), need to be performed with cleaning gobies to measure their foraging preferences. Alternative causes should also be considered. For example, if cleaning gobies inflict very small costs on their clients when cheating, the cleaner control tactics displayed by clients of cleaner wrasses would be largely unnecessary, as might be the advertisement of (honest) cleaning service. In addition, punishing a cleaner that misbehaves is useful to a client only if the cleaner changes its behaviour in subsequent interactions with the punisher. The cognitive ability required is considerable, as the cleaner needs to recall the act that led to punishment as well as the identity of the individual that retaliated. Cleaner wrasses, which have been described as having a 'Machiavellian intelligence' (Bshary and Würth, 2001), are able to do this (Tebbich *et al.*, 2002; Bshary and Grutter, 2002a), but we do not know whether cleaning gobies can.

Finally, the apparently limited role played by cleaning gobies in determining client abundance or community structure needs to be confirmed (or refuted). Interested researchers should note in this context that the earliest cleaner wrasse removal experiments invariably concluded that cleaner wrasses had little or no effect on client populations (Youngbluth, 1968; Gorlick *et al.*, 1987; Grutter, 1996, 1997). Yet, later efforts with longer-running and better-replicated field experiments concluded the opposite (Bshary, 2003; Grutter *et al.*, 2003).

Given that so much of our recent understanding of cleaning behaviour has been influenced by research on cleaner wrasses, should we study cleaning gobies at all? Our answer is an emphatic yes! Work on cleaning gobies has revealed that cleaner wrasses are not a universal model to which all cleanerfish conform. This comparison has given, and will continue to provide, deep insights into the evolution of cleaning interactions. We have only begun to unravel the mysteries of cleaning gobies.

References

Allen, G.R. and D.R. Robertson. 1994. *Fishes of the Tropical Eastern Pacific*. University of Hawaii Press, Honolulu.

Arnal, C. and I.M. Côté. 2000. Diet of broadstripe cleaning gobies on a Barbadian reef. *Journal of Fish Biology* 57: 1075-1082.

Arnal, C., I.M. Côté and S. Morand. 2001. Why clean and be cleaned? The importance of client ectoparasites and mucus in a marine cleaning symbiosis. *Behavioral Ecology and Sociobiology* 51: 1-7.

Arnal, C., O. Verneau and Y. Desdevises. 2006. Phylogenetic relationships and evolution of cleaning behaviour in the family Labridae: importance of body colour pattern. *Journal of Evolutionary Biology* 19: 755-763.

Arnal, C., I.M. Côté, P. Sasal and S. Morand. 2000. Cleaner-client interactions on a Caribbean reef: influence of correlates of parasitism. *Behavioral Ecology and Sociobiology* 47: 353-358.

Bansemer, C., A.S. Grutter and R. Poulin. 2002. Geographic variation in the behaviour of the cleaner fish *Labroides dimidiatus* (Labridae). *Ethology* 108: 353-366.

Begon, M., J.L. Harper and C.R.Townsend. 1990. *Ecology: Individuals, Populations, and Communities*. Blackwell Scientific, Cambridge, MA. 2nd Edition.

Bohlke, J.E. and J.E. McCosker. 1973. Two additional West Atlantic gobies (genus *Gobiosoma*) that remove ectoparasites from other fishes. *Copeia* 1973: 609-610.

Bshary, R. 2001. The cleaner fish market. In: *Economics in Nature*, R. Noë, J.A.R.A.M. van Hooff and P. Hammerstein (eds.), Cambridge University Press, Cambridge, pp. 146-172.

Bshary, R. 2003. The cleaner wrasse, *Labroides dimidiatus*, is a key organism for reef fish diversity at Ras Mohammed National Park, Egypt. *Journal of Animal Ecology* 72: 169-176.

Bshary, R. and M. Würth. 2001. Cleaner fish *Labroides dimidiatus* manipulate client reef fish by providing tactile stimulation. *Proceedings of the Royal Society of London Series B-Biological Sciences* 268: 1495-1501.

Bshary, R. and D. Schaffer. 2002. Choosy reef fish select cleaner fish that provide high-quality service. *Animal Behaviour* 63: 557-564.

Bshary, R. and A.S. Grutter. 2002a. Asymmetric cheating opportunities and partner control in a cleaner fish mutualism. *Animal Behaviour* 63: 547-555.

Bshary, R. and A.S. Grutter. 2002b. Experimental evidence that partner choice is a driving force in the payoff distribution among cooperators or mutualists: the cleaner fish case. *Ecology Letters* 5: 130-136.

Bshary, R. and A.S. Grutter. 2005. Punishment and partner switching cause cooperative behaviour in a cleaning mutualism. *Biology Letters* 1: 396-399.

Bshary, R., R.F. Oliveira, T.S.F. Oliveira and A.V.M. Canario. 2007. Do cleaning organisms reduce the stress response of client reef fish? *Frontiers in Zoology* 4: 21-29.

Chambers, S.D. and P.C. Sikkel. 2002. Diel emergence patterns of ecologically important, fish-parasitic, gnathiid isopod larvae on Caribbean coral reefs. *Caribbean Journal of Science* 38: 37-43.

Cheney, K.L. and I.M. Côté. 2001. Are Caribbean cleaning symbioses mutualistic? Costs and benefits of visiting cleaning stations to longfin damselfish. *Animal Behaviour* 62: 927-933.

Cheney, K.L. and I.M. Côté. 2003a. The ultimate effect of being cleaned: does ectoparasite removal have reproductive consequences for damselfish clients? *Behavioral Ecology* 14: 892-896.

Cheney, K.L. and I.M. Côté. 2003b. Habitat choice in adult longfin damselfish: territory characteristics and relocation times. *Journal of Experimental Marine Biology and Ecology* 287: 1-12.

Cheney, K.L. and I.M. Côté. 2003c. Do ectoparasites determine cleaner fish abundance? Evidence on two spatial scales. *Marine Ecology-Progress Series* 263: 189-196.

Cheney, K.L. and I.M. Côté. 2005. Mutualism or parasitism? The variable outcome of cleaning symbioses. *Biology Letters* 1: 162-165.

Cheney, K.L., A.S. Grutter and N.J. Marshall. 2008. Facultative mimicry: cues for colour change and colour accuracy in a coral reef fish. *Proceedings of the Royal Society B-Biological Sciences* 275: 117-122.

Colin, P. 1975. *The Neon Gobies: The Comparative Biology of the Gobies of the Genus* Gobiosoma, *Subgenus* Elacatinus *(Pisces: Gobiidae) in the Tropical Western North Atlantic Ocean.* T.H.F. Publications, Neptune City.

Côté, I.M. 2000. Evolution and ecology of cleaning symbioses in the sea. *Oceanography and Marine Biology* 38: 311-355.

Côté, I.M. and P.P. Molloy. 2003. Temporal variation in cleanerfish and client behaviour: Does it reflect ectoparasite availability? *Ethology* 109: 487-499.

Côté, I.M., C. Arnal and J.D. Reynolds. 1998. Variation in posing behaviour among fish species visiting cleaning stations. *Journal of Fish Biology* 53 (Supplement A): 256-266.

Darcy, G.H., E. Maisel and J.C. Ogden. 1974. Cleaning preferences of gobies *Gobiosoma evelynae* and *Gobiosoma prochilos* and juvenile wrasse *Thalassoma bifasciatum*. *Copeia* 1974: 375-378.

Davies, A.J., J.C. Eiras and R.T.E. Austin. 1994. Investigations into the transmission of *Haemogregarina bigemina* Laveran and Mesnil, 1901 (Apicomplexa, Adeleorina) between intertidal fishes in Portugal. *Journal of Fish Diseases* 17: 283-289.

Eibl-Eibesfeldt, I. 1955. Über Symbiosen, Parasitismus und andere besondere zwischenartliche Beziehungen tropischer Meeresfische. *Zeitschrift für Tierpsychologie* 12: 203-219.

Eschmeyer, W.N. 1998. *Catalog of Fishes*. California Academy of Sciences, San Francisco.

Feder, H.M. 1966. Cleaning symbiosis in the marine environment. In: *Symbioses*, S.M. Henry (ed.), Academic Press, New York, pp. 327-380.

Francini-Filho, R.B. and I. Sazima. 2008. A comparative study of cleaning activity of two reef fishes at Fernando de Noronha Archipelago, tropical West Atlantic. *Environmental Biology of Fishes* 83: 213-220.

Francihi-Filho, R.B., R.L. Moura and I. Sazima. 2000. Cleaning by the wrasse *Thalassoma noronhanum*, with two records of predation by its grouper client *Cephalopholis fulva*. *Journal of Fish Biology* 56: 802-809.

Gorlick, D.L. 1980. Ingestion of host fish surface mucus by the Hawaiian cleaning wrasse, *Labroides phthirophagus* (Labridae), and its effect on host species preference. *Copeia* 1980: 863-868.

Gorlick, D.L., P.D. Atkins and G.S. Losey. 1978. Cleaning stations as water holes, garbage dumps, and sites for evolution of reciprocal altruism. *American Naturalist* 112: 341-353.

Gorlick, D.L., P.D. Atkins and G.S. Losey. 1987. Effect of cleaning by *Labroides dimidiatus* (Labridae) on an ectoparasite population infecting *Pomacentrus vaiuli* (Pomacentridae) at Enewetak Atoll. *Copeia* 1987: 41-45.

Grove, J.S. and R.J. Lavenberg. 1997. *The Fishes of the Galapagos Islands*. Stanford University Press, Stanford.

Grutter, A.S. 1996. Experimental demonstration of no effect by the cleaner fish *Labroides dimidiatus* (Cuvier and Valenciennes) on the host fish *Pomacentrus moluccensis* (Bleeker). *Journal of Experimental Marine Biology and Ecology* 196: 285-298.

Grutter, A.S. 1997. Effect of the removal of cleaner fish on the abundance and species composition of reef fish. *Oecologia* 111: 137-143.

Grutter, A.S. 1999. Cleaner fish really do clean. *Nature* 398: 672-673.

Grutter, A.S. 2004. Cleaner fish use tactile dancing behavior as a preconflict management strategy. *Current Biology* 14: 1080-1083.

Grutter, A.S. and R. Bshary. 2003. Cleaner wrasse prefer client mucus: support for partner control mechanisms in cleaning interactions. *Proceedings of the Royal Society of London Series B-Biological Sciences* 270: S242-S244.

Grutter, A.S. and R. Bshary. 2004. Cleaner fish, *Labroides dimidiatus*, diet preferences for different types of mucus and parasitic gnathiid isopods. *Animal Behaviour* 68: 583-588.

Grutter, A.S., J.M. Murphy and J.H. Choat. 2003. Cleaner fish drives local fish diversity on coral reefs. *Current Biology* 13: 64-67.

Guimaraes, R.Z.P., J.L. Gasparini and L.A. Rocha. 2004. A new cleaner goby of the genus *Elacatinus* (Teleostei : Gobiidae), from Trindade Island, off Brazil. *Zootaxa* 770: 1-8.

Hobson, E.S. 1968. Predatory behaviour of some shore fishes in the Gulf of California. *U.S. Fish and Wildlife Service Research Report* 73: 1-92.

Hobson, E.S. 1971. Cleaning symbiosis among California inshore fishes. *Fishery Bulletin* 69: 491-524.

Hobson, E.S. 1976. Rock wrasse, *Halichoeres semicinctus*, as a cleaner fish. *California Fish and Game* 62: 73-78.

Humann, P. 1993. *Reef Fish Identification: Galapagos*. New World Publications, Jacksonville, FL.

Johnson, W.S. and P. Ruben. 1988. Cleaning behavior of *Bodianus rufus*, *Thalassoma bifasciatum*, *Gobiosoma evelynae*, and *Periclimenes pedersoni* along a depth gradient at Salt River submarine canyon, St Croix. *Environmental Biology of Fishes* 23: 225-232.

Limbaugh, C. 1961. Cleaning symbiosis. *Scientific American* 205: 42-49.

Losey, G.S. 1971. Communication between fishes in cleaning symbiosis. In: *Aspects of the Biology of Symbioses*, T.C. Cheng (ed.), University Park Press, Baltimore, pp. 45-76.

Losey, G.S. 1972. Ecological importance of cleaning symbiosis. *Copeia* 1972: 820-833.

Losey, G.S. 1974. Cleaning symbiosis in Puerto Rico with comparison to tropical Pacific. *Copeia* 1974: 960-970.

McCourt, R.M. and D.A. Thomson. 1984. Cleaning behavior of the juvenile Panamic sergeant major, *Abudefduf troschelii* (Gill), with a resume of cleaning associations in the Gulf of California and adjacent waters. *California Fish and Game* 70: 234-239.

McFarland, D. 1985. *Animal Behaviour: Psychobiology, Ethology and Evolution.* Pitman, London.

Michael, SW. 1993. *Reef Sharks and Rays of the World.* Sea Challengers, Monterey, CA.

Noë, R. 1990. A veto game played by baboons: a challenge to the use of the Prisoner's Dilemma as a paradigm for reciprocity and cooperation. *Animal Behaviour* 39: 78-90.

Noë, R. 2001. Biological markets: partner choice as a driving force behind the evolution of mutualism. In: *Economics in Nature,* R. Noë, J.A.R.A.M van Hooff and P. Hammerstein (eds.), Cambridge University Press, Cambridge, pp. 146-172.

Noë, R. 2006. Digging for the roots of trading. In: *Cooperation in Primates and Humans: Mechanisms and Evolution,* P.M. Kappeler and C.P. van Schaik (eds.), Springer-Verlag, Berlin, pp. 233-261.

Noë, R. and R. Hammerstein. 1994. Biological markets — Supply and demand determine the effect of partner choice in cooperation, mutualism and mating. *Behavioral Ecology and Sociobiology* 35: 1-11.

Noë, R., C.P. van Schaik and J. van Hooff. 1991. The market effect — an explanation for pay-off asymmetries among collaborating animals. *Ethology* 87: 97-118.

Potts, G.W. 1973a. Cleaning symbiosis among British fish with special reference to *Crenilabrus melops* (Labridae). *Journal of the Marine Biological Association of the United Kingdom* 53: 1-10.

Potts, G.W. 1973b. Ethology of *Labroides dimidiatus* (Cuv and Val) (Labridae, Pisces) on Aldabra. *Animal Behaviour* 21: 250-291.

Randall, J.E. 1955. Fishes of the Gilbert Islands. *Atoll Research Bulletin* 47: 143-144.

Randall, J.E. 1958. A review of the genus *Labroides,* with description of two new species and notes on ecology. *Pacific Science* 12: 327-347.

Rocha, L.A., I.L. Rosa and B.M. Feitoza. 2000. Sponge-dwelling fishes of northeastern Brazil. *Environmental Biology of Fishes* 59: 453-458.

Rüber, L., J.L. van Tassell and R. Zardoya. 2003. Rapid speciation and ecological divergence in the American seven-spined gobies (Gobiidae, Gobiosomatini) inferred from a molecular phylogeny. *Evolution* 57: 1584-1598.

Sargent, R.C. and G.E. Wagenbach. 1975. Cleaning behavior of shrimp, *Periclimenes anthophilus* Holthuis and Eibl-Eibesfeldt (Crustacea, Decapoda, Natantia). *Bulletin of Marine Science* 25: 466-472.

Sazima, I. and R.L. Moura. 2000. Shark (*Carcharhinus perezi*), cleaned by the goby (*Elacatinus randalli*), at Fernando de Noronha Archipelago, Western South Atlantic. *Copeia* 2000: 297-299.

Sazima, I., R.L. Moura and R.S. Rosa. 1997. *Elacatinus figaro* sp. n. (Perciformes: Gobiidae), a new cleaning goby from the western South Atlantic. *Aqua Journal of Ichthyology and Aquatic Biology* 2: 33-38.

Sazima, I., R.L. Moura and C. Sazima. 1999. Cleaning activity of juvenile angelfish, *Pomacanthus paru*, on the reefs of the Abrolhos Archipelago, western South Atlantic. *Environmental Biology of Fishes* 56: 399-407.

Sazima, I., A. Carvalho and C. Sazima. 2008. A new cleaner species of *Elacatinus* (Actinopterygii: Gobiidae) from the Southwestern Atlantic. *Zootaxa* 1932: 27-32.

Sazima, I., C. Sazima, R.B Francini and R.L. Moura. 2000. Daily cleaning activity and diversity of clients of the barber goby, *Elacatinus figaro*, on rocky reefs in southeastern Brazil. *Environmental Biology of Fishes* 59: 69-77.

Sikkel, P.C., K.L. Cheney and I.M. Côté. 2004. In situ evidence for ectoparasites as a proximate cause of cleaning interactions in reef fish. *Animal Behaviour* 68: 241-247.

Sikkel, P.C., C.S. Schaumburg and J.K. Mathenia. 2006. Diel infestation dynamics of gnathiid isopod larvae parasitic on Caribbean reef fish. *Coral Reefs* 25: 683-689.

Soares, M.C., S.C. Cardoso and I.M. Côté. 2007. Client preferences by Caribbean cleaning gobies: food, safety or something else? *Behavioral Ecology and Sociobiology* 61: 1015-1022.

Soares, M.C., I.M. Côté, S.C. Cardoso and R. Bshary. 2008a. The cleaning goby mutualism: a system without punishment, partner switching or tactile stimulation. *Journal of Zoology* 276: 306-312.

Soares, M.C., R. Bshary, S.C. Cardoso and I.M. Côté. 2008b. The meaning of jolts by fish clients of cleaning gobies. *Ethology* 114: 209-214.

Soares, M.C., R. Bshary, S.C. Cardoso and I.M. Côté. 2008c. Does competition for clients increase service quality in cleaning gobies? *Ethology* 114: 625-632.

Streelman, J.T. and P.D. Danley. 2003. The stages of vertebrate evolutionary radiation. *Trends in Ecology and Evolution* 18: 126-131.

Stummer, L.E., J.A. Weller, M.L. Johnson and I.M. Côté. 2004. Size and stripes: how fish clients recognize cleaners. *Animal Behaviour* 68: 145-150.

Taylor, M.S. and L. Akins. 2007. Two new species of *Elacatinus* (Teleostei: Gobiidae) from the Mexican coast of the Gulf of Mexico. *Zootaxa* 1425: 45-51.

Taylor, M.S. and M.E. Hellberg. 2005. Marine radiations at small geographic scales: speciation in neotropical reef gobies (*Elacatinus*). *Evolution* 59: 374-385.

Tebbich, S., R. Bshary and A.S. Grutter. 2002. Cleaner fish *Labroides dimidiatus* recognise familiar clients. *Animal Cognition* 5: 139-145.

Thompson, J.N. 1994. *The Coevolutionary Process*. University of Chicago Press, Chicago, IL.

Trivers, R.L. 1971. Evolution of reciprocal altruism. *Quarterly Review of Biology* 46: 35-57.

White, J.W., C.J. Grigsby and R.R. Warner. 2007. Cleaning behavior is riskier and less profitable than an alternative strategy for a facultative cleaner fish. *Coral Reefs* 26: 87-94.

Whiteman, E.A. and I.M. Côté. 2002. Cleaning activity of two Caribbean cleaning gobies: intra- and interspecific comparisons. *Journal of Fish Biology* 60: 1443-1458.

Whiteman, E.A. and I.M. Côté. 2004a. Individual differences in microhabitat use in a Caribbean cleaning goby: a buffer effect in a marine species? *Journal of Animal Ecology* 73: 831-840.

Whiteman, E.A. and I.M. Côté. 2004b. Dominance hierarchies in group-living cleaning gobies: causes and foraging consequences. *Animal Behaviour* 67: 239-247.

Whiteman, E.A., I.M. Côté and J.D. Reynolds. 2002. Do cleaning stations affect the distribution of territorial reef fishes? *Coral Reefs* 21: 245-251.

Wickler, W. 1963. Zum Problem der Signalbildung, am Beispiel der Verhaltens-Mimikry zwischen *Aspidontus* und *Labroides*. *Zeitschrift für Tierpsychologie* 20: 657-679.

Wicksten, M.K. 1998. Behaviour of cleaners and their client fishes at Bonaire, Netherlands Antilles. *Journal of Natural History* 32: 13-30.

Youngbluth, M.J. 1968. Aspects of the ecology and ethology of the cleaning fish, *Labroides phthirophagus* Randall. *Zeitschrift für Tierpsychologie* 25: 915-932.

CHAPTER

4.4

The Partnership between Gobiid Fishes and Burrowing Alpheid Shrimps

Ilan Karplus[1] and Andrew Richard Thompson[2]

INTRODUCTION

The association between non-burrowing gobiid fishes and burrowing alpheid shrimps is a mutualistic, co-evolved partnership. The shrimp constructs and maintains a burrow which is used by the goby as a temporary shelter during the day, a permanent resting place over night and a safe place for breeding. The goby provides the shrimp with a tactile based alarm system, warning it against approaching predators or fish that dig in the sediment and may block its burrow entrance. A detailed review of the information available on goby-shrimp associations was published twenty years ago (Karplus, 1987). The aim of the present review is to focus mainly on new findings that cover a wide range of topics including goby-shrimp taxonomy,

Authors' addresses: [1] Aquaculture Research Unit, the Institute of Animal Sciences, The Volcani Research Center, P.O. Box 6, Bet-Dagan 20250, Israel. E-mail: karplus@agri.gov.il

[2] NOAA Fisheries Service, Southwest Fisheries Science Center, 8604 La Jolla Shores Drive, La Jolla, CA 92037-1508, USA. E-mail: andrew.thompson@noaa.gov

biogeography, ecology, behavior, partner specificity, phylogeography and evolution. However, in order to expose the reader to the full picture of these fascinating associations, older studies will be briefly reviewed at the beginning of each section. In some areas of research (e.g. goby and shrimp reproduction) few or no studies were carried out over the last twenty years. In these cases a more detailed overview based on older studies will be presented.

SYSTEMATICS OF GOBIES AND SHRIMPS

The majority of studies addressing the goby-shrimp partnership focus mostly on the description of new gobies without describing their behavior, ecology or interrelationship with their shrimp partners. Historically, description of associated gobies lagged behind the description of other coral reef fishes, due to the habit of the gobies to shelter inside the shrimp-constructed burrow when disturbed. These species were often not collected with traditional collecting techniques such as large scale indiscriminate poisoning. The advent of SCUBA diving and selective fish collecting using a variety of techniques reviewed in Karplus (1987) resulted in the continuous discovery of many species of associated gobies, particularly in the tropical Indo-Pacific.

Close to forty new species of gobies associated with shrimps were described over the last twenty years (Table 4.4.1). Over 120 species of associated gobies belonging to more than twenty different genera are now known to live in association with burrowing alpheid shrimps (Karplus, 1987; Table 4.4.1). Previous and current debates concerning the generic affinity of associated gobies (e.g. Ikeda *et al.*. 1995; Chen and Fang, 2003; Shibukawa and Suzuki, 2004; Iwata *et al.*, 2007; Randall, 2007b; Randall and Chen, 2007), contributed to the common occurrence of synonymy. Throughout this text we adopted the taxonomic classification presented at the FishBase website (Froese and Pauly, 2008).

Although the nature of the relationship between gobies and shrimps is not known for some species, many species of gobies interact either obligately or facultatively with shrimps. The obligate species belong to twelve genera that consist exclusively of gobies that associate with shrimps (but see below for a possible exception). The two most speciose genera are the Indo-Pacific genera *Amblyeleotris* and *Cryptocentrus*, each containing over thirty described species. *Amblyeleotris* species share a pair of sensory papillae at the center of the lower lip, and the presence of both transverse and

Table 4.4.1 New species of gobies associated with shrimps (and reported depth distribution) described over the last 20 years (1987-2007).

Fish Species	*Distribution and Depth*	*Shrimp Partner*	*References*
Amblyeleotris arcupinna	Papua New Guinea (23 m)	*Alpheus bellulus*	Mohlmann and Munday, 1999
Amblyeleotris bellicauda	New Caledonia Papua New Guinea (3 m)	not described	Randall, 2004
Amblyeleotris biguttata	Solomon Islands New Caledonia (11-50 m)	not described	Randall, 2004
Amblyeleotris downingi	Arabian Gulf (13-17 m)	*Alpheus bellulus* *Alpheus ochrostriatus*	Randall, 1994
Amblyeleotris ellipse	New Caledonia Papua New Guinea	not described	Randall, 2004
Amblyeleotris harrisorum	Christmas Island (32 m)	not described	Mohlmann and Randall, 2002
Amblyeleotris katherine	Mariana and Marshall Islands, American Samoa Cook and Society Islands (3-38 m)	*Alpheus ochrostriatus*	Randall, 2004
Amblyeleotris ***marquesas***	Marquesas Islands (20-25 m)	*Alpheus randalli*	Mohlmann and Randall, 2002
Amblyeleotris melanocephala	Okinoshima and Okinawa Islands, Japan	not described	Aonuma *et al.*, 2000
Amblyeleotris nasuii	Ryukyu Islands, Japan (35-40 m)	not described	Aonuma and Yoshino, 1996
Amblyeleotris neumanni	New Britain Papua New Guinea (26 m)	*Alpheus rapacida* and additional non described species	Randall and Earle, 2006
Amblyeleotris rubrimarginata	Great Barrier Reef New Caledonia, (3-26 m)	not described	Mohlmann and Randall, 2002
Amblyeleotris stentaeniata	New Caledonia (10-20 m)	*Alpheus bellulus*	Randall, 2004
Amblyeleotris triguttata	Red Sea, Gulf of Oman Arabian Gulf (1.5-17 m)	*Alpheus bellulus*	Randall, 1994
Ambleyleotris yanoi	Ryukyu Islands, Japan (3-30 m)	*Alpheus randalli*	Aonuma and Yoshino, 1996
Ctenogobiops formosa	Taiwan	not described	Randall *et al.*, 2003

Contd...

Contd...

Fish Species	*Distribution and Depth*	*Shrimp Partner*	*References*
Ctenogobiops mitodes	South China Sea, Marshall Islands, Solomon Islands, Papua New-Guinea (2-21m)	not described	Randall *et al.*, 2007
Ctenogobiops phaeostictus	Papua New-Guinea (10 m)	expected to be associated with a shrimp	Randall *et al.*, 2007
Flabelligobius smithi	Taiwan (100 m)	expected to be associated with a shrimp	Chen and Fang, 2003
Stonogobiops pentafasciata	Kochi prefecture, Japan (18-40 m)	*Alpheus randalli* *Alpheus bellulus*	Iwata and Hirata, 1994
Stonogobiops yasha	Ryukyu Islands, Japan Palau Islands New Caledonia and Bali, Indonesia (15-40 m)	*Alpheus randalli*	Yoshino and Shimada, 2001
Tomiyamichthys alleni	Japan and Indonesia	not described	Iwata *et al.* 2000
Tomiyamichthys tanyspilus	Island of Flores, Indonesia (4 m)	Alpheus cf. rapacida	Randall and Chen, 2007
Vanderhorstia attenuata	Solomon Islands (38-48 m)	expected to be associated with a shrimp	Randall, 2007b
Vanderhorstia auronotata	Molluca Islands, Indonesia	expected to be associated with a shrimp	Randall, 2007b
Vanderhorstia bella	Fiji Islands (8 m)	not described	Greenfield and Longenecker, 2005
Vanderhorstia belloides	Papua New Guinea (21m)	expected to be associated with a shrimp	Randall, 2007b
Vandrhorstia dorsomacula	Papua New Guinea (25m)	*Alpheus rapacida*	Randall, 2007b
Vanderhorstia flavilineata	Papua New Guinea (15-30 m)	not described	Allen and Munday, 1995
Vanderhorstia hiramatsui	Japan, (30m)	not described	Iwata *et al.*, 2007
Vanderhorstia kizakura	Japan (45m)	not described	Iwata *et al.*, 2007
Vanderhorstia longimanus*	Indonesia (118 m)	Expected to be associated with a shrimp	Weber, 1909 Iwata *et al.*, 2007

Contd...

Contd...

Fish Species	*Distribution and Depth*	*Shrimp Partner*	*References*
Vanderhorstia macropteryx*	Japan (20-110 m)	not described	Franz, 1910 Ikeda *et al.*, 1995
Vanderhorstia nannai	Palau and Philippines (0-15 m)	not described	Winterbottom *et al.*, 2005
Vanderhorstia nobilis	Philippines, Sulawesi, Indonesia (5-25 m)	not described	Allen and Randall, 2006
Vanderhorsta opercularis	Red Sea (27-40m)	*Alpheus bellulus* *Alpheus ochrostriatus*	Randall, 2007a
Vanderhorstia papilio	Ryuku Islands, Japan (45 m)	not described	Shibukawa and Suzuki, 2004
Vanderhorstia puncticeps	Japan (60-123) China (76-98)	expected to be associated with a shrimp	Deng and Xiong, 1980 Iwata *et al.*, 2007
Vanderhorstia rapa	Japan (25-50 m)	not described	Deng and Xiong, 1980; Iwata *et al.*, 2007

* not described before associated with shrimp

longitudinal cheek papillae rows (Hoese and Steene, 1978; Winterbottom, 2002). The genus *Cryptocentrus* also has both transverse and longitudinal papillae rows, but differs from *Amblyeleotris* in having fewer dorsal fin rays, cycloid scales only (rather than ctenoid), and differences in papillae pattern and head shape (Chen *et al.*, 1998; Mohlman and Randall, 2002). The genera *Vanderhorstia* and *Ctenogobiops* consist of an intermediate number of species. *Ctenogobiops* was recently reviewed by Randall *et al.* (2003) and Thacker *et al.* (2010); it shares with *Vanderhorstia* characters of fin ray counts, scalation, head and tongue shape, and sensory pore pattern. The two genera are distinguished only by caudal fin length: longer and more tapered in *Vanderhorstia* as compared to *Ctenogobiops* (Shibukawa and Suzuki, 2004). Genera such as the Atlantic *Nes* and the Indo-Pacific *Lotilia* contain only one described, associated species. Additional genera that contain obligatory shrimp gobies are *Psilogobius, Flabelligobius, Tomiyamichthys, Myersina* and *Mahidolia.*

At present there are only five reports of facultative associations between gobies and burrowing alpheid shrimps. *Bathygobius curacao* (Karplus, 1992) and *Ctenogobius saepepallens* (Randall *et al.*, 2005 ; Kramer *et al.* 2009) were both reported in a facultative partnership with *Alpheus floridanus* from the Western Atlantic. Three additional facultative fish partners were reported from the Indo-Pacific: *Vireosa hanae* (Yanagisawa, 1978), *Acentrogobius pflaumi* (Yanagisawa, 1978) and *Cryptocentrus singapurensis*

(Palomar, 2001). Facultative gobies apparently belong to genera that include only a single or few associated species and a large number of free living gobies. For example, the facultative Atlantic *Bathygobius curacao* (Karplus, 1992), is the only associated goby in a genus which contains 36 free living species. Similarly, the facultative Indo-Pacific *Acentrogobius pflaumi* (Yanagisawa, 1978), is the only associated goby in a large genus that contains an additional 21 free living species. However, the facultative fish partner *Cryptocentrus singapurensis* belongs to a genus previously known to contain only obligatory fish partners.

At present there has not been a comprehensive analysis of the evolutionary relationship among the different genera and species of obligatory, facultative and free living species (but see evolution section of this review). Such an analysis based on morphological traits and molecular markers could be very important for the understanding of the evolution of the goby-shrimp partnership.

Our understanding of the taxonomy of goby-associated alpheid shrimps was poor twenty years ago (Karplus, 1987), and remains confused and incomplete to this day. Currently, there are fourteen described species (*Alpheus bellulus, A. bisincisus, A. brevicristatus, A. brevirostris, A. djeddensis, A. djiboutensis, A. fenneri, A. floridanus, A. crassimanus, A. macellarius, A. malabaricus, A. randalli, A. rapacida* and *A. rapax* Karplus, 1987; Talwar and Jhingram, 1991; Table 4.4.2). There are three additional species (*A. purpurilenticularis, A. ochrostriatus* and *A. rubromaculatus*) that have not been formally described but their names were given to Karplus by Miya after a detailed morphological study. Photos of the latter three species and their color description were presented by Karplus *et al.* (1981). These three associated shrimps were first described from the Red-Sea and later reported in several popular and scientific publications (Debelius, 1999; Randall, 1994, 2004) from different localities in the Indo-Pacific. There is urgent need for the formal description of these three species as their current names are invalid without proper description. In addition to the described and semi-described alpheid species, numerous shrimp "types" were noted by several researchers. These types were differentiated by distinct colorations in studies carried out in the Seychelles (Polunin and Lubbock, 1977; 7 types), the Great Barrier Reef (Cummins, 1979; 4 types), Palau (Sakaue *et al.*, 2005; 3 types) and Thailand (Nakasone and Manthachitra, 1986; 1 type).

Elucidating alpheid shrimp taxonomy is made difficult by two main factors. First, goby-associated shrimps are difficult to collect even in comparison with their goby partners. Whereas gobies sometimes wander away from a burrow, shrimp always retreat into the burrow when disturbed

Table 4.4.2 New species of shrimps associated with gobies described during the last 20 years (1987-2007)* and earlier described species of shrimps not reported before as living with gobies**.

Shrimp species	*Distribution and Depth*	*Fish partner*	*References*
*Alpheus djeddensis***	South Pacific	*Ctenogobiops feroculus*	Thompson, 2005
*Alpheus fenneri**	Sulawesi, Indonesia (6 m)	*Amblyeleotris fontannesii*	Bruce, 1994
*Alpheus macellarius**	Philippines	*Cryptocentrus octafasciatus,*	Palomar *et al.*, 2004, 2005
		Cryptocentrus singapurensis	Chace, 1988

and are virtually impossible to excavate. Although techniques have been developed to catch these elusive shrimps (Karplus and Vercheson, 1978; Karplus, 1987), collection is still a time-intensive process. Second, alpheid shrimps are known to exhibit a high degree of cryptic speciation (Knowlton, 1993; Thompson *et al.*, 2005; Matthews, 2006) and are morphologically variable within a species. To clarify species distinctions it is necessary to not rely solely on museum work with dead specimens but to also make careful field observations (such as live coloration, goby partner identity and preferred microhabitat) and conduct laboratory studies on live specimens (Banner and Banner, 1982; Anker, 2000). There is a real need for the description of associated alpheid shrimp species and analysis of their relatedness and relationship with free living species. This analysis can be achieved by combining classical morphological, field observations and molecular techniques.

BIOGEOGRAPHY

Over the past two decades several studies have helped to further our understanding of how patterns of goby and shrimp species richness vary around the world. For example, Randall *et al.* (2005, 2007) examined the distribution and described three new species of gobies in the genus *Ctenogobiops*. Recent reviews and species descriptions of the genera *Amblyeleotris* (Aonuma and Yoshino, 1996; Chen *et al.*, 1998; Mohlmann and Randall, 2002; Denou and Aonuma, 2007), *Cryptocentrus* (Winterbottom, 2002; Koichi *et al.*, 2005), *Flabelligobius* (Chen and Fang, 2003; Koichi *et al.*, 2005), *Myersina* (Winterbottom, 2002), *Stonogobiops* (Winterbottom, 2002), *Tomiyamichthys* (Iwata *et al.*, 2000; Randall and Chen, 2007) and

Vanderhorstia (Randall, 2007) have also expanded our knowledge of the distribution of Indo-Pacific associated gobies. Although the rate at which new species of associated gobies are being discovered indicates that there is still much work to be done to fully elucidate global patterns of species richness of associated gobies and shrimps, much progress has been made in this direction since this association was reviewed by Karplus (1987).

Despite the recent research on the distribution of different goby genera, at present there has not been a synthetic, quantitative examination of the relationship between latitude and longitude on variability in species richness of associated gobies and shrimps. To evaluate how species richness changes with global location, we compiled species lists from several locations in the Indo-Pacific. Our primary source for site-specific fish species information was the FishBase website (Froese and Pauly, 2008), but we also reviewed species accounts from the recent literature (e.g. Randall, 2007; Randall and Chen, 2007). Because species lists are not necessarily comprehensive in the Fish Base database, we restricted our analysis to locations that met at least one of the following conditions: 1) specific studies on shrimp gobies had been conducted at the location (Red Sea, Seychelles, Maldives), 2) comprehensive surveys of fish diversity were made at the location (Red Sea, Taiwan, Guam, Indonesia, Papua New Guinea), and/or 3) we had personally collected associated gobies and shrimps at the location (Red Sea, Society Islands, Fiji, Papua New Guinea, Taiwan). If we identified associated gobies that were not included in the fishbase.org database for a particular site, we added them to the list. It is important to note that the species lists generated for this analysis incorporate those species that are confirmed as being present at a location and that absence of a species does not necessarily indicate that it is not present at a given locality. We restricted our analysis to include only species of gobies that depend obligately on shrimps.

We tested the hypothesis that goby and shrimp diversity is maximal near Indonesia and decreases away from this location (Bellwood and Hughes 2001). Despite the caveats associated with our data, results reveal a clear pattern: species richness of associated gobies peaks near Indonesia and decreases away from this center of diversity (Figure 1-A, Table 4.4.3A). Indeed, we identified 46 species of associated gobies that are recorded in Indonesia (Table 4.4.1A). By contrast, islands west of Indonesia are reported to contain 16 (Maldives) and 13 species (Seychelles). Taiwan, which is approximately 3000 kilometers north of Indonesia contains a reported 22 associated species. Diversity also drops as distance increases eastward from Indonesia as records indicate that Papua New Guinea contains 36 species,

Table 4.4.3 **A.** Species of gobies that are obligately associated with alpheid shrimps from several locations in the Indo-Pacific. Species were primarily identified using Froese and Pauly (2008). Superscript numbers indicate species that were identified by the following sources: [1]Mous *et al.*, 2005; [2]photographed and/or collected and deposited in the Section of Fishes of the Natural History Museum of Los Angeles County by Andrew Thompson; [3]Allen and Munday, 1999, Checklist of the Fishes of Kimbe Bay, Papua New Guinea (www.walindi.com/fish), [7]Randall, 2007, [8]Iwata *et al.*, 2000, [9]Randall and Chen, 2007, [10]Randall *et al.*, 2007. **B.** Species of shrimps which are obligately associated with gobies from several locations in the Indo-Pacific. All sites were visited by Andrew Thompson and/or Ilan Karplus except Hawaii. Superscript numbers indicate the source of shrimp species information: [2]same as above; [4]same shrimp as identified by Masayoshi and Shiratori, 2003; [5]Karplus *et al.*, 1981, [6]unknown shrimp, [11]Preston, 1978 and Nelson, 2005.

A

Indonesia	Papua New Guinea	Taiwan	Maldives	Seychelles	Red Sea	Fiji	Society Islands	Hawaii
Amblyeleotris arcupinna [1]	*Amblyeleotris arcupinna* [3]	*Amblyeleotris fasciata*	*Amblyeleotris aurora*	*Amblyeleotris aurora*	*Amblyeleotris diagonalis*	*Amblyeleotris guttata* [2]	*Amblyeleotris katherine* [2]	*Psilogobius mainlandi*
Amblyeleotris aurora	*Amblyeleotris diagonalis*	*Amblyeleotris fontanesii*	*Amblyeleotris diagonalis*	*Amblyeleotris diagonalis*	*Amblyeleotris fasciatus*	*Amblyeleotris periophthalma* [2]	*Ctenogobiops fercoculus* [2]	
Amblyeleotris diagonalis	*Amblyeleotris fontanesii*	*Amblyeleotris guttata*	*Amblyeleotris periophthalma*	*Amblyeleotris periophthalma*	*Amblyeleotris steinitzi*	*Amblyeleotris steinitzi* [2]	*Mahidolia mystacina*	
Amblyeleotris fasciata	*Amblyeleotris guttata*	*Amblyeleotris periophthalma*	*Amblyeleotris steinitzi*	*Amblyeleotris sungami*	*Amblyeleotris sungami*	*Amblyeleotris randalli*	*Vanderhorstia ornatissima*	
Amblyeleotris fontanesii	*Amblyeleotris gymnocephala*	*Amblyeleotris steinitzi*	*Amblyeleotris wheeleri*	*Cryptocentrus cryptocentrus*	*Amblyeleotris triguttata*	*Amblyeleotris wheeleri*		
Amblyeleotris guttata	*Amblyeleotris novaecaledoniae*	*Amblyeleotris wheeleri*	*Cryptocentrus fasciatus*	*Cryptocentrus fasciatus*	*Amblyeleotris wheeleri*	*Ctenogobiops aurocingulus*		
Amblyeleotris gymnocephala	*Amblyeleotris periophthalma*	*Cryptocentrus albidorsus*	*Cryptocentrus strigilliceps*	*Ctenogobiops crocineus*	*Cryptocentrus caeruleopunctatus*	*Ctenogobiops crocineus* [2]		
Amblyeleotris periophthalma	*Amblyeleotris randalli*	*Cryptocentrus cryptocentrus*	*Ctenogobiops crocineus*	*Ctenogobiops feroculus*	*Cryptocentrus cryptocentrus*	*Ctenogobiops ferocululs* [2]		
Amblyeleotris latifasciata [1]	*Amblyeleotris rhyax*	*Cryptocentrus nigrocellatus*	*Ctenogobiops feroculus*	*Stonogobiops dracula*	*Cryptocentrus fasciatus*	*Ctenogobiops mitodes* [10]		
Amblyeleotris randalli	*Amblyeleotris steinitzi*	*Cryptocentrus strigilliceps*	*Flabelligobius latruncularia*	*Stonogobiops nematodes*	*Cryptocentrus lutheri*	*Ctenogobiops pomastictus* [2]		
Amblyeleotris steinitzi	*Amblyeleotris wheeleri*	*Ctenogobiops aurocingulus*	*Stonogobiops dracula*	*Vanderhorstia delagoae*	*Ctenogobiops crocineus*	*Cryptocentrus koumansi*		
Amblyeleotris wheeleri	*Amblyeleotris yanoi*	*Ctenogobiops crocineus*	*Tomiyamichthys latruncularius* [9]	*Vanderhorstia ornatissima*	*Ctenogobiops feroculus*	*Cryptocentrus strigilliceps* [2]		
Amblyeleotris yanoi	*Cryptocentrus cinctus*	*Ctenogobiops feroculus*	*Tomiyamichthys praealta* [9]	*Vanderhorstia praealta*	*Flabelligobius latruncularia*	*Lotilia graciliosa*		
Cryptocentrus albidorsus	*Cryptocentrus cyanotaenia*	*Ctenogobiops formosa*	*Vanderhorstia ambanoro*		*Tomiyamichthys fourmanoiri* [9]	*Vanderhorstia bella*		
Cryptocentrus caeruleomaculatus	*Cryptocentrus fasciatus*	*Ctenogobiops pomastictus*	*Vanderhorstia ornatissima*		*Vanderhorstria delagoae*			
Cryptocentrus cinctus	*Cryptocentrus inexplicatus* [2]	*Ctenogobiops tangaroai*	*Vanderhorstia praealta*		*Vanderhorstia mertensi*			
Cryptocentrus cyanotaenia	*Cryptocentrus leptocephalus*	*Flabelligobius smithi*			*Vanderhorstria opercularis* [7]			
Cryptocentrus fasciatus	*Cryptocentrus leucostictus*	*Lotilia graciliosa*						
Cryptocentrus inexplicatus	*Cryptocentrus octofasciatus* [3]	*Mahidolia mystacina*						
Cryptocentrus leptocephalus	*Cryptocentrus singapurensis*	*Myersina yangii*						
Cryptocentrus leucostictus	*Cryptocentrus strigilliceps*	*Tomiyamichthys smithi* [9]						
Cryptocentrus maudae	*Cryptocentrus sp. A*	*Vanderhorstia mertensi*						
Cryptocentrus nigrocellatus	*Ctenogobiops aurocingulus*							
Cryptocentrus octofasciatus [1]	*Ctenogobiops crocineus*							
Cryptocentrus strigilliceps	*Ctenogobiops feroculus*							
Cryptocentrus sp. 2 [1]	*Ctenogobiops mitodes* [10]							
Cryptocentrus sp. 3 [1]	*Ctenogobiops phaeostictus* [10]							
Cryptocentrus sp. 4 [1]	*Ctenogobiops pomastictus*							
Ctenogobiops aurocingulus	*Ctenogobiops tangaroai*							
Ctenogobiops crocineus	*Lotilia graciliosa*							
Ctenogobiops feroculus	*Mahidolia mystacina*							
Ctenogobiops mitodes [10]	*Myersina lachneri* [3]							
Ctenogobiops pomastictus	*Stonogobiops xanthorhinica*							
Ctenogobiops tangaroai	*Tomiyamichthys oni*							
Ctenogobiops sp. [2]	*Vanderhorstia ambanoro*							
Lotilia graciliosa	*Vanderhorstia lanceolata*							
Mahidolia mystacina	*Vanderhorstia mertensi*							
Myersina nigrivirgata [1]	*Vanderhorstia ornatissima*							
Myersina filifer								
Myersina papuanus								
Stonogobiops nematodes								
Stonogobiops xanthorhinica								
Tomiyamichthys alleni [8]								
Tomiyamichthys oni								
Tomiyamichthys tanyspilus [9]								
Vanderhorstia ambanoro								
Vanderhorstia lanceolata								

B

Papua New Guinea	Taiwan	Red Sea	Fiji	Society Islands	Hawaii
Alpheus bellulus[2]	*Alpheus bellulus*[2]	*Alpheus bellulus*[5]	*Alpheus djeddensis*[2]	*Alpheus djeddensis*[2]	*Alpheus rapax*[11]
Alpheus brevicristatus[2]	*Alpheus djeddensis*[2]	*Alpheus djiboutensis*[5]	*Alpheus ochrostriatus*[2]	*Alpheus ochrostriatus*[2]	*Alpheus rapicida*[11]
Alpheus djeddensis[2]	*Alpheus djiboutensis*[2]	*Alpheus ochrostriatus*[5]	*Alpheus rapax*[2]	*Alpheus rapax*[2]	
Alpheus ochrostriatus[2]	*Alpheus ochrostriatus*[2]	*Alpheus rapax*[5]	*Alpheus sp. 5 pink*[2]		
Alpheus randalli[2]	*Alpheus rubromaculatus*[2]	*Alpheus rubromaculatus*[5]			
Alpheus rapax[2]		*Alpheus* sp.[5, 6]			
Alpheus sp. 1[2,4]					
Alpheus sp. 5 pink[2,4]					
Alpheus sp. 7 white[2,4]					
Alpheus sp. A[2,6]					

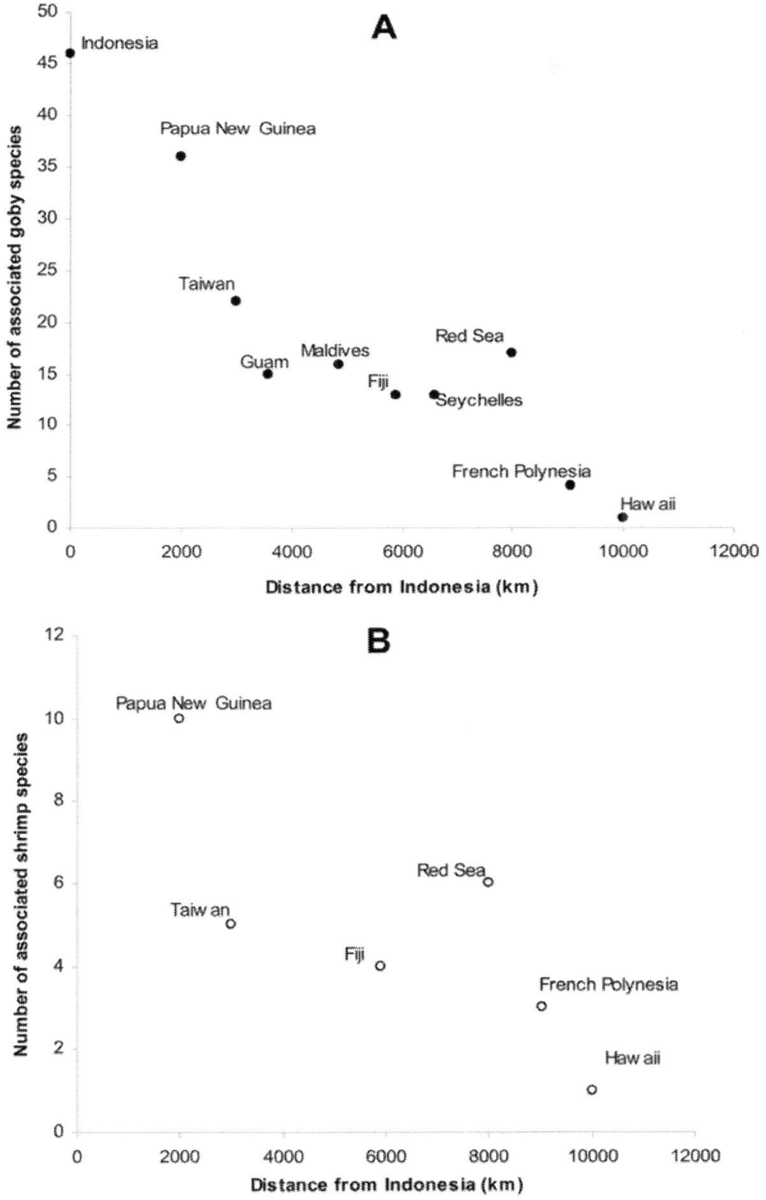

Fig. 4.4.1 Species richness of associated gobies and shrimps in the Indo-Pacific. A. Minimum number of obligately associated shrimp-goby species from ten locations in the Indo-Pacific. B. Minimum number of shrimp species that are obligately associated with gobies from six locations in the Indo-Pacific.

Guam 15 species, Fiji 13 species, and French Polynesia four species. The most distant site, Hawaii, harbors only one species of associated goby.

Patterns of species richness appear to be similar for obligately associated shrimps. Because there are few published, comprehensive site-specific surveys for alpheid shrimps in the Indo-Pacific, we present unpublished data from our own research on Papua New Guinea, Taiwan, Fiji, and the Society Islands (Thompson and Thacker, unpublished data) and published reports from the Red Sea (summarized in Karplus 1987). In addition, we include information on shrimp species from Hawaii where the list is likely complete. We collected and/or surveyed ten associated shrimps in Papua New Guinea, three in Taiwan, three in Fiji, and three in French Polynesia. In Hawaii, two species of associated shrimps are reported (Nelson, 2005). While these surveys were not necessarily comprehensive across all habitat types in these locations, the pattern of decreasing diversity away from Indonesia is similar to that shown by associated gobies (Fig. 4.4.1B, Table 4.4.3B). As mentioned in the Systematics section of this review, better elucidation of shrimp taxonomy may further clarify how species richness varies across coral reefs worldwide.

Our finding that goby and shrimp diversity peaks near Indonesia and declines eastward and westward parallels studies of other coral reef taxa (Bellwood *et al.*, 2005). A number of hypothesis exist to explain the notably high species diversity of marine organisms in the Indo-West Pacific (IWP) (reviewed in Palumbi, 1996, 1997). The center of origin model postulates that species arose in the IWP region in the west, and dispersed eastward. The center of accumulation model holds that speciation took place in outlying archipelagos such as Hawaii, the Marquesas Islands or the Society Islands, followed by dispersal westward into the IWP. The center of overlap model would indicate no consistent pattern for speciation or dispersal, because IWP diversity would simply be an artifact of overlapping biogeographic provinces in which speciation and dispersal occurred throughout. Although it is difficult to definitively state that one or more of these hypothesis explains the gradient in associated goby and shrimp species richness, Thompson *et al.* (2005) showed that decreasing within-site genetic diversity for *Ctenogobiops feroculus* and *Alpheus djeddensis* in an eastward direction is consistent with dispersal away from the center of diversity (see goby-shrimp phylogeography chapter in this review). Further molecular work on more goby-shrimp taxa could further elucidate causal patterns of the biogeographic gradient in the species richness of associated gobies and shrimps.

Whereas most of the research on goby-shrimp partnerships focused on obligate, Indo-Pacific genera, associated gobies and shrimps are also found in the Atlantic. Five species of gobies (*Bathygobius curacao*, *Gobionellus saepepallens*, *Gobionellus stigmalophius*, and *Nes longus*) are known to associate with burrowing alpheid shrimp in the western Atlantic. Of these species only *Nes longus* maintains an obligate relationship with its shrimp partner (Karplus, 1992; Randall *et al.*, 2005; Kramer *et al.* 2009). Thus it appears that although the total number of goby species that associate with shrimp is much lower in the Atlantic than the Indo-Pacific, the proportion of species that interact facultatively with shrimp may be higher in the Atlantic. Additional research on the relative number of facultative and obligate species in different parts of the world is needed to confirm this pattern.

ECOLOGY

Demographic Impacts of the Mutual Partnership

Although it is well established that many species of gobies and alpheid shrimps may participate in this mutualism, the exact nature of the benefits provided by each partner are not fully understood. From a demographic perspective, the mutual benefits are reflected through either an augmentation in rates of birth or a reduction in rates of death (Thompson *et al.*, 2006). Given the abundance of predators on coral reefs, it is generally believed that the primary function of the goby-shrimp mutualism is to directly reduce the rates of death for both partners (Karplus, 1987; Nelson, 2005). From the perspective of the goby, the shrimp provides a burrow within which the fish seeks shelter from predators, thus reducing goby mortality. Rates of death are also expected to be alleviated for the shrimp as gobies warn shrimp when active outside the burrows of the presence of predators through tactile communication (Karplus, 1979, 1987; Preston, 1978). Therefore, shrimp are believed to be less vulnerable to predators when paired with a goby.

A laboratory experiment was carried out to determine the impact of the partnership on rates of growth of both gobies and shrimp forming an obligatory partnership (Thompson, 2003). Recently recruited gobies (*Ctenogobiops feroculus*) and shrimp (*Alpheus djeddensis*) from a coral reef lagoon in Moorea, French Polynesia were placed into aquaria containing optimal substrate (70% sand, 30% rubble; Thompson, 2004). Individuals were initially measured, weighed and randomly assigned to one of four treatments: goby alone, shrimp alone, goby plus small shrimp, or goby plus large shrimp. Individuals in each treatment were fed TetraMin flake food and brine shrimp *ad libitum*. After 24 days, individuals were reweighed and it was determined whether growth rates differed among treatments. Results showed that shrimp growth was significantly greater in treatments where a goby was present (Figure 4.4.2A). By contrast, there was no difference in goby growth between treatments where shrimp were present or absent (Figure 4.4.2B). Thompson (2003) suggested that shrimp growth rates are reduced without gobies because shrimp curtail foraging when separated from their partner. Reduced rates of foraging are known to retard growth (von Bertalanffy, 1938), which, in turn, reduces reproductive potential. Gobies, however, continue foraging even if shrimp are absent and hence, in the absence of predators, there is no impact on goby growth rate by shrimp. Although this experiment is suggestive of a demographic effect, the results must be treated with some caution because it is difficult to perform an artifact-free laboratory experiment on the goby-shrimp system. For example, although Thompson (2003) carefully used sediments from the lagoon (where these goby and shrimp species co-occur), that contained some natural food, the primary source of food in the experiment was TetraMin flakes that differed in structure, composition and location from natural food. Further, although the experimental, the goby and shrimp seemed to behave normally (despite the absence of predators) with the shrimp constructing burrows that were occupied by both partners, one cannot rule out that the shrimp occasionally left the burrows wander to and feed over the entire aquarium as shrimp often do in captivity (Karplus, pers. obs.). Despite the caveats inherent to Thompson's (2003) findings that the goby impacted the growth rate of shrimp a subsequent field study carried out in Hawaii on *A. rapax* (Nelson, 2005) seemed to corroborate Thompson's (2003) results. In this study a shrimp deprived of its fish partner *Psilogobius mainlandi* revealed a marked reduction in its foraging activity compared with an associated shrimp. In addition, field observations in the Red Sea (Magnus, 1967; Karplus, 1987) and the Great Barrier Reef (Cummins, 1979) suggested that shrimp deprived of their goby shifted to a subterranean existence, very likely that the growth of these shrimp was reduced. Field experiments contrasting the growth of marked associated shrimp deprived of their goby and in its presence could reveal if gobies indeed affect the growth of shrimps.

The impact of the shrimp on goby mortality was assessed through a field experiment in which marked gobies were released into areas of a lagoon in Moorea containing resident *Ctenogobiops feroculus* and *Alpheus djeddensis* (Thompson, 2005). Prior to goby release, the location of residents was mapped for a period of approximately two weeks. Upon goby addition, introduced gobies competed with residents for burrows and 20 of the 48

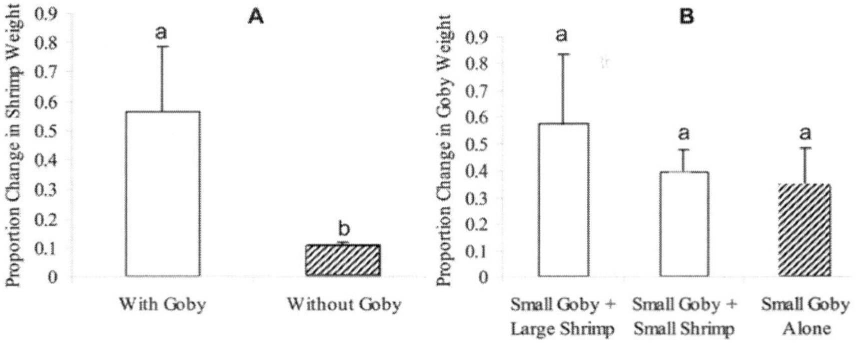

Fig. 4.4.2 The influence of gobies on the growth rate of their shrimp partners and vice versa (Error bars depict 1 SE). **A.** Proportion change in *Alpheus djeddensis* weight after 24 days in the presence and absence of a *Ctenogobiops feroculus*. **B.** Proportion change in the weight of *Ctenogobiops feroculus* after 24 days when paired with a large (mean total length = 53.7 mm) *A. djeddensis*, small (mean total length = 23.4 mm) *A. djeddensis* or when the goby was alone (after Thompson, 2003).

invaders were observed in burrows that previously contained a resident. Of the remaining 28, none were observed even following extensive searching of the areas surrounding release. Thompson (2005) interpreted this finding to mean that gobies that failed to pair with shrimp were consumed by predators.

The direct impact of the goby on the shrimp survival is difficult to evaluate because shrimp deprived of their goby greatly reduce their activity outside the burrows. Long term monitoring of marked shrimp with and without gobies in the field and of the burrow openings of isolated shrimp are needed to establish whether shrimp suffer from greater predation in the absence of their fish partner.

Diet and Feeding Behavior

Little research has been conducted on the diet of various species of associated gobies and shrimps subsequent to Karplus's (1987) review. Briefly, gobies of the genera *Amblyeleotris*, *Cryptocentrus*, *Ctenogobiops* and *Vanderhorstia* feed by picking organisms directly from the sand as well as taking mouthfuls of sand and filtering organisms through their gill rakers (Magnus, 1967; Hoese and Allen, 1976; Cummins, 1979). Gobies typically pick through sediment around the burrow entrance and have been observed to pick through sediment extracted from a burrow by a shrimp (Hoese and Steene, 1978; A. Thompson, pers. obs.). In addition, some species such as *Amblyeleotris japonica* in Japan (Yanagisawa, 1982), and *Stonogobiops xanthorhinica* in Papua New Guinea (A. Thompson, pers. obs.) were observed to ascend approximately 1-5 cm in the water column and apparently pick at planktonic food items.

Analysis of *A. japonica* gut contents showed that these gobies consumed primarily corophiid amphipods and other small species of crustaceans (Yanagisawa, 1982). Similarly, Cummins (1979) stated that stomach contents of shrimp-associated gobies found near One Tree Reef in Australia of the genera *Amblyeleotris*, *Cryptocentrus*, *Ctenogobiops* and *Vanderhorstia*, consisted largely of small invertebrates (i.e., amphipods, copepods, bivalves, and worms), as well as some algae.

Goby-associated shrimps appear to be opportunistic feeders that consume interstitial matter exhumed during burrow excavation and material such as algae that is taken when shrimp are outside the burrow. Shrimps in Hawaii (Moehring, 1972), One Tree Reef (Cummins, 1979) and the Philippines (Palomar *et al.*, 2004) were all observed taking algae into their burrows. A recent study using stable isotope analysis of the gut contents of *Alpheus macellarius* in the Philippines demonstrated that organic matter from the sediment and sea grass constitute the majority of the diet of this shrimp. Furthermore, there was a positive relationship between sediment organic content and the density of this species (Palomar *et al.*, 2004). In addition, at least one species of shrimp forages by cleaning its goby partner. *Alpheus djiboutensis* was documented inside its burrow placing its first set of chelae on *Cryptocentrus cryptocentrus* and repeatedly moving its second set of chelae from the fish to its mouth (Karplus *et al.*, 1972a).

Habitat Specificity

Several studies indicate that the distribution of goby and shrimp species is influenced by habitat parameters. For example, Yanagisawa (1984) demonstrated that *Amblyeleotris japonica* and *Alpheus bellulus* in Japan are found primarily in locations with a mixture of sand, pebbles, coral debris and shell fragments, but are rarely found in pure-sand habitats. Karplus *et al.* (1981) showed that four species of goby-associated shrimps in the Red Sea inhabit discrete microhabitats that are characterized by depth and sediment grain size.

More recently, studies in Australia and French Polynesia quantified habitat associations of shrimp-associated gobies. Thompson (2004) found that there were non linear relationships between the proportion of sand, rubble, coral and coral pavement within plots and the density of *Ctenogobiops feroculus* that paired with *Alpheus djeddensis* throughout the relatively shallow lagoons on the north shore of Moorea, French Polynesia.

Specifically, he found that maximal goby densities occurred in locations with approximately 70% sand and 30% rubble, and that densities tended to be low in sites comprised entirely of sand, rubble, coral, or pavement (Fig. 4.4.3). A laboratory experiment demonstrated that this mixture of sand and rubble is optimal for burrow construction by *A. djeddensis* as this shrimp was unable to construct burrows in pure-sand habitats and relied on the rubble to buttress tunnels (Thompson 2004; see also section 5 on burrow structure). Thompson (2004) also found that the three species of associated gobies and shrimps that are found in Moorea were completely allopatric and segregated by habitat as *C. feroculus* and *A. djeddensis* were found in shallow, sandy lagoons, *Vanderhorstia ornatissima* and *Alpheus brevicristatus* occurred only in silty, highly protected areas, and *Amblyeleotris fasciata* and *Alpheus ochrostriatus* lived exclusively on the outer reef crest. Similarly, Syms and Jones (2004) showed that habitat composition dictated the distribution of multiple shrimp-associated gobies at Lizard Island in Australia's Great Barrier Reef. Here, *Vanderhorstia ornatissima* was found

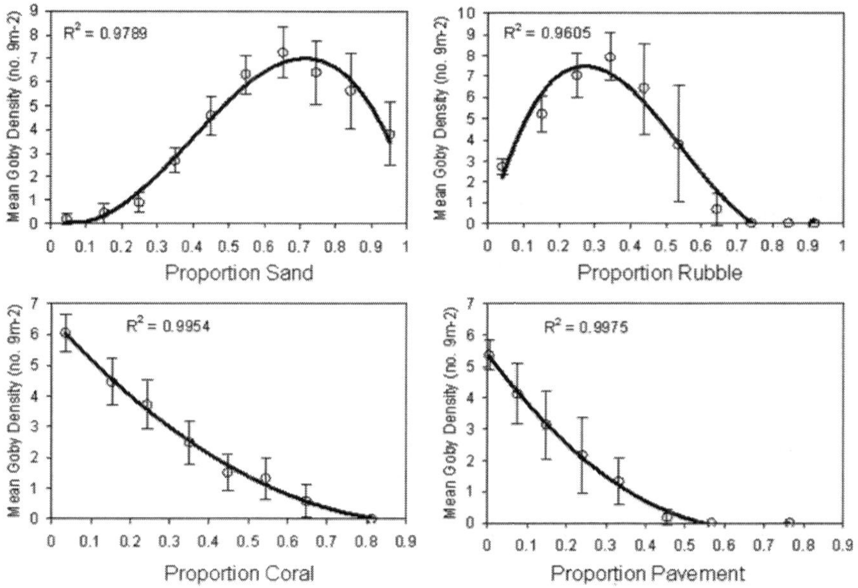

Fig. 4.4.3 *Ctenogobiops feroculus* habitat and density. Scatter plots and best-fit trend lines depicting relationships between habitat and goby density using data calculated at 0.1 intervals bins of each habitat type. Error bars represent 2 SE. Trend lines were used to fit values taken from each bin and were characterized by the following equations: $y = (-)51.86x^3 + 60.89x^2 - 7.62x + 0.28$ (sand), $y = 92.28x^3 - 152.29x^2 + 62.29x - 0.06$ (rubble), $y = 7.27x^2 - 13.87x + 6.50$ (coral), $y = 12.84x^2 - 16.89x + 5.40$ (pavement), where y = mean goby density and x = proportion habitat (after Thompson, 2004).

exclusively in the central lagoon and on front reef slopes, *Ctenogobiops pomastictus* and *C. aurocingulus* on reef flats, sheltered back reefs, and near the lagoon entrance, *Cryptocentrus cinctus* on the sheltered back reef, and *Amblyeleotris steinitizi* on the sheltered back reef and near the lagoon entrance.

Population Dynamics

Population dynamics of multiple, sympatric species of gobies and shrimps were studied through a series of observational and manipulative experiments on One Tree Reef by Cummins (1979). First, censuses of goby and shrimp abundances at three locations demonstrated that total population size and relative abundance of particular goby and shrimp species changed significantly through time. Next, Cummins (1979) followed eight different burrows that contained four species of shrimps. Over three observation periods spanning a period of five months, the species of goby occupying a burrow changed at least once in five of the burrows. Finally, Cummins removed gobies from 14 burrows and found that 12 were recolonized within three weeks. Of the 12 recolonizations, 10 were by the same species of goby that was initially removed, one was by a goby that was observed to commonly associate with that species of shrimp in the field, and one was colonized by a goby that was rarely observed with that species of shrimp. The latter association, however, was temporary; subsequent observations demonstrated that this goby abandoned the burrow (see also section below on partner specificity).

The influence of biotic factors on shrimp/goby population dynamics were explored recently in Moorea by Thompson (2005) and in Hawaii by Nelson (2005). Thompson conducted several observational and field experiments to elucidate the interplay between predation and intraspecific competition on *Ctenogobiops feroculus* populations. First, surveys of sites with habitat appropriate for *C. feroculus*/*Alpheus djeddenis* [a combination of rubble and sand; see Thompson (2004)] but with variable densities of piscivorous predators showed that there was no correlation between *C. feroculus* and predator densities. Predator density, however, correlated negatively with the proportion of large *C. feroculus* and positively with the proportion of small *C. feroculus* paired with large shrimp. Next, an experiment in which tagged *C. feroculus* were added to several plots demonstrated that 1) these gobies competed intraspecifically for burrows, 2) large gobies outcompeted smaller gobies, and 3) gobies preferred pairing with large as opposed to small *A. djeddensis*. Finally, Thompson (2005) experimentally augmented predator densities by placing roof tiles that attracted ambush predators such as spotted sandperch (*Parapercis millepunctata*) on several plots. Although goby mortality increased significantly in predator-addition sites relative to control plots, goby abundance did not change in the predator addition plots. Rather, gobies that were consumed were replaced quickly by recruits. As such, the proportion of large gobies per site decreased following predator addition. Predation, however, seemed to not impact *A. djeddensis* as the size structure of shrimp remained unchanged in response to elevated predator densities. Therefore, the proportion of small gobies paired with large shrimp increased following predator addition. Nelson (2005) found a very similar impact of predators on *Psilogobius mainlandi* and *Alpheus rapax* in Hawaii. Rather than attract predators, Nelson (2005) utilized predator exclusion cages to reduce goby-shrimp exposure to predators. He found that while goby densities did not change after five months of predator exclusion, there were more than twice as many large gobies in exclusion plots as compared to control plots.

BEHAVIOR

Burrow Structure, Construction and Dynamics

The three dimensional subterranean structure of the burrows of associated alpheid shrimps was revealed with the application of polyester and epoxy resin casts (Karplus, 1987). Overall, burrows were shallow (not more than half a meter deep), and in close contact with hard objects such as coral and rock boulders. Due to the tendency of sand to collapse, hard objects were used to support the subterranean burrow structure. Sections of the burrows leading under rocks or corals had chamber-like enlargements whereas casts retrieved from sediment lacking supporting objects had an even diameter at different points (Fig. 4.4.4B, C).

The effect of the substratum on burrow structure has been demonstrated for *Alpheus crassimanus* (Farrow, 1971). Whereas burrows of this species that were located in coarse substratum with hard objects had an irregular structure, those located in muddy/silty substratum had a regular, dichotomous branching pattern.

Three basic types of burrow openings have been described for Red Sea associated shrimps including asymmetrical tube, symmetrical tube, and funnel-like types (Fig. 4.4.4A). Whereas the burrow openings of all associated shrimps are reinforced by coral and shell fragments, the subterranean burrow sections of only some of the shrimp species are reinforced with hard substrates (Cummins, 1979; Karplus, 1987). In contrast to the lack of species specificity in the structure of the subterranean burrows, some specificity was revealed in the number (ranging from one to

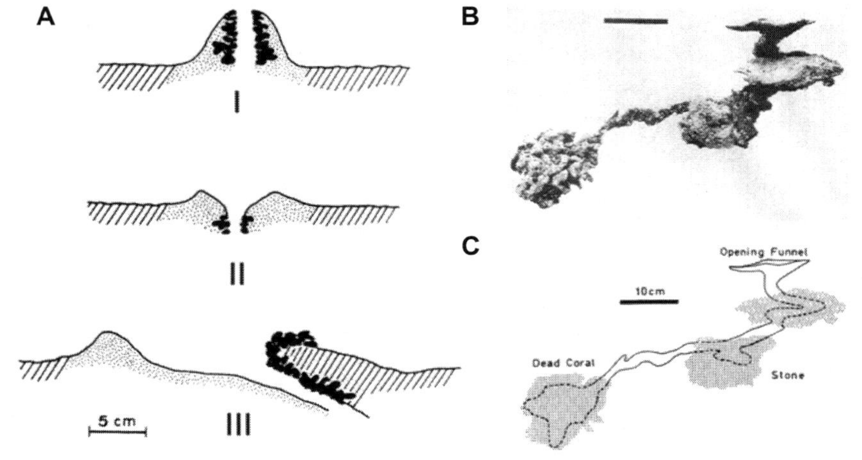

Fig. 4.4.4 The burrows of alpheid shrimps associated with gobies. **A.** Three different types of burrow openings: **I.** Tube-like symmetrical opening of *Alpheus rapax*. **II.** Funnel-like symmetrical opening of *Alpheus rubromaculatus*. **III.** Asymmetrical opening of *Alpheus djiboutensis*. **B.** Epoxy resin cast of the burrow of *Alpheus djiboutensis*. **C.** Drawing indicating the same burrow structure relation to large objects in the sediment (after Karplus *et al.*, 1974).

six) and structure of the burrow openings of Red Sea (Karplus *et al.*, 1974) and Great Barrier Reef associated shrimps (Cummins, 1979).

The behavior of several species of goby-associated shrimps during burrow construction has been described in aquaria (Harada, 1969; Karplus *et al.*, 1972a) and in the sea (Luther, 1958a; Macnae and Kalk, 1962; Magnus, 1967; Farrow, 1971; Yanagisawa, 1984). Three different digging techniques have been revealed: 1) Digging with the first pair of chelae into a vertical sand wall and twisting until the sand collapses; 2) Digging with the second pair of chelae and the third and fourth pairs of periopods; and 3) Digging with the pleopods, mostly practiced inside the burrow. The transport of sand and small stones from inside the burrow is carried out by the first pair of chelae that are held together to form a flat spade broadened by rows of long hairs fringing the chelae margins. Occasionally, large, flat objects such as a shell are grasped by the first pair of chelae and lifted up to reinforce the burrow entrance or used like a tool for more efficient transfer of sediment (Magnus, 1967; Karplus *et al.*, 1972a).

With very few exceptions, associated gobies rely exclusively on shrimp to construct burrows. There is only a single report from Aldabra Atoll of a non-identified goby that participated with *Alpheus crassimanus* in burrow construction. Here, the goby enlarged upper parts of the burrow by removing mouthfuls of mud from within the burrow and ejecting them at the periphery (Farrow, 1971). By contrast, aquarium-based observations of several species of associated gobies such as *Nes longus* (Weiler, 1976), *Cryptocentrus caeruleopunctatus* (Magnus, 1967), *Amblyeleotris japonica* (Harada, 1969), and *Ctenogobiops feroculus* Thompson (2004) indicated that these gobies are unable to construct a burrow. When isolated from shrimp, these species formed a shallow depression by splashing sand that only provided minimal shelter. It seems that in almost all associations the burrow is constructed and maintained solely by the shrimp.

Recently, Palomar *et al.* (2005) studied in the Philippines the behavior of male *Alpheus macellarius* associated with *Cryptocentrus octafasciatus* in the laboratory under four sediments treatments that differed in the level of gravel content (0, 5, 15, and 25% gravel) for five consecutive days. Burrowing was mainly conducted during the first day and the early mornings of subsequent days, thus implying a greater priority for the shrimp to construct and maintain its burrows relative to other activities. Gravel content did not significantly influence burrowing behavior, but marked variations were noted in burrowing success, burrow structure and stability. Shrimp in 15% and 25% gravel substrates produced larger, more complex burrows that lasted longer than those of shrimp in 0% and 5% gravel substrate. The importance of a high percentage of rubble in the sediment for optimal burrow construction was also reported for *A. djeddensis* paired with *Ctenogobiops feroculus* (Thompson, 2004).

Daily changes in the position of burrow openings have been reported for several species and types of associated shrimps (Karplus, 1987). Monitoring daily changes (i.e., distance and angle) revealed maximal daily shifts of the opening ranging from 25 to 160 cm; however, the spatial location of entrances was confined to a rather limited zone of up to several square meters (Fig. 4.4.5B). Changes in burrow structure are probably restricted to the upper, shallow parts of the burrow, whereas the deeper parts that often lead under and between large boulders likely remain stable (Karplus *et al.*, 1974).

Species-specific trends were found in the daily shifts of the location of burrow openings of several species and types of shrimp in the northern Red Sea (Karplus *et al.*, 1974) and the Great Barrier Reef (Cummins, 1979). Three different, non-exclusive mechanisms have been suggested to underlie the shifting of burrow opening of different associations. First, Magnus (1967) suggested that the daily changes in the position of the opening for *Alpheus rapax* and *A. rapacida* (both associated with *Cryptocentrus caeruleopunctatus* and *Vanderhorstia delagoae*) resulted from the feeding

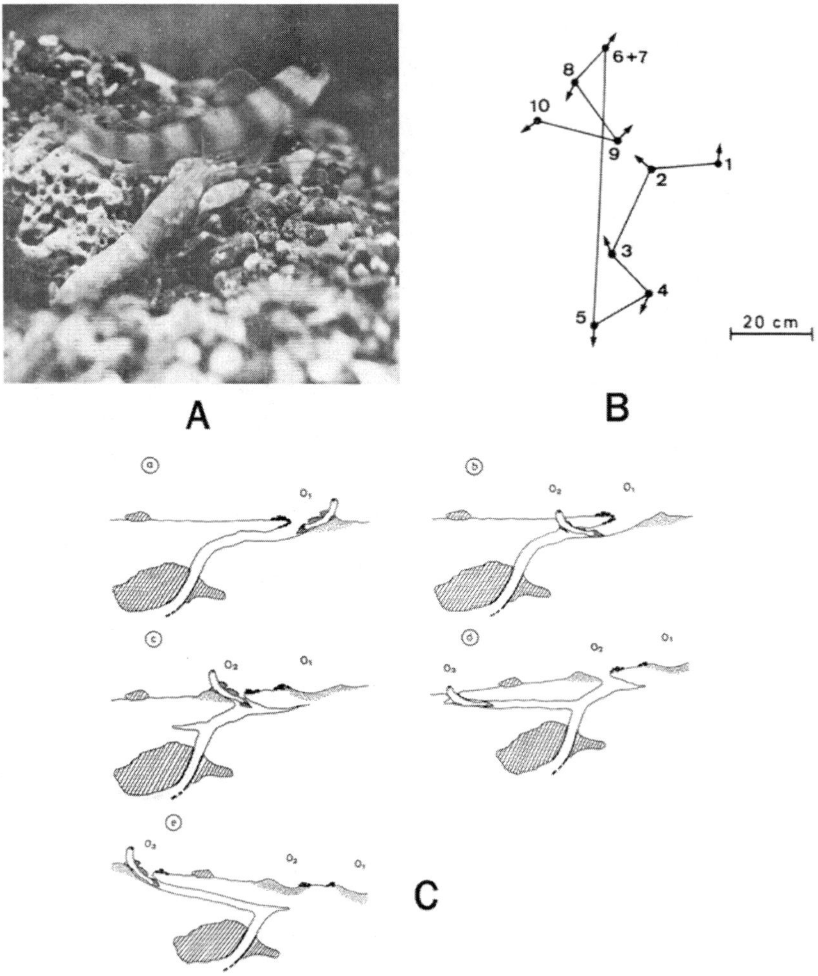

Fig. 4.4.5 Multidirectional shifts of the burrow opening due to the combined activities of goby and shrimp. A. *Amblyeleotris steinitzi* associated with *Alpheus purpurilenticularis*. B. Daily shift in position of burrow opening. Numbers indicate consecutive days. Arrows indicate location of sand pile in front of burrow opening. C. Dynamics of the upper burrow system. (a) Goby positioned in front of burrow opening (O1); (b) formation of new burrow opening (O2) by goby; (c) opening O1 collapsed, new opening functional; (d) formation of additional opening (O3) by goby; (e) opening O1 and O2 collapsed, opening O3 functional. (after Karplus *et al.*, 1974).

activity of the shrimps. These shrimps feed on non-disturbed sediment overlying the burrow and thus continuously shift the opening backwards. Second, Cummins (1979) suggested that the fixed changes in the position of associated shrimp types from the Great Barrier Reef were due to the digging activity of the shrimps. In this case, both the upper and lower parts of the burrows were reinforced by coral fragments. Therefore, all parts of the burrows were stable and the shrimps only alternately cleared or blocked them with sediment, thus reforming the opening at the same position. Third, Karplus *et al.* (1974) suggested that the irregular, multidirectional shifts in burrow entrances resulted from the combined activities of the goby *Amblyeleotris steinitzi* and the shrimp *A. purpurilenticularis* (Fig. 4.4.5A). Here, the goby wedges its head through the substratum to create a new opening and the shrimp follows and enlarges it while the old opening rapidly collapses (Fig. 4.4.5C).

Daily shifts in the location of burrow openings may be important for several reasons. First, changing the position of the burrow can continuously expose shrimps to non-disturbed sediment. Samples of sediments excavated from the burrows of *Alpheus macellarius* associated with *Cryptocentrus octafasciatus* contained significantly less organic carbon than samples taken from undisturbed sediments in proximity of the burrows (Palomar, 2001). Because shrimps feed opportunistically on detritus and other items found in the substrate (Palomar *et al.*, 2005), this may be an important source of nutrition for the shrimps and may compensate for the limited exposure to non-disturbed sediment within the relatively shallow burrows. Second, relocation of burrow entrances may reduce territorial conflicts among gobies. Conflicts between neighbors may be resolved by forming new openings further away from a nearby, dominant neighbor. Third, spacing of burrows may alleviate territorial conflicts between neighboring shrimps. Intense fighting between associated shrimps during paired staged encounters in the laboratory has been reported for *Alpheus rapax* and *A. rapacida* (both paired with *Psilogobius mainlandi*) in Hawaii (Moehring, 1972) and between same-sex *Alpheus djeddensis* in Moorea (Thompson, 2003). Fourth, during the reproductive season of gobies, shifting an opening towards potential mates could reduce the exposure to predation when individuals travel between burrows to mate (Karplus *et al.*, 1974).

Activity Rhythms

All studied species of associated gobies and shrimps have demonstrated a strict diurnal activity rhythm. Both partners remain in their burrows at night, and the burrow entrances are often closed due to wave action or are intentionally sealed off by the shrimp (Yanagisawa, 1984; Karplus,

1987). Activity is renewed by the goby around sunrise as it wedges its head through the sediment. The goby is followed very soon after by its shrimp partner who enlarges the opening and resumes its activity outside the burrow. Activity is terminated around sunset by the goby as it retreats into the burrow. Behavioral changes towards the end of the day were observed in several shrimp species as they remained outside their burrows for shorter amounts of time and reinforced the entrances with coral and shell fragments. Termination of the daily activity is more synchronous than its onset, probably due to the fact that the end is triggered by low light levels and the onset more by an endogenous rhythm as levels of light may not be perceived accurately by the goby and shrimp inside their burrows (Karplus, 1987).

Variation in the daily activity rhythms of several species of associated shrimps from different localities often revealed a marked increase in activity in the late afternoon (Karplus, 1987; Palomar, 2001; Nelson, 2005). However, maximal activity of several shrimp types in the Great Barrier Reef occurred at noon (Cummins, 1979). In the Northern Red Sea three species of associated shrimps (*Alpheus djiboutensis, A. rapax* and *A. purpurilenticularis*) displayed marked variation in mean time spent outside the burrow during the day that was concomitant with a change in the type of activity performed. Shrimps activity in the morning mostly consisted of short exits with chelae loaded with sediment cleared from burrows that collapsed during the night. In the afternoon, by contrast, exits were lengthier, and the shrimps excavated non-disturbed sediment, that was subsequently transported into the burrow, probably for nocturnal feeding (Karplus, 1987).

The activity of *Alpheus floridanus* was contrasted in southern Florida while living in association with an obligatory partner (*Nes longus*), a facultative partner (*Bathygobius curacao*) and while free living (Karplus, 1992). The shrimp spent five fold more time outside the burrow when associated with the obligatory partner than when associated with the facultative one due to both more exits per time and increased time spent outside the burrow per exit (Fig. 4.4.6). There was no difference in the types of activities carried out over the daily cycle for shrimp with obligatory and facultative partners. Shrimp associated with the obligatory partner, however, spent more time picking at plants and digging in sediment than shrimp associated with the facultative one. Shrimp with the facultative goby engaged almost entirely in burrow maintenance (e.g. dumping of sediment). Free living shrimp spent much less time (only about 2%) outside their burrows than any of the associated shrimp.

The activity of *Alpheus rapax* outside its burrows in Hawaii while it was associated with its obligate goby partner, *Psilogobius mainlandi*, or

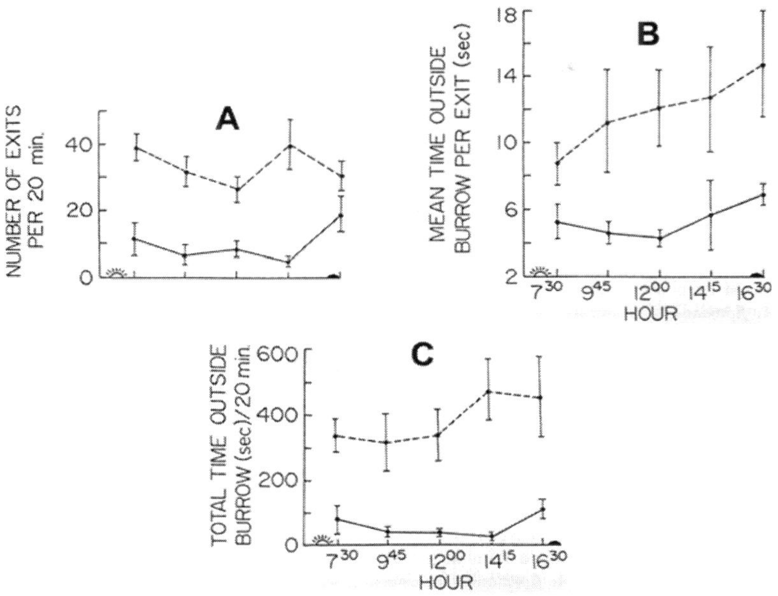

Fig. 4.4.6 Activity of *Alpheus floridanus* outside its burrow when associated with *Nes longus* (obligatory partnership-broken line) and *Bathygobius curio* (facultative partnership-continuous line) (after Karplus, 1992).

while it was deprived of its fish partner was contrasted by Nelson (2005). Whereas deprived shrimp spent only 7% of the time outside the burrow (for an average duration of 3 seconds), those with a goby were outside 54% of the time (for an average of 14 seconds). Deprived shrimp were almost entirely (98% of the exits) occupied with burrow maintenance activities such as clearing the burrow of sediment and manipulating rubble around the burrow entrance, whereas associated individuals carried out burrow maintenance in only 53% of their exits and spent the rest of their time conducting other activities such as foraging.

Interspecific Communication

The restrictive conditions for tactile communication (i.e., close proximity of the communicating parties) are perfectly met in the partnership between gobies and alpheid shrimps. As long as the shrimp stays outside the burrow it maintains a continuous antennal contact with its fish partner (Fig. 4.4.10). The goby warns the shrimp of danger by two different types of warning signals:

(1) **Tail-flicks:** rapid, small lateral fin-flicks which are only produced when the shrimp is outside the burrow and often induce the shrimp to enter its burrow or "freeze" at the entrance.

(2) **Head first entries:** rapid head first entries of the goby into the burrow produced when the shrimp is either present or not outside the burrow the shrimp always enters the burrow when the goby enters the burrow head first.

The production of warning signals by the goby *Amblyeleotris steinitzi* was studied in a shallow lagoon in the northern Red Sea during the afternoon when its shrimp partner, *Alpheus purpurilenticularis*, spent most of its time outside of the burrow (Karplus, 1979). The goby generated an average of 7.4 tail-flicks per hour. Signals were produced in series (spaced less than 5 s apart) with numbers varying from 1 to 9 and a mean of 1.7 signals per series. Tail-flicks were emitted in response to all approaching large fishes irrespective of their feeding habits, whereas all small fish were ignored. The goby was selective in its response towards approaching medium sized fishes, signaling in response to piscivorus fishes and fishes digging in the sediments that can block the burrow entrances (e.g. goatfish) but ignoring medium sized fishes that feed on algae or coral (e.g. surgeon fish and butterfly fish). In a controlled experiment using live fishes presented to the goby inside a perspex box at a fixed distance from its burrow the differential signaling response to a predatory vs. a non predatory fish was confirmed. The shrimp retreated into the burrow in response to approximately 60% of the single signals but to 90% of series composed of these same signals. Some series of signals that did not induce the shrimp's retreat seem to have been made in situations of little danger (e.g. an intruder already leaving the burrow vicinity). Head first entries by the goby were performed at a rate of 0.3 times per hour and always caused a rapid retreat of the shrimp into its burrow. The same intruding fish that induced the warning signals also induced at close range the goby's head first retreat.

Production of tail-flick warning signals by the goby *Amblyeleotris steinitzi* in response to the approach of differently sized two-dimensional predator models was investigated in the northern Red Sea (Karplus and Ben-Tuvia, 1979). The distance between the model and the goby at which the goby entered its burrow head first was defined as the critical point. A negative correlation was found between the number of tail-flicks and the distance from the critical point. The occurrence of dynamic fields of interspecific communication around the burrow opening has been suggested. As soon as a specific distance (typical for each type of intruding fish) is crossed and the shrimp is outside the burrow, the goby begins to flick its tail, signaling to the shrimp to enter its burrow. The rate of signaling increases as the intruder approaches the burrow with the alert goby remaining outside the burrow and carrying out activities such as foraging. However, as soon as the intruder reaches the critical point, the goby performs a head first entry either in the presence or absence of the shrimp partner, thus terminating all activities outside the burrow by both the goby and shrimp.

Due to the speed and complexity of the goby-shrimp tactile communication, detailed analyses can only be made through the use of video or film recordings. (Karplus *et al.*, 1979; Karplus, 1987). Quantification of films revealed five parameters of the tail-flick signal of *Amblyeleotris steinitzi* (e.g. duration and amplitude). Specifically, the communicative tail-flicks are graded and coded signals that are conspicuously different from other tail movements. The tail-flick warning signal of this species is shorter than one second in duration and had a maximal amplitude of less than 1 cm (Fig. 4.4.7). A principle component factor analysis followed by a stepwise regression analysis revealed the impact of 18 variables (e.g. shrimp chelae empty or loaded with sediment, the speed of the shrimp and its direction) on the shrimp retreat response. The structure of the preceding and actual warning signals and the area of contact of the antenna accounted for approximately a third of the differences in the shrimp's retreat response. The most important signal variable is the total distance traversed (TDT) by the caudal fin.

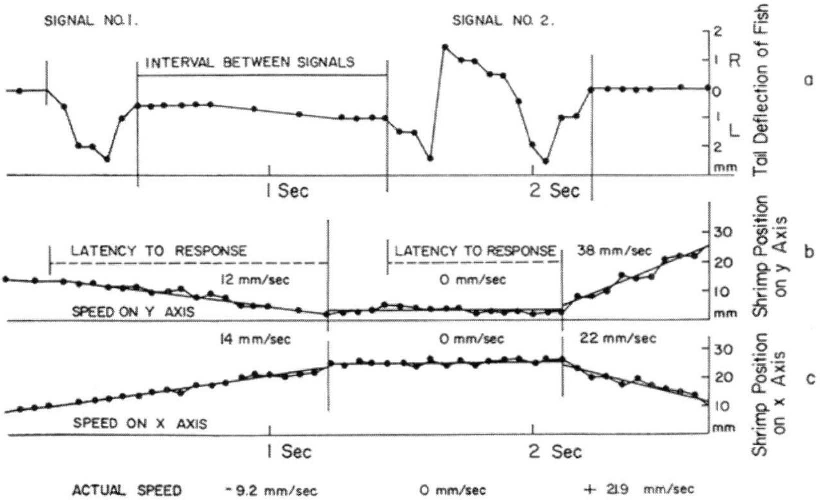

Fig. 4.4.7 Graphical presentation of the interaction between the tail-flick warning signal of *Amblyeleotris steinitzi* and the response of its shrimp partner *Alpheus purpurilenticularis* (after Karplus *et al.*, 1979).

The feedback mechanism by which the goby regulates the intensity of the tail-flick warning signals, according to the shrimp's response is one of the most complex aspects of this communication system. If the shrimp does not respond to its tail-flicks, the goby increases the intensity of its signal (measured by TDT). By contrast, upon the rapid retreat of the shrimp the fish decreases the signal intensity. The goby's regulation of the intensity of the warning signal according to the shrimp's response is possible because the mean time interval between two consecutive signals of a series (1.6 ± 1.1 s) is significantly longer than the mean latency of the shrimp's response (0.5 ± 0.4 s).

When a goby retreats head first into a burrow, the shrimp inevitably also retreats and the movement of both partners is so rapid that only a cloud of sand can be perceived by the observer. Film analysis revealed that during the head first entry the goby, which initially was facing away from the burrow entrance, bends laterally its head, almost touching its tail and then enters the burrow head first. The shrimp latency of response to a head first entry is much shorter (0.16 s) compared with its mean latency of response to a tail-flick (0.51 ± 0.36 s). Similarly, the mean speed of retreat following a head first entry (93.7 ± 54.9 mm/s) is five times more rapid than the mean retreat speed following a tail-flick (16.5 ± 7.2 mm/s). Because the head first retreat takes place under circumstances of extreme danger such as a direct predator attack, any delay in the shrimp response may cause it to be killed. Therefore, there was likely a strong selection for rapid response.

The body site of contact between the shrimp and goby differs according to the position of the shrimp relative to the goby (Karplus *et al.*, 1979). When the shrimp was behind the goby, both its antennae pointed forward and one of them touched the goby's caudal fin. When the shrimp was parallel to the goby, one antenna was bent sideways touching the second dorsal fin, while the other still pointed forward. When the shrimp was further away from the burrow than the goby, one antenna pointed backwards and touched the goby's pectoral fin, while the other was directed forward. Warning signals are given with various fins including the caudal, second dorsal, anal and pectoral fins according to the area of antennal contact. In each case, however, the caudal fin is involved in signaling (Fig. 4.4.8). The possibility of signaling using several fins enables the shrimp to move further away from the goby while still being protected through the warning system.

The importance of the antennal contact for the transmission of information in the goby-shrimp partnership was revealed in observations

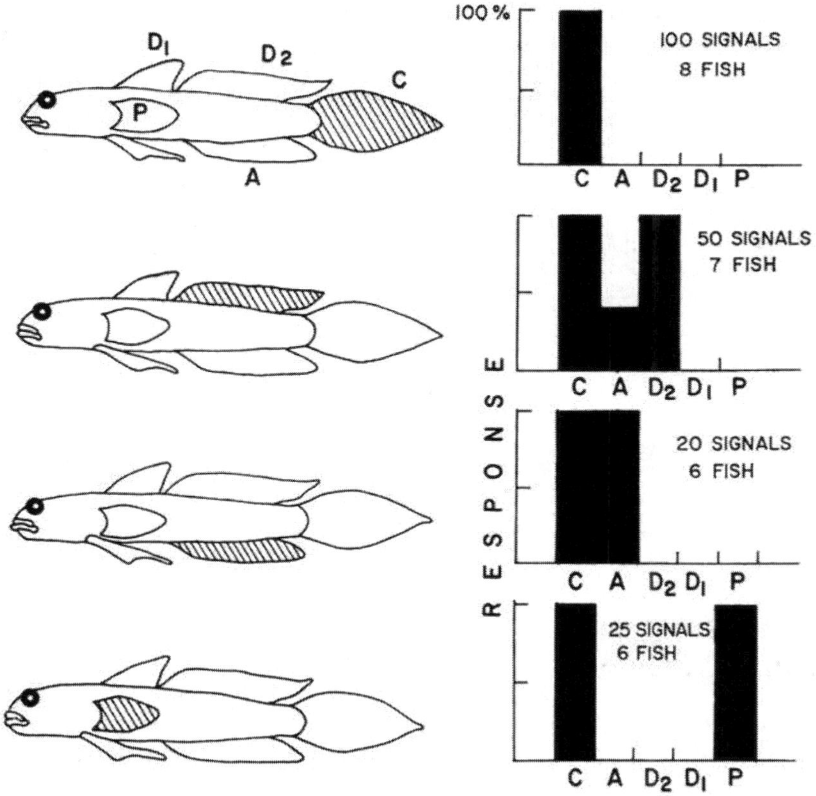

Fig. 4.4.8 Fins of *Amblyeleotris steinitzi* taking part in signaling (left side) and the areas of antennal contact (right side-stippled). C = caudal fin; A = anal fin; D2 = second dorsal fin; D1 = first dorsal fin; P = pectoral fin (after Karplus *et al.*, 1979).

on *A. djiboutensis* with a single partially ablated antenna (Karplus *et al.*, 1972a). Without the antennal contact the shrimp did not respond to the head first retreat of its partner *Cryptocentrus cryptocentrus* which would normally result in its rapid withdrawal.

Differences in the response of two shrimp species found in Hawaii, *Alpheus rapax* and *A. rapacida*, that both pair with *Psilogobius mainlandi* is probably related to differences in length of their antenna (Preston, 1978). On the one hand, *A. rapax*, which has long antennae, can dig at a relatively far distance from the goby and still maintain contact with the fish. *Alpheus rapax* is capable of differentiating between generalized goby movements and tail flick signals (Preston, 1978). On the other hand, the relatively short antennaed *A. rapacida* maintains more body contact with the goby and usually cannot distinguish between generalized and specialized movements of the goby.

Over the last twenty years research on interspecific communication has focused mainly on Atlantic species. This work has contributed to our understanding of the possible evolution of these associations by studying both obligatory and facultative partners (Karplus, 1992; Randall *et al.*, 2005). In a field study carried out in southern Florida, the facultative shrimp partner *Alpheus floridanus* maintained antennal contact with both its obligatory fish partner, *Nes longus*, and its facultative partner, *Bathytgobius curacao*. Occasionally, however, contact was not maintained when the partners were within a distance of several centimeters and the goby's body was located between the left and right shrimp's antenna (Karplus, 1992). In contrast, no antennal or incidental contact was maintained by the same shrimp species when associated with another species of facultative fish partner, *Ctenogobius saepepallens*, in Belize (Randall *et al.*, 2005). In obligatory Indo-Pacific associations the antennal contact between goby and shrimp is continuous when they are close to each other. Only when the goby is positioned at the far end of a long furrow (leading in a straight line to the burrow opening) does the shrimp venture out of its burrow without antennal contact. However, also under these circumstances the shrimp is still protected by the goby since the same furrow that is used by the shrimp when leaving the burrow, is also used by the goby as its escape route, such that the goby collides with the shrimp when rushing into the burrow.

Obligatory and facultative fish partners in southern Florida differed markedly in the types of warning signals they produced to warn their shrimp partner, *Alpheus floridanus* (Karplus, 1992). Whereas the facultative fish partner *Bathygobius curacao* produced only head first entries, the obligatory partner *Nes longus* produced both head first entries and rapid tail-flicks. The latter warning signals resemble those described for obligatory Indo-Pacific fish partners. The generation of warning signals by *Nes longus* associated with *A. floridanus* was studied in the laboratory with a film camera and a motor driven three dimensional barracuda model. Preliminary analysis of the video revealed an increase in the rate of signaling as the model approached the critical point, followed by a head first retreat of the goby (Karplus, unpublished data). One of the important differences between the obligatory partner, *Nes longus*, and the facultative *Bathygobius curacao* is related to the way the goby positions itself at the burrow entrance. Whereas *Nes longus* always positioned itself in the activity zone of the shrimp with its tail directed towards the opening, *Bathygobius caracao* positioned itself in the shrimp activity zone during only 40% of the observations and its tail typically was not directed towards the entrance (Fig. 4.4.9). Therefore, the shrimp was likely much more aware of the obligate than the facultative goby. Aborted exits (i.e., exits of shrimp terminated by a retreat when no contact is made with the goby) were five times more common in associations with the facultative fish partner and often occurred when the goby was in the vicinity of the burrow. When associated with an obligate partner, aborted exits only occurred when the goby was absent from the entrance area (Karplus, 1992).

The study of the communication between *Lotilia graciliosa* and the shrimp *A. rubromaculatus* (Fig. 4.4.10C) is of special interest because of the extreme partner specificity in this obligatory association and the habit of this goby to hover above the burrow opening (most associated gobies perch on the sediment in front of the burrows). The difficulty in studying this shy goby was overcome by installing a remote controlled video camera in front of the burrow that was connected to a monitor and recorder on shore. In this obligatory, Indo-Pacific association, the shrimp maintains continuous antennal contact with the goby while outside the burrow. The goby does not produce any tail-flicks but warns its shrimp exclusively by head first and tail first burrow entries. The frequency of these signals is much higher than in gobies that produce both tail-flick and burrow entry warning signals. Both goby and shrimp stay close to the burrow opening. Shrimp activities are restricted to burrow maintenance and some feeding near the burrow entrance with little introduction of sediment for nocturnal feeding. The greater reliance of this shrimp on subterraneous food may be related to its deeper burrow. The unique partner specificity in this association is probably related to a different interspecific communication system (Karplus *et al.*, 2002; Karplus, in preparation).

Reproduction

At present only a single study in Japan addressed the important issues of reproduction and the early formation of the goby-shrimp partnership (Yanagisawa, 1982, 1984). Yangisawa monitored over extended periods of time marked individuals of *Amblyeleotris japonica* and its shrimp partner *Alpheus bellulus*. In this relatively high-latitude system, which is characterized by distinct seasonal fluctuation in water temperature, male gobies form pairs with females from May to September by moving cautiously over the substratum and rarely venturing more than 3 m from their burrows. Females were rather passive and sometimes refuse to pair with wandering

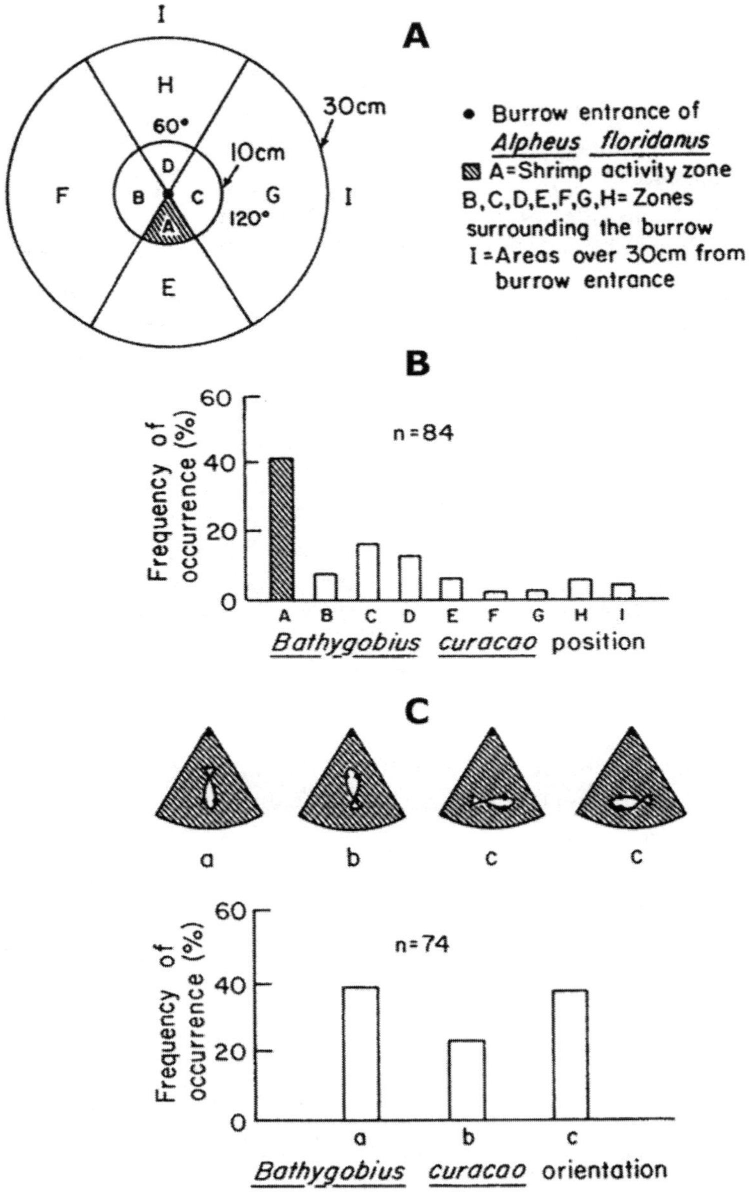

Fig. 4.4.9 **A.** Shematic view of the area surrounding the burrow entrance of *Alpheus floridanus.* **B.** Position of *Bathygobius curacao* at the burrow entrance. **C.** Orientation of *Bathygobius curacao* to the burrow entrance while in the shrimp activity zone (after Karplus, 1992).

Fig. 4.4.10 Associations between gobies and alpheid shrimps from the Indo-Pacific region. **A.** *Stonogobiops xanthorhinica* and *Alpheus randalli*, Japan, (Photo: T. Hirata). **B.** *Amblyeleotris wheeleri* and *Alpheus ochrostriatus*, Japan, (Photo: T. Hirata). **C.** *Lotilia graciliosa* and *Alpheus rubromaculatus* (Photo: P. Nahke). **D.** *Cryptocentrus maudae* and *Alpheus* sp. Bali, Indonesia, (Photo: R. Steene). Note the bright color of both gobies and shrimps and the antennal contact between the partners.

Color image of this figure appears in the color plate section at the end of the book.

males. Paired males were sometimes attacked and replaced by other males. The strong competition in this species may be accounted for by the fact that only a small percentage of females were gravid at one time while most adult males were apparently sexually active throughout the breeding season. At most only about 7% of all shrimp burrows contained pairs of gobies during July and August, and males typically paired with a female only once during the breeding season. Established pairs were maintained for several days. A 77-mm female was observed to lay in captivity an egg mass containing about 20,000 eggs. After spawning, females either leave the burrow or position themselves at the entrance while the males spend from four to seven days inside the burrow taking care of the eggs until they hatch.

Most *Alpheus bellulus* burrows are inhabited by a stable, heterosexual pair. The size of the two genders is positively correlated, although in adult pairs females are slightly larger than males. The proportion of ovigerous females was highest from mid July to mid August. The number of eggs carried by a female was positively correlated with her size and its maximal recorded number was close to 4500. Juvenile shrimp settle on the substrate from late July to early October. They mature and reproduce within one year of settlement. At the onset of their benthic lives they are single, but gradually form pairs, with 50% already paired four to six months after settlement but prior to sexual maturation. The shrimp probably acquire mates subterraneously when their burrows merge.

Yanagisawa (1982, 1984) suggested that the post larval shrimp starts digging a small burrow as soon as it settles to the bottom. Following settlement to the benthic environment, a juvenile goby exploring the bottom may encounter such a burrow and form a partnership with its occupant. The establishment of the association as early as possible after settlement is essential to avoid predation (e.g. a shrimp with a carapace length of 1.7 mm was observed associated with a goby 8.7 mm standard length). When the burrows of two shrimp merge and they form a pair, the smaller of the two gobies is forced to abandon the burrow by the larger one. Many of the evicted gobies are lost to predation, resulting in strong density-dependent mortality of juvenile gobies.

Partner Specificity

The occurrence of several different species of associated alpheid shrimps and gobiid fishes in the same area poses several questions with respect to the degree of partner specificity, its regulating mechanisms, and functions. Early studies on partner specificity simply noted the occurrence or non-occurrence of certain species of gobies and shrimps in the same burrow (Harada, 1972; Polunin and Lubbock, 1977; Nakasone and Manthachitra, 1986). Using occurrence vs. non-occurrence as the sole criteria for specificity, however, could be misleading since the association composition could result from random independent distribution of both species over burrows. The first quantitative study on partner specificity was carried out by Cummins (1979) on six species of gobies and four types of shrimps in the sandy lagoon of One Tree Reef in the Great Barrier Reef, Australia. Chi-square analysis was used to determine whether the distribution of goby species across types of shrimp differed significantly from random. Three species of gobies were associated both as juveniles and adults with a single "preferred" type of shrimp, while in one species of *Amblyeleotris* only the adults were partner specific. One species of goby was equally and significantly more associated with two types of shrimps and another species was associated at random with all four types of shrimps.

The most detailed study to date on partner specificity utilized a combination of field and laboratory studies to discern patterns among Red Sea goby-shrimp associations (Karplus *et al.*, 1972b, 1974, 1981; Karplus,

1981). The following three questions were addressed by analyzing the composition of over 750 associations: (1) Does a species of goby occur together with a species of shrimp in the same burrow?; (2) Is the number of co-occurrences significantly different (more or less) than would be expected from a random distribution of both species over burrows?; and (3) What is the strength of the association between two species, measured using Pielou's (1969) correlation coefficient? Three different types of specificity were found in shallow water where four different species of shrimps occupied different microhabitats: **Type I**) Co-occurrences of goby and shrimp species in the same burrow, with a significant positive correlation coefficient; **Type II**) Co-occurrences of goby and shrimp species in the same burrow, with a significant negative correlation coefficient; and **Type III**) no co-occurrence of goby and shrimp species in the same burrow, and a negative correlation coefficient. In the shallow water each shrimp species had a strong association with a single species of goby of a different genus. In deep water each of the species of shrimp occupied the same habitat and no partner specificity was revealed (Fig. 4.4.11).

Laboratory observations revealed that both gobies and shrimps exhibit strong, negative phototactic responses as well as strong, positive thigmotaxis (touch-based orientation) (Karplus *et al.*, 1972b). Laboratory experiments also revealed a mutual attraction between certain species of gobies and shrimps based on different sensory modalities. The gobies are visually attracted to specific species of shrimp which have distinct colors and patterns (Fig. 4.4.10). Shrimps, by contrast, are chemically attracted to specific gobies (Fig. 4.4.12; Karplus *et al.*, 1972b; Karplus, 1981). The three major mechanisms that regulate partner specificity are the attraction of the goby to the shrimp's burrow, the formation of the tactile communication system and the mutual attraction of goby–shrimp pairs. The presence or absence of each of these mechanisms determines the type of specificity of a specific goby and shrimp. In **Type I** partnerships, the goby is attracted to the shrimp's burrow, the tactile communication is formed and the goby and shrimp are mutually attracted. These associations are common, stable and have been observed for long periods of time (i.e., months). In **Type II** partnerships, the goby is attracted to the shrimp's burrow, the tactile communication is formed but there is no mutual attraction. These associations are rare and unstable lasting for relative short periods of time (i.e., days or a few weeks). In **Type III** partnerships, the goby is attracted to the shrimp's burrow, the tactile communication system is not formed and the goby and shrimp are not mutually attracted. These associations are not found in the sea and cannot be formed in the laboratory.

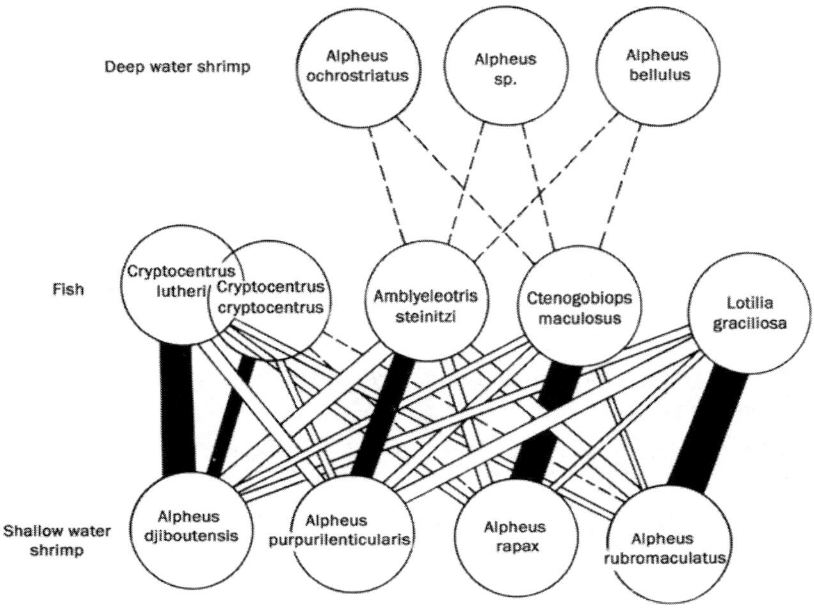

Fig. 4.4.11 Diagramatic presentation of the strength of the associations between gobiid fishes and alpheid shrimps in the northern Red Sea (Eilat Nature Reserve): black lines- more frequent than random co-occurrences of goby and shrimp in the same burrows, a significant positive correlation coefficient, with the width of the line representing the strength of the association; white lines-less frequent than random co-occurrences of goby and shrimp in the same burrows, a significant negative correlation coefficient; stippled lines- independent co-occurrences of goby and shrimp in the same burrows (after Karplus *et al.*,1981).

The shrimp plays an active role in determining its partner's identity. Shrimps deprived of their goby partners often block the burrow openings and leave only a small aperture. Gobies seeking shelter in a shrimp's burrow insert their tail into that opening. The shrimp readily touches the goby's tail with its antenna. If the goby is a preferred species, the opening is enlarged and the goby is allowed to enter (Karplus *et al.*, 1981). If the goby is not preferred by the shrimp, an association may not form. For example, in the laboratory it was not possible to form an association between *Alpheus purpurilenticularis* and *Cryptocentrus lutheri* which are not found to be associated in the sea. During an entire month the shrimp avoided any antennal contact with this species of goby. Replacement of *C. lutheri* by the "preferred" goby (*Amblyeleotris steinitzi*) resulted in the immediate formation of the antennal contact and response of the shrimp to the goby's

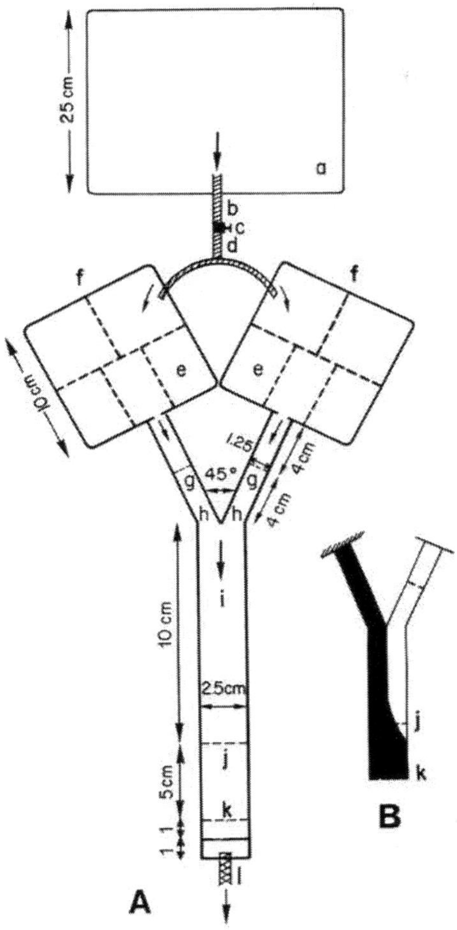

Fig. 4.4.12 **A.** Apparatus for testing the chemical attraction between gobies and their shrimp partners; **B.** the flow pattern of stained water within the Y-channel; a-water reservoir; b-connecting plastic tube; c-regulation clamp; d-bifurcation clamp; e-cells for holding gobies or shrimp; f-perforated partitions; g-partition regulating water level; h-arm of Y channel; i-stem of Y-channel; j-removable net partition; k-non-removable net partition; l-exit of water (after Karplus, 1981).

tail-flick warning signals. The actual formation of a specific association is probably much more complex than suggested. In addition to the attraction to the burrow and the shrimp, the goby may also be attracted to a specific microhabitat and a species-specific burrow entrance structure (Fig. 4.4.4A).

Interspecific competition among gobies over shrimp burrows may also affect the identity of goby-shrimp partners.

Preference of the goby for some traits of the burrow such as size and stability may affect partner specificity. Species of large gobies were associated at One Tree Reef, Australia with large types of shrimp that construct large burrows (Cummins, 1979). At Palau, mature males of the dominant goby species *Cryptocentrus singapurensis* select the burrows of *Alpheus macellarius* out of five types of shrimp probably because *A. macellarius* construct very stable burrows which may enhance spawning success. Juveniles and females of this species of goby occupied the burrows of all five shrimp types (Sakaue *et al.*, 2005).

The most intriguing and least understood aspect of partner specificity is probably the way it is related to the goby-shrimp communication. Polunin and Lubbock (1977) suggested that a species-specific communication system between goby and shrimp species would enhance communication efficacy but reduce the number of available hosts whereas a non-specific communication system could have lower transmission efficiency but a higher number of potential partners. Although species-specific differences in communication systems of several Red Sea associations were found (Karplus, 1976, 1987, 2005), more detailed studies are required to better understand the role interspecific communication plays in the regulation of partner specificity. Some insight may be gained from the obligatory association between *Lotilia graciliosa* and *Alpheus rubromaculatus* (Fig. 4.4.10C). In this association, extreme partner specificity occurs together with a unique communication system based entirely on head first and tail first burrow entries and lack of tail-flicks that is common in other species (Karplus, 1987, 2005; Karplus *et al.* 2002).

GOBY-SHRIMP PHYLOGEOGRAPHY

A recent study examined dispersal dynamics of associated gobies and shrimp. Thompson *et al.* (2005) used mitochondrial DNA to examine the phylogeography of the fierce shrimp goby, *Ctenogobiops feroculus* and the shrimp, *Alpheus djeddensis*. Both of these species have planktonic larvae and hence harbor the potential to disperse significant distances from their place of birth (Caley *et al.*, 1996). Female shrimp brood eggs that are released into the water column upon hatching (A. Thompson, unpublished data). Although the reproductive mechanisms have not been studied specifically for *C. feroculus*, plankton tows within a lagoon in Moorea collected larvae of this species (A. Thompson and C. Thacker, unpublished data). Thompson *et al.* (2005) tested the hypothesis that both *C. ferocułus* and *A. djeddensis* exhibit similar patterns of gene flow and colonization dynamics.

Thompson *et al.*'s (2005) study focused on the central-eastern South Pacific. Between 11 and 27 individual *Ctenogobiops feroculus* and *Alpheus djeddensis* were collected from Fiji, Rarotonga (Cook Islands), Rangiroa (Tuamotu Islands), Moorea, Raitea, Huahine and Bora Bora (Society Islands). *Ctenogobiops feroculus* and *A. djeddensis* were the only mutualistic goby and shrimp taxa found in the lagoons of islands east of Fiji. In Fiji, however, *A. djeddensis* paired with *Ambleleotris steinitzi* as well as *C. feroculus*. Genetic analyses based on 997 and 348 base-pairs of mitochondrial cytochrome *b* from *C. feroculus* and *A. djeddensis*, respectively, provided several important findings. First, shrimp grouped into multiple clades unrelated to geography. This result was interpreted to mean that the shrimp recognized as *A. djeddensis* may actually have been comprised of more than one, cryptic species. Cryptic speciation is known to be common in alpheid shrimps (Mathews, 2006) and this result underscores the need for more research on the taxonomy of alpheid shrimp (see Systematics section of this review). Gobies and shrimp from the most common clade (two-thirds of the sampled individuals fell into this clade), exhibited very similar genetic patterns. First, genetic signatures were significantly different between Fiji and all other sample locations, but did not differ significantly among any other populations. Second, isolation-by-distance slopes (a measure of genetic versus geographic differentiation for pairs of sample locations) were almost identical for gobies and shrimp (Figure 4.4.13). Isolation by distance slopes are often used to infer mean dispersal distances for aquatic species (Palumbi, 2003); thus it is probably that *C. feroculus* and *A. djeddensis* have very similar dispersal potentials.

The finding that these species exhibit similar patterns of dispersal has implications for the ability of these species to colonize habitats. Because *Ctenogobiops feroculus* and *Alpheus djeddensis* pair exclusively with each other in the eastern South Pacific, successful colonization of empty habitats in this region likely depends on the near-simultaneous arrival of individuals from both species. If one species disperses a greater distance than the other, then mortality rates of the long-distance disperser may be greater than the short-distance disperser if the former settles in locations where its partner is not present. Hence, congruent dispersal distances may depict the optimal strategy for pairs of gobies and shrimp to colonize new habitats.

Indeed, Thompson *et al.* (2005) found that intra-site genetic variability was lower for both species in the eastern than in the western islands

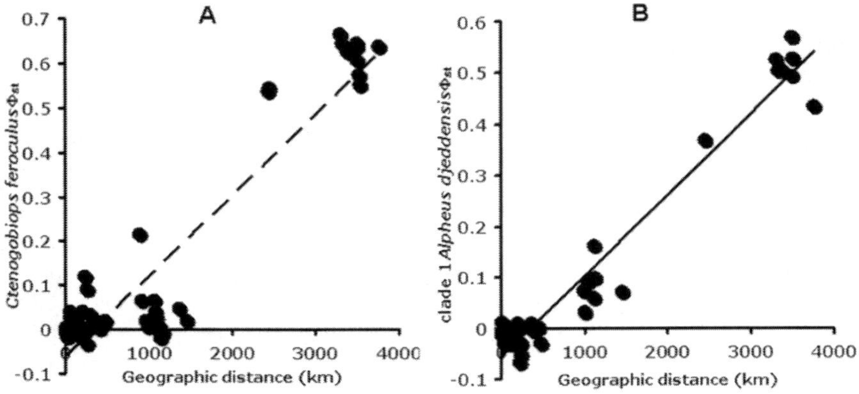

Fig. 4.4.13 Isolation by distance plots for A. *C. feroculus* (y = 0.0002x – 0.06) and B. clade 1 *A. djeddensis* (y = 0.0002x – 0.06). Geographic distances are based on straight-line separation between sample locations (after Thompson *et al.*, 2005).

and that there was a significant, positive relationship between goby and shrimp inter-site genetic variability among islands (Fig. 4.4.14). This result was interpreted to mean that eastern populations were founded more recently than those in the west. A potential reason for eastward population expansion is that the eastern islands are volcanic whereas many in the west are part of continental plates. Volcanic islands such as Moorea are characterized by narrow, shallow lagoons surrounded by steep-sloped reef crests. On continental plates, by contrast, the drop off from a lagoon to the reef crest is much less steep. Thus, lagoons on volcanic islands are extremely vulnerable to drops in sea level, and there is strong evidence that these lagoons dried out during periods of low sea level associated with the last ice-age (approximately 14 000 years ago). Therefore, Thompson *et al.*'s (2005) results suggest that *Ctenogobiops feroculus* and *Alpheus djeddensis* colonized eastern Pacific islands by dispersing in tandem from western islands that were more likely to persist during periods of low sea level.

EVOLUTION

Complex, obligatory associations may have evolved from loose, facultative partnerships. Such a facultative relationship was described for *Ptereleotris hanae* and *Alpheus bellulus* in Japan. This hovering goby occasionally takes shelter in shrimp burrows thereby warning the shrimp in case of danger by

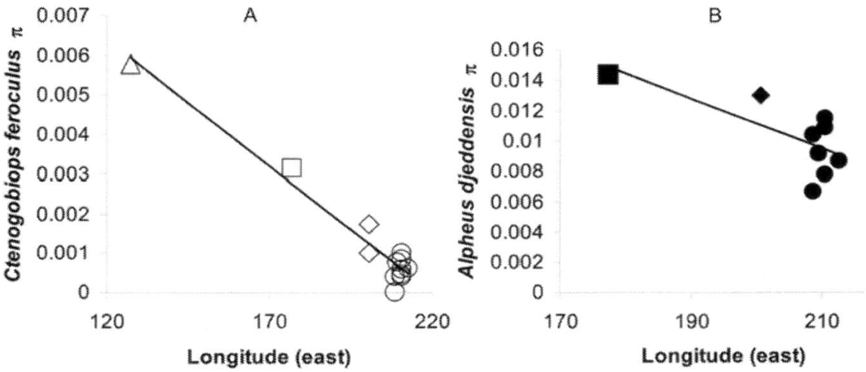

Fig. 4.4.14 Nucleotide (π) diversities as a function of longitude for A. *Ctenogobiops feroculus* (r^2 = 0.95, y = –0.00006 + 0.014) and B. *A. djeddensis* (r^2 = 0.53, y = –0.00016x + 0.044). Triangles depict samples from Okinawa, squares from Fiji, diamonds from the Cook Islands, and circles from French Polynesia (after Thompson *et al.*, 2005).

its head first entry (Yanagisawa, 1978). A similar relationship was recently described in the western Atlantic between the facultative non-hovering *Ctenogobius saepepallens* and *A. floridanus* (Randall *et al.*, 2005; Kramer *et al.*, 2009). Antennal contacts were gradually established by the shrimp with gobies entering their burrows. The poor vision of the subterranean shrimp (Luther, 1958; Magnus, 1967) probably facilitated the evolution of this tactile communication system between shrimp and goby. An interspecifc comparison of the visual system of free living alpheids and alpheid shrimp associated with gobies is still lacking. Causual observations in aquaria on associated Indo-Pacific shrimp suddenly isolated from their fish partner indicated that they were largely disoriented and did not respond to the approach of a rod. In the field, isolated shrimp walked straight into a jar that was hold by a diver (Karplus, pers. observ.). The initial purpose of the antennal contact was probably only to inform the shrimp of the goby's presence at the entrance, with the goby retreat into the burrow being the only warning signal. Such a relationship has been described for *Bathygobius curacao* associated with *A. floridanus (*Karplus, 1992). Concomitant with the formation of the antennal contact with the goby the shrimp may have begun to spend more time outside its burrow (Karplus, 1992). Magnus (1967) and Preston (1978) both suggested that the specialized tail-flick warning signals evolved from intentional movements of the goby related with its retreat into the burrow. An intermediate phase in the formation of the ritualized tail-flicks is the wide amplitude tail-beat. The tail-beat is still the dominant warning signal in the Hawaiian shrimp goby *Psilogobius mainlandi*, produced in cases of higher danger than the tail-flick. In *Amblyeleotris steinitzi*, the tail-beat signal is seldom emitted and only under conditions of heavy body contact. In this species the tail-flick is the dominant signal—effective, economic, graded and coded, a signal produced at a low amplitude and short duration.

The obligatory associations between burrowing alpheid shrimp and non-burrowing gobiid fishes are among the few marine associations that very likely have co-evolved. The complex tactile communication system between gobies and shrimps, including the unique feed-back mechanism, has clearly co-evolved. Similarly the mutual attraction between gobies and shrimps based on different sensory modalities could only evolve with these two organisms in close association. The bright colors of many of the different shrimp species and types (Fig. 4.4.10) are probably used for their selection by their "preferred" gobiid partner since the shrimps themselves are almost blind. Additional aspects of the associated goby and shrimp biology, such as their intervened activity rhythms and their synchronized breeding season reflect a complex co-evolutionary history. There is no doubt that the many diverse interrelated and elaborate adaptations which resulted from the association between gobies and shrimp provide both of them with advantages over their free-living relatives that are reflected in their richness in species, wide distribution and abundance.

Given the complex nature of the interactions between gobies and shrimps, an important evolutionary question is whether all associated species evolved from a single common ancestor or if the interaction evolved independently on multiple occasions. Preliminary phylogenetic analyses based on the mitochondrial ND2 gene indicate that the interaction evolved more than once. Within a consensus tree derived from parsimony analysis that included 27 species of non associated gobies and 7 genera of shrimp gobies, species in the genera *Amblyeleotris*, *Ctenogobiops*, and *Vanderhorstia* formed a well-defined clade that separated from a clade containing species within *Cryptocentrus*, *Tomiyamichthys*, *Stonogobiops* and *Mahidolia* (Thacker and Thompson, in preparation). Similar to the shrimp gobies, preliminary results based on the mitochondrial CO1 gene using seven species of non associated shrimps that were analyzed by Williams *et al.* (2001) and ten species of goby-associated alpeheids indicated that goby-associate alpheid shrimps are not monophyletic. Rather, a non-associated species, *Alpheus cristulifrons* was nested within one of the clades of associated shrimp

(Thacker and Thompson, in preparation). Further analysis is needed to better define the evolution of both gobies and shrimp.

CONCLUSIONS AND FUTURE RESEARCH

There has been considerable progress in the study of goby-shrimp associations over the last twenty years. Close to forty new species of associated gobies have been described. A better understanding of the dynamic feeding behavior of associated shrimp was gained by combining research with stable isotopes with stomach content analysis. The use of remote controlled video cameras allowed the reliable study of goby-shrimp communication systems. In addition, carefully controlled observational and manipulative experiments examined the population dynamics of gobies and shrimp in multiple geographic locations. Recent application of mitochondrial DNA analysis elucidated the so far completely unknown dispersion potential, colonization patterns and evolution of both gobies and shrimps.

Our study of the goby-shrimp partnership is only at its very beginning. Only the morphology is known for most described species and nothing about their behavior and ecology. It is currently also clear that goby-shrimp associations are far from being uniform differing in many aspects of their co-evolved biology. Throughout the text of this review many areas that need further research have been emphasized. However, some of the most intriguing and complex questions that should be addressed are related to the mechanism, function and evolution of partner specificity. Such research should be interdisciplinary, combining taxonomical, molecular, behavioral and ecological studies.

ACKNOWLEDGEMENTS

Many thanks are due to Profs. L. Fishelson, J.E. Randall, C. Thacker and Dr. A. Barki for their valuable comments on the manuscript. We wish to thank D. Karplus for the preparations of the figures. We also wish to thank R. Steene and T. Hirata for allowing us to use their photos. We are grateful to the following publishers: Elsevier, Wiley-Blackwell, Springer Science + Business Media and Balban for allowing us to use again our figures first published in their journals. We are very grateful to H. Debelius for allowing us to use the photo of *Lotilia graciliosa* (photo Rippel/IKAN).

REFERENCES

Allen, G.R. and J.E. Randall. 2006. *Vanderhorstia nobilis*, a new species of shrimpgoby from Indonesia and the Philippines. *Aqua. Journal of Ichthyology and Aquatic Biology* 12: 39-44.

Allen, G.R. and P. Munday. 1995. Description of four new gobies (Gobiidae) from the western Pacific ocean. *Revue francaise d'Aquariologie* 22: 99-104.

Allen, G.R. and P. Munday. 1999. Checklist of the Fishes of Kimbe Bay, Papua New Guinea (www.walindi.com/fish).

Anker, A. 2000. Taxonomical problems of the goby-associated species of *Alpheus* (Decapoda, Alpheidae). I.O.P. Diving News 1-7.

Aonuma, Y. and T. Yoshino. 1996. Two new species of the genus *Amblyeleotris* (Pisces: Gobiidae) from the Ryukyu Islands, Japan. *Ichthyogical Research* 43: 161-168.

Aonuma, Y., A. Iwata and T. Yoshino. 2000. A new species of the genus *Amblyeleotris* (Pisces: Gobiidae) from Japan. *Ichthyological Research* 47: 113-117.

Banner, D.M. and A.H. Banner. 1982. The alpheid shrimps of Australia. Part III: The remaining alpheids, principally the genus *Alpheus*, and the family Ogyrididae. *Records of the Australian Museum* 34: 1-357.

Bellwood, D.R. and T.P. Huges. 2001. Regional-scale assembly rules and biodiversity of coral reefs. *Science* 292: 1532-1534.

Bellwood, D.R., T.P. Hughes, S.R. Connolly and J. Tanner. 2005. Environmental and geometric constraints on Indo-Pacific coral reef biodiversity. Ecology Letters 8: 643-651.

Bruce, A.J. 1994. *Alpheus fenneri* sp. nov. and *A. williamsi* sp. nov., two new Indo-West Pacific alpheid shrimps of the brevirostris species group. *The Beagle, Records of the Museums and art Galleries of the Northern Territory* 11: 15-28.

Caley, M.J., M.H. Carr, M.A. Hixon, T.P. Hughes, G.P. Jones and B.A. Menge. 1996. Recuritment and the local dynamics of open marine populations. *Annual Review of Ecology and Systematics* 27: 477-500.

Chace, F.A. Jr. 1988. The Caridean shrimps (Crustacea: Decapoda) of the Albatross Philippine expedition, 1907-1910, Part 5: Family Alpheidae. *Smithsonian Contributions to Zoology* 466: 1-99.

Chen, I-S. and L-S. Fang. 2003. A new marine goby of genus *Flabelligobius* (Teleostei: Gobiidae) from Taiwan. *Ichthyological Research* 50: 333-338.

Chen, J-P., I-S. Chen and K-T. Shao. 1998. Review of the marine gobiid genus *Amblyeleotris* (Pisces: Gobiidae) with seven new records from Taiwan. *Zoological Studies* 37: 110-118.

Cummins, R.A. 1979. *Ecology of Gobiid Fishes Associated with Alpheid Shrimps at One Tree Reef.* PhD Thesis. University of Sydney, Sydney, Australia. 252 pp.

Debelius, H. 1999. *Crustacea Guide of the World.* IKAN, Jahr-Verlag, Hamburg, 321 pp.

Farrow, G.E. 1971. Back-reef and lagonal environments of Aldabra atoll distinguished by their crustacean burrows. *Symposium Zoological Society London* 28: 455-500.

Franz, V. 1910. Die japanische Knochenfische der Sammlungen Haberer und Doflein. In Beitrage zur Naturgeschichte Ostasiens. Abhandlungen der Mathematisch-Physikalischen Klasse der Koniglich Bayerischen Akademie der Wissenschaften Serie A Supplement 1 *abha*1-135.

Froese, R. and D. Pauly. (eds.). 2008. FishBase. World Wide Web electronic publication. www.fishbase.org, version (12/2008).

Greenfield, D.W. and K.R. Longenecker. 2005. *Vanderhorstia bella*, a new goby from Fiji (Teleosei: Gobiidae). *Proceedings California Academy of Sciences* 56: 619-623.

Harada, E. 1969. On the interspecific association of a snapping shrimp and gobioid fishes. *Publications of the Seto Marine Biological Laboratory* 16: 315-334.

Harada, E. 1972. On the interspecific association between snapping shrimps and gobioid fishes, observed in coastal areas of Nansei Islands. *Biological Magazine Okinawa* 9: 1-8.

Hoese, D.F. and R. Steene. 1978. *Amblyeleotris randalli*, a new species of gobiid fish living in association with alpheid shrimps. *Records of the Western Australian Museum* 6: 379-389.

Hoese, D.F. and G.R. Allen. 1978. *Signigobius biocellatus*, a new genus and species of sand dwelling coral reef gobiid fish from the western tropical Pacific. *Japanese Journal of Ichthyology* 23: 199-207.

Ikeda, Y., T. Nakabo and W. Hiramatsu. 1995. Redescription of *Vanderhorstia macropteryx* (Perciformes: Gobiidae) with designation of a neotype. *Japanese Journal of Ichthyology* 42: 303-310.

Iwata, A. and T. Hirata. 1994. A new gobiid fish, *Stonogobiops pentafasciata*, from Kashiwajima Island, Kochi Prefecture, Japan. *Japanese Journal of Ichthyology* 41: 189-193.

Iwata, A., N. Ohnishi and T. Hirata. 2000. *Tomiyamichthys alleni*: a new species of Gobiidae from Japan and Indonesia. *Copeia* 2000: 771-776.

Iwata, A., K. Shibukawa, and N. Ohnishi. 2007. Three new species of the shrimp-associated goby genus *Vanderhorstia* (Perciformes: Gobiidae: Gobiinae) from Japan, with re-descriptions of two related congeners. *Bulletin of the National Science Museum. Series A Supplement* 1: 185-205.

Karplus, I. 1976. *Comparative studies in behaviour and distribution of the associations of gobiid fishes and alpheid shrimps in the Gulf of Elat*. PhD thesis, The Hebrew University. 123 pp.

Karplus, I. 1979. The tactile communication between *Cryptocentrus steinitzi* (Pisces, Gobiidae) and *Alpheus purpurilenticularis* (Crustacea, Alpheidae). *Zeitschrift für Tierpsychologie* 49: 173-196.

Karplus, I. 1981. Goby-shrimp partner specificity. II. The behavioural mechanisms regulating partner specificity. *Journal Experimental Marine Biology and Ecology* 51: 21-35.

Karplus, I. 1987. The association between gobiid fishes and burrowing alpheid shrimps. *Oceanography and Marine Biology Annual Review* 25: 507-562.

Karplus, I. 1992. Obligatory and facultative goby-shrimp partnerships in the western tropical Atlantic. *Symbiosis* 12: 275-291.

Karplus, I. 2005. Comparative studies of interspecific communication in goby-shrimp associations. The 7th International Indo-Pacific Fish Conference (abstract).

Karplus, I. and A. Vercheson. 1978. A method for collecting live alpheid shrimp and their symbiotic gobiid fish partners. *Crustaceana* 34: 220-221.

Karplus, I. and S. Ben-Tuvia. 1979. Warning signals of *Cryptocentrus steinitzi* (Pisces, Gobiidae) and predator models. *Zeitschrift fiir Tierpsychologie* 51: 225-232.

Karplus, I., R. Szlep and M. Tsurnamal. 1972a. Associative behavior of the fish *Cryptocentrus cryptocentrus* (Gobiidae) and the pistol shrimp *Alpheus djiboutensis* (Alpheidae) in artificial burrows. *Marine Biology* 15: 95-104.

Karplus, I., M. Tsurnamal and R. Szlep. 1972b. Analysis of the mutual attraction in the association of the fish *Cryptocentrus cryptocentrus* (Gobiidae) and the shrimp *Alpheus djiboutensis* (Alpheidae). *Marine Biology* 17: 275-283.

Karplus, I., R. Szlep and M. Tsurnamal. 1974. The burrows of alpheid shrimp associated with gobiid fish in the northern Red Sea. *Marine Biology* 24: 259-268.

Karplus, I., R. Szlep and M. Tsurnamal. 1981. Goby-shrimp partner specificity. I. Distribution in the northern Red Sea and partner specificity. *Journal of Experimental Marine Biology and Ecology* 51: 1-19.

Karplus, I., L. Shaish and R. Ogorek. 2002. The association between *Lotilia graciliosa* and *Alpheus rubromaculatus*. The 8th Colloquium Crustacean Decpoda Mediterranea, Corfu Island, Greece (abstract).

Karplus, I., M. Tsurnamal, R. Szlep and D. Algom. 1979. Film analysis of the tactile communication between *Cryptocentrus steinitzi* (Pisces, Gobiidae) and *Alpheus purpurilenticularis* (Crustacea, Alpheidae). *Zeitschrift für Tierpsychologie* 49: 337-351.

Knowlton, N. 1993. Sibling species in the sea. *Annual Review of Ecology and Systematics* 24: 189-216.

Kramer, A., J.L. Van Tassell and R.A. Patzner. 2009. A comparative study of two goby shrimp associations in the Caribbean Sea. *Symbiosis* 49: 137-141.

Luther, W. 1958. Symbiose von Fischen (Gobiidae) mit einem Krebs (*Alpheus djiboutensis*) im Roten Meer. *Zeitschrift für Tierpsychologie* 15: 175-177.

Macnae, W. and M. Kalk. 1962. The fauna and flora of sand flats at Inhaca Island, Mocambique. *Journal of Animal Ecology* 31: 93-128.

Magnus, D.B.E. 1967. Zur Ökologie sedimentbewohnender *Alpheus*-Garnelen (Decapoda, Natantia) des Roten Meeres. *Helgoländer Wissenschaftliche Meeresuntersuchungen* 15: 506-522.

Masayoshi, H. and T. Shiratori. 2003. Gobies of Japanese Waters. TBS, Japan.

Mathews, L.M. 2006. Cryptic biodiversity and phylogeographical patterns in a snapping shrimp species complex. *Molecular Ecology* 15: 4049-4063.

Moehring, J.L. 1972. *Communication Systems of a Goby-Shrimp Symbiosis*. Ph.D. thesis, University of Hawaii. Hawaii. 271 pp.

Mohlmann, M.S. and P.L. Munday. 1999. *Amblyeleotris arcupinna*, a new species of shrimp goby from Papua New Guinea. *Revue francaise d'Aquariologie* 26: 55-58.

Mohlmann, M.S. and J.E. Randall. 2002. Three new species of gobiid fishes of the genus *Amblyeleotris* from the central and western Pacific. *Raffles Buletin of Zoology* 50: 215-226.

Mous, P.J., D. Nainggolan and L.M. Barmawi. 2005. Report on a rapid ecological assessment of the Raja Ampat Islands, Papua, Eastern Indonesia held October 30–November 22, 2002. The Nature Conservancy — Southeast Asia Center for Marine Protected Areas, Sanur, Bali, Indonesia (www.coraltrianglecenter.org/).

Nakasone, Y. and V. Manthachitra. 1986. A preliminary report on the associations between gobies and alpheid shrimps in the Sichang island, The Gulf of Thailand. *Galaxea* 5: 157-162.

Nelson, R.P. 2005. *A Behavioral Study of the Hawaiin Goby Shrimp Relationship and the Effects of Predation on the System.* M.Sc. Thesis, University of Hawaii.

Palomar, N.E. 2001. *The Population Structure and Behavior of the Burrowing Shrimp Alpheus macellarius as influenced by Sediment Conditions and Seagrass.* M.Sc. Thesis, University of the Philippines, Diliman, Quezon City. 63 pp.

Palomar, N.E., M.A. Juinio-Meñez and I. Karplus. 2004. Feeding habits of the burrowing shrimp *Alpheus macellarius. Journal of the Marine Biological Association of the UK* 84: 1199-1202.

Palomar, N.E., M.A. Juinio-Meñez and I. Karplus. 2005. Behavior of the burrowing shrimp *Alpheus macellarius* in varying gravel substrate conditions. *Journal of Ethology* 23: 173-180.

Palumbi, S.R. 1996. What can molecular genetics contribute to marine biogeography? An urchin's tale. *Journal of Experimental Marine Biology and Ecology* 203: 75-92.

Palumbi, S.R. 1997. Molecular biogeography of the Pacific. *Coral Reefs* 16 (Supplement): S47-S52.

Palumbi, S.R. 2003. Population genetics, demographic connectivity, and the design of marine reserves. *Ecological Applications* 13: S146-S158.

Pielou, E.C. 1969. *An Introduction to Mathematical Ecology.* John Wiley & Sons Inc., New York, 286 pp.

Polunin, N.V.C. and R. Lubbock. 1977. Prawn-associated gobies (Teleostei: Gobiidae) from the Seychelles, western Indian Ocean: systematics and ecology. *Journal of Zoology (London)* 183: 63-101.

Preston, J.L. 1978. Communication systems and social interactions in a goby-shrimp symbiosis. *Animal Behaviour* 26: 791-802.

Randall, J.E. 1994. A new genus and six new gobiid fishes (Perciformes: Gobiidae) from Arabian waters. *Fauna of Arabia* 14: 317-340.

Randall, J.E. 2004. Five new shrimp gobies of the genus *Amblyeleotris* from islands of Oceania. *Aqua, Journal of Ichthyology and Aquatic Biology* 8: 61-78.

Randall, J.E. 2007a. *Vanderhorstia opercularis*, a new shrimp goby from the northern Red Sea. *Electronic Journal of Ichthyology* 1: 18-25.

Randall, J.E. 2007b. Descriptions of four new shrimp gobies of the genus *Vanderhorstia* from the western Pacific. *Aqua* 12: 89-100.

Randall, J.E. and J.L. Earle. 2006. *Amblyeleotris neumanni*, a new species of shrimp goby from New Britain. *Aqua, Journal of Ichthyology and Aquatic Biology* 11: 19-24.

Randall, J.E. and I.-S. Chen. 2007. *Tomiyamichthys tanyspilus*, a new species of Gobiid fish from Indonesia. *Zoological Studies* 46: 651-655.

Randall, J.E., K.T. Shao and J.P. Chen. 2003. A review of the Indo-Pacific gobiid fish genus *Ctenogobiops*, with descriptions of two new species. *Zoological Studies* 42: 506-515.

Randall, J.E., P.S. Lobel and C.W. Kennedy. 2005. Comparative ecology of the gobies *Nes longus* and *Ctenogobius saepepallens*, both symbiotic with the snapping shrimp *Alpheus floridanus*. *Environmental Biology of Fishes* 74: 119-127.

Randall, J.E., K-T. Shao and J.P. Chen. 2007. Two new shrimp gobies of the genus *Ctenogobiops* (Perciformes: Gobiidae), from the western Pacific. *Zoological Studies* 46: 26-34.

Sakaue, J., T. Asahida,, H. Terashima and H. Ida. 2005. What type of alpheid does the goby *Cryptocentrus singapurensis* prefer ? The 7th International Indo-Pacific Fish Conference, Taiwan (abstract).

Senou, H. and Y. Aonuma. 2007. A new shrimp goby of the genus *Amblyeleotris* (Perciformes: Gobiidae) from the Ogasawara Islands, Japan. *Bulletin of the National Museum of Natural Sciences, Series A Supplement* 1: 101-107.

Shibukawa, K. and T. Suzuki. 2004. *Vanderhorstia papilio*, a new shrimp-associated goby from the Ryukyu Islands, Japan (Perciformes: Gobiidae: Gobiinae), with comments on the limits of the genus. *Ichthyological Research* 51: 113-119.

Syms, C. and G.P. Jones. 2004. Habitat structure, disturbance and the composition of sand-dwelling goby assemblages in a coral reef lagoon. *Marine Ecology Progress Series* 268: 221-230.

Talwar, P.K. and A.G. Jhingram. 1991. *Inland Fishes of India and Adjacent Countries.* Vol. 2. Balkema, Rotterdam.

Thacker, C.E. 2003. Molecular phylogeny of the gobioid fishes (Teleostei: Perciformes: Gobioidei). *Molecular Phylogenetics and Evolution* 26: 354-368.

Thacker, C.E., A.R. Thompson, T.C. Adam and J.P. Chen. 2010. Phylogeny and character evolution in the Indo-Pacific genus *Ctenogobiops* (Gobiiformes: Gobiidae). *Ichthyological Research* 57: 231-239.

Thompson, A.R. 2002. Habitat-use patterns of the fierce shrimp goby (*Ctenogobiops feroculus*) at multiple scales: a geospatial approach. In: *Proceedings of the 9th International Coral Reef Symposium* 1: 509-514. Ministry of the Environment, the Indonesian Institute of Sciences and the International Society for Reef Studies, Bali, Indonesia.

Thompson, A.R. 2003. *Population Ecology of Marine Mutualists.* Ph.D. dissertation. University of California, Santa Barbara.

Thompson, A.R. 2004. Habitat and mutualism affect the distribution and abundance of a shrimp-associated goby. *Marine Freshwater Research* 55: 105-113.

Thompson, A.R. 2005. Dynamics of demographically open mutualists: immigration, intraspecific competition, and predation impact goby populations. *Oecologia* 143: 61-69.

Thompson, A.R., C.E. Thacker and E.Y. Shaw. 2005. Phylogeography of marine mutualists: Parallel patterns of genetic structure between obligate goby and shrimp partners. *Molecular Ecology* 14: 3557-3572.

Thompson, A.R., R.J. Schmitt and R.M. Nisbet. 2006. Dynamics of mutualistic populations with open life histories. *Journal of Animal Ecology* 75: 1239-1251.

von Bertalanffy, L. 1938. A quantitative theory of organic growth (Inquiries on growth laws. II). *Human Biology* 10: 181-213.

Weber, M. 1909. Diagnosen neuer Fische der Siboga-Expedition. *Notes Leyden Museum* 31: 143-169.

Weiler, D.A. 1976. *Burrow-dwelling Fishes in a Back-reef Area and their Relation to Sediment Grain Size.* M.Sc. Thesis. University of Puerto Rico, Mayaguez, Puerto Rico. 65 pp.

Winterbottom, R.A. 2002. A redescription of *Cryptocentrus crocatus* Wongratana, a redefinition of *Myersina* Herre (Acanthopterygii: Gobiidae), a key to the species, and comments on relationships. *Ichthyological Research* 49: 69-75.

Winterbottom, R., A. Iwata and T. Kozawa. 2005. *Vanderhorstia nannai*, a new species of burrow associated goby from Palau and the Philippines (Pisces: Gobiidae). *Aqua, Journal of Ichthyology and Aquatic Biology* 9: 109-114.

Yanagisawa, Y. 1978. Studies on the interspecific relationship between gobiid fish and snapping shrimp. I. Gobiid fishes associated with snapping shrimps in Japan. *Publications of the Seto Marine Biological Laboratory* 24: 269-325.

Yanagisawa, Y. 1982. Social behaviour and mating system of the gobiid fish *Amblyeleotris japonica. Japanese Journal of Ichthyology* 28: 401-422.

Yanagisawa, Y. 1984. Studies on the interspecific relationship between gobiid fish and snapping shrimp. II. Life history and pair formation of snapping shrimp *Alpheus bellulus. Publications of the Seto Marine Biological Laboratory* 29: 93-116.

Yoshino, T. and K. Shimada. 2001. *Stonogobiops yasha*, a new shrimp-associated goby from Japan. *Ichthyological Research* 48: 404-408.

Xu, C.S., S. Deng, G. Xiong and H. Zhan. 1980. Two new fishes from East China sea. *Oceanologia et Limnologia Sinica* 11: 179-184. (In Chinese with English abstract.)

CHAPTER

4.5

Mudskippers: Front Runners in the Modern Invasion of Land

Atsushi Ishimatsu[1] and Tomas T. Gonzales[2]

WHAT ARE THE MUDSKIPPERS?

Only a small number of extant fish groups show adaptations to amphibious lifestyle (Graham, 1997). Among them, mudskippers stand out as one of the largest groups that volitionally emerge from water and spend a significant portion of time on mudflat surface. The term "mudskippers" typically refers to four genera of oxudercine gobies (Gobiidae: Oxudercinae) including *Boleophthalmus*, *Periophthalmodon*, *Periophthalmus*, and *Scartelaos*. Species of *Periophthalmus* are the most terrestrial of the four genera: *Periophthalmus* mudskippers have been observed to remain emerged even during high tide, perching on rocks and stems of mangrove trees and other halophytes (Ikebe and Oishi, 1996, 1997; Takita *et al.*, 1999; Zhang *et al.*, 2003). The degree of terrestriality of *Periophthalmodon* mudskippers seems to vary between

Authors' addresses: [1]Institute for East China Sea Research, Nagasaki University, Tairamachi 1551-7, Nagasaki 851-2213, Japan. E-mail: a-ishima@nagasaki-u.ac.jp

[2]Center for General Education, AMA International University, Bahrain, P.O. Box 18041, Manama, Kingdom of Bahrain. E-mail: t2gonzales@yahoo.com

species. Zhang *et al.* (2003) reported that *Pn. septemradiatus* often spends more than 30 min on an exposed mudflat, whilst *Pn. schlosseri* stays at the water's edge during both low and high tide. In addition, Milward (1974) inferred that *Pn. schlosseri* remains inside its burrow when the burrow openings are covered by high tides. *Boleophthalmus* and *Scartelaos* species are more aquatic; neither ventures far from water pools on the mudflat and both appear to reside in their burrows during high tide (Milward, 1974; Zhang *et al.*, 2000; Lee *et al.*, 2005). More recently, Polgar and Crosa (2009) reported on the habitat characteristics of seven Malaysian mudskippers (*B. boddarti, B. pectinirostris, Pn. schlosseri, Ps. chrysospilos, Ps. gracilis, Ps. variabilis*, and *S. histophorus*). Their analysis basically confirmed earlier findings of different affinities to terrestrial environment among the mudskipper genera, but also indicated possible influence of predation pressure on mudskipper's distribution on tidal flats. In spite of the alleged burrow confinement of more aquatic mudskippers during high tide, observations of underwater activities of mudskippers are often obscured by generally high turbidity of water inundating the mudflats, and therefore direct confirmation of the supposition is difficult. Males of all mudskipper species probably reside in burrows during egg care (see section on Reproduction).

Murdy (1989) published a cladistic analysis of oxudercine gobies, and verified 5 species belonging to the genus *Boleophthalmus*, 3 to *Periophthalmodon*, 12 to *Periophthalmus*, and 4 to *Scartelaos*. To our best knowledge, five new *Periophthalmus* species have been reported thereafter (*Ps. magnuspinnatus,* Lee *et al.*, 1995; *Ps. spilotus,* Murdy and Takita, 1999; *Ps. walailakae*, Darumas and Tantichodok, 2002; *Ps. darwini*, Larson and Takita, 2004; *Ps. takita*, Jaafar and Larson, 2008). In addition, Jaafar *et al.* (2009) claimed that *Ps. novemradiatus* includes two species (i.e. *Ps. novemradiatus* and *Ps. variabilis*). This brings the total number of *Periophthalmus* species to 18*. No new species have been reported for the other three genera. The distribution of mudskippers centers mainly in the tropical Indo-Pacific. The known localities of *Boleophthalmus* and *Scartelaos* species are similar, ranging from the Red Sea throughout the East Asian countries, and south to the tropical regions of Australia. The distribution of *Periophthalmodon* species is more restricted to Southeast Asia and mid-northern Australia. *Periophthalmus* species show the widest distribution among mudskippers, from west coast of Africa to Polynesian islands (Murdy, 1989).

* *Ps. murdyi*, once claimed as a new species (Larson and Takita 2004), was subsequently re-examined and judged as a junior synonym of *Ps. novaeguineaensis* (Jaafar and Larson, 2008).

Mudflats are one of the most productive ecosystems of the earth (Willmer *et al.*, 2005), but at the same time pose severe challenges to inhabitants due to extreme environmental fluctuations driven by tidal cycles (e.g. daily cycles of emersion and submersion, changes in salinity and temperature, sedimentation and resuspension of mud particles, high current velocity, and oxygen shortages within the sediment, Little, 2000). Therefore, only those animals that have evolved adaptive machinery for this harsh environment can flourish on mudflats. Mudskippers show remarkable capabilities for air breathing (Graham, 1997; Martin and Bridges, 1999), osmoregulation (Evans *et al.*, 1999), terrestrial locomotion (Harris, 1960; Swanson and Gibb, 2004; Pace and Gibb, 2009) and feeding (Sponder and Lauder, 1981), and aerial vision (Sayer, 2005), as well as high tolerance to thermal stress (Sayer, 2005) and environmental ammonia (Ip *et al.*, 2004a). These support a wide variety of activities displayed by mudskippers on exposed mudflats, including feeding, territorial defense, courtship, and burrowing behaviours. In contrast to ample information available on surface activities, the subsurface life of mudskippers in burrows has been little studied. The most notable yet crucial feature of their subsurface activities is reproduction in burrows (Ishimatsu *et al.*, 2007, 2009). It has been known that mudskippers excavate burrows, and use them as a refuge from predators, for protection from desiccation and temperature extremes, and for reproduction (Stebbins and Kalk, 1961; Kobayashi *et al.*, 1971; Brillet, 1976; Clayton and Vaughan, 1986. Although burrow sequestration affords egg protection, the severe hypoxia of burrow water (Ishimatsu *et al.*, 1998a, b, 2000, 2007) would not permit embryonic development. Mudskippers have evolved a highly sophisticated strategy to ensure embryonic development and larval hatching by integrating their aerial respiratory capability into reproduction (Fig. 4.5.1, see section of Reproduction).

Mudskippers could give useful insights into our understanding of land invasion by ancient vertebrates, despite their distant taxonomic position from those animals that gave rise to ancestral tetrapods, a large difference in body size (body length of mudskippers ranging from only a few to 20-30 cm, in contrast to 15-40 cm of the skull size alone of basal tetrapodomorph fish, Clack, 2006), distinctly different skull and skeletal anatomy, and different environmental and ecological settings of the present and the past. Vertebrate invasion to land has arguably taken place in tropical freshwaters (Zimmer, 1998), but recent fossil evidence has pointed out that ancestral vertebrates of tetrapods inhabited both freshwater and marine regions, and

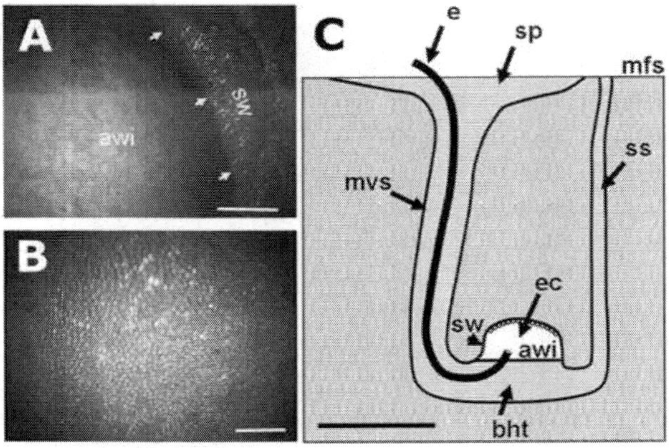

Fig. 4.5.1 View of the air–water interface (awi; A) and the inner top surface (B) of an egg chamber in a *Periophthalmodon schlosseri* burrow. A monolayer of spawned eggs was confirmed only on the inner top surface (B), not on the side wall (sw; A). The image in A was taken using an endoscope CCD camera, which was raised above the air–water interface and bent obliquely downward toward the interface. Arrows in A indicate the edge of the air–water interface. Scales in A and B represent 3 and 2 cm, respectively (from Ishimatsu *et al.*, 2009). Diagram of the position of the endoscope inside a burrow (C). awi, air–water interface; bht: bottom horizontal tunnel, e endoscope, ec egg chamber, mfs mudflat surface; mvs: main vertical shaft; ss: side shaft; sp: surface pool; sw: side wall of the egg chamber. Scale 30 cm.

Color image of this figure appears in the color plate section at the end of the book.

that land invasion by some of these animals might have occurred in marine habitats (Zimmer, 1998; Schultze, 1999; Clack, 2006, 2007).

RESPIRATION

Mudskippers lack specialized accessory respiratory organs, as found in the majority of freshwater air-breathing fishes (Graham, 1976). Instead, their gas exchange surface is comprised of the epithelia lining the buccal, pharyngeal and opercular cavities, the gills, and the skin. The lack of specialized respiratory organs is not unique in mudskippers but prevalent among marine air-breathing, amphibious fishes such as blennies. This led Graham (1976) to infer that "marine forms, in respiring aerially with no more than gills, modified buccal or pharyngeal epithelium, or possibly a vascularized skin, are invariably relegated to the ranks of the least specialized air-breathing fishes". The suprabranchial chamber, which was reported to be present in both *Periophthalmodon schlosseri* (Yadav *et al.*, 1990) and another oxudercine *Pseudapocryptes lanceolatus* (Yadav and Singh, 1989,

= *Pss. elongatus*, the latter is the valid scientific name of the species, see Ferraris, 1995) appears to be no more than a simple expansion of the buccopharyngeal cavity (Gonzales and Ishimatsu, unpublished observation) and, in structural complexity, is nothing like the suprabranchial chambers of the freshwater air-breathing fishes (Graham, 1997). The relative importance of gas exchange through the mouth (the buccal, opercular cavity and the gills) and body surface varies among mudskipper species and with environment (air and water), the former usually accounting for 25 to 60% of the total O_2 uptake (Graham, 1997).

The bucco-opercular cavity (we define as the spaces in the buccal, pharyngeal and opercular cavities combined) of mudskippers can accommodate a large volume of air when distended. For example, the mouth gas volume amounts to 16% of body volume in *Boleophthalmus pectinirostris* (Graham *et al.*, 2007) and *Periophthalmodon schlosseri* (Aguilar *et al.*, 2000), 18% in *Periophthalmus modestus* (Ishimatsu *et al.*, 2007), and 14% in *Scartelaos histophorus* (Graham *et al.*, 2007), in contrast to only 2 to 4% reported for non-oxudercine air-breathing gobies (Gee and Gee, 1991). Another air-breathing goby, *Odontamblyopus lacepedii* (Amblyopinae) was found to hold a somewhat larger volume (6% of body mass, Gonzales *et al.*, 2006), but this is still less than half the values found for mudskippers. The large volume of the respiratory organ would be beneficial for aquatic air-breathing fishes in that it can reduce the frequency of surfacing, but may not be of great respiratory significance for amphibious air-breathing fishes because air is readily available when emerged. Rather, it might have developed as an adaptation to air storage in their burrows (see below). To our experience, mudskippers never surface for air when mudflats are covered by water. There is little trabeculation in the inner surface of the bucco-opercular cavity of the mudskippers to increase respiratory surface area.

Mudskippers retain the basic gill plan seen in purely aquatic teleosts. They have four pairs of gills, each of which bears two rows of gill filaments (holobranch), and the filaments carry secondary lamellae on either surface. Similar with the general trend seen in other air-breathing fishes, gills diminish in size with increasing reliance on air breathing among mudskippers (Tamura and Moriyama, 1976; Graham, 1997, Graham *et al.*, 2007): gill filaments of *Periophthalmodon* and *Periophthalmus* are short, sparsely spaced, and twisted (*Periophthalmus*) or frequently bifurcated (*Periophthalmodon*, Low *et al.*, 1988; Gonzales, 2007; Gonzales *et al.*, 2011). The secondary lamellae of *Pn. schlosseri* are covered by thick epithelia, and frequently fused together (Schöttle, 1931; Low *et al.*, 1988; Wilson *et al.*, 1999; Graham *et al.*, 2007), to the extent that led Wilson *et al.* (1999) to conclude that the gill lamellae of this fish are generally ineffective for gas exchange. Yadav *et al.* (1990) reported interlamellar fusion in *Pn. schlosseri* but only at the filament tips (body length of their fish ranging from 1 to 6 cm). The difference between these two studies could be due to developmental changes in gill morphology or misidentification of the species used by Yadav *et al.* (1990). The first possibility is supported by the observation by Schöttle (1931) that the gill anatomy of small (body length 1.5 cm) *Pn. schlosseri* resembles that in *Ps. chrysospilos* and *Ps. vulgaris* (= *argentilineatus*, Murdy, 1989), both of which have free secondary lamellae, whereas the lamellae are extensively fused together in larger specimens (size not stated). Parallel developmental changes in gill anatomy were reported by Brauner *et al.* (2004) for *Arapaima gigas*. In *A. gigas*, gill filaments in small individuals are similar to other water-breathing fishes but in adults, fishes change to a rod-like structure in which interlamellar space is filled partially and then eventually completely with proliferating mitochondria-rich cells (also called "chloride cells"). Possible misidentification by Yadav *et al.*, (1990) was proposed by Clayton (1993), who concluded that the species used by those authors was in fact *Ps. minutus* on the basis of body size and gill morphology.

It is well known that in animals, partial pressure of CO_2 (PCO_2) of the blood rises with increasing reliance on aerial respiration: arterial PCO_2 typically ranges from 1 to 4 mmHg in purely water-breathing animals, but is in the range of 30 to 40 mmHg in most air-breathing animals, with bimodally breathing animals showing intermediate values (Randall *et al.*, 1981; Truchot, 1987; Maina, 1998). The relatively high arterial PCO_2 (10-15 mmHg) of *Pn. schlosseri* (Kok *et al.*, 1998; Ishimatsu *et al.*, 1999) is in line with the supposition that the gills play only a limited role in gas exchange. However, it is curious that a more aquatic mudskipper, *Boleophthalmus boddarti*, which possesses much more developed gills (Low *et al.*, 1988) and presumably high ability of aquatic CO_2 excretion, still shows similarly high arterial PCO_2 even when totally submerged (12-13 mmHg, Kok *et al.*, 1998).

Most aquatic air-breathing fishes experience an accumulation of CO_2 and the resulting acidosis in their body fluid when placed in air due to their limited capacity to excrete CO_2 in air (Graham, 1997). In contrast, mudskippers (and most other amphibious fishes) can successfully excrete CO_2 irrespective of the respiratory medium, and thereby maintain their body fluid pH in both air and in water (Ishimatsu *et al.*, 1999). This seems to be accounted for at least in part by the difference in air ventilation mechanism: many aquatic air-breathing fishes depend on water for air ventilation of their accessory respiratory organs (Liem, 1987), and therefore, when placed out of water, might not be able to excrete CO_2 at the necessary rate. Obviously, mudskippers do not need water for aerial ventilation of the bucco-pharyngeal cavity. But this water-independent ventilatory mechanism and the position of the main respiratory surface inside the mouth entails reductions in percentage renewal of the mouth gas. In *Pn. schlosseri*, approximately 55% of the mouth gas is replenished in each ventilatory cycle (Aguilar *et al.*, 2000). This percentage is much smaller compared to nearly 100% renewal of the mouth gas volume reported for the aquatic facultative air-breathing eel goby, *Odontamblyopus lacepedii*, which vertically raises its anterior body for air breathing at burrow water surface during low tide (Gonzales *et al.*, 2006). Almost complete renewal of the respiratory gas in the air-breathing organs is also confirmed in several aquatic air-breathing fishes that use diverticulum derived from the buccal cavity for aerial respiration (*Channa*, Ishimatsu *et al.*, 1981; Liem 1984, *Osphronemus* and *Ctenopoma*, Peters, 1978).

The gas transport system of *Periophthalmodon schlosseri* is evidently more efficient in air than in water. The fish can enhance oxygen uptake to repay oxygen debt after exhaustive exercise only in air, but not in water (Takeda *et al.*, 1999). Furthermore, arterial PO_2, O_2 content and O_2 saturation all significantly decreased in 1.5 h after forced submersion in normoxic water (Ishimatsu *et al.*, 1999). Kok *et al.*, (1998) reported similar reductions in arterial O_2 level in the same species in response to forced submersion, although the same procedure did not result in decreased arterial PO_2 in *B. boddarti*.

BLOOD CIRCULATION

After the classical study on respiratory and circulatory anatomy of gobies by Schöttle (1931), little has been added to the knowledge of the circulatory anatomy in mudskippers. This is probably due to the small size of the mudskippers, which makes it difficult to apply modern techniques such as corrosion casting, and the fact that Schöttle's diagrams on *Boleophthalmus*, *Periophthalmodon* and *Periophthalmus* showed little deviation of their circulatory system from typical teleostean configuration. In contrast, the circulatory system of freshwater air-breathing fishes shows various types of circulatory modifications, some of which appear to progress toward separation of systemic and respiratory circuits (Satchell, 1976; Olson, 1994; Graham, 1997). However, some papers describe anomalous circulatory patterns in mudskippers, which requires re-inspection (see below).

In mudskippers, the entire volume of blood ejected from the heart is directed to the gills and, after traversing the gills, perfuses the systemic circuit (Schöttle, 1931). This has been confirmed for *B. boddarti*, *Pn. schlosseri* (referred to as *Ps. schlosseri*) and three *Periophthalmus* species (*Ps. chrysospilos*, *Ps. vulgaris* = *Ps. argentilineatus*, and *Ps. koelreuteri fuscatus* = *Ps. kalolo*, Murdy, 1989). There appears to be no vascular shunt bypassing the gills, as found in several freshwater air-breathing fishes (e.g. lungfishes and swamp eels, Satchell, 1976; Olson, 1994; Graham, 1997). However, there are a few papers that reported vascular connections that differ from this basic pattern. For instance, Schöttle (1931) reported that the ventral aorta of *B. viridis* (this scientific name is no longer valid, possibly *Scartelaos histophorus*) extends further anteriorly after giving off the first afferent branchial arteries, and bifurcates to become the hyomandibular artery, which supplies mainly the respiratory surfaces of the buccopharyngeal and opercular cavities. Similarly, Niva *et al.* (1981) reported the presence of the bucco-opercular artery branching off from the first afferent branchial artery in *B. boddarti*, which is not shown in Schöttle's figure for the same species. In addition, Das (1934) described that the first afferent branchial artery in a non-mudskipper oxudercine *Pseudapocryptes elongatus* sends off the bucco-opercular artery at about one third of its origin from the ventral aorta, which supplies the lining of the buccal and opercular cavities. If these observations are true, the respiratory surfaces would be perfused by both pre- and post-branchial blood. In general, the pre-branchial blood has a higher pressure than the post-branchial blood, which makes it unlikely for the same vascular bed to be perfused by these two types of blood. These ambiguities need to be clarified by using modern methods.

In spite of the general retention of teleostean circulatory configuration at macroscopic levels, more subtle circulatory modifications have been reported in the bucco-opercular cavity, the skin, and the gills of mudskippers. The most obvious is vascularization of the inner epithelia of the bucco-opercular cavity and the skin. Table 4.5.1 compares the respiratory capillary density of the inner epithelia of the bucco-opercular cavity in a purely water-breathing goby, *Acanthogobius hasta*, a facultative air-breathing eel goby, *Odontamblyopus lacepedii*, and two species of mudskipper (*Periophthalmus modestus* and *Periophthalmodon schlosseri*, Gonzales, 2007; Gonzales *et al.*, 2011; see also Gonzales *et al.*, 2008). Increasing air-breathing capability is reflected by the progressively higher density of the capillaries in these epithelia culminating in *Pn. schlosseri*, which has nearly the same capillary density on its tongue, palate, and the inner operculum. Zhang (2000) studied morphological adaptations of oxudercine gobies and

Table 4.5.1 Blood capillary density (per mm) within 0 - 10 μm diffusion distance in the bucco-opercular epithelia of mudflat-burrowing gobies

Species	*Tongue*	*Palate*	*Operculum*	*Reference*
Non-air breather				
Acanthogobius hasta	0^{a}	0^{a}	0^{a}	Gonzales *et al.*, 2008
Facultative air breather				
Odontamblyopus lacepedii	0^{a}	0.3 ± 0.3^{a}	14.5 ± 3.0^{b}	Gonzales *et al.*, 2008
Amphibious air breather				
Periophthalmus modestus	1.6 ± 0.5^{a}	1.9 ± 1.6^{b}	59.1 ± 8.5^{c}	Gonzales *et al.*, 2008
Periophthalmodon schlosseri	47.8 ± 5.6^{b}	50.8 ± 12.0^{c}	71.7 ± 10.3^{c}	Gonzales *et al.*, 2011

Values are mean ± s.d. of 3 individuals. Different letters within column denote significant difference ($P < 0.05$).

demonstrated rich vascularization of the inner epithelium of the operculum in *Boleophthalmus*, *Oxuderces*, *Periophthalmodon*, *Periophthalmus*, *Pseudapocryptes*, and *Scartelaos*. In addition, rich vascularization has been demonstrated on the gill arches and leading edges of gill filament in *Pn. schlosseri* (Wilson *et al.*, 1999; Gonzales, 2007).

The mudskipper skin is in general richly vascularized, although capillary density in the outermost layer of the epidermis varies between species and among body parts. Capillaries are distributed in higher density and with shorter diffusion distance over more exposed parts of the body (i.e. head, outer surface of the operculum, and dorsolateral trunk surface) than in ventral surfaces which are in contact with mud when mudskippers emerge (Park *et al.*, 2006; Zhang *et al.*, 2000, 2003). In *Boleophthalmus* and to a lesser extent in *Scartelaos*, the entire body surface except ventral surface is studded with characteristic papillary protrusions (dermal bulge). An artery and a vein run through the dermis into each of the dermal bulge at its center, and the artery sends off radially arranged capillaries (Schöttle, 1931; Zhang *et al.*, 2000; Park *et al.*, 2003a). In *Periophthalmodon* or *Periophthalmus*, no such papilla-like structure has been found. Another characteristic feature of mudskipper skin is the presence of large, vacuolated cells (often referred to as 'swollen cells') in the middle layer of the epidermis, which has been reported in all four genera (for ultrastructure of these cells, see Suzuki, 1992; Yokoya and Tamura, 1992). The functional role of the vacuolated cell layer has been speculated by some authors to be for water conservation (Yokoya and Tamura, 1992), although the presence of such cell layer in the non-amphibious aquatic air-breathing goby (*Odontamblyopus lacepedii*) contradicts this supposition (Park *et al.*, 2003b).

Periophthalmodon schlosseri shows a specialized vascular arrangement in the secondary lamellae, wherein the blood space of the secondary lamellae consists of 5-8 parallel blood vessels, rather than forming a lacuna as seen in most teleosts (see Olson, 2002; Evans *et al.*, 2005), and the internal diameter of these lamellar vessels is larger than the size of red blood cells (Gonzales *et al.*, 2011). Similar channelization of lamellar blood space has been reported in a few freshwater air-breathing fishes (e.g. *Amia calva*, Olson, 1981; *Anabas testudineus*, Olson *et al.*, 1986). Such structural modification is expected to reduce vascular resistance of the gills, and, together with the interlamellar fusion and thickening of the lamellar epithelia (see above), would ensure sustained blood input into the systemic and aerial gas exchange vasculatures during air breathing. It is currently unknown whether other mudskippers have similar vascular specialization in their lamellae.

Respiratory capillaries of the bucco-opercular cavity and the trunk are believed to originate from the systemic arteries such that systemic and respiratory capillary beds are placed in parallel (Schöttle, 1931), although there seems to have been no direct demonstration of this vascular connection. In this case, the aerial respiratory capillaries are supplied with arterial blood having the same oxygen content as systemic capillaries would receive, which may be considered to be physiologically ineffective. Oxygenated blood from the respiratory epithelia is drained by systemic veins, and therefore mixed with deoxygenated systemic venous blood before returning to the heart.

Nearly no physiological data are available as to cardiovascular performance of mudskippers under various conditions. Garey (1962) was the first to demonstrate lowering of heart rate upon submersion (water oxygen level not stated) in a mudskipper (*Periophthalmus australis* = *Periophthalmodon freycineti*, Murdy 1989). We also have shown in *Pn. schlosseri* that heart rate rapidly decreased in response to forced submersion in normoxic seawater but rose to a higher than pre-submersion level when the fish took the first air breath upon re-emersion. Blood pressure was depressed during submersion and did not show any overshoot upon re-emersion (Ishimatsu *et al.*, 1999). Kok *et al.* (1998) reported a similar decrease in heart rate during submersion in *Pn. schlosseri*, but not in *B. boddarti*.

OSMOREGULATION

Mudskippers are highly euryhaline, tolerating a wide range of environmental salinities (Gordon *et al.*, 1965; Lee *et al.*, 1987; Ip *et al.*, 1991; Lee *et al.*,

1991). Early studies have unequivocally demonstrated the mudskipper's outstanding capacity for osmoregulation. However, these studies subjected mudskippers to different salinities for several days or longer, which may not be ecologically relevant considering that they are usually exposed to a certain salinity level for only several hours in their natural mudflat habitat. Rather, mudskippers often experience rapid changes in salinity each day with tide. Therefore, it seems more useful to study physiological adjustment to salinity changes on hourly basis and in this regard, the study of Sakamoto *et al.* (2000) bears more relevance to the mudskipper's natural life adaptation. These authors demonstrated that in *Periophthalmus modestus*, the apical crypts of the ion-transporting chloride cells were rapidly closed in response to freshwater transfer, and re-opened after returning to seawater, without any changes in the total number of the cells. These morphological changes in the apical crypts appear to be triggered at least in part by environmental Ca^{2+} (Sakamoto and Ando, 2002). The apical crypt of the chloride cells is the principal site of ion transport with the external media (Evans *et al.*, 2005) and therefore the rapid morphological changes in this structure are thought to be an adaptive mechanism for life in a fluctuating salinity environment. The highest density of chloride cells in *Ps. modestus* occurs in the skin under the pectoral fin followed by the ventral trunk (Yokota *et al.*, 1997), which correlates with its highly terrestrial mode of life. In aquatic fish, the chloride cells are located primarily in the gills, where ion exchange takes place with continual respiratory water current (Evans *et al.*, 2005). In highly terrestrial mudskippers such as *Ps. modestus*, ion exchange would be hampered in air, if the localization of chloride cells were similar to that of the aquatic fishes. The position of the cells on the ventral body permits osmoregulation whenever a small puddle of water is contacted.

Water drinking rate of *Periophthalmus vulgaris* (= *Ps. argentilineatus*) was determined by Dall and Milward (1969). Even though one might expect a low drinking rate in mudskippers due to reduced surface area and thick epithelium of the gills as stated above, the drinking rate and sodium efflux measured in full-strength seawater were not exceptionally low as compared with other aquatic species. Takahashi *et al.* (2006a,b) investigated cell turnover and its endocrinological regulation in the anterior intestine in *Ps. modestus*, which may play an important role in regulating the permeability of the gastrointestinal tract of the species. Their results showed increases in intestinal apoptosis during seawater acclimation, but enhanced cell proliferation in freshwater-acclimated fish. It was suggested that both cortisol and prolactin are involved in these morphological changes. In laboratory multiple-choice experiments, Gordon *et al.*, (1985) demonstrated that *Ps.*

cantonensis (= *Ps. modestus*) had no preference for any particular salinity (ranging from 25 to 100% seawater), but avoided exposure to freshwater.

AMMONIA EXCRETION

As one of the end products of metabolism, ammonia is present in both terrestrial and aquatic environments (Eddy, 2005). Ammonia, although potentially toxic, is highly soluble in water and ammonia excretion through the gills is the principal method used by fishes (Randall and Tsui, 2002; Randall and Ip, 2006). Therefore, water availability and unlimited access to it is important.

Due to their active terrestrial lifestyle and constant exposure to ambient burrow water, mudskippers are thought to often face with high levels of ammonia. Similar with some air-breathing fishes, mudskippers exhibit high tolerance to environmental ammonia (Graham, 1997; Ip *et al.*, 2004a). Peng *et al.* (1998) showed that *Periophthalmodon schlosseri* can maintain a low steady level of internal ammonia during exposure for several days to an elevated concentration of external ammonia that can be lethal to other fishes. During aerial exposure, both *Pn. schlosseri* and *Boleophthalmus boddarti* minimize ammonia buildup by reducing the rate of protein and amino acid catabolism (Lim *et al.*, 2001). It has also been reported that *Pn. schlosseri* can actively excrete ammonia through the gills during emersion (Chew *et al.*, 2007), thus maintaining low levels of ammonia in the body. *Pn. schlosseri* has a unique gill morphology (see section on Respiration) wherein the tightly arranged and sometimes fused lamellae form fenestrae that trap water in between adjacent lamellae (Wilson *et al.*, 1999). The interlamellar water may have implications not only for preventing branchial desiccatory stress but also for providing a fluid medium to eliminate ammonia. Chew *et al.* (2007) measured the ammonia level of water trapped in the branchial surface of *Pn. schlosseri*, and found a surprisingly high concentration of ammonia after 24 h of fish emersion. In addition, individuals injected with ammonium acetate eliminated more ammonia in the branchial fluid than control individuals given sodium acetate (Chew *et al.*, 2007). Randall *et al.* (2004) hypothesized that *Pn. schlosseri* can maintain tolerable ammonia level in the tissue through active ammonium ion transport across the chloride cells via an apical $Na^+/H^+(NH_4^+)$ exchanger and a basolateral $Na^+/K^+(NH_4^+)$ ATPase or a $Na^+/K^+(NH_4^+)/2Cl^-$ cotransporter. To prevent the inward flux of exogenous ammonia, acidification of the branchial fluid and burrow water through CO_2 and H^+ efflux from the body has been proposed (Randall *et al.*, 2004; Chew *et al.*, 2007). The acidified water will supposedly trap excreted ammonium ions and reduce the formation of toxic ammonia (Randall *et al.*, 2004). These results suggest that branchial water acts as a sink for ammonia during emergence and that the lamellae serve as excretory pathway for the outward diffusion of ammonium ions from the body. Previous studies demonstrated that ammonia volatilization in air accounts for a minor contribution (<1-3%) of the total ammonia excretion (Wilson *et al.*, 1999). Ip *et al.* (2004b) also demonstrated that *Pn. schlosseri* is capable of modulating postprandial ammonia excretion in anticipation of the increase in ammonia concentration as a result of feeding activity. Although these studies have demonstrated high physiological capacities of mudskippers against ammonia toxicity, it should be noted that most mudskippers stay out of water for up to 10-20 min only and frequently visit their burrows or water pools during their activities on the mudflat (Milward, 1974). The high ammonia tolerance of mudskippers might rather be of greater biological significance during burrow confinement during high tide or overwintering (in temperate species).

FEEDING

Of the four mudskipper genera, species of *Periophthalmodon* and *Periophthalmus* are purely carnivorous (Sponder and Lauder, 1981; Kruitwagen *et al.*, 2007; Kutschera *et al.*, 2008), whereas *Boleophthalmus* species are exclusively herbivorous (Yang *et al.*, 2003). Although not much is known about feeding habit of *Scartelaos* mudskippers, Milward (1974) reported them to be omnivorous.

Terrestrial life demands fundamentally different mechanism for feeding. Aquatic fishes usually capture prey using suctioning (Ferry-Graham and Lauder, 2001; Grubich, 2001; Wainwright *et al.*, 2007), however, this mechanism is ineffective in air due to the much lower density and viscosity of the air compared to water. In addition to exploiting aquatic prey, mudskippers are compelled to forage terrestrially because of their propensities to emerge for extended periods and engage in taxing activities over the mud surface. Similar with the acute aerial vision reported for amphibious carnivorous blennies (Graham, 1970; Nieder, 2001), *Periophthalmus* can likewise effectively locate and follow their prey visually (Kutschera *et al.*, 2008)*. Terrestrial carnivorous feeding, however, does not only require visual acuity in air but also a strategy to subdue or dismember

*Functional significance of the peculiar olfactory organ structure recently described for *Ps. barbarus* is currently unclear (Kuciel *et al.*, 2010 Acta Zoologica (Stockholm) doi: 10.1111/j.1463-6395.2010.00495.x).

live preys before ingestion. Van Wassenbergh *et al.* (2006) discovered that when the freshwater amphibious eel catfish, *Channallabes apus*, forages on land, it lifts the anterior part of its body and bends its head downward onto the prey, followed by alternate closing and opening of mouth until the prey is firmly secured between the jaws. This elaborate feeding scheme will probably ensure that the prey is bodily infirmed so as not to inflict injury into the oral structures of the fish, although Van Wassenbergh *et al.* (2006) claimed that head bending will prevent prey escape and facilitate capture. A similar terrestrial feeding kinematics has been reported in *Periophthalmus koelreuteri* (= *Ps. kalolo*, Sponder and Lauder, 1981). Swallowing is accomplished by coordinated movement of the pharyngeal jaws and the buccal cavity is filled with air during feeding (Sponder and Lauder, 1981).

Although feeding action of *Boleophthalmus* has not been studied in detail, our observation suggests that *B. pectinirostris* fills its bucco-opercular cavities with water by drinking at a burrow or a nearby water pool prior to feeding (Ishimatsu *et al.*, unpublished). Foraging consists of a characteristic lateral sweeping of the head to scrape off diatoms from the mudflat surface. Yang *et al.* (2003) stated that "*B. pectinirostris* feed by 'drinking', filling their mouths with water, which is then discharged through the operculae". Diatoms with maximum length of 50 μm were selectively ingested (Yang *et al.*, 2003). Gill rakers of *B. pectinirostris* show a characteristic morphology. Those on the 1st, 2nd and anterior (rostral) side of the 3rd gill arches are made up of coarsely arranged plates, whereas those on the posterior (opercular) side of the 3rd and both sides of the 4th gill arches are much densely packed (Fig. 4.5.2). Yang (1996, as cited in Yang *et al.*, 2003) found that the distance between adjacent gill rakers was approximately 50-60 μm. Following foraging, *B. pectinirostris* lifts its mouth off the surface and then does a 'gargling' movement, an action which is likely involved in sieving diatoms from mud particles (Ishimatsu *et al.*, unpublished). The herbivorous feeding of *Boleophthalmus* probably does not require similarly acute aerial vision as suspected for *Periophthalmus*.

REPRODUCTION

One of the most remarkable adaptations to mudflat environment shown by mudskippers is their use of air for reproduction. It has been long known that mudskippers deposit eggs in burrows they excavate in mudflats, but how their embryos can develop remains an enigma because of extremely low dissolved oxygen concentration of burrow water (Gordon, 1995; Ishimatsu *et al.* 1998a, b, 2000, 2007). Our previous findings of the presence of oxygen-

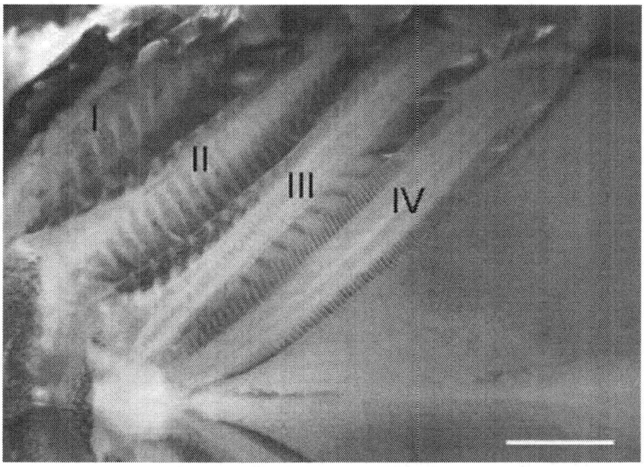

Fig. 4.5.2 Dorsal view of the gill rakers of *Boleophthalmus pectinirostris*. Roman numerals indicate gill arch number. Note different morphology of the gill rakers on the posterior (opercular) side of III and on both sides of IV (Aketagawa and Ishimatsu, unpublished). Scale 2 mm.

containing gas in burrows of the Malaysian mudskipper *Periophthalmodon schlosseri* shed new light on this issue (Ishimatsu *et al.*, 1998a).

Subsequent to the findings on *Pn. schlosseri*, we conducted a 10-year field study on reproduction of the Japanese mudskipper, *Periophthalmus modestus*. This has resulted in much better understanding of how this mudskipper secures oxygen for developing embryos, and how hatching occurs in the mudflat burrows (Ishimatsu *et al.*, 2007). Burrow-guarding males of *Ps. modestus* deposit mouthfuls of fresh air into the egg chamber during each low tide. As a result, oxygen pressure (PO_2) of the egg-chamber gas continually increased until burrow openings were covered by incoming tide. During high tide, the male ceases air deposition and this leads to a gradual decline in egg-chamber PO_2, which can be attributed mostly to O_2 consumption by developing embryos (Etou *et al.*, 2007, Fig. 4.5.3).

The air deposition behaviour during low tide appears to be governed by burrow-gas PO_2 and tidal cycle (Ishimatsu *et al.*, 2007). The burrow-guarding male responded to artificial reductions of burrow-gas PO_2 by overtly increasing its air deposition frequency, indicating that the ability of the fish to sense burrow-gas PO_2 affects its behaviour. However, when plotted against burrow-gas PO_2, the frequency of the induced air-adding behaviour lies above the relationship obtained for the natural cycles of air deposition, as if the fish somehow "recognizes" the time left before the next submersion of the burrow opening. In fact, the data obtained from burrows

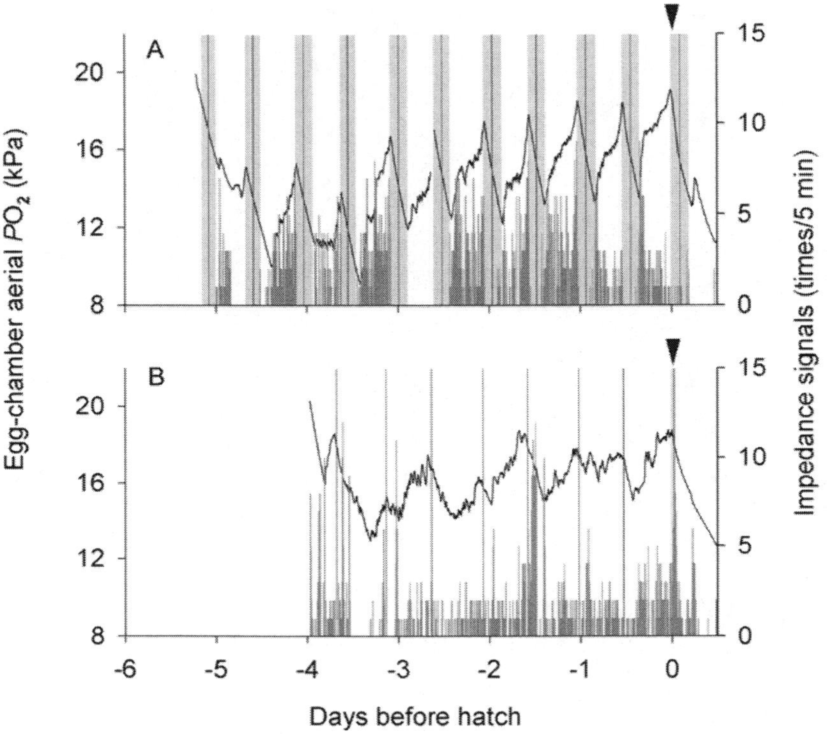

Fig. 4.5.3 Continuous records of *Periophthalmus modestus* egg-chamber air PO_2 (black line) and the timing of the male's egg-chamber visits (indicated by impedance signals, grey spikes) in relation to the tidal cycle. Data acquisition began with the day of instrumentation and continued until egg hatching, initiated by the male on a rising tide (blue triangle). The beginning of day zero is set to the time when a burrow was covered by an incoming tide, following which egg hatching occurred. The initial high PO_2 reflects the opening of the egg chamber to air during instrumentation. (A) Record for a burrow regularly covered (blue bars) and uncovered by the tidal oscillation (red lines show the highest tide). (Note: No impedance data for day –3 due to a technical problem.) (B) Record for a burrow located sufficiently high on the mudflat to be continually exposed to air until day zero when the rising tide covered the burrow (blue bar) and hatching occurred (blue triangle, from Ishimatsu *et al.*, 2007).

Color image of this figure appears in the color plate section at the end of the book.

excavated in high locations of mudflats suggest that the fish modulate air-adding behaviour according to the tidal oscillation at their specific location. These burrows are located sufficiently high on the mudflat to remain uncovered by the highest neap tides. Nevertheless burrow-gas PO_2 reflected tidal cycle, of which the peak PO_2 in the egg chamber coincided with the time of highest tide (Fig. 4.5.3).

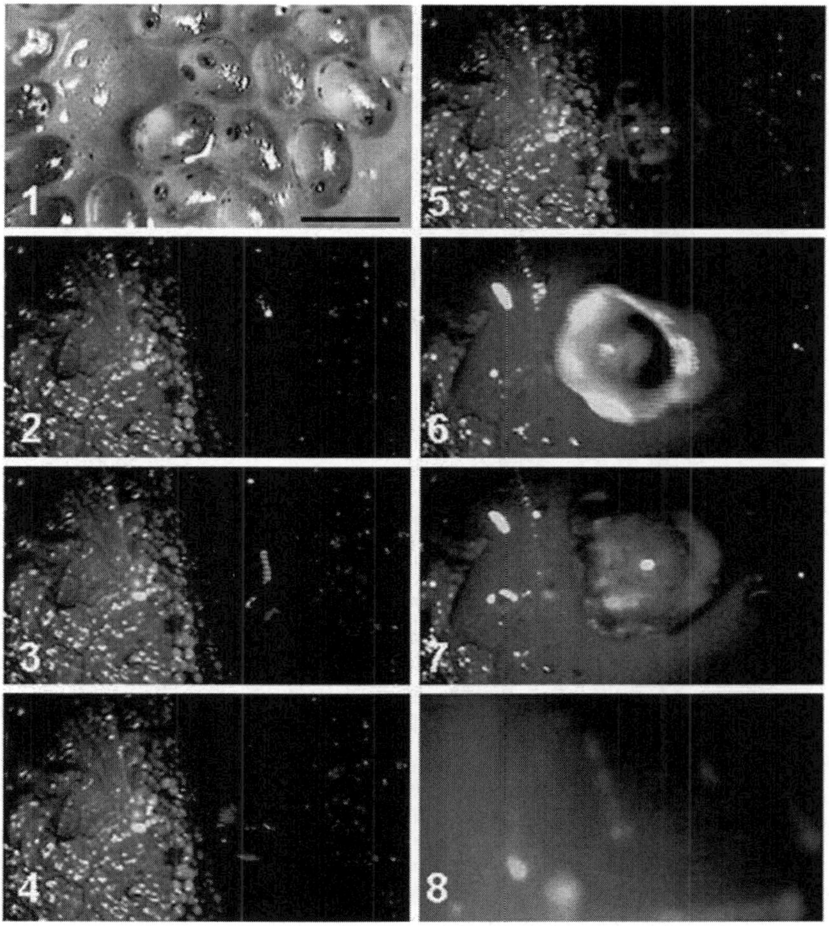

Fig. 4.5.4 Interior view of a *Periophthalmus modestus* egg chamber. (1) Eggs on the mud wall of a laboratory-incubated chamber (scale bar 1 mm). (2 – 7) Endoscope video frames documenting hatch induction behaviour (June 19, 2002). Vertical view is down, through the egg chamber (note glistening monolayer of eggs on the mud wall), to the burrow water surface. The male appears at the surface with its mouth open, and becomes progressively larger as it moves closer to the camera. Removing mouthfuls of air (3, 4, 6, 7) raises the water level, which immerses the eggs, causing them to hatch. (8) Hatched larvae [average total length 2.84 mm (Kobayashi *et al.*, 1972)] swimming in the flooded egg chamber 48 min after frame 7 (from Ishimatsu *et al.*, 2007).

Color image of this figure appears in the color plate section at the end of the book.

When egg development is complete, the male removes the egg-chamber air (by regulping the air) and releases it outside the burrow. This air-removal activity floods the egg chamber and induces egg hatching (Fig. 4.5.4).

Ps. modestus eggs develop normally when incubated in a temperature-controlled, humidified air-filled chamber in the laboratory, but need to be immersed (either in freshwater or seawater) to trigger hatching. The aerial-egg development and water-induced hatching demonstrated for *Ps. modestus* agree with observations of Brillet (1976) who reported that the eggs of *Ps. sobrinus* (= *P. argentilineatus*, Murdy, 1989) required both air incubation and subsequent submersion for hatching and died if incubated in water or kept in air but not immersed after development.

Air storage in burrows appears to be common among mudskippers, and has been reported for all the four genera (*Boleophthalmus*, A. Toba and A. Ishimatsu, in preparation; *Periophthalmodon*, Ishimatsu *et al.*, 1998a, 2009 *Periophthalmus*, Ishimatsu *et al.*, 2007; *Scartelaos*, Lee *et al.*, 2005) and also for a non-mudskipper oxudercine goby, *Oxuderces dentatus* (Dr. Toru Takita, Professor Emeritus of Nagasaki University, unpublished). We have recently found that both males and females of *B. pectinirostris* maintain mudflat burrows throughout the year (in both spawning and non-spawning seasons), and deposit air therein, although burrow gas volume decreases during winter months (A. Toba and A. Ishimatsu, in preparation). Likewise, Lee *et al.* (2005) observed air deposition by both males and females of *S. histophorus* that shared burrows on mudflats in Australia. This was also observed in an artificial burrow system in the laboratory. Both *Boleophthalmus* (Zhang *et al.*, 2000) and *Scartelaos* (Lee *et al.*, 2005) are thought to reside in their burrows during high tide. These findings support the hypothesis that air deposition by mudskippers serves the need for respiration by adult fish during burrow confinement when mudflat surface is inundated by high tide.

We presume that aerial egg incubation in burrow is prevalent among mudskippers, even though this has so far been confirmed for only two species, *Periophthalmus modestus* (Ishimatsu *et al.*, 2007) and *Periophthalmodon schlosseri* (Ishimatsu *et al.*, 2009). Hong *et al.* (2007) injected a dye into the burrows of *Boleophthalmus pectinirostris*, and found that the top inner wall of the egg chambers, where eggs were deposited, was not stained. We have also found for the same species that the redox-potential of the top inner wall of egg chambers was significantly higher than the values recorded in the other parts of the burrow, indicating a more oxygenated condition on the egg-chamber ceiling (A. Toba and A. Ishimatsu, in preparation). Both of these findings strongly suggest the presence of air in egg chambers of this mudskipper. Configuration of *B. pectinirostris* burrows is deeper, much more complex and variable than that of *Periophthalmus* and *Pn. schlosseri* burrows, therefore direct observation of the egg chambers has not been successfully done (the structure may vary among populations, see Hong *et al.*, 2007). Hong *et al.* (2007) also found that the egg chambers are present only in the burrows excavated by the male *B. pectinirostris*. A brochure providing information about the conservation of endangered species on Okinawa Island, Japan, states that eggs of *Scartelaos histophorus* are attached to the top inner wall of egg chambers. Whether or not air is present in the egg chambers of this species is not stated (Okinawa Prefecture, 2000). Aerial egg incubation might also occur in non-mudskipper gobies dwelling on mudflats. The eel goby *Odontamblyopus lacepedii*, which is sympatric with *Ps. modestus*, has recently been shown to be a facultative air breather (Gonzales *et al.*, 2006). *Odontamblyopus*, however, is not amphibious and is restricted to its extensive mudflat burrow where, during low tide, it breathes air to sustain metabolism. The eel goby's reproductive behaviour is not known, however, fish that spawned in captivity attached their eggs along the upper, inner side of the submerged clay pipes in their holding tank (Dotsu and Takita, 1967), implying that burrow spawning may occur.

FUTURE RESEARCH THEMES

During the course of preparing this review, we recognized that there are still many gaps in our knowledge of mudskipper biology. The following list outlines several of the most important open questions:

1. Oxudercine non-mudskipper gobies need to be investigated for their potential amphibious nature. Possible candidates include species of *Apocryptes*, *Oxuderces*, *Pseudapocryptes* and *Zappa*. Graham (1997) classified *Apocryptes bato* (Oxudercinae) as amphibious based on Murdy's (1989) account, but we were unable to confirm this in Murdy (1989) or in Hora (1936). Zhang (2000) reported that "*Oxuderces dentatus* sometimes moves on the exposed mudflats for a short distance when moving from one water pool to another, exposing the whole or a part of body to air". Takita (personal communication) recently found that *Pseudapocryptes elongatus* emerges from water during low tide and feeds on a mudflat at the Sungai Pinang in Balik Pulau, Penang, Malaysia (Fig. 4.5.5). On the contrary, Hora (1935) stated that "it (*Pss. elongatus*) lives in small, tidal pools, flooded at spring tide only, and in its natural haunts rarely comes to the surface for aerial respiration". Finally, Roberts (1978) described that "hundreds of *Zappa confluentus* (reported as *Pss. confluentus*) flipped about on an exposed mudflat, but never balanced on their pelvic fins or assumed an upright position". Even though these are

Fig. 4.5.5 *Pseudapocryptes elongatus* feeding on a mudflat along Sungai Pinang, Penang Island, Malaysia (by the courtesy of Dr. T. Takita).

no more than preliminary observations, we are of the opinion that possible amphibious nature of non-mudskipper oxudercine gobies needs to be examined in both field and the laboratory.*

2. Some details about the mechanics of burrow spawning by mudskippers remain unknown. We confirmed that there is air in the burrows of the courting male *Ps. modestus*. On the basis of this finding, we speculate that the male fills the egg chamber with air prior to mating and that both spawning and fertilization occur in air, although these have not been observed (Ishimatsu *et al.*, 2007). The presence of air in burrows of *Periophthalmodon schlosseri* where no eggs were found in the egg chambers (Ishimatsu *et al.*, 2009) may also represent air deposition prior to egg laying. In support of the possibility of aerial spawning and fertilization, Shimizu *et al.* (2006) reported aerial egg deposition and fertilization in the amphibious rockhopper blenny, *Andamia tetradactyla*. Direct video recording of burrow spawning must answer this question.
3. Potential ovoviviparity of *Periophthalmus vulgaris* (= *argentilineatus*, Murdy, 1989) reported by Harms (1934) is unlikely considering burrow egg deposition shown for many species of mudskippers, but nevertheless needs rigorous investigation to verify or disprove the description "The oldest stage, which I could observe in the ovary, was already a young fish with optic vesicles, heart, spinal cord,

*Recently, Polgar *et al.* (2010) analysed habitat conditions of oxdercine gobies in Papua New Guinea including *Z. confluentus*, but did not describe its behaviour (Polgar *et al.* (2010) J. Fish Biol. doi:10.1111/j.1095-8649.2010.02807.x).

notochord and somites. These animals moved already weakly within their egg membrane..." (translation by the authors).

4. Nitrogen metabolism of mudskipper embryos also warrants investigation. During early development of fish, ammonia is continuously formed as yolk proteins and amino acids are catabolized. Fish embryos excrete ammonia into surrounding water, but also have a system to convert ammonia to glutamine and urea and to store ammonia in the yolk to prevent ammonia accumulation in embryonic tissues (Steele *et al.*, 2001; Essex-Fraser *et al.*, 2005; Wright, 2007). Mudskipper embryos remain out of water throughout their entire development (one week in *Ps. modestus*), and thus are more prone to ammonia buildup. Therefore it is expected that mudskipper embryos have a high capacity for ammonia detoxification and/or high tolerance to elevated tissue ammonia.
5. The mechanism for rapid escape of the hatched larvae from hypoxic burrows needs to be studied. Our laboratory test revealed that larvae would succumb if they are kept in hypoxic water simulating typical burrow-water conditions (Etou *et al.*, 2007). Whether or not the guarding male is involved in transporting larvae out of the burrow remains to be determined.
6. During winter, mudskippers in temperate regions remain in the mud for several months, as in Japan and Korea (Baeck *et al.*, 2008). It is not known how overwintering mudskippers maintain their metabolism or whether they occur/hibernate in water-filled burrows or bury themselves in mud or even in an air-filled chamber. Baeck *et al.* (2008) reported both *Periophthalmus magnuspinnatus* and *Ps. modestus* possess burrows for wintering.
7. One of the peculiarities of oxudercine anatomy is the position of the anterior nostril. In these gobies, a tissue flap droops down over the upper jaw with the anterior nostril opening at tip. This is in fact one of the diagnostic features of the whole oxudercine group, not exclusively of the mudskipper genera (Murdy, 1989). The migration of the anterior nostril down to the level of the upper lip might have some functional connection with feeding on the substratum (similar positioning of the anterior nostril is seen in other aquatic benthic/demersal gobies, e.g. several species of *Luciogobius*, *Odontamblyopus*, and *Taenioides*). It may be merely a coincidence, but the anterior nostril position in these gobies is somewhat reminiscent of that in ancestral tetrapodomorph fish (Zhu and Ahlberg, 2004), even

though the nasal structure of mudskippers and early tetrapods are clearly not homologous.

ACKNOWLEDGMENTS

We are grateful to Ms. Mizuri Murata for her help in preparing the manuscript.

REFERENCES

Aguilar, N.M., A. Ishimatsu, K. Ogawa and K.H. Khoo. 2000. Aerial ventilatory responses of the mudskipper, *Periophthalmodon schlosseri*, to altered aerial and aquatic respiratory gas concentrations. *Comparative Biochemistry and Physiology* A 120: 285-292.

Baeck, G.W., T. Takita and Y.H. Yoon. 2008. Lifestyle of Korean mudskipper *Periophthalmus magnuspinnatus* with reference to a congeneric species *Periophthalmus modestus. Ichthyological Research* 55: 43-52.

Brauner, C.J., V. Matey, J.M. Wilson, N.J. Bernier and A.L. Val. 2004. Transition in organ function during the evolution of air-breathing: Insights from *Arapaima gigas*, an obligate air-breathing teleost from the Amazon. *Journal of Experimental Biology* 207: 1433-1438.

Brillet, C. 1976. Structure du terrier, reproduction et comportement des jeunes chez le poisson amphibie *Periophthalmus sobrinus* Eggert. *Revue d'Ecologie: La Terre et la Vie* 30: 465-483.

Chew, S.F., M.Y. Sim, Z.C. Phua, W.P. Wong and Y.K. Ip. 2007. Active ammonia excretion in the giant mudskipper, *Periophthalmodon schlosseri* (Pallas), during emersion. *Journal of Experimental Zoology* A 307: 357-369.

Clack, J.A. 2006. The emergence of early tetrapods. *Palaeogeography, Palaeoclimatology, Palaeoecology* 232: 167-189.

Clack, J.A. 2007. Devonian climate change, breathing, and the origin of the tetrapod stem group. *Integrative and Comparative Biology* 47: 510-523.

Clayton, D.A. 1993. Mudskippers. *Oceanography and Marine Biology: An Annual Review* 31: 507-577.

Clayton, D.A. and T.C. Vaughan. 1986. Territorial acquisition in the mudskipper *Boleophthalmus boddarti* (Teleostei, Gobiidae) on the mudflats of Kuwait. *Journal of Zoology, London* 209: 501-519.

Dall, W. and N.E. Milward. 1969. Water intake, gut absorption and sodium fluxes in amphibious and aquatic fishes. *Comparative Biochemistry and Physiology* 30: 247-260.

Darumas, U. and P. Tantichodok. 2002. A new species of mudskipper (Gobiidae: Oxudercinae) from Southern Thailand. *Phuket Marine Biological Center Research Bulletin* 64: 99-106.

Das, B.K. 1934. The habits and structure of *Pseudapocryptes lanceolatus*, a fish in the first stages of structural adaptation to aerial respiration. *Proceedings of the Royal Society of London* 115: 422-431.

Dotsu, Y. and T. Takita. 1967. Induced spawning by hormone operation, egg development and larva of blind gobioid fish, *Odontamblyopus rubicundus*. *Bulletin of the Faculty of Fisheries, Nagasaki University* 23: 135-144. (In Japanese.)

Eddy, F.B. 2005. Ammonia in estuaries and effects on fish. *Journal of Fish Biology* 67: 1495-1513.

Essex-Fraser, P.A., S.L. Steele, N.J. Bernier, B.W. Murray, E.D. Stevens and P.A. Wright. 2005. Expression of four glutamine synthetase genes in the early stages of development of rainbow trout (*Oncorhynchus mykiss*) in relation to nitrogen excretion. *Journal of Biological Chemistry* 280: 20268-20273.

Etou, A., T. Takeda, Y. Yoshida and A. Ishimatsu. 2007. Oxygen consumption during embryonic development of the mudskipper (*Periophthalmus modestus*): Implication for the aerial development in burrows. In: *Fish Respiration and Environment*, M.N. Fernandes, F.T. Rantin, M.L. Glass and B.G. Kapoor (eds.), Science Publishers, Enfield, pp. 83-91.

Evans, D.H., J.B. Claiborne and G.A. Kormanik. 1999. Osmoregulation, acid-base regulation, and nitrogen excretion. In: *Intertidal Fishes: Life in Two Worlds*, M.H. Horn, K.L.M. Martin and M.A. Chotkowski (eds.), Academic Press, San Diego, pp. 79-96.

Evans, D.H., P.M. Piermarini and K.P. Choe. 2005. The multifunctional fish gill: Dominant site of gas exchange, osmoregulation, acid-base regulation, and excretion of nitrogenous waste. *Physiological Reviews* 85: 97-177.

Ferraris, C.J. Jr. 1995. On the validity of the name *Pseudapocryptes lanceolatus* (Bloch and Schneider, 1801) (Osteichthyes: Percomorpha: Gobiidae). *Copeia* 1995: 984

Ferry-Graham, L.A. and G.V. Lauder. 2001. Aquatic prey capture in ray-finned fishes: a century of progress and new directions. *Journal of Morphology* 248: 99-119.

Garey, W.F. 1962. Cardiac responses of fishes in asphyxic environments. *Biological Bulletin* 122: 362-368.

Gee, J.H. and P.A. Gee. 1991. Reactions of gobioid fishes to hypoxia: buoyancy control and aquatic surface respiration. *Copeia* 1991: 17-28.

Gonzales, T.T. 2007. *Respiratory Adaptation of the Intertidal Burrow-dwelling Eel Goby*, Odontamblyopus lacepedii (Gobiidae: Amblyopinae). Ph.D. Thesis, Nagasaki University.

Gonzales, T.T., M. Katoh and A. Ishimatsu. 2006. Air breathing of aquatic burrow-dwelling eel goby, *Odontamblyopus lacepedii* (Gobiidae: Amblyopinae). *Journal of Experimental Biology* 209: 1085-1092.

Gonzales, T.T., M. Katoh and A. Ishimatsu. 2008. Respiratory vasculatures of the intertidal air-breathing eel goby, *Odontamblyopus lacepedii* (Gobiidae: Amblyopinae). *Environmental Biology of Fishes* 82: 341-351.

Gonzales, T.T., M. Katoh, M.A. Ghaffar and A. Ishimatsu. 2011. Gross and fine anatomy of the respiratory vasculature of the mudskipper, *Periophthalmodon schlosseri* (Gobiidae: Oxudercinae). *Journal of Morphology* DOI: 10.1002/Jmor.10944.

Gordon, M.S. 1995. Functional evidence from living vertebrates. In: *Invasion of the Land*, M.S. Gordon and E.C. Olson (eds.), Columbia University Press, New York, pp. 216-250.

Gordon, M.S., D.J. Gabaldon and A.Y.-w Yip. 1985. Exploratory observations on microhabitat selection within the intertidal zone by the Chinese mudskipper fish *Periophthalmus cantonensis*. *Marine Biology* 85: 209-215.

Gordon, M.S., J. Boëtius, I. Boëtius, D.H. Evans, R. McCarthy and L.C. Oglesby. 1965. Salinity adaptation in the mudskipper fish *Periophthalmus sobrinus*. *Hvalrdets Skrifter* 48: 85-93.

Graham, J.B. 1970. Preliminary studies on the biology of the amphibious clinid *Mnierpes macrocephalus*. *Marine Biology* 5: 136-140.

Graham, J.B. 1976. Respiratory adaptations of marine air-breathing fishes. In: *Respiration of Amphibious Vertebrates*, G.M. Hughes (ed.), Academic Press, London, pp. 165-187.

Graham, J.B. 1997. *Air-breathing Fishes: Evolution, Diversity and Adaptation*. Academic Press, San Diego.

Graham, J.B., H.J. Lee and N.C. Wegner. 2007. Transition from water to land in an extant group of fishes: Air breathing and the acquisition sequence of adaptations for amphibious life in oxudercine gobies. In: *Fish Respiration and Environment*, M.N. Fernandes, F.T. Rantin, M.L. Glass and B.G. Kapoor (eds.), Science Publishers, Enfield, pp. 255-288.

Grubich, J.R. 2001. Prey capture in actinopterygian fishes: a review of suction feeding motor patterns with new evidence from an elopomorph fish, *Megalops atlanticus*. *American Zoologist* 41: 1258-1265.

Harms, J.W. 1934. *Wandlungen des Artgefüges unter Natürlichen and Künstlichen Umweltbedingungen*. Johann Ambrosius Barth, Leipzig.

Harris, V.A. 1960. On the locomotion of the mudskipper *Periophthalmus koelreuteri* (Pallas): (Gobiidae). *Proceedings of the Royal Society of London* 134: 107-135.

Hong, W., S. Chen, Q. Zhang and Q. Wang. 2007. Reproductive ecology of the mudskipper *Boleophthalmus pectinirostris*. *Acta Oceanologica Sinica* 26: 72-81.

Hora, S.L. 1935. Physiology, bionomics, and evolution of the air-breathing fishes of India. *Transactions of the National Institute of Science of India* 1: 1-16.

Hora, S.L. 1936. Ecology and bionomics of the gobioid fishes of the Gangetic Delta. *Comptes Rendus du XIIe Congrès International de Zoologie, Lisbon* 1935: 841-865.

Ikebe, Y. and T. Oishi. 1996. Correlation between environmental parameters and behaviour during high tides in *Periophthalmus modestus*. *Journal of Fish Biology* 49: 139-147.

Ikebe, Y. and T. Oishi. 1997. Relationships between environmental factors and diel and annual changes of the behaviors during low tides in *Periophthalmus modestus*. *Zoological Science* 14: 49-55.

Ip, Y.K., C.G.L. Lee, W.P. Low and T.J. Lam. 1991. Osmoregulation in the mudskipper. *Boleophthalmus boddaerti* I. Responses of branchial cation activated and anion stimulated adenosine triphosphatases to changes in salinity. *Fish Physiology and Biochemistry* 9: 63-68.

Ip, Y.K., S.F. Chew, J.M. Wilson and D.J. Randall. 2004a. Defences against ammonia toxicity in tropical air-breathing fishes exposed to high concentrations of environmental ammonia: A review. *Journal of Comparative Physiology* B 174: 565-575.

Ip, Y.K., C.K. Lim, S.L.M. Lee, W.P. Wong and S.F Chew. 2004b. Postprandial increases in nitrogenous excretion and urea synthesis in the giant mudskipper *Periophthalmodon schlosseri*. *Journal of Experimental Biology* 207: 3015-3023.

Ishimatsu, A. and Y. Itazawa. 1981. Ventilation of the air-breathing organ in the snakehead *Channa argus*. *Japanese Journal of Ichthyology* 28: 276-282.

Ishimatsu, A., T. Takeda, T. Kanda, S. Oikawa and K.H. Khoo. 2000. Burrow environment of mudskippers in Malaysia. *Journal of Bioscience* 11: 17-28.

Ishimatsu, A., T. Takeda, Y. Tsuhako, T.T. Gonzales and K.H. Khoo. 2009. Direct evidence for aerial egg deposition in the burrows of the Malaysian mudskipper, *Periophthalmodon schlosseri*. *Ichthyological Research* 56: 417-420.

Ishimatsu, A., Y. Yoshida, N. Itoki, T. Takeda, H.J. Lee and J.B. Graham. 2007. Mudskippers brood their eggs in air but submerge them for hatching. *Journal of Experimental Biology* 210: 3946-3954.

Ishimatsu, A., Y. Hishida, T. Takita, T. Kanda, S. Oikawa, T. Takeda and K.H. Khoo. 1998a. Mudskippers store air in their burrows. *Nature* 391: 237-238.

Ishimatsu, A., K.H. Khoo and T. Takita. 1998b. Deposition of air in burrows of tropical mudskippers as an adaptation to the hypoxic mudflat environment. *Science Progress* 81: 289-297.

Ishimatsu, A., N.M. Aguilar, K. Ogawa, Y. Hishida, T. Takeda, S. Oikawa, T. Kanda and K.H. Khoo. 1999. Arterial blood gas levels and cardiovascular function during varying environmental conditions in a mudskipper, *Periophthalmodon schlosseri*. *Journal of Experimental Biology* 202: 1753-1762.

Jaafar, Z. and H.K. Larson. 2008. A new species of mudskipper, *Periophthalmus takita* (Teleostei: Gobiidae: Oxudercinae), from Australia, with a key to the genus. *Zoological Science* 25: 946-952.

Jaafar, Z., M. Perrig and L.M. Chou. 2009. *Periophthalmus variabilis* (Teleostei: Gobiidae: Oxudercinae), a valid species of mudskipper, and a re-diagnosis of *Periophthalmus novemradiatus*. *Zoological Science* 26: 309-314.

Kobayashi, T., Y. Dotsu and T. Takita. 1971. Nest and nesting behavior of the mudskipper, *Periophthalmus cantonensis* in Ariake Sound. *Bulletin of the Faculty of Fisheries, Nagasaki University* 32: 27-40. (In Japanese.)

Kobayashi, T., Y. Dotsu and N. Miura. 1972. Egg development and rearing experiments of the larvae of the mudskipper, *Periophthalmus cantonensis* in Ariake Sound. *Bulletin of the Faculty of Fisheries, Nagasaki University* 33: 49-62. (In Japanese.)

Kok, W.K., C.B. Lim, T.J. Lam and Y.K. Ip. 1998. The mudskipper *Periophthalmodon schlosseri* respires more efficiently on land than in water and vice versa for *Boleophthalmus boddaerti*. *Journal of Experimental Zoology* 280: 86-90.

Kruitwagen, G., I. Nagelkerken, B.R. Lugendo, H.B. Pratap and S.E. Wendelaar Bonga. 2007. Influence of morphology and amphibious lifestyle on the feeding ecology of the mudskipper *Periophthalmus argentilineatus*. *Journal of Fish Biology* 71: 39-52.

Kutschera, U., H. Burghagen and J-P. Ewert. 2008. Prey-catching behaviour in mudskippers and toads: A comparative analysis. *OnLine Journal of Biological Sciences* 8: 41-43.

Larson, H.K. and T. Takita. 2004. Two new species of *Periophthalmus* (Teleostei: Gobiidae: Oxudercinae) from northern Australia, and a re-diagnosis of *Periophthalmus novaeguineaensis*. *The Beagle, Records of the Museums and Art Galleries of the Northern Territory* 20: 175-185.

Lee, C.G.L., W.P. Low and Y.K. Ip. 1987. Na^+, K^+ and volume regulation in the mudskipper, *Periophthalmus chrysospilos*. *Comparative Biochemistry and Physiology* A 87: 439-448.

Lee, C.G.L., P.W. Low, T.J. Lam, A.D. Munro and Y.K. Ip. 1991. Osmoregulation in the mudskipper, *Boleophthalmus boddaerti* II. Transepithelial potential and hormonal control. *Fish Physiology and Biochemistry* 9: 69-75.

Lee, H.J., C.A. Martinez, K.J. Hertzberg, A.L. Hamilton and J.B. Graham. 2005. Burrow air phase maintenance and respiration by the mudskipper *Scartelaos histophorus* (Gobiidae: Oxudercinae). *Journal of Experimental Biology* 208: 169-177.

Lee, Y.-J., Y. Choi and B.-S. Ryu. 1995. A taxonomic revision of the genus *Periophthalmus* (Pisces: Gobiidae) from Korea with description of a new species. *Korean Journal of Ichthyology* 7: 120-127.

Liem, K.F. 1984. The muscular basis of aquatic and aerial ventilation in the air-breathing teleost fish *Channa*. *Journal of Experimental Biology* 113: 1-18.

Liem, K.F. 1987. Functional design of the air ventilation apparatus and overland excursions by teleosts. *Fieldiana Zoology New Series* 37: 1-29.

Lim, C.B., S.F. Chew, P.M. Anderson and Y.K. Ip. 2001. Reduction in the rates of protein and amino acid catabolism to slow down the accumulation of endogenous ammonia: A strategy potentially adopted by mudskippers (*Periophthalmodon schlosseri* and *Boleophthalmus boddaerti*) during aerial exposure in constant darkness. *Journal of Experimental Biology* 204: 1605-1614.

Little, C. 2000. *The Biology of Soft Shores and Estuaries*. Oxford University Press, Oxford.

Low, W.P., D.J.W. Lane and Y.K. Ip. 1988. A comparative study of terrestrial adaptations of the gills in three mudskippers—*Periophthalmus chrysospilos*, *Boleophthalmus boddaerti*, and *Periophthalmodon schlosseri*. *Biological Bulletin* 175: 434-438.

Maina, J.N. 1998. *The Gas Exchangers: Structure, Function, and Evolution of the Respiratory Processes.* Springer, Berlin.

Martin, K.L.M. and C.R. Bridges. 1999. Respiration in water and air. In: *Intertidal Fishes: Life in Two Worlds,* M.H. Horn, K.L.M. Martin and M.A. Chotkowski (eds.), Academic Press, San Diego, pp. 54-78.

Milward, N.E. 1974. *Studies on the Taxonomy, Ecology, and Physiology of Queensland Mudskippers.* Ph.D. Thesis, University of Queensland.

Murdy, E.O. 1989. A taxonomic revision and cladistic analysis of the oxudercine gobies (Gobiidae: Oxudercinae). *Records of the Australian Museum Suppl.* 11: 1-93.

Murdy, E.O. and T. Takita. 1999. *Periophthalmus spilotus,* a new species of mudskipper from Sumatra (Gobiidae: Oxudercinae). *Ichthyological Research* 46: 367-370.

Nieder, J. 2001. Amphibious behaviour and feeding ecology of the four-eyed blenny (*Dialommus fuscus,* Labrisomidae) in the intertidal zone of the island of Santa Cruz (Galapagos, Ecuador). *Journal of Fish Biology* 58: 755-767.

Niva, B., J. Ojha and J.S.D. Munshi. 1981. Morphometrics of the respiratory organs of an estuarine goby, *Boleophthalmus boddaerti. Japanese Journal of Ichthyology* 27: 316-326.

Okinawa Prefecture. 2000. *Tokage-haze no hanashi (An account of* Scartelaos histophorus). Okinawa Sekkei Service, Okinawa. (In Japanese.)

Olson, K.R. 1981. Morphology and vascular anatomy of the gills of a primitive air-breathing fish, the bowfin (*Amia calva*). *Cell and Tissue Research* 218: 499-517.

Olson, K.R. 1994. Circulatory anatomy in bimodally breathing fish. *American Zoologist* 34: 280-288.

Olson, K.R. 2002. Vascular anatomy of the fish gill. *Journal of Experimental Zoology* 293: 214-231.

Olson, K.R., J.S.D. Munshi, T.K. Ghosh and J. Ojha. 1986. Gill microcirculation of the air-breathing climbing perch, *Anabas testudineus* (Bloch): Relationships with the accessory respiratory organs and systemic circulation. *American Journal of Anatomy* 176: 305-320.

Pace, C.M. and A.C. Gibb. 2009. Mudskipper pectoral fin kinematics in aquatic and terrestrial environments. *Journal of Experimental Biology* 212: 2279-2286.

Park, J.Y., I.S. Kim and Y.J. Lee. 2006. A study on the vascularization and structure of the epidermis of the air-breathing mudskipper, *Periophthalmus magnuspinnatus* (Gobiidae, Teleostei), along different parts of the body. *Journal of Applied Ichthyology.* 22: 62-67.

Park, J.-Y., Y.-J. Lee, I.-S. Kim and S.-Y. Kim. 2003a. A comparative study of the regional epidermis of an amphibious mudskipper fish, *Boleophthalmus pectinirostris* (Gobiidae, Pisces). *Folia Zoologica* 52: 431-440.

Park, J.-Y., Y.-J. Lee, I.-S. Kim and S.-Y. Kim. 2003b. Morphological and cytochemical study on the skin of Korean eel goby, *Odontamblyopus lacepedii* (Pisces, Gobiidae). *Korean Journal of Biological Science* 7: 43-47.

Peng, K.W., S.F. Chew, C.B. Lim, S.S.L. Kuah, W.K. Kok and Y.K. Ip. 1998. The mudskippers *Periophthalmodon schlosseri* and *Boleophthalmus boddaerti* can tolerate environmental NH_3 concentrations of 446 and 36 μM, respectively. *Fish Physiology and Biochemistry* 19: 59-69.

Peters, H.M. 1978. On the mechanism of air ventilation in Anabantoids (Pisces: Teleostei). *Zoomorphologie* 89: 93-123.

Polgar, G. and G. Crosa. 2009. Multivariate characterization of the habitats of seven species of Malaysian mudskippers (Gobiidae: Oxudercinae). *Marine Biology* 156: 1475-1486.

Randall, D.J. and T.K.N. Tsui. 2002. Ammonia toxicity in fish. *Marine Pollution Bulletin* 45: 17-23.

Randall, D.J. and Y.K. Ip. 2006. Ammonia as a respiratory gas in water and air-breathing fishes. *Respiratory Physiology and Neurobiology* 154: 216-225.

Randall, D.J., W.W. Burggren, A.P. Farrell and M.S. Haswell. 1981. *The Evolution of Air Breathing in Vertebrates*. Cambridge University Press, Cambridge.

Randall, D.J., Y.K. Ip, S.F. Chew and J.M. Wilson. 2004. Air breathing and ammonia excretion in the giant mudskipper, *Periophthalmodon schlosseri*. *Physiological and Biochemical Zoology* 77: 783-788.

Roberts, T. 1978. An Ichthyological survey of the Fly River in Papua New Guinea with description of new species. *Smithsonian Contributions to Zoology* 281: 1-72.

Sakamoto, T. and M. Ando. 2002. Calcium ion triggers rapid morphological oscillation of chloride cells in the mudskipper, *Periophthalmus modestus*. *Journal of Comparative Physiology* B 172: 435-439.

Sakamoto, T., S. Yokota and M. Ando. 2000. Rapid morphological oscillation of mitochondrion-rich cell in estuarine mudskipper following salinity changes. *Journal of Experimental Zoology* 286: 666-669.

Satchell, G.H. 1976. The circulatory system of air-breathing fish. In: *Respiration of Amphibious Vertebrates*, G.M. Hughes (ed.). Academic Press, London, pp. 105-123.

Sayer, M.D.J. 2005. Adaptations of amphibious fish for surviving life out of water. *Fish and Fisheries* 6: 186-211.

Schöttle, E. 1931. Morphologie und Physiologie der Atmung bei wasser-, schlamm- und landlebenden Gobiiformes. *Zeitschrift für Wissenschaftliche Zoologie* 140: 1-114.

Schultze, H.-P. 1999. The fossil record of the intertidal zone. In: *Intertidal Fishes: Life in Two Worlds*, M.H. Horn, K.L.M. Martin and M.A. Chotkowski (eds.), Academic Press, San Diego, pp. 373-392.

Shimizu, N., Y. Sakai, H. Hashimoto and K. Gushima. 2006. Terrestrial reproduction by the air-breathing fish *Andamia tetradactyla* (Pisces; Blenniidae) on supralittoral reefs. *Journal of Zoology* 269: 357-364.

Sponder, D.L. and G.V. Lauder. 1981. Terrestrial feeding in the mudskipper *Periophthalmus* (Pisces: Teleostei): A cineradiographic analysis. *Journal of Zoology, London* 193: 517-530.

Stebbins, R.C. and M. Kalk. 1961. Observations on the natural history of the mudskipper, *Periophthalmus sobrinus*. *Copeia* 1961: 18-27.

Steele, S.L., T.D. Chadwick and P.A. Wright. 2001. Ammonia detoxification and localization of urea cycle enzyme activity in embryos of the rainbow trout (*Oncorhynchus mykiss*) in relation to early tolerance to high environmental ammonia levels. *Journal of Experimental Biology* 204: 2145-2154.

Suzuki, N. 1992. Fine structure of the epidermis of the mudskipper, *Periophthalmus modestus* (Gobiidae). *Japanese Journal of Ichthyology* 38: 379-396.

Swanson, B.O. and A.C. Gibb. 2004. Kinematics of aquatic and terrestrial escape responses in mudskippers. *Journal of Experimental Biology.* 207: 4037-4044.

Takahashi, H., T. Sakamoto and K. Narita. 2006a. Cell proliferation and apoptosis in the anterior intestine of an amphibious, euryhaline mudskipper (*Periophthalmus modestus*). *Journal of Comparative Physiology* B 176: 463-468.

Takahashi, H., A. Takahashi and T. Sakamoto. 2006b. In vivo effects of thyroid hormone, corticosteroids and prolactin on cell proliferation and apoptosis in the anterior intestine of the euryhaline mudskipper (*Periophthalmus modestus*). *Life Sciences* 79: 1873-1880.

Takeda, T., A. Ishimatsu, S. Oikawa, T. Kanda, Y. Hishida and K.H. Khoo. 1999. Mudskipper *Periophthalmodon schlosseri* can repay oxygen debts in air but not in water. *Journal of Experimental Zoology* 284: 265-270.

Takita, T., Agusnimar and B.A. Ali. 1999. Distribution and habitat requirements of oxudercine gobies (Gobiidae: Oxudercinae) along the Straits of Malacca. *Ichthyological Research* 46: 131-138.

Tamura, O. and T. Moriyama. 1976. On the morphological feature of the gill of amphibious and air-breathing fishes. *Bulletin of the Faculty of Fisheries, Nagasaki University* 41: 1-8.

Truchot, J.P. 1987. *Comparative Aspects of Extracellular Acid-Base Balance.* Springer-Verlag, Berlin.

Van Wassenbergh, S., A. Herrel, D. Adriaens, F. Huysentruyt, S. Devaere and P. Aerts. 2006. A catfish that can strike its prey on land. *Nature* 440: 881.

Wainwright, P., A.M. Carroll, D.C. Collar, S.W. Day, T.E. Higham and R.A. Holzman. 2007. Suction feeding mechanics, performance, and diversity in fishes. *Integrative and Comparative Biology* 47: 96-106.

Willmer, P., G. Stone and I. Johnson. 2005. *Environmental Physiology of Animals*, Blackwell, Massachusetts. 2nd edition.

Wilson, J.M., T.W.K. Kok, D.J. Randall and W.A. Wogl. 1999. Fine structure of the gill epithelium of the terrestrial mudskipper, *Periophthalmodon schlosseri*. *Cell and Tissue Research* 298: 345-356.

Wright, P.A. 2007. Ionic, osmotic and nitrogenous waste regulation. In: *Primitive Fishes* (*Fish Physiology* Vol. 26), D.J. McKenzie, A.P. Farrell and C.J. Brauner (eds.), Academic Press, Amsterdam, pp. 283-318.

Yadav, A.N. and B.R. Singh. 1989. Gross structure and dimensions of the gill in an air-breathing estuarine goby, *Pseudapocryptes lanceolatus*. *Japanese Journal of Ichthyology* 36: 252-259.

Yadav, A.N., M.S. Prasad and B.R. Singh. 1990. Gross structure of the respiratory organs and dimensions of the gill in the mudskipper, *Periophthalmodon schlosseri* (Bleeker). *Journal of Fish Biology* 37: 383-392.

Yang, K.Y. 1996. *The Feeding Ecology of the Mudskipper* Boleophthalmus pectinirostris *(Pisces: Periophthalmidae) at the Mai Po Marshes Nature Reserve, Hong Kong*. Master's Thesis, University of Hong Kong.

Yang, K.Y., S.Y. Lee and G.A. Williams. 2003. Selective feeding by the mudskipper (*Boleophthalmus pectinirostris*) on the microalgal assemblage of a tropical mudflat. *Marine Biology* 143: 245-256.

Yokota, S., K. Iwata, Y. Fujii and M. Ando. 1997. Ion transport across the skin of the mudskipper *Periophthalmus modestus*. *Comparative Biochemistry and Physiology* A 118: 903-910.

Yokoya, S. and O.S. Tamura. 1992. Fine structure of the skin of the amphibious fishes, *Boleophthalmus pectinirostris* and *Periophthalmus cantonensis*, with special reference to the location of blood vessels. *Journal of Morphology* 214: 287-297.

Zhang, J. 2000. *A Comparative Study on the Morphological Adaptations in Oxudercine Gobies to their Terrestrial Life*. Ph.D. Thesis, Nagasaki University.

Zhang, J., T. Taniguchi, T. Takita and A.B. Ali. 2000. On the epidermal structure of *Boleophthalmus* and *Scartelaos* mudskippers with reference to their adaptation to terrestrial life. *Ichthyological Research* 47: 359-366.

Zhang, J., T. Taniguchi, T. Takita and A.B. Ali. 2003. A study on the epidermal structure of *Periophthalmodon* and *Periophthalmus* mudskippers with reference to their terrestrial adaptation. *Ichthyological Research* 50: 310-317.

Zhu, M. and P.E. Ahlberg. 2004. The origin of the internal nostril of tetrapods. *Nature* 432: 94-97.

Zimmer, C. 1998. *At the Water's Edge*. Touchstone, New York.

Index

B

C

D

E

F

H

The Editors

Robert A. Patzner (Salzburg, 1945) is professor of Marine Biology at the University of Salzburg in Austria since 1981. His earlier work focused on the reproduction of hagfish (Myxinoidea) that included year-long research experiences in both Norway and Japan (University of Tokyo). Later work dealt with the reproduction of Mediterranean blennies and included several short stays at marine biological stations in Italy, France and Croatia. In recent years his research addressed the ecology of small benthic fishes – gobies and blennies – as well as limnology. He has published several hundred papers in scientific journals and some books, e.g. on Mediterranean fishes and invertebrates, methods for teaching marine biology and in 2009 "The Biology of Blennies". As a high school teacher at the University of Salzburg in marine biology and limnology he guided student excursions to different places of the Mediterranean, North Sea, Red Sea and Papua New Guinea. He is an underwater photographer and collaborator of FishBase. More details can be found at: http://www.uni-salzburg.at/orgbiol/robert.patzner.

James L. Van Tassell (New York, 1945) received his doctorate in systematics and evolutionary biology from the City University of New York's program at the American Museum of Natural History in 1998 and is currently a Research Associate-in-Residence at the museum. During his tenure as a high school teacher (1970-2000); at H. F. Carey High School in Franklin Square, NY; he developed courses in Ecology, Marine biology, a science research program for gifted students, and an Ecology Club which was recognized by the United Nations as one of the top ten youth organizations in the world. Evenings and vacation periods were occupied by his research on the fishes of the Canary Islands. His studies of the shore fish populations and the effect of tourism on the marine environment lead to the establishment of the first underwater preserve in Spain. Since 1984 he has focused his attention on the systematics of gobiid fishes in the Americas and eastern Atlantic; developed a web site on the systematics, biogeography, and breeding of gobiid fishes (http://gobiidae.com); and published over 30 scientific papers. He worked in the Canary Islands for 26 years and traveled extensively to other areas in the east Pacific, Central America, Caribbean, and coastal zones in the US. Currently he is working with D. Ross Robertson on the development of a website for the identification of all shorefishes in the greater Caribbean region.

Marcelo Kovačić (Split, 1966) is senior curator of vertebrates at the Natural History Museum Rijeka in Croatia. Since he was first empolyed in the museum in 1992 his main interest was in ichthyology and marine biology. His researches address taxonomy and ecology of small marine and freshwater fishes, especially of family Gobiidae. He has published about fifty papers in scientific journals, mostly on Mediterranean and European gobies. The published educational works include also four multimedial interactive CD-ROMs or DVD-ROMs on vertebrates (Mammals, Birds, Amphibians and Reptiles, Sharks and Rays). He gives lectures on Marine Vertebrates at the University of Pula. SCUBA diver since high school, he managed to turn his childhood hobbies, fishing and diving, in the adult career.

B.G. Kapoor was formerly professor and Head of Zoology in Jodhpur University (India). Dr Kapoor has co-edited 22 books published by Science Publishers, Enfield, NH, USA. The most recent ones are Fish Defenses (Volume 1), 2009 (with Giacomo Zaccone, José Meseguer Peñalver and Alfonsa García-Ayala); The Biology of Blennies, 2009 (with Robert A. Patzner, Emanuel J. Gonçalves and Philip A. Hastings); Development of Non-Teleost Fishes, 2009 (with Yvette W. Kunz and Carl A. Luer); Fish Defenses (Volume 2), 2009 (with Giacomo Zaccone, C. Perrière and A. Mathis); Fish Locomotion: An Eco-ethological Perspective (with Paolo Domenici), Biology of Subterranean Fishes, 2010 (with Eleonora Trajano and Maria Elina Bichuette); Biological Clock in Fish, 2010 (with Ewa Kulczykowska and Wlodzimierz Popek). Dr Kapoor has also co-edited The Senses of Fish: Adaptations for the Reception of Natural Stimuli, 2004 (with G. von der Emde and J. Mogdans); and co-authored Ichthyology Handbook, 2004 (with B. Khanna), both from Springer, Heidelberg. He has also been a contributor in books from Academic Press, London (1969, 1975 and 2001). His E-mail is: bhagatgopal.kapoor@rediffmail.com

Color Plate Section

Section 1: Systematics

Chapter 1.2

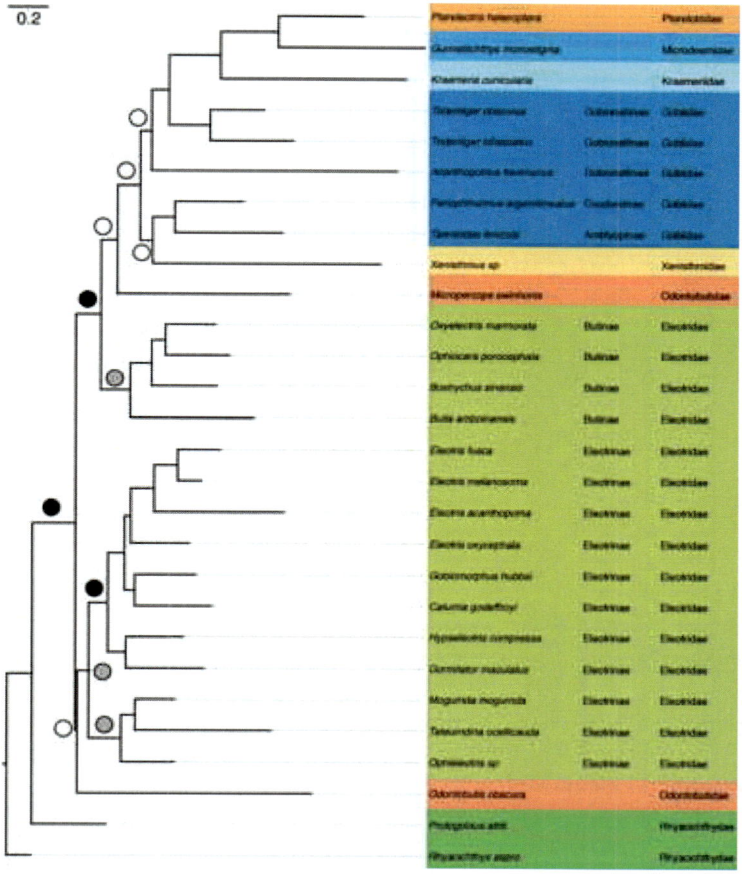

Fig. 1.2.2 Maximum likelihood reanalysis of the Akihito *et al.* (2000) data set using *Rhyacichthys* as outgroup. The analysis was based on a 1140 bp alignment of the complete *cob* gene and analyzed with Garli v0.96 (Zwickl, 2006) using default settings (-lnL = 14990.30) and visualized with FigTree v1.2.1 (Rambaut, 2006). Circles indicate bootstrap support (BS) for selected branches based on 1,000 replicates; white = BS < 50%, grey = 50 ≤ BS < 70, black = BS ≥ 70.

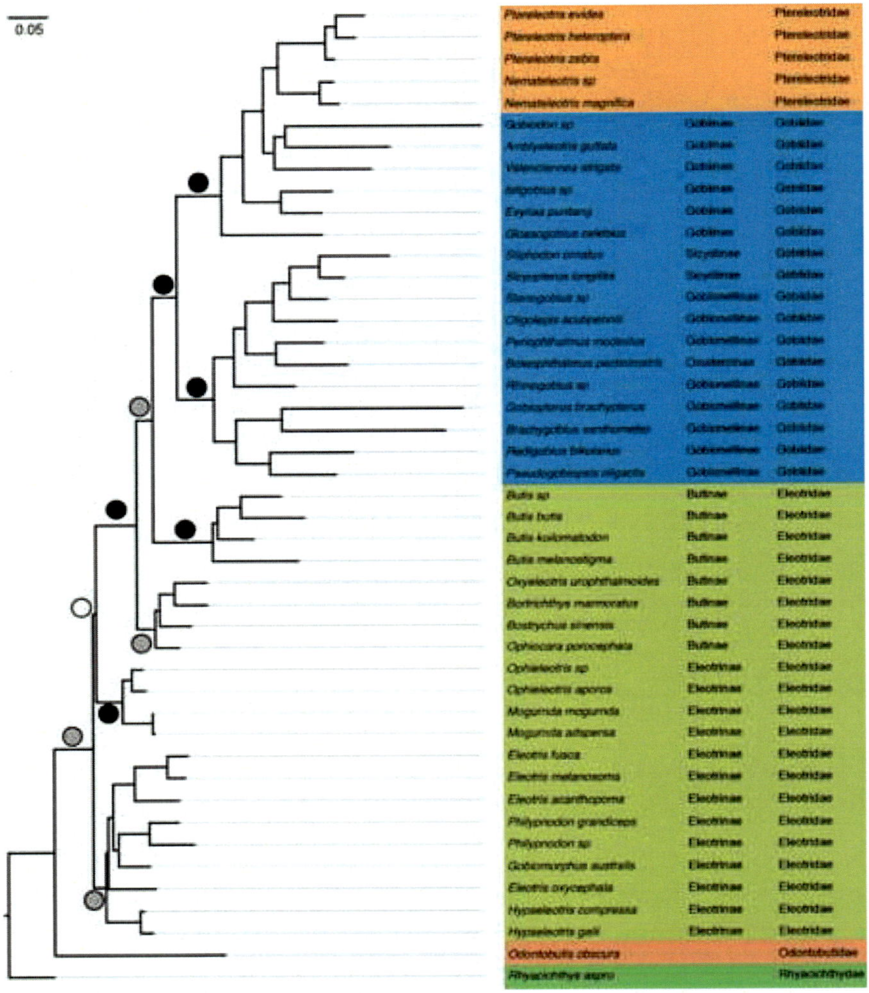

Fig. 1.2.3 Maximum likelihood reanalysis of the Wang *et al.* (2001) data set using *Rhyacichthys* as outgroup. The analysis was based on a 905 bp alignment of 12S rRNA and analyzed with Garli v0.96 (Zwickl, 2006) using default settings (-lnL = 11108.60) and visualized with FigTree v1.2.1 (Rambaut, 2006). Note, the original analyses by Wang *et al.* (2001) was based on 12S rRNA and $tRNA^{Val}$. However, the $tRNA^{Val}$ data are not available from GenBank. Circles indicate bootstrap support (BS) for selected branches based on 1,000 replicates; white = BS < 50%, grey = 50 ≤ BS < 70, black = BS ≥ 70.

A

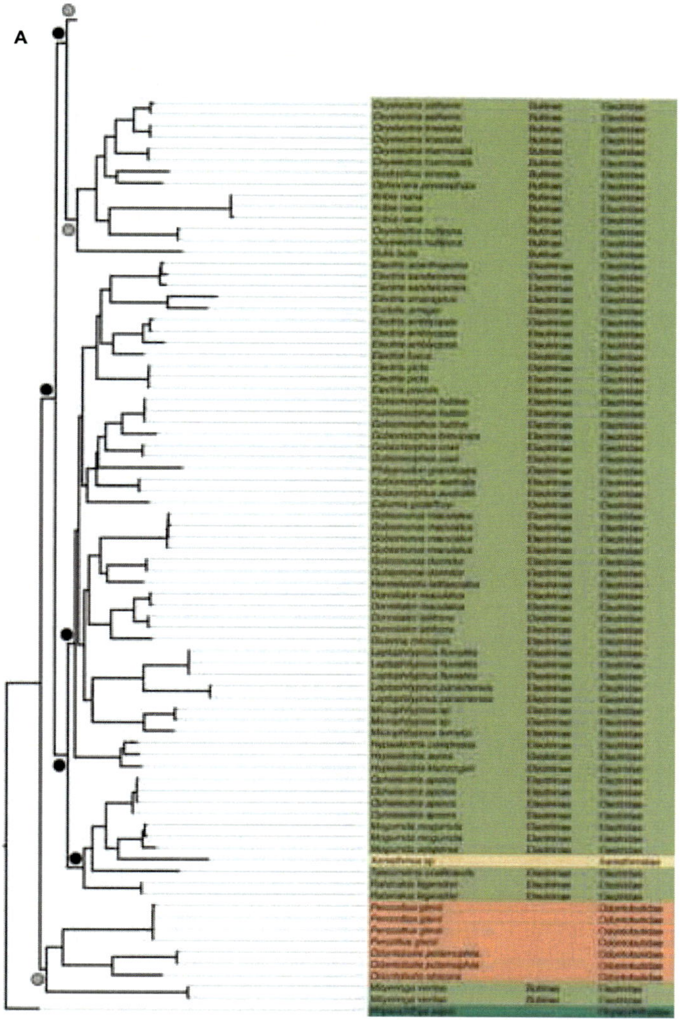

Fig. 1.2.4 Contd..

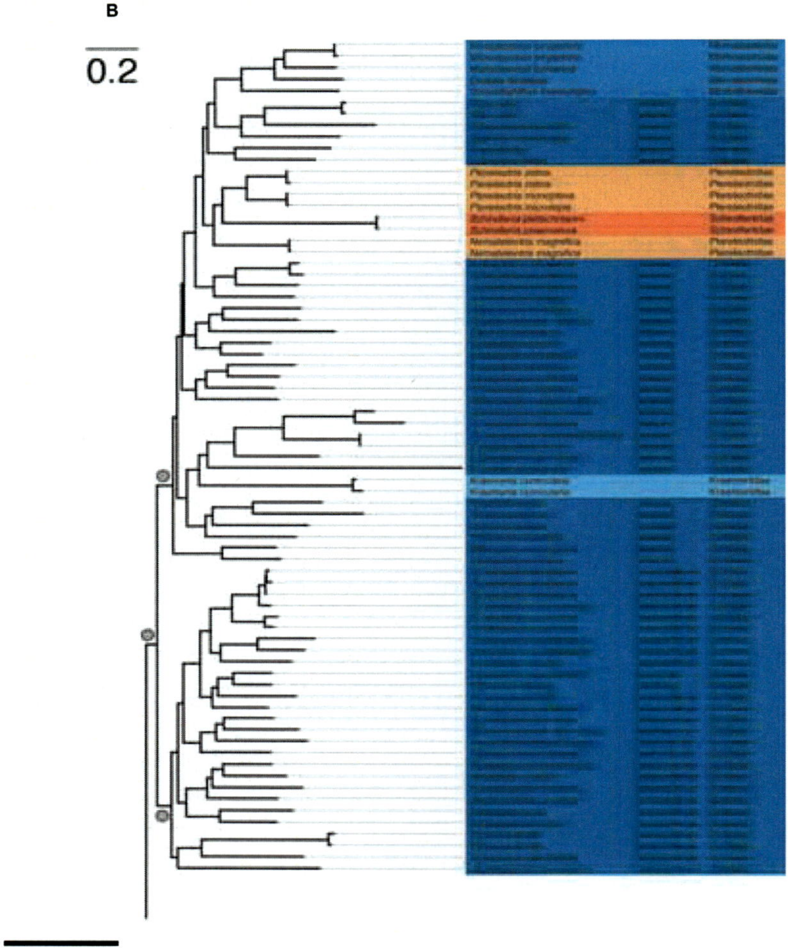

Fig. 1.2.4 Maximum likelihood reanalysis of the Thacker (2009) data set using *Rhyacichthys* as outgroup. The analysis was based on a 4647 bp alignment of concatenated *cox1*, *nad1*, *nad2*, and *cob* sequences and analyzed with Garli v0.96 (Zwickl, 2006) using default settings (-lnL = 196497.43) and visualized with FigTree v1.2.1 (Rambaut, 2006). Circles indicate bootstrap support (BS) for selected branches based on 1,000 replicates; white = BS < 50%, grey = 50 ≤ BS < 70, black = BS ≥ 70.

Chapter 1.4

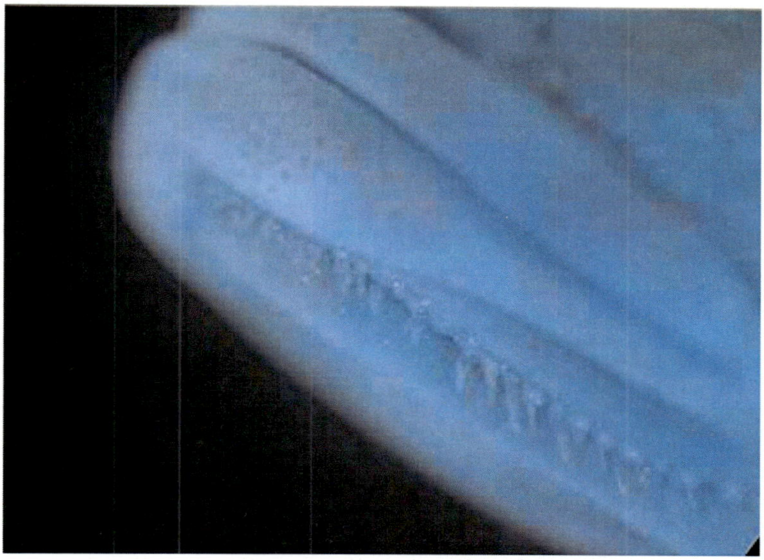

Fig. 1.4.4 Barbel-like projection of sensory papilla (ventral part of lower jaw) of *Neodontobutis hainanensis*.

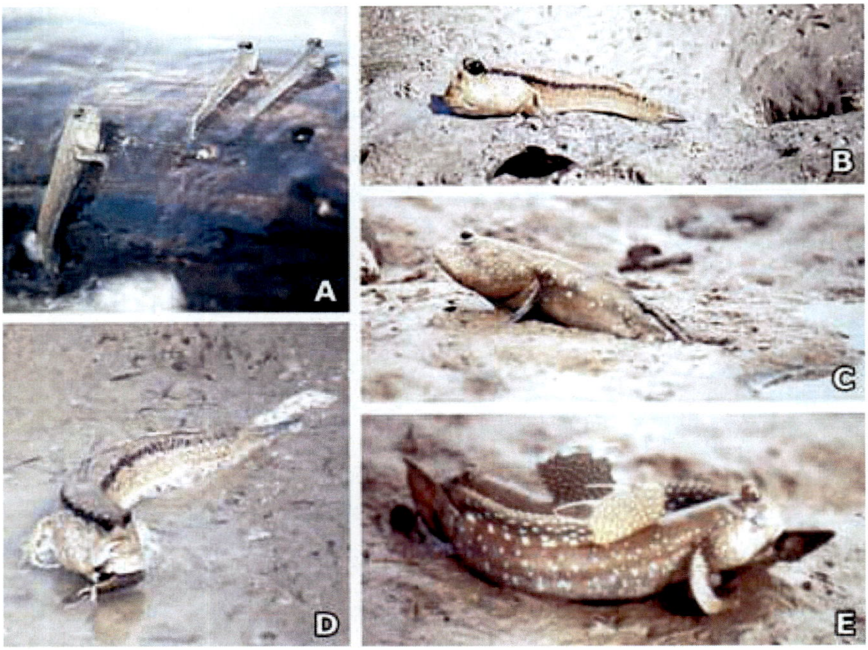

Fig. 1.7.1(A) *Periophthalmus chrysospilos* climbing log (image by E.O. Murdy)

(B) *Periophthalmodon schlosseri* at burrow entrance (image by E.O. Murdy)

(C) *Boleophthalmus dussumieri* entering burrow tail first (image by E.O. Murdy)

(D) *Periophthalmodon schlosseri* eating fiddler crab (image by E.O. Murdy)

(E) *Boleophthalmus dussumieri* in courting display at low tide on the mudflat (image by E.O. Murdy)

Chapter 1.8

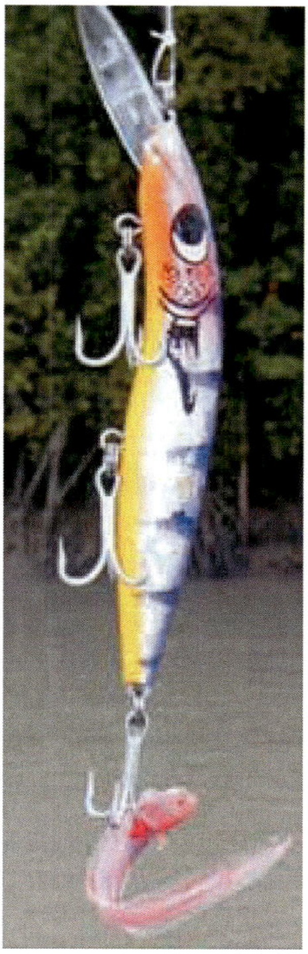

Fig. 1.8.1 *Caragobius rubristriatus* snagged on a fishing lure in Northern Territory, Australia, displaying the pinkish coloration common to many amblyopines. (Image by T. Robb and provided courtesy of M. McGrouther, Australian Museum).

Chapter 1.9

Fig. 1.9.4 Some Sicydiinae species (all male) showing the different genera. A = *Cotylopus rubripinnis*; B = *Akihito vanuatu*; C = *Lentipes kaaea*; D = *Sicydium plumieri*; E = *Sicyopterus lagocephalus*; F = *Sicyopus chloe*; G = *Stiphodon rutilaureus*. Photos P. Keith and E. Vigneux.

Section 2: Zoogeography

Chapter 2.1

Fig. 2.1.1 Representative members of the Tribe Gobiosomatini. **(A)** *Aboma etheostoma* **(B)** *Barbulifer ceuthoecus* **(C)** *Chriolepis zebra* **(D)** *Elacatinus evelynae* **(E)** *Enypnias seminudus* **(F)** *Evermannichthys spongicola* **(G)** *Ginsburgellus novemlineatus* **(H)** *Gobiosoma bosc* **(I)** *Gobulus birdsong* **(J)** *Gymneleotris seminudus* **(K)** *Psilotris batracoides* **(M)** *Tigrigobius multifasciatus* female **(N)** *Tigrigobius gemmatum* **(O)** *Tigrigobius multifasciatus* male **(P)** *Akko brevis* **(Q)** *Bollmannia litura* **(R)** *Microgobius brevispinis*. Photographs by J. Van Tassell; plate composed by R. Patzner.

Chapter 2.3

Fig. 2.3.2 **A.** *Psammogobius knysnaensis* (Nenga River mouth at Coffee Bay), **B.** *Caffrogobius caffer* (Tide Pools at Hole in Wall, South Africa), **C.** *Caffrogobius gilchristi* (Bomvu River mouth close to Coffee Bay, South Africa), **D.** *Caffrogobius nudiceps* (Kowie River mouth, South Africa), **E.** *Hetereleotris zonata* (Tide Pools at Hole in Wall, South Africa). Photos U. Schliewen.

Chapter 2.5

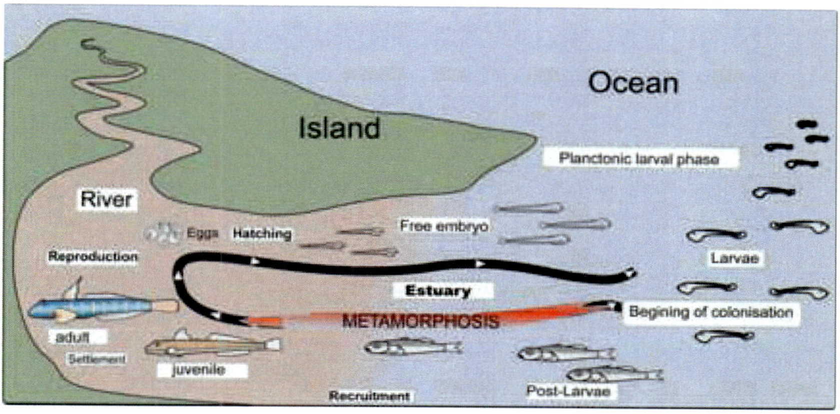

Fig. 2.5.1 Amphidromy in freshwater gobies. © P. Torres and P. Keith.

Section 4: Ecology

Chapter 4.1

Fig. 4.1.2 Live pictures of some paedomorphic gobioids, showing the characteristic body morphology and colouration pattern. (**A**) *A. minuta*, (**B**) *C. linearis*, (**C**) *P. ferreri* and (**D**) *L. petersii* (photographs by kind permission of Dr. Mat Vestjens).

Chapter 4.2

Fig. 4.2.1 Ten Indo-Pacific gobiid species as examples of the various life-styles of coral reefs gobies. A: *Amblygobius albimaculatus* sifting sand in a seagrass meadow. B: a pair of *Valenciennea puellaris* at the entrance of their self-dug burrow. C: *Amblyeleotris sungami* which lives in a mutualistic relationship with a species of *Alpheus* snapping shrimp. D: *Gnatholepis anjerensis* on the edge between coral rock and sand bottom. E: *Asterropteryx semipunctata* between algae on coral rubble. F: *Trimma avidori* swimming upside down in a small coral rock cave. G: *Eviota zebrina* inhabits live corals and coral rock. H: *Bryaninops natans* is an obligate associate of *Acropora* corals. I: *Paragobiodon echinocephalus* in the pocilloporid coral *Stylophora pistillata*. J: *Gobiodon histrio* among branches of *Acropora gemmifera*. Photographs by J. Herler.

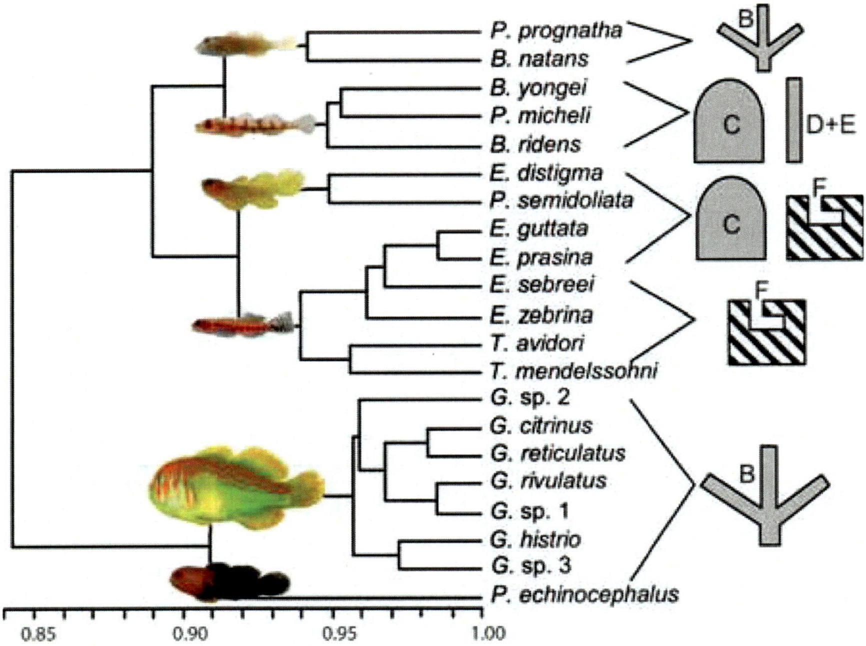

Fig. 4.2.5 Cluster analysis (Bray Curtis – similarity) of nine proportional morphometric variables (all expressed as percent of standard length) of 21 reef gobies from the Gulf of Aqaba, northern Red Sea (modified from Herler, 2007). The most frequented microhabitats are indicated (B = branching corals, C = massive corals, D = *Millepora dichotoma*, E = *Cirripathes* sp., F = coral rock). Fish genera abbreviations: B = *Bryaninops*, E = *Eviota*, G = *Gobiodon*, P = *Paragobiodon* (*echinocephalus*), *Pleurosicya* (*prognatha* and *micheli*), and *Priolepis* (*semidoliata*), and T = *Trimma* respectively.

Chapter 4.3

Fig. 4.3.2 A Cleaning goby *Elacatinus* sp. inspecting the buccal area of a tiger grouper *Mycteroperca tigris* on a Little Cayman reef. Photo: Isabelle Côté. B Gnathiid isopod larva – the most ubiquitous ectoparasite found in stomach contents of cleaning gobies. Photo courtesy of Paul Sikkel.

Chapter 4.4

Fig. 4.4.10 Associations between gobies and alpheid shrimps from the Indo-Pacific region. **A.** *Stonogobiops xanthorhinica* and *Alpheus randalli*, Japan, (Photo: T. Hirata). **B.** *Amblyeleotris wheeleri* and *Alpheus ochrostriatus*, Japan, (Photo: T. Hirata). **C.** *Lotilia graciliosa* and *Alpheus rubromaculatus* (Photo: P. Nahke). **D.** *Cryptocentrus maudae* and *Alpheus* sp. Bali, Indonesia, (Photo: R. Steene). Note the bright color of both gobies and shrimps and the antennal contact between the partners.

Chapter 4.5

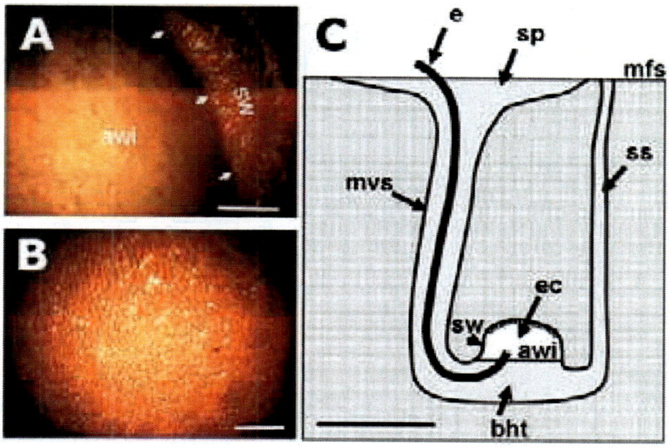

Fig. 4.5.1 View of the air–water interface (awi; A) and the inner top surface (B) of an egg chamber in a *Periophthalmodon schlosseri* burrow. A monolayer of spawned eggs was confirmed only on the inner top surface (B), not on the side wall (sw; A). The image in A was taken using an endoscope CCD camera, which was raised above the air–water interface and bent obliquely downward toward the interface. Arrows in A indicate the edge of the air–water interface. Scales in A and B represent 3 and 2 cm, respectively (from Ishimatsu *et al.*, 2009). Diagram of the position of the endoscope inside a burrow (C). awi, air–water interface; bht, bottom horizontal tunnel; e, endoscope; ec, egg chamber; mfs, mudflat surface; mvs, main vertical shaft; ss, side shaft; sp, surface pool; sw, side wall of the egg chamber. Scale 30 cm

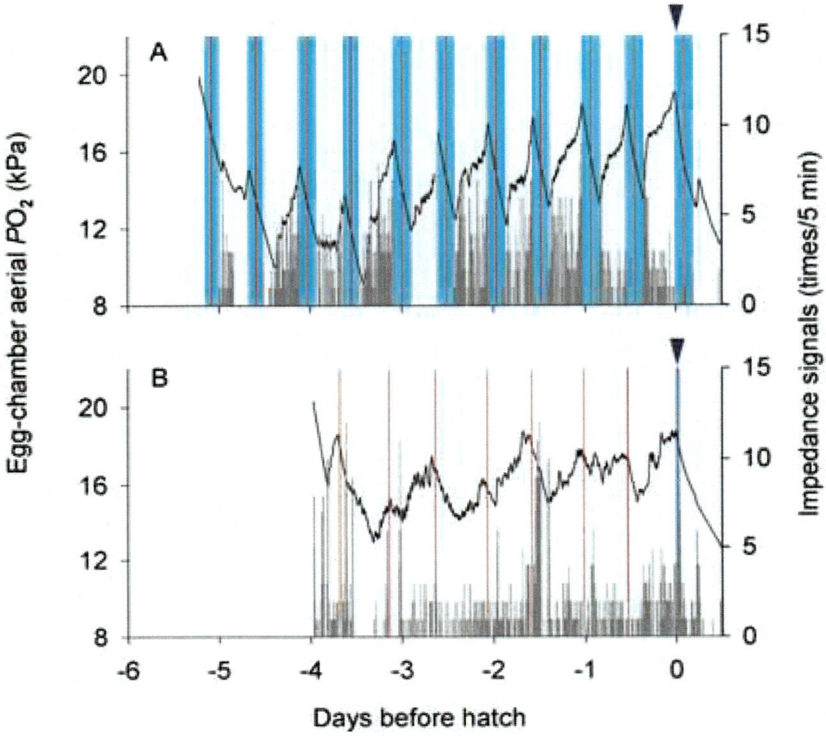

Fig. 4.5.3 Continuous records of *Periophthalmus modestus* egg-chamber air PO_2 (black line) and the timing of the male's egg-chamber visits (indicated by impedance signals, grey spikes) in relation to the tidal cycle. Data acquisition began with the day of instrumentation and continued until egg hatching, initiated by the male on a rising tide (blue triangle). The beginning of day zero is set to the time when a burrow was covered by an incoming tide, following which egg hatching occurred. The initial high PO_2 reflects the opening of the egg chamber to air during instrumentation. (A) Record for a burrow regularly covered (blue bars) and uncovered by the tidal oscillation (red lines show the highest tide). (Note: No impedance data for day –3 due to a technical problem.) (B) Record for a burrow located sufficiently high on the mudflat to be continually exposed to air until day zero when the rising tide covered the burrow (blue bar) and hatching occurred (blue triangle, from Ishimatsu *et al.*, 2007).

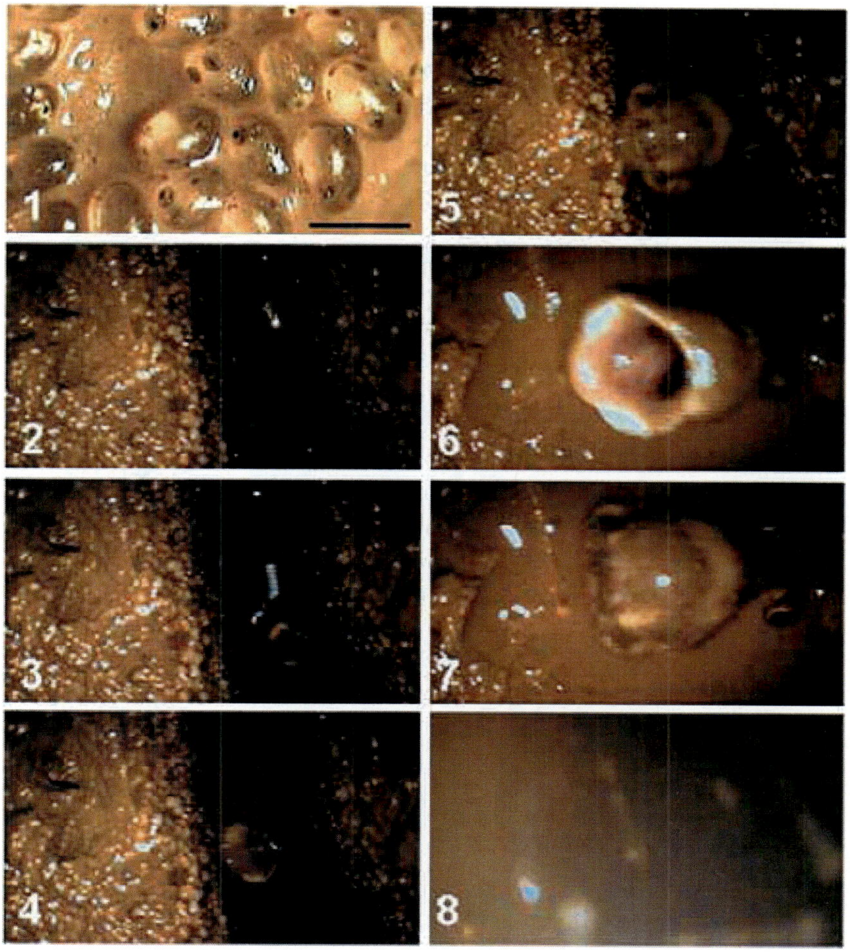

Fig. 4.5.4 Interior view of a *Periophthalmus modestus* egg chamber. (1) Eggs on the mud wall of a laboratory-incubated chamber (scale bar 1 mm). (2–7) Endoscope video frames documenting hatch induction behaviour (June 19, 2002). Vertical view is down, through the egg chamber (note glistening monolayer of eggs on the mud wall), to the burrow water surface. The male appears at the surface with its mouth open, and becomes progressively larger as it moves closer to the camera. Removing mouthfuls of air (3, 4, 6, 7) raises the water level, which immerses the eggs, causing them to hatch. (8) Hatched larvae [average total length 2.84 mm (Kobayashi *et al.*, 1972)] swimming in the flooded egg chamber 48 min after frame 7 (from Ishimatsu *et al.*, 2007).